Further Engineering Mathematics

2

Companion volume

K.A. Stroud, *Engineering Mathematics*

Related titles

John Berry and Patrick Wainwright, *Foundation Mathematics for Engineers*

Keith L. Watson, *Foundation Science for Engineers*

***Foundations of Engineering* Series**

J. Cain and R. Hulse, *Structural Mechanics*

G.E. Drabble, *Dynamics*

R.G. Powell, *Electromagnetism*

P. Silvester, *Electric Circuits*

John Simonson, *Thermodynamics*

Martin Widden, *Fluid Mechanics*

Further Engineering Mathematics

Programmes and Problems

K.A. Stroud

formerly Principal Lecturer
Department of Mathematics, Coventry University

THIRD EDITION

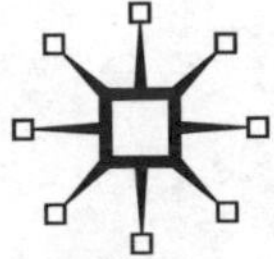

First edition 1986
Reprinted twice
Second edition 1990
Reprinted seven times
Third edition 1996

Published by
PALGRAVE
Houndmills, Basingstoke, Hampshire RG21 6XS and
175 Fifth Avenue, New York, N. Y. 10010
Companies and representatives throughout the world

PALGRAVE is the new global academic imprint of
St. Martin's Press LLC Scholarly and Reference Division and
Palgrave Publishers Ltd (formerly Macmillan Press Ltd).

ISBN 0–333–65741–1

This book is printed on paper suitable for recycling and made from fully managed and sustained forest sources.

A catalogue record for this book is available from the British Library.

10 9 8 7 6
05 04 03 02 01 00

Printed in Malaysia

Contents

Preface to the First Edition

The purpose of this book is essentially to provide a sound second year course in Mathematics appropriate to studies leading to B.Sc. Engineering Degrees and other qualifications of a comparable level. The emphasis throughout is on techniques and applications, supported by sufficient formal proofs to warrant the methods being employed.

The structure of the text and the techniques used follow closely those of the author's first year book, *Engineering Mathematics—Programmes and Problems*, to which this further book is a companion volume and a continuation of the highly successful learning strategies devised. As with the previous work, the text is based on a series of self-instructional programmes arising from extensive research and rigid evaluation in a variety of relevant courses and, once again, the individualised nature of the development makes the book eminently suitable both for general class use and for personal study.

Each of the course programmes guides the student through the development of a particular topic, with numerous worked examples to demonstrate the techniques and with increased responsibility passing to the student as mastery is achieved. Revision exercises are provided where appropriate and each programme terminates with a *Revision Summary* of the main points covered, a *Test Exercise* based directly on the work of the programme and a set of *Further Problems* which provides opportunity for the additional practice that is essential for ensured success. The ability to work at one's own pace throughout is of utmost importance in maintaining motivation and in achieving mastery.

In several instances, the topic of a programme is a direct extension of basic work covered in *Engineering Mathematics* and where this is so, the title page of the programme carries a brief reference to the relevant programme in the first year treatment. This clearly directs the student to worthwhile revision of the prerequisites assumed in the further development of the subject matter.

A complete set of Answers to all problems and a detailed Index are provided at the end of the book.

Grateful acknowledgement is made of the constructive suggestions and cooperation received from many quarters both in the development of the original programmes and in the final preparation of the text.

Recognition must also be made of the many sources from which examples have been gleaned over the years and which contribute in no small measure to the success of the work.

Finally my sincere appreciation is due to the publishers for their patience, advice and ready cooperation in the preparation of the text for publication.

K.A. Stroud

Preface to the Second Edition

Since the first publication of *Further Engineering Mathematics* as core material for a typical second year engineering degree course, requests have been received from time to time for the inclusion of further topics to cover the particular requirements of individual syllabuses.

Some limit, inevitably, has to be placed on the physical size of the text, but it has been possible at least to include a programme on *Linear Optimisation* (*Linear Programming*) which was one of the subjects most frequently required.

The treatment of the additional material follows the structure of the rest of the book and the emphasis is largely on the practical use of the *simplex method* for the solution of both maximisation and minimisation problems.

The opportunity has also been taken to amend and clarify a number of minor points in the existing text and my thanks are due to those correspondents who have undertaken to write with constructive comment. Such feedback is always welcome.

K.A.S.

Preface to the Third Edition

With the new edition of *Further Engineering Mathematics*, the opportunity has been taken to incorporate a number of minor revisions and amendments to the previous text.

The format of the pages has been changed and the publishers have undertaken the complete resetting of the text to result in a more open presentation of the material and to facilitate the learning process still further.

Once again, my sincere thanks are due to all those correspondents who have kindly written with constructive comment concerning the book and to the publishers for their continued support, advice and cooperation throughout the preparation, production and marketing of the work.

K.A.S.

Hints on Using the Book

This book contains twenty lessons, each of which has been written in such a way as to make learning more effective and more interesting. It is almost like having a personal tutor, for you proceed at your own rate of learning and any difficulties you may have are cleared before you have the chance to practise incorrect ideas or techniques.

You will find that each programme is divided into sections called frames. When you start a programme, begin at frame 1. Read each frame carefully and carry out any instructions or exercise which you are asked to do. In almost every frame, you are required to make a response of some kind, testing your understanding of the information in the frame, and you can immediately compare your answer with the correct answer given in the next frame. To obtain the greatest benefit, you are strongly advised to cover up the following frame until you have made your response. When a series of dots occurs, you are expected to supply the missing word, phrase, or number. At every stage, you will be guided along the right path. There is no need to hurry: read the frames carefully and follow the directions exactly. In this way, you must learn.

At the end of each programme, you will find a short Test Exercise. This is set directly on what you have learned in the lesson: the questions are straightforward and contain no tricks. To provide you with the necessary practice, a set of Further Problems is also included: do as many of these problems as you can. Remember that in mathematics, as in many other situations, practice makes perfect—or more nearly so.

Even if you feel you have done some of the topics before, work steadily through each programme: it will serve as useful revision and fill in any gaps in your knowledge that you may have.

Useful Background Information

I. Algebraic Identities

$(a+b)^2 = a^2 + 2ab + b^2 \qquad (a+b)^3 = a^3 + 3a^2b + 3ab^2 + b^3$
$(a-b)^2 = a^2 - 2ab + b^2 \qquad (a-b)^3 = a^3 - 3a^2b + 3ab^2 - b^3$
$(a+b)^4 = a^4 + 4a^3b + 6a^2b^2 + 4ab^3 + b^4$
$(a-b)^4 = a^4 - 4a^3b + 6a^2b^2 - 4ab^3 + b^4$
$a^2 - b^2 = (a-b)(a+b)$
$a^3 + b^3 = (a+b)(a^2 - ab + b^2)$
$a^3 - b^3 = (a-b)(a^2 + ab + b^2)$

II. Trigonometrical Identities

(1) $\sin^2\theta + \cos^2\theta = 1; \qquad \sec^2\theta = 1 + \tan^2\theta;$
$\operatorname{cosec}^2\theta = 1 + \cot^2\theta$

(2) $\sin(A+B) = \sin A\cos B + \cos A\sin B$
$\sin(A-B) = \sin A\cos B - \cos A\sin B$
$\cos(A+B) = \cos A\cos B - \sin A\sin B$
$\cos(A-B) = \cos A\cos B + \sin A\sin B$

$$\tan(A+B) = \frac{\tan A + \tan B}{1 - \tan A\tan B}$$

$$\tan(A-B) = \frac{\tan A - \tan B}{1 + \tan A\tan B}$$

(3) Let $A = B = \theta$. $\quad \therefore\ \sin 2\theta = 2\sin\theta\cos\theta$

$$\cos 2\theta = \cos^2\theta - \sin^2\theta$$
$$= 1 - 2\sin^2\theta$$
$$= 2\cos^2\theta - 1$$
$$\tan 2\theta = \frac{2\tan\theta}{1 - \tan^2\theta}$$

(4) Let $\theta = \dfrac{\phi}{2}$ $\quad \therefore\ \sin\phi = 2\sin\dfrac{\phi}{2}\cos\dfrac{\phi}{2}$

$$\cos\phi = \cos^2\frac{\phi}{2} - \sin^2\frac{\phi}{2}$$

$$= 1 - 2\sin^2\frac{\phi}{2}$$

$$= 2\cos^2\frac{\phi}{2} - 1$$

$$\tan\phi = \frac{2\tan\dfrac{\phi}{2}}{1 - \tan^2\dfrac{\phi}{2}}$$

(5) $\sin C + \sin D = 2\sin\dfrac{C+D}{2}\cos\dfrac{C-D}{2}$

$\sin C - \sin D = 2\cos\dfrac{C+D}{2}\sin\dfrac{C-D}{2}$

$\cos C + \cos D = 2\cos\dfrac{C+D}{2}\cos\dfrac{C-D}{2}$

$\cos D - \cos C = 2\sin\dfrac{C+D}{2}\sin\dfrac{C-D}{2}$

(6) $2\sin A\cos B = \sin(A+B) + \sin(A-B)$

$2\cos A\sin B = \sin(A+B) - \sin(A-B)$

$2\cos A\cos B = \cos(A+B) + \cos(A-B)$

$2\sin A\sin B = \cos(A-B) - \cos(A+B)$

(7) Negative angles: $\sin(-\theta) = -\sin\theta$

$\cos(-\theta) = \cos\theta$

$\tan(-\theta) = -\tan\theta$

(8) Angles having the same trig. ratios:

(i) Same sine: θ and $(180° - \theta)$

(ii) Same cosine: θ and $(360° - \theta)$, i.e. $(-\theta)$

(iii) Same tangent: θ and $(180° + \theta)$

(9) $a\sin\theta + b\cos\theta = A\sin(\theta + \alpha)$

$a\sin\theta - b\cos\theta = A\sin(\theta - \alpha)$

$a\cos\theta + b\sin\theta = A\cos(\theta - \alpha)$

$a\cos\theta - b\sin\theta = A\cos(\theta + \alpha)$

where: $\begin{cases} A = \sqrt{(a^2 + b^2)} \\ \alpha = \tan^{-1}\dfrac{b}{a} \qquad (0^\circ < \alpha < 90^\circ) \end{cases}$

III. Standard Curves

(1) *Straight line*:

Slope, $m = \dfrac{dy}{dx} = \dfrac{y_2 - y_1}{x_2 - x_1}$

Angle between two lines, $\tan\theta = \dfrac{m_2 - m_1}{1 + m_1 m_2}$

For parallel lines, $m_2 = m_1$
For perpendicular lines, $m_1 m_2 = -1$

Equation of a straight line (slope $= m$)

(i) Intercept c on real y-axis: $y = mx + c$
(ii) Passing through (x_1, y_1): $y - y_1 = m(x - x_1)$
(iii) Joining (x_1, y_1) and (x_2, y_2): $\dfrac{y - y_1}{y_2 - y_1} = \dfrac{x - x_1}{x_2 - x_1}$

(2) *Circle*:

Centre at origin, radius r: $x^2 + y^2 = r^2$
Centre (h, k), radius r: $(x - h)^2 + (y - k)^2 = r^2$
General equation: $x^2 + y^2 + 2gx + 2fy + c = 0$
with centre $(-g, -f)$: radius $= \sqrt{(g^2 + f^2 - c)}$
Parametric equation: $x = r\cos\theta$, $y = r\sin\theta$

(3) *Parabola*:

Vertex at origin, focus $(a, 0)$: $y^2 = 4ax$
Parametric equations: $x = at^2$, $y = 2at$

(4) *Ellipse*:

Centre at origin, foci $(\pm\sqrt{[a^2 - b^2]}, 0)$: $\dfrac{x^2}{a^2} + \dfrac{y^2}{b^2} = 1$

where $a =$ semi major axis, $b =$ semi minor axis

Parametric equations: $x = a\cos\theta$, $y = b\sin\theta$

(5) *Hyperbola*:

Centre at origin, foci $(\pm\sqrt{[a^2 + b^2]}, 0)$: $\dfrac{x^2}{a^2} - \dfrac{y^2}{b^2} = 1$

Parametric equations: $x = a\sec\theta$, $y = b\tan\theta$

Rectangular hyperbola:

Centre at origin, vertex $\pm\left(\frac{a}{\sqrt{2}}, \frac{a}{\sqrt{2}}\right)$: $xy = \frac{a^2}{2} = c^2$

where $c = \frac{a}{\sqrt{2}}$ i.e. $xy = c^2$

Parametric equations: $x = ct$, $y = c/t$

IV. Laws of Mathematics

(1) *Associative laws* – for addition and multiplication

$$a + (b + c) = (a + b) + c$$
$$a(bc) = (ab)c$$

(2) *Commutative laws* – for addition and multiplication

$$a + b = b + a$$
$$ab = ba$$

(3) *Distributive laws* – for multiplication and division

$$a(b + c) = ab + ac$$
$$\frac{b + c}{a} = \frac{b}{a} + \frac{c}{a} \quad (\text{ provided } a \neq 0)$$

Programme 1

Theory of Equations

PART 1

Function Notation

1

If $f(x)$ represents the function $a_1x^n + a_2x^{n-1} + \ldots + a_{n+1}$ and $f(a)$ denotes the value of the function when x is replaced by the value a, then if

$$f(x) = 2x^3 - 3x^2 + 5x - 6$$
$$f(1) = 2 - 3 + 5 - 6 = -2$$
$$f(-1) = -2 - 3 - 5 - 6 = -16$$

and $f(2.5) = \ldots\ldots\ldots$

2

19

for $f(2.5) = 2(2.5)^3 - 3(2.5)^2 + 5(2.5) - 6$
$= 31.25 - 18.75 + 12.5 - 6 = \underline{19}$

Calculation of function values by 'nesting'

If the function is a polynomial, the calculation of function values is more efficiently accomplished by the method of 'nesting'. This entails re-writing the function in such a way that the ensuing calculation can be carried out as a series operation and requires no recording of intermediate partial results—and thus avoids a possible source of human error in practice.

For example, we can write $f(x) = 2x^3 - 3x^2 + 5x - 6$ in the form
$f(x) = [(2x - 3)x + 5]x - 6$

Then, starting with the innermost brackets, we can, for example, obtain $f(4)$ thus:

evaluate $2 \times 4 - 3$	i.e.	5
multiply the result by 4	i.e.	20
add on 5	i.e.	25
multiply the result by 4	i.e.	100
subtract 6	i.e.	94

In practice, the partial results are not written down. They are included here simply by way of explanation.

On to the next frame.

3

The method of 'nesting' is particularly suitable for use with a simple computer or calculator, as the calculation proceeds in a chain formation.

To express a polynomial in 'nested' form, we simply proceed as follows:

$$f(x) = x^3 + 4x^2 - 3x + 7$$

Starting at the left-hand end

(a) First write the variable x and the next coefficient in brackets, thus

$$(x + 4)$$

(b) Now multiply by x and add the next coefficient, and enclose in brackets

$$[(x + 4)x - 3]$$

(c) Repeat the process: multiply by x and add the next coefficient

$$f(x) = [(x + 4)x - 3]x + 7$$

So, using this result and starting with the innermost brackets,

$$f(-2) = \ldots\ldots\ldots$$

4

21

Similarly, if $f(x) = 5x^4 + 2x^3 - 4x^2 - x + 3$, we start off with

$$f(x) = \{[(5x + 2)x - 4]\ldots\}$$

Note that we include the coefficient 5 with the initial x and that, at each stage, we multiply by x and add the next coefficient.

So, finish it off. $f(x) = \ldots\ldots\ldots$

5

$$\boxed{f(x) = \{[(5x + 2)x - 4]x - 1\}x + 3}$$

From this we can obtain $f(2) = \ldots\ldots\ldots$

6

$$f(2) = 81$$

Finally, one more example. $f(x) = 2x^4 - 5x^3 + 3x^2 - 4x + 6$

In nested form $f(x) = \dots\dots\dots$

Hence $f(3) = \dots\dots\dots$

7

$$f(x) = \{[(2x - 5)x + 3]x - 4\}x + 6$$
$$f(3) = 48$$

Numerical representation of a function and interpolation

Normally, a function is defined algebraically as in the previous examples and function values for particular values of the independent variable can be obtained by direct substitution.

Sometimes, however, a function is represented by a set of function values, as occurs when a set of readings is compiled as the result of an experiment or practical test.

x	$f(x)$
1	4
2	14
3	40
4	88
5	164
6	274

Intermediate function values, e.g. for $x = 2.5$, are obtained by *interpolation*. The value of $f(2.5)$ will clearly lie between 14 and 40, the function values of $x = 2$ and $x = 3$.

Purely as an estimate $f(2.5) = \dots\dots\dots$ What do you suggest?

8

27

1. Linear interpolation

If you gave the result as 27, you no doubt argued that $x = 2.5$ is midway between $x = 2$ and $x = 3$, and that therefore $f(2.5)$ would be midway between 14 and 40, i.e. 27. This is the simplest form of interpolation, but there is no evidence that there is a linear relationship between x and $f(x)$, and the result is therefore suspect.

Of course, we could have determined the function value at $x = 2.5$ by other means, such as

....

9

by drawing the graph of $f(x)$ against x

2. Graphical interpolation

We could, indeed, plot the graph of $f(x)$ against x and, from it, estimate the value of $f(x)$ at $x = 2.5$.

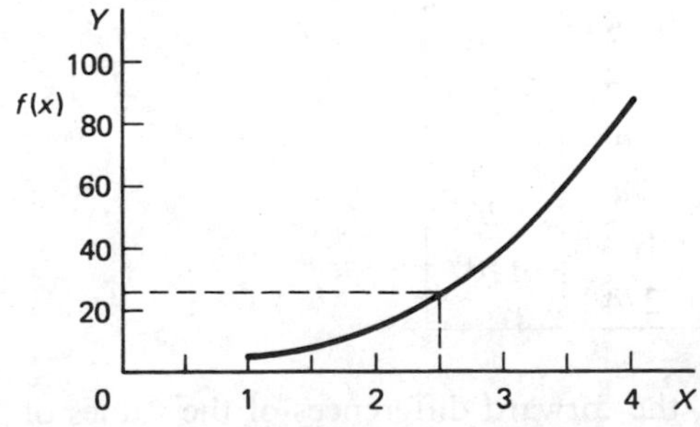

This method is also approximate and time consuming.

$$f(2.5) \approx 26$$

A much more reliable method is set out in the next frame.

10

3. Gregory–Newton interpolation formula using forward finite differences

x	$f(x)$
$\vdots$	$\vdots$
x_0	$f(x_0)$
x_1	$f(x_1)$
$\vdots$	$\vdots$

$$\Delta f_0 = f(x_1) - f(x_0)$$

For each pair of consecutive function values, $f(x_0)$ and $f(x_1)$, in the table, the *forward difference* Δf_0 is calculated by subtracting $f(x_0)$ from $f(x_1)$. This difference is written in a third column of the table, midway between the lines carrying $f(x_0)$ and $f(x_1)$.

x	$f(x)$	Δf
1	4	
		10
2	14	
		26
3	40	
$\vdots$	$\vdots$	

Complete the table, which then becomes ....

11

x	$f(x)$	Δf
1	4	
		10
2	14	
		26
3	40	
		48
4	88	
		76
5	164	
		110
6	274	

We now form a fourth column, the forward differences of the values of Δf, denoted by $\Delta^2 f$ and again written midway between the lines of Δf. These are the second forward differences of $f(x)$.

So the table then becomes ....

12

x	$f(x)$	Δf	$\Delta^2 f$
1	4		
		10	
2	14		16
		26	
3	40		22
		48	
4	88		28
		76	
5	164		34
		110	
6	274		

A further column can now be added in like manner, giving the third differences, denoted by $\Delta^3 f$, so that we then have ....

13

x	$f(x)$	Δf	$\Delta^2 f$	$\Delta^3 f$
1	4			
		10		
2	14		16	
		26		6
3	40		22	
		48		6
4	88		28	
		76		6
5	164		34	
		110		
6	274			

Note that the table has now been completed, for the third differences are constant and all subsequent differences would be zero.

Now we will see how to deal with it. So move on.

14

To find $f(2.5)$

h: $x_0 \rightarrow$, x_p, $x_1 \rightarrow$

x	$f(x)$	Δf	$\Delta^2 f$	$\Delta^3 f$
1	4			
		10		
2	14		16	
		26		6
3	40		22	
		48		6
4	88		28	
		76		6
5	164		34	
		110		
6	274			

We have to find $f(2.5)$. Therefore denote $\left.\begin{array}{l} x = 2 \text{ as } x_0 \\ x = 3 \text{ as } x_1 \end{array}\right\} x = 2.5 \text{ as } x_p$

Let h = the constant range between successive values of x, i.e. $h = x_1 - x_0$

Express $(x_p - x_0)$ as a fraction of h, i.e. $p = \dfrac{x_p - x_0}{h}$, $\quad 0 < p < 1$

Therefore, in the case above, $h = 1$ and $p = \dfrac{2.5 - 2.0}{1} = 0.5.$

All we now use from the table is the set of values underlined by the broken line, corresponding to x_0.

So we have

$p = \ldots\ldots;\ \ f_0 = \ldots\ldots;\ \ \Delta f_0 = \ldots\ldots;\ \ \Delta^2 f_0 = \ldots\ldots;\ \ \Delta^3 f_0 = \ldots\ldots$

15

$p = 0.5; \quad f_0 = 14; \quad \Delta f_0 = 26; \quad \Delta^2 f_0 = 22; \quad \Delta^3 f_0 = 6$

Now at long last, we are ready to deal with the *Gregory–Newton forward difference interpolation formula*

$$f_p = f_0 + p\Delta f_0 + \frac{p(p-1)}{1 \times 2}\Delta^2 f_0 + \frac{p(p-1)(p-2)}{1 \times 2 \times 3}\Delta^3 f_0 + \ldots$$

This is sometimes written in operator form

$$f_p = \left\{1 + p\Delta + \frac{p(p-1)}{1 \times 2}\Delta^2 + \frac{p(p-1)(p-2)}{1 \times 2 \times 3}\Delta^3 + \ldots\right\} f_0$$

Which you no doubt recognise as the binomial expansion of

$$f_p = (1 + \Delta)^p \times f_0$$

Substituting the values in the above example gives

$$f(2.5) = f_p = \ldots\ldots\ldots$$

16

24.625

for $f_p = 14 + 0.5(26) + \dfrac{0.5(-0.5)}{1 \times 2}(22) + \dfrac{0.5(-0.5)(-1.5)}{1 \times 2 \times 3}(6)$

$= 14 + 13 - 2.75 + 0.375$

$= 27.375 - 2.75 = 24.625$

Comparing the results of the three methods we have discussed

(a) Linear interpolation $f(2.5) = 27$

(b) Graphical interpolation $f(2.5) = 26$

(c) Gregory–Newton formula $f(2.5) = 24.625$—the true value

Example 2

x	$f(x)$
2	14
4	88
6	274
8	620
10	1174

It is required to determine the value of $f(x)$ at $x = 5.5$
In this case

$x_0 = \ldots\ldots\ldots; \quad x_1 = \ldots\ldots\ldots; \quad h = \ldots\ldots\ldots; \quad p = \ldots\ldots\ldots$

17

$$x_0 = 4;\ x_1 = 6;\ h = 2;\ p = 0.75$$

since $\quad h = x_1 - x_0 = 6 - 4 = 2$

$$p = \frac{x_p - x_0}{h} = \frac{5.5 - 4}{2} = \frac{1.5}{2} = 0.75$$

First compile the table of forward differences

. . . .

18

	x	$f(x)$	Δf	$\Delta^2 f$	$\Delta^3 f$
	2	14			
			74		
$x_0 \longrightarrow$	4	88		112	
			186		48
$x_1 \longrightarrow$	6	274		160	
			346		48
	8	620		208	
			554		
	10	1174			

The Gregory–Newton forward difference interpolation formula is

$$f_p = (1 + \Delta)^p \times f_0$$

i.e. $\quad f_p = \ldots\ldots\ldots$

19

$$f_p = \left\{1 + p\Delta + \frac{p(p-1)}{1 \times 2}\Delta^2 + \frac{p(p-1)(p-2)}{1 \times 2 \times 3}\Delta^3 \ldots\right\} f_0$$
$$= f_0 + p\Delta f_0 + \frac{p(p-1)}{1 \times 2}\Delta^2 f_0 + \frac{p(p-1)(p-2)}{1 \times 2 \times 3}\Delta^3 f_0 + \ldots$$

So, substituting the relevant values from the table, gives

$$f(5.5) = f_p = \ldots\ldots\ldots$$

20

214.4

for

	x	$f(x)$	Δf	$\Delta^2 f$	$\Delta^3 f$
	2	14			
			74		
$x_0 \longrightarrow$	4	88		112	
$x_p \longrightarrow$			186		48
$x_1 \longrightarrow$	6	274		160	
			346		48
	8	620		208	
			554		
	10	1174			

$$f(5.5) = f_p = 88 + 0.75(186) + \frac{0.75(-0.25)}{1 \times 2}(160) + \frac{0.75(-0.25)(-1.25)}{1 \times 2 \times 3}(48)$$

$$= 88 + 139.5 - 15 + 1.875 = 214.375$$

$$\therefore \underline{f(5.5) = 214.4}$$

Finally, one more.

Example 3 Determine the value of $f(-1)$ from the set of function values.

x	−4	−2	0	2	4	6	8
$f(x)$	541	55	1	−53	−155	31	1225

Complete the working and then check with the next frame.

21

$$\boxed{f(-1) = 10}$$

Here is the working; method as before.

	x	$f(x)$	Δf	$\Delta^2 f$	$\Delta^3 f$	$\Delta^4 f$
	-4	541				
			-486			
$x_0 \longrightarrow$	-2	55		432		
$x_p \longrightarrow$			-54		-432	
$x_1 \longrightarrow$	0	1		0		384
			-54		-48	
	2	-53		-48		384
			-102		336	
	4	-155		288		384
			186		720	
	6	31		1008		
			1194			
	8	1225				

$x_0 = -2; \quad x_1 = 0; \quad x_p = -1; \qquad \therefore h = 2; \quad p = \frac{1}{2}$

$$f_p = f_0 + p\Delta f_0 + \frac{p(p-1)}{1 \times 2}\Delta^2 f_0 + \frac{p(p-1)(p-2)}{1 \times 2 \times 3}\Delta^3 f_0$$

$$+ \frac{p(p-1)(p-2)(p-3)}{1 \times 2 \times 3 \times 4}\Delta^4 f_0$$

$$= 55 + \frac{1}{2}(-54) + \frac{\frac{1}{2}(-\frac{1}{2})}{1 \times 2}(0) + \frac{\frac{1}{2}(-\frac{1}{2})(-\frac{3}{2})}{1 \times 2 \times 3}(-48) + \frac{\frac{1}{2}(-\frac{1}{2})(-\frac{3}{2})(-\frac{5}{2})}{1 \times 2 \times 3 \times 4}(384)$$

$$= 55 - 27 + 0 - 3 - 15 = 10 \qquad \therefore \underline{f_p = f(-1) = 10}$$

Now to something rather different.

22

Remainder theorem

If $x = a$ is substituted in the polynomial $f(x)$, the value of $f(a)$ so obtained is the remainder that would result from dividing $f(x)$ by $(x - a)$.

For, if $q(x)$ is the quotient from such division, then

$$f(x) = (x - a) \cdot q(x) + r \qquad \text{where } r \text{ is the remainder.}$$

Substituting $x = a$, we obtain $\underline{f(a) = r}$

This is the basic form of the remainder theorem.

$$\text{For example, if} \quad f(x) = 5x^3 - 8x^2 - 7x + 6$$
$$= [(5x - 8)x - 7]x + 6$$
$$\text{then} \quad f(3) = 48$$

i.e. if $5x^3 - 8x^2 - 7x + 6$ were divided by $(x - 3)$, the remainder would be 48.

In the same way, $f(2) = \ldots\ldots\ldots$

23

0

Therefore, if $f(x) = 5x^3 - 8x^2 - 7x + 6$ were divided by $(x - 2)$, the remainder would be zero.

Therefore $(x - 2)$ is $\ldots\ldots\ldots$

24

a factor of $5x^3 - 8x^2 - 7x + 6$

The basic form of the remainder theorem can be extended to include quadratic divisors.

If $f(x)$, a polynomial of degree n, is divided by $(x - a)(x - b)$, the quotient $q(x)$ will be a polynomial of degree $(n - 2)$ with a remainder of the form $(rx + s)$.

$$f(x) = (x - a)(x - b) \cdot q(x) + (rx + s)$$

Substituting $x = a \qquad f(a) = ra + s$

$x = b \qquad f(b) = rb + s$

If we solve these two equations we obtain

$$r = \frac{f(a) - f(b)}{a - b} \quad \text{and} \quad s = \frac{a \cdot f(b) - b \cdot f(a)}{a - b} \qquad a \neq b$$

so that, the remainder

$$rx + s = \frac{[f(a) - f(b)]x + a \cdot f(b) - b \cdot f(a)}{a - b} \qquad a \neq b$$

Example Determine the remainder resulting from dividing

$$x^3 + 2x^2 - 4x - 7 \text{ by } (x + 1)(x - 2)$$

Remainder $= \ldots\ldots\ldots$

25

$$\boxed{x - 1}$$

For $f(x) = x^3 + 2x^2 - 4x - 7 = [(x + 2)x - 4]x - 7$

Divisor $= (x + 1)(x - 2) \quad \therefore a = -1,\ b = 2$

$$f(a) = f(-1) = -2; \quad f(b) = f(2) = 1$$

$$\therefore rx + s = \frac{[f(a) - f(b)]x + a \cdot f(b) - b \cdot f(a)}{a - b}$$

$$= \frac{[-2 - 1]x + (-1)1 - 2(-2)}{-1 - 2} = x - 1$$

The remainder is therefore $(x - 1)$

Let us pursue this a little further. Move on.

26

Square divisors

The method we have used breaks down when $a = b$, because of the denominator $(a - b)$. When that occurs, we proceed as follows.

$$\text{Let } b = a \quad \therefore \text{divisor} = (x - a)^2$$

Then $f(x) = (x - a)^2 \cdot q(x) + (rx + s)$

Differentiating with respect to x, gives

$$f'(x) = 2(x - a) \cdot q(x) + (x - a)^2 \cdot q'(x) + r$$

Putting $x = a$ in these two, we have

$$f(a) = ra + s$$

$$f'(a) = r \qquad \therefore s = f(a) - a \cdot f'(a)$$

The remainder, $(rx + s)$, is therefore $x \cdot f'(a) + f(a) - a \cdot f'(a)$

Therefore, if $f(x)$ is divided by $(x - a)^2$, the remainder $(rx + s)$ is given by $x \cdot f'(a) + f(a) - a \cdot f'(a)$

Example Find the remainder when $f(x) = x^4 - 3x^3 - 7x^2 + 10x + 6$ is divided by $x^2 - 6x + 9$.

In nesting form, $f(x) = \dots\dots\dots$

27

$$\boxed{f(x) = \{[(x-3)x-7]x+10\}\,x+6}$$

$$f(x) = x^4 - 3x^3 - 7x^2 + 10x + 6$$

$$f'(x) = \ldots\ldots\ldots$$

$$= \ldots\ldots\ldots \text{ in nesting form}$$

28

$$\boxed{\begin{aligned} f'(x) &= 4x^3 - 9x^2 - 14x + 10 \\ &= [(4x-9)x-14]x+10 \end{aligned}}$$

The divisor is $x^2 - 6x + 9$ i.e. $(x-3)^2$ $\quad \therefore a = 3$

$$\therefore f(a) = f(3) = \ldots\ldots\ldots; \quad f'(a) = f'(3) = \ldots\ldots\ldots$$

29

$$\boxed{f(a) = f(3) = -27; \quad f'(a) = f'(3) = -5}$$

$$\left.\begin{aligned} f(a) &= f(3) = 3r + s = -27 \\ f'(a) &= f'(3) = r = -5 \end{aligned}\right\} \quad \therefore r = -5; \quad s = -12$$

$$\therefore \text{Remainder} = rx + s = -5x - 12 = \underline{-(5x+12)}$$

Move on to the next frame.

30

Here are two more examples by way of practice—the more the better.

Exercise Determine the remainder that results when

1. $x^4 + 5x^3 + 5x^2 - 4x - 2$ is divided by $(x^2 - 1)$
2. $x^4 + 4x^3 + 3x^2 - 2x - 5$ is divided by $(x^2 + 4x + 4)$.

When you have completed both, check the results with the next frame.

31

1. $(x + 4)$	2. $(2x - 1)$

Here is the working

1. $f(x) = x^4 + 5x^3 + 5x^2 - 4x - 2 = \{[(x + 5)x + 5]x - 4\}x - 2$

   Divisor $= (x^2 - 1) = (x - 1)(x + 1) \quad \therefore a = 1;\ b = -1$

   $f(a) = f(1) = 5; \qquad f(b) = f(-1) = 3$

   $$\therefore rx + s = \frac{[f(a) - f(b)]x + a \cdot f(b) - b \cdot f(a)}{a - b}$$

   $$= \frac{[5 - 3]x + 3 - (-1)5}{1 - (-1)} = \frac{2x + 8}{2}$$

   $\therefore$ Remainder $= x + 4$

2. $f(x) = x^4 + 4x^3 + 3x^2 - 2x - 5 = \{[(x + 4)x + 3]x - 2\}x - 5$

   $f'(x) = 4x^3 + 12x^2 + 6x - 2 = [(4x + 12)x + 6]x - 2$

   Divisor $= x^2 + 4x + 4 = (x + 2)^2 \quad \therefore a = -2$

   $f(a) = f(-2) = -5; \qquad f'(a) = f'(-2) = 2$

   $$\left.\begin{array}{l} f(a) = r(-2) + s = -5 \\ f'(a) = r = 2 \end{array}\right\} \quad \therefore r = 2; \quad s = -1$$

   $\therefore$ Remainder $= 2x - 1$

Now we will move on and consider polynomial equations.

Polynomial equations

32

Formation of an equation with given roots

If $x = 3$ and $x = 5$ are the roots of a quadratic equation, then $x - 3 = 0$ and $x - 5 = 0$ and hence the equation is $(x - 3)(x - 5) = 0$ which multiplies up to form $x^2 - 8x + 15 = 0$.

Similarly if the roots of an equation are 1, –2 and 4, then the equation is ....

33

$$\boxed{x^3 - 3x^2 - 6x - 8 = 0}$$

since
$$(x - 1)(x + 2)(x - 4) = 0$$
$$(x - 1)(x^2 - 2x - 8) = 0$$
$$\underline{x^3 - 3x^2 - 6x - 8 = 0}$$

In general, the equation with roots $\alpha, \beta, \gamma \ldots$ is ....

34

$$\boxed{(x - \alpha)(x - \beta)(x - \gamma) \ldots = 0}$$

So, if the roots are known to be 2, 1 + j3, 1 – j3, the equation is

....

35

$$x^3 - 4x^2 + 14x - 20 = 0$$

for $(x-2)(x-[1+j3])(x-[1-j3]) = 0$

$(x-2)([x-1]-j3)([x-1]+j3) = 0$

$(x-2)([x-1]^2+9) = 0$

$(x-2)(x^2-2x+10) = 0 \qquad \therefore \underline{x^3 - 4x^2 + 14x - 20 = 0}$

Relation between the coefficients and roots of a polynomial equation

Let α, β, γ be the roots of $x^3 + px^2 + qx + r = 0$. Then, writing the expression $x^3 + px^2 + qx + r$ in terms of α, β, γ gives

$$x^3 + px^2 + qx + r \equiv \ldots\ldots\ldots$$

36

$$(x-\alpha)(x-\beta)(x-\gamma)$$

i.e. $x^3 + px^2 + qx + r = (x-\alpha)(x-\beta)(x-\gamma)$

$= (x^2 - [\alpha+\beta]x + \alpha\beta)(x-\gamma)$

$= x^3 - (\alpha+\beta)x^2 + \alpha\beta x - \gamma x^2 + (\alpha+\beta)\gamma x - \alpha\beta\gamma$

$= x^3 - (\alpha+\beta+\gamma)x^2 + (\alpha\beta+\beta\gamma+\gamma\alpha)x - \alpha\beta\gamma$

$\therefore$ equating coefficients

(i) $\alpha+\beta+\gamma = \ldots\ldots\ldots$

(ii) $\alpha\beta+\beta\gamma+\gamma\alpha = \ldots\ldots\ldots$

(iii) $\alpha\beta\gamma = \ldots\ldots\ldots$

37

(i) $-p$; (ii) q; (iii) $-r$

This, of course, applies to a cubic equation. Let us extend this to a more general equation.

So on to the next frame.

38 In general, if $\alpha_1, \alpha_2, \alpha_3 \ldots \alpha_n$ are roots of the equation

$$p_0x^n + p_1x^{n-1} + p_2x^{n-2} + \ldots + p_{n-1}x + p_n = 0$$

then sum of the roots $= -\dfrac{p_1}{p_0}$

sum of products of the roots, two at a time $= \dfrac{p_2}{p_0}$

sum of products of the roots, three at a time $= -\dfrac{p_3}{p_0}$

sum of products of the roots, n at a time $= (-1)^n \cdot \dfrac{p_n}{p_0}$

These results are often popping up, so make a note of them.

So for the equation $3x^4 + 2x^3 + 5x^2 + 7x - 4 = 0$, if $\alpha, \beta, \gamma, \delta$ are the four roots, then

(i) $\alpha + \beta + \gamma + \delta = \ldots\ldots\ldots$

(ii) $\alpha\beta + \beta\gamma + \gamma\delta + \delta\alpha + \delta\beta + \gamma\alpha = \ldots\ldots\ldots$

(iii) $\alpha\beta\gamma + \beta\gamma\delta + \gamma\delta\alpha + \alpha\beta\delta = \ldots\ldots\ldots$

(iv) $\alpha\beta\gamma\delta = \ldots\ldots\ldots$

39

(i) $-\dfrac{2}{3}$; (ii) $\dfrac{5}{3}$; (iii) $-\dfrac{7}{3}$; (iv) $-\dfrac{4}{3}$

Now for a problem or two on the same topic.

Example 1 Solve the equation $x^3 - 8x^2 + 9x + 18 = 0$ given that the sum of two of the roots is 5.

Using the same approach as before, if α, β, γ are the roots, then

(i) $\alpha + \beta + \gamma = \ldots\ldots\ldots$

(ii) $\alpha\beta + \beta\gamma + \gamma\alpha = \ldots\ldots\ldots$

(iii) $\alpha\beta\gamma = \ldots\ldots\ldots$

40

(i) 8; (ii) 9; (iii) -18

So we have $\alpha+\beta+\gamma=8$ Let $\alpha+\beta=5$

$\therefore 5+\gamma=8 \quad \therefore \underline{\gamma=3}$

Also $\alpha\beta\gamma=-18 \quad \alpha\beta(3)=-18 \quad \therefore \alpha\beta=-6$

$\alpha+\beta=5 \quad \therefore \beta=5-\alpha \quad \therefore \alpha(5-\alpha)=-6$

$\alpha^2-5\alpha-6=0 \quad \therefore (\alpha-6)(\alpha+1)=0 \quad \therefore \alpha=-1 \text{ or } 6$

$\therefore \beta=6 \text{ or } -1$

Roots are $x=-1,\ 3,\ 6$

41

Example 2 Solve the equation $2x^3+3x^2-11x-6=0$ given that the three roots form an arithmetic sequence.

Let us represent the roots by $(a-k)$, a, $(a+k)$

Then the sum of the roots $=3a=\ldots\ldots\ldots$

and the product of the roots $=a(a-k)(a+k)=\ldots\ldots\ldots$

42

$$3a = -\frac{3}{2}; \quad a(a+k)(a-k) = \frac{6}{2} = 3$$

$\therefore a = -\frac{1}{2} \qquad -\frac{1}{2}\left(\frac{1}{4} - k^2\right) = 3 \qquad \therefore k = \pm\frac{5}{2}$

If $k = \frac{5}{2} \qquad a = -\frac{1}{2}; \qquad a - k = -3; \qquad a + k = 2$

If $k = -\frac{5}{2} \qquad a = -\frac{1}{2}; \qquad a - k = 2; \qquad a + k = -3$

$\therefore$ required roots are $-3, -\frac{1}{2}, 2$

Here is a similar one.

Example 3 Solve the equation $x^3 + 3x^2 - 6x - 8 = 0$ given that the three roots are in geometric sequence.

This time, let the roots be $\frac{a}{k}$, a, ak

Then $\frac{a}{k} + a + ak =$ .... and $\left(\frac{a}{k}\right)(a)(ak) =$ ....

43

sum of roots $= -3$; product of roots $= 8$

It then soon follows that the roots are ...., ..., ...

44

$-4, \quad 2, \quad -1$

The working rests on the relationships between the roots and the coefficients, i.e. if α, β, γ are the roots of the cubic equation

$$ax^3 + bx^2 + cx + d = 0$$

then (i) $\alpha + \beta + \gamma =$ ....

(ii) $\alpha\beta + \beta\gamma + \gamma\alpha =$ ....

(ii) $\alpha\beta\gamma =$ ....

45

(i) $-\frac{b}{a}$; (ii) $\frac{c}{a}$; (iii) $-\frac{d}{a}$

Now on to the next stage.

Transformation of Equations 46

Let us now consider the formation of equations given other types of conditions. There is no hard and fast rule, but an example or two will show the general approach.

Example 1 Form the equation whose roots are the reciprocals of the roots of the equation

$$5x^4 - 2x^3 + 4x^2 + 7x + 3 = 0$$

All we do here is to write $y = \frac{1}{x}$, i.e. put $x = \frac{1}{y}$

Then $$\frac{5}{y^4} - \frac{2}{y^3} + \frac{4}{y^2} + \frac{7}{y} + 3 = 0$$

Multiplying throughout by y^4 and reversing the order of the terms gives

$$\underline{3y^4 + 7y^3 + 4y^2 - 2y + 5 = 0}$$

Example 2 Form the equation whose roots are those of the equation

$$x^4 - 4x^3 + 8x + 3 = 0$$

each diminished by 1. Hence solve the original equation.

This time we substitute $y = x - 1$, i.e. $x = y + 1$
The equation, when simplified, then becomes ....

47

$$\boxed{y^4 - 6y^2 + 8 = 0}$$

for $(y+1)^4 - 4(y+1)^3 + 8(y+1) + 3 = 0$

$$(y^4 + 4y^3 + 6y^2 + 4y + 1) - 4(y^3 + 3y^2 + 3y + 1) + 8(y+1) + 3 = 0$$
$$y^4 + 4y^3 + 6y^2 + 4y + 1 - 4y^3 - 12y^2 - 12y - 4 + 8y + 8 + 3 = 0$$

$$\therefore \underline{y^4 - 6y^2 + 8 = 0}$$

This happens to be a quadratic in y^2 and can be solved in the usual way to give $y^2 =$ ....

48

$$\boxed{y^2 = 2 \text{ or } y^2 = 4}$$

$\therefore y = \pm\sqrt{2}$ and $y = \pm 2$, and reverting to the original equation,

$$x = \dots\dots\dots$$

49

$$\boxed{1 \pm \sqrt{2},\ 3,\ -1}$$

One more.

Example 3 Form the equation whose roots are those of the equation

$$2x^3 + 5x^2 - 4x + 1 = 0$$

each increased by 2.

Work through it and check the results with the next frame.

50

$$\boxed{2y^3 - 7y^2 + 13 = 0}$$

For, putting $y = x + 2$, i.e. $x = y - 2$, the equation becomes

$$2(y-2)^3 + 5(y-2)^2 - 4(y-2) + 1 = 0$$

$$2(y^3 - 6y^2 + 12y - 8) + 5(y^2 - 4y + 4) - 4(y-2) + 1 = 0$$

$$2y^3 - 12y^2 + 24y - 16 + 5y^2 - 20y + 20 - 4y + 8 + 1 = 0$$

$$\therefore \underline{2y^3 - 7y^2 + 13 = 0}$$

Reciprocal Equations

51

A *reciprocal equation* is an equation that is unaltered when the variable is replaced by its reciprocal.
For example, if we consider the equation

$$ax^3 + bx^2 + cx + d = 0 \qquad \text{(i)}$$

and replace x by $\dfrac{1}{x}$, then

$$\frac{a}{x^3} + \frac{b}{x^2} + \frac{c}{x} + d = 0$$

$$\therefore\ dx^3 + cx^2 + bx + a = 0 \qquad \text{(ii)}$$

If (i) and (ii) are identical, then $a = d$ and $b = c$ so that the original equation is of the form

$$ax^3 + bx^2 + bx + a = 0$$

where the coefficients are symmetrical about the centre.

Note that if $x = k$ is a root of the equation, so also is $x = \dfrac{1}{k}$ and the roots must occur in pairs. This leads us straight into two important points.

(a) If the reciprocal equation, $f(x) = 0$, is of *odd* degree and therefore has an odd number of roots, one of these must be its own reciprocal, i.e. $x = 1$ (or -1). Hence $(x - 1)$ or $(x + 1)$ is a factor of $f(x)$. The equation can then be expressed as the product of $(x - 1)$, or $(x + 1)$, and an expression of even degree equated to zero.

(b) If the reciprocal equation is of *even* degree, we use the substitution $x + \dfrac{1}{x} = t$ from which it follows that $x^2 + \dfrac{1}{x^2} = \ldots\ldots\ldots$

52

$$\boxed{x^2 + \frac{1}{x^2} = t^2 - 2}$$

for $x + \dfrac{1}{x} = t \qquad \therefore\ x^2 + 2 + \dfrac{1}{x^2} = t^2 \qquad \therefore\ x^2 + \dfrac{1}{x^2} = t^2 - 2$

So $\quad x + \dfrac{1}{x} = t$

$$x^2 + \frac{1}{x^2} = t^2 - 2$$

Proceeding on much the same lines, we can establish that

$$x^3 + \frac{1}{x^3} = \ldots\ldots\ldots$$

53

$$x^3 + \frac{1}{x^3} = t^3 - 3t$$

Since $\quad x + \dfrac{1}{x} = t$, then $\left(x + \dfrac{1}{x}\right)^3 = t^3$

$$x^3 + 3x + \frac{3}{x} + \frac{1}{x^3} = t^3$$

$$\left(x^3 + \frac{1}{x^3}\right) + 3\left(x + \frac{1}{x}\right) = t^3 \qquad \therefore\ x^3 + \frac{1}{x^3} = t^3 - 3t$$

So we have
$$x + \frac{1}{x} = t$$

$$x^2 + \frac{1}{x^2} = t^2 - 2$$

$$x^3 + \frac{1}{x^3} = t^3 - 3t$$

Make a note of these: we shall be using them straight away.

Example 1 To solve $6x^4 + 5x^3 - 38x^2 + 5x + 6 = 0$.

First we divide through by the power of x in the middle term, i.e. x^2

$$6x^2 + 5x - 38 + \frac{5}{x} + \frac{6}{x^2} = 0$$

Now pair the terms from each end of the expression

$$6\left(x^2 + \frac{1}{x^2}\right) + 5\left(x + \frac{1}{x}\right) - 38 = 0$$

Then, making the substitutions $x + \dfrac{1}{x} = t$ and $x^2 + \dfrac{1}{x^2} = t^2 - 2$ we can simplify and solve for t, getting $t = \dots\dots\dots$

54

$$t = \tfrac{5}{2} \quad \text{or} \quad -\tfrac{10}{3}$$

since $6(t^2 - 2) + 5t - 38 = 0$

$$6t^2 + 5t - 50 = 0$$

$$(3t + 10)(2t - 5) = 0 \qquad \therefore\ t = \frac{5}{2} \quad \text{or} \quad -\frac{10}{3}$$

So (i) $t = \dfrac{5}{2} \quad \therefore\ x + \dfrac{1}{x} = \dfrac{5}{2} \quad \therefore\ (2x - 1)(x - 2) = 0$

$$\therefore\ x = \frac{1}{2} \quad \text{or} \quad 2$$

(ii) $t = -\dfrac{10}{3} \quad \therefore\ x = \dots\dots\dots$

55

$$x = -\tfrac{1}{3} \quad \text{or} \quad -3$$

The given quartic equation, therefore, has the roots

$$x = -3,\ -\tfrac{1}{3},\ \tfrac{1}{2},\ 2$$

Example 2 Solve the equation $x^5 - 6x^4 + 7x^3 + 7x^2 - 6x + 1 = 0$.

We see at once that this is a reciprocal equation because ....

56

the coefficients are symmetrical about the centre

In that case, we know that .... or .... is a factor of $f(x)$

57

$(x-1)$ or $(x+1)$

If $(x-1)$ is a factor of $f(x)$, then $f(1)=0$. Try it.

$$f(1)=1-6+7+7-6+1=4 \quad \therefore (x-1) \text{ is not a factor.}$$

If $(x+1)$ is a factor of $f(x)$, then $f(-1)=0$. Try that.

$$f(-1)=-1-6-7+7+6+1=0 \quad \therefore (x+1) \text{ is a factor of } f(x)$$

Dividing therefore by $(x+1)$

$$f(x)=(x+1)(\ldots\ldots\ldots)$$

58

$x^4-7x^3+14x^2-7x+1$

$\therefore\ (x+1)(x^4-7x^3+14x^2-7x+1)=0$

$\therefore\ \underline{x=-1}$ or $x^4-7x^3+14x^2-7x+1=0$

The next step is therefore to ....

59

divide through by the power of x in the middle term, i.e. x^2

We then have $\quad x^2-7x+14-\dfrac{7}{x}+\dfrac{1}{x^2}=0$

Now, pairing up the terms from each end and applying the substitutions

$$x+\frac{1}{x}=t \quad \text{and} \quad x^2+\frac{1}{x^2}=t^2-2$$

we can solve the equation for t and hence obtain the roots of the original equation in x.

The required roots are ....

$$x = -1; \quad \tfrac{1}{2}(3 \pm \sqrt{5}); \quad 2 \pm \sqrt{3}$$ 60

Here is the working: check it.

$$\left(x^2 + \frac{1}{x^2}\right) - 7\left(x + \frac{1}{x}\right) + 14 = 0$$

$$\therefore (t^2 - 2) - 7t + 14 = 0 \qquad t^2 - 7t + 12 = 0$$

$$\therefore t = 3 \quad \text{or} \quad t = 4$$

For $t = 3$, $\quad x + \dfrac{1}{x} = 3 \quad \therefore\ x^2 - 3x + 1 = 0 \quad \therefore x = \dfrac{1}{2}(3 \pm \sqrt{5})$

For $t = 4$, $\quad x + \dfrac{1}{x} = 4 \quad \therefore x^2 - 4x + 1 = 0 \quad \therefore x = 2 \pm \sqrt{3}$

Therefore, the five roots of the original equation are

$$\underline{x = -1,\ \tfrac{1}{2}(3 \pm \sqrt{5}),\ 2 \pm \sqrt{3}}$$

Repeated roots 61

The cubic equation $x^3 - x^2 - 8x + 12 = 0$ can be written in the form $(x - 2)(x - 2)(x + 3) = 0$ and therefore has a repeated root, $x = 2$, and a third root, $x = -3$. It is said to have a repeated root $x = 2$ of *multiplicity* 2 (since there are two of them).

The quartic equation $x^4 + x^3 - 9x^2 + 11x - 4 = 0$ can be written as $(x - 1)(x - 1)(x - 1)(x + 4) = 0$ and therefore has .... and a ....

---

repeated root ($x = 1$) of multiplicity 3; fourth root $x = -4$ 62

Let us now see how we can detect repeated roots when the function $f(x)$ of the equation $f(x) = 0$ is not conveniently factorised ready for us.

On then to the next frame.

63 Determination of repeated roots of multiplicity 2

If $x = k$ is a repeated root of the equation $f(x) = 0$, it will also be a root of the equation $f'(x) = 0$, where $f'(x)$ is the first derivative of $f(x)$.

For, if
$$f(x) = (x - k)(x - k)F(x) = 0 \quad \text{(i)}$$
$$= (x - k)^2 \cdot F(x) = 0$$
then
$$f'(x) = (x - k)^2 \cdot F'(x) + 2(x - k) \cdot F(x) = 0$$
$$= (x - k)\{(x - k) \cdot F'(x) + 2F(x)\} = 0 \quad \text{(ii)}$$

We can see that $x = k$ is a repeated root of (i)

and that $x = k$ is a single root of (ii)

Let us illustrate this with an example.

Example 1 If $x^3 + x^2 - 16x + 20 = 0$ is known to have a repeated root $x = k$, then $x = k$ will also be a root of the equation

....

64

$$\boxed{3x^2 + 2x - 16 = 0}$$

$\therefore$ Substituting $x = k$ in the two equations, we have

	$k^3 + k^2 - 16k + 20 = 0$	(i)
	$3k^2 + 2k - 16 = 0$	(ii)
(i) $\times$ 3	$3k^3 + 3k^2 - 48k + 60 = 0$	(iii)
(ii) $\times$ k	$3k^3 + 2k^2 - 16k = 0$	(iv)
Subtract	$\therefore k^2 - 32k + 60 = 0$	(v)
(v) $\times$ 3	$3k^2 - 96k + 180 = 0$	
(ii)	$3k^2 + 2k - 16 = 0$	
Subtract	$\therefore -98k + 196 = 0 \quad \therefore k = 2$	

$\therefore$ $\underline{x = 2 \text{ is a repeated root of } x^3 + x^2 - 16x + 20 = 0}$

Example 2 Given that $x^3 + 5x^2 + 3x - 9 = 0$ has a repeated root $x = k$, determine the value of k.

This is the same as before. The two initial equations we need are

....

....

65

$$x^3 + 5x^2 + 3x - 9 = 0$$
$$3x^2 + 10x + 3 = 0$$

Now we substitute $x = k$ in both equations and proceed to eliminate the highest power of k at each stage until we arrive at

$$k = \ldots\ldots\ldots$$

66

$$k = -3$$

Here it is

	$k^3 + 5k^2 + 3k - 9 = 0$	(i)
	$3k^2 + 10k + 3 = 0$	(ii)
(i) $\times 3$	$3k^3 + 15k^2 + 9k - 27 = 0$	(iii)
(ii) $\times k$	$3k^3 + 10k^2 + 3k = 0$	(iv)
Subtract	$\therefore\ 5k^2 + 6k - 27 = 0$	(v)
(ii) $\times 5$	$15k^2 + 50k + 15 = 0$	(vi)
(v) $\times 3$	$15k^2 + 18k - 81 = 0$	(vii)
Subtract	$\therefore\ 32k + 96 = 0 \qquad \therefore \underline{k = -3}$	

With a cubic equation that has a repeated root, the method is always the same, so that an expression for the repeated root $x = k$ can be obtained in terms of the coefficients.

In general, if $x^3 + ax^2 + bx + c = 0$ is known to have a repeated root $x = k$, then $3x^2 + 2ax + b = 0$ also has a root $x = k$.

	$k^3 + ak^2 + bk + c = 0$	(i)
	$3k^2 + 2ak + b = 0$	(ii)
(i) $\times 3$	$3k^3 + 3ak^2 + 3bk + 3c = 0$	(iii)
(ii) $\times k$	$3k^3 + 2ak^2 + bk = 0$	(iv)

If we now substract and then continue the elimination process, we arrive at a formula for k in terms of the coefficients.

$$k = \ldots\ldots\ldots$$

Finish it off.

67

$$k = \frac{9c - ab}{2a^2 - 6b}$$

For, subtracting (iv) from (iii) gives

$$ak^2 + 2bk + 3c = 0 \qquad \text{(v)}$$

(ii) $\times a$ $\quad 3ak^2 + 2a^2k + ab = 0 \qquad$ (vi)

(v) $\times 3$ $\quad 3ak^2 + 6bk + 9c = 0 \qquad$ (vii)

Subtract $\quad \therefore (2a^2 - 6b)k + ab - 9c = 0$

$$\therefore k = \frac{9c - ab}{2a^2 - 6b}$$

WARNING: This result involving the coefficients *a, b, c,* applies only when it is known that the cubic equation $x^3 + ax^2 + bx + c = 0$ has a repeated root $x = k$. It must not be used for cubic equations generally, but is useful on occasions. Make a note of it then, but Beware!

68 Complex roots

The roots of some polynomial equations with real coefficients are complex. For example, $x^3 - 4x^2 - 6x - 20 = 0$ can be factorised into

$$(x - 2)(x^2 - 2x + 10) = 0$$

Hence $x = 2$ or $x^2 - 2x + 10 = 0$ giving $x = 1 \pm \text{j}3$.

In this case, the three roots are

$$x = 2; \quad x = 1 + \text{j}3; \quad x = 1 - \text{j}3$$

Note that where complex roots occur, they always do so in *conjugate pairs*, otherwise the coefficients themselves of the original equation would be complex.

And that brings us to the end of this particular programme on Theory of Equations. There will be more to come in Programme 2, but, for the time being, check down the Revision Summary that follows and revise anything that you think needs it. Then work through the Test Exercise, which is straightforward and set on the topics with which we have been dealing.

REVISION SUMMARY

69

1. *Function evaluation* — nesting

$$f(x) = ax^3 + bx^2 + cx + d$$
$$= [(ax + b)x + c]x + d$$

2. *Interpolation* (a) Linear

(b) Graphical

(c) Gregory–Newton forward differences

$$\Delta f_0 = f(x_1) - f(x_0)$$

$$h \updownarrow \quad \begin{array}{ll} x_0 & f(x_0) = f_0 \\ x_p \rightarrow & f_p \\ x_1 & f(x_1) = f_1 \end{array}$$

$$f_p = f_0 + p\Delta f_0 + \frac{p(p-1)}{2!}\Delta^2 f_0 + \frac{p(p-1(p-2)}{3!}\Delta^3 f_0 + \dots$$
$$= \left\{1 + p\Delta + \frac{p(p-1)}{2!}\Delta^2 + \frac{p(p-1)(p-2)}{3!}\Delta^3 + \dots\right\} f_0$$
$$= (1 + \Delta)^p f_0$$

3. *Remainder theorem*

If $f(x) = a_1x^n + a_2x^{n-1} + \dots + a_nx + a_{n+1}$

then $f(k) =$ the remainder resulting from dividing $f(x)$ by $(x - k)$.

If $f(k) = 0$, then $(x - k)$ is a factor of $f(x)$.

If $f(x)$ is divided by $(x - a)(x - b)$, the remainder $(rx + s)$ is given

by $$r = \frac{f(a) - f(b)}{a - b}; \quad s = \frac{a \cdot f(b) - b \cdot f(a)}{a - b}$$

If $a = b$, i.e. if $f(x)$ is divided by $(x - a)^2$, the remainder $(rx + s)$ is

given by $$x \cdot f'(a) + f(a) - a \cdot f'(a)$$

4. *Equation with given roots*

If $\alpha, \beta, \gamma \dots$ are roots, then the equation is

$$(x - \alpha)(x - \beta)(x - \gamma)\dots = 0$$

5. *Relation between coefficients and roots*

If α, β, γ are the roots of $x^3 + px^2 + qx + r = 0$

$$\alpha + \beta + \gamma = -p$$
$$\alpha\beta + \beta\gamma + \gamma\alpha = q$$
$$\alpha\beta\gamma = -r$$

In general, if $\alpha, \beta, \gamma \ldots$ are roots of the equation

$$p_0x^n + p_1x^{n-1} + p_2x^{n-2} + \ldots + p_{n-1}x + p_n = 0$$

then sum of the roots $= -\dfrac{p_1}{p_0}$

sum of the products of roots, taken two at a time $= -\dfrac{p_2}{p_0}$

sum of the products of roots, taken three at a time $= -\dfrac{p_3}{p_0}$

sum of products of roots, n at a time $= (-1)^n \dfrac{p_n}{p_0}$

6. *Transformation of equations*

$$f(x) = ax^4 + bx^3 + cx^2 + dx + c = 0$$

If the new roots are the old roots increased by k, i.e. $y = x + k$, substitute $x = y - k$ and simplify

$$f(y) = a(y - k)^4 + b(y - k)^3 + c(y - k)^2 + d(y - k) + c = 0 \quad \text{etc.}$$

7. *Reciprocal equations* — unaltered when the variable is replaced by its reciprocal. Coefficients are symmetrical about the centre.

If $x = k$ is a root, so also is $x = \dfrac{1}{k}$.

(a) *Odd degree*: If the reciprocal equation $f(x) = 0$ is of odd degree

$$x = 1 \quad \text{or} \quad x = -1 \quad \text{is a root.}$$

(b) *Even degree*: If the reciprocal equation $f(x) = 0$ is of even degree

$$\text{substitute } x + \frac{1}{x} = t$$

$$x^2 + \frac{1}{x^2} = t^2 - 2$$

$$x^3 + \frac{1}{x^3} = t^3 - 3t \quad \text{etc.}$$

8. *Repeated roots* — roots of the same value.

If $x = k$ is a repeated root of $f(x) = 0$ it is also a root of $f'(x) = 0$.

For a cubic $x^3 + ax^2 + bx + c = 0$, if $x = k$ is a repeated root

$$\text{then } k = \frac{9c - ab}{2a^2 - 6b}.$$

9. *Complex roots*, when they occur, do so in conjugate pairs.

Now on to the Test Exercise.

TEST EXERCISE I

70

1. Rewrite the function $f(x) = 3x^4 + 5x^3 - 4x^2 + x - 2$ in 'nested' form and hence evaluate $f(2.5)$ giving the result to four significant figures.

2. A polynomial function is defined by the following set of function values.

x	2	4	6	8	10
$y = f(x)$	−7.00	9.00	97.0	305	681

   Use the Gregory–Newton formula to evaluate $f(4.8)$.

3. Without dividing in full, determine the remainder which occurs when $x^4 - x^3 - 7x^2 + 3x + 5$ is divided by $(x^2 - 2x - 3)$.

4. Solve the equation $x^3 - 5x^2 - 8x + 12 = 0$, given that the sum of two of the roots is 7.

5. Form the equation whose roots are those of the equation $x^3 + x^2 + 9x + 9 = 0$ each increased by 2.

6. Solve the equation $6x^4 - 5x^3 - 38x^2 - 5x + 6 = 0$.

7. If the equation $4x^3 + 8x^2 - 35x - 75 = 0$ is known to have a repeated root $x = k$, determine the value of k.

FURTHER PROBLEMS I

1. Find the values of the constants p and q such that the function $f(x) = 2x^3 + px^2 + qx + 6$ may be exactly divisible by $(x - 2)(x + 1)$.

2. If $f(x) = 4x^4 + px^3 - 23x^2 + qx + 11$ and when $f(x)$ is divided by $2x^2 + 7x + 3$ the remainder is $3x + 2$, determine the values of p and q.

3. If one root of the equation $x^3 - 2x^2 - 9x + 18 = 0$ is the negative of another, determine the three roots.

4. Solve the equation $x^3 - 7x^2 - 21x + 27 = 0$, given that the roots form a geometric sequence.

5. Form the equation whose roots exceed by 3 the roots of the equation $x^3 - 4x^2 + x + 6 = 0$.

6. If the equation $4x^3 - 4x^2 - 5x + 3 = 0$ is known to have two roots whose sum is 2, solve the equation.

7. Solve the equation $x^3 - 10x^2 + 8x + 64 = 0$, given that the product of two of the roots is the negative of the third.

8. Form the equation whose roots exceed those of the equation $2x^3 - 3x^2 - 11x + 6 = 0$ by 2.

9. If α, β, γ are the roots of the equation $x^3 + px^2 + qx + r = 0$, prove that $\alpha^2 + \beta^2 + \gamma^2 = p^2 - 2q$.

10. Solve the reciprocal equation $x^4 + 6x^3 - 25x^2 + 6x + 1 = 0$.

11. Solve the equation $x^5 + 3x^4 - 11x^3 - 11x^2 + 3x + 1 = 0$.

12. The equation $4x^3 + 8x^2 - 11x + 3 = 0$ is known to have a repeated root $x = k$. Determine the value of k and hence solve the equation.

13.

x	4	5	6	7	8	9	10
$f(x)$	−10	12	56	128	234	380	572

For the function $f(x)$ indicated, apply the Gregory–Newton interpolation method to evaluate

(a) $f(4.5)$ (b) $f(6.4)$.

14.

x	2	4	6	8	10	12
$f(x)$	−9	35	231	675	1463	2691

For the function defined in the table above, evaluate (a) $f(2.6)$ and (b) $f(7.2)$.

15. A function $f(x)$ is defined by the following table

x	−4	−2	0	2	4	6	8
$f(x)$	277	51	1	−17	−147	−533	−1319

Determine the values of (a) $f(-3)$ and (b) $f(1.6)$.

Programme 2

Theory of Equations

PART 2

Cubic Equations

1

The cubic equation $ax^3 + bx^2 + cx + d = 0$ has three different sets of solutions depending on the values of the coefficients.

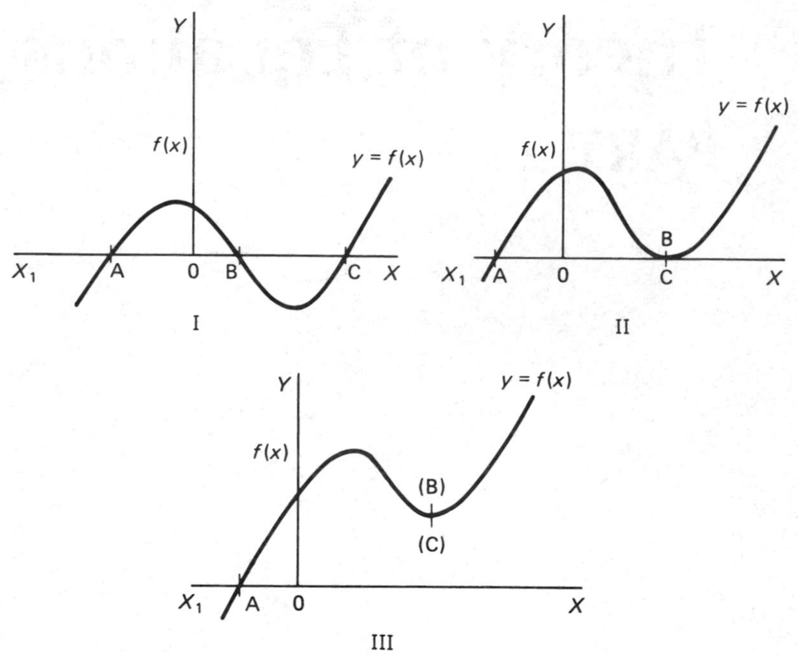

In I, the equation $f(x) = 0$ has *three real and different* roots, given by the values of x at A, B and C.

In II, the equation $f(x) = 0$ has *three real roots* at A, B and C, *two of which are identical*, i.e. at B and C.

In III, the equation $f(x) = 0$ has *one real root* at A and two complex roots corresponding to B and C—these being a *pair of conjugate complex roots*.

Whatever the form of the roots of a cubic equation, one root at least is always ....

2

real

If the one real root, $x = k$, can be found, $f(x)$ can be factorised into $(x-k)(ax^2+px+q)$ by long division, so that the equation $f(x) = 0$ becomes $(x-k)(ax^2+px+q) = 0$.
$\therefore$ $\underline{x = k}$ or $ax^2+px+q = 0$ which can always be solved by factors or formula.

Example 1 To solve $x^3 - x^2 + 2x + 4 = 0$.

First we seek a linear factor by using the remainder theorem

$$f(x) = x^3 - x^2 + 2x + 4 = [(x-1)x+2]x+4$$
$$f(1) = \ldots\ldots\ldots$$

3

$f(1) = 6$

$\therefore$ $(x-1)$ is not a factor of $f(x)$. However $f(-1) = 0$
so that ....

4

$(x+1)$ is a factor of $f(x)$

If we now divide $f(x)$ by $(x+1)$, we obtain the remaining quadratic factor which is ....

5

$x^2 - 2x + 4$

So we now have $(x+1)(x^2-2x+4) = 0$ giving $x = -1$, $x = 1 \pm \mathrm{j}3$.

$\therefore$ The required roots are $\underline{x = -1;\ x = 1 + \mathrm{j}3;\ x = 1 - \mathrm{j}3}$

Example 2 To solve $x^3 + 3x^2 - 6x + 2 = 0$.

Proceeding as before,

(i) find a linear factor of $f(x)$ by the remainder theorem
(ii) by long division, find the resulting quadratic factor
(iii) so determine the three roots of the particular cubic.

By trial and error, substituting $x = 1, -1, 2, -2$, etc., we use the remainder theorem and find the linear factor of $f(x)$ which is ....

6

$$\boxed{x-1}$$

for $f(x) = x^3 + 3x^2 - 6x + 2$ i.e. $[(x+3)x - 6]x + 2$

$f(1) = 0 \qquad \therefore (x-1)$ is a factor of $f(x)$.

Dividing out to find the quadratic factor, and then solving in the normal way, we obtain $\underline{x = 1; \quad x = -2 + \sqrt{6}; \quad x = -2 - \sqrt{6}}$

Example 3 Solve $2x^3 - x^2 - 9x + 2 = 0$.

This again is all very simple, the final results being ....

7

$$\boxed{x = -2; \quad \frac{5 \pm \sqrt{17}}{4} \quad \text{i.e.} -2; \quad 2.2808; \quad 0.2192}$$

Note: The real root of a cubic equation is not always a simple whole number to be found by the remainder theorem. Also there is no general formula by which the three roots can be evaluated. Other means, therefore, have to be found, *so let us start a new frame.*

8

Tartaglia's solution for the real root

In the sixteenth century, Tartaglia discovered that the real root of the cubic equation $x^3 + ax + b = 0$, where $a > 0$, is given by

$$x = \left\{-\frac{b}{2} + \sqrt{\frac{a^3}{27} + \frac{b^2}{4}}\right\}^{1/3} + \left\{-\frac{b}{2} - \sqrt{\frac{a^3}{27} + \frac{b^2}{4}}\right\}^{1/3}$$

That looks pretty formidable, but it is a good deal easier than it appears. Notice that $\frac{b}{2}$ and $\sqrt{\frac{a^3}{27} + \frac{b^2}{4}}$ occur twice and it is convenient to evaluate these first and then substitute the results in the main expression for x.

Example 1 Find the real root of $x^3 + 2x + 5 = 0$.

Here, $a = 2,\ b = 5 \qquad \therefore \frac{b}{2} = 2.5$

$$\sqrt{\frac{a^3}{27} + \frac{b^2}{4}} = \sqrt{\frac{8}{27} + \frac{25}{4}} = \sqrt{6.5463} = 2.5586$$

Then $x = (-2.5 + 2.5586)^{1/3} + (-2.5 - 2.5586)^{1/3}$

$= 0.3884 - 1.7166 = 1.3282 \qquad \underline{x = -1.328}$

Once we have a real root, the equation can be reduced to a quadratic and the remaining two roots determined.

Example 2 Determine the real root of $2x^3 + 3x - 4 = 0$.

This is first written $x^3 + 1.5x - 2 = 0 \quad \therefore a = 1.5,\ b = -2$

Now you can evaluate $\frac{b}{2}$ and $\sqrt{\frac{a^3}{27} + \frac{b^2}{4}}$ and so determine

$x = \ldots\ldots\ldots$

9

$$x = 0.8796$$

for $\frac{b}{2} = -1$ and $\sqrt{\frac{a^3}{27} + \frac{b^2}{4}} = 1.0607$ and the rest follows.

You will already have noticed the restriction in Tartaglia's method that $a > 0$, i.e. the coefficient of the x term must be positive. What if it is negative? Well, there are other ways of tackling the problem, as we shall see, so let us make a fresh start.

10

Trigonometrical solution of equations of the form $x^3 + px + q = 0$

We have already noted that every cubic equation has at least one real root. Let us consider the three standard cases.

(a) *Three real and different roots*

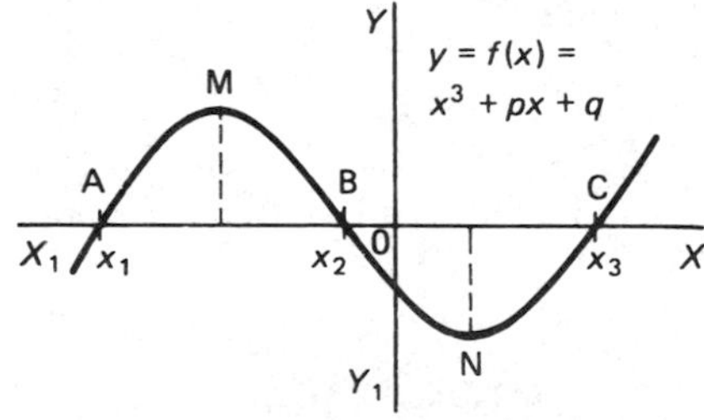

Roots x_1, x_2, x_3 are given at points A, B, C, as shown. The turning points, M and N, occur where $\frac{dy}{dx} = 0$, i.e. where

$$3x^2 + p = 0 \quad \therefore x = \pm\sqrt{\frac{-p}{3}}$$

Therefore, for the two values of x to be real, *p must be negative.*

$$\therefore \text{ At M and N, } y = \pm\left(\frac{-p}{3}\right)\sqrt{-\frac{p}{3}} \pm p\sqrt{-\frac{p}{3}} + q$$

$$= q \pm \frac{2p}{3}\sqrt{-\frac{p}{3}}$$

$$\therefore y = q + \frac{2p}{3}\sqrt{-\frac{p}{3}} \quad \text{and} \quad y = q - \frac{2p}{3}\sqrt{-\frac{p}{3}}$$

If these y-values of M and N are of opposite signs, as in the figure, the product of the ordinates must be negative, that is

....

11

$$\boxed{4p^3 + 27q^2 < 0}$$

for $\left(q + \frac{2p}{3}\sqrt{-\frac{p}{3}}\right)\left(q - \frac{2p}{3}\sqrt{-\frac{p}{3}}\right) < 0$

$$\therefore\ q^2 - \frac{4p^2}{9}\left(-\frac{p}{3}\right) < 0 \qquad \text{i.e. } \underline{4p^3 + 27q^2 < 0}$$

So, if (a) p is negative
and (b) $4p^3 + 27q^2 < 0$
then $x^3 + px + q = 0$ will have three real and different roots.

Note: The expression $4p^3 + 27q^2$ is called the *discriminant* and is akin to the expression $b^2 - 4ac$ in the solution of a quadratic equation.

(b) *Three real roots, two identical*

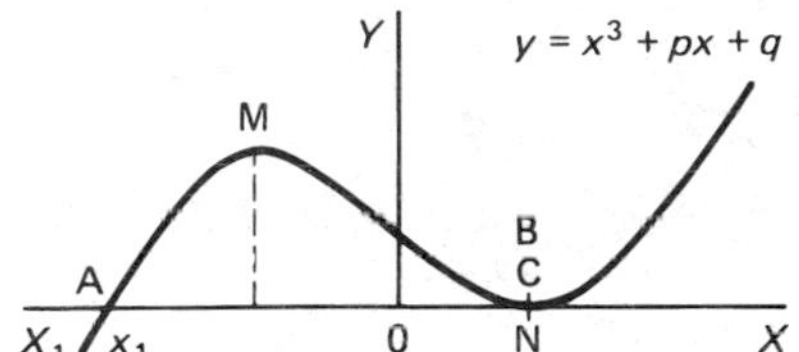

In this case, the y-value of one of the turning points is zero. The discriminant is therefore ....

12

$$\boxed{\text{zero} \quad \text{i.e.} \quad 4p^3 + 27q^2 = 0}$$

(c) *One real root and two complex roots*

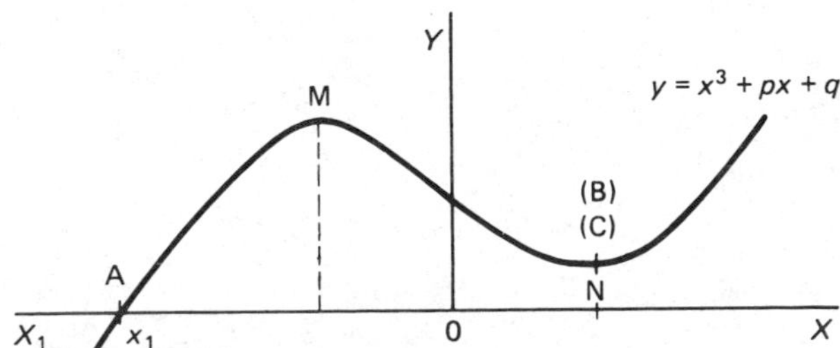

Here the y-values of the turning points M and N are of the same sign (both positive or both negative). The product of the ordinates is therefore positive. Therefore ....

13

$$\boxed{4p^3 + 27q^2 > 0}$$

So to solve a particular equation of the form $x^3 + px + q = 0$, we first investigate the value of the discriminant $4p^3 + 27q^2$. This will tell us the type of roots the equation has.

If $4p^3 + 27q^2 < 0$ ....

If $4p^3 + 27q^2 = 0$ ....

If $4p^3 + 27q^2 > 0$ ....

14

$4p^3 + 27q^2 < 0$ —3 real and different roots
$4p^3 + 27q^2 = 0$ —3 real roots, 2 identical
$4p^3 + 27q^2 > 0$ —1 real root, 2 conjugate complex roots

Example 1 Investigate the roots of $x^3 - 5x - 2 = 0$.

If we compare this with the general form $x^3 + px + q = 0$, then the value of the discriminant is ....

15

$$\boxed{-392}$$

for $4p^3 + 27q^2 = 4(-5)^3 + 27(-2)^2 = -392$ i.e. < 0

Therefore, the equation has ....

16

three real and different roots

That is true enough, but now we have to find them. We do this by using the trigonometrical substitution $x = \lambda \cos\theta$.

Then $\qquad x^3 - 5x \qquad - 2 = 0$

becomes $\qquad \lambda^3 \cos^3\theta - 5\lambda\cos\theta \quad - 2 = 0$

We now make use of the identity $\cos 3\theta = 4\cos^3\theta - 3\cos\theta$.
So we have the two equations

$$\lambda^3 \cos^3\theta - 5\lambda\cos\theta - 2 = 0$$

$$4\cos^3\theta - 3\cos\theta - \cos 3\theta = 0$$

Comparing coefficients of powers of $\cos\theta$

$$\frac{\lambda^3}{4} = \dots\dots\dots$$

17

$$\boxed{\frac{\lambda^3}{4} = \frac{-5\lambda}{-3} = \frac{-2}{-\cos 3\theta}}$$

That is, $\dfrac{\lambda^3}{4} = \dfrac{5\lambda}{3} = \dfrac{2}{\cos 3\theta}$

From the first two, we have

$$\frac{\lambda^3}{4} = \frac{5\lambda}{3} \quad \therefore\ 3\lambda^3 = 20\lambda \quad \therefore\ \lambda = 0 \quad \text{or} \quad \lambda^2 = \frac{20}{3}$$

$\lambda = 0$ would give x permanently zero and is therefore neglected.

$$\therefore \quad \lambda^2 = \frac{20}{3} \qquad \lambda = \pm\sqrt{\frac{20}{3}} = \pm 2.5820$$

Taking $\lambda = 2.5820$ and using the second relationship

$$\frac{5(2.5820)}{3} = \frac{2}{\cos 3\theta} \qquad \therefore\ \cos 3\theta = 0.46476$$

$$\theta = \dots\dots\dots;\ \dots\dots\dots;\ \dots\dots\dots$$

18

$$\theta = 20° \, 46'; \quad 99° \, 14'; \quad 140° \, 46'$$

for $\cos 3\theta = 0.46476 \therefore 3\theta = 62° \, 18'; \quad 297° \, 42'; \quad 422° \, 18'$

$\therefore \quad \theta = 20° \, 46'; \quad 99° \, 14'; \quad 140° \, 46'$

Finally, since $\lambda = 2.5820$ and $x = \lambda \cos\theta$
then $x = \ldots\ldots\ldots; \ \ldots\ldots\ldots; \ \ldots\ldots\ldots$

19

$$x = 2.4143; \quad -0.4143; \quad -2.0000$$

If we had taken the negative value for λ, we should have had different values for θ, but the final values for x would have been the same.

Example 2 Solve the equation $4x^3 - 3x + 1 = 0$.

We first divide through by 4 to give unity coefficient for x^3

$$\therefore x^3 - \frac{3}{4}x + \frac{1}{4} = 0 \qquad \text{i.e. } p = -\frac{3}{4}; \quad q = \frac{1}{4}$$

The discriminant $4p^3 + 27q^2 = \ldots\ldots\ldots$

20

$$0$$

So we know that the equation has ……… roots

21

3 real roots, two of which are identical

As before, we put $x = \lambda\cos\theta$ in the equation and compare the result with the identity $\cos 3\theta = 4\cos^3\theta - 3\cos\theta$.

$$\lambda^3\cos^3\theta - \frac{3}{4}\lambda\cos\theta + \frac{1}{4} = 0$$

$$4\cos^3\theta - 3\cos\theta - \cos 3\theta = 0$$

Therefore, comparing coefficients ………

22

$$\boxed{\frac{\lambda^3}{4} = \frac{\lambda}{4} = -\frac{1}{4\cos 3\theta}}$$

$$\therefore \frac{\lambda^3}{4} = \frac{\lambda}{4} \quad \therefore \lambda^3 = \lambda \quad \therefore \lambda = 0 \quad \text{or} \quad \lambda^2 = 1 \quad \therefore \text{ Use } \lambda = \pm 1.$$

$$\frac{\lambda}{4} = -\frac{1}{4\cos 3\theta} \qquad \therefore \text{ using } \lambda = 1, \qquad \cos 3\theta = -1$$

From this we obtain 3 values for θ and hence use $x = \lambda \cos\theta$ to obtain

$x = \ldots\ldots\ldots; \ \ldots\ldots\ldots; \ \ldots\ldots\ldots$

23

$$\boxed{x = -1; \quad 0.5; \quad 0.5}$$

for $\cos 3\theta = -1 \quad \therefore 3\theta = 180°, 540°, 900°$

$\therefore \theta = 60°, 180°, 300°$

$\therefore x = (1)\cos 60°, \quad (1)\cos 180°, \quad (1)\cos 300°$

$\underline{x = 0.5; \ -1; \ 0.5}$

Now for an example that is rather different.

Example 3 Solve the equation $x^3 - 4x + 6 = 0$.

In this case, $p = -4$, $q = 6 \quad \therefore 4p^3 + 27q^2 = 716$ i.e. > 0

The equation therefore has ………

24

one real and two complex roots (conjugate pair)

Working as before, we put $x = \lambda \cos\theta$ giving

$$\lambda^3 \cos^3\theta - 4\lambda\cos\theta + 6 = 0$$

and, as usual, $4\cos^3\theta - 3\cos\theta - \cos 3\theta = 0$

Equating coefficients $\dfrac{\lambda^3}{4} = \dfrac{4\lambda}{3} = -\dfrac{6}{\cos 3\theta}$

From the first pair, $3\lambda^3 = 16\lambda \quad \therefore \lambda = 0$ or $\lambda^2 = \dfrac{16}{3} \quad \therefore \lambda = 2.3094$

From the second pair, $\dfrac{4(2.3094)}{3} = -\dfrac{6}{\cos 3\theta}$

$$\therefore \cos 3\theta = \frac{-18}{4(2.3094)} = -1.9484$$

and there is no real angle with a cosine of -1.9484! and we seem to have come to a sticky end.

We therefore return to the beginning and substitute $x = \lambda \cosh\theta$ and use the corresponding hyperbolic identity

$$\cosh 3\theta = 4\cosh^3\theta - 3\cosh\theta$$

Doing that we find $\lambda = \dots\dots\dots$

25

$$\boxed{\lambda = 2.3094}$$

for $$x^3 - 4x + 6 = 0$$

$$\therefore\ \lambda^3 \cosh^3\theta - 4\lambda\cosh\theta + 6 = 0$$

$$4\cosh^3\theta - 3\cosh\theta - \cosh 3\theta = 0$$

$$\therefore\ \frac{\lambda^3}{4} = \frac{4\lambda}{3} = -\frac{6}{\cosh 3\theta}$$

$$\lambda^3 = \frac{16\lambda}{3} \quad \therefore\ \lambda = 0 \text{ or } \lambda^2 = \frac{16}{3} \quad \therefore\ \lambda = \pm 2.3094$$

This, you will notice, is the same value for λ as we had before.

Now for the second part, we have

$$\frac{4\lambda}{3} = -\frac{6}{\cosh 3\theta} \quad \therefore\ \cosh 3\theta = -\frac{18}{4(2.3094)} = -1.9486$$

$$\therefore\ \frac{e^{3\theta} + e^{-3\theta}}{2} = -1.9486 \quad \therefore\ e^{3\theta} + \frac{1}{e^{3\theta}} = -3.8972$$

$$\therefore\ (e^{3\theta})^2 + 3.8972(e^{3\theta}) + 1 = 0$$

which is a quadratic in $e^{3\theta}$, from which we get

$$e^{3\theta} = \ldots\ldots\ldots$$

26

$$\boxed{e^{3\theta} = -0.2762 \quad \text{or} \quad -3.6210}$$

If we now take $e^{3\theta} = -0.2762$, this can be written

$$e^{3\theta} = 0.2762\,e^{j\pi};\ \ 0.2762\,e^{j3\pi};\ \text{ or }\ 0.2762\,e^{j5\pi}$$

$$\therefore\ e^{\theta} = (0.2762)^{1/3}\,e^{j\pi/3};\ (0.2762)^{1/3}\,e^{j\pi};\ \text{or } (0.2762)^{1/3}\,e^{j5\pi/3}$$

$$= \ldots\ldots\ldots;\ \ldots\ldots\ldots;\ \text{or }\ \ldots\ldots\ldots$$

27

$$e^{\theta} = 0.6512\,e^{j\pi/3}; \quad 0.6512\,e^{j\pi}; \quad \text{or} \quad 0.6512\,e^{j5\pi/3}$$

$$\begin{aligned} e^{-\theta} &= (0.6512)^{-1}e^{-j\pi/3};\ (0.6512)^{-1}e^{-j\pi};\ \text{or}\ (0.6512)^{-1}e^{-j5\pi/3} \\ &= 1.5356\,e^{-j\pi/3};\ 1.5356\,e^{-j\pi};\ \text{or}\ 1.5356\,e^{-j5\pi/3} \end{aligned}$$

Now $e^{j\theta} = \cos\theta + j\sin\theta$ and $e^{-j\theta} = \cos\theta - j\sin\theta$

$$\begin{aligned} e^{j\pi/3} &= \dots\dots\dots; \quad e^{-j\pi/3} = \dots\dots\dots \\ e^{j\pi} &= \dots\dots\dots; \quad e^{-j\pi} = \dots\dots\dots \\ e^{j5\pi/3} &= \dots\dots\dots; \quad e^{-j5\pi/3} = \dots\dots\dots \end{aligned}$$

28

$$\begin{aligned} e^{j\pi/3} &= 0.5 + j0.8660; \quad e^{-j\pi/3} = 0.5 - j0.8660 \\ e^{j\pi} &= -1 \quad ; \quad e^{-j\pi} = -1 \\ e^{j5\pi/3} &= 0.5 - j0.8660; \quad e^{-j5\pi/3} = 0.5 + j0.8660 \end{aligned}$$

So, making use of these intermediate results, we can write

$$\begin{aligned} 2\cosh\theta = e^{\theta} + e^{-\theta} &= 0.6512(0.5 + j0.8660) + 1.5365(0.5 - j0.8660) \\ &= 1.0934 - j0.7659 \end{aligned}$$

$$\therefore \underline{\cosh\theta = 0.5467 - j0.3830}$$

or $2\cosh\theta = e^{\theta} + e^{-\theta} = 0.6512(-1) + 1.5356(-1) = -2.1868$

$$\therefore \underline{\cosh\theta = -1.0934}$$

or $2\cosh\theta = e^{\theta} + e^{-\theta} = 0.6512(0.5 - j0.8660) + 1.5356(0.5 + j0.8660)$

$$= 1.0934 + j0.7659$$

$$\therefore \underline{\cosh\theta = 0.5467 + j0.3830}$$

We therefore have three values for $\cosh\theta$ and since $x = \lambda\cosh\theta$ with $\lambda = 2.3094$ then

$$x = \dots\dots\dots; \ \dots\dots\dots; \ \dots\dots\dots$$

29

$$x = 1.2625 - j0.8845; \quad -2.5251; \quad 1.2625 + j0.8845$$

i.e. $\quad x = -2.5251; \quad 1.2625 \pm j0.8845$

i.e. one real root and two complex roots that are a conjugate pair.

Note The substitution $x = \lambda \cosh \theta$ is necessary in the equation $x^3 + px + q = 0$ when $4p^3 + 27q^2 > 0$ and *p is negative*, as in the example we have just worked.

When $4p^3 + 27q^2 > 0$ and *p is positive*, we use the substitution $x = \lambda \sinh \theta$ as in the next example.

Example 4 Solve the equation $x^3 + 3x + 3 = 0$.

In this case, $p = 3$, $q = 3$ and $4p^3 + 27q^2 = 351$, i.e. > 0. *p is positive.*
$\therefore$ We substitute $x = \lambda \sinh \theta$ and use the identity

$$\sinh 3\theta = 4\sinh^3 \theta + 3\sinh \theta$$

Proceeding as before, we arrive at the stage

$$e^{3\theta} = \ldots\ldots\ldots$$

30

$$\boxed{e^{3\theta} = 0.3028}$$

Here is the working to date

$$\lambda^3 \sinh^3\theta + 3\lambda\sinh\theta + 3 = 0$$

$$4\sinh^3\theta + 3\sinh\theta - \sinh 3\theta = 0$$

$$\frac{\lambda^3}{4} = \lambda \quad \therefore\ \lambda = 0 \quad \text{or} \quad \lambda^2 = 4 \quad \therefore\ \lambda = \pm 2 \qquad \text{Use } \lambda = 2.$$

$$\therefore\ 2 = -\frac{3}{\sinh 3\theta} \qquad \therefore\ \sinh 3\theta = -1.5$$

$$\therefore\ \frac{e^{3\theta} - e^{-3\theta}}{2} = -1.5 \qquad (e^{3\theta})^2 + 3(e^{3\theta}) - 1 = 0$$

which is a quadratic, giving $e^{3\theta} = 0.3028$ or -3.3028.

Now we continue as in the previous example.

$$e^{3\theta} = 0.3028, \text{ i.e. } 0.3028\,e^{j2\pi};\ 0.3028\,e^{j4\pi};\ 0.3028\,e^{j6\pi}$$

$$\therefore\ e^{\theta} = 0.6715\,e^{j2\pi/3};\ 0.6715\,e^{j4\pi/3};\ 0.6715\,e^{j2\pi}$$

$$= 0.6715(-0.5 + j0.8660);\ 0.6715(-0.5 - j0.8660);\ 0.6715(1)$$

$$\therefore\ e^{\theta} = -0.3358 + j0.5815; \quad -0.3358 - j0.5815; \quad 0.6715$$

Taking reciprocals of $e^{3\theta}$ we evaluate $e^{-3\theta}$ and then get

$$e^{-\theta} = -0.7446 - j1.2896; \quad -0.7446 + j1.2896; \quad 1.4892$$

Having evaluated e^{θ} and $e^{-\theta}$, we can determine three values for $\sinh\theta$ and hence, with $\lambda = 2$, we obtain the three required roots

$$x = \ldots\ldots\ldots;\ \ldots\ldots\ldots;\ \ldots\ldots\ldots$$

31

$$x = -0.8178; \quad 0.4088 \pm j\,1.8712$$

Summary of procedures to solve $x^3 + px + q = 0$

(a) $4p^3 + 27q^2 < 0$ — 3 real and different roots.

Put $x = \lambda \cos\theta$ and use $\cos 3\theta = 4\cos^3\theta - 3\cos\theta$.

(b) $4p^3 + 27q^2 = 0$ — 3 real roots, 2 identical

Put $x = \lambda \cos\theta$ and use $\cos 3\theta = 4\cos^3\theta - 3\cos\theta$.

(c) $4p^3 + 27q^2 > 0$ — 1 real root, 2 complex (conjugate pair)

(i) If p is negative
put $x = \lambda \cosh\theta$ and use $\cosh 3\theta = 4\cosh^3\theta - 3\cosh\theta$

(ii) If p is positive
put $x = \lambda \sinh\theta$ and use $\sinh 3\theta = 4\sinh^3\theta + 3\sinh\theta$.

Make a note of these for future use; they are quite important.

32

Reduced form of a cubic equation

You will have realised that all the previous cubic equations that we have solved have been of a special kind in that ....

33

in every case, the function $f(x)$ has not included a term in x^2

This is called the *reduced form* of a cubic and, in many practical cases, a general form of a cubic, $x^3 + ax^2 + bx + c = 0$, has first to be re-written in this form, $y^3 + py + q = 0$, before solution can be carried out. Just how this is done, the next section of work will show.

34

Transforming a cubic to reduced form

In every case, an equation of the form

$$x^3 + ax^2 + bx + c = 0$$

can be converted into the reduced form $y^3 + py + q = 0$ by the substitution $x = y - \frac{a}{3}$.

An example will demonstrate the method.

Example 1 Express $f(x) = x^3 + 6x^2 - 4x + 5 = 0$ in reduced form.

Substitute $x = y - \frac{a}{3}$ i.e. $x = y - \frac{6}{3} = y - 2$. Put $x = y - 2$.

The equation then becomes

$$(y-2)^3 + 6(y-2)^2 - 4(y-2) + 5 = 0$$

$$(y^3 - 3y^2 2 + 3y4 - 8) + 6(y^2 - 4y + 4) - 4(y-2) + 5 = 0$$

which simplifies to.........

35

$$\boxed{y^3 - 16y + 29 = 0}$$

This is the reduced form of the given equation $f(x) = 0$ and could be solved as in the previous examples.

Example 2 To solve the equation $x^3 + 3x^2 - 4x - 1 = 0$.

First we express the equation in reduced form. To do this with this particular equation, we substitute.........

36

$$\boxed{x = y - \frac{a}{3} \quad \text{i.e.} \quad x = y - 1}$$

The reduced form of the cubic equation is then

....

37

$$y^3 - 7y + 5 = 0$$

for, putting $x = y - 1$ in $x^3 + 3x^2 - 4x - 1 = 0$, we have

$$(y-1)^3 + 3(y-1)^2 - 4(y-1) - 1 = 0$$

$$(y^3 - 3y^2 + 3y - 1) + 3(y^2 - 2y + 1) - 4(y-1) - 1 = 0$$

which simplifies to $\underline{y^3 - 7y + 5 = 0}$

To go on and solve this, we first investigate the discriminant

$$4p^3 + 27q^2 = \ldots\ldots\ldots$$

38

-697 i.e. < 0

Therefore we know that the three roots of the equation are

....

39

all real and different

From our list of procedures, we therefore substitute $y = \ldots\ldots\ldots$ and use the identity ....

40

$$y = \lambda\cos\theta; \quad \cos 3\theta = 4\cos^3\theta - 3\cos\theta$$

Continuing then as in our previous examples, we eventually get

$$y = \ldots\ldots\ldots;\ \ldots\ldots\ldots;\ \ldots\ldots\ldots$$

41

$$y = 2.1659; \quad 0.7830; \quad -2.9489$$

Here is the working

$$y^3 - 7y + 5 = 0$$

$$\lambda^3\cos^3\theta - 7\lambda\cos\theta + 5 = 0$$

$$4\cos^3\theta - 3\cos\theta - \cos 3\theta = 0$$

....

42

$$\frac{\lambda^3}{4} = \frac{7\lambda}{3} = -\frac{5}{\cos 3\theta}$$

$$\frac{\lambda^3}{4} = \frac{7\lambda}{3} \quad \therefore \lambda = 0 \text{ or } \lambda^2 = \frac{28}{3} \quad \therefore \lambda = \pm 3.0551$$

Using $\lambda = 3.0551$, $\cos 3\theta = -\dfrac{15}{7(3.0551)} = -0.7014$

$$\therefore 3\theta = 134^\circ\ 32';\quad 225^\circ\ 28';\quad 494^\circ\ 32'$$

$$\therefore \theta = 44^\circ\ 51';\quad 75^\circ\ 9';\quad 164^\circ\ 51'$$

$$\therefore y = 3.0551 \cos 44^\circ\ 51';\quad 3.0551 \cos 75^\circ\ 9';\quad 3.0551 \cos 164^\circ\ 51'$$
$$= 2.1659\quad ;\quad 0.7830\quad ;\quad -2.9489$$

So $x = \ldots\ldots\ldots;\ \ldots\ldots\ldots;\ \ldots\ldots\ldots$

43

$$x = 1.166;\quad -0.217;\quad -3.949$$

since the original substitution was $x = y - 1$.

The essential step is to write down the reduced form of the given cubic equation $x^3 + ax^2 + bx + c = 0$ and, in every case, we do this by substituting $x = \ldots\ldots\ldots$

44

$$x = y - \frac{a}{3}$$

Many practical problems give rise to fourth degree equations, so we now turn our attention to the solution of quartic equations. Let us start the new section in the next frame.

Quartic Equations

45

$$x^4 + ax^3 + bx^2 + cx + d = 0$$

The first step in solving a quartic equation is to write the equation in *reduced form*, i.e. to eliminate the term in x^3. We accomplish this by a method very like that for dealing with a cubic equation.

Reduced form of a quartic equation $x^4 + ax^3 + bx^2 + cx + d = 0$

Substitution of $x = y - \frac{a}{4}$ removes the term in x^3.

Example $x^4 - 8x^3 + 3x^2 - x + 2 = 0.$

Here, $a = -8$ $\therefore$ we substitute $x = y + 2$ and the equation then becomes

. . . .

46

$$\boxed{y^4 - 21y^2 - 53y - 36 = 0}$$

for $(y+2)^4 - 8(y+2)^3 + 3(y+2)^2 - (y+2) + 2 = 0$

$$(y^4 + 8y^3 + 24y^2 + 32y + 16) - 8(y^3 + 6y^2 + 12y + 8) + 3(y^2 + 4y + 4)$$
$$-(y+2) + 2 = 0$$

which simplifies to $\underline{y^4 - 21y^2 - 53y - 36 = 0}$

So, to obtain the reduced form

(i) of a cubic, we substitute $x = y - \frac{a}{3}$

(ii) of a quartic, we substitute $x = y - \frac{a}{4}$

Having written the quartic in reduced form, how do we then proceed to solve it?

On to the next frame.

47

Solution of a quartic equation in reduced form

Consider the equation in reduced form

$$y^4 + py^2 + qy + r = 0$$

We express this as a product of two quadratic factors. Since there is no term in y^3, this product can be written

$$(y^2 + ay + b)(y^2 - ay + c) = 0$$

Multiplying these two factors together, gives ....

48

$$\boxed{y^4 - (a^2 - b - c)y^2 - a(b - c)y + bc = 0}$$

Comparing this with the reduced equation

$$y^4 - (a^2 - b - c)y^2 - a(b - c)y + bc = 0$$

$$y^4 \qquad + \quad py^2 \qquad + \quad qy + r = 0$$

$$p = b + c - a^2; \qquad q = a(c - b); \qquad r = bc$$

$$\left.\begin{aligned} b + c &= p + a^2 \\ b - c &= -\frac{q}{a} \end{aligned}\right\} \therefore b = \frac{1}{2}\left(p + a^2 - \frac{q}{a}\right); \quad c = \frac{1}{2}\left(p + a^2 + \frac{q}{a}\right)$$

Also, $bc = r$ $\quad \therefore \frac{1}{4}\left(p + a^2 - \frac{q}{a}\right)\left(p + a^2 + \frac{q}{a}\right) = r$

$$\therefore (p + a^2)^2 - \frac{q^2}{a^2} = 4r$$

which simplifies into ....

49

$$\boxed{a^6 + 2pa^4 + (p^2 - 4r)a^2 - q^2 = 0}$$

This is a cubic in a^2 and, having found a, the values of b and c can be determined, and finally the four roots of the equation.

Example 1 To solve $x^4 + 4x^3 - x^2 - 10x + 6 = 0$.

In reduced form this becomes ....

50

$$\boxed{y^4 - 7y^2 + 12 = 0}$$

for, substituting $x = y - \frac{a}{4}$ i.e. $x = y - 1$, the original equation becomes

$$(y-1)^4 + 4(y-1)^3 - (y-1)^2 - 10(y-1) + 6 = 0$$
$$(y^4 - 4y^3 + 6y^2 - 4y + 1) + 4(y^3 - 3y^2 + 3y - 1) - (y^2 - 2y + 1)$$
$$-10(y-1) + 6 = 0$$
$$\therefore\ y^4 - 7y^2 + 12 = 0$$

It happens that this is a quadratic in y^2. That makes it easy. We can now find y and hence x. $x = \ldots\ldots\ldots$

51

$$\boxed{x = -3,\ 1,\ -1 \pm \sqrt{3}}$$

Before we embark on a further example, let us revise the steps in the general procedure.

(i) Express the given equation in reduced form $y^4 + py^2 + qy + r = 0$.
(ii) Express the function as the product of two quadratic factors

$$(y^2 + ay + b)(y^2 - ay + c) = 0$$

(iii) Comparing coefficients gives $b = \frac{1}{2}\left(p + a^2 - \frac{q}{a}\right)$

$$c = \frac{1}{2}\left(p + a^2 + \frac{q}{a}\right)$$
$$bc = r$$

(iv) From these, form a cubic equation in a^2.
(v) Solve this for one value of a and hence evaluate b and c.
(vi) Finally solve the quadratic equations for y and hence determine values for x.

It all sounds a little involved, so let us apply the method to a new example.

52

Example 2 To solve $x^4 - 8x^3 + 24x^2 + 8x - 25 = 0$.

First express this is reduced form, by substituting $x = y - \frac{a}{4}$ i.e. $x = y + 2$. The reduced equation is then

....

53

$$\boxed{y^4 + 40y + 39 = 0}$$

We compare coefficients with the general form of the reduced equation

$$y^4 + py^2 + qy + r = 0$$

In this case, $p = 0, \quad q = 40, \quad r = 39$

We now have to write $y^4 + 40y + 39 = 0$ as the product of two quadratic factors. These will take the form

....

54

$$\boxed{(y^2 + ay + b)(y^2 - ay + c) = 0}$$

Our next task is to evaluate a, b and c.

$$b = \frac{1}{2}\left(p + a^2 - \frac{q}{a}\right) = \frac{1}{2}\left(a^2 - \frac{40}{a}\right)$$

$$c = \frac{1}{2}\left(p + a^2 + \frac{q}{a}\right) = \frac{1}{2}\left(a^2 + \frac{40}{a}\right)$$

$$bc = r \quad \therefore \ \frac{1}{4}\left(a^2 - \frac{40}{a}\right)\left(a^2 + \frac{40}{a}\right) = 39$$

which gives $a^6 - 156a^2 - 1600 = 0$ which is cubic in a^2

$$(a^2)^3 - 156(a^2) - 1600 = 0$$

This now has to be solved for a^2. It is just a case of taking one step at a time.

Let $z = a^2$. $\quad \therefore \ z^3 - 156z - 1600 = 0$

Now we go through the cubic routine, with $p = -156$ and $q = -1600$.

The discriminant $4p^3 + 27q^2 = \ldots\ldots\ldots$

55

$$4p^3 + 27q^2 = 53\,934\,000 \qquad \text{i.e.} > 0$$

The three required roots are therefore of the type

. . . .

56

1 real and 2 complex (conjugate pair)

In our case $\quad z^3 - 156z - 1600 = 0$

$$z^3 + pz + q = 0$$

p is negative. Therefore we substitute $z =$ . . . . and use the identity . . . .

57

$$z = \lambda \cosh\theta; \qquad \cosh 3\theta = 4\cosh^3\theta - 3\cosh\theta$$

$$\therefore \lambda^3\cosh^3\theta - 156\lambda\cosh\theta - 1600 = 0$$

$$4\cosh^3\theta - 3\cosh\theta - \cosh 3\theta = 0$$

Comparing coefficients $\quad \dfrac{\lambda^3}{4} = \dfrac{156\lambda}{3} = \dfrac{1600}{\cosh 3\theta}$

and therefore $\lambda =$ . . . . ; $\quad \cosh 3\theta =$ . . . .

58

$$\lambda = 14.42; \quad \cosh 3\theta = 2.1335$$

for $\dfrac{\lambda^3}{4} = \dfrac{156\lambda}{3} \quad \therefore \lambda = 0 \quad \text{or} \quad \lambda^2 = \dfrac{4(156)}{3} = 208 \quad \therefore \lambda = \pm 14.422$

Using $\lambda = 14.422, \quad \cosh 3\theta = \dfrac{3(1600)}{156(14.422)} = 2.1335$

So $\dfrac{e^{3\theta} + e^{-3\theta}}{2} = 2.1335 \qquad e^{3\theta} =$ . . . .

59

$$e^{3\theta} = 4.0181 \quad \text{or} \quad 0.2489$$

since $e^{3\theta} + e^{-3\theta} = 4.2670 \qquad (e^{3\theta})^2 - 4.2670(e^{3\theta}) + 1 = 0$
whence $e^{3\theta} = 4.0181$ or 0.2489. We will use $e^{3\theta} = 4.0181$.

$$\therefore\ e^{\theta} = 4.0181^{1/3} = 1.5898 \quad \therefore\ e^{-\theta} = 0.6290$$

$$\cosh\theta = \tfrac{1}{2}(e^{\theta} + e^{-\theta}) = \tfrac{1}{2}(1.5898 + 0.6290) = 1.1094$$

Remembering that $\lambda = 14.422$, we can now work back through the various substitutions we have made.

$$a^2 = z = \lambda\cosh\theta; \quad b = \frac{1}{2}\left(a^2 - \frac{40}{a}\right); \quad c = \frac{1}{2}\left(a^2 + \frac{40}{a}\right)$$

$(y^2 + ay + b)(y^2 - ay + c) = 0$ to obtain y; and finally $x = y + 2$.
The required four roots are therefore

$$x = \dots\dots\dots$$

60

$$x = \pm 1; \qquad 4 \pm j3$$

for $a^2 = z = 14.422(1.1094) = 16.000 \quad \therefore\ a = \pm 4$

Using $a = 4$

$$b = \frac{1}{2}\left(16 - \frac{40}{4}\right) = 3; \quad c = \frac{1}{2}\left(16 + \frac{40}{4}\right) = 13.$$

$\therefore\ (y^2 + 4y + 3)(y^2 - 4y + 13) = 0 \qquad (y+1)(y+3)(y^2 - 4y + 13) = 0$

$$\therefore\ y = -1,\ -3,\quad 2 \pm j3$$

Finally, $x = y + 2 \qquad \therefore\ \underline{x = 1,\ -1,\quad 4 \pm j3}$

Numerical Solution of Equations

Newton–Raphson iterative method 61

Consider the graph of $y = f(x)$ as shown. Then the x-value at the point A where the graph crosses the x-axis, gives a solution of the equation $f(x) = 0$.

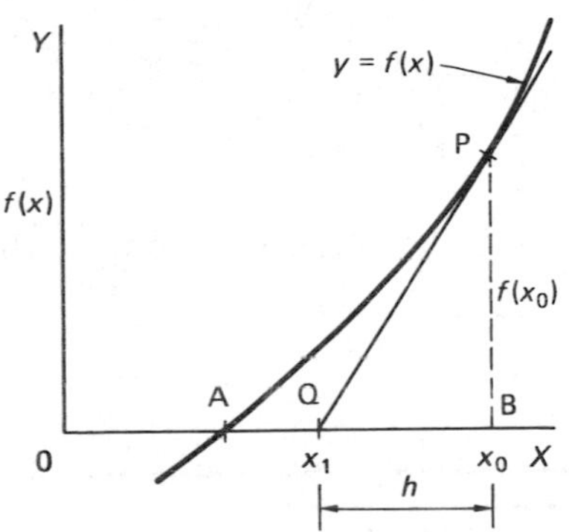

If P is a point on the curve near to A, then $x = x_0$ is an approximate value of the root of $f(x) = 0$, the error of the approximation being given by AB.

Let PQ be the tangent to the curve at P, crossing the x-axis at $Q(x_1, 0)$. Then $x = x_1$ is a better approximation to the required root.

From the diagram, $\dfrac{PB}{QB} = \left[\dfrac{dy}{dx}\right]_P$ i.e. the value of the differential coefficient of y at the point P, $x = x_0$.

$$\therefore \frac{PB}{QB} = f'(x_0) \quad \text{and} \quad PB = f(x_0)$$

$$\therefore QB = \frac{PB}{f'(x_0)} = \frac{f(x_0)}{f'(x_0)} = h(\text{say})$$

$$x_1 = x_0 - h \qquad \therefore x_1 = x_0 - \frac{f(x_0)}{f'(x_0)}$$

If we begin, therefore, with an approximate value (x_0) of the root, we can determine a better approximation (x_1). Naturally, the process can then be repeated to improve the result still further.
Let us see this in operation.

On to the next frame.

62

Example 1 The equation $x^3 - 3x - 4 = 0$ is known to have a root at approximately $x = 2$. Find a better approximation to the root.

We have $f(x) = x^3 - 3x - 4 \qquad \therefore f'(x) = 3x^2 - 3$
If the first approximation is $x_0 = 2$, then

$$f(x_0) = f(2) = -2 \quad \text{and} \quad f'(x_0) = f'(2) = 9$$

A better approximation x_1 is given by $x_1 = x_0 - \dfrac{f(x_0)}{f'(x_0)}$

$$x_1 = 2 - \frac{(-2)}{9} = 2.22$$

$$\therefore \underline{x_0 = 2; \qquad x_1 = 2.22}$$

If we now start from x_1 we can get a better approximation still by repeating the process. $\quad x_2 = x_1 - \dfrac{f(x_1)}{f'(x_1)}$

Here $x_1 = 2.22 \qquad f(x_1) = \dots\dots\dots \quad f'(x_1) = \dots\dots\dots$

63

$$\boxed{f(x_1) = 0.281; \quad f'(x_1) = 11.785}$$

Then $x_2 = \dots\dots\dots$

64

$$\boxed{x_2 = 2.196}$$

since $x_2 = 2.22 - \dfrac{0.281}{11.79} = \underline{2.196}$
Using $x_2 = 2.196$ as a starter value, we can continue the process until successive results agree to the desired degree of accuracy.

$$x_3 = \dots\dots\dots$$

65

$$\boxed{x_3 = 2.196}$$

for $f(x_2) = f(2.196) = 0.002026; \quad f'(x_2) = f'(2.196) = 11.467$

$$\therefore x_3 = x_2 - \frac{f(x_2)}{f'(x_2)} = 2.196 - \frac{0.00203}{11.467} = 2.196 \text{ (to 4 sig. fig.)}$$

The required solution is therefore $\underline{x = 2.196}$ to 4 significant figures.

The process is simple but effective and can be repeated again and again. Each repetition, or *iteration*, gives a result nearer to the required root $x = x_A$.

In general $x_{n+1} = \dots\dots\dots$

$$x_{n+1} = x_n - \frac{f(x_n)}{f'(x_n)}$$ 66

Tabular display of results

In practice, it is neater, and less prone to error, if the intermediate results are entered into a table as they are obtained—particularly as most of the actual calculation will be carried out on some form of calculator.

For the previous example, we should have

$$f(x) = x^3 - 3x - 4 \qquad f'(x) = 3x^2 - 3 \qquad x_{n-1} = x_n - \frac{f(x_n)}{f'(x_n)}$$

n	x_n	$f(x_n)$	$f'(x_n)$	$h = \frac{f(x_n)}{f'(x_n)}$	$x_{n+1} = x_n - h$
0	2	−2	9	−0.22	2.22
1	2.22	0.281	11.785	0.0238	2.1962
2	2.1962	0.00203	11.467	0.00018	2.19602

$\therefore$ Root is $\underline{x = 2.196}$ to 4 significant figures

Now let us have another example.

Example 2 The equation $x^3 + 2x^2 - 5x - 1 = 0$ has a root near to 67
$x = 1.4$. Use the Newton–Raphson method to find the root to four significant figures.

Applying the nesting technique

$$f(x) = x^3 + 2x^2 - 5x - 1 = [(x + 2)x - 5]x - 1$$

$$f'(x) = 3x^2 + 4x - 5 = (3x + 4)x - 5$$

$$x_{n+1} = x_n - h = x_n - \frac{f(x_n)}{f'(x_n)}$$

n	x_n	$f(x_n)$	$f'(x_n)$	$h = \frac{f(x_n)}{f'(x_n)}$	$x_{n+1} = x_n + h$
0	1.4	−1.336	6.48	−0.206	1.606
1	1.61				

Now taking $x_1 = 1.61$, complete the next line of the table.

68 Here it is

n	x_n	$f(x_n)$	$f'(x_n)$	$h=\dfrac{f(x_n)}{f'(x_n)}$	$x_{n+1}=x_n-h$
0	1.4	−1.336	6.48	−0.206	1.606
1	1.61	0.3075	9.2163	0.0334	1.5766
2	1.577				

Continue with the next line.

69

n	x_n	$f(x_n)$	$f'(x_n)$	$h=\dfrac{f(x_n)}{f'(x_n)}$	$x_{n+1}=x_n-h$
:					
2	1.577	0.01075	8.7688	0.001226	1.5758
3	1.5758				

Repeat the process once again.

70 Finally, the table is complete

n	x_n	$f(x_n)$	$f'(x_n)$	$h=\dfrac{f(x_n)}{f'(x_n)}$	$x_{n+1}=x_n-h$
0	1.4	−1.336	6.48	−0.206	1.606
1	1.61	0.3075	9.2163	0.0334	1.5766
2	1.577	0.01075	8.7688	0.00123	1.5758
3	1.5758	0.00023	8.7526	0.000027	1.57577

To the required 4 significant figures, the last two results for x_{n+1} agree. Therefore the required root is 1.576 to 4 significant figures.

Now here is another. Set up the table in the same way and continue the iterations until two successive entries in the right-hand column agree to the required level of accuracy.

Example 3 The equation $2x^3-7x^2-x+12=0$ has a root near to $x=1.5$. Determine the root correct to 4 significant figures.

This time we have

$$f(x)=2x^3-7x^2-x+12=[(2x-7)x-1]x+12$$
$$f'(x)=6x^2-14x-1 \qquad =(6x-14)x-1$$

Complete the work and check with the next frame.

71

n	x_n	$f(x_n)$	$f'(x_n)$	$h = \dfrac{f(x_n)}{f'(x_n)}$	$x_{n+1} = x_n - h$
0	1.5	1.5	−8.5	−0.176	1.676
1	1.676	0.0769	−7.610	−0.0101	1.6861
2	1.6861	0.000307	−7.5478	−0.000041	1.68614

Therefore, to 4 significant figures, $\underline{x = 1.686.}$

First approximations The whole process hinges on knowing a 'starter' value as first approximation. If we are not given a hint, this information can be found by either

(a) applying the remainder theorem to the function
(b) drawing a sketch graph of the function.

Example 4 Find the real root of the equation $x^3 + 5x^2 - 3x - 4 = 0$ correct to three significant figures.

Application of the remainder theorem involves substituting $x = 0$, $x = \pm 1$, $x = \pm 2$, etc. until two adjacent values give a change in sign.

$f(x) = x^3 + 5x^2 - 3x - 4 = [(x + 5)x - 3]x - 4$
$f(0) = -4; \quad f(1) = -1; \quad f(-1) = 3$

The sign changes from $f(0)$ to $f(-1)$. There is thus a root between $x = 0$ and $x = -1$.

Therefore choose $x = -0.6$ as the first approximation and then proceed as before.

Complete the table and obtain the root

$x = \ldots\ldots\ldots$

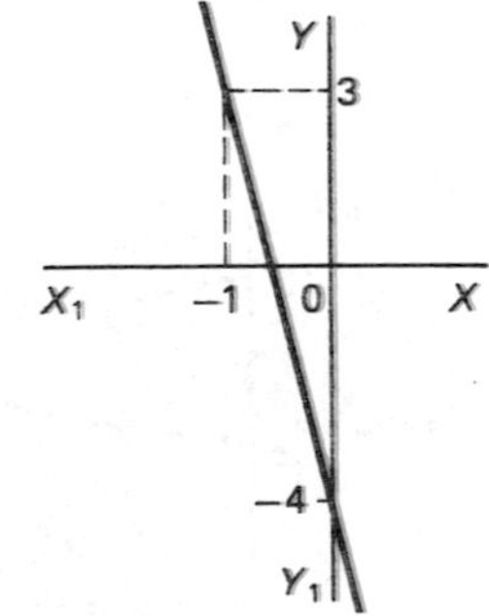

72

$$f(x) = x^3 + 5x^2 - 3x - 4 = [(x+5)x - 3]x - 4$$
$$f'(x) = 3x^2 + 10x - 3 \qquad = (3x + 10)x - 3$$

n	x_n	$f(x_n)$	$f'(x_n)$	$h = \dfrac{f(x_n)}{f'(x_n)}$	$x_{n+1} = x_n -$
0	−0.6	−0.616	−7.92	0.0778	−0.678
1	−0.678	0.0208	−8.401	−0.00248	−0.6755
2	−0.6755	−0.00023	−8.386	0.000027	−0.6755

$\underline{x = -0.676}$

73 *Example 5* Solve the equation $e^x + x - 2 = 0$ giving the root to 4 significant figures.

It is sometimes more convenient to obtain a first approximation to the required root from a sketch graph of the function, or by some other graphical means.

In this case, the equation can be re-written as $e^x = 2 - x$ and we therefore sketch graphs of $y = e^x$ and $y = 2 - x$.

x	0.2	0.4	0.6	0.8	1.0
e^x	1.22	1.49	1.82	2.23	2.72

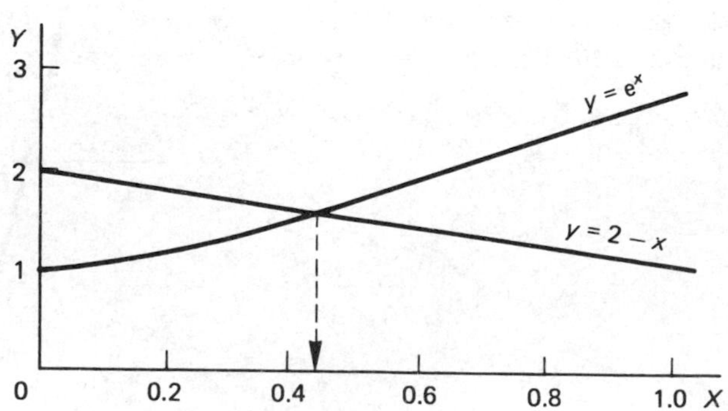

Approximate root $x = 0.4$

$$f(x) = e^x + x - 2 \qquad f'(x) - e^x + 1$$

$x = \ldots\ldots\ldots$ Finish it off.

74

$x = 0.4429$

n	x_n	$f(x_n)$	$f'(x_n)$	$h = \dfrac{f(x_n)}{f'(x_n)}$	$x_{n+1} = x_n - h$
0	0.4	−0.108	2.49	−0.0434	0.4434
1	0.443	0.000372	2.557	0.000145	0.44286
2	0.4429	0.000117	2.5572	0.0000458	0.44285

Therefore, to 4 significant figures, $\underline{x = 0.4429}$

Note There are times when the normal application of the Newton–Raphson method fails to converge to the required root. This is particularly so when $f'(x_0)$ is very small, so before we leave this useful section of work, let us consider the difficulty.

On to the next frame.

75 Modified Newton–Raphson method

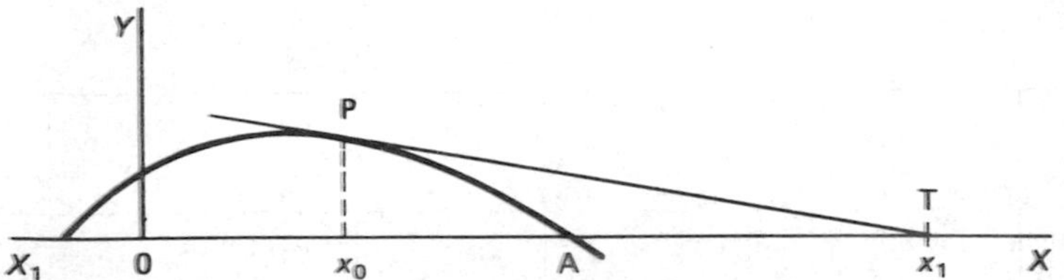

If the slope of the curve at $x = x_0$ is small, the value of the second approximation $x = x_1$ may be further from the exact root at A than the first approximation. That is, if $f'(x)$ is small, the method, as it stands, fails to converge.

If $x = x_0$ is an approximate root of $f(x) = 0$ and $x = x_0 + h$ is an exact root, then $f(x_0 + h) = 0$.

By Taylor's series

$$f(x_0 + h) = f(x_0) + h \cdot f'(x_0) + \frac{h^2}{2!} f''(x_0) + \ldots + \frac{h^n}{n!} f^n(x_0) + \ldots$$

(a) If we consider the first two terms only, then

$$0 \approx f(x_0) + h \cdot f'(x_0) \qquad \therefore\ h \approx -\frac{f(x_0)}{f'(x_0)}$$

and the new approximation $x = x_1 = x_0 + h$ is given by

$$x_1 = x_0 - \frac{f(x_0)}{f'(x_0)}$$

and this, of course, is the relationship we have been using and which is seen to fail when $f'(x_0)$ is small.

(b) If we consider the first three terms, then

$$0 \approx f(x_0) + h \cdot f'(x_0) + \frac{h^2}{2!} f''(x_0)$$

$$\therefore\ h^2 \cdot f''(x_0) + 2h \cdot f'(x_0) + 2f(x_0) \approx 0$$

This is a quadratic in h giving

$$h = \frac{-2f'(x_0) \pm \sqrt{4\{f'(x_0)\}^2 - 8f(x_0) \cdot f''(x_0)}}{2f''(x_0)}$$

Since $f'(x_0)$ is small

$$h \approx \pm \frac{\sqrt{-8f(x_0) \cdot f''(x_0)}}{2f''(x_0)} = \pm\sqrt{\frac{-2f(x_0)}{f''(x_0)}}$$

Then $\quad x_1 = x_0 + h \qquad \therefore x_1 = x_0 \pm \sqrt{\frac{-2f(x_0)}{f''(x_0)}}$

We use this result *only* when $f'(x_0)$ is found to be very small. Having found x_1 from x_0, we revert for further iterations to the normal relationship $x_{n+1} = x_n - \dfrac{f(x)}{f'(x)}$

Note this.

76

Example 6 The equation $x^3 - 1.3x^2 + 0.4x - 0.03 = 0$ is known to have a root near $x = 0.7$. Determine the root to 3 significant figures.

We start off in the usual way.

$$f(x) = x^3 - 1.3x^2 + 0.4x - 0.03 = [(x - 1.3)x + 0.4]x - 0.03$$
$$f'(x) = 3x^2 - 2.6x + 0.4 \qquad = (3x - 2.6)x + 0.4$$

and complete the first line of the normal table.

n	x_n	$f(x_n)$	$f'(x_n)$	$h = \dfrac{f(x_n)}{f'(x_n)}$	$x_{n+1} = x_n - h$
0	0.7				

Complete just the first line of values.

77

We have

n	x_n	$f(x_n)$	$f'(x_n)$	$h = \dfrac{f(x_n)}{f'(x_n)}$	$x_{n+1} = x_n - h$
0	0.7	−0.044	0.05	−0.88	1.58

We notice at once that

(a) The value of x_1 is well away from the approximate value (0.7) of the root.

(b) The value of $f'(x_0)$ is small, i.e. 0.05.

To obtain x_1 we therefore make a fresh start, using the modified relationship $x_1 = \dots\dots\dots$

78

$$x_1 = x_0 \pm \sqrt{\frac{-2f(x_0)}{f''(x_0)}}$$

$$\begin{aligned} f(x) &= x^3 - 1.3x^2 + 0.4x - 0.03 = [(x - 1.3)x + 0.4]x - 0.03 \\ f'(x) &= 3x^2 - 2.6x + 0.4 \qquad\qquad = (3x - 2.6)x + 0.4 \\ f''(x) &= 6x - 2.6 \end{aligned}$$

n	x_0	$f(x_0)$	$f''(x_0)$	$h = \sqrt{\dfrac{-2f(x_0)}{f''(x_0)}}$	$x_1 = x_0 \pm h$
0	0.7	−0.044			

Complete the line.

79

n	x_0	$f(x_0)$	$f''(x_0)$	$h = \sqrt{\dfrac{-2f(x_0)}{f''(x_0)}}$	$x_1 = x_0 \pm h$
0	0.7	–0.044	1.6	0.2345	0.9345

Note that in the expression $x_1 = x_0 \pm h$, we chose the positive sign since at $x_0 = 0.7$, $f(x_0)$ is negative and the slope $f'(x_0)$ is positive.

Having established that $x_1 = 0.9345$, we now revert to the usual $x_{n+1} = x_n - \dfrac{f(x_n)}{f'(x_n)}$ for the rest of the calculation. Complete the table therefore and obtain the required root.

80

The table finally looks like this

n	x_n	$f(x_n)$	$f'(x_n)$	$h = \dfrac{f(x_n)}{f'(x_n)}$	$x_{n+1} = x_n - h$
0	0.7	–0.044	(0.05)	—	(0.9345)
1	0.935	0.0249	0.5917	0.0421	0.8929
2	0.893	0.00264	0.4705	0.00561	0.8874
3	0.8874	0.000046	0.4552	0.00010	0.8873

Therefore, to 3 significant figures, the required root is $\underline{x = 0.887}$

Note that we use the modified method only to find x_1. After that, the normal relationship is used.

Now there remains the Test Exercise which covers the work of the programme. First check the Revision Summary and revise any special points that need it.

81 REVISION SUMMARY

1. *Cubic Equations*

Every cubic equation has at least one real root.

Tartaglia's solution for the real root of $x^3 + ax + b = 0 \quad a > 0$

$$x = \left\{-\frac{b}{2} + \sqrt{\frac{a^3}{27} + \frac{b^2}{4}}\right\}^{1/3} + \left\{-\frac{b}{2} - \sqrt{\frac{a^3}{27} + \frac{b^2}{4}}\right\}^{1/3}$$

Trigonometrical solution of $x^3 + ax^2 + bx + c = 0$.

Express in *reduced form* by substituting $x = y - \frac{a}{3}$ to give

$$y^3 + py + q = 0$$

Discriminant: $4p^3 + 27q^2$

(a) $4p^3 + 27q^2 < 0$ — 3 real and different roots

Put $y = \lambda \cos\theta$ and use $\cos 3\theta = 4\cos^3\theta - 3\cos\theta$.

(b) $4p^3 + 27q^2 = 0$ — 3 real roots, 2 of which identical

Put $y = \lambda \cos\theta$ and use $\cos 3\theta = 4\cos^3\theta - 3\cos\theta$.

(c) $4p^3 + 27q^2 > 0$ — 1 real root, 2 complex (conjugate pair)

(i) If p is *negative*

put $y = \lambda \cosh\theta$ and use $\cosh 3\theta = 4\cosh^3\theta - 3\cosh\theta$.

(ii) If p is *positive*

put $y = \lambda \sinh\theta$ and use $\sinh 3\theta = 4\sinh^3\theta + 3\sinh\theta$.

2. *Quartic Equations*

$$x^4 + ax^3 + bx^2 + cx + d = 0$$

Express in *reduced form* by substituting $x = y - \frac{a}{4}$ to give

$$y^4 + py^2 + qy + r = 0$$

Then as $\quad (y^2 + ay + b)(y^2 - ay + c) = 0$

where $b = \frac{1}{2}\left(p + a^2 - \frac{q}{a}\right)$

$c = \frac{1}{2}\left(p + a^2 + \frac{q}{a}\right)$

$$bc = r \qquad \therefore (p + a^2)^2 - \frac{q^2}{a^2} = 4r$$

This gives a cubic in a^2. Determine one value of a and then b and c can be found.

3. *Numerical Solution of Equations*

Newton–Raphson iterative method

If $x = x_0$ is an approximate value of a root of the equation $f(x) = 0$, a better approximation $x = x_1$ is given by

$$x_1 = x_0 - \frac{f(x_0)}{f'(x_0)}$$

The process can be repeated and in general

$$x_{n+1} = x_n - \frac{f(x_n)}{f'(x_n)}$$

Modified Newton–Raphson method

If $f'(x_0)$ is very small, the normal method fails and gives a second approximation x_1 further from the true root than x_0. In that case, we obtain x_1 from the relationship

$$x_1 = x_0 \pm \sqrt{\frac{-2f(x_0)}{f''(x_0)}}$$

We revert to the normal relationship for subsequent iterations.

82 TEST EXERCISE II

1. Using Tartaglia's solution, find the real root of the equation $2x^3 + 4x - 5 = 0$ giving the result to 4 significant figures.
2. Solve the equation $x^3 - 6x - 4 = 0$.
3. Re-write the equation $x^3 + 6x^2 + 9x + 4 = 0$ in reduced form and hence determine the three roots.
4. Solve the equation $x^4 - 4x^3 + 8x + 3 = 0$.
5. Express the equation $x^4 - 4x^3 + 4x^2 + 4x - 5 = 0$ in reduced form and hence determine the four roots.
6. Show that the equation $x^3 + 3x^2 - 4x - 6 = 0$ has a root between $x = 1$ and $x = 2$, and use the Newton–Raphson iterative method to evaluate this root to 4 significant figures.

FURTHER PROBLEMS II

1. Find the real root of the equations

   (a) $x^3 + 4x + 3 = 0$ (b) $5x^3 + 2x - 1 = 0$.

2. Solve the following equations

   (a) $x^3 - 5x + 1 = 0$ (b) $x^3 + 2x - 3 = 0$

   (c) $x^3 - 4x + 1 = 0$.

3. Express the following in reduced form and determine the roots

   (a) $x^3 + 6x^2 + 9x + 5 = 0$

   (b) $8x^3 + 20x^2 + 6x - 9 = 0$

   (c) $4x^3 - 9x^2 + 42x - 10 = 0$.

4. Re-write the following in reduced form and hence solve the equations

   (a) $x^4 + 8x^3 + 18x^2 + 8x - 35 = 0$

   (b) $x^4 - 12x^3 + 49x^2 - 78x + 40 = 0$

   (c) $x^4 - 8x^3 + 15x^2 + 4x - 20 = 0$

   (d) $x^4 - 8x^3 + 25x^2 - 36x + 18 = 0$

   (e) $x^4 - 12x^3 + 57x^2 - 126x + 104 = 0$

   (f) $x^4 - 4x^3 + 11x^2 - 14x + 10 = 0$.

5. Express the following equations in reduced form and make the appropriate substitutions to determine the roots

(a) $x^4 - 8x^3 + 17x^2 + 2x - 24 = 0$

(b) $x^4 - 12x^3 + 46x^2 - 28x - 87 = 0$

(c) $x^4 - 4x^3 + 9x^2 + 16x - 52 = 0$.

6. Use the Newton–Raphson iterative method to solve the following.

(a) Show that a root of the equation $x^3 + 3x^2 + 5x + 9 = 0$ occurs between $x = -2$ and $x = -3$. Evaluate the root to four significant figures.

(b) Show graphically that the equation $e^{2x} = 25x - 10$ has two real roots and find the larger root correct to four significant figures.

(c) Verify that the equation $x - \cos x = 0$ has a root near to $x = 0.8$ and determine the root correct to three significant figures.

(d) Obtain graphically an approximate root of the equation $2 \ln x = 3 - x$. Evaluate the root correct to four significant figures.

(e) Verify that the equation $x^4 + 5x - 20 = 0$ has a root at approximately $x = 1.8$. Determine the root correct to five significant figures.

(f) Show that the equation $x + 3 \sin x = 2$ has a root between $x = 0.4$ and $x = 0.6$. Evaluate the root to four significant figures.

(g) The equation $2 \cos x = e^x - 1$ has a real root between $x = 0.8$ and $x = 0.9$. Evaluate the root correct to four significant figures.

(h) The equation $20x^3 - 22x^2 + 5x - 1 = 0$ has a root at approximately $x = 0.6$. Determine the value of the root correct to four significant figures.

Programme 3

Partial Differentiation

Prerequisites: Engineering Mathematics (fourth edition)
Programmes 9, 10, 11

Small Increments

1

Taylor's theorem for one independent variable

Taylor's theorem expands $f(x+h)$ in terms of $f(x)$, powers of h and successive derivatives of $f(x)$, and can be stated as

$$f(x+h) = f(x) + hf'(x) + \frac{h^2}{2!}f''(x) + \ldots + \frac{h^n}{n!}f^n(x) + \ldots$$

where $f^n(x)$ denotes the nth derivative of $f(x)$. You will also, no doubt, remember that, by putting $x = 0$ in the result and then letting $h = x$, we obtain Maclaurin's series

$$f(x) = f(0) + hf'(0) + \frac{h^2}{2!}f''(0) + \ldots + \frac{h^n}{n!}f^n(0) + \ldots$$

Taylor's theorem for two independent variables

If we consider $z = f(x, y)$ where z is a function of two independent variables x and y, then, in general, increases in x and y will produce a combined increase in z.

So, if $z = f(x, y)$ then $z + \delta z = f(x+h,\ y+k)$

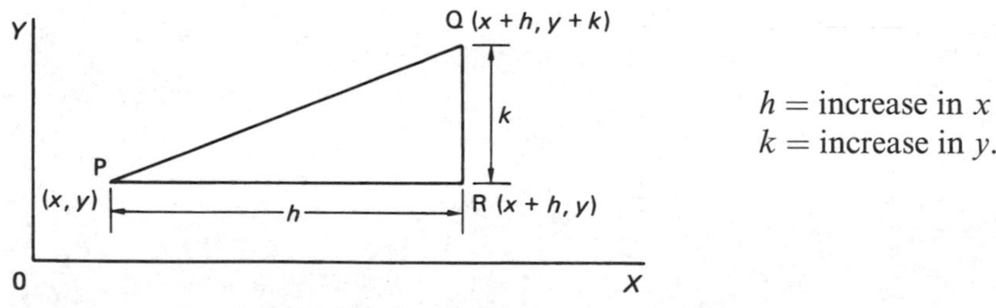

h = increase in x
k = increase in y.

For R: $f(x+h, y) = f(x, y) + hf'_x(x, y) + \dfrac{h^2}{2!}f''_{xx}(x, y) + \ldots$ (1)

where $f'_x(x, y)$ denotes $\dfrac{\partial}{\partial x}f(x, y)$; $f''_{xx}(x, y)$ denotes $\dfrac{\partial^2}{\partial x^2}f(x, y)$ etc.

From R to Q: $(x+h)$ is constant; y changes to $(y+k)$

$$\therefore f(x+h, y+k) = f(x+h, y) + kf'_y(x+h, y) + \frac{k^2}{2!}f''_{yy}(x+h, y) + \ldots$$

(2)

To express (2) in terms of $f(x,y)$ we can substitute result (1) for the first term $f(x+h,y)$ and similar expressions which we shall obtain for $f'_y(x+h,y)$, $f''_{yy}(x+h,y)$ and so on.

If we differentiate (1) with respect to y, we have

$$f'_y(x+h,y) = \dots\dots\dots$$

2

$$f'_y(x+h,y) = f'_y(x,y) + hf''_{yx}(x,y) + \frac{h^2}{2!}f'''_{yxx}(x,y) + \dots$$

If we now differentiate this result again with respect to y

$$f''_{yy}(x+h,y) = \dots\dots\dots$$

3

$$f''_{yy}(x+h,y) = f''_{yy}(x,y) + hf'''_{yyx}(x,y) + \frac{h^2}{2!}f^{\text{iv}}_{yyxx}(x,y) + \dots$$

Then our previous expansion (2), i.e.

$$f(x+h,y+k) = f(x+h,y) + kf'_y(x+h,y) + \frac{k^2}{2!}f''_{yy}(x+h,y) + \dots$$

now becomes

$$\begin{aligned} f(x+h,y+k) = {} & f(x,y) + hf'_x(x,y) + \frac{h^2}{2!}f''_{xx}(x,y) + \dots \\ & + k\left\{f'_y(x,y) + hf''_{yx}(x,y) + \frac{h^2}{2!}f'''_{yxx}(x,y) + \dots\right\} \\ & + \frac{k^2}{2!}\left\{f''_{yy}(x,y) + hf'''_{yyx}(x,y) + \frac{h^2}{2!}f^{\text{iv}}_{yyxx}(x,y) + \dots\right\} \\ & + \dots \end{aligned}$$

Rearranging the terms by collecting together all the first derivatives, and then all the second derivatives, and so on, we get

$$f(x+h,y+k) = \dots\dots\dots$$

4

$$f(x+h, y+k) = f(x, y) + \{hf'_x(x, y) + kf'_y(x, y)\}$$
$$+ \frac{1}{2!}\{h^2 f''_{xx}(x, y) + 2hkf''_{yx}(x, y) + k^2 f''_{yy}(x, y)\} + \ldots$$

Then, if $z = f(x, y)$, $h = \delta x$, $k = \delta y$, this can be written

$$z + \delta z = z + \left\{h\frac{\partial z}{\partial x} + k\frac{\partial z}{\partial y}\right\} + \frac{1}{2!}\left\{h^2\frac{\partial^2 z}{\partial x^2} + 2hk\frac{\partial^2 z}{\partial y\,\partial x} + k^2\frac{\partial^2 z}{\partial y^2}\right\} + \ldots$$

Subtracting z from each side

$$\delta z = \frac{\partial z}{\partial x}\delta x + \frac{\partial z}{\partial y}\delta y + \frac{1}{2!}\left\{\frac{\partial^2 z}{\partial x^2}(\delta x)^2 + 2\frac{\partial^2 z}{\partial y\,\partial x}(\delta x\,\delta y) + \frac{\partial^2 z}{\partial y^2}(\delta y)^2\right\} + \ldots$$

Since δx and δy are small, the expression in the brackets is of the next order of smallness and can be discarded for our purposes. Therefore, we arrive at the result

$$\text{If} \quad z = f(x, y) \quad \text{then} \quad \underline{\delta z = \frac{\partial z}{\partial x}\delta x + \frac{\partial z}{\partial y}\delta y}$$

As already explained above, this result is, in fact, an approximation since the smaller terms in the series have been neglected. For practical purposes, however, the result can be used as stated.

Be sure to make a note of the result, for it is the foundation of much that follows.

5

$$z = f(x, y); \qquad \delta z = \frac{\partial z}{\partial x}\delta x + \frac{\partial z}{\partial y}\delta y$$

The following diagram illustrates the result

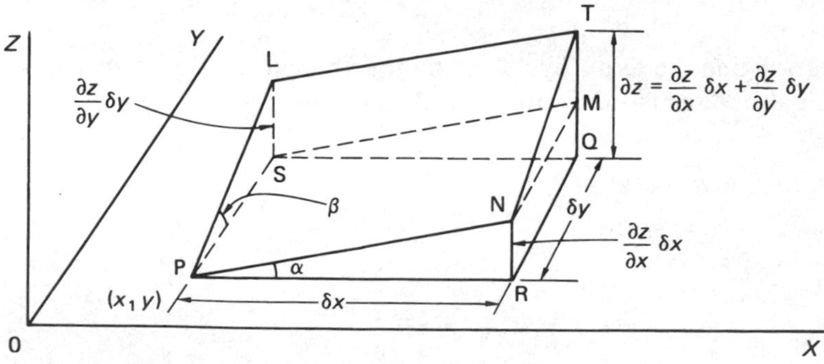

$\dfrac{\partial z}{\partial x}$ is the slope of PN $\quad\therefore$ RN $= \dfrac{\partial z}{\partial x}\delta x =$ QM

$\dfrac{\partial z}{\partial y}$ is the slope of PL $\quad\therefore$ SL $= \dfrac{\partial z}{\partial y}\delta y =$ MT

QT = QM + MT $\quad\therefore \delta z = \dfrac{\partial z}{\partial x}\delta x + \dfrac{\partial z}{\partial y}\delta y$

This is the total increment of $z = f(x, y)$ from P to Q.

It is worth noting at this stage that the result can be extended to the case of three independent variables, i.e. if $u = f(x, y, z)$

$$\delta u = \frac{\partial u}{\partial x}\delta x + \frac{\partial u}{\partial y}\delta y + \frac{\partial u}{\partial z}\delta z$$

One or two straightforward applications will lay the foundations for future development.

Example 1 A rectangular box has sides measured as 30 mm, 40 mm and 60 mm. If these measurements are liable to be in error by ± 0.5 mm, ± 0.8 mm and ± 1.0 mm respectively, calculate the length of the diagonal of the box and the maximum possible error in the result.

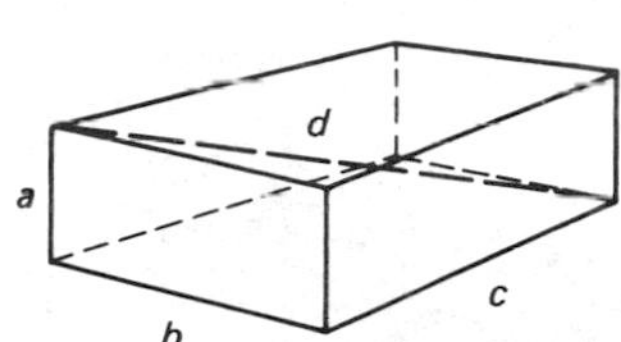

First build up an expression for the diagonal d in terms of the sides, a, b and c.

$d = \ldots\ldots\ldots$

6

$$\boxed{d = (a^2 + b^2 + c^2)^{\frac{1}{2}}}$$

Then $\quad \delta d = \dfrac{\partial d}{\partial a}\delta a + \dfrac{\partial d}{\partial b}\delta b + \dfrac{\partial d}{\partial c}\delta c$

We now determine the partial differential coefficients and obtain an expression for δd, but all in terms of a, b and c. Do not yet insert numerical values.

$\delta d = \ldots\ldots\ldots$

7

$$\delta d = \frac{1}{\sqrt{a^2 + b^2 + c^2}}\{a\delta a + b\delta b + c\delta c\}$$

Now, substituting the values $a = 30$, $b = 40$, $c = 60$
$\delta a = \pm 0.5$, $\delta b = \pm 0.8$, $\delta c = \pm 1.0$
the calculated length of the diagonal = ...
the maximum possible error = ...

8

diagonal $= \sqrt{a^2 + b^2 + c^2} = 78.10$ mm
maximum error $= \pm 1.37$ mm

for $\delta d = \dfrac{1}{78.10}\{30(\pm 0.5) + 40(\pm 0.8) + 60(\pm 1.0)\}$

Greatest error when the signs are the same

$$\therefore \delta d = \frac{1}{78.10}\{\pm (15 + 32 + 60)\} = \pm 1.37 \text{ mm}$$

Rates of change

If $z = f(x, y)$, then we have seen that $\delta z = \dfrac{\partial z}{\partial x}\delta x + \dfrac{\partial z}{\partial y}\delta y$

Dividing through by δt, $\quad \dfrac{\delta z}{\delta t} = \dfrac{\partial z}{\partial x}\dfrac{\delta x}{\delta t} + \dfrac{\partial z}{\partial y}\dfrac{\delta y}{\delta t}$

Then if $\delta t \to 0$, $\quad \dfrac{\mathrm{d}z}{\mathrm{d}t} = \dfrac{\partial z}{\partial x}\dfrac{\mathrm{d}x}{\mathrm{d}t} + \dfrac{\partial z}{\partial y}\dfrac{\mathrm{d}y}{\mathrm{d}t}$

Note the result. Then on to an example.

Example The base radius r of a right circular cone is increasing at the rate of 1.5 mm/s while the perpendicular height is decreasing at 6.0 mm/s. Determine the rate at which the volume V is changing when $r = 12$ mm and $h = 24$ mm.

Find an expression for $\dfrac{\mathrm{d}V}{\mathrm{d}t}$ in terms of r and h which is ....

9

$$\boxed{\frac{dV}{dt} = \frac{2\pi rh}{3} \cdot \frac{dr}{dt} + \frac{\pi r^2}{3} \cdot \frac{dh}{dt}}$$

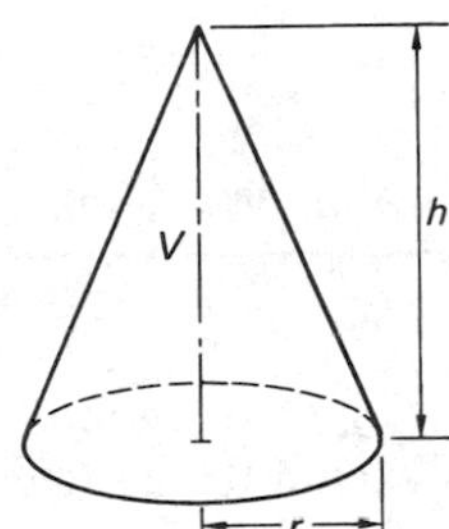

$$V = \frac{1}{3}\pi r^2 h; \qquad \frac{dV}{dt} = \frac{\partial V}{\partial r} \cdot \frac{dr}{dt} + \frac{\partial V}{\partial h} \cdot \frac{dh}{dt}$$

$$\frac{\partial V}{\partial r} = \frac{2\pi rh}{3}; \qquad \frac{\partial V}{\partial h} = \frac{\pi r^2}{3}$$

$$\therefore \frac{dV}{dt} = \frac{2\pi rh}{3} \cdot \frac{dr}{dt} + \frac{\pi r^2}{3} \cdot \frac{dh}{dt}$$

Finally, we insert the numerical values

$$r = 12; \quad h = 24; \quad \frac{dr}{dt} = 1.5; \quad \frac{dh}{dt} = -6.0 \quad (h \text{ is decreasing})$$

$$\frac{dV}{dt} = 288\pi - 288\pi = 0$$

$\therefore$ At the instant when $r = 12$ mm and $h = 24$ mm,

the volume is unchanging.

Implicit functions

The same initial result, $\delta z = \frac{\partial z}{\partial x}\delta x + \frac{\partial z}{\partial y}\delta y$ enables us to determine the differential coefficient of an implicit function $f(x, y) = 0$, i.e. in a case where y is not defined explicitly in terms of x.

If $f(x, y) = 0$ is an implicit function, we let $z = f(x, y)$.

Then, as before, $$\delta z = \frac{\partial z}{\partial x}\delta x + \frac{\partial z}{\partial y}\delta y$$

Dividing through by δx, $$\frac{\delta z}{\delta x} = \frac{\partial z}{\partial x} + \frac{\partial z}{\partial y} \cdot \frac{\delta y}{\delta x}$$

Then, if $\delta x \to 0$, $$\frac{dz}{dx} = \frac{\partial z}{\partial x} + \frac{\partial z}{\partial y} \cdot \frac{dy}{dx}$$

But $z = 0 \quad \therefore \frac{dz}{dx} = 0$ $$\therefore \frac{\partial z}{\partial x} + \frac{\partial z}{\partial y} \cdot \frac{dy}{dx} = 0$$

$$\therefore \frac{dy}{dx} = -\left(\frac{\partial z}{\partial x} \Big/ \frac{\partial z}{\partial y}\right)$$

So, if $x^2 - xy - y^2 = 0$, $\frac{dy}{dx} = \ldots\ldots\ldots$

10

$$\boxed{\frac{dy}{dx} = \frac{2x - y}{x + 2y}}$$

Putting $z = x^2 - xy - y^2$, $\dfrac{\partial z}{\partial x} = 2x - y$ and $\dfrac{\partial z}{\partial y} = -x - 2y$
The rest follows immediately.

Now on to the next frame.

11

The work so far, important though it is, is largely by way of revision of the more basic ideas of partial differentiation. We now extend these same ideas to further applications.

Change of Variables

If $z = f(x, y)$ and x and y are themselves functions of two new independent variables, u and v, then we need expressions for $\dfrac{\partial z}{\partial u}$ and $\dfrac{\partial z}{\partial v}$.

Yet again, we start from the result we established at the beginning of this programme.

If $z = f(x, y)$ then $\delta z = \dfrac{\partial z}{\partial x}\delta x + \dfrac{\partial z}{\partial y}\delta y$

Dividing in turn by δu and δv

$$\frac{\delta z}{\delta u} = \frac{\partial z}{\partial x}\cdot\frac{\delta x}{\delta u} + \frac{\partial z}{\partial y}\cdot\frac{\delta y}{\delta u}$$

$$\frac{\delta z}{\delta v} = \frac{\partial z}{\partial x}\cdot\frac{\delta x}{\delta v} + \frac{\partial z}{\partial y}\cdot\frac{\delta y}{\delta v}$$

Then, as $\delta u \to 0$ and $\delta v \to 0$, these become

$$\underline{\frac{\partial z}{\partial u} = \frac{\partial z}{\partial x}\cdot\frac{\partial x}{\partial u} + \frac{\partial z}{\partial y}\cdot\frac{\partial y}{\partial u}}$$

$$\underline{\frac{\partial z}{\partial v} = \frac{\partial z}{\partial x}\cdot\frac{\partial x}{\partial v} + \frac{\partial z}{\partial y}\cdot\frac{\partial y}{\partial v}}$$

Example 1 If $z = x^2 - y^2$ and $x = r\cos\theta$ and $y = r\sin\theta$, then

$$\frac{\partial z}{\partial r} = \frac{\partial z}{\partial x}\cdot\frac{\partial x}{\partial r} + \frac{\partial z}{\partial y}\cdot\frac{\partial y}{\partial r}$$

and

$$\frac{\partial z}{\partial \theta} = \frac{\partial z}{\partial x}\cdot\frac{\partial x}{\partial \theta} + \frac{\partial z}{\partial y}\cdot\frac{\partial y}{\partial \theta}$$

We now need the various partial derivatives

$$\frac{\partial z}{\partial x} = \dots\dots\dots; \quad \frac{\partial x}{\partial r} = \dots\dots\dots; \quad \frac{\partial y}{\partial r} = \dots\dots\dots$$

$$\frac{\partial z}{\partial y} = \dots\dots\dots; \quad \frac{\partial x}{\partial \theta} = \dots\dots\dots; \quad \frac{\partial y}{\partial \theta} = \dots\dots\dots$$

12

$$\frac{\partial z}{\partial x} = 2x; \quad \frac{\partial x}{\partial r} = \cos\theta; \quad \frac{\partial y}{\partial r} = \sin\theta$$

$$\frac{\partial z}{\partial y} = -2y; \quad \frac{\partial x}{\partial \theta} = -r\sin\theta; \quad \frac{\partial y}{\partial \theta} = r\cos\theta$$

Substituting in the two equations and simplifying

$$\frac{\partial z}{\partial r} = \dots\dots\dots; \quad \frac{\partial z}{\partial \theta} = \dots\dots\dots$$

13

$$\frac{\partial z}{\partial r} = 2x\cos\theta - 2y\sin\theta; \quad \frac{\partial z}{\partial \theta} = -(2xr\sin\theta + 2yr\cos\theta)$$

Finally, we can express x and y in terms of r and θ as given, so that, after tidying up, we obtain

$$\frac{\partial z}{\partial r} = \dots\dots\dots; \quad \frac{\partial z}{\partial \theta} = \dots\dots\dots$$

14

$$\frac{\partial z}{\partial r} = 2r(\cos^2\theta - \sin^2\theta); \quad \frac{\partial z}{\partial \theta} = -4r^2\sin\theta\cos\theta$$

Of course, we could express these as

$$\underline{\frac{\partial z}{\partial r} = 2r\cos 2\theta} \quad \text{and} \quad \underline{\frac{\partial z}{\partial \theta} = -2r^2\sin 2\theta}$$

From these results, we can, if necessary, find the second partial derivatives in the normal manner.

$$\frac{\partial^2 z}{\partial r^2} = \frac{\partial}{\partial r}\left(\frac{\partial z}{\partial r}\right) = \frac{\partial}{\partial r}(2r\cos 2\theta) = \underline{2\cos 2\theta}$$

Similarly $\dfrac{\partial^2 z}{\partial \theta^2} = \dots\dots\dots$ and $\dfrac{\partial^2 z}{\partial r\partial \theta} = \dots\dots\dots$

15

$$\frac{\partial^2 z}{\partial\theta^2} = -4r^2\cos 2\theta; \quad \frac{\partial^2 z}{\partial r\partial\theta} = -4r\sin 2\theta$$

for $\dfrac{\partial^2 z}{\partial\theta^2} = \dfrac{\partial}{\partial\theta}\left(\dfrac{\partial z}{\partial\theta}\right) = \dfrac{\partial}{\partial\theta}(-2r^2\sin 2\theta) = \underline{-4r^2\cos 2\theta}$

and $\dfrac{\partial^2 z}{\partial r\partial\theta} = \dfrac{\partial}{\partial r}\left(\dfrac{\partial z}{\partial\theta}\right) = \dfrac{\partial}{\partial r}(-2r^2\sin 2\theta) = \underline{-4r\sin 2\theta}$

Example 2 If $z = f(x, y)$ and $x = \frac{1}{2}(u^2 - v^2)$ and $y = uv$, show that

$$u\frac{\partial z}{\partial v} - v\frac{\partial z}{\partial u} = 2\left(x\frac{\partial z}{\partial y} - y\frac{\partial z}{\partial x}\right)$$

Although this is much the same as the last example, there is, at least, one difference. In this case, we are not told the precise nature of $f(x, y)$. We must remember that z is a function of x and y and, therefore, of u and v. With that in mind, we set off with the usual two equations.

$$\frac{\partial z}{\partial u} = \dots\dots\dots$$

$$\frac{\partial z}{\partial v} = \dots\dots\dots$$

16

$$\frac{\partial z}{\partial u} = \frac{\partial z}{\partial x}\cdot\frac{\partial x}{\partial u} + \frac{\partial z}{\partial y}\cdot\frac{\partial y}{\partial u}$$

$$\frac{\partial z}{\partial v} = \frac{\partial z}{\partial x}\cdot\frac{\partial x}{\partial v} + \frac{\partial z}{\partial y}\cdot\frac{\partial y}{\partial v}$$

From the given information

$$\frac{\partial x}{\partial u} = \dots\dots\dots; \quad \frac{\partial y}{\partial u} = \dots\dots\dots$$

$$\frac{\partial x}{\partial v} = \dots\dots\dots; \quad \frac{\partial y}{\partial v} = \dots\dots\dots$$

17

$$\frac{\partial x}{\partial u} = u; \qquad \frac{\partial y}{\partial u} = v$$
$$\frac{\partial x}{\partial v} = -v; \qquad \frac{\partial y}{\partial v} = u$$

Whereupon $\dfrac{\partial z}{\partial u} = \ldots\ldots\ldots$

$\dfrac{\partial z}{\partial v} = \ldots\ldots\ldots$

18

$$\frac{\partial z}{\partial u} = u\frac{\partial z}{\partial x} + v\frac{\partial z}{\partial y}$$

$$\frac{\partial z}{\partial v} = -v\frac{\partial z}{\partial x} + u\frac{\partial z}{\partial y}$$

If we now multiply the first of these by $(-v)$ and the second by u and add the two equations, we get the desired result.

$$-v\frac{\partial z}{\partial u} = -uv\frac{\partial z}{\partial x} - v^2\frac{\partial z}{\partial y}$$

$$u\frac{\partial z}{\partial v} = -uv\frac{\partial z}{\partial x} + u^2\frac{\partial z}{\partial y}$$

Adding $$u\frac{\partial z}{\partial v} - v\frac{\partial z}{\partial u} = -2uv\frac{\partial z}{\partial x} + (u^2 - v^2)\frac{\partial z}{\partial y}$$

$$= -2y\frac{\partial z}{\partial x} + 2x\frac{\partial z}{\partial y}$$

$$\therefore\ u\frac{\partial z}{\partial v} - v\frac{\partial z}{\partial u} = 2\left(x\frac{\partial z}{\partial y} - y\frac{\partial z}{\partial x}\right)$$

With the same given data, i.e.

$$z = f(x, y) \text{ with } x = \frac{1}{2}(u^2 - v^2) \text{ and } y = uv$$

we can now show that $\dfrac{\partial^2 z}{\partial u^2} + \dfrac{\partial^2 z}{\partial v^2} = (u^2 + v^2)\left(\dfrac{\partial^2 z}{\partial x^2} + \dfrac{\partial^2 z}{\partial y^2}\right)$.

In determining the second partial derivatives, keep in mind that z is a function of u and v and that both of these variables also occur in $\dfrac{\partial z}{\partial x}$ and $\dfrac{\partial z}{\partial y}$.

$$\frac{\partial^2 z}{\partial u^2} = \ldots\ldots\ldots$$

19

$$\boxed{\frac{\partial^2 z}{\partial u^2} = u^2\frac{\partial^2 z}{\partial x^2} + 2uv\frac{\partial^2 z}{\partial x \partial y} + v^2\frac{\partial^2 z}{\partial y^2}}$$

for $\dfrac{\partial}{\partial u} = \left(u\dfrac{\partial}{\partial x} + v\dfrac{\partial}{\partial y}\right)$ and $\dfrac{\partial}{\partial v} = \left(-v\dfrac{\partial}{\partial x} + u\dfrac{\partial}{\partial y}\right)$

$$\therefore \frac{\partial^2 z}{\partial u^2} = \frac{\partial}{\partial u}\left(u\frac{\partial z}{\partial x} + v\frac{\partial z}{\partial y}\right) = u\frac{\partial}{\partial u}\left(\frac{\partial z}{\partial x}\right) + v\frac{\partial}{\partial u}\left(\frac{\partial z}{\partial y}\right)$$

$$= u\left(u\frac{\partial}{\partial x} + v\frac{\partial}{\partial y}\right)\frac{\partial z}{\partial x} + v\left(u\frac{\partial}{\partial x} + v\frac{\partial}{\partial y}\right)\frac{\partial z}{\partial y}$$

$$= u^2\frac{\partial^2 z}{\partial x^2} + uv\frac{\partial^2 z}{\partial x \partial y} + uv\frac{\partial^2 z}{\partial x \partial y} + v^2\frac{\partial^2 z}{\partial y^2}$$

$$\therefore \frac{\partial^2 z}{\partial u^2} = u^2\frac{\partial^2 z}{\partial x^2} + 2uv\frac{\partial^2 z}{\partial x \partial y} + v^2\frac{\partial^2 z}{\partial y^2} \qquad (1)$$

Likewise, $\dfrac{\partial^2 z}{\partial v^2} = \dfrac{\partial}{\partial v}\left(\dfrac{\partial z}{\partial v}\right) = \dfrac{\partial}{\partial v}\left(-v\dfrac{\partial z}{\partial x} + u\dfrac{\partial z}{\partial y}\right)$

$$= \ldots\ldots\ldots$$

20

$$\frac{\partial^2 z}{\partial v^2} = v^2\frac{\partial^2 z}{\partial x^2} - 2uv\frac{\partial^2 z}{\partial x \partial y} + u^2\frac{\partial^2 z}{\partial y^2}$$

for $\quad \dfrac{\partial^2 z}{\partial v^2} = \dfrac{\partial}{\partial v}\left(-v\dfrac{\partial z}{\partial x} + u\dfrac{\partial z}{\partial y}\right)$

$$= -v\left(-v\frac{\partial}{\partial x} + u\frac{\partial}{\partial y}\right)\frac{\partial z}{\partial x} + u\left(-v\frac{\partial}{\partial x} + u\frac{\partial}{\partial y}\right)\frac{\partial z}{\partial y}$$

$$= v^2\frac{\partial^2 z}{\partial x^2} - uv\frac{\partial^2 z}{\partial x \partial y} - uv\frac{\partial^2 z}{\partial x \partial y} + u^2\frac{\partial^2 z}{\partial y^2}$$

$$\therefore \quad \frac{\partial^2 z}{\partial v^2} = v^2\frac{\partial^2 z}{\partial x^2} - 2uv\frac{\partial^2 z}{\partial x \partial y} + u^2\frac{\partial^2 z}{\partial y^2} \qquad (2)$$

Adding together results (1) and (2), we get

. . . .

21

$$\frac{\partial^2 z}{\partial u^2} + \frac{\partial^2 z}{\partial v^2} = (u^2 + v^2)\left(\frac{\partial^2 z}{\partial x^2} + \frac{\partial^2 z}{\partial y^2}\right)$$

and that is it.

Now, for something slightly different, move on to the next frame.

Inverse Functions

22

If $z = f(x, y)$ and x and y are functions of two independent variables u and v defined by $u = g(x, y)$ and $v = h(x, y)$, we can theoretically solve these two equations to obtain x and y in terms of u and v. Hence we can determine $\dfrac{\partial x}{\partial u}, \dfrac{\partial x}{\partial v}, \dfrac{\partial y}{\partial u}, \dfrac{\partial y}{\partial v}$ and then $\dfrac{\partial z}{\partial x}$ and $\dfrac{\partial z}{\partial y}$ as required.

In practice, however, the solution of $u = g(x, y)$ and $v = h(x, y)$ may well be difficult or even impossible by normal means. The following example shows how we can get over this difficulty.

Example 1 If $z = f(x, y)$ and $u = e^x \cos y$ and $v = e^{-x} \sin y$, we have to find $\dfrac{\partial x}{\partial u}, \dfrac{\partial x}{\partial v}, \dfrac{\partial y}{\partial u}, \dfrac{\partial y}{\partial v}$.

We start off once again with our standard relationships

$$\delta u = \frac{\partial u}{\partial x}\delta x + \frac{\partial u}{\partial y}\delta y \qquad (1)$$

$$\delta v = \frac{\partial v}{\partial x}\delta x + \frac{\partial v}{\partial y}\delta y \qquad (2)$$

Now $u = e^x \cos y$ and $v = e^{-x} \sin y$

So $\dfrac{\partial u}{\partial x} = \dots\dots\dots \qquad \dfrac{\partial u}{\partial y} = \dots\dots\dots$

$\dfrac{\partial v}{\partial x} = \dots\dots\dots \qquad \dfrac{\partial v}{\partial y} = \dots\dots\dots$

23

$$\frac{\partial u}{\partial x} = e^x \cos y; \qquad \frac{\partial u}{\partial y} = -e^x \sin y$$
$$\frac{\partial v}{\partial x} = -e^{-x} \sin y; \qquad \frac{\partial v}{\partial y} = e^{-x} \cos y$$

Substituting in equations (1) and (2) above, we have

$$\delta u = e^x \cos y\,\delta x - e^x \sin y\,\delta y \qquad (3)$$

$$\delta v = -e^{-x} \sin y\,\delta x + e^{-x} \cos y\,\delta y \qquad (4)$$

Eliminating δy from (3) and (4), we get

$$\delta x = \dots\dots\dots$$

24

$$\delta x = \frac{e^{-x}\cos y}{\cos 2y}\delta u + \frac{e^{x}\sin y}{\cos 2y}\delta v$$

for (3) $\times\, e^{-x}\cos y$: $\quad e^{-x}\cos y\,\delta u = \cos^2 y\,\delta x - \sin y\cos y\,\delta y$

(4) $\times\, e^{x}\sin y$: $\quad e^{x}\sin y\,\delta v = -\sin^2 y\,\delta x + \sin y\cos y\,\delta y$

Adding: $\quad e^{-x}\cos y\,\delta u + e^{x}\sin y\,\delta v = (\cos^2 y - \sin^2 y)\,\delta x$

$$\therefore\ \delta x = \frac{e^{-x}\cos y}{\cos 2y}\delta u + \frac{e^{x}\sin y}{\cos 2y}\delta v$$

But
$$\delta x = \frac{\partial x}{\partial u}\delta u + \frac{\partial x}{\partial v}\delta v$$

$$\therefore\ \underline{\frac{\partial x}{\partial u} = \frac{e^{-x}\cos y}{\cos 2y}} \quad \text{and} \quad \underline{\frac{\partial x}{\partial v} = \frac{e^{x}\sin y}{\cos 2y}}$$

which are, of course, two of the expressions we have to find.

Starting again with equations (3) and (4), we can obtain

$\delta y = \ldots\ldots\ldots$

25

$$\delta y = \frac{e^{-x}\sin y}{\cos 2y}\delta u + \frac{e^{x}\cos y}{\cos 2y}\delta v$$

for (3) $\times\, e^{-x}\sin y$: $\quad e^{-x}\sin y\,\delta u = \sin y\cos y\,\delta x - \sin^2 y\,\delta y$

(4) $\times\, e^{x}\cos y$: $\quad e^{x}\cos y\,\delta v = -\sin y\cos y\,\delta x + \cos^2 y\,\delta y$

Adding: $\quad e^{-x}\sin y\,\delta u + e^{x}\cos y\,\delta v = (\cos^2 y - \sin^2 y)\,\delta y$

$$\therefore\ \delta y = \frac{e^{-x}\sin y}{\cos 2y}\delta u + \frac{e^{x}\cos y}{\cos 2y}\delta v$$

But, $\delta y = \ldots\ldots\ldots$ Finish it off.

26

$$\delta y = \frac{\partial y}{\partial u}\delta u + \frac{\partial y}{\partial v}\delta v$$
$$\therefore \frac{\partial y}{\partial u} = \frac{e^{-x}\sin y}{\cos 2y} \quad \text{and} \quad \frac{\partial y}{\partial v} = \frac{e^{x}\cos y}{\cos 2y}$$

So, collecting our four results together

$$\frac{\partial x}{\partial u} = \frac{e^{-x}\cos y}{\cos 2y}; \quad \frac{\partial x}{\partial v} = \frac{e^{x}\sin y}{\cos 2y}$$

$$\frac{\partial y}{\partial u} = \frac{e^{-x}\sin y}{\cos 2y}; \quad \frac{\partial y}{\partial v} = \frac{e^{x}\cos y}{\cos 2y}$$

We can tackle most similar problems in the same way, but it is more efficient to investigate a general case and to streamline the results. Let us do that.

27

General case

If $z = f(x, y)$ with $u = g(x, y)$ and $v = h(x, y)$, then we have

$$\delta u = \frac{\partial u}{\partial x}\delta x + \frac{\partial u}{\partial y}\delta y \qquad (1)$$

$$\delta v = \frac{\partial v}{\partial x}\delta x + \frac{\partial v}{\partial y}\delta y \qquad (2)$$

We now solve these for δx and δy. Eliminating δy, we have

$(1) \times \dfrac{\partial v}{\partial y}$:
$$\frac{\partial v}{\partial y}\delta u = \frac{\partial v}{\partial y}\cdot\frac{\partial u}{\partial x}\delta x + \frac{\partial v}{\partial y}\cdot\frac{\partial u}{\partial y}\delta y$$

$(2) \times \dfrac{\partial u}{\partial y}$:
$$\frac{\partial u}{\partial y}\delta v = \frac{\partial u}{\partial y}\cdot\frac{\partial v}{\partial x}\delta x + \frac{\partial u}{\partial y}\cdot\frac{\partial v}{\partial y}\delta y$$

Subtracting:
$$\frac{\partial v}{\partial y}\delta u - \frac{\partial u}{\partial y}\delta v = \left(\frac{\partial u}{\partial x}\cdot\frac{\partial v}{\partial y} - \frac{\partial v}{\partial x}\cdot\frac{\partial u}{\partial y}\right)\delta x$$

$$\therefore \delta x = \frac{\dfrac{\partial v}{\partial y}\delta u - \dfrac{\partial u}{\partial y}\delta v}{\dfrac{\partial u}{\partial x}\cdot\dfrac{\partial v}{\partial y} - \dfrac{\partial v}{\partial x}\cdot\dfrac{\partial u}{\partial y}}$$

Starting afresh from (1) and (2) and eliminating δx, we have

$$\delta y = \ldots\ldots\ldots$$

28

$$\delta y = \frac{\dfrac{\partial u}{\partial x}\delta v - \dfrac{\partial v}{\partial x}\delta u}{\dfrac{\partial u}{\partial x}\cdot\dfrac{\partial v}{\partial y} - \dfrac{\partial v}{\partial x}\cdot\dfrac{\partial u}{\partial y}}$$

The two results so far are therefore

$$\delta x = \frac{\dfrac{\partial v}{\partial y}\delta u - \dfrac{\partial u}{\partial y}\delta v}{\dfrac{\partial u}{\partial x}\cdot\dfrac{\partial v}{\partial y} - \dfrac{\partial v}{\partial x}\cdot\dfrac{\partial u}{\partial y}} \quad \text{and} \quad \delta y = \frac{\dfrac{\partial u}{\partial x}\delta v - \dfrac{\partial v}{\partial x}\delta u}{\dfrac{\partial u}{\partial x}\cdot\dfrac{\partial v}{\partial y} - \dfrac{\partial v}{\partial x}\cdot\dfrac{\partial u}{\partial y}}$$

You will notice that the denominator is the same in each case and that it can be expressed in determinant form

$$\frac{\partial u}{\partial x}\cdot\frac{\partial v}{\partial y} - \frac{\partial v}{\partial x}\cdot\frac{\partial u}{\partial y} = \begin{vmatrix} \dfrac{\partial u}{\partial x} & \dfrac{\partial v}{\partial x} \\ \dfrac{\partial u}{\partial y} & \dfrac{\partial v}{\partial y} \end{vmatrix}$$

This determinant is called the *Jacobian* of u, v with respect to x, y and is denoted by the symbol J

i.e. $$\mathrm{J} = \begin{vmatrix} \dfrac{\partial u}{\partial x} & \dfrac{\partial v}{\partial x} \\ \dfrac{\partial u}{\partial y} & \dfrac{\partial v}{\partial y} \end{vmatrix} \quad \text{and is often written as} \quad \frac{\partial(u, v)}{\partial(x, y)}$$

So, $$\mathrm{J} = \frac{\partial(u, v)}{\partial(x, y)} = \begin{vmatrix} \dfrac{\partial u}{\partial x} & \dfrac{\partial v}{\partial x} \\ \dfrac{\partial u}{\partial y} & \dfrac{\partial v}{\partial y} \end{vmatrix}$$

Out last two results can therefore be written

$\delta x = \ldots\ldots\ldots;\qquad \delta y = \ldots\ldots\ldots$

29

$$\delta x = \frac{\dfrac{\partial v}{\partial y}\delta u - \dfrac{\partial u}{\partial y}\delta v}{J}; \quad \delta y = \frac{\dfrac{\partial u}{\partial x}\delta v - \dfrac{\partial v}{\partial x}\delta u}{J}$$

We can now get a number of useful relationships

(a) *If v is kept constant,* $\delta v = 0$ $\qquad \therefore \delta x = \dfrac{\partial v}{\partial y}\delta u \Big/ J$

Dividing by δu and letting $\delta u \to 0$ $\qquad \dfrac{\partial x}{\partial u} = \dfrac{\partial v}{\partial y}\Big/ J$

Similarly $\qquad \dfrac{\partial y}{\partial u} = -\dfrac{\partial v}{\partial x}\Big/ J$

(b) *If u is kept constant,* $\delta u = 0$ $\qquad \therefore \delta x = -\dfrac{\partial u}{\partial y}\delta v \Big/ J$

Dividing by δv and letting $\delta v \to 0$ $\qquad \dfrac{\partial x}{\partial v} = -\dfrac{\partial u}{\partial y}\Big/ J$

Similarly $\qquad \dfrac{\partial y}{\partial v} = \dfrac{\partial u}{\partial x}\Big/ J$

So, at this stage, we had better summarise the results.

Summary

If $z = f(x, y)$ and $u = g(x, y)$ and $v = h(x, y)$ then

$$\frac{\partial x}{\partial u} = \frac{\partial v}{\partial y}\Big/ J \qquad \frac{\partial x}{\partial v} = -\frac{\partial u}{\partial y}\Big/ J$$

$$\frac{\partial y}{\partial u} = -\frac{\partial v}{\partial x}\Big/ J \qquad \frac{\partial y}{\partial v} = \frac{\partial u}{\partial x}\Big/ J$$

where, in each case

$$J = \frac{\partial(u, v)}{\partial(x, y)} = \begin{vmatrix} \dfrac{\partial u}{\partial x} & \dfrac{\partial v}{\partial x} \\ \dfrac{\partial u}{\partial y} & \dfrac{\partial v}{\partial y} \end{vmatrix}$$

Let us put this into practice by doing again the same example that we started with (Example 1), but by the new method. First of all, however, make a note of the important summary listed above for future reference.

30 *Example 1A*

If $z=f(x,y)$ and $u=e^x\cos y$ and $v=e^{-x}\sin y$, find the derivatives $\dfrac{\partial x}{\partial u}, \dfrac{\partial x}{\partial v}, \dfrac{\partial y}{\partial u}, \dfrac{\partial y}{\partial v}$.

$$u=e^x\cos y \qquad v=e^{-x}\sin y$$

$$\frac{\partial u}{\partial x}=e^x\cos y \qquad \frac{\partial v}{\partial x}=-e^{-x}\sin y$$

$$\frac{\partial u}{\partial y}=-e^x\sin y \qquad \frac{\partial v}{\partial y}=e^{-x}\cos y$$

$$J=\frac{\partial(u,v)}{\partial(x,y)}=\begin{vmatrix}\dfrac{\partial u}{\partial x} & \dfrac{\partial v}{\partial x}\\ \dfrac{\partial u}{\partial y} & \dfrac{\partial v}{\partial y}\end{vmatrix}=\begin{vmatrix}e^x\cos y & -e^{-x}\sin y\\ -e^x\sin y & e^{-x}\cos y\end{vmatrix}$$

$$=(e^x\cos y)(e^{-x}\cos y)-(-e^x\sin y)(-e^{-x}\sin y)$$

$$=\cos^2 y - \sin^2 y = \cos 2y$$

Then $\dfrac{\partial x}{\partial u}=\dfrac{\partial v}{\partial y}\Big/J=\dfrac{e^{-x}\cos y}{\cos 2y}$; $\quad \dfrac{\partial x}{\partial v}=-\dfrac{\partial u}{\partial y}\Big/J=\dfrac{e^x\sin y}{\cos 2y}$

$$\frac{\partial y}{\partial u}=-\frac{\partial v}{\partial x}\Big/J=\frac{e^{-x}\sin y}{\cos 2y}; \qquad \frac{\partial y}{\partial v}=\frac{\partial u}{\partial x}\Big/J=\frac{e^x\cos y}{\cos 2y}$$

which is a lot shorter than our first approach.

Move on for a further example.

---

31 *Example 2*

If $z=f(x,y)$ with $u=x^2-y^2$ and $v=xy$, find expressions for $\dfrac{\partial x}{\partial u}$, $\dfrac{\partial x}{\partial v}, \dfrac{\partial y}{\partial u}, \dfrac{\partial y}{\partial v}$.

First we need

$$\frac{\partial u}{\partial x}=\dots\dots\dots; \quad \frac{\partial u}{\partial y}=\dots\dots\dots; \quad \frac{\partial v}{\partial x}=\dots\dots\dots; \quad \frac{\partial v}{\partial y}=\dots\dots\dots$$

32

$$\boxed{\frac{\partial u}{\partial x}=2x; \quad \frac{\partial u}{\partial y}=-2y; \quad \frac{\partial v}{\partial x}=y; \quad \frac{\partial v}{\partial y}=x}$$

Then we calculate J which, in this case, is ....

33

$$\boxed{J = 2(x^2 + y^2)}$$

$$\text{for} \quad J = \frac{\partial(u, v)}{\partial(x, y)} = \begin{vmatrix} \frac{\partial u}{\partial x} & \frac{\partial v}{\partial x} \\ \frac{\partial u}{\partial y} & \frac{\partial v}{\partial y} \end{vmatrix} = \begin{vmatrix} 2x & y \\ -2y & x \end{vmatrix} = 2x^2 + 2y^2$$

Finally, we have the four relationships

$$\frac{\partial x}{\partial u} = \frac{\partial v}{\partial y} \Big/ J = \ldots\ldots\ldots; \qquad \frac{\partial x}{\partial v} = -\frac{\partial u}{\partial y} \Big/ J = \ldots\ldots\ldots$$

$$\frac{\partial y}{\partial u} = -\frac{\partial v}{\partial x} \Big/ J = \ldots\ldots\ldots; \qquad \frac{\partial y}{\partial v} = \frac{\partial u}{\partial x} \Big/ J = \ldots\ldots\ldots$$

34

$$\frac{\partial x}{\partial u} = \frac{x}{2(x^2 + y^2)}; \qquad \frac{\partial x}{\partial v} = -\frac{y}{x^2 + y^2}$$

$$\frac{\partial y}{\partial u} = \frac{-y}{2(x^2 + y^2)}; \qquad \frac{\partial y}{\partial v} = \frac{x}{x^2 + y^2}$$

And that is all there is to it.

If we know the details of the function $z = f(x, y)$ then we can go one stage further and use the results $\frac{\partial x}{\partial u}, \frac{\partial x}{\partial v}, \frac{\partial y}{\partial u}, \frac{\partial y}{\partial v}$ to find $\frac{\partial z}{\partial u}$ and $\frac{\partial z}{\partial v}$.

Let us see this in a further example.

Example 3

If $z = 2x^2 + 3xy + 4y^2$ and $u = x^2 + y^2$ and $v = x + 2y$, determine

(a) $\frac{\partial x}{\partial u}, \frac{\partial x}{\partial v}, \frac{\partial y}{\partial u}, \frac{\partial y}{\partial v}$ (b) $\frac{\partial z}{\partial u}$ and $\frac{\partial z}{\partial v}$.

Section (a) is just like the previous example. Complete that on your own.

35

$$\boxed{\frac{\partial x}{\partial u} = \frac{1}{2x - y}; \quad \frac{\partial x}{\partial v} = \frac{-y}{2x - y}; \quad \frac{\partial y}{\partial u} = \frac{-1}{2(2x - y)}; \quad \frac{\partial y}{\partial v} = \frac{x}{2x - y}}$$

for, if $u = x^2 + y^2$ and $v = x + 2y$

$$\frac{\partial u}{\partial x} = 2x; \quad \frac{\partial u}{\partial y} = 2y; \quad \frac{\partial v}{\partial x} = 1; \quad \frac{\partial v}{\partial y} = 2$$

$$\mathrm{J} = \frac{\partial(u, v)}{\partial(x, y)} = \begin{vmatrix} \dfrac{\partial u}{\partial x} & \dfrac{\partial v}{\partial x} \\ \dfrac{\partial u}{\partial y} & \dfrac{\partial v}{\partial y} \end{vmatrix} = \begin{vmatrix} 2x & 1 \\ 2y & 2 \end{vmatrix} = 4x - 2y = 2(2x - y)$$

Then $\dfrac{\partial x}{\partial u} = \dfrac{\partial v}{\partial y} \Big/ \mathrm{J} = 2 \Big/ 2(2x - y) = \dfrac{1}{2x - y}$

$$\frac{\partial x}{\partial v} = -\frac{\partial u}{\partial y} \Big/ \mathrm{J} = -2y \Big/ 2(2x - y) = \frac{-y}{2x - y}$$

$$\frac{\partial y}{\partial u} = -\frac{\partial v}{\partial x} \Big/ \mathrm{J} = -1 \Big/ 2(2x - y) = \frac{-1}{2(2x - y)}$$

$$\frac{\partial y}{\partial v} = \frac{\partial u}{\partial x} \Big/ \mathrm{J} = 2x \Big/ 2(2x - y) = \frac{x}{2x - y}$$

$$\therefore \frac{\partial x}{\partial u} = \frac{1}{2x - y}; \quad \frac{\partial x}{\partial v} = \frac{-y}{2x - y}; \quad \frac{\partial y}{\partial u} = \frac{-1}{2(2x - y)}; \quad \frac{\partial y}{\partial v} = \frac{x}{2x - y}$$

Now for part (b).

Since z is also a function of u and v, the expressions for $\dfrac{\partial z}{\partial u}$ and $\dfrac{\partial z}{\partial v}$ are

$$\frac{\partial z}{\partial u} = \ldots\ldots\ldots$$

$$\frac{\partial z}{\partial v} = \ldots\ldots\ldots$$

36

$$\boxed{\begin{aligned}\frac{\partial z}{\partial u} &= \frac{\partial z}{\partial x}\cdot\frac{\partial x}{\partial u} + \frac{\partial z}{\partial y}\cdot\frac{\partial y}{\partial u}\\ \frac{\partial z}{\partial v} &= \frac{\partial z}{\partial x}\cdot\frac{\partial x}{\partial v} + \frac{\partial z}{\partial y}\cdot\frac{\partial y}{\partial v}\end{aligned}}$$

The only remaining items of information we need are the expressions for $\dfrac{\partial z}{\partial x}$ and $\dfrac{\partial z}{\partial y}$ which we obtain from $z = 2x^2 + 3xy + 4y^2$

$$\frac{\partial z}{\partial x} = 4x + 3y \quad \text{and} \quad \frac{\partial z}{\partial y} = 3x + 8y$$

Using these and the previous set of derivatives, we now get

$$\frac{\partial z}{\partial u} = \ldots\ldots\ldots; \quad \frac{\partial z}{\partial v} = \ldots\ldots\ldots$$

37

$$\boxed{\frac{\partial z}{\partial u} = \frac{5x - 2y}{2(2x - y)}; \qquad \frac{\partial z}{\partial v} = \frac{3x^2 + 4xy - 3y^2}{2x - y}}$$

for $\dfrac{\partial z}{\partial u} = \dfrac{\partial z}{\partial x}\cdot\dfrac{\partial x}{\partial u} + \dfrac{\partial z}{\partial y}\cdot\dfrac{\partial y}{\partial u}$

$$\therefore \frac{\partial z}{\partial u} = (4x + 3y)\left\{\frac{1}{2x - y}\right\} + (3x + 8y)\left\{\frac{-1}{2(2x - y)}\right\}$$

$$= \frac{5x - 2y}{2(2x - y)} \qquad \therefore \underline{\frac{\partial z}{\partial u} = \frac{5x - 2y}{2(2x - y)}}$$

and $\dfrac{\partial z}{\partial v} = \dfrac{\partial z}{\partial x}\cdot\dfrac{\partial x}{\partial v} + \dfrac{\partial z}{\partial y}\cdot\dfrac{\partial y}{\partial v}$

$$\therefore \frac{\partial z}{\partial v} = (4x + 3y)\left\{\frac{-y}{2x - y}\right\} + (3x + 8y)\left\{\frac{x}{2x - y}\right\}$$

$$= \frac{3x^2 + 4xy - 3y^2}{2x - y} \qquad \therefore \underline{\frac{\partial z}{\partial v} = \frac{3x^2 + 4xy - 3y^2}{2x - y}}$$

They are all done in the same general way.

Now on to the next topic.

Stationary Values of a Function

38

You will doubtless remember that in earlier work you established the characteristics of *turning points* on a plane curve and derived the conditions that enable these critical points to be calculated.

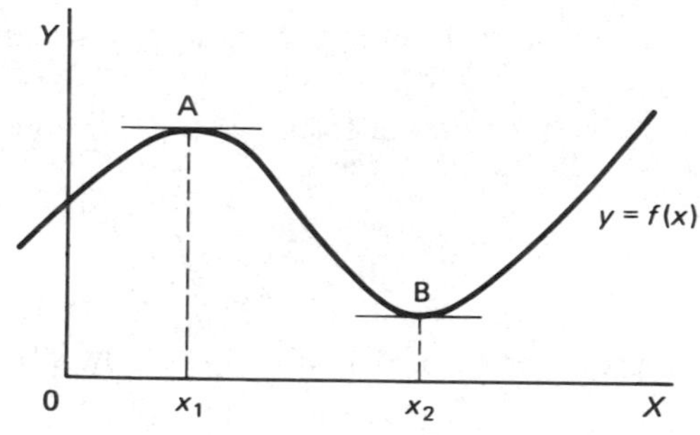

At A and B $\quad \dfrac{dy}{dx} = 0$

For maximum $\quad \dfrac{d^2y}{dx^2}$ is negative $(x = x_1)$

For minimum $\quad \dfrac{d^2y}{dx^2}$ is positive $(x = x_2)$

We now progress to the application of these same considerations to three dimensions, where $z = f(x, y)$. The function is now represented by a surface and stationary values of the function $z = f(x, y)$ occur when the tangent plane to the surface at a point $P(a, b)$ is parallel to the plane $z = 0$, i.e. to the xy-plane.

Let us take a closer look at this.

39

Maximum and minimum values

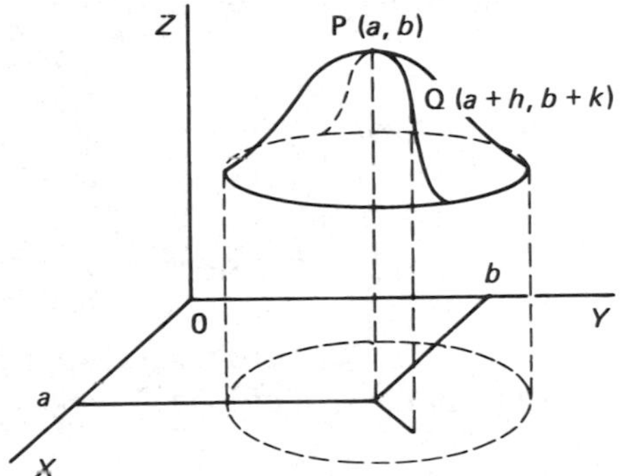

A function $z = f(x, y)$ is said to have *maximum* value at $P(a, b)$ if $f(a, b)$ is greater than the value at a near-by point $Q(a + h, b + k)$ for all values of h and k however small, positive or negative, i.e. in all directions from P.

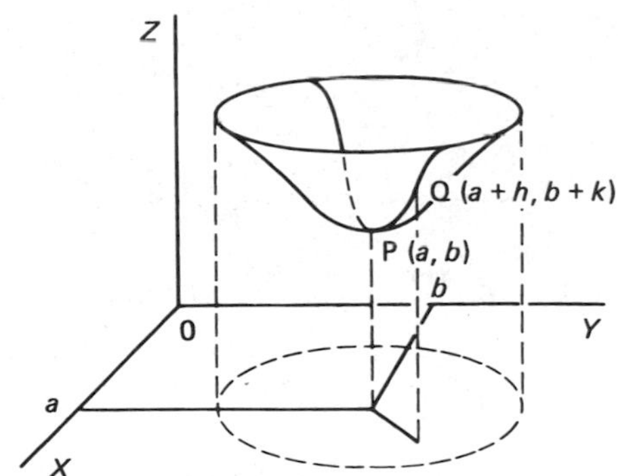

Similarly, $z = f(x, y)$ is said to have a *minimum* value at $P(a, b)$ if $f(a, b)$ is less than the value at a neighbouring point $Q(a + h, b + k)$ in any direction from P.

To establish maximum and minimum values, we must therefore investigate the sign of the value of $f(a + h, b + k) - f(a, b)$.

If $f(a + h, b + k) - f(a, b) < 0$ we have a maximum value at $P(a, b)$.

If $f(a + h, b + k) - f(a, b) > 0$ we have a minimum value at $P(a, b)$.

To pursue this further, we turn, once again, to Taylor's theorem.

$$f(a+h,b+k)=f(a,b)+h\frac{\partial f}{\partial x}+k\frac{\partial f}{\partial y}$$
$$+\frac{1}{2!}\left(h^2\frac{\partial^2 f}{\partial x^2}+2hk\frac{\partial^2 f}{\partial x\partial y}+k^2\frac{\partial^2 f}{\partial y^2}\right)+\dots$$

$$\therefore f(a+h,b+k)-f(a,b)=h\frac{\partial f}{\partial x}+k\frac{\partial f}{\partial y}$$
$$+\frac{1}{2!}\left(h^2\frac{\partial^2 f}{\partial x^2}+2hk\frac{\partial^2 f}{\partial x\partial y}+k^2\frac{\partial^2 f}{\partial y^2}\right)+\dots$$

For very small values of h and k, the third and subsequent terms are of a higher order of smallness and the sign of the left-hand side is determined by the sign of $h\frac{\partial f}{\partial x}+k\frac{\partial f}{\partial y}$.

For a stationary value of z at (a,b),

$$f(a+h,b+k)-f(a,b)=\dots\dots\dots$$

40

0

$$h\frac{\partial f}{\partial x}+k\frac{\partial f}{\partial y}=0$$

and since h and k are small independent increments

$$h\frac{\partial f}{\partial x}=0 \quad\text{and}\quad k\frac{\partial f}{\partial y}=0$$

$\therefore$ For $z=f(x,y)$ to have a stationary value

$$\frac{\partial f}{\partial x}=0 \quad\text{and}\quad \frac{\partial f}{\partial y}=0$$

Example 1 Determine the values of x and y at which stationary values of $z=5xy-6x^2-y^2+7x-2y$ occur.

All we need to do is to obtain expressions for $\frac{\partial z}{\partial x}$ and $\frac{\partial z}{\partial y}$; equate each to zero; and solve the pair of simultaneous equations obtained.

In which case $x=\dots\dots\dots$; $y=\dots\dots\dots$

41

$$\boxed{x = -4; \qquad y = -11}$$

for $z = 5xy - 6x^2 - y^2 + 7x - 2y \quad \therefore \quad \dfrac{\partial z}{\partial x} = 5y - 12x + 7$

$$\therefore \quad \frac{\partial z}{\partial y} = 5x - 2y - 2$$

Then solving $\left.\begin{aligned} 12x - 5y - 7 &= 0 \\ 5x - 2y - 2 &= 0 \end{aligned}\right\}$ gives $\underline{x = -4; \; y = -11}$

Although a stationary value occurs at $(-4, -11)$ we have no evidence as to whether it is a maximum or minimum value. Let us investigate further.

From the previous definitions

$f(a, b)$ will be a maximum value if $f(a+h, b+k) - f(a, b) < 0$

$f(a, b)$ will be a minimum value if $f(a+h, b+k) - f(a, b) > 0$.

We have already seen that at a stationary value $h\dfrac{\partial f}{\partial x} + k\dfrac{\partial f}{\partial y} = 0$.

Then Taylor's theorem becomes

$$f(a+h, b+k) - f(a, b) = \frac{1}{2!}\left(h^2\frac{\partial^2 f}{\partial x^2} + 2hk\frac{\partial^2 f}{\partial x \partial y} + k^2\frac{\partial^2 f}{\partial y^2}\right) + \ldots$$

where subsequent terms are of higher orders of h and k and are neglected.

The expression in the brackets on the right-hand side can be written as

$$\frac{1}{\dfrac{\partial^2 f}{\partial x^2}}\left\{\left(h\frac{\partial^2 f}{\partial x^2} + k\frac{\partial^2 f}{\partial x \partial y}\right)^2 + k^2\left(\frac{\partial^2 f}{\partial x^2}\cdot\frac{\partial^2 f}{\partial y^2} - \left[\frac{\partial^2 f}{\partial x \partial y}\right]^2\right)\right\}$$

Take a moment and square out the brackets and confirm that this is so.

42 So $h^2\dfrac{\partial^2 f}{\partial x^2}+2hk\dfrac{\partial^2 f}{\partial x\partial y}+k^2\dfrac{\partial^2 f}{\partial y^2}$

$$=\frac{1}{\dfrac{\partial^2 f}{\partial x^2}}\left\{\left(h\frac{\partial^2 f}{\partial x^2}+k\frac{\partial^2 f}{\partial x\partial y}\right)^2+k^2\left(\frac{\partial^2 f}{\partial x^2}\cdot\frac{\partial^2 f}{\partial y^2}-\left[\frac{\partial^2 f}{\partial x\partial y}\right]^2\right)\right\}$$

Now $\left(h\dfrac{\partial^2 f}{\partial x^2}+k\dfrac{\partial^2 f}{\partial x\partial y}\right)^2$ being a square, is always positive and if $\dfrac{\partial^2 f}{\partial x^2}\cdot\dfrac{\partial^2 f}{\partial y^2}>\left[\dfrac{\partial^2 f}{\partial x\partial y}\right]^2$ the second term will also be positive. In that case the sign of the whole expression is given by that of $\dfrac{\partial^2 f}{\partial x^2}$ at the front.

Furthermore, if $\dfrac{\partial^2 f}{\partial x^2}\cdot\dfrac{\partial^2 f}{\partial y^2}>\left(\dfrac{\partial^2 f}{\partial x\partial y}\right)^2$ i.e. $\dfrac{\partial^2 f}{\partial x^2}\cdot\dfrac{\partial^2 f}{\partial y^2}-\left(\dfrac{\partial^2 f}{\partial x\partial y}\right)^2>0$ this can be so only if $\dfrac{\partial^2 f}{\partial x^2}$ and $\dfrac{\partial^2 f}{\partial y^2}$ have the same sign.

Therefore, for $f(a,b)$ to be a maximum, $\dfrac{\partial^2 f}{\partial x^2}$ and $\dfrac{\partial^2 f}{\partial y^2}$ are both negative

and for $f(a,b)$ to be a minimum, $\dfrac{\partial^2 f}{\partial x^2}$ and $\dfrac{\partial^2 f}{\partial y^2}$ are both positive.

So, to determine whether a known stationary value is a maximum or a minimum value, we must find the second derivatives $\dfrac{\partial^2 f}{\partial x^2}$, $\dfrac{\partial^2 f}{\partial y^2}$ and $\dfrac{\partial^2 f}{\partial x\partial y}$.

Then

(a) If $\dfrac{\partial^2 f}{\partial x^2}\cdot\dfrac{\partial^2 f}{\partial y^2}-\left(\dfrac{\partial^2 f}{\partial x\partial y}\right)^2>0$, the stationary value is a true maximum or minimum value.

(b) In that case

(i) if $\dfrac{\partial^2 f}{\partial x^2}$ and $\dfrac{\partial^2 f}{\partial y^2}$ are both *negative*, $f(a,b)$ is a *maximum*

(ii) if $\dfrac{\partial^2 f}{\partial x^2}$ and $\dfrac{\partial^2 f}{\partial y^2}$ are both *positive*, $f(a,b)$ is a *minimum*.

Make a careful note of the conclusions (a) and (b): then let us apply them.

43

Example 1 Investigate stationary values of the function

$$z = x^2 + xy + y^2 + 5x - 5y + 3$$

First, to find where stationary values, if any, occur, we have

$$\frac{\partial z}{\partial x} = 0 \text{ and } \frac{\partial z}{\partial y} = 0$$

This is so when $x = \ldots\ldots\ldots;\ y = \ldots\ldots\ldots$

44

$$\boxed{x = -5; \qquad y = 5}$$

for $\dfrac{\partial z}{\partial x} = 2x + y + 5$ and $\dfrac{\partial z}{\partial y} = x + 2y - 5$

Solving $\left.\begin{aligned} 2x + \ y + 5 &= 0 \\ x + 2y - 5 &= 0 \end{aligned}\right\}$ $\qquad x = -5,\ y = 5$

This tells us that, if a stationary value exists, it occurs at $(-5, 5)$.

Next, we investigate the value of $\left(\dfrac{\partial^2 z}{\partial x^2}\right)\left(\dfrac{\partial^2 z}{\partial y^2}\right) - \left(\dfrac{\partial^2 z}{\partial x \partial y}\right)^2$. If this is greater than zero at $(-5, 5)$, it will confirm that a stationary value does in fact occur at $(-5, 5)$.

Check whether this is so.

45

$$\boxed{\text{yes.} \quad \left(\frac{\partial^2 z}{\partial x^2}\right)\left(\frac{\partial^2 z}{\partial y^2}\right) - \left(\frac{\partial^2 z}{\partial x \partial y}\right)^2 > 0}$$

since $\qquad \dfrac{\partial^2 z}{\partial x^2} = 2; \quad \dfrac{\partial^2 z}{\partial y^2} = 2; \quad \dfrac{\partial^2 z}{\partial x \partial y} = 1.$

This confirms that $(-5, 5)$ is either a maximum or a minimum.

To decide which it is, we note that $\dfrac{\partial^2 z}{\partial x^2}$ and $\dfrac{\partial^2 z}{\partial y^2}$ are both *positive*.

$\therefore$ at $(-5, 5)$, z is a $\ldots\ldots\ldots$

46

minimum

Of course, to find the actual minimum value of z we could substitute $x = -5$ and $y = 5$ in the expression for z. That is really all there is to it. Another example.

Example 2 Determine the stationary values, if any, of the function

$$z = x^3 - 6xy + y^3$$

The four steps in the routine are:

(a) Find $\dfrac{\partial z}{\partial x}$ and $\dfrac{\partial z}{\partial y}$ and solve the equations $\dfrac{\partial z}{\partial x} = 0$ and $\dfrac{\partial z}{\partial y} = 0$.

(b) Verify that $\left(\dfrac{\partial^2 z}{\partial x^2}\right)\left(\dfrac{\partial^2 z}{\partial y^2}\right) - \left(\dfrac{\partial^2 z}{\partial x \partial y}\right)^2 > 0$.

(c) Note the sign of $\dfrac{\partial^2 z}{\partial x^2}$ and $\dfrac{\partial^2 z}{\partial y^2}$ to distinguish between max. and min.

(d) Evaluate the maximum or minimum value of z.

In this example, possible stationary values occur at ....

47

$(0, 0)$ and $(2, 2)$

for $z = x^3 - 6xy + y^3 \quad \therefore \dfrac{\partial z}{\partial x} = 3x^2 - 6y \qquad \dfrac{\partial z}{\partial y} = -6x + 3y^2$

$$\frac{\partial z}{\partial x} = 0 \text{ and } \frac{\partial z}{\partial y} = 0 \quad \therefore x^2 - 2y = 0 \text{ and } -2x + y^2 = 0$$

$$\therefore (x^2 - y^2) + 2(x - y) = 0 \quad \therefore (x - y)(x + y + 2) = 0$$

$$\therefore y = x \quad \text{or} \quad y = -(x + 2)$$

(a) If $y = -(x + 2)$, $\quad x^2 + 2(x + 2) = 0 \quad \therefore x^2 + 2x + 4 = 0$

for which there is no real value of x.

(b) If $y = x$, $\quad x^2 - 2x = 0 \quad \therefore x = 0 \quad \text{or} \quad x = 2$

$\qquad y = 0 \quad \text{or} \quad y = 2$

$\therefore$ There are possible stationary values at (0, 0) and (2, 2)

Next we confirm that $\left(\dfrac{\partial^2 z}{\partial x^2}\right)\left(\dfrac{\partial^2 z}{\partial y^2}\right) - \left(\dfrac{\partial^2 z}{\partial x \partial y}\right)^2 > 0$ Result ....

48

No stationary value at (0, 0); Stationary value at (2, 2)

$$\frac{\partial z}{\partial x} = 3x^2 - 6y \qquad \therefore \frac{\partial^2 z}{\partial x^2} = 6x$$

$$\frac{\partial^2 z}{\partial x \partial y} = -6$$

$$\frac{\partial z}{\partial y} = -6x + 3y^2 \qquad \therefore \frac{\partial^2 z}{\partial y^2} = 6y$$

$\therefore$ at (0, 0) $$\left(\frac{\partial^2 z}{\partial x^2}\right)\left(\frac{\partial^2 z}{\partial y^2}\right) - \left(\frac{\partial^2 z}{\partial x \partial y}\right)^2 = (0)(0) - 36 < 0$$

$\therefore$ No max. or min. at (0, 0)

At (2, 2) $$\left(\frac{\partial^2 z}{\partial x^2}\right)\left(\frac{\partial^2 z}{\partial y^2}\right) - \left(\frac{\partial^2 z}{\partial x \partial y}\right)^2 = (12)(12) - 36 > 0$$

$\therefore$ There is a stationary value at (2, 2)

We see that at (2, 2) both $\frac{\partial^2 z}{\partial x^2}$ and $\frac{\partial^2 z}{\partial y^2}$ are positive. Therefore the stationary value at (2, 2) is a ....

49

minimum

Finally, the minimum value of z is ....

50

-8

Therefore, $z_{min} = -8$ and occurs at (2, 2)

Before doing a further example, let us consider one other aspect of stationary values.

On to a new frame.

51

Saddle Point

In the last example, when we substituted the coordinates (0, 0) in the expression $\left(\frac{\partial^2 z}{\partial x^2}\right)\left(\frac{\partial^2 z}{\partial y^2}\right) - \left(\frac{\partial^2 z}{\partial x \partial y}\right)^2$ we found that this did not satisfy the condition that for a maximum or minimum value

$$\left(\frac{\partial^2 z}{\partial x^2}\right)\left(\frac{\partial^2 z}{\partial y^2}\right) - \left(\frac{\partial^2 z}{\partial x \partial y}\right)^2 > 0$$

In fact, if $\frac{\partial z}{\partial x} = 0$ and $\frac{\partial z}{\partial y} = 0$

and $$\left(\frac{\partial^2 z}{\partial x^2}\right)\left(\frac{\partial^2 z}{\partial y^2}\right) - \left(\frac{\partial^2 z}{\partial x \partial y}\right)^2 < 0$$

this is an indication of a form of stationary value described as a *saddle point* as shown at P below.

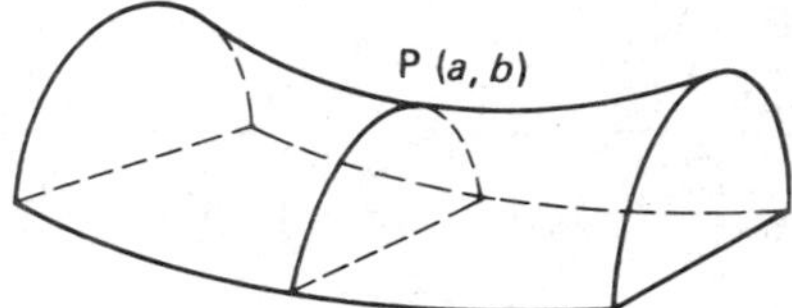

A saddle point is, in effect, a combined maximum and minimum configuration in different directions. Its name is obvious from the shape.

Add this then to the list of conditions for stationary values that we have built up.

52

At this stage, one naturally asks, what is implied if

$$\left(\frac{\partial^2 z}{\partial x^2}\right)\left(\frac{\partial^2 z}{\partial y^2}\right) - \left(\frac{\partial^2 z}{\partial x \partial y}\right)^2 = 0$$

In such a case, further detailed study of the function is necessary. In all problems that we shall meet, stationary values will refer to maxima, minima, or saddle points.

Now for an example to see it all in practice.

Example 3 Determine the stationary values of

$z = 5xy - 4x^2 - y^2 - 2x - y + 5.$

Possible stationary values (or turning points) occur where

$$\frac{\partial z}{\partial x} = 0 \quad \text{and} \quad \frac{\partial z}{\partial y} = 0, \text{ i.e. at } \ldots\ldots\ldots$$

53

$$\boxed{x = 1, \;\; y = 2}$$

$$\frac{\partial z}{\partial x} = 5y - 8x - 2 \qquad \frac{\partial z}{\partial y} = 5x - 2y - 1$$

$$\therefore \left.\begin{aligned} 8x - 5y + 2 &= 0 \\ 5x - 2y - 1 &= 0 \end{aligned}\right\} \text{ gives } x = 1, \; y = 2$$

Therefore, the only possible stationary value occurs at (1, 2).

Next we substitute these x and y values in

$$\left(\frac{\partial^2 z}{\partial x^2}\right)\left(\frac{\partial^2 z}{\partial y^2}\right) - \left(\frac{\partial^2 z}{\partial x \partial y}\right)^2 \quad \text{and find } \ldots\ldots\ldots$$

54

$$\boxed{\left(\frac{\partial^2 z}{\partial x^2}\right)\left(\frac{\partial^2 z}{\partial y^2}\right) - \left(\frac{\partial^2 z}{\partial x \partial y}\right)^2 < 0}$$

$$\text{for } \frac{\partial^2 z}{\partial x^2} = -8; \; \frac{\partial^2 z}{\partial y^2} = -2; \; \frac{\partial^2 z}{\partial x \partial y} = 5$$

$$\therefore \left(\frac{\partial^2 z}{\partial x^2}\right)\left(\frac{\partial^2 z}{\partial y^2}\right) - \left(\frac{\partial^2 z}{\partial x \partial y}\right)^2 = (-8)(-2) - 25 = -9 \quad \text{i.e.} < 0$$

The stationary value at (1, 2) is therefore a ....

55

$$\boxed{\text{saddle point}}$$

Example 4 Determine stationary values of $z = x^3 - 3x + xy^2$ and their nature.

We go through the same routine as before.

First find $\frac{\partial z}{\partial x}$ and $\frac{\partial z}{\partial y}$ and solve $\frac{\partial z}{\partial x} = 0$ and $\frac{\partial z}{\partial y} = 0$.

Possible stationary values therefore occur at ....

56

$$x = 0,\ y = \pm\sqrt{3}; \qquad x = \pm 1, y = 0$$

$$\frac{\partial z}{\partial x} = 3x^2 - 3 + y^2 \qquad \frac{\partial z}{\partial y} = 2xy \qquad \therefore x = 0 \text{ or } y = 0$$

If $x = 0$, $y^2 = 3$ $\quad \therefore y = \pm\sqrt{3} \qquad x = 0,\ y = \pm\sqrt{3}$

If $y = 0$, $3x^2 = 3$ $\quad \therefore x = \pm 1 \qquad x = \pm 1, y = 0.$

Now we need the second derivatives and the usual tests. Finish if off. The nature of the stationary values:

$(0, \sqrt{3})$ ....; $(0, -\sqrt{3})$ ....

$(1, 0)$ ....; $(-1, 0)$ ....

57

$(0, \sqrt{3})$ saddle point; $\quad (0, -\sqrt{3})$ saddle point

$(1, 0)$ minimum; $\quad (-1, 0)$ maximum

$$\frac{\partial^2 z}{\partial x^2} = 6x; \quad \frac{\partial^2 z}{\partial y^2} = 2x; \quad \frac{\partial^2 z}{\partial x \partial y} = 2y$$

$$\left(\frac{\partial^2 z}{\partial x^2}\right)\left(\frac{\partial^2 z}{\partial y^2}\right) - \left(\frac{\partial^2 z}{\partial x \partial y}\right)^2$$

$(0, \sqrt{3})$	$(0)(0) - 12$	i.e. < 0	$\therefore$ saddle point
$(0, -\sqrt{3})$	$(0)(0) - 12$	i.e. < 0	$\therefore$ saddle point
$(1, 0)$	$(6)(2) - 0$	i.e. > 0	minimum
$(-1, 0)$	$(-6)(-2) - 0$	i.e. > 0	maximum

and that just about does everything.

Substitution of $(1, 0)$ and $(-1, 0)$ in $z = x^3 - 3x + xy^2$ gives the minimum and maximum values of z. $z_{\text{min}} = -2$; $z_{\text{max}} = 2$. The value of z at each of the saddle points is zero.

Lagrange Undetermined Multipliers

58

In many practical situations, the independent variables are, in fact, often required to conform to a given constraint. That is, we are required to determine the points at which stationary values occur, with the added condition that the points lie on a pre-described curve.

Functions with two independent variables

Let us assume that we have to determine stationary points of the function

$$u = f(x, y) \qquad (1)$$

with the variables x and y constrained by the relation

$$\phi(x, y) = 0 \qquad (2)$$

As we saw previously, at stationary points, the total differential reduces to zero, i.e.

$$\frac{\partial u}{\partial x}\delta x + \frac{\partial u}{\partial y}\delta y = 0 \qquad (3)$$

Also, since $\phi(x, y) = 0$

$$\frac{\partial \phi}{\partial x}\delta x + \frac{\partial \phi}{\partial y}\delta y = 0 \qquad (4)$$

If we multiply each term in (4) by a multiplier λ and then add (4) to (3), we have

....

59

$$\boxed{\left(\frac{\partial u}{\partial x} + \lambda\frac{\partial \phi}{\partial x}\right)\delta x + \left(\frac{\partial u}{\partial y} + \lambda\frac{\partial \phi}{\partial y}\right)\delta y = 0}$$

Since δx and δy are independent increments

$$\frac{\partial u}{\partial x} + \lambda\frac{\partial \phi}{\partial x} = 0 \qquad (5)$$

and

$$\frac{\partial u}{\partial y} + \lambda\frac{\partial \phi}{\partial y} = 0 \qquad (6)$$

Then equations (5) and (6) and the original constraint (2) give us three relationships from which the values of x and y at the stationary points—and also the value of λ if required—can be found. Quite often, the actual value of λ is not important.

Let us see how it works in a simple example.

60 *Example 1* Find the stationary points of the function $u = x^2 + y^2$ subject to the constraint $x^2 + y^2 + 2x - 2y + 1 = 0$.

In this case,

$$u = x^2 + y^2$$
$$\phi = x^2 + y^2 + 2x - 2y + 1$$

We need to know

$$\frac{\partial u}{\partial x} = \dots\dots\dots; \quad \frac{\partial u}{\partial y} = \dots\dots\dots; \quad \frac{\partial \phi}{\partial x} = \dots\dots\dots; \quad \frac{\partial \phi}{\partial y} = \dots\dots\dots$$

61

$$\frac{\partial u}{\partial x} = 2x; \quad \frac{\partial u}{\partial y} = 2y; \quad \frac{\partial \phi}{\partial x} = 2x + 2; \quad \frac{\partial \phi}{\partial y} = 2y - 2$$

Then we form and solve

$$\frac{\partial u}{\partial x} + \lambda \frac{\partial \phi}{\partial x} = 0$$

$$\frac{\partial u}{\partial y} + \lambda \frac{\partial \phi}{\partial y} = 0$$

together with $\phi = x^2 + y^2 + 2x - 2y + 1 = 0$

which gives $x = \dots\dots\dots; \; y = \dots\dots\dots; \; \lambda = \dots\dots\dots$

62

$$x = -1 \pm \frac{\sqrt{2}}{2}; \quad y = 1 \mp \frac{\sqrt{2}}{2}; \quad \lambda = \sqrt{2} - 1$$

$$\frac{\partial u}{\partial x} + \lambda \frac{\partial \phi}{\partial x} = 0 \quad \therefore 2x + \lambda(2x + 2) = 0 \quad \therefore x + \lambda(x + 1) = 0$$

$$\frac{\partial u}{\partial y} + \lambda \frac{\partial \phi}{\partial y} = 0 \quad \therefore 2y + \lambda(2x - 2) = 0 \quad \therefore y + \lambda(y - 1) = 0$$

$$\therefore \frac{x}{y} = \frac{-\lambda(x + 1)}{-\lambda(y - 1)} \quad \therefore xy - x = xy + y \quad \therefore y = -x$$

Substituting this in ϕ

$$x^2 + x^2 + 2x + 2x + 1 = 0 \qquad 2x^2 + 4x + 1 = 0$$

$$\therefore \underline{x = -1 \pm \frac{\sqrt{2}}{2}}$$

But $y = -x$ $\qquad \therefore \underline{y = 1 \mp \frac{\sqrt{2}}{2}}$

To find λ, we have $x + \lambda(x + 1) = 0 \quad \therefore \underline{\lambda = \sqrt{2} - 1}$

As we have already said, we do not really need to find the value of λ.

On to the next.

63

Functions with three independent variables

The argument is very much the same as before.

To find stationary points of the function $u = f(x, y, z)$ (1)

subject to the constraint $\phi(x, y, z) = 0.$ (2)

Again we have, at stationary points

$$\frac{\partial u}{\partial x}\delta x + \frac{\partial u}{\partial y}\delta y + \frac{\partial u}{\partial z}\delta z = 0 \tag{3}$$

and since $\phi(x, y, z) = 0$

then
$$\frac{\partial \phi}{\partial x}\delta x + \frac{\partial \phi}{\partial y}\delta y + \frac{\partial \phi}{\partial z}\delta z = 0 \tag{4}$$

Multiplying each term in (4) by λ and adding (4) to (3), we have

. . . .

64

$$\boxed{\left(\frac{\partial u}{\partial x} + \lambda\frac{\partial \phi}{\partial x}\right)\delta x + \left(\frac{\partial u}{\partial y} + \lambda\frac{\partial \phi}{\partial y}\right)\delta y + \left(\frac{\partial u}{\partial z} + \lambda\frac{\partial \phi}{\partial z}\right)\delta z = 0}$$

from which
$$\frac{\partial u}{\partial x} + \lambda\frac{\partial \phi}{\partial x} = 0 \tag{5}$$

$$\frac{\partial u}{\partial y} + \lambda\frac{\partial \phi}{\partial y} = 0 \tag{6}$$

$$\frac{\partial u}{\partial z} + \lambda\frac{\partial \phi}{\partial z} = 0 \tag{7}$$

Equations (5), (6), (7), together with the constraint (2) provide all the information to determine x, y, z, and, if necessary, λ.

65

Example 2 To find the stationary points of the function

$$u = x^2 + 2y^2 + z$$

subject to the constraint $\phi(x, z) = x^2 - z^2 - 2 = 0.$

So $\frac{\partial u}{\partial x} = \ldots\ldots\ldots;\quad \frac{\partial u}{\partial y} = \ldots\ldots\ldots;\quad \frac{\partial u}{\partial z} = \ldots\ldots\ldots$

$\frac{\partial \phi}{\partial x} = \ldots\ldots\ldots;\quad \frac{\partial \phi}{\partial y} = \ldots\ldots\ldots;\quad \frac{\partial \phi}{\partial z} = \ldots\ldots\ldots$

66

$$\frac{\partial u}{\partial x} = 2x; \quad \frac{\partial u}{\partial y} = 4y; \quad \frac{\partial u}{\partial z} = 1$$

$$\frac{\partial \phi}{\partial x} = 2x; \quad \frac{\partial \phi}{\partial y} = 0; \quad \frac{\partial \phi}{\partial z} = -2z$$

Now compile the equations

$$\frac{\partial u}{\partial x} + \lambda\frac{\partial \phi}{\partial x} = 0; \qquad \frac{\partial u}{\partial v} + \lambda\frac{\partial \phi}{\partial v} = 0; \qquad \frac{\partial u}{\partial z} + \lambda\frac{\partial \phi}{\partial z} = 0$$

and, together with the constraint $\phi = x^2 - z^2 - 2 = 0$, establish that stationary points occur at

. . . .

67

$$\left(\tfrac{3}{2},\ 0,\ -\tfrac{1}{2}\right) \text{ and } \left(-\tfrac{3}{2},\ 0,\ -\tfrac{1}{2}\right)$$

for $\dfrac{\partial u}{\partial x} + \lambda\dfrac{\partial \phi}{\partial x} = 0 \quad \therefore\ 2x + \lambda 2x = 0 \qquad \therefore\ \lambda = -1$

$$\frac{\partial u}{\partial y} + \lambda\frac{\partial \phi}{\partial y} = 0 \qquad 4y + \lambda(0) = 0 \qquad \therefore\ y = 0$$

$$\frac{\partial u}{\partial z} + \lambda\frac{\partial \phi}{\partial z} = 0 \qquad 1 - \lambda 2z = 0 \qquad \therefore\ z = \frac{1}{2\lambda} = -\frac{1}{2}$$

$\phi = x^2 - z^2 - 2 = 0 \quad \therefore\ x^2 - \frac{1}{4} - 2 = 0 \quad \therefore\ x = \pm\frac{3}{2}$.

Therefore, stationary points at $\underline{\left(\tfrac{3}{2},\ 0,\ -\tfrac{1}{2}\right) \text{ and } \left(-\tfrac{3}{2},\ 0,\ -\tfrac{1}{2}\right)}$.

The method of Lagrange multipliers does not lend itself easily to give a distinction between the various types of stationary points. In many practical applications, however, whether a result is a maximum or a minimum value will be apparent from the physical consideration of the problem.

Let us finish with one further example.

So move on.

68

Example 3 A hot water storage tank is a vertical cylinder surmounted by a hemispherical top of the same diameter. The tank is designed to hold 400 m^3 of liquid. Determine the total height and the diameter of the tank if the surface heat loss is to be a minimum.

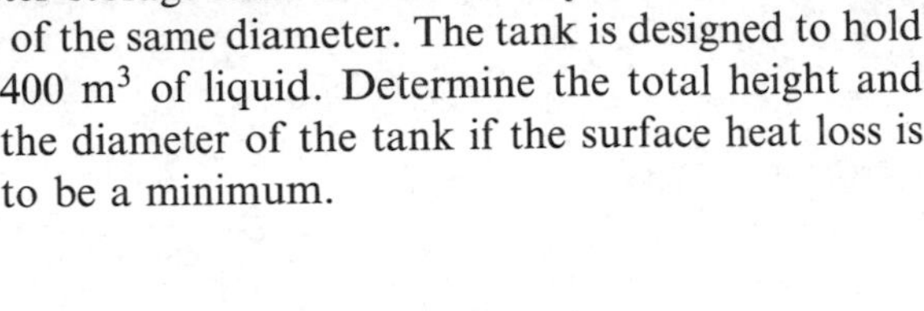

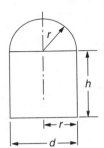

We first write down the function for the total surface area. A.

$$A = \dots\dots\dots$$

69

$$\boxed{A = 3\pi r^2 + 2\pi rh}$$

This is the function which has to be a minimum. The constraint in this problem is that ....

70

$$\boxed{\text{the volume is } 400 \text{ m}^3}$$

So we have

$$A = 3\pi r^2 + 2\pi rh \qquad (1)$$

constraint $\quad V = \pi r^2 h + \frac{2}{3}\pi r^3 = 400$

i.e. $\quad V = \pi r^2 h + \frac{2}{3}\pi r^3 - 400 = 0 \qquad (2)$

We now want $\dfrac{\partial A}{\partial r} = \dots\dots\dots; \quad \dfrac{\partial A}{\partial h} = \dots\dots\dots$

$$\frac{\partial V}{\partial r} = \dots\dots\dots; \quad \frac{\partial V}{\partial h} = \dots\dots\dots$$

71

$$\frac{\partial A}{\partial r} = 6\pi r + 2\pi h; \quad \frac{\partial V}{\partial r} = 2\pi rh + 2\pi r^2$$

$$\frac{\partial A}{\partial h} = 2\pi r; \quad \frac{\partial V}{\partial h} = \pi r^2$$

Now we form $\quad \dfrac{\partial A}{\partial r} + \lambda\dfrac{\partial V}{\partial r} = 0$

and $\quad \dfrac{\partial A}{\partial h} + \lambda\dfrac{\partial V}{\partial h} = 0$

and, with the constraint, $V = \pi r^2 h + \frac{2}{3}\pi r^3 - 400 = 0$,

we eventually obtain $r = \ldots\ldots\ldots$ and $h = \ldots\ldots\ldots$

Finish it off and hence find the total height and the diameter.

72

$$r = 4.243 \text{ m}; \quad h = 4.243 \text{ m}$$

Check the working

$$\frac{\partial A}{\partial r} + \lambda\frac{\partial V}{\partial r} = 0 \quad \therefore\ 6\pi r + 2\pi h + \lambda(2\pi rh + 2\pi r^2) = 0 \qquad (3)$$

$$\frac{\partial A}{\partial h} + \lambda\frac{\partial V}{\partial h} = 0 \quad \therefore\ 2\pi r + \lambda\pi r^2 = 0 \qquad (4)$$

From (4) $\quad \lambda = -\dfrac{2}{r} \quad$ Substitute this in (3)

$$6\pi r + 2\pi h - \frac{2}{r}(2\pi rh + 2\pi r^2) = 0$$

$$\therefore\ 6r + 2h - 4h - 4r = 0 \qquad \therefore\ \underline{h = r}$$

Also $\quad \pi r^2 h + \frac{2}{3}\pi r^3 = 400 \qquad \therefore\ \frac{5}{3}\pi r^3 = 400 \qquad \therefore\ \underline{r = 4.243}$

$\therefore$ Total height $= h + r = 8.49$ m; Diameter $= 8.49$ m

That brings us to the end of this particular programme and to the usual Revision Summary that follows. Check through it and be sure that you are happy with its contents. You can always revise any section should you feel that is necessary. Then you will find the Test Exercise straightforward—no tricks.

REVISION SUMMARY

73

1. *Small increments*

$$z = f(x, y) \qquad \delta z = \frac{\partial z}{\partial x}\delta x + \frac{\partial z}{\partial y}\delta y$$

$$u = f(x, y, z) \quad \delta u = \frac{\partial u}{\partial x}\delta x + \frac{\partial u}{\partial y}\delta y + \frac{\partial u}{\partial z}\delta z$$

2. *Rates of change*

$$z = f(x, y) \qquad \frac{\mathrm{d}z}{\mathrm{d}t} = \frac{\partial z}{\partial x}\cdot\frac{\mathrm{d}x}{\mathrm{d}t} + \frac{\partial z}{\partial y}\cdot\frac{\mathrm{d}y}{\mathrm{d}t}$$

3. *Implicit functions*

$$f(x, y) = 0 \qquad \frac{\mathrm{d}y}{\mathrm{d}x} = -\left(\frac{\partial z}{\partial x}\bigg/\frac{\partial z}{\partial y}\right)$$

4. *Change of variables*

$z = f(x, y)$ x and y are functions of u and v

$$\frac{\partial z}{\partial u} = \frac{\partial z}{\partial x}\cdot\frac{\partial x}{\partial u} + \frac{\partial z}{\partial y}\cdot\frac{\partial y}{\partial u}$$

$$\frac{\partial z}{\partial v} = \frac{\partial z}{\partial x}\cdot\frac{\partial x}{\partial v} + \frac{\partial z}{\partial y}\cdot\frac{\partial y}{\partial v}$$

5. *Inverse functions*

$$z = f(x, y) \quad u = g(x, y) \quad v = h(x, y)$$

$$\frac{\partial x}{\partial u} = \frac{\partial v}{\partial y}\bigg/\mathrm{J}; \qquad \frac{\partial x}{\partial v} = -\frac{\partial u}{\partial y}\bigg/\mathrm{J}$$

$$\frac{\partial y}{\partial u} = -\frac{\partial v}{\partial x}\bigg/\mathrm{J}; \qquad \frac{\partial y}{\partial v} = \frac{\partial u}{\partial x}\bigg/\mathrm{J}$$

where

$$\mathrm{J} = \frac{\partial(u, v)}{\partial(x, y)} = \begin{vmatrix} \dfrac{\partial u}{\partial x} & \dfrac{\partial v}{\partial x} \\ \dfrac{\partial u}{\partial y} & \dfrac{\partial v}{\partial y} \end{vmatrix}$$

6. *Stationary points*

$z=f(x,y)$ (a) $\dfrac{\partial z}{\partial x}=0$ and $\dfrac{\partial z}{\partial y}=0$

(b) $\left(\dfrac{\partial^2 z}{\partial x^2}\right)\left(\dfrac{\partial^2 z}{\partial y^2}\right)-\left(\dfrac{\partial^2 z}{\partial x \partial y}\right)^2 > 0$ for max. or min.

< 0 for saddle point

(c) $\dfrac{\partial^2 z}{\partial x^2}$ and $\dfrac{\partial^2 z}{\partial y^2}$ both *negative* for maximum

$\dfrac{\partial^2 z}{\partial x^2}$ and $\dfrac{\partial^2 z}{\partial y^2}$ both *positive* for minimum.

7. *Lagrange multipliers*

Two independent variables

$u=f(x,y)$ with constraint $\phi(x,y)=0$

Solve
$$\frac{\partial u}{\partial x}+\lambda\frac{\partial \phi}{\partial x}=0$$
$$\frac{\partial u}{\partial y}+\lambda\frac{\partial \phi}{\partial y}=0$$
with
$$\phi(x,y)=0.$$

Three independent variables

$u=f(x,y,z)$ with constraint $\phi(x,y,z)=0$

Solve
$$\frac{\partial u}{\partial x}+\lambda\frac{\partial \phi}{\partial x}=0$$
$$\frac{\partial u}{\partial y}+\lambda\frac{\partial \phi}{\partial y}=0$$
$$\frac{\partial u}{\partial z}+\lambda\frac{\partial \phi}{\partial z}=0$$
with
$$\phi(x,y,z)=0.$$

TEST EXERCISE III 74

1. If $z = \dfrac{xy}{x - y}$, show that

(a) $x\dfrac{\partial z}{\partial x} + y\dfrac{\partial z}{\partial y} = z$

(b) $x^2\dfrac{\partial^2 z}{\partial x^2} - y^2\dfrac{\partial^2 z}{\partial y^2} = 0$

(c) $z\dfrac{\partial^2 z}{\partial x \partial y} = 2\dfrac{\partial z}{\partial x}\cdot\dfrac{\partial z}{\partial y}$.

2. Two sides of a triangular plate are measured as 125 mm and 160 mm, each to the nearest millimetre. The included angle is quoted as $60° \pm 1°$. Calculate the length of the remaining side and the maximum possible error in the result.

3. If $2x^2 + 4xy + 3y^2 = 1$, obtain expressions for $\dfrac{dy}{dx}$ and $\dfrac{d^2y}{dx^2}$.

4. If $u = x^2 + y^2$ and $v = 4xy$, determine

$$\frac{\partial x}{\partial u}, \frac{\partial x}{\partial v}, \frac{\partial y}{\partial u}, \frac{\partial y}{\partial v}.$$

5. Determine the position and nature of the stationary points of the function $z = 2x^2y^2 + 4xy^2 - 4y^3 + 16y + 5$.

6. A rectangular storage tank is to have a capacity of 1.0 m^3. If the tank is closed and the top is made of metal half as thick as the sides and base, use Lagrange's method of undetermined multipliers to determine the dimensions of the tank for the total amount of metal used in its construction to be a minimum.

FURTHER PROBLEMS III

1. If $z = 2x^2 - 3y$ with $u = x^2 \sin y$ and $v = 2y \cos x$, determine expressions for $\dfrac{\partial z}{\partial u}$ and $\dfrac{\partial z}{\partial v}$.

2. If $u = x^2 + e^{-3y}$ and $v = 2x + e^{3y}$, determine $\dfrac{\partial x}{\partial u}, \dfrac{\partial x}{\partial v}, \dfrac{\partial y}{\partial u}, \dfrac{\partial y}{\partial v}$.

3. If $z = f(x, y)$ where $x = uv$ and $y = u^2 - v^2$, show that

   (a) $2x\dfrac{\partial z}{\partial x} + 2y\dfrac{\partial z}{\partial y} = u\dfrac{\partial z}{\partial u} + v\dfrac{\partial z}{\partial v}$

   (b) $2\dfrac{\partial z}{\partial y} = \dfrac{1}{u^2 + v^2}\left\{u\dfrac{\partial z}{\partial u} - v\dfrac{\partial z}{\partial v}\right\}$.

4. If $V = f(x, y)$ and $x = r\cos\theta$ and $y = r\sin\theta$, show that

$$\frac{\partial^2 V}{\partial x^2} + \frac{\partial^2 V}{\partial y^2} = \frac{\partial^2 V}{\partial r^2} + \frac{1}{r}\frac{\partial V}{\partial r} + \frac{1}{r^2}\frac{\partial^2 V}{\partial \theta^2}.$$

5. If $z = \cosh 2x \sin 3y$ and $u = e^x(1 + y^2)$ and $v = 2ye^{-x}$, determine expressions for $\dfrac{\partial x}{\partial u}, \dfrac{\partial x}{\partial v}, \dfrac{\partial y}{\partial u}, \dfrac{\partial y}{\partial v}$, and hence find $\dfrac{\partial z}{\partial u}$ and $\dfrac{\partial z}{\partial v}$.

6. If $z = f(u, v)$ where $u = \frac{1}{2}(x^2 - y^2)$ and $v = xy$, prove that

$$\frac{\partial^2 z}{\partial x^2} - \frac{\partial^2 z}{\partial y^2} = 2u\left(\frac{\partial^2 z}{\partial u^2} - \frac{\partial^2 z}{\partial v^2}\right) + 4v\frac{\partial^2 z}{\partial u \partial v} + 2\frac{\partial z}{\partial u}.$$

7. Locate the stationary points of the following functions. Determine the nature of the points and calculate the critical function values.

   (a) $z = y^2 + xy + x^2 + 4y - 4x + 5$

   (b) $z = y^2 + xy + 2x + 3y + 6$

   (c) $z = 3xy - 6y^2 - 3x^2 + 6y + 6x + 7$.

8. Find the stationary points of the function

$$z = (x^2 + y^2)^2 - 8(x^2 - y^2)$$

   and determine their nature.

9. Verify that the function $z = (x + y - 1)/(x^2 + 2y^2 + 2)$ has stationary values at (2, 1) and $(-\frac{2}{3}, -\frac{1}{3})$ and determine their nature.

10. Locate stationary points of the function

$$z = 4x^2 + 10xy + 4y^2 - x^2y^2$$

and determine their nature.

11. Find the stationary points of the following functions and determine their nature.

(a) $z = x(x^2 - 3) + 3y(x - 1)^2 + 18y^2(2y - 3)$

(b) $z = x^2y^2 - x^2 - y^2$.

12. A metal channel is formed by turning up the sides of width x of a rectangular sheet of metal through an angle θ. If the sheet is 200 mm wide, determine the values of x and θ for which the cross-section of the channel will be a maximum.

13. A container is in the form of a right circular cylinder of length l and diameter d, with equal conical ends of the same diameter and height h. If V is the fixed volume of the container, find the dimensions l, h and d, for minimum surface area.

14. A solid consists of a cylinder of length l and diameter d, surmounted at one end by a cone of vertex angle 2θ and base diameter d, and at the other end by a hemisphere of the same diameter. If the volume V of the solid is 50 cm^3, determine the dimensions l, d and θ, so that the total surface area shall be a minimum.

15. A rectangular solid of maximum volume is to be cut from a solid sphere of radius r. Determine the dimensions of the solid so formed and its volume.

16. Use Lagrange's method of undetermined multipliers to obtain the stationary values of the following functions u, subject in each case to the constraint ϕ.

(a) $u = x^2y^2z^2$ $\qquad \phi = x^2 + y^2 + z^2 - 4 = 0$

(b) $u = x^2 + y^2$ $\qquad \phi = 4x^2 + 6xy + 4y^2 = 9$

(c) $u = x^2 + y^2 + z^2$ $\qquad \phi = 3x - 2y + z - 4 = 0$.

Programme 4

Integral Functions

Prerequisites: Engineering Mathematics (fourth edition)
Programmes 15, 16, 17

Integral Functions

1 Some functions are most conveniently defined in the form of integrals and we shall deal with one or two of these in the present programme.

The gamma function

The gamma function $\Gamma(x)$ is defined by the integral

$$\Gamma(x) = \int_0^\infty t^{x-1}e^{-t}\,dt \tag{1}$$

and is convergent for $x > 0$.

From (1): $\Gamma(x+1) = \int_0^\infty t^x e^{-t}\,dt$

Integrating by parts

$$\Gamma(x+1) = \left[t^x\left(\frac{e^{-t}}{-1}\right)\right]_0^\infty + x\int_0^\infty e^{-t}t^{x-1}\,dt$$

$$= \{0 - 0\} + x\Gamma(x)$$

$$\therefore \Gamma(x+1) = x\Gamma(x) \tag{2}$$

This is a fundamental recurrence relation for gamma functions. It can also be written as $\Gamma(x) = (x-1)\Gamma(x-1)$

With it we can derive a number of other results.

For instance, when $x = n$, a positive integer ≥ 1, then

$$\begin{aligned}\Gamma(n+1) &= n\Gamma(n) &\quad \text{But } \Gamma(n) &= (n-1)\Gamma(n-1)\\ &= n(n-1)\Gamma(n-1) & \Gamma(n-1) &= (n-2)\Gamma(n-2)\\ &= n(n-1)(n-2)\Gamma(n-2) && \\ &\quad ------ && \\ &= n(n-1)(n-2)(n-3)\ldots 1\Gamma(1) = n!\Gamma(1) &&\end{aligned}$$

But, from the original definition $\Gamma(1) = \ldots\ldots\ldots$

2

$$\boxed{\Gamma(1) = 1}$$

for $\quad \Gamma(1) = \int_0^\infty t^0 e^{-t} dt = \Big[-e^{-t}\Big]_0^\infty = 0 + 1 = 1$

Therefore, we have $\underline{\Gamma(1) = 1}$ (3)

and $\quad \underline{\Gamma(n+1) = n!}$ provided n is a positive integer.

$$\therefore \Gamma(7) = \ldots\ldots\ldots$$

3

$$\boxed{\Gamma(7) = 720}$$

for $\quad \Gamma(7) = \Gamma(6+1) = 6! = 720.$

Knowing $\Gamma(7) - 720$, $\quad \Gamma(8) = \ldots\ldots\ldots$ and $\Gamma(9) = \ldots\ldots\ldots$

4

$$\boxed{\Gamma(8) = 5040; \qquad \Gamma(9) = 40\,320}$$

since

$$\Gamma(8) = \Gamma(7+1) = 7\Gamma(7) = 7(720) \;\; = \underline{5040}$$
$$\Gamma(9) = \Gamma(8+1) = 8\Gamma(8) = 8(5040) = \underline{40\,320}$$

We can also use the recurrence relation in reverse

$$\Gamma(x+1) = x\Gamma(x) \quad \therefore \Gamma(x) = \frac{\Gamma(x+1)}{x} \tag{4}$$

For example, given that $\Gamma(7) = 720$, we can determine $\Gamma(6)$

$$\Gamma(6) = \frac{\Gamma(6+1)}{6} = \frac{\Gamma(7)}{6} = \frac{720}{6} = 120$$

and then $\quad \Gamma(5) = \ldots\ldots\ldots$

5

$$\boxed{\Gamma(5) = 24}$$

$$\Gamma(5) = \frac{\Gamma(5+1)}{5} = \frac{\Gamma(6)}{5} = \frac{120}{5} = 24.$$

So far, we have used the original definition

$$\Gamma(x) = \int_0^\infty t^{x-1}e^{-t}dt$$

for cases where x is a positive integer n.

What happens when $x = \frac{1}{2}$? We will investigate.

$$\Gamma\left(\tfrac{1}{2}\right) = \int_0^\infty t^{-1/2}\,e^{-t}dt$$

Putting $t = u^2$, $dt = 2u\,du$, then

$$\Gamma\left(\tfrac{1}{2}\right) = \ldots\ldots\ldots$$

6

$$\boxed{\Gamma\left(\tfrac{1}{2}\right) = 2\int_0^\infty e^{-u^2}du}$$

for $\Gamma\left(\frac{1}{2}\right) = \displaystyle\int_0^\infty u^{-1}e^{-u^2}2u\,du = 2\int_0^\infty e^{-u^2}\,du.$

Unfortunately, $\displaystyle\int_0^\infty e^{-u^2}\,du$ cannot easily be determined by normal means. It is, however, important, so we have to find a way of getting round the difficulty.

Evaluation of $\displaystyle\int_0^\infty e^{-x^2}dx$

Let $I = \displaystyle\int_0^\infty e^{-x^2}dx$, then also $I = \displaystyle\int_0^\infty e^{-y^2}dy$

$$\therefore I^2 = \left(\int_0^\infty e^{-x^2}dx\right)\left(\int_0^\infty e^{-y^2}dy\right) = \int_0^\infty\int_0^\infty e^{-(x^2+y^2)}\,dx\,dy$$

$\delta a = \delta x\,\delta y$ represents an element of area in the xy-plane and the integration with the stated limits covers the whole of the first quadrant.

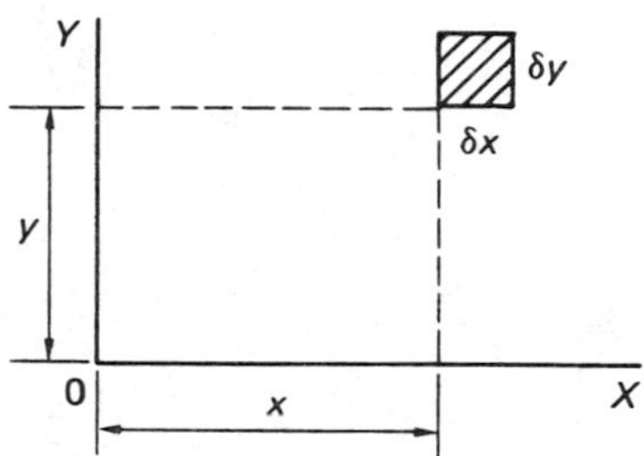

Converting to polar coordinates, the element of area $\delta a = r\,\delta\theta\,\delta r$. Also,

$x^2 + y^2 = r^2$ $\quad\quad \therefore\ e^{-(x^2+y^2)} = e^{-r^2}$.

For the integration to cover the same region as before,

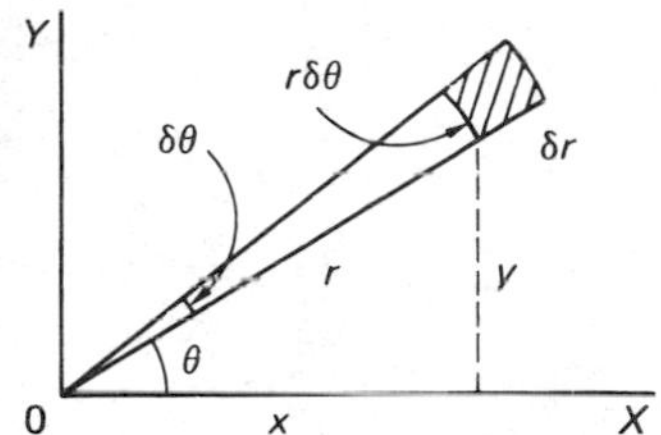

the limits of r are $r = 0$ to $r = \infty$

the limits of θ are $\theta = 0$ to $\theta = \pi/2$.

$$\therefore\ I^2 = \int_0^{\pi/2}\int_0^{\infty} e^{-r^2} r\,dr\,d\theta = \int_0^{\pi/2}\left[-\frac{e^{-r^2}}{2}\right]_0^{\infty} d\theta$$

$$= \int_0^{\pi/2}\left(\frac{1}{2}\right) d\theta = \left[\frac{\theta}{2}\right]_0^{\pi/2} = \frac{\pi}{4} \quad \therefore\ I = \frac{\sqrt{\pi}}{2}$$

$$\therefore \int_0^{\infty} e^{-x^2}\,dx = \frac{\sqrt{\pi}}{2} \qquad (5)$$

This result opens the way for others, so make a note of it and then move on to the next frame.

7

Before that diversion, we had established that

$$\Gamma\left(\tfrac{1}{2}\right) = 2\int_0^\infty e^{-u^2}\,du$$

We now know that $\displaystyle\int_0^\infty e^{-u^2}\,du = \frac{\sqrt{\pi}}{2}$ $\qquad \therefore \underline{\Gamma\left(\tfrac{1}{2}\right) = \sqrt{\pi}}$

From this, using the recurrence relation $\Gamma(x+1) = x\Gamma(x)$, we can obtain the following.

$$\Gamma\left(\tfrac{3}{2}\right) = \tfrac{1}{2}\ \Gamma\left(\tfrac{1}{2}\right) = \tfrac{1}{2}(\sqrt{\pi}) \qquad \therefore \Gamma\left(\tfrac{3}{2}\right) = \frac{\sqrt{\pi}}{2}$$

$$\Gamma\left(\tfrac{5}{2}\right) = \tfrac{3}{2}\ \Gamma\left(\tfrac{3}{2}\right) = \tfrac{3}{2}\left(\frac{\sqrt{\pi}}{2}\right) \qquad \therefore \Gamma\left(\tfrac{5}{2}\right) = \frac{3\sqrt{\pi}}{4}$$

$$\Gamma\left(\tfrac{7}{2}\right) = \ldots\ldots\ldots$$

8

$$\boxed{\Gamma\left(\tfrac{7}{2}\right) = \frac{15\sqrt{\pi}}{8}}$$

$$\text{for } \Gamma\left(\tfrac{7}{2}\right) = \Gamma\left(\tfrac{5}{2}+1\right) = \tfrac{5}{2}\,\Gamma\left(\tfrac{5}{2}\right) = \frac{5}{2}\left(\frac{3\sqrt{\pi}}{4}\right) = \underline{\frac{15\sqrt{\pi}}{8}}$$

Using the recurrence relation in reverse, i.e. $\Gamma(x) = \dfrac{\Gamma(x+1)}{x}$, we can also obtain

$$\Gamma\left(-\tfrac{3}{2}\right) = \frac{\Gamma\left(-\frac{1}{2}\right)}{-\frac{3}{2}} = \frac{\Gamma\left(\frac{1}{2}\right)}{\left(-\frac{3}{2}\right)\left(-\frac{1}{2}\right)} = \underline{\tfrac{4}{3}\sqrt{\pi}}$$

Negative values of x

Since $\Gamma(x) = \dfrac{\Gamma(x+1)}{x}$, then as $x \to 0$, $\Gamma(x) \to \infty$ $\therefore$ $\Gamma(0) = \infty$.

The same result occurs for all negative integral values of x—which does not follow from the original definition, but which is obtainable from the recurrence relation.

For at $x = -1$, $\quad \Gamma(-1) = \dfrac{\Gamma(0)}{-1} = \infty$

$x = -2$, $\quad \Gamma(-2) = \dfrac{\Gamma(-1)}{-2} = \infty$ etc.

Also, at $x = -\frac{1}{2}$, $\quad \Gamma\left(-\frac{1}{2}\right) = \dfrac{\Gamma\left(\frac{1}{2}\right)}{-\frac{1}{2}} = -2\sqrt{\pi}$

and at $x = -\frac{3}{2}$, $\quad \Gamma\left(-\frac{3}{2}\right) = \dfrac{\Gamma\left(-\frac{1}{2}\right)}{-\frac{3}{2}} = \dfrac{4}{3}\sqrt{\pi}$

Similarly $\Gamma\left(-\frac{5}{2}\right) = \ldots\ldots\ldots$

and $\Gamma\left(-\frac{7}{2}\right) = \ldots\ldots\ldots$

9

$$\Gamma(-\tfrac{5}{2}) = -\frac{8}{15}\sqrt{\pi}; \quad \Gamma(-\tfrac{7}{2}) = \frac{16}{105}\sqrt{\pi}$$

So we have

(a) For n a positive integer

$$\Gamma(n+1) = n\Gamma(n) = n!$$

$$\Gamma(1) = 1; \quad \Gamma(0) = \infty; \quad \Gamma(-n) = \pm\infty$$

(b) $\Gamma(\tfrac{1}{2}) = \sqrt{\pi}; \qquad \Gamma(-\tfrac{1}{2}) = -2\sqrt{\pi}$

$$\Gamma(\tfrac{3}{2}) = \frac{\sqrt{\pi}}{2}; \qquad \Gamma(-\tfrac{3}{2}) = \frac{4}{3}\sqrt{\pi}$$

$$\Gamma(\tfrac{5}{2}) = \frac{3\sqrt{\pi}}{4}; \qquad \Gamma(-\tfrac{5}{2}) = -\frac{8}{15}\sqrt{\pi}$$

$$\Gamma(\tfrac{7}{2}) = \frac{15\sqrt{\pi}}{8}; \qquad \Gamma(-\tfrac{7}{2}) = \frac{16}{105}\sqrt{\pi}$$

This is quite a useful list. Make a note of it for future use.

10

Graph of $y = \Gamma(x)$

Values of $\Gamma(x)$ for a range of positive values of x are available in tabulated form in various sets of mathematical tables. These, together with the results established above, enable us to draw the graph of $y = \Gamma(x)$.

x	0	0.5	1.0	1.5	2.0	2.5	3.0	3.5	4.0
$\Gamma(x)$	∞	1.772	1.000	0.886	1.000	1.329	2.000	3.323	6.000

x	−0.5	−1.5	−2.5	−3.5
$\Gamma(x)$	−3.545	2.363	−0.945	0.270

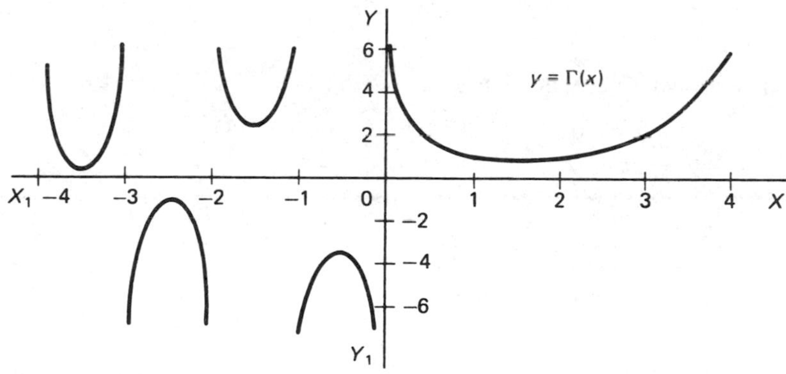

11

Revision

Let us now revise the main points before we move on to some examples.

The definition of $\Gamma(x)$ is that $\Gamma(x) = \ldots\ldots\ldots$

12

$$\boxed{\Gamma(x) = \int_0^\infty t^{x-1} e^{-t} \mathrm{d}t}$$

The recurrence relation states that

$$\Gamma(x+1) = \ldots\ldots\ldots$$

13

$$\boxed{\Gamma(x+1) = x\Gamma(x)}$$

When x is a positive integer, i.e. $x = n$, then

$$\Gamma(n+1) = \ldots\ldots\ldots$$

14

$$\boxed{\Gamma(n+1) = n!}$$

Then we have a number of specific results

$$\Gamma(1) = \ldots\ldots\ldots; \quad \Gamma(0) = \ldots\ldots\ldots; \quad \Gamma\left(\tfrac{1}{2}\right) = \ldots\ldots\ldots$$

15

$$\boxed{\Gamma(1) = 1; \quad \Gamma(0) = \infty; \quad \Gamma\left(\tfrac{1}{2}\right) = \sqrt{\pi}}$$

and finally, for all negative integral values of n

$$\Gamma(n) = \ldots\ldots\ldots$$

16

$$\boxed{\Gamma(n) = \pm\infty}$$

Listing them together, we have

$$\Gamma(x) = \int_0^\infty t^{x-1} e^{-t} \mathrm{d}t$$

$$\Gamma(x+1) = x\Gamma(x)$$

$$\Gamma(n+1) = n! \qquad \text{for } n \text{ a positive integer}$$

$$\Gamma(1) = 1; \quad \Gamma(0) = \infty; \quad \Gamma\left(\tfrac{1}{2}\right) = \sqrt{\pi}$$

$$\Gamma(n) = \pm\infty \quad \text{for } n \text{ a negative integer.}$$

17

Now for a few examples

Example 1 Evaluate $\int_0^\infty x^7 e^{-x} dx$.

We recognise this as the standard form of the gamma function

$$\Gamma(x) = \int_0^\infty t^{x-1} e^{-t} dt \qquad \text{with the variables changed.}$$

It is often convenient to write the gamma function as

$$\Gamma(v) = \int_0^\infty x^{v-1} e^{-x} dx$$

Our example then becomes

$$I = \int_0^\infty x^7 e^{-x} dx = \int_0^\infty x^{v-1} e^{-x}\, dx \qquad \text{where } v = \dots\dots\dots$$

18

$$\boxed{v = 8}$$

$$\therefore I = \Gamma(v) = \Gamma(8) = \dots\dots\dots$$

19

$$\boxed{\Gamma(8) = 7! = 5040}$$

i.e. $\int_0^\infty x^7\, e^{-x}\, dx = \Gamma(8) = 7! = \underline{5040}$

Example 2 Evaluate $\int_0^\infty x^3\, e^{-4x}\, dx$.

If we compare this with $\Gamma(v) = \int_0^\infty x^{v-1}\, e^{-x}\, dx$, we must reduce the power of e to a single variable, i.e. put $y = 4x$, and we use this substitution to convert the whole integral into the required form.

$$y = 4x \qquad \therefore dy = 4\, dx \qquad \text{Limits remain unchanged.}$$

The integral now becomes ....

20

$$I=\int_0^\infty \left(\frac{y}{4}\right)^3 e^{-y}\,\frac{dy}{4}$$

$$\therefore\ I=\frac{1}{4^4}\int_0^\infty y^3\,e^{-y}\,dy=\frac{1}{4^4}\Gamma(v) \qquad \text{where } v=\ldots\ldots\ldots$$

21

$$v=4$$

Since $\displaystyle\int_0^\infty y^{v-1}e^{-y}dy=\int_0^\infty y^3e^{-y}dy \qquad \therefore\ v=4$

$$\therefore\ I=\frac{1}{4^4}\,\Gamma(4)=\ldots\ldots\ldots$$

22

$$I=\frac{3}{128}$$

for $I=\dfrac{1}{256}\Gamma(4)=\dfrac{1}{256}(3!)=\dfrac{6}{256}=\dfrac{3}{128}$

One more.

Example 3 Evaluate $\displaystyle\int_0^\infty x^{1/2}\,e^{-x^2}\,dx$.

The substitution here is to put ....

23

$$y=x^2$$

Work through it as before. When you have completed it, check with the next frame.

24

Here is the working.

$y = x^2 \quad \therefore \mathrm{d}y = 2x\,\mathrm{d}x \quad$ Limits $x = 0,\ y = 0; \quad x = \infty,\ y = \infty.$

$x = y^{1/2} \quad \therefore x^{1/2} = y^{1/4}$

$$\therefore I = \int_0^\infty y^{1/4}\,\mathrm{e}^{-y}\,\mathrm{d}y/2x = \int_0^\infty \frac{y^{1/4}\,\mathrm{c}^{-y}\,\mathrm{d}y}{2y^{1/2}}$$

$$= \frac{1}{2}\int_0^\infty y^{-1/4}\,\mathrm{e}^{-y}\,\mathrm{d}y$$

$$= \frac{1}{2}\int_0^\infty y^{v-1}\,\mathrm{e}^{-y}\mathrm{d}y \quad \text{where } v = \frac{3}{4} \qquad \therefore I = \frac{1}{2}\Gamma\left(\frac{3}{4}\right)$$

From tables, $\Gamma(0.75) = 1.2254$

$$\therefore I = 0.613$$

Now we will move on to another set of functions closely related to gamma functions.

Let us start a new frame.

25

The beta function

The beta function B(*m*, *n*) is defined by

$$B(m,n) = \int_0^1 x^{m-1}(1-x)^{n-1}\,dx \tag{1}$$

which converges for $m > 0$ and $n > 0$.

Putting $(1-x) = u \quad \therefore x = 1-u \quad \therefore dx = -\,du$
Limits: when $x = 0,\ u = 1$; when $x = 1,\ u = 0$

$$\therefore B(m,n) = -\int_0^1 (1-u)^{m-1}\,u^{n-1}\,du = \int_0^1 (1-u)^{m-1}\,u^{n-1}du$$

$$= \int_0^1 u^{n-1}(1-u)^{m-1}du = B(n,m)$$

$$\therefore \underline{B(m,n) = B(n,m)} \tag{2}$$

Alternative form of the beta function

We had $\qquad B(m,n) = \int_0^1 x^{m-1}(1-x)^{n-1}\,dx$

If we put $x = \sin^2\theta$, the result then becomes ....

26

$$\boxed{B(m,n) = 2\int_0^{\pi/2} \sin^{2m-1}\theta\ \cos^{2n-1}\theta\ d\theta}$$

for if $x = \sin^2\theta,\quad dx = 2\sin\theta\ \cos\theta\ d\theta$.

When $x = 0,\ \theta = 0$; when $x = 1,\ \theta = \pi/2$. $1 - x = 1 - \sin^2\theta = \cos^2\theta$

$$\therefore B(m,n) = 2\int_0^{\pi/2} \sin^{2m-2}\theta\cos^{2n-2}\theta\sin\theta\ \cos\theta\ d\theta$$

$$\therefore \underline{B(m,n) = 2\int_0^{\pi/2} \sin^{2m-1}\theta\cos^{2n-1}\theta\ d\theta} \tag{3}$$

Make a note of this result. We shall need to use it later.

27

Reduction formulae

In Programme 17 of *Engineering Mathematics* (fourth edition) we established useful reduction formulae relating to integrals of powers of sines and cosines, particularly when the integral limits are 0 and $\pi/2$.

(a) $$\int_0^{\pi/2} \sin^n x\,dx = \frac{n-1}{n}\int_0^{\pi/2} \sin^{n-2} x\,dx \quad \text{i.e. } S_n = \frac{n-1}{n} S_{n-2} \qquad (4)$$

(b) $$\int_0^{\pi/2} \cos^n x\,dx = \frac{n-1}{n}\int_0^{\pi/2} \cos^{n-2} x\,dx \quad \text{i.e. } C_n = \frac{n-1}{n} C_{n-2} \qquad (5)$$

A third reduction formula for products of powers of sines and cosines is

(c) $$\int_0^{\pi/2} \sin^m x \cos^n x\,dx = \frac{m-1}{m+n}\int_0^{\pi/2} \sin^{m-2} x \cos^n x\,dx$$

If we denote $\int_0^{\pi/2} \sin^m x \cos^n x\,dx$ by $I_{m,\,n}$, the last result can be written

$$I_{m,\,n} = \frac{m-1}{m+n} I_{m-2,\,n} \qquad (6)$$

Alternatively, $\int_0^{\pi/2} \sin^m x \cos^n x\,dx$ can be expressed as

$$\frac{n-1}{m+n}\int_0^{\pi/2} \sin^m x \cos^{n-2} x\,dx$$

i.e. $$I_{m,\,n} = \frac{n-1}{m+n} I_{m,\,n-2} \qquad (7)$$

Now $\mathrm{B}(m,n) = 2\int_0^{\pi/2} \sin^{2m-1}\theta \cos^{2n-1}\theta\,d\theta$ and if we apply (6) to the integral, we have

$$\int_0^{\pi/2} \sin^{2m-1}\theta \cos^{2n-1}\theta\,d\theta = \frac{(2m-1)-1}{(2m-1)+(2n-1)}\int_0^{\pi/2} \sin^{2m-3}\theta \cos^{2n-1}\theta\,d\theta$$

$$= \frac{m-1}{m+n-1}\int_0^{\pi/2} \sin^{2m-3}\theta \cos^{2n-1}\theta\,d\theta$$

Now, using (7) with the right-hand integral

$$\int_0^{\pi/2} \sin^{2m-1}\theta \, \cos^{2n-1}\theta \, d\theta = \frac{m-1}{m+n-1} \cdot \frac{(2n-1)-1}{(2m-3)+(2n-1)}$$

$$\times \int_0^{\pi/2} \sin^{2m-3}\theta \cos^{2n-3}\theta \, d\theta$$

$$= \frac{m-1}{m+n-1} \cdot \frac{n-1}{m+n-2}$$

$$\times \int_0^{\pi/2} \sin^{2m-3}\theta \times \cos^{2n-3}\theta \, d\theta$$

$$\therefore \text{B}(m,n) = \frac{(m-1)(n-1)}{(m+n-1)(m+n-2)} \cdot 2\int_0^{\pi/2} \sin^{2m-3}\theta \, \cos^{2n-3}\theta \, d\theta$$

i.e. $$\underline{\text{B}(m,n) = \frac{(m-1)(n-1)}{(m+n-1)(m+n-2)}\text{B}(m-1,n-1)} \qquad (8)$$

This is obviously a reduction formula for $\text{B}(m,n)$ and the process can be repeated as required.

For example $\text{B}(4, 3)$ = ...

28

$$\text{B}(4,3) = \frac{(3)(2)}{(6)(5)}\frac{(2)(1)}{(4)(3)}\text{B}(2,1)$$

For, applying (8)

$$\text{B}(4,3) = \frac{(3)(2)}{(6)(5)}\,\text{B}(3,2) = \frac{(3)(2)}{(6)(5)}\frac{(2)(1)}{(4)(3)}\,\text{B}(2,1)$$

Now we must evaluate B(2, 1) for we can go no further in the reduction process, since, from the definition of $\text{B}(m, n)$, m and n must be

....

29

$$\boxed{>0}$$

But $B(2, 1) = 2\int_0^{\pi/2} \sin^3\theta \cos\theta \, d\theta = 2\left[\frac{\sin^4\theta}{4}\right]_0^{\pi/2} = \frac{1}{2}$

$$\therefore \ B(4,3) = \frac{(3)(2)}{(6)(5)}\frac{(2)(1)}{(4)(3)}\frac{1}{2}$$

$$= \frac{(3)(2)(1)\times(2)(1)}{(6)(5)(4)(3)(2)(1)} = \underline{\frac{(3!)(2!)}{(6!)}}$$

Similarly, $B(5, 3) = \ldots\ldots\ldots\ldots\ldots$

30

$$\boxed{B(5,3) = \frac{(4!)(2!)}{(7!)}}$$

for $\quad B(5,3) = \frac{(4)(2)}{(7)(6)}B(4,2) = \frac{(4)(2)}{(7)(6)}\frac{(3)(1)}{(5)(4)}B(3,1)$

$$B(3,1) = 2\int_0^{\pi/2} \sin^5\theta \cos\theta \, d\theta = 2\left[\frac{\sin^6\theta}{6}\right]_0^{\pi/2} = \frac{1}{3}$$

$$\therefore B(5,3) = \frac{(4)(2)}{(7)(6)}\frac{(3)(1)}{(5)(4)}\frac{1}{3}\frac{(2)}{(2)} = \underline{\frac{(4!)(2!)}{(7!)}}$$

In general $\qquad \underline{B(m,n) = \frac{(m-1)!(n-1)!}{(m+n-1)!}} \qquad (9)$

Note that $\qquad B(k,1) = 2\int_0^{\pi/2} \sin^{2k-1}\theta \cos\theta \, d\theta$

$$= 2\int_0^{\pi/2} \sin^{2k-1}\theta \, d(\sin\theta)$$

$$= 2\left[\frac{\sin^{2k}\theta}{2k}\right]_0^{\pi/2} = \frac{1}{k} \qquad \therefore \underline{B(k,1) = \frac{1}{k}}$$

$$\underline{\therefore B(k,1) = B(1,k) = \frac{1}{k}} \qquad (10)$$

We can also use the trigonometrical definition (3) to evaluate $B\left(\frac{1}{2},\frac{1}{2}\right)$

$$B\left(\tfrac{1}{2},\tfrac{1}{2}\right) = \ldots\ldots\ldots$$

31

$$\boxed{B\left(\tfrac{1}{2}, \tfrac{1}{2}\right) = \pi}$$

$$\text{for } B(m, n) = 2\int_0^{\pi/2} \sin^{2m-1}\theta \, \cos^{2n-1}\theta \, d\theta$$

$$\therefore B\left(\tfrac{1}{2}, \tfrac{1}{2}\right) = 2\int_0^{\pi/2} \sin^0\theta \, \cos^0\theta \, d\theta$$

$$= 2\int_0^{\pi/2} 1 \, d\theta = 2\left[\theta\right]_0^{\pi/2} = \pi \qquad (11)$$

Now let us summarise our various results so far. *Next frame.*

32 *Revision*

$$B(m, n) = \int_0^1 x^{m-1}(1-x)^{n-1} \, dx \qquad m > 0, \; n > 0$$

$$B(m, n) = B(n, m)$$

$$B(m, n) = 2\int_0^{\pi/2} \sin^{2m-1}\theta \, \cos^{2n-1}\theta \, d\theta$$

$$B(m, n) = \frac{(m-1)(n-1)}{(m+n-1)(m+n-2)} B(m-1, \; n-1)$$

$$B(m, n) = \frac{(m-1)!(n-1)!}{(m+n-1)!} \qquad m \text{ and } n \text{ positive integers}$$

$$B(k, 1) = B(1, k) = \frac{1}{k} \qquad \therefore B(1, 1) = 1$$

$$B\left(\tfrac{1}{2}, \tfrac{1}{2}\right) = \pi$$

Be sure that you are familiar with all these. We shall be using them all in due course.

33

Relation between the gamma and beta functions

If m and n are positive integers

$$B(m,n) = \frac{(m-1)!(n-1)!}{(m+n-1)!}$$

Also, we have previously established that, for n a positive integer,

$$n! = \Gamma(n+1)$$

$\therefore\ (m-1)! = \Gamma(m)$ and $(n-1)! = \Gamma(n)$

and also $(m+n-1)! = \Gamma(m+n)$

$$\therefore\ B(m,n) = \frac{(m-1)!(n-1)!}{(m+n-1)!} = \frac{\Gamma(m)\Gamma(n)}{\Gamma(m+n)} \qquad (12)$$

The relation $B(m,n) = \dfrac{\Gamma(m)\Gamma(n)}{\Gamma(m+n)}$ holds good even when m and n are not necessarily integers.

We will prove this in the next frame, so move on.

34

Proof that $\quad B(m,n) = \dfrac{\Gamma(m)\Gamma(n)}{\Gamma(m+n)}$

$$\text{Let } \Gamma(m) = \int_0^\infty x^{m-1}\,e^{-x}\,dx \quad\text{and}\quad \Gamma(n) = \int_0^\infty y^{n-1}\,e^{-y}\,dy$$

$$\therefore\ \Gamma(m)\Gamma(n) = \int_0^\infty x^{m-1}\,e^{-x}\,dx \int_0^\infty y^{n-1}\,e^{-y}\,dy$$

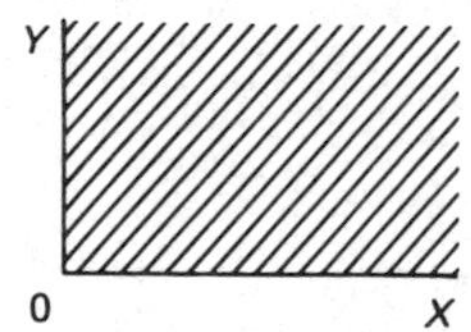

$$= \int_0^\infty\int_0^\infty x^{m-1}\,y^{n-1}\,e^{-(x+y)}\,dx\,dy$$

Note that the integration is carried out over the first quadrant of the xy-plane.

Putting $x = u^2$ and $y = v^2$ $\quad dx = 2u\,du$ and $dy = 2v\,dv$

$$\therefore\ \Gamma(m)\Gamma(n) = 4\int_0^\infty\int_0^\infty u^{2m-2}\,v^{2n-2}\,e^{-(u^2+v^2)}uv\,du\,dv$$

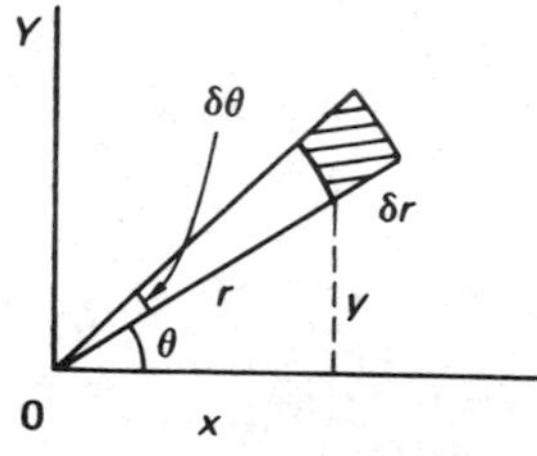

$$= 4\int_0^\infty\int_0^\infty u^{2m-1}\,v^{2n-1}\,e^{-(u^2+v^2)}\,du\,dv$$

If we now convert to polar coordinates, $u = r\cos\theta$; $\quad v = r\sin\theta$; $du\,dv = r\,dr\,d\theta$

$u^2+v^2 = r^2 \qquad 0 < r < \infty; \qquad 0 < \theta < \pi/2$

$$\therefore\ \Gamma(m)\Gamma(n) = 4\int_0^{\pi/2}\int_0^\infty r^{2m-1}\cos^{2m-1}\theta\, r^{2n-1}\sin^{2n-1}\theta\, e^{-r^2}\,r\,dr\,d\theta$$

$$= 4\int_0^{\pi/2}\int_0^\infty r^{2m+2n-2}\,e^{-r^2}\cos^{2m-1}\theta\,\sin^{2n-1}\theta\,r\,dr\,d\theta$$

Then, writing $w = r^2 \qquad \therefore\ dw = 2r\,dr$

$$\Gamma(m)\Gamma(n) = 2\int_0^\infty w^{m+n-1}\,e^{-w}\,dw\int_0^{\pi/2}\sin^{2n-1}\theta\,\cos^{2m-1}\theta\,d\theta$$

$$= \Gamma(m+n)\times B(m,n)$$

$$\therefore\ B(m,n) = \frac{\Gamma(m)\Gamma(n)}{\Gamma(m+n)} \qquad (13)$$

So $B(\frac{3}{2},\frac{1}{2}) = \ldots\ldots\ldots$

35

$$\mathrm{B}\left(\tfrac{3}{2}, \tfrac{1}{2}\right) = \tfrac{\pi}{2}$$

for $\mathrm{B}\left(\tfrac{3}{2}, \tfrac{1}{2}\right) = \dfrac{\Gamma\left(\tfrac{3}{2}\right)\Gamma\left(\tfrac{1}{2}\right)}{\Gamma(2)} = \dfrac{\sqrt{\pi}/2 \times \sqrt{\pi}}{1} = \dfrac{\pi}{2}$

Now for some examples.

36

Application of gamma and beta functions

The use of gamma and beta functions in the evaluation of definite integrals depends largely on the ability to change the variables to express the integral in the basic form of the beta function $\int_0^1 x^{m-1}(1-x)^{n-1}\,\mathrm{d}x$ or its trigonometrical form $2\int_0^{\pi/2} \sin^{2m-1}\theta\,\cos^{2n-1}\theta\,\mathrm{d}\theta$.

Example 1 Evaluate $I = \int_0^1 x^5(1-x)^4\,\mathrm{d}x$.

Compare this with $\mathrm{B}(m, n) = \int_0^1 x^{m-1}(1-x)^{n-1}\,\mathrm{d}x$

Then $m - 1 = 5 \quad \therefore m = 6$ and $n - 1 = 4 \quad \therefore n = 5$

$$\therefore I = \mathrm{B}(6, 5) = \ldots\ldots\ldots$$

37

$$\mathrm{I} = \mathrm{B}(6, 5) = \frac{5!\,4!}{10!} = \frac{1}{1260}$$

Example 2 Evaluate $I = \int_0^1 x^4\sqrt{1-x^2}\,\mathrm{d}x$.

Comparing this with $\mathrm{B}(m, n) = \int_0^1 x^{m-1}(1-x)^{n-1}\,\mathrm{d}x$

we see that we have x^2 in the root, instead of a single x.

Therefore, put $x^2 = y \quad \therefore x = y^{\frac{1}{2}} \quad \mathrm{d}x = \tfrac{1}{2}y^{-\frac{1}{2}}\,\mathrm{d}y$

The limits remain unchanged. $\quad \therefore I = \ldots\ldots\ldots$

38

$$I = \tfrac{1}{2}\mathrm{B}\left(\tfrac{5}{2}, \tfrac{3}{2}\right)$$

for $\quad I = \displaystyle\int_0^1 y^2(1-y)^{\frac{1}{2}}\,\frac{1}{2}y^{-\frac{1}{2}}\,\mathrm{d}y = \frac{1}{2}\int_0^1 y^{\frac{3}{2}}(1-y)^{\frac{1}{2}}\,\mathrm{d}y$

$$m - 1 = \tfrac{3}{2} \quad \therefore m = \tfrac{5}{2} \quad \text{and} \quad n - 1 = \tfrac{1}{2} \quad \therefore n = \tfrac{3}{2}$$

$$\therefore I = \tfrac{1}{2}\mathrm{B}\left(\tfrac{5}{2}, \tfrac{3}{2}\right)$$

Expressing this in gamma functions

$$I = \ldots\ldots\ldots$$

39

$$I = \frac{1}{2}\,\frac{\Gamma\left(\frac{5}{2}\right)\Gamma\left(\frac{3}{2}\right)}{\Gamma(4)}$$

From our previous work on gamma functions

$$\Gamma\left(\tfrac{3}{2}\right) = \frac{\sqrt{\pi}}{2};\quad \Gamma\left(\tfrac{5}{2}\right) = \frac{3\sqrt{\pi}}{4};\quad \Gamma(4) = 3!$$

$$\therefore I = \ldots\ldots\ldots$$

40

$$I = \frac{\pi}{32}$$

for $\quad I = \dfrac{1}{2}\cdot\dfrac{(3\sqrt{\pi}/4)(\sqrt{\pi}/2)}{3!} = \dfrac{\pi}{32}.$

Now you can work through this one in much the same way. There are no tricks.

Example 3 $\quad$ Evaluate $I = \displaystyle\int_0^3 \frac{x^3\,\mathrm{d}x}{\sqrt{3-x}}$.

You need to compare this with $\mathrm{B}(m, n) = \displaystyle\int_0^1 x^{m-1}(1-x)^{n-1}\,\mathrm{d}x$ so bring everything up on to the top line and then make the necessary change in the variables. Finish it off and then compare the results with the next frame.

41

$$I = \frac{864\sqrt{3}}{35} = 42.76$$

Here is the working; see whether you agree.

$$I = \int_0^3 \frac{x^3\,\mathrm{d}x}{\sqrt{3-x}} = \int_0^3 x^3(3-x)^{-\frac{1}{2}}\,\mathrm{d}x = 3^{-\frac{1}{2}}\int_0^3 x^3\left(1-\frac{x}{3}\right)^{-\frac{1}{2}}\mathrm{d}x$$

Put $\dfrac{x}{3} = y$, i.e. $x = 3y \quad \therefore\ \mathrm{d}x = 3\,\mathrm{d}y$

Limits: $x = 0,\ y = 0;\quad x = 3,\ y = 1$

$$\therefore\ I = 27\sqrt{3}\int_0^1 y^3(1-y)^{-\frac{1}{2}}\,\mathrm{d}y \qquad m-1=3 \quad \therefore\ m=4$$

$$n-1 = -\tfrac{1}{2} \quad \therefore\ n = \tfrac{1}{2}$$

$$\therefore\ I = 27\sqrt{3}\ \ B(4,\tfrac{1}{2}) = 27\sqrt{3}\ \frac{\Gamma(4)\Gamma(\frac{1}{2})}{\Gamma\ (9/2)}$$

Now $\quad \Gamma\left(\dfrac{1}{2}\right) = \sqrt{\pi};\ \Gamma\left(\dfrac{9}{2}\right) = \dfrac{105\sqrt{\pi}}{16};\ \Gamma(4) = 3!$

$$\therefore\ I = 27\sqrt{3}\times 6\times\sqrt{\pi}\times\frac{16}{105\sqrt{\pi}} = \frac{864\sqrt{3}}{35} = \underline{42.76}$$

Example 4 Evaluate $I = \displaystyle\int_0^{\pi/2} \sin^5\theta\cos^4\theta\,\mathrm{d}\theta$.

$$\mathrm{B}(m,n) = 2\int_0^{\pi/2} \sin^{2m-1}\theta\ \cos^{2n-1}\theta\ \mathrm{d}\theta$$

$$\therefore\ 2m-1 = 5 \quad \therefore\ m = 3;\quad 2n-1 = 4 \quad \therefore\ n = 5/2$$

$$\therefore\ I = \tfrac{1}{2}\,\mathrm{B}(3, 5/2) = \dots\dots\dots$$

Finish it off.

42

$$I = \frac{8}{315}$$

$$I = \frac{1}{2}\,\mathrm{B}(3, 5/2) = \frac{1}{2} \cdot \frac{\Gamma(3)\Gamma(5/2)}{\Gamma(11/2)}$$

$$= \frac{1}{2} \cdot \frac{2!(3\sqrt{\pi})/4}{(945\sqrt{\pi})/32} = \frac{3\sqrt{\pi}}{4} \cdot \frac{32}{945\sqrt{\pi}} = \underline{\frac{8}{315}}$$

Finally, one more.

Example 5 Evaluate $I = \int_0^{\pi/2} \sqrt{\tan\theta}\,\mathrm{d}\theta$.

Somehow, we need to turn this into the form

$$\mathrm{B}(m, n) = 2\int_0^{\pi/2} \sin^{2m-1}\theta\, \cos^{2n-1}\theta\, \mathrm{d}\theta$$

So off you go; express the result in gamma functions

$$I = \dots\dots\dots$$

43

$$\boxed{I = \tfrac{1}{2} \cdot \frac{\Gamma(\frac{3}{4})\Gamma(\frac{1}{4})}{\Gamma(1)}}$$

for
$$I = \int_0^{\pi/2} \sqrt{\tan\theta}\,\mathrm{d}\theta = \int_0^{\pi/2} \sin^{\frac{1}{2}}\theta \cos^{-\frac{1}{2}}\theta\,\mathrm{d}\theta$$

$$\therefore\ 2m - 1 = \frac{1}{2} \quad \therefore\ m = \frac{3}{4}; \quad 2n - 1 = -\frac{1}{2} \quad \therefore\ n = \frac{1}{4}$$

$$\therefore\ I = \frac{1}{2}\,\mathrm{B}\left(\frac{3}{4}, \frac{1}{4}\right) = \frac{1}{2} \cdot \frac{\Gamma(\frac{3}{4})\Gamma(\frac{1}{4})}{\Gamma(1)}$$

and, unless we have appropriate tables to evaluate $\Gamma(\frac{3}{4})$ and $\Gamma(\frac{1}{4})$, we cannot proceed much further.

Here is part of a table that may be useful

x	$\Gamma(x)$
0.25	3.6256
0.50	1.7725
0.75	1.2254
1.00	1.0000
1.25	0.9064
1.50	0.8862
1.75	0.9191
2.00	1.0000
2.25	1.1330
2.50	1.3293

x	$\Gamma(x)$
2.75	1.6084
3.00	2.0000
3.25	2.5493
3.50	3.3234
3.75	4.4230
4.00	6.0000
4.25	8.2851
4.50	11.6318
4.75	16.5862
5.00	24.0000

Now we can evaluate the integral of our example

$$I = \ldots\ldots\ldots$$

44

$$\boxed{I = 2.2214}$$

for $\Gamma(0.25) = 3.6256$ and $\Gamma(0.75) = 1.2254$

$$\therefore I = \frac{1}{2} \cdot \frac{(1.2254)(3.6256)}{1.0000} = \underline{2.2214}$$

Duplication formula for gamma functions

We already know that, when n is a positive integer

$$\Gamma(n) = (n-1)!$$

A useful formula enables us to calculate the gamma functions for values of n halfway between the integers. This is the *duplication formula* which can be stated as

$$\underline{\Gamma(n + \tfrac{1}{2}) = \frac{\Gamma(2n)\sqrt{\pi}}{2^{2n-1}\Gamma(n)}} \qquad (14)$$

Thus, to find $\Gamma(3.5)$ $\quad \Gamma(n) = \Gamma(3) = 2!$

$$\Gamma(2n) = \Gamma(6) = 5!$$

$$\therefore \Gamma(3.5) = \Gamma(3 + \tfrac{1}{2}) = \frac{5!\sqrt{\pi}}{2^5\, 2!} = \underline{3.3234}$$

The formula is quoted here without proof, but it is useful to have on occasions.

So $\qquad \Gamma(6.5) = \ldots\ldots\ldots$

45

$$\boxed{\Gamma(6.5) = 287.9}$$

$$\Gamma(6.5) = \Gamma(6 + \tfrac{1}{2}) = \frac{\Gamma(12)\sqrt{\pi}}{2^{11}\Gamma(6)}$$

$$\Gamma(6) = 5!;\quad \Gamma(12) = 11!;\quad 2^{11} = 2048$$

$$\therefore\ \Gamma(6.5) = \frac{11!\sqrt{\pi}}{2048 \times 5!} = \underline{287.9}$$

Now let us consider another function represented by an integral.

On then to the next frame.

46

The error function

The *error function* is defined by

$$\text{erf}(x) = \frac{2}{\sqrt{\pi}}\int_0^x e^{-t^2}\, dt$$

and occurs in statistics and various studies in physics, and heat conduction. This integral cannot be evaluated directly and values of $\text{erf}(x)$ for different values of x are obtained from tables.

Where the limits of $\displaystyle\int_a^b e^{-x^2}\, dx$ are zero or $\pm\infty$, however, a result is possible.

The integral $I = \displaystyle\int_0^\infty e^{-x^2}\, dx$ we have already considered earlier in this programme when dealing with gamma functions and we established then that

$$\int_0^\infty e^{-x^2}\, dx = \dots\dots\dots$$

47

$$\int_0^{\infty} e^{-x^2}\, dx = \frac{\sqrt{\pi}}{2}$$

Graph of $y = e^{-x^2}$

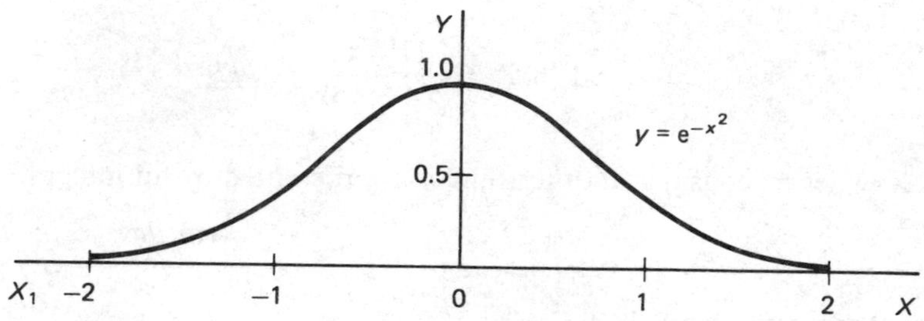

The graph of $y = e^{-x^2}$ is symmetrical about the y-axis. Therefore, the total area under the curve

$$= \int_{-\infty}^{\infty} e^{-x^2}\, dx = 2\int_0^{\infty} e^{-x^2}\, dx = \underline{\sqrt{\pi}}$$

The integral $\int_0^x e^{-t^2}\, dt$ denotes the area shaded, i.e. the area under the curve $y = e^{-t^2}$ between $t = 0$ and $t = x$.

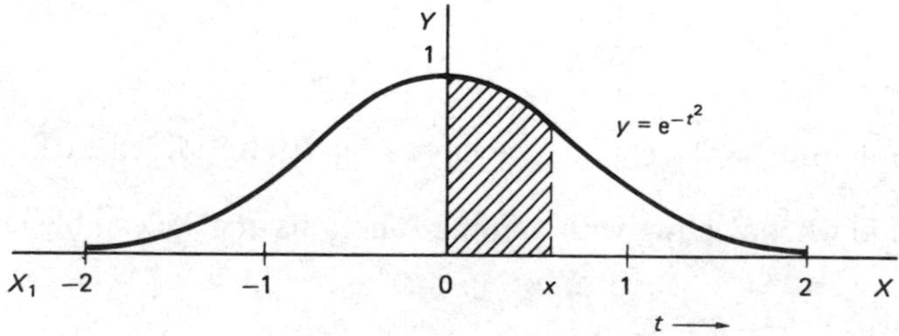

Therefore, if $x \to \infty$, $\int_0^x e^{-t^2}\, dt \to$ ....

48

$$\frac{\sqrt{\pi}}{2}$$

If we therefore introduce a multiplying factor $\frac{2}{\sqrt{\pi}}$ to the integral, then $\frac{2}{\sqrt{\pi}}\int_0^\infty e^{-t^2}\,dt$ will have a value of 1.

The error function $\text{erf}(x) = \frac{2}{\sqrt{\pi}}\int_0^x e^{-t^2}\,dt$ will tend to the value 1 as $x \to \infty$.

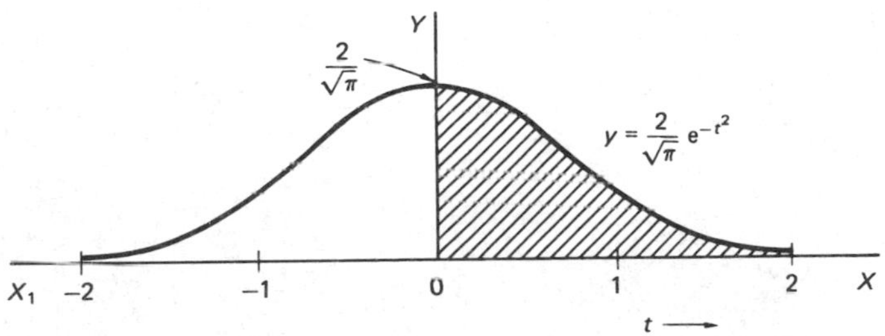

The graph of $y = \text{erf}(x)$

The graph of $y = \text{erf}(x)$ is symmetrical about the y-axis. For positive values of x, the values of $\text{erf}(x)$ are as follows.

x	0	0.2	0.4	0.6	0.8	1.0
$\text{erf}(x)$	0	0.223	0.428	0.604	0.742	0.843

x	1.2	1.4	1.6	1.8	2.0
$\text{erf}(x)$	0.910	0.952	0.976	0.989	0.995

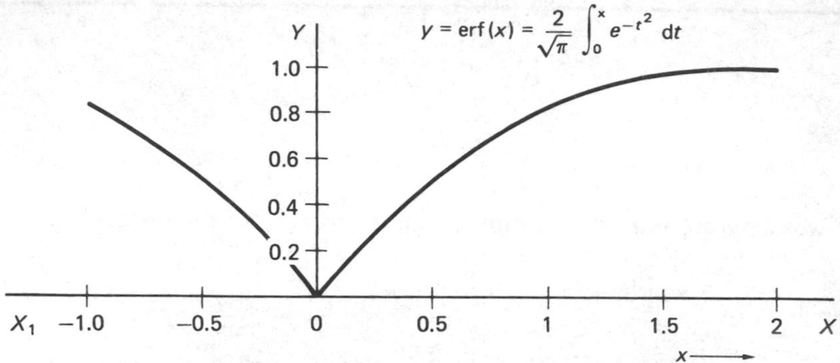

In statistics, the integral $\frac{2}{\sqrt{\pi}}\int_0^x e^{-t^2}\,dt$ is called the *probability integral.*

Now let us consider a new set of integral functions.

49

Elliptic functions

The use of *elliptic functions* provides a means of evaluating a further range of definite integrals, provided that the integrals can be converted by various appropriate substitutions into certain standard forms.

If an integrand is a rational function of x and of $\sqrt{P(x)}$ where $P(x)$ is a polynomial in x of degree 3 or 4, then the integral is said to be *elliptic*.

For example, $\displaystyle\int_0^1 \frac{\mathrm{d}x}{\sqrt{(1-2x^2)(4-3x^2)}}$ is an elliptic function. The name is derived from such an integral occurring in the determination of the arc length of part of an ellipse.

Standard forms of elliptic functions

(a) *Of the first kind*

$$F(k, \phi) = \int_0^{\phi} \frac{\mathrm{d}\theta}{\sqrt{1 - k^2 \sin^2 \theta}} \qquad (1)$$

where $0 \leq \phi \leq \pi/2$ and $0 < k < 1$.

(b) *Of the second kind*

$$E(k, \phi) = \int_0^{\phi} \sqrt{1 - k^2 \sin^2 \theta}\, \mathrm{d}\theta \qquad (2)$$

where $0 \leq \phi \leq \dfrac{\pi}{2}$ and $0 < k < 1$.

Make a careful note of these two standard forms: then we can apply them to some examples.

50

Example 1 Evaluate $\int_0^{\pi/2} \sqrt{4-\sin^2\theta}\,d\theta$

Taking out a factor 4 to reduce the first term to 1

$$I = 2\int_0^{\pi/2} \sqrt{1-\frac{1}{4}\sin^2\theta}\,d\theta$$

The integral now agrees with the standard form, where $k^2 = \frac{1}{4}$, i.e. $k = \frac{1}{2}$ and $\phi = \pi/2$.

$$\therefore\ I = \ldots\ldots\ldots$$

51

$$\boxed{I = 2E(\tfrac{1}{2}, \pi/2)}$$

Complete elliptic functions

In each of the cases (1) and (2) listed above, if $\phi = \pi/2$, the integral is said to be *complete* and then

$$F(k, \pi/2) \quad \text{is denoted by} \quad K(k)$$

and

$$E(k, \pi/2) \quad \text{is denoted by} \quad E(k).$$

Values of the functions $F(k, \phi)$, $E(k, \phi)$, $K(k)$ and $E(k)$ for various values of k and ϕ are available in published tables. The method, therefore, rests on making suitable substitutions in a given integral to transform the integrand into one of the standard forms stated above.

Incidentally, the result of Example 1 above, i.e. $I = 2E(\frac{1}{2}, \pi/2)$ could also be written as

$$I = \ldots\ldots\ldots$$

52

$$\boxed{I = 2E(\tfrac{1}{2})}$$

since, in this case, $\phi = \pi/2$.

From tables, we find that $E(\frac{1}{2}) = 1.4675 \quad \therefore \underline{I = 2.935}$

Example 2 Evaluate $I = \int_0^{\pi/6} \frac{d\theta}{\sqrt{1 - 4\sin^2\theta}}$.

At first sight, this seems to be in standard form, but notice that the value of k^2 is 4, i.e. $k = 2$—and this does not comply with the requirement that $0 < k < 1$. We therefore proceed as follows.

$$I = \int_0^{\pi/6} \frac{d\theta}{\sqrt{1 - 4\sin^2\theta}} \qquad \text{Put } 4\sin^2\theta = \sin^2\psi$$

$$\text{i.e. } 2\sin\theta = \sin\psi$$

$$\therefore 2\cos\theta \, d\theta = \cos\psi \, d\psi \quad \therefore d\theta = \frac{\cos\psi \, d\psi}{2\cos\theta}$$

Also, for the new limits, when $\theta = 0$, ; $\psi = \dots\dots\dots$

and when $\theta = \pi/6$, $\psi = \dots\dots\dots$

53

$$\boxed{\theta = 0, \ \psi = 0; \quad \theta = \pi/6, \ \psi = \pi/2}$$

$$\therefore I = \int_0^{\pi/2} \frac{1}{\sqrt{1 - \sin^2\psi}} \cdot \frac{\cos\psi \, d\psi}{2 \cos\theta}$$

We now transform the $\cos\theta$

$$\sin\theta = \tfrac{1}{2}\sin\psi \quad \therefore 1 - \cos^2\theta = \tfrac{1}{4}\sin^2\psi \quad \therefore \cos\theta = \sqrt{1 - \tfrac{1}{4}\sin^2\psi}$$

$$\therefore I = \frac{1}{2}\int_0^{\pi/2} \frac{1}{\cos\psi} \cdot \frac{\cos\psi \, d\psi}{\sqrt{1 - \frac{1}{4}\sin^2\psi}}$$

$$= \frac{1}{2}\int_0^{\pi/2} \frac{d\psi}{\sqrt{1 - \frac{1}{4}\sin^2\psi}} \quad \text{which is now in standard form}$$

$$\therefore I = \dots\dots\dots$$

54

$$I = \tfrac{1}{2}F(\tfrac{1}{2}, \pi/2) = \tfrac{1}{2}K(\tfrac{1}{2})$$

From the appropriate tables, $K(\tfrac{1}{2}) = 1.6858 \quad \therefore \underline{I = 0.8429}$

Now for another

Example 3 Evaluate $I = \displaystyle\int_0^{\pi/3} \frac{d\theta}{\sqrt{3 - 4\sin^2\theta}}.$

The first step is to ....

55

take out a factor 3 to reduce the first term to 1

$$\therefore I = \frac{1}{\sqrt{3}}\int_0^{\pi/3} \frac{d\theta}{\sqrt{1 - \frac{4}{3}\sin^2\theta}}$$

Next, we see that $k^2 > 1$. Therefore, we put ....

56

$$\tfrac{4}{3}\sin^2\theta = \sin^2\psi$$

$$\frac{2}{\sqrt{3}}\sin\theta = \sin\psi \quad \therefore \frac{2}{\sqrt{3}}\cos\theta \, d\theta = \cos\psi \, d\psi \quad \therefore d\theta = \frac{\sqrt{3}\cos\psi \, d\psi}{2\cos\theta}$$

Then, so far, we have $I =$ ....

57

$$I = \frac{1}{\sqrt{3}} \int_{\theta=0}^{\theta=\pi/3} \frac{1}{\sqrt{1-\sin^2\psi}} \cdot \frac{\sqrt{3}\cos\psi\, d\psi}{2\cos\theta}$$

$$\frac{2}{\sqrt{3}}\sin\theta = \sin\psi$$

Limits: when $\theta = 0, \ \psi = 0$

$$\theta = \frac{\pi}{3}, \ \frac{2}{\sqrt{3}}\sin\theta = \frac{2}{\sqrt{3}} \cdot \frac{\sqrt{3}}{2} = 1 \qquad \therefore \psi = \pi/2$$

Also $\cos\theta = \sqrt{1-\sin^2\theta} = \sqrt{1-\frac{3}{4}\sin^2\psi}$

$$\therefore I = \dots\dots\dots$$

58

$$I = \frac{1}{2}\int_0^{\pi/2} \frac{d\psi}{\sqrt{1-\frac{3}{4}\sin^2\psi}}$$

which is now in standard form with $k = \dfrac{\sqrt{3}}{2}$ and $\phi = \pi/2$

$$\therefore I = \frac{1}{2}F\left(\frac{\sqrt{3}}{2}, \ \pi/2\right) = \frac{1}{2}K\left(\frac{\sqrt{3}}{2}\right)$$

From tables $K\left(\dfrac{\sqrt{3}}{2}\right) = 2.1565 \qquad \underline{\therefore I = 1.078}$

Now, what about this one?

Example 4 Evaluate $I = \displaystyle\int_0^{\pi/2} \frac{d\theta}{\sqrt{1+4\sin^2\theta}}.$

The trouble here is the *plus* sign in the denominator. Were it a minus sign as in Example 2, the integral could be converted into standard form and would present no difficulty.

In this case, the key is to put $\theta = \pi/2 - \psi$, i.e. $\sin\theta = \cos\psi$. Expressing the integral in terms of ψ, we have

$$I = \dots\dots\dots$$

59

$$I = \int_{\pi/2}^{0} \frac{-\mathrm{d}\psi}{\sqrt{5 - 4\sin^2 \psi}}$$

for $\quad \theta = \pi/2 - \psi \qquad \therefore \ \mathrm{d}\theta = -\mathrm{d}\psi$

$$1 + 4\sin^2 \theta = 1 + 4(1 - \cos^2 \theta) = 5 - 4\cos^2 \theta = 5 - 4\sin^2 \psi$$

Limits: when $\theta = 0$, $\psi = \pi/2$; when $\theta = \pi/2$, $\psi = 0$ and the expression above immediately follows.

Move on.

60

So we have $\quad I = \int_{\pi/2}^{0} \frac{-\mathrm{d}\psi}{\sqrt{5 - 4\sin^2 \psi}}$

The minus sign in the numerator can be absorbed by ....

61

changing the order of the limits

$$\therefore I = \int_{0}^{\pi/2} \frac{\mathrm{d}\psi}{\sqrt{5 - 4\sin^2 \psi}}$$

Finally, taking out a factor 5 from the denominator, the integral becomes

$$I = \frac{1}{\sqrt{5}} \int_{0}^{\pi/2} \frac{\mathrm{d}\psi}{\sqrt{1 - \frac{4}{5}\sin^2 \psi}}$$

and this can then be written ....

62

$$I = \frac{1}{\sqrt{5}} F\left(\frac{2}{\sqrt{5}}, \frac{\pi}{2}\right) = \frac{1}{\sqrt{5}} K\left(\frac{2}{\sqrt{5}}\right)$$

From tables $K\left(\frac{2}{\sqrt{5}}\right) = K(0.8944) = 2.2435 \quad \therefore \underline{I = 1.003}$

Alternative forms of elliptic functions

(a) *Of the first kind*

$$F(k, x) = \int_0^x \frac{\mathrm{d}u}{\sqrt{(1-u^2)(1-k^2u^2)}} \tag{3}$$

where $0 \leq x \leq 1$ and $0 < k < 1$.

(b) *Of the second kind*

$$E(k, x) = \int_0^x \sqrt{\frac{1-k^2u^2}{1-u^2}}\, \mathrm{d}u \tag{4}$$

where $0 \leq x \leq 1$ and $0 < k < 1$.

Note these two new forms and then we can deal with a few examples. As before, it is a case of transforming the given integrand into the required form by suitable substitutions.

63

Example 1 Evaluate $I = \int_0^{1/\sqrt{2}} \sqrt{\frac{4-3u^2}{1-u^2}}\, \mathrm{d}u.$

Here we remove a factor 4 from the numerator to reduce the first term to 1.

$$I = 2\int_0^{1/\sqrt{2}} \sqrt{\frac{1-\frac{3}{4}u^2}{1-u^2}}\, \mathrm{d}u$$

This is now in standard form with $k =$ .... and $x =$ ....

64

$$k = \frac{\sqrt{3}}{2}; \quad x = \frac{1}{\sqrt{2}}$$

$$\therefore I = 2E\left(\frac{\sqrt{3}}{2}, \frac{1}{\sqrt{2}}\right) = 2(0.7282) \text{ from tables}$$

$$\therefore I = 1.4564$$

Example 2 Evaluate $I = \int_0^{1/2} \frac{\mathrm{d}u}{\sqrt{5 - 6u^2 + u^4}}$.

Factorising the denominator, gives $I = \ldots\ldots\ldots$

65

$$I = \int_0^{1/2} \frac{\mathrm{d}u}{\sqrt{(1 - u^2)(5 - u^2)}}$$

Taking out a factor 5

$$I = \frac{1}{\sqrt{5}} \int_0^{1/2} \frac{\mathrm{d}u}{\sqrt{(1 - u^2)(1 - \frac{1}{5}u^2)}}$$

which is in standard form with $k = 1/\sqrt{5}$ and $x = 1/2$

$$\therefore I = \ldots\ldots\ldots$$

66

$$I = \frac{1}{\sqrt{5}} F\left(\frac{1}{\sqrt{5}}, \frac{1}{2}\right)$$

In some tables, k is quoted as $\sin\theta$, i.e. $\sin\theta = \frac{1}{\sqrt{5}}$ $\therefore \theta = 26^\circ\ 34'$

and x is quoted as $\sin\phi$, i.e. $\sin\phi = \frac{1}{2}$ $\therefore \phi = 30^\circ$.

Then $F(1/\sqrt{5}, 1/2) = 0.528$ $\therefore \underline{I = 0.236}$

Now move on for Example 3.

67

Example 3 Evaluate $I = \int_0^{\sqrt{3}/4} \sqrt{\frac{2 - x^2}{1 - 4x^2}}\, dx$.

We have to convert this into the form $\int \sqrt{\frac{1 - k^2 u^2}{1 - u^2}}\, du$, so first concentrate on the denominator. Any suggestions?

68

Put $4x^2 = u^2$ i.e. $2x = u$

$$4x^2 = u^2 \quad \therefore 2x = u \quad \therefore 2\,dx = du$$

Limits: when $x = 0$, $u = 0$ and when $x = \sqrt{3}/4$, $u = \sqrt{3}/2$

Also $2 - x^2 = 2 - u^2/4$

The integral now becomes ....

69

$$I = \int_0^{\sqrt{3}/2} \sqrt{\frac{2 - u^2/4}{1 - u^2}} \cdot \frac{du}{2}$$

Finally, taking out the factor 2 in the numerator

$$I = \ldots\ldots\ldots$$

70

$$I = \frac{1}{\sqrt{2}} \int_0^{\sqrt{3}/2} \frac{\sqrt{1 - u^2/8}}{1 - u^2} du$$

i.e. $k^2 = \frac{1}{8}$ $\quad \therefore k = \frac{\sqrt{2}}{4}$ $\quad$ and $\quad x = \frac{\sqrt{3}}{2}$

So $\quad I = \ldots\ldots\ldots$

71

$$I = \frac{1}{\sqrt{2}} E\left(\frac{\sqrt{2}}{4}, \frac{\sqrt{3}}{2}\right)$$

Then $\sin\theta = \frac{\sqrt{2}}{4}$ $\quad \therefore \theta = 20° \; 42'$ and $\sin\phi = \frac{\sqrt{3}}{2}$ $\quad \therefore \phi = 60°$

From tables, $\quad E\left(\frac{\sqrt{2}}{4}, \frac{\sqrt{3}}{2}\right) = 1.029 \quad \therefore \underline{I = 0.728}$

So it is all just a question of manipulation to transform the given integral into the required standard forms, and then of reference to the appropriate tables.

REVISION SUMMARY

72

1. *Gamma functions*

(a) $$\Gamma(x) = \int_0^\infty t^{x-1}e^{-t}\,dt \qquad x > 0$$

$$\Gamma(x+1) = x\Gamma(x)$$

(b) If $x = n$, a positive integer,

$$\Gamma(n+1) = n!$$
$$\Gamma(1) = 1$$
$$\Gamma(0) = \infty \quad \Gamma(-n) = \pm\infty$$

(c) $$\int_0^\infty e^{-x^2}\,dx = \frac{\sqrt{\pi}}{2}.$$

(d) $\Gamma(\frac{1}{2}) = \sqrt{\pi}$ $\qquad \Gamma(\frac{3}{2}) = \dfrac{\sqrt{\pi}}{2}$

$\Gamma(\frac{5}{2}) = \dfrac{3\sqrt{\pi}}{4}$ $\qquad \Gamma(\frac{7}{2}) = \dfrac{15\sqrt{\pi}}{8}$

$\Gamma(-\frac{1}{2}) = -2\sqrt{\pi}$ $\qquad \Gamma(-\frac{3}{2}) = \dfrac{4\sqrt{\pi}}{3}.$

(e) Duplication formula $\Gamma(n+\frac{1}{2}) = \dfrac{\Gamma(2n)\sqrt{\pi}}{2^{2n-1}\cdot\Gamma(n)}.$

2. *Beta functions*

(a) $$B(m,n) = \int_0^1 x^{m-1}(1-x)^{n-1}\,dx \qquad m > 0;\ n > 0$$

$$B(m,n) = B(n,m)$$

$$B(m,n) = 2\int_0^{\pi/2} \sin^{2m-1}\theta\cos^{2n-1}\theta\,d\theta$$

(b) $$B(m,n) = \frac{(m-1)(n-1)}{(m+n-1)(m+n-2)}B(m-1,\ n-1)$$

$$B(k,1) = B(1,k) = \frac{1}{k}$$

$$B(1,1) = 1; \qquad B(\tfrac{1}{2},\tfrac{1}{2}) = \pi$$

$$B(m,n) = \frac{\Gamma(m)\cdot\Gamma(n)}{\Gamma(m+n)}.$$

(c) m and n positive integers

$$B(m,n) = \frac{(m-1)!(n-1)!}{(m+n-1)!}.$$

3. *Error function*

(a)
$$\operatorname{erf}(x) = \frac{2}{\sqrt{\pi}}\int_0^x e^{-t^2}\,dt$$

(b)
$$\int_0^\infty e^{-x^2}\,dx = \frac{\sqrt{\pi}}{2}$$

$$\int_{-\infty}^\infty e^{-x^2}\,dx = \sqrt{\pi}; \qquad \int_{-\infty}^\infty e^{-x^2/2}\,dx = \sqrt{2\pi}.$$

4. *Elliptic functions*

(a) *Standard forms*

(i) of the first kind: $F(k, \phi) = \displaystyle\int_0^\phi \frac{d\theta}{\sqrt{1 - k^2\sin^2\theta}}$

(ii) of the second kind: $E(k, \phi) = \displaystyle\int_0^\phi \sqrt{1 - k^2\sin^2\theta}\,d\theta$

In each case, $0 \le \phi \le \pi/2$; $0 < k < 1$.

(b) *Complete elliptic integrals* $\phi = \dfrac{\pi}{2}$

$$F\left(k, \frac{\pi}{2}\right) = K(k)$$

$$E\left(k, \frac{\pi}{2}\right) = E(k).$$

(c) *Alternative forms of elliptic functions*

(i) of the first kind: $F(k, x) = \displaystyle\int_0^x \frac{du}{\sqrt{(1 - u^2)(1 - k^2u^2)}}$

(ii) of the second kind: $E(k, x) = \displaystyle\int_0^x \sqrt{\frac{1 - k^2u^2}{1 - u^2}}\,du$

In each case $0 \le x \le 1$; $0 < k < 1$.

TEST EXERCISE IV

73

1. Evaluate (a) $\dfrac{\Gamma(6)}{3\Gamma(4)}$ (b) $\dfrac{\Gamma(1.5)}{\Gamma(2.5)}$ (c) $\dfrac{\Gamma(-\frac{1}{2})}{\Gamma(\frac{1}{2})}$

(d) $\displaystyle\int_0^{\infty} x^5 \, \mathrm{e}^{-x} \, \mathrm{d}x$ (e) $\displaystyle\int_0^{\infty} x^6 \, \mathrm{e}^{-4x^2} \, \mathrm{d}x$.

2. Determine (a) $\displaystyle\int_0^1 x^5(2-x)^4 \mathrm{d}x$

(b) $\displaystyle\int_0^{\pi/2} \sin^7\theta \cos^3\theta \, \mathrm{d}\theta$

(c) $\displaystyle\int_0^{\pi/8} \sin^2 4\theta \cos^5 4\theta \, \mathrm{d}\theta$.

3. Express the following in elliptic functions

(a) $\displaystyle\int_0^{\pi/4} \frac{\mathrm{d}\theta}{\sqrt{1-2\sin^2\theta}}$

(b) $\displaystyle\int_0^{\sqrt{3}/2} \frac{\mathrm{d}u}{\sqrt{4-5u^2+u^4}}$.

FURTHER PROBLEMS IV

1. Evaluate (a) $\dfrac{\Gamma(5)}{2\Gamma(3)}$; (b) $\dfrac{\Gamma(\frac{1}{2})}{\Gamma(-\frac{1}{2})}$; (c) $\dfrac{\Gamma(2.5)}{\Gamma(3.5)}$;

   (d) $\displaystyle\int_0^{\infty} x^4 e^{-x}\,dx$; (e) $\displaystyle\int_0^{\infty} x^8 e^{-2x}\,dx$.

2. Determine (a) $\displaystyle\int_0^{\infty} x^3 e^{-x}\,dx$; (b) $\displaystyle\int_0^{\infty} x^4 e^{-3x}\,dx$

   (c) $\displaystyle\int_0^{\infty} x^2 e^{-2x^2}\,dx$; (d) $\displaystyle\int_0^{\infty} \sqrt{x}\cdot e^{-\sqrt{x}}\,dx$.

3. If m and n are positive constants, show that $\displaystyle\int_0^{\infty} x^m e^{-ax^n}\,dx$ can be expressed in the form $\dfrac{1}{n\cdot a^{(m+1)/n}}\Gamma\left(\dfrac{m+1}{n}\right)$.

4. Evaluate the following

   (a) $\displaystyle\int_0^{1/2} x^4(1-2x)^3\,dx$

   (b) $\displaystyle\int_0^{1/\sqrt{2}} x^2\sqrt{1-2x^2}\,dx$

   (c) $\displaystyle\int_0^{\pi/2} \sin^5\theta\cos^4\theta\,d\theta$

   (d) $\displaystyle\int_0^{\pi/2} \sin\theta\sqrt{\cos^5\theta}\,d\theta$

   (e) $\displaystyle\int_0^{\pi/4} \sin^3 2\theta\cos^6 2\theta\,d\theta$

   (f) $\displaystyle\int_0^{1/3} x^2\sqrt{1-9x^2}\,dx$.

5. Express the following in elliptic functions

   (a) $\displaystyle\int_0^{\pi/2} \sqrt{1+4\sin^2\theta}\,d\theta$

   (b) $\displaystyle\int_0^{\pi/2} \frac{d\theta}{\sqrt{\cos\theta}}$

(c) $\displaystyle\int_0^1 \sqrt{\frac{4-x^2}{1-x^2}}\,dx$

(d) $\displaystyle\int_0^2 \frac{dx}{\sqrt{(9-x^2)(16-x^2)}}$

(e) $\displaystyle\int_0^2 \frac{dx}{\sqrt{(4-x^2)(5-x^2)}}$

(f) $\displaystyle\int_0^{\pi/6} \frac{d\theta}{\sqrt{\sin^2\theta + 2\cos^2\theta}}$

(g) $\displaystyle\int_{\pi/4}^{\pi/3} \frac{d\theta}{\sqrt{\sin^2\theta + 2\cos^2\theta}}$.

6. Using the substitution $x = \tan\theta$ prove that the integral

$\displaystyle\int_0^1 \frac{dx}{\sqrt{(1+x^2)(1+4x^2)}}$ can be expressed in the form

$$\frac{1}{2}\int_0^{\pi/4} \frac{d\theta}{\sqrt{1-\frac{3}{4}\cos^2\theta}}.$$

Hence, using $\theta = \frac{\pi}{2} - \phi$, evaluate the integral in terms of elliptic functions.

7. Evaluate the following

(a) $\displaystyle\int_0^{0.5} \frac{dx}{\sqrt{3-4x^2+x^4}}$

(b) $\displaystyle\int_{0.5}^{1.0} \frac{dx}{\sqrt{3-4x^2+x^4}}$

(c) $\displaystyle\int_0^{\pi/2} \frac{d\theta}{\sqrt{25+9\sin^2\theta}}$

(d) $\displaystyle\int_0^{\pi/3} \frac{d\theta}{\sqrt{4+3\sin^2\theta}}.$

Programme 5

Power Series Solution of Differential Equations

Prerequisites: Engineering Mathematics (fourth edition)
Programmes 13, 14

Higher Differential Coefficients

1

If $y = \sin x$ $\quad \dfrac{dy}{dx} = \cos x = \sin\left(x + \dfrac{\pi}{2}\right)$

$$\frac{d^2y}{dx^2} = -\sin x = \sin(x + \pi) = \sin\left(x + \frac{2\pi}{2}\right)$$

$$\frac{d^3y}{dx^3} = -\cos x = \sin\left(x + \frac{3\pi}{2}\right) \quad \text{etc.}$$

We see a pattern developing.

In general $\quad \dfrac{d^ny}{dx^n} = \sin\left(x + \dfrac{n\pi}{2}\right)$.

Before we go further, it is useful to introduce a shorthand notation, where

$$y_1 = \frac{dy}{dx}, \quad y_2 = \frac{d^2y}{dx^2}, \quad y_3 = \frac{d^3y}{dx^3} \quad \text{and} \quad y_n = \frac{d^ny}{dx^n} \quad \text{etc.}$$

The results above can therefore be written

If $y = \sin x \quad \therefore y_1 = \cos x = \sin\left(x + \dfrac{\pi}{2}\right)$

$$y_2 = -\sin x = \sin\left(x + \frac{2\pi}{2}\right)$$

$$y_3 = -\cos x = \sin\left(x + \frac{3\pi}{2}\right)$$

and, in general, $\underline{y_n = \sin\left(x + \dfrac{n\pi}{2}\right)}$

It is therefore possible to write down any particular differential coefficient of $\sin x$ without calculating all the previous derivatives. For example

$$\frac{d^7y}{dx^7} = y_7 = \sin\left(x + \frac{7\pi}{2}\right) = -\cos x$$

Similarly, starting with $y = \cos x$, we can determine an expression for the nth differential coefficient of y which is ....

2

$$y_n = \cos\left(x + \frac{n\pi}{2}\right)$$

For $y = \cos x \quad \therefore y_1 = -\sin x = \cos\left(x + \frac{\pi}{2}\right)$

$$y_2 = -\cos x = \cos\left(x + \frac{2\pi}{2}\right)$$

$$y_3 = \sin x \quad = \cos\left(x + \frac{3\pi}{2}\right) \quad \text{etc.}$$

$$\therefore y_n = \cos\left(x + \frac{n\pi}{2}\right)$$

Many of the standard functions can be treated in a similar manner. For example, if $y = e^{ax}$, then $y_n = \ldots\ldots\ldots$

3

$$y_n = a^n e^{ax}$$

since $y = e^{ax}$, $\quad y_1 = a\,e^{ax}$, $\quad y_2 = a^2\,e^{ax}$, $\quad y_3 = a^3\,e^{ax}$, $\quad$ etc.

In general, $y_n = a^n\,e^{ax}$.

With no great effort, we can now write down expressions for the following

If $y = \sin ax$, $\quad y_n = \ldots\ldots\ldots$

If $y = \cos ax$, $\quad y_n = \ldots\ldots\ldots$

4

$$y = \sin ax, \qquad y_n = a^n \sin\left(ax + \frac{n\pi}{2}\right)$$

$$y = \cos ax, \qquad y_n = a^n \cos\left(ax + \frac{n\pi}{2}\right)$$

Now one more. $\qquad$ If $y = \ln x$, $\quad y_n = \ldots\ldots\ldots$

5

$$y_n = (-1)^{n-1} \cdot \frac{(n-1)!}{x^n}$$

for $y = \ln x \quad \therefore y_1 = \dfrac{1}{x}$

$$y_2 = -\frac{1}{x^2}$$

$$y_3 = \frac{2}{x^3}$$

$$y_4 = -\frac{3!}{x^4} \qquad \therefore y_n = (-1)^{n-1} \cdot \frac{(n-1)!}{x^n}$$

We already know that, if $y = \ln x$, $\dfrac{dy}{dx} = y_1 = \dfrac{1}{x} = x^{-1}$.
Therefore, if the result obtained for y_n is to be valid for $n = 1$, then

$$y_1 = (-1)^0 \cdot \frac{0!}{x} = \frac{0!}{x}$$

But $y_1 = x^{-1}$ $\qquad \therefore 0! = \ldots\ldots\ldots$

6

$$0! = 1$$

Now let us consider the differential coefficients of $\sinh ax$ and $\cosh ax$.

Next frame.

7

If $y = \sinh ax$,

$$y_1 = a \cosh ax$$
$$y_2 = a^2 \sinh ax$$
$$y_3 = a^3 \cosh ax \qquad \text{etc.}$$

Since $\sinh ax$ is not periodic, we cannot proceed as we did with $\sin ax$. We need to find a general statement for y_n containing terms in $\sinh ax$ and in $\cosh ax$, such that, when n is even, the term in $\cosh ax$ disappears and, when n is odd, the term in $\sinh ax$ disappears.

This we can do by writing y_n in the form

$$y_n = \frac{a^n}{2}\{[1 + (-1)^n] \sinh ax + [1 - (-1)^n] \cosh ax\}$$

In very much the same way, we can determine the nth differential coefficient of $y = \cosh ax$ as $\ldots\ldots\ldots$

8

$$y_n = \frac{a^n}{2}\{[1-(-1)^n]\sinh ax + [1+(-1)^n]\cosh ax\}$$

Finally, let us deal with $y = x^a$

$$y = x^a \quad \therefore \quad y_1 = ax^{a-1}$$
$$y_2 = a(a-1)\,x^{a-2}$$
$$y_3 = a(a-1)(a-2)\,x^{a-3}$$
$$\cdots\cdots\cdots$$
$$\therefore \quad y_n = a(a-1)(a-2)\ldots(a-\overline{n-1})\,x^{a-n}$$
$$\therefore \quad y_n = \frac{a!}{(a-n)!}\,x^{a-n}$$

So, collecting our results together, we have

$y = x^a$	$y_n = \frac{a!}{(a-n)!}x^{a-n}$
$y = e^{ax}$	$y_n = a^n e^{ax}$
$y = \sin ax$	$y_n = a^n \sin\left(ax + \frac{n\pi}{2}\right)$
$y = \cos ax$	$y_n = a^n \cos\left(ax + \frac{n\pi}{2}\right)$
$y = \sinh ax$	$y_n = \frac{a^n}{2}\{[1+(-1)^n]\sinh ax + [1-(-1)^n]\cosh ax\}$
$y = \cosh ax$	$y_n = \frac{a^n}{2}\{[1-(-1)^n]\sinh ax + [1+(-1)^n]\cosh ax\}$

Make a note of these, as a set, and then move on to the next frame.

9

Exercise Determine the following differential coefficients

1. $y = \sin 4x$ $\quad y_5 = \ldots\ldots\ldots$
2. $y = e^{x/2}$ $\quad y_8 = \ldots\ldots\ldots$
3. $y = \cosh 3x$ $\quad y_{12} = \ldots\ldots\ldots$
4. $y = \cos(x\sqrt{2})$ $\quad y_{10} = \ldots\ldots\ldots$
5. $y = x^8$ $\quad y_6 = \ldots\ldots\ldots$
6. $y = \sinh 2x$ $\quad y_7 = \ldots\ldots\ldots$

Finish them all; then check with the next frame.

10

Here are the solutions

1. $y_5 = 4^5 \sin\left(4x + \dfrac{5\pi}{2}\right) = 1024 \sin\left(4x + \dfrac{\pi}{2}\right) = \underline{1024 \cos 4x}$

2. $y_8 = \left(\frac{1}{2}\right)^8 e^{x/2} = \dfrac{1}{256} e^{x/2} = \underline{e^{x/2}/256}$

3. $y_{12} = \dfrac{3^{12}}{2}\{0 \sinh 3x + 2 \cosh 3x\} = \underline{3^{12} \cosh 3x}$

4. $y_{10} = (\sqrt{2})^{10} \cos\left(x\sqrt{2} + \dfrac{10\pi}{2}\right)$

   $= 32 \cos(x\sqrt{2} + 5\pi) = \underline{-32 \cos(x\sqrt{2})}$

5. $y_6 = \dfrac{8!}{2!} x^2 = \underline{20\,160\, x^2}$

6. $y_7 = \dfrac{2^7}{2}\left\{[1 + (-1)^7] \sinh 2x + [1 - (-1)^7] \cosh 2x\right\}$

   $= \underline{2^7 \cosh 2x}$

Leibnitz theorem—*n*th differential coefficient of a product of two functions.

If $y = uv$, where u and v are functions of x, then

$$y_1 = uv_1 + vu_1 \quad \text{where} \quad v_1 = \frac{dv}{dx} \quad \text{and} \quad u_1 = \frac{du}{dx}$$

$$\text{and} \quad y_2 = uv_2 + v_1u_1 + vu_2 + u_1v_1 = u_2v + 2u_1v_1 + uv_2$$

If we differentiate the last result and collect up like terms, we obtain

$$y_3 = \ldots\ldots\ldots$$

$$y_3 = u_3v + 3u_2v_1 + 3u_1v_2 + uv_3$$

A further stage of differentiation would give

$$y_4 = u_4v + 4u_3v_1 + 6u_2v_2 + 4u_1v_3 + uv_4$$

These results can therefore be written

$$y\ = uv$$
$$y_1 = u_1v + uv_1$$
$$y_2 = u_2v + 2u_1v_1 + uv_2$$
$$y_3 = u_3v + 3u_2v_1 + 3u_1v_2 + uv_3$$
$$y_4 = u_4v + 4u_3v_1 + 6u_2v_2 + 4u_1v_3 + uv_4$$

Notice that in each case

(i) the subscript of u decreases regularly by 1
(ii) the subscript of v increases regularly by 1
(iii) the numerical coefficients are the normal binomial coefficients.

The expression for the nth differential coefficient can therefore be written as

$$y_n = u_nv + nu_{n-1}v_1 + \frac{n(n-1)}{1 \times 2}u_{n-2}v_2 + \frac{n(n-1)(n-2)}{1 \times 2 \times 3}u_{n-3}v_3 + \dots$$

$$= u_nv + nu_{n-1}v_1 + \frac{n(n-1)}{2!}u_{n-2}v_2 + \frac{n(n-1)(n-2)}{3!}u_{n-3}v_3 + \dots$$

i.e. $y_n = u_nv + {}^nC_1\,u_{n-1}v_1 + {}^nC_2\,u_{n-2}v_2 + \dots + {}^nC_{n-1}u_1v_{n-1} + uv_n$

where ${}^nC_r = \dfrac{n!}{r!(n-r)!}$

If $y = uv$ $\qquad$ $\underline{y_n = \sum_{r=0}^{n} {}^nC_r\,u_{n-r}v_r}$

This is *Leibnitz theorem*. We shall certainly be using it often in the work ahead, so make a note of it for future reference. Then we can see it in use.

12

Choice of function for u and v

Of the product function $y = uv$ the function taken as

(a) u is the one whose nth differential coefficient can readily be obtained
(b) v is the one whose derivatives reduce to zero after a small number of stages of differentiation.

Example 1 To find y_n when $y = x^3 e^{2x}$.

Here we choose $v = x^3$— whose fourth derivative is zero
$u = e^{2x}$— since we know that the nth derivative

$$u_n = \ldots\ldots\ldots$$

13

$$\boxed{u_n = 2^n e^{2x}}$$

Leibnitz theorem:

$$y_n = u_n v + n u_{n-1} v_1 + \frac{n(n-1)}{2!} u_{n-2} v_2 + \frac{n(n-1)(n-2)}{3!} u_{n-3} v_3 + \ldots$$

$v = x^3; \quad v_1 = 3x^2; \quad v_2 = 6x; \quad v_3 = 6; \quad v_4 = 0$

$u = e^{2x}; \quad u_n = 2^n e^{2x}$

$$\therefore y_n = \ldots\ldots\ldots$$

14

$$\boxed{y_n = e^{2x}\, 2^{n-3}\{8x^3 + 12nx^2 + n(n-1)\,6x + n(n-1)(n-2)\}}$$

Example 2 If $x^2 y_2 + x y_1 + y = 0$, show that

$$x^2 y_{n+2} + (2n+1)\, x y_{n+1} + (n^2+1)\, y_n = 0.$$

We take the given equation $x^2 y_2 + x y_1 + y = 0$ and differentiate n times, treating each term in turn.

If $w = x^2 y_2$ $\quad w_n = \ldots\ldots\ldots$
If $w = x y_1$ $\quad w_n = \ldots\ldots\ldots$
If $w = y$ $\quad w_n = \ldots\ldots\ldots$

15

$$w = x^2 y_2 \qquad \therefore \quad w_n = y_{n+2}\, x^2 + n y_{n+1}\, 2x + \frac{n(n-1)}{2!}\, y_n 2 + 0 \dots$$
$$w = x y_1 \qquad \therefore \quad w_n = y_{n+1}\, x + n y_n 1 + 0 + \dots$$
$$w = y \qquad \therefore \quad w_n = y_n$$

Then $[x^2 y_2 + x y_1 + y]_n = 0$ becomes

....

16

$$x^2 y_{n+2} + (2n+1) x y_{n+1} + (n^2+1) y_n = 0$$

which is what we had to show.

Example 3 Differentiate n times

$$(1+x^2) y_2 + 2 x y_1 - 5y = 0.$$

The result ....

17

$$(1+x^2)\,y_{n+2}+2(n+1)\,xy_{n+1}+(n^2+n-5)\,y_n=0$$

for, by Leibnitz theorem

$$\left\{y_{n+2}(1+x^2)+ny_{n+1}\,2x+\frac{n(n-1)}{2!}\,y_n 2\right\}+2\{xy_{n+1}+ny_n\cdot 1\}-5y_n=0$$

$$(1+x^2)\,y_{n+2}+2(n+1)\,xy_{n+1}+\{n(n-1)+2n-5\}\,y_n=0$$

$$\underline{(1+x^2)\,y_{n+2}+2(n+1)\,xy_{n+1}+(n^2+n-5)\,y_n=0}$$

We shall be using Leibnitz theorem in the rest of this programme, so let us move on to see some of its applications.

Power Series Solutions

18 Second-order linear differential equations with constant coefficients of the form $a\dfrac{\mathrm{d}^2y}{\mathrm{d}x^2}+b\dfrac{\mathrm{d}y}{\mathrm{d}x}+cy=0$ can be solved by algebraic methods giving solutions in terms of the normal elementary functions.

In general, equations of the form $\dfrac{\mathrm{d}^2y}{\mathrm{d}x^2}+P\,(x)\dfrac{\mathrm{d}y}{\mathrm{d}x}+Q\,(x)\,y=0$ where $P\,(x)$ and $Q\,(x)$ are functions of x, cannot be solved in this way. However, it is often possible to obtain solutions in the form of infinite series of powers of x—and the next section of work investigates some of the methods which make this possible.

1. Leibnitz–Maclaurin method

As the title suggests, for this we need to be familiar with Leibnitz theorem and with Maclaurin's series.

Leibnitz theorem states that, if $y=uv$, where u and v are functions of x, then $\quad y_n=\ldots\ldots\ldots$

19

$$y_n = u_n v + n u_{n-1} v_1 + \frac{n(n-1)}{2!} u_{n-2} v_2 + \ldots + \frac{n(n-1)\ldots(n-r+1)}{r!} u_{n-r} v_r \ldots$$

where u_r and v_r denote $\dfrac{d^r u}{dx^r}$ and $\dfrac{d^r v}{dx^r}$ respectively.

Maclaurin's series for $y = f(x)$ can be stated as

$y = \ldots\ldots\ldots$

20

$$y = y(0) + x y_1(0) + \frac{x^2}{2!} y_2(0) + \ldots + \frac{x^r}{r!} y_r(0) + \ldots$$

where $y_r(0)$ denotes the value of $\dfrac{d^r y}{dx^r}$ at $x = 0$.

Having revised those two expansions, we will now see the method by an example.

On to the next frame.

21

Example 1 Find the power series solution of the equation

$$x\frac{d^2y}{dx^2} + \frac{dy}{dx} + xy = 1.$$

The equation can be written

$$xy_2 + y_1 + xy = 1$$

In the first product term xy_2, treat y_2 as u and x as v. Then, differentiating the equation n times by Leibnitz theorem, gives

$\ldots\ldots\ldots$

22

$$(xy_{n+2} + n \cdot 1 \cdot y_{n+1}) + y_{n+1} + (xy_n + n \cdot 1 \cdot y_{n-1}) = 0$$
$$\text{i.e.} \quad xy_{n+2} + (n+1)y_{n+1} + xy_n + ny_{n-1} = 0$$

At $x = 0$, this becomes

$$(n+1)\,[y_{n+1}]_0 + n\,[y_{n-1}]_0 = 0$$

$$\therefore\ [y_{n+1}]_0 = -\frac{n}{n+1}\,[y_{n-1}]_0 \qquad n \geq 1$$

This relationship is called a *recurrence relation.*

We can now substitute $n = 1, 2, 3, \ldots$ and get a set of relationships between the various coefficients.

$$\begin{aligned} n &= 1 & (y_2)_0 &= -\tfrac{1}{2}(y)_0 \\ n &= 2 & (y_3)_0 &= -\tfrac{2}{3}(y_1)_0 \\ n &= 3 & (y_4)_0 &= -\tfrac{3}{4}(y_2)_0 = \left(-\tfrac{3}{4}\right)\left(-\tfrac{1}{2}\right)(y)_0 \end{aligned}$$

Continuing in the same way,

$$(y_5)_0 = \ldots\ldots\ldots$$
$$(y_6)_0 = \ldots\ldots\ldots$$
$$(y_7)_0 = \ldots\ldots\ldots$$
$$(y_8)_0 = \ldots\ldots\ldots$$

23

$$\begin{aligned} n &= 4 & (y_5)_0 &= -\tfrac{4}{5}(y_3)_0 = \left(-\tfrac{4}{5}\right)\left(-\tfrac{2}{3}\right)(y_1)_0 \\ n &= 5 & (y_6)_0 &= -\tfrac{5}{6}(y_4)_0 = \left(-\tfrac{5}{6}\right)\left(-\tfrac{3}{4}\right)\left(-\tfrac{1}{2}\right)(y)_0 \\ n &= 6 & (y_7)_0 &= -\tfrac{6}{7}(y_5)_0 = \left(-\tfrac{6}{7}\right)\left(-\tfrac{4}{5}\right)\left(-\tfrac{2}{3}\right)(y_1)_0 \\ n &= 7 & (y_8)_0 &= -\tfrac{7}{8}(y_6)_0 = \left(-\tfrac{7}{8}\right)\left(-\tfrac{5}{6}\right)\left(-\tfrac{3}{4}\right)\left(-\tfrac{1}{2}\right)(y)_0 \end{aligned}$$

Notice that, by this means, the values of all the differential coefficients at $x = 0$ can be expressed in terms of $(y)_0$ and $(y_1)_0$.

If we now substitute these values for $(y_r)_0$ in the Maclaurin series

$$y = (y)_0 + x(y_1)_0 + \frac{x^2}{2!}(y_2)_0 + \frac{x^3}{3!}(y_3)_0 + \ldots \frac{x^r}{r!}(y_r)_0 + \ldots$$

we obtain ....

24

$$y = (y)_0 + x(y_1)_0 + \frac{x^2}{2!}\left(-\frac{1}{2}\right)(y)_0 + \frac{x^3}{3!}\left(-\frac{2}{3}\right)(y_1)_0 + \frac{x^4}{4!}\left(-\frac{3}{4}\right)\left(-\frac{1}{2}\right)(y)_0 + \frac{x^5}{5!}\left(-\frac{4}{5}\right)\left(-\frac{2}{3}\right)(y_1)_0 + \frac{x^6}{6!}\left(-\frac{5}{6}\right)\left(-\frac{3}{4}\right)\left(-\frac{1}{2}\right)(y)_0 + \dots\dots\dots$$

Simplifying, this gives

$$y = (y)_0\left\{1 - \frac{x^2}{2^2} + \frac{x^4}{2^2 \times 4^2} - \frac{x^6}{2^2 \times 4^2 \times 6^2} + \dots\right\}$$

$$+(y_1)_0\left\{x - \frac{x^3}{3^2} + \frac{x^5}{3^2 \times 5^2} + \dots\right\}$$

The values of $(y)_0$ and $(y_1)_0$ provide the two arbitrary constants for the second-order equation and are obtained from the given initial conditions.

For example, if at $x = 0$, $y = 2$ and $\frac{dy}{dx} = 1$, then the relevant particular solution is ....

25

$$y = 2\left\{1 - \frac{x^2}{2^2} + \frac{x^4}{2^2 \times 4^2} - \frac{x^6}{2^2 \times 4^2 \times 6^2} + \dots\right\} + \left\{x - \frac{x^3}{3^2} + \frac{x^5}{3^2 \times 5^2} + \dots\right\}$$

since at $x = 0$, $y = 2$ i.e $(y)_0 = 2$

$$\frac{dy}{dx} = 1 \quad \text{i.e.} \quad (y_1)_0 = 1.$$

To be a valid solution, the series obtained must converge. Application of the ratio test will normally indicate any restrictions on the values that x may have.

The process therefore involves the following main steps:

(a) Differentiate the given equation n times, using Leibnitz theorem.
(b) Rearrange the result to obtain the recurrence relation at $x = 0$.
(c) Determine the values of the differential coefficients at $x = 0$, usually in terms of $(y)_0$ and $(y_1)_0$.
(d) Substitute in the Maclaurin expansion for $y = f(x)$.
(e) Simplify the result where possible and apply boundary conditions if provided.

That is all there is to it. Let us go through the various steps with another example.

Example 2 Determine a series solution of the equation

$$\frac{d^2y}{dx^2} + x\frac{dy}{dx} + y = 0.$$

The equation can be written $y_2 + xy_1 + y = 0$

(a) Differentiate n times using Leibnitz theorem, which gives

....

26

$$y_{n+2} + xy_{n+1} + (n+1)y_n = 0$$

for
$$y_2 + xy_1 + y = 0$$
$$\therefore\ y_{n+2} + \{xy_{n+1} + n \cdot 1 \cdot y_n\} + y_n = 0$$

$$\therefore\ y_{n+2} + xy_{n+1} + (n+1)y_n = 0.$$

(b) Determine the recurrence relation at $x = 0$, which is

....

27

$$y_{n+2} = -(n+1)\,y_n$$

(c) Now taking $n = 0, 1, 2, 3, 4, 5$, determine the differential coefficients at $x = 0$ in terms of $(y)_0$ and $(y_1)_0$. List them, as we did before, in table form.

28

$$\begin{array}{rll}
n=0 & (y_2)_0 = -(y)_0 & = -(y)_0 \\
1 & (y_3)_0 = -2\,(y_1)_0 & = -2\,(y_1)_0 \\
2 & (y_4)_0 = -3\,(y_2)_0 = (-3)[-(y)_0] & = 3\,(y)_0 \\
3 & (y_5)_0 = -4\,(y_3)_0 = (-4)[-2(y_1)_0] & = 2 \times 4\,(y_1)_0 \\
4 & (y_6)_0 = -5\,(y_4)_0 = (-5)[-3(y_2)_0] & = -3 \times 5\,(y)_0 \\
5 & (y_7)_0 = -6\,(y_5)_0 = (-6)[-4(y_3)_0] & = -2 \times 4 \times 6\,(y_1)_0
\end{array}$$

(d) Substitute these expressions for the differential coefficients in terms of $(y)_0$ and $(y_1)_0$ in Maclaurin's expansion

$$y = (y)_0 + x\,(y_1)_0 + \frac{x^2}{2!}(y_2)_0 + \frac{x^3}{3!}(y_3)_0 + \frac{x^4}{4!}(y_4)_0 + \ldots$$

Then $y = \ldots\ldots\ldots$

29

$$y = (y)_0 + x(y_1)_0 + \frac{x^2}{2!}(-y)_0 + \frac{x^3}{3!}(-2y_1)_0 + \frac{x^4}{4!}(3y)_0 + \frac{x^5}{5!}(8y_1)_0$$
$$+ \frac{x^6}{6!}(-15y)_0 + \frac{x^7}{7!}(-48y_1)_0 + \ldots$$

Collecting now the terms in $(y)_0$ and $(y_1)_0$, we finally obtain

$$y = (y)_0\left\{1 - \frac{x^2}{2} + \frac{x^4}{2 \times 4} - \frac{x^6}{2 \times 4 \times 6} + \ldots\right\}$$

$$+(y_1)_0\left\{x - \frac{x^3}{3} + \frac{x^5}{3 \times 5} - \frac{x^7}{3 \times 5 \times 7} + \ldots\right\}$$

They are all done in very much the same way. Here is another.

Example 3 Solve the equation $\frac{d^2y}{dx^2} + \frac{dy}{dx} + 2xy = 0$ given that at $x = 0$, $y = 0$ and $\frac{dy}{dx} = 1$.

First write the equation as $y_2 + y_1 + 2xy = 0$, differentiate n times by Leibnitz theorem and obtain the recurrence relation at $x = 0$, which is ....

30

$$y_{n+2} = -\{y_{n+1} + 2ny_{n-1}\} \quad n \geq 1$$

for $\quad y_2 + y_1 + 2xy = 0$

$$\therefore y_{n+2} + y_{n+1} + 2xy_n + n2y_{n-1} = 0$$

At $x = 0$, $\quad y_{n+2} + y_{n+1} + 2ny_{n-1} = 0$

$$\therefore \underline{y_{n+2} = -\{y_{n+1} + 2ny_{n-1}\}}$$

Since we have a term in y_{n-1}, then n must start at 1 to give $(y)_0$. Therefore the recurrence relation applies for $n \geq 1$.

We now take $n = 1, 2, 3, \ldots$ to obtain the relationships between the coefficients up to $(y_6)_0$. Complete the table and check with the next frame.

31

$n=1$	$(y_3)_0 = -\{(y_2)_0 + 2(y)_0\}$
$n=2$	$(y_4)_0 = -\{(y_3)_0 + 4(y_1)_0\}$
$n=3$	$(y_5)_0 = -\{(y_4)_0 + 6(y_2)_0\}$
$n=4$	$(y_6)_0 = -\{(y_5)_0 + 8(y_3)_0\}$

We therefore have expressions for $(y_3)_0, (y_4)_0, (y_5)_0, (y_6)_0$, but what about $(y_2)_0$?

If we refer to the initial conditions, we know that at $x=0$, $y=0$ and $y_1 = 1$. $\therefore\ (y)_0 = 0$ and $(y_1)_0 = 1$.
We can find $(y_2)_0$ by reference to the given equation itself, for

$$y_2 + y_1 + 2xy = 0$$

Therefore, at $x=0$, $(y_2)_0 + (y_1)_0 = 0 \quad \therefore\ (y_2)_0 = -(y_1)_0 = -1$.

So now we have
$$\begin{aligned}(y)_0 &= 0\\ (y_1)_0 &= 1\\ (y_2)_0 &= -1\\ (y_3)_0 &= -\{(y_2)_0 + 2(y)_0\} = -\{(-1)+0\} = 1\\ (y_4)_0 &= -\{(y_3)_0 + 4(y_1)_0\} = -\{(1)+4\} = -5\\ (y_5)_0 &= -\{(y_4)_0 + 6(y_2)_0\} = -\{(-5)-6\} = 11\\ (y_6)_0 &= -\{(y_5)_0 + 8(y_3)_0\} = -\{(11)+8\} = -19\end{aligned}$$

The required series solution is therefore

$$y = \ldots\ldots\ldots$$

32

$$y = x - \frac{x^2}{2!} + \frac{x^3}{3!} - \frac{5x^4}{4!} + \frac{11x^5}{5!} - \frac{19x^6}{6!} + \ldots$$

since $y = (y)_0 + x(y_1)_0 + \frac{x^2}{2!}(y_2)_0 + \frac{x^3}{3!}(y_3)_0 + \frac{x^4}{4!}(y_4)_0 + \ldots$

$$= 0 + x(1) + \frac{x^2}{2!}(-1) + \frac{x^3}{3!}(1) + \frac{x^4}{4!}(-5) + \frac{x^5}{5!}(11) + \frac{x^6}{6!}(-19)$$

$$\therefore\ y = x - \frac{x^2}{2!} + \frac{x^3}{3!} - \frac{5x^4}{4!} + \frac{11x^5}{5!} - \frac{19x^6}{6!} + \ldots$$

33

One more of the same kind.

Example 4 Determine the general series solution of the equation

$$(x^2+1)y_2 + xy_1 - 4y = 0.$$

As usual, establish the recurrence relation at $x = 0$, which is

....

34

$$\boxed{y_{n+2} = (4 - n^2)y_n}$$

Then, starting with $n = 0$, determine expressions for $(y_r)_0$ as far as $r = 7$.

35

$n = 0$	$(y_2)_0 = 4(y)_0$	$= 4(y)_0$
$n = 1$	$(y_3)_0 = 3(y_1)_0$	$= 3(y_1)_0$
$n = 2$	$(y_4)_0 = 0$	$= 0$
$n = 3$	$(y_5)_0 = -5(y_3)_0$	$= -15(y_1)_0$
$n = 4$	$(y_6)_0 = -12(y_4)_0$	$= 0$
$n = 5$	$(y_7)_0 = -21(y_5)_0$	$= (-21)(-15)(y_1)_0$
$n = 6$	$(y_8)_0 = -32(y_6)_0$	$= 0$

Now substitute in Maclaurin's expansion and simplify the result.

$y = $ ....

36

$$y = A(1 + 2x^2) + B\left\{x + \frac{x^3}{2} - \frac{x^5}{8} + \frac{x^7}{16} + \dots\right\}$$

for $y = (y)_0 + x(y_1)_0 + \dfrac{x^2}{2!}(y_2)_0 + \dfrac{x^3}{3!}(y_3)_0 + \dfrac{x^4}{4!}(y_4)_0 + \dots$

$$= (y)_0 + x(y_1)_0 + \frac{x^2}{2!}4(y)_0 + \frac{x^3}{3!}3(y_1)_0 + \frac{x^4}{4!}(0) + \frac{x^5}{5!}(-15)(y_1)_0$$

$+$ etc.

$$= (y)_0\{1 + 2x^2\} + (y_1)_0\left\{x + \frac{x^3}{2} - \frac{x^5}{8} + \frac{x^7}{16} + \dots\right\}$$

Putting $(y)_0 = A$ and $(y_1)_0 = B$, we have the result stated.

Now to something slightly different.

37

2. Frobenius' method

In each of the previous examples, we established the solution as a power series in integral powers of x. Such a solution is not always possible and a more general method is to assume a trial solution of the form

$$y = x^c\{a_0 + a_1x + a_2x^2 + a_3x^3 + \ldots + a_rx^r + \ldots\}$$

where a_0 is the first coefficient that is not zero.

The type of equation that can be solved by this method is of the form

$$y_2 + Py_1 + Qy = 0$$

where P and Q are functions of x.

However, certain conditions have to be satisfied.

(a) If the functions P and Q are such that both are finite when x is put equal to zero, $x = 0$ is called an *ordinary point* of the equation.
(b) If xP and x^2Q remain finite at $x = 0$, then $x = 0$ is called a *regular point* of the equation.

In both of these cases, the method of Frobenius can be applied.

(c) If, however, P and Q do not satisfy either of these conditions stated in (a) or (b), then $x = 0$ is called a *singular point* of the equation and the method of Frobenius cannot be applied.

Solution of differential equations by the method of Frobenius

To solve a given equation, we have to find the coefficients $a_0, a_1, a_2, \ldots$ and also the index c in the trial solution. Basically, the steps in the method are as follows

(a) Differentiate the trial series as required.
(b) Substitute the results in the given differential equation.
(c) Equate coefficients of corresponding powers of x on each side of the equation.

The following examples will demonstrate the method—so move on.

38

Example 1 Find a series solution for the equation

$$2x\frac{d^2y}{dx^2}+\frac{dy}{dx}+y=0.$$

The equation can be written as $2xy_2+y_1+y=0$.

Assume a solution of the form

$$y=x^c\{a_0+a_1x+a_2x^2+a_3x^3+\ldots+a_rx^r+\ldots\} \qquad a_0\neq 0.$$

$$\therefore\ y=a_0x^c+a_1x^{c+1}+a_2x^{c+2}+\ldots+a_rx^{c+r}+\ldots$$

Differentiating term by term, we get

$$y_1=\ldots\ldots\ldots$$

39

$$y_1=a_0cx^{c-1}+a_1(c+1)x^c+a_2(c+2)x^{c+1}+\ldots+a_r(c+r)x^{c+r-1}+\ldots$$

Repeating the process one stage further, we have

$y_2=\ldots\ldots\ldots$ (give yourself plenty of room)

40

$$y_2=a_0c(c-1)x^{c-2}+a_1c(c+1)x^{c-1}+a_2(c+1)(c+2)x^c+\ldots$$
$$+a_r(c+r-1)(c+r)x^{c+r-2}+\ldots$$

So far, we have $\quad 2xy_2+y_1+y=0$

$$y=a_0x^c+a_1x^{c+1}+a_2x^{c+2}+\ldots+a_rx^{c+r}+\ldots$$

$$y_1=a_0cx^{c-1}+a_1(c+1)x^c+a_2(c+2)x^{c+1}+\ldots$$
$$+a_r(c+r)x^{c+r-1}+\ldots$$
$$y_2=a_0c(c-1)x^{c-2}+a_1c(c+1)x^{c-1}+a_2(c+1)(c+2)x^c+\ldots$$
$$+a_r(c+r-1)(c+r)x^{c+r-2}+\ldots$$

Considering each term of the equation in turn

$$2xy_2=2a_0c(c-1)x^{c-1}+2a_1c(c+1)x^c+2a_2(c+1)(c+2)x^{c+1}$$
$$+\ldots+a_r(c+r-1)(c+r)x^{c+r-1}+\ldots$$
$$y_1=a_0cx^{c-1}+a_1(c+1)x^c+a_2(c+2)x^{c+1}+\ldots$$
$$+a_r(c+r)x^{c+r-1}+\ldots$$
$$y=a_0x^c+a_1x^{c+1}+\ldots+a_rx^{c+r}+\ldots$$

Adding these three lines to form the left-hand side of the equation, we can equate the total coefficient of each power of x to zero, since the right-hand side is zero.

$[x^{c-1}]$ gives $\ldots\ldots\ldots$

41

$$[x^{c-1}]: \qquad 2a_0c(c-1) + a_0c = 0$$
$$\therefore\ a_0c(2c-1) = 0$$

So, $[x^{c-1}]$ gives $\qquad a_0c(2c-1) = 0 \qquad (1)$

Similarly, $[x^c]$ gives ....

42

$$2a_1c\,(c+1) + a_1(c+1) + a_0 = 0$$

Simplifying, this becomes

$$a_1(2c^2 + 3c + 1) + a_0 = 0$$

i.e. $\qquad a_1(c+1)(2c+1) + a_0 = 0 \qquad (2)$

Also $[x^{c+1}]$ gives ....

43

$$2a_2(c+1)(c+2) + a_2(c+2) + a_1 = 0$$

and this simplifies straight away to

$$a_2(c+2)(2c+3) + a_1 = 0 \qquad (3)$$

Note that the coefficient of x^c involves all three lines of the expressions and, from then on, a general relationship can be obtained for $x^{c+r}, r \geq 0$.

In the expression for $2xy_2$ and y_1 we have terms in x^{c+r-1}. If we replace r by $(r+1)$, we shall obtain the corresponding terms in x^{c+r}.

In the series for $2xy_2$, this is $2a_{r+1}(c+r)(c+r+1)x^{c+r}$

In the series for y_1 this is $a_{r+1}(c+r+1)x^{c+r}$

In the series for y, this is a_rx^{c+r}

Therefore, equating the total coefficient of x^{c+r} to zero, we have

....

44

$$2a_{r+1}(c+r)(c+r+1) + a_{r+1}(c+r+1) + a_r = 0$$

and this tidies up to

$$a_{r+1}\{(c+r+1)(2c+2r+1)\} + a_r = 0 \qquad (4)$$

Make a note of results (1), (2), (3) and (4): we shall return to them in due course. *Then move on.*

45

Indicial equation

Equation (1), formed from the coefficient of the lowest power of x, is called the *indicial equation* from which the values of c can be obtained. In the present example $a_0c(2c-1) = 0$

$$\therefore c = \dots\dots\dots$$

46

$$c = 0 \text{ or } \tfrac{1}{2}, \text{since } a_0 \neq 0, \text{ by definition}$$

Both values of c are valid, so that we have two possible solutions of the given equation. We will consider each in turn.

(a) *Using* $c = 0$

(2) gives $a_1(1)(1) + a_0 = 0 \qquad \therefore a_1 = -a_0$

Similarly

(3) gives ………

47

$$\boxed{a_2(2)(3) + a_1 = 0}$$

$$a_1 = -a_0 \quad \text{and} \quad a_2 = -\frac{a_1}{2 \times 3} = \frac{a_0}{2 \times 3}$$

and from (4) $$a_{r+1} = \frac{-a_r}{(r+1)(2r+1)} \qquad r \geq 0$$

From the combined series, the term in x^c and all subsequent terms involve all three lines and the coefficient of the general term can be used.

So we have $a_1 = -a_0$ and $a_{r+1} = \dfrac{-a_r}{(r+1)(2r+1)}$ for $r = 0, 1, 2 \ldots$

$$\therefore a_2 = \frac{-a_1}{2 \times 3} = \frac{a_0}{2 \times 3}$$

$$a_2 = \frac{-a_2}{3 \times 5} = \frac{-a_0}{(2 \times 3)\,(3 \times 5)}$$

$$a_4 = \frac{-a_3}{4 \times 7} = \frac{a_0}{(2 \times 3 \times 4)\,(3 \times 5 \times 7)} \qquad \text{etc.}$$

$$\therefore y = x^0\{a_0 - a_0 x + \frac{a_0}{2 \times 3}x^2 - \frac{a_0}{(2 \times 3)\,(3 \times 5)}x^3 + \ldots\}$$

$$\therefore y = a_0\{1 - x + \frac{x^2}{(2)\,(3)} - \frac{x^3}{(2 \times 3)\,(3 \times 5)} + \frac{x^4}{(2 \times 3 \times 4)\,(3 \times 5 \times 7)} + \ldots\}$$

Now we go through the same steps using our second value for c, i.e. $c = \frac{1}{2}$. *Next frame.*

48

(b) *Using* $c = \frac{1}{2}$

Our equations relating the coefficients were

$$a_0 c(2c - 1) = 0 \quad \text{which gave} \quad c = 0 \quad \text{or} \quad c = \tfrac{1}{2} \qquad (1)$$

$$a_1(c+1)(2c+1) + a_0 = 0 \qquad (2)$$

$$a_2(c+2)(2c+3) + a_1 = 0 \qquad (3)$$

$$a_{r+1}(c+r+1)(2c+2r+1) + a_r = 0 \qquad (4)$$

Putting $c = \frac{1}{2}$ in (2) gives ....

49

$$\boxed{a_1 = -\frac{a_0}{3}}$$

Similarly (3) gives $\qquad a_2 = -\frac{a_1}{10} = \frac{a_0}{3 \times 10}$

and from the general relationship, (4), we have ....

50

$$\boxed{a_{r+1} = \frac{-a_r}{(r+1)(2r+3)}}$$

So $\qquad a_1 = -\frac{a_0}{3}$

$$a_2 = -\frac{a_1}{2 \times 5} = \frac{a_0}{(1 \times 2)\,(3 \times 5)}$$

$$a_3 = -\frac{a_2}{3 \times 7} = \frac{-a_0}{(1 \times 2 \times 3)\,(3 \times 5 \times 7)}$$

$$a_4 = -\frac{a_3}{4 \times 9} = \frac{a_0}{(1 \times 2 \times 3 \times 4)(3 \times 5 \times 7 \times 9)} \qquad \text{etc.}$$

$$y = x^c\{a_0 + a_1 x + a_2 x^2 + a_3 x^3 + \ldots + a_r x^r + \ldots\}$$

i.e. $y =$ ....

51

$$y = x^{\frac{1}{2}}\left\{a_0 - \frac{a_0}{3}x + \frac{a_0}{(1\times 2)\,(3\times 5)}x^2 - \frac{a_0}{(1\times 2\times 3)\,(3\times 5\times 7)}x^3 + \ldots\right.$$

$$\text{i.e.}\quad y = a_0 x^{\frac{1}{2}}\left\{1 - \frac{x}{(1\times 3)} + \frac{x^2}{(1\times 2)\,(3\times 5)} - \frac{x^3}{(1\times 2\times 3)\,(3\times 5\times 7)} + \ldots\right.$$

Since a_0 is an arbitrary constant in each solution, its values may well be different. If we denote the first solution by $u(x)$ and the second by $v(x)$, then

$$u = A\left\{1 - x + \frac{x^2}{(2\times 3)} - \frac{x^3}{(2\times 3)(3\times 5)} + \frac{x^4}{(2\times 3\times 4)(3\times 5\times 7)} + \ldots\right.$$

$$\text{and } v = Bx^{\frac{1}{2}}\left\{1 - \frac{x}{(1\times 3)} + \frac{x^2}{(1\times 2)(3\times 5)} - \frac{x^3}{(1\times 2\times 3)(3\times 5\times 7)} + \ldots\right\}$$

The general solution $y = u + v$ is therefore ....

52

$$y = A\left\{1 - x + \frac{x^2}{(2\times 3)} - \frac{x^3}{(2\times 3)(3\times 5)} + \ldots\right\} + Bx^{\frac{1}{2}}\left\{1 - \frac{x}{(1\times 3)}\right.$$
$$\left. + \frac{x^2}{(1\times 2)(3\times 5)} - \frac{x^3}{(1\times 2\times 3)(3\times 5\times 7)} + \ldots\right\}$$

The method may seem somewhat lengthy, but we have set it out in detail. It is a straightforward routine. Here is another example with the same steps.

Example 2 Find the series solution for the equation

$$3x^2y_2 - xy_1 + y - xy = 0.$$

We proceed in just the same way as in the previous example.

Assume $y = x^c\{a_0 + a_1x + a_2x^2 + a_3x^3 + \ldots + a_rx^r + \ldots\}$

$$\text{i.e.}\quad y = a_0x^c + a_1x^{c+1} + a_2x^{c+2} + \ldots + a_rx^{c+r} + \ldots$$

$$\therefore y_1 = a_0cx^{c-1} + a_1(c+1)x^c + a_2(c+2)x^{c+1} + \ldots$$
$$+ a_r(c+r)x^{c+r-1} + \ldots$$

and $y_2 =$ ....

53

$$y_2 = a_0c(c-1)x^{c-2} + a_1(c+1)cx^{c-1} + a_2(c+2)(c+1)x^c + \dots + a_r(c+r)(c+r-1)x^{c+r-2} + \dots$$

Now we build up the terms in the given equation.

$$3x^2y_2 = 3a_0c(c-1)x^c + 3a_1(c+1)cx^{c+1} + 3a_2(c+2)(c+1)x^{c+2} + \dots + 3a_r(c+r)(c+r-1)x^{c+r} + \dots$$

$$-xy_1 = -a_0cx^c - a_1(c+1)x^{c+1} - a_2(c+2)x^{c+2} + \dots - a_r(c+r)x^{c+r} + \dots$$

$$y = a_0x^c + a_1x^{c+1} + a_2x^{c+2} + \dots + a_rx^{c+r} + \dots$$

$$-xy = -a_0x^{c+1} - a_1x^{c+2} + \dots - a_rx^{c+r+1} \dots$$

The *indicial equation*, i.e. equating the coefficient of the lowest power of x to zero, gives the values of c. Thus, in this case

$$c = \dots\dots\dots$$

54

$$c = 1 \quad \text{or} \quad \tfrac{1}{3}$$

for the lowest power is x^c and the coefficient of x^c equated to zero gives

$$3a_0c(c-1) - a_0c + a_0 = 0$$

$$\therefore a_0(3c^2 - 4c + 1) = 0 \quad \therefore (3c-1)(c-1) = 0 \quad \text{since} \quad a_0 \neq 0$$

$$\therefore c = 1 \quad \text{or} \quad \tfrac{1}{3}$$

The coefficient of the general term, i.e. x^{c+r} gives

$$3a_r(c+r)(c+r-1) - a_r(c+r) + a_r - a_{r-1} = 0$$

$$\therefore a_r = \dots\dots\dots$$

55

$$a_r = \frac{a_{r-1}}{3(c+r)^2 - 4(c+r) + 1} = \frac{a_{r-1}}{(c+r-1)(3c+3r-1)}$$

(a) *Using* $c = 1$ the recurrence relation becomes

$$a_r = \frac{a_{r-1}}{r(3r+2)}$$

$$\therefore\ r = 1 \qquad a_1 = \frac{a_0}{1 \times 5}$$

$$r = 2 \qquad a_2 = \frac{a_1}{2 \times 8} = \frac{a_0}{(1 \times 2)\,(5 \times 8)}$$

$$r = 3 \qquad a_3 = \frac{a_2}{3 \times 11} = \frac{a_0}{(1 \times 2 \times 3)\,(5 \times 8 \times 11)}$$

Our first solution is therefore

$$y = \ldots\ldots\ldots$$

56

$$y = x^1\left\{a_0 - \frac{a_0 x}{(1 \times 5} + \frac{a_0 x^2}{(1 \times 2)\,(5 \times 8)} + \frac{a_0 x^3}{(1 \times 2 \times 3)(5 \times 8 \times 11)} + \ldots\right\}$$

$$\therefore\ y = Ax\left\{1 + \frac{x}{1 \times 5} + \frac{x^2}{(1 \times 2)(5 \times 8)} + \frac{x^3}{(1 \times 2 \times 3)(5 \times 8 \times 11)} + \ldots\right\}$$

(b) *For the second solution*, we put $c = \frac{1}{3}$. The recurrence relation then becomes

$$a_r = \ldots\ldots\ldots$$

57

$$a_r = \frac{a_{r-1}}{r(3r-2)}$$

Therefore we can now determine the coefficients for $r = 1, 2, 3, \ldots$ and complete the second solution.

$$y = \ldots\ldots\ldots$$

58

$$y = Bx^{\frac{1}{3}}\left\{1 + x + \frac{x^2}{2 \times 4} + \frac{x^3}{(2 \times 3)(4 \times 7)} + \frac{x^4}{(2 \times 3 \times 4)(4 \times 7 \times 10)} + \dots\right\}$$

for $a_1 = \dfrac{a_0}{1 \times 1}$; $\quad a_2 = \dfrac{a_1}{2 \times 4} = \dfrac{a_0}{(1 \times 2)(2 \times 4)}$

$$a_3 = \frac{a_2}{3 \times 7} = \frac{a_0}{(2 \times 3)(4 \times 7)}$$

$$a_4 = \frac{a_3}{4 \times 10} = \frac{a_0}{(2 \times 3 \times 4)\,(4 \times 7 \times 10)}$$

$$\therefore\ y = a_0x^{\frac{1}{3}}\left\{1 + x + \frac{x^2}{2 \times 4} + \frac{x^3}{(2 \times 3)(4 \times 7)} + \frac{x^4}{(2 \times 3 \times 4)\,(4 \times 7 \times 10)} + \dots\right\}$$

Therefore, the general solution is

$$y = \dots\dots\dots$$

59

$$y = Ax\left\{1 + \frac{x}{1 \times 5} + \frac{x^2}{(1 \times 2)(5 \times 8)} + \frac{x^3}{(1 \times 2 \times 3)(5 \times 8 \times 11)} + \dots\right\}$$

$$+ Bx^{\frac{1}{3}}\left\{1 + x + \frac{x^2}{2 \times 4} + \frac{x^3}{(2 \times 3)(4 \times 7)} + \frac{x^4}{(2 \times 3 \times 4)(4 \times 7 \times 10)} + \dots\right\}$$

Example 3 Find the series solution for the equation

$$\frac{d^2y}{dx^2} - y = 0 \quad \text{i.e.} \quad y_2 - y = 0.$$

As usual, we start off with the assumed solution

$$y = x^c\{a_0 + a_1 x + a_2 x^2 + \dots + a_r x^r + \dots\}$$

$$\text{i.e.} \quad y = a_0 x^c + a_1 x^{c+1} + a_2 x^{c+2} + \dots + a_r x^{c+r} + \dots$$

$$\therefore y_1 = a_0 c x^{c-1} + a_1(c+1)x^c + a_2(c+2)x^{c+1} + \dots$$
$$+ a_r(c+r)x^{c+r-1} + \dots$$

$$y_2 = a_0 c(c-1)x^{c-2} + a_1(c+1)cx^{c-1} + a_2(c+2)(c+1)x^c + \dots$$
$$+ a_r(c+r)(c+r-1)x^{c+r-2} + \dots$$

These three expansions are required regularly, so make a note of them.

60

Now we build up the terms in the left-hand side of the equation.

$$y_2 = a_0 c(c-1)x^{c-2} + a_1(c+1)cx^{c-1} + a_2(c+2)(c+1)x^c + \dots$$
$$+ a_r(c+r)(c+r-1)x^{c+r-2} + \dots$$
$$y = a_0 x^c + a_1 x^{c+1} + \dots + a_r x^{c+r} + \dots$$

The term in x^{c+r} in the first of these expansions is

. . . .

61

$$a_{r+2}(c+r+2)(c+r+1)x^{c+r}$$

since replacing r by $(r+2)$ in $a_r(c+r)(c+r+1)x^{c+r-2}$ gives this result.

Then $y_2 - y = \dots\dots\dots$

62

$$y_2 - y = a_0 c(c-1)x^{c-2} + a_1(c+1)cx^{c-1} + [a_2(c+2)(c+1) - a_0]x^c + \ldots + [a_{r+2}(c+r+2)(c+r+1) - a_r]x^{c+r} + \ldots$$

We now equate each coefficient in turn to zero, since the right-hand side of the equation is zero. The coefficient of the lowest power of x gives the *indicial equation* from which we obtain the values of c.

So, in this case, $c = \ldots\ldots\ldots$

63

$$c = 0 \quad \text{or} \quad 1$$

For the term in x^{c-1}, we have

$[x^{c-1}]$: $a_1(c+1)c = 0.$ With $c = 1$, $a_1 = 0$.
But with $c = 0$, a_1 is indeterminate, since any value of a_1 combined with the zero value of c, would make the product zero.

$[x^c]$: $a_2(c+2)(c+1) - a_0 = 0 \quad \therefore a_2 = \dfrac{a_0}{(c+1)(c+2)}$

For the general term

$[x^{c+r}]$: $\ldots\ldots\ldots$

64

$$a_{r+2} = \frac{a_r}{(c+r+1)(c+r+2)}$$

for $a_{r+2}(c+r+2)(c+r+1) - a_r = 0$. Hence the result above.

From the indicial equation, $c = 0$ or $c = 1$.

(a) When $c = 0$ $\quad a_1$ is indeterminate

$$a_2 = \frac{a_0}{2}$$

In general $\quad a_{r+2} = \dfrac{a_r}{(r+1)(r+2)}$

$r = 1 \quad \therefore a_3 = \dfrac{a_1}{2 \times 3}$

$r = 2 \quad a_4 = \dfrac{a_2}{3 \times 4} = \dfrac{a_0}{4!}$

Therefore, one solution is ....

65

$$y = x^0\left\{a_0 + a_1 x + \frac{a_0}{2!}x^2 + \frac{a_1}{3!}x^3 + \frac{a_0}{4!}x^4 \dots\right\}$$

i.e. $y = a_0\left\{1 + \dfrac{x^2}{2!} + \dfrac{x^4}{4!} + \dots\right\} + a_1\left\{x + \dfrac{x^3}{3!} + \dfrac{x^5}{5!} + \dots\right\}$

a_0 and a_1 are arbitrary constants depending on the boundary conditions.

$$\therefore y = A\left\{1 + \frac{x^2}{2!} + \frac{x^4}{4!} + \dots\right\} + B\left\{x + \frac{x^3}{3!} + \frac{x^5}{5!} + \dots\right\}$$

(b) Similarly,

when $c = 1 \quad a_1 = 0$

$$a_2 = \frac{a_0}{2 \times 3}$$

$$a_{r+2} = \dots\dots\dots$$

66

$$\boxed{a_{r+2} = \frac{a_r}{(r+2)(r+3)}}$$

$$\therefore a_1 = 0$$

$$a_2 = \frac{a_0}{3!}$$

$$r = 1 \qquad a_3 = \frac{a_1}{3 \times 4} = 0$$

$$r = 2 \qquad a_4 = \frac{a_2}{4 \times 5} = \frac{a_0}{5!}$$

$$r = 3 \qquad a_5 = \frac{a_3}{5 \times 6} = 0 \qquad \text{etc.}$$

A second solution with $c = 1$, is therefore

$$y = \ldots\ldots\ldots$$

67

$$y = a_0\left\{x + \frac{x^3}{3!} + \frac{x^5}{5!} + \dots\right\}$$

and, since a_0 is an arbitrary constant

$$y = C\left\{x + \frac{x^3}{3!} + \frac{x^5}{5!} + \frac{x^7}{7!} + \dots\right\}$$

Note: This is not, in fact, a separate solution, since it already forms the second series in the solution for $c = 0$ obtained previously. Therefore, the first solution, with its two arbitrary constants, A and B, gives the general solution. This happens when the two values of c differ by an integer.

Make a note of the following:

If the two values of c, i.e. c_1 and c_2, differ by an integer, and if $c = c_1$ results in a_1 being indeterminate, then this value of c gives the general solution.
The solution resulting from $c = c_2$ is then merely a multiple of one of the series forming the first solution.

Our last problem was an example of this.

So far, we have met two distinct cases concerning the two roots $c = c_1$ and $c = c_2$ of the indicial equation.

(a) *If c_1 and c_2 differ by a quantity NOT an integer* then two independent solutions, $y = u(x)$ and $y = v(x)$, are obtained. The general solution is then $y = Au + Bv$.
(b) *If c_1 and c_2 differ by an integer*, i.e. $c_2 = c_1 + n$, and if one coefficient (a_r) is indeterminate when $c = c_1$, the complete general solution is given by using this value of c. Using $c = c_1 + n$ gives a series which is a simple multiple of one of the series in the first solution.

Make a note of these two points in your record book. Then move on.

68

There is a third category to be added to (a) and (b) above.

(c) If the roots $c = c_1$ and $c = c_1 + n$, of the indicial equation differ by an integer and one coefficient (a_r) becomes infinite when $c = c_1$, the series is re-written with a_0 replaced by $k(c - c_1)$.

Putting $c = c_1$ in the new series z and that of $\frac{\partial z}{\partial c}$ after differentiating, gives two independent solutions.

Add this to the previous two. Then we will see how it works in practice.

69

Example 4 Find the series solution of the equation

$$xy_2 + (2 + x)y_1 - 2y = 0.$$

Using $y = x^c(a_0 + a_1x + a_2x^2 + a_3x^3 + \ldots + a_rx^r + \ldots)$

and its first two derivatives, the expansions for

$$xy_2 = \ldots\ldots\ldots$$
$$2y_1 = \ldots\ldots\ldots$$
$$xy_1 = \ldots\ldots\ldots$$
$$-2y = \ldots\ldots\ldots$$

Method as before.

70

$$xy_2 = a_0c(c-1)x^{c-1} + a_1(c+1)cx^c + a_2(c+2)(c+1)x^{c+1} + \ldots$$
$$+ a_r(c+r)(c+r-1)x^{c+r-1} + \ldots$$
$$2y_1 = 2a_0cx^{c-1} + 2a_1(c+1)x^c + 2a_2(c+2)x^{c+1} + 2a_3(c+3)x^{c+2}$$
$$+ \ldots + 2a_r(c+r)x^{c+r-1} + \ldots$$
$$xy_1 = a_0cx^c + a_1(c+1)x^{c+1} + a_2(c+2)c^{c+2} + \ldots$$
$$+ a_r(c+r)x^{c+r} + \ldots$$
$$-2y = -2a_0x^c - 2a_1x^{c+1} - 2a_2x^{c+2} - 2a_3x^{c+3} + \ldots$$
$$- 2a_rx^{c+r} + \ldots$$

From which, the indicial equation is ....

71

$$\boxed{a_0(c^2 + c) = 0}$$

i.e. equating the coefficient of the lowest power of x, (x^{c-1}), to zero.

$$a_0 \neq 0 \qquad \therefore \underline{c = 0 \text{ or } -1}$$

Also, from the expansions, the total coefficient of x^c gives

$$a_1 = \ldots\ldots\ldots$$

72

$$a_1 = \frac{-a_0(c-2)}{(c+1)(c+2)}$$

From the terms in x^c, all four expansions are involved, so we can form the recurrence relation from the coefficient of x^{c+r}.

$$a_{r+1} = \ldots\ldots\ldots$$

73

$$a_{r+1} = \frac{-a_r(c+r-2)}{(c+r+1)(c+r+2)}$$

for $a_{r+1}(c+r+1)\,(c+r)\ + 2a_{r+1}(c+r+1) + a_r(c+r) - 2a_r = 0$

$$a_{r+1}(c+r+1)\,(c+r+2)\ + a_r(c+r-2) = 0$$

$$\therefore a_{r+1} = \frac{-a_r(c+r-2)}{(c+r+1)\,(c+r+2)} \qquad r \geq 0$$

$$\therefore\ a_2 = \ldots\ldots\ldots$$

74

$$a_2 = \frac{a_0(c-1)(c-2)}{(c+1)(c+2)^2(c+3)}$$

and, from the recurrence relation, when $r = 2$

$$a_3 = \ldots\ldots\ldots$$

75

$$a_3 = \frac{-a_0 c(c-1)(c-2)}{(c+1)(c+2)^2(c+3)^2(c+4)}$$

$$\therefore\ y = a_0 x^c \left\{1 - \frac{c-2}{(c+1)(c+2)}x + \frac{(c-1)(c-2)}{(c+1)(c+2)^2(c+3)}x^2 - \frac{c(c-1)(c-2)}{(c+1)(c+2)^2(c+3)^2(c+4)}x^3 + \dots\right\}$$

From the indicial equation above, the values of c and 0 and (–1).
Putting $c = 0$, we have one solution

$$y = u = \dots\dots\dots$$

76

$$y = u = a_0\left\{1 + x + \frac{x^2}{6}\right\}$$

Note that coefficients after the x^2 term are zero, because of the factor c in the numerator.
Putting $c = -1$, we soon find that ....

77

coefficients become infinite, because of the factor $(c+1)$ in the denominator.

Therefore, we substitute $a_0 = k(c - c_1) = k(c - [-1]) = k(c+1)$.

$$\therefore\ y = k(c+1)x^c \left\{1 - \frac{c-2}{(c+1)(c+2)}x + \frac{(c-1)(c-2)}{(c+1)(c+2)^2(c+3)}x^2 - \frac{c(c-1)(c-2)}{(c+1)(c+2)^2(c+3)^2(c+4)}x^3 + \dots\right\}$$

$$= kx^c \left\{(c+1) - \frac{c-2}{c+2}x + \frac{(c-1)(c-2)}{(c+2)^2(c+3)}x^2 - \frac{c(c-1)(c-2)}{(c+2)^2(c+3)^2(c+4)}x^3 \dots\right\}$$

Now, putting $c = -1$,

$$y = \dots\dots\dots$$

78

$$y = kx^{-1}\left\{3x + 3x^2 + \frac{x^3}{2}\right\}$$

All subsequent terms are zero, since the numerators all contain a factor $(c+1)$.

$$\therefore\ y = v = \left\{3 + 3x + \frac{x^2}{2}\right\} \text{ is a solution.}$$

A solution is also given by $\dfrac{\partial y}{\partial c} = 0.$

So, starting from

$$y = kx^c\left\{(c+1) - \frac{c-2}{c+2}x + \frac{(c-1)(c-2)}{(c+2)^2(c+3)}x^2 - \frac{c(c-1)(c-2)}{(c+2)^2(c+3)^2(c+4)}x^3 \ldots\right\}$$

$$\frac{\partial y}{\partial c} = kx^c \ln x\left\{(c+1) - \frac{c-2}{c+2}x + \frac{(c-1)(c-2)}{(c+2)^2(c+3)}x^2 - \frac{c(c-1)(c-2)}{(c+2)^2(c+3)^2(c+4)}x^3 \ldots\right\}$$

$$+kx^c\frac{\partial}{\partial c}\left\{(c+1) - \frac{c-2}{c+2}x + \frac{(c-1)(c-2)}{(c+2)^2(c+3)}x^2 + \ldots\right\}$$

We now have to determine the partial derivative of each term.

$$\frac{\partial}{\partial c}(c+1) = 1$$

$$\frac{\partial}{\partial c}\left\{\frac{c-2}{c+2}\right\} = \ldots\ldots\ldots$$

79

$$\boxed{\frac{\partial}{\partial c}\left\{\frac{c-2}{c+2}\right\} = \frac{4}{(c+2)^2}}$$

Now we have to differentiate $\dfrac{(c-1)(c-2)}{(c+2)^2(c+3)}$

Let $t = \dfrac{(c-1)(c-2)}{(c+2)^2(c+3)}$ $\therefore$ $\ln t = \ln(c-1) + \ln(c-2) - 2\ln(c+2)$

$$- \ln(c+3)$$

$$\therefore \frac{1}{t}\frac{\partial t}{\partial c} = \frac{1}{c-1} + \frac{1}{c-2} - \frac{2}{c+2} - \frac{1}{c+3}$$

$$\therefore \frac{\partial t}{\partial c} = \frac{(c-1)(c-2)}{(c+2)^2(c+3)}\left\{\frac{1}{c-1} + \frac{1}{c-2} - \frac{2}{c+2} - \frac{1}{c+3}\right\}$$

$\therefore$ when $c = -1$, $\dfrac{\partial}{\partial c}(c+1) = 1$

$$\frac{\partial}{\partial c}\left\{\frac{c-2}{c+2}\right\} = 4$$

$$\frac{\partial}{\partial c}\left\{\frac{(c-1)(c-2)}{(c+2)^2(c+3)}\right\} = \dots\dots\dots$$

80

$$\boxed{-10}$$

Therefore, when $c = -1$,

$$\frac{\partial y}{\partial c} = kx^{-1}\ln x\left\{0 + 3x + 3x^2 + \frac{x^3}{2} + \ldots\right\}$$

$$+kx^{-1}\{1 - 4x - 10x^2 + \ldots\}$$

$\therefore$ Another solution is

$$y = w = C\left\{\ln x\left(3 + 3x + \frac{x^2}{2} + \ldots\right) + x^{-1}(1 - 4x - 10x^2 + \ldots)\right\}$$

Now we have a problem, for we seem to have three separate series solutions for a second-order differential equation.

(a) $y = u = A\left(1 + x + \dfrac{x^2}{6}\right)$

(b) $y = v = B\left(3 + 3x + \dfrac{x^2}{2}\right)$

(c) $y = w = C\left\{\ln x\left(3 + 3x + \dfrac{x^2}{2} + \ldots\right) + x^{-1}(1 - 4x - 10x^3 + \ldots)\right\}$

But (b) is clearly a simple multiple of (a) and thus not a distinct solution. So finally, we have just (a) and (c).

i.e. $y = u = A\left(1 + x + \dfrac{x^2}{6}\right)$

and $y = w = B\left\{\ln x\left(3 + 3x + \dfrac{x^2}{2} + \ldots\right) + x^{-1}(1 - 4x - 10x^3 + \ldots)\right\}$

The complete solution is then $\underline{y = u + w}$

Finally we have just one more variation, so move on.

81 *Example 5* Solve the equation $xy_2 + y_1 - xy = 0$.

Start off as before and build up expansions for the terms in the left-hand side of the equation.

$$xy_2 = \ldots\ldots\ldots$$

$$y_1 = \ldots\ldots\ldots$$

$$-xy = \ldots\ldots\ldots$$

82

$$xy_2 = a_0c(c-1)x^{c-1} + a_1(c+1)cx^c + a_2(c+2)(c+1)x^{c+1} + \dots$$
$$+ a_r(c+r)(c+r-1)x^{c+r-1} + \dots$$
$$y_1 = a_0cx^{c-1} + a_1(c+1)x^c + a_2(c+2)x^{c+1} + \dots$$
$$+ a_r(c+r)x^{c+r-1} + \dots$$
$$-xy = \quad -a_0x^{c+1} - a_1x^{c+2} - \dots$$
$$- a_rx^{c+r+1} + \dots$$

The indicial equation, therefore, gives $c = \dots\dots\dots$

83

$$c = 0 \text{ (twice)}$$

since $a_0\,\{c(c-1)+c\} = 0 \quad a_0 \neq 0 \quad \therefore c^2 = 0 \quad \therefore \underline{c = 0 \text{ (twice)}}$

Coefficient of x^c gives $\dots\dots\dots$

84

$$a_1 = 0$$

$[x^c]$: $\quad a_1(c^2+c+c+1) = 0 \quad \therefore a_1(c+1)^2 = 0 \quad \therefore \underline{a_1 = 0}$

$[x^{c+1}]$: This involves all three expansions and from this point, we can use the general recurrence relation.

$[x^{c+r-1}]$: $\quad a_r\{(c+r)(c+r-1)+(c+r)\} - a_{r-2} = 0$

$$\therefore a_r(c+r)^2 = a_{r-2} \quad \therefore \underline{a_r = \frac{a_{r-2}}{(c+r)^2}}$$

$\therefore y = \dots\dots\dots$

85

$$y = x^c\left\{a_0 + \frac{a_0}{(c+2)^2}x^2 + \frac{a_0}{(c+2)^2(c+4)^2}x^4 + \dots\right\}$$

i.e. $$y = a_0x^c\left\{1 + \frac{x^2}{(c+2)^2} + \frac{x^4}{(c+2)^2(c+4)^2} + \dots\right\}$$

$\therefore$ *When* $c = 0$

$$\underline{y = u = A\left\{1 + \frac{x^2}{2^2} + \frac{x^4}{2^2 \times 4^2} + \dots\right\}} \qquad (1)$$

This is one solution. Another is given by $v = \dfrac{\partial y}{\partial c}$

$$\frac{\partial y}{\partial c} = a_0x^c \ln x\left\{1 + \frac{x^2}{(c+2)^2} + \frac{x^4}{(c+2)^2(c+4)^2} + \dots\right\}$$

$$+a_0x^c\frac{\partial}{\partial c}\left\{1 + \frac{x^2}{(c+2)^2} + \frac{x^4}{(c+2)^2(c+4)^2} + \dots\right\}$$

Now $\dfrac{\partial}{\partial c}(1) = 0$; $\quad \dfrac{\partial}{\partial c}\left\{\dfrac{1}{(c+2)^2}\right\} = \dfrac{-2}{(c+2)^3}$

Let $t = \dfrac{1}{(c+2)^2(c+4)^2}$ $\quad \therefore \ln t = -2\ln(c+2) - 2\ln(c+4)$

$$\therefore \frac{1}{t}\frac{\partial t}{\partial c} = \frac{-2}{c+2} - \frac{2}{c+4} \quad \therefore \frac{\partial t}{\partial c} = \frac{-2}{(c+2)^2(c+4)^2}\left\{\frac{1}{c+2} + \frac{1}{c+4}\right\}$$

$$\therefore \frac{\partial y}{\partial c} = a_0x^c \ln x\left\{1 + \frac{x^2}{(c+2)^2} + \frac{x^4}{(c+2)^2(c+4)^2} + \dots\right\}$$

$$+a_0x^c\left\{0 - \frac{2x^2}{(c+2)^3} - \frac{4x^4(c+3)}{(c+2)^3(c+4)^3} + \dots\right\}$$

$\therefore$ *When* $c = 0$

$$y = v = \dots\dots\dots$$

86

$$y = v = B\left\{\ln x\left(1 + \frac{x^2}{2^2} + \frac{x^4}{2^2 \times 4^2} + \ldots\right) - \frac{x^2}{2^2} - \frac{3x^4}{2^3 \times 4^2} + \ldots\right\} \quad (2)$$

So our two solutions are $y = u$ (at 1) and $y = v$ (at 2). The complete solution is therefore $\underline{y = u + v}$

87

Summary

Let us now summarise the four types of procedures in the method of Frobenius that we have covered.

(a) Assume a series of the form

$$y = x^c(a_0 + a_1x + a_2x^2 + \ldots + a_rx^r + \ldots)$$

(b) Indicial equation gives $c = c_1$ and $c = c_2$.

(c) *Case 1.* c_1 and c_2 differ by a quantity *not an integer*. Substitute $c = c_1$ and $c = c_2$ in the series for y.

(d) *Case 2.* c_1 and c_2 differ by *an integer* and make a coefficient *indeterminate* with $c = c_1$. Substitution of $c = c_1$ gives the complete solution.

(e) *Case 3.* c_1 and c_2 $(c_1 < c_2)$ differ by *an integer* and make a coefficient *infinite* for $c = c_1$. Replace a_0 by $k\,(c - c_1)$. Put $c = c_1$ in the new series for y and for $\dfrac{\partial y}{\partial c}$.

(f) *Case 4.* c_1 and c_2 *equal*. Substitute $c = c_1$ in the series for y and for $\dfrac{\partial y}{\partial c}$. Make the substitution after differentiating.

Make a note of this summary for future reference.

Bessel's Equation

88 A second-order differential equation that occurs frequently in branches of technology is of the form

$$x^2y_2 + xy_1 + (x^2 - v^2)y = 0$$

where v is a real constant.

Starting with $y = x^c(a_0 + a_1x + a_2x^2 + a_3x^3 + \ldots + a_rx^r + \ldots)$ and proceeding as before, we obtain

$$c = \pm v \quad \text{and} \quad a_1 = 0$$

The recurrence relation is $\quad a_r = \dfrac{a_{r-2}}{v^2 - (c+r)^2} \quad$ for $r \geq 2$.

It follows that $a_1 = a_3 = a_5 = a_7 = \ldots = 0$

and that $\quad a_2 = \ldots\ldots\ldots; \quad a_4 = \ldots\ldots\ldots; \quad a_6 = \ldots\ldots\ldots$

89

$$a_2 = \frac{a_0}{v^2 - (c+2)^2}; \quad a_4 = \frac{a_0}{[v^2 - (c+2)^2]\,[v^2 - (c+4)^2]};$$

$$a_6 = \frac{a_0}{[v^2 - (c+2)^2]\,[v^2 - (c+4)^2]\,[v^2 - (c+6)^2]}$$

$\therefore$ *When* $c = +v \quad a_2 = \ldots\ldots\ldots; \quad a_4 = \ldots\ldots\ldots$

$a_6 = \ldots\ldots\ldots; \quad a_r = \ldots\ldots\ldots$

90

$$a_2 = \frac{-a_0}{2^2(v+1)}; \qquad a_4 = \frac{a_0}{2^4 \times 2(v+1)(v+2)};$$

$$a_6 = \frac{-a_0}{2^6 \times 3!(v+1)(v+2)(v+3)};$$

$$a_r = \frac{(-1)^{r/2} a_0}{2^r \times (r/2)!(v+1)(v+2)\dots(v+r/2)} \text{ for } r \text{ even}$$

The resulting series solution is therefore

$y = u = \dots\dots\dots$

91

$$y = u = Ax^v \left\{ 1 - \frac{x^2}{2^2(v+1)} + \frac{x^4}{2^4 \times 2!(v+1)(v+2)} - \frac{x^6}{2^6 \times 3!(v+1)(v+2)(v+3)} + \dots \right\}$$

This is valid provided v is not a negative integer.

Similarly, *when* $c = -v$

$$y = w = Bx^{-v} \left\{ 1 + \frac{x^2}{2^2(v-1)} + \frac{x^4}{2^4 \times 2!(v-1)(v-2)} + \frac{x^6}{2^6 \times 3!(v-1)(v-2)(v-3)} + \dots \right\}$$

This is a valid provided v is not a positive integer.

Except for these two restrictions, the complete solution of Bessel's equation is therefore $\underline{y = u + w}$ with the two arbitrary constants A and B.

92

Bessel functions

It is convenient to present the two results obtained above in terms of gamma functions, remembering that for $x > 0$

$$\Gamma(x+1) = x\Gamma(x)$$

$$\Gamma(x+2) = (x+1)\Gamma(x+1) = (x+1)x\Gamma(x)$$

$$\Gamma(x+3) = (x+2)\Gamma(x+2) = (x+2)(x+1)x\Gamma(x), \text{ etc.}$$

If, at the same time, we assign to the arbitrary constant a_0 the value $\dfrac{1}{2^v\Gamma(v+1)}$, then we have, for $c = v$

$$a_2 = \frac{a_0}{v^2-(c+2)^2} = \frac{a_0}{(v-c-2)(v+c+2)} = \frac{a_0}{-2(2v+2)}$$

$$= \frac{-1}{2^2(v+1)} \cdot \frac{1}{2^v\Gamma(v+1)} = \underline{\frac{-1}{2^{v+2}(1!)\Gamma(v+2)}}$$

Similarly $a_4 = \dots\dots\dots$

93

$$\boxed{a_4 = \frac{1}{2^{v+4}(2!)\Gamma(v+3)}}$$

for $a_4 = \dfrac{a_2}{v^2-(c+4)^2} = \dfrac{a_2}{(v-c-4)(v+c+4)} = \dfrac{a_2}{-4(2v+4)}$

$$= \frac{-1}{2^3(v+2)} \cdot \frac{-1}{2^{v+2}(1!)\Gamma(v+2)} = \underline{\frac{1}{2^{v+4}(2!)\Gamma(v+3)}}$$

and $a_6 = \dots\dots\dots$

94

$$a_6 = \frac{-1}{2^{v+6}(3!)\Gamma(v+4)}$$

We can see the pattern taking shape.

$$a_r = \frac{(-1)^{r/2}}{2^{v+r}\left(\frac{r}{2}!\right)\Gamma\left(v+\frac{r}{2}+1\right)} \quad \text{for } r \text{ even.} \qquad \therefore \text{ Put } r = 2k$$

The result then becomes

$$a_{2k} = \ldots\ldots\ldots$$

95

$$a_{2k} = \frac{(-1)^k}{2^{v+2k}(k!)\Gamma(v+k+1)} \qquad k = 1, 2, 3, \ldots$$

Therefore, we can write the new form of the series for y as

$$y = x^v\left\{\frac{1}{2^v\Gamma(v+1)} - \frac{x^2}{2^{v+2}(1!)\Gamma(v+2)} + \frac{x^4}{2^{v+4}(2!)\Gamma(v+3)} + \ldots\right\}$$

This is called the *Bessel function of the first kind of order v* and is denoted by $J_v(x)$.

$$\therefore\ J_v(x) = \left(\frac{x}{2}\right)^v\left\{\frac{1}{\Gamma(v+1)} - \frac{x^2}{2^2(1!)\Gamma(v+2)} + \frac{x^4}{2^4(2!)\Gamma(v+3)}\ldots\right\}$$

This is valid provided v is not ....

96

a negative integer

— otherwise, some of the terms would become infinite.

If we take the other value for c, i.e. $c = -v$, the corresponding result becomes

$$J_{-v}(x) = \ldots\ldots\ldots$$

97

$$J_{-v}(x) = \left(\frac{x}{2}\right)^{-v}\left\{\frac{1}{\Gamma(1-v)} - \frac{x^2}{2(1!)\Gamma(2-v)} + \frac{x^4}{2^2(2!)\Gamma(3-v)}\cdots\right\}$$

provided v is not a positive integer.

In general terms

$$J_v(x) = \left(\frac{x}{2}\right)^{v}\sum_{k=0}^{\infty}\frac{(-1)^k x^{2k}}{2^{2k}(k!)\Gamma(v+k+1)}$$

$$J_{-v}(x) = \left(\frac{x}{2}\right)^{-v}\sum_{k=0}^{\infty}\frac{(-1)^k x^{2k}}{2^{2k}(k!)\Gamma(k-v+1)}$$

The convergence of the series for all values of x can be established by the normal ratio test.

$J_v(x)$ and $J_{-v}(x)$ are two independent solutions of the original equation. Hence, the complete solution is

$$\underline{y = A \times J_v(x) + B \times J_{-v}(x)}$$

Make a note of the expressions for $J_v(x)$ and $J_{-v}(x)$. Then on to the next frame.

98

Some Bessel functions are commonly used and are worthy of special mention. This arises when v is a positive integer, denoted by n.

$$\therefore J_n(x) = \left(\frac{x}{2}\right)^n \sum_{k=0}^{\infty} \frac{(-1)^k x^{2k}}{2^{2k}(k!)\Gamma(n+k+1)}$$

From our work on gamma functions, $\Gamma(k+1) = k!$ for $k = 0, 1, 2, \ldots$

$$\therefore \Gamma(n+k+1) = (n+k)!$$

and the result above then becomes

$$J_n(x) = \ldots\ldots\ldots$$

99

$$\boxed{J_n(x) = \left(\frac{x}{2}\right)^n \sum_{k=0}^{\infty} \frac{(-1)^k x^{2k}}{2^{2k}(k!)(n+k)!}}$$

We have seen that $J_v(x)$ and $J_{-v}(x)$ are two solutions of Bessel's equation. When v and $-v$ are not integers, the two solutions are independent of each other. Then $y = A \times J_v(x) + B \times J_{-v}(x)$.

When, however, $v = n$ (integer), then $J_n(x)$ and $J_{-n}(x)$ are not independent, but are related by $J_{-n}(x) = (-1)^n J_n(x)$. This can be shown by referring once again to our knowledge of gamma functions.

$$\Gamma(x+1) = x\Gamma(x) \qquad \therefore \Gamma(x) = \frac{\Gamma(x+1)}{x}$$

and for negative integral values of x, or zero, $\Gamma(x)$ is infinite.

From the previous result,

$$J_{-v}(x) = \left(\frac{x}{2}\right)^{-v} \sum_{k=0}^{\infty} \frac{(-1)^k x^{2k}}{2^{2k}(k!)\Gamma(k-v+1)} \qquad k = 0, 1, 2, \ldots$$

Let us consider the gamma function $\Gamma(k-v+1)$ in the denominator and let v approach closely to a positive integer n.

Then $\quad \Gamma(k-v+1) \to \Gamma(k-n+1)$.

When $k-n+1 \leq 0$, i.e. when $k \leq (n-1)$, then $\Gamma(k-n+1)$ is infinite.

The first finite value of $\Gamma(k-n+1)$ occurs for $k = n$.

When values of $\Gamma(k-v+1)$ are infinite the coefficients of $J_{-v}(x)$ are

$$\ldots\ldots\ldots$$

100

zero

The series, therefore, starts at $k = n$

$$\therefore J_{-n}(x) = \left(\frac{x}{2}\right)^{-n} \sum_{k=n}^{\infty} \frac{(-1)^k x^{2k}}{2^{2k}(k!)\Gamma(k-n+1)}$$

$$= \sum_{k=n}^{\infty} \frac{(-1)^k x^{2k-n}}{2^{2k-n}(k!)\Gamma(k-n+1)} \qquad \text{Put } k = p+n$$

$$= \sum_{p=0}^{\infty} \frac{(-1)^{p+n} x^{2p+n}}{2^{2p+n}(k!)(k-n)!}$$

$$= (-1)^n \sum_{p=0}^{\infty} \frac{(-1)^p x^{2p+n}}{2^{2p+n}(p!)(p+n)!}$$

$$= (-1)^n \left(\frac{x}{2}\right)^n \sum_{p=0}^{\infty} \frac{(-1)^p x^{2p}}{2^{2p}(p!)(p+n)!}$$

$$= (-1)^n \left(\frac{x}{2}\right)^n \sum_{k=0}^{\infty} \frac{(-1)^k x^{2k}}{2^{2k}(k!)(k+n)!}$$

$$\therefore \underline{J_{-n}(x) = (-1)^n J_n(x)}$$

So, after all that, the series for $J_n(x) = \dots\dots\dots$

101

$$J_n(x) = \left(\frac{x}{2}\right)^n \left\{\frac{1}{n!} - \frac{1}{(n+1)!}\left(\frac{x}{2}\right)^2 + \frac{1}{(2!)(n+2)!}\left(\frac{x}{2}\right)^4 - \dots\dots\dots\right\}$$

From this we obtain two commonly used functions

$$J_0(x) = \dots\dots\dots$$

102

$$J_0(x) = 1 - \frac{1}{(1!)^2}\left(\frac{x}{2}\right)^2 + \frac{1}{(2!)^2}\left(\frac{x}{2}\right)^4 - \frac{1}{(3!)^2}\left(\frac{x}{2}\right)^6 + \dots$$

and $J_1(x) = \dots\dots\dots$

103

$$J_1(x) = \frac{x}{2}\left\{1 - \frac{1}{(1!)(2!)}\left(\frac{x}{2}\right)^2 + \frac{1}{(2!)(3!)}\left(\frac{x}{2}\right)^4 + \dots\right\}$$

Bessel functions for a range of values of n and x are tabulated in published lists of mathematical data. Of these, $J_0(x)$ and $J_1(x)$ are most commonly used.

104

Graphs of Bessel functions $J_0(x)$ and $J_1(x)$

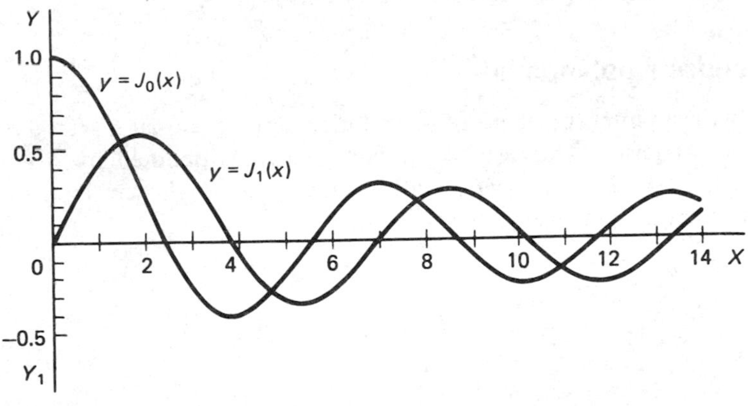

Legendre's Equation

105

Another equation of special interest in engineering applications is Legendre's equation of the form

$$(1-x^2)y_2 - 2xy_1 + k(k+1)y = 0$$

where k is a real constant.

This may be solved by the Frobenius method as before. In this case, the indicial equation gives $c = 0$ and $c = 1$, and the two corresponding solutions are

(a) $c = 0: y = a_0\left\{1 - \dfrac{k(k+1)}{2!}x^2 + \dfrac{k(k-2)(k+1)(k+3)}{4!}x^4 \dots\right\}$

(b) $c = 1: y = a_1\left\{x - \dfrac{(k-1)(k+2)}{3!}x^3\right.$

$$\left. + \frac{(k-1)(k-3)(k+2)(k+4)}{5!}x^5 \dots\right\}$$

where a_0 and a_1 are the usual arbitrary constants

Legendre's polynomials

When k is an integer (n), one of the solution series terminates after a finite number of terms. The resulting polynomial in x, denoted by $P_n(x)$, is called a *Legendre polynomial*, with a_0 or a_1 being chosen so that the polynomial has unit value when $x = 1$.

For example $\quad P_2(x) = \dots\dots\dots$

106

$$P_2(x) = \frac{1}{2}(3x^2 - 1)$$

for, in $P_2(x)$, $n = k = 2$

$$\therefore\ y = a_0\left\{1 - \frac{2 \times 3}{2!}x^2 + 0 + 0 + \dots\right\}$$

$$= a_0\{1 - 3x^2\}$$

The constant a_0 is then chosen to make $y = 1$ when $x = 1$

i.e. $\quad 1 = a_0(1 - 3) \qquad \therefore\ a_0 = -\frac{1}{2}$

$$\therefore\ P_2(x) = -\frac{1}{2}(1 - 3x^2) = \underline{\frac{1}{2}(3x^2 - 1)}$$

Similarly $\qquad P_3(x) = \dots\dots\dots$

107

$$P_3(x) = \frac{1}{2}(5x^3 - 3x)$$

Here $\quad n = k = 3$

$$\therefore\ y = a_1\left\{x - \frac{2 \times 5}{3!}x^3 + 0 + 0 + \dots\right\}$$

$$= a_1\left\{x - \frac{5x^3}{3}\right\}$$

$y = 1$ when $x = 1 \qquad \therefore\ a_1\left(1 - \frac{5}{3}\right) = 1 \qquad \therefore\ a_1 = -\frac{3}{2}$

$$\therefore\ P_3(x) = -\frac{3}{2}\left(x - \frac{5x^3}{3}\right) = \underline{\frac{1}{2}(5x^3 - 3x)}$$

Relatively few differential equations can be solved by the normal analytical means and solution by power series provides a powerful tool in many situations.

The main points that we have covered are listed, as usual, in the Revision Summary that follows. Note any sections that may need further attention and refer back to the relevant parts of the programme, if necessary. There will then be no trouble with the Test Exercise.

108 REVISION SUMMARY

1. *Higher derivatives*

y	y_n
x^a	$\dfrac{a!}{(a-n)!}x^{a-n}$
e^{ax}	$a^n e^{ax}$
$\sin ax$	$a^n \sin\left(ax + \dfrac{n\pi}{2}\right)$
$\cos ax$	$a^n \cos\left(ax + \dfrac{n\pi}{2}\right)$
$\sinh ax$	$\dfrac{a^n}{2}\{[1+(-1)^n]\sinh ax + [1-(-1)^n]\cosh ax\}$
$\cosh ax$	$\dfrac{a^n}{2}\{[1-(-1)^n]\sinh ax + [1+(-1)^n]\cosh ax\}$

2. *Leibnitz theorem* — nth derivative of a product of functions.

If $y = uv$

$$y_n = u_n v + n u_{n-1} v_1 + \frac{n(n-1)}{2!} u_{n-2} v_2 + \frac{n(n-1)(n-2)}{3!} u_{n-3} v_3 \dots$$

$$\dots + \frac{n(n-1)(n-2)\dots(n-r+1)}{r!} u_{n-r} v_r + \dots$$

i.e. $\quad y_n = \displaystyle\sum_{r=0}^{\infty} {}^nC_r u_{n-r} v_r.$

3. *Power series solution of second-order differential equations*

(a) *Leibnitz–Maclaurin method*

(i) Differentiate the equation n times by Leibnitz theorem.

(ii) Put $x = 0$ to establish a recurrence relation.

(iii) Substitute $n = 1, 2, 3, \ldots$ to obtain $y_1, y_2, y_3, \ldots$ at $x = 0$.

(iv) Substitute in Maclaurin's series and simplify where possible.

(b) *Frobenius method*

Assume a series solution of the form

$$y = x^c\{a_0 + a_1x + a_2x^2 + \ldots + a_rx^r + \ldots\} \quad a_0 \neq 0$$

(i) Differentiate the assumed series to find y_1 and y_2.

(ii) Substitute in the equation.

(iii) Equate coefficients of corresponding powers of x on each side of the equation — usually written with zero on the right-hand side.

(iv) Coefficient of the lowest power of x gives the *indicial equation* from which values of c are obtained, $c = c_1$ and $c = c_2$.

Case 1: c_1 and c_2 differ by a quantity *not an integer*. Substitute $c = c_1$ and $c = c_2$ in the series for y.

Case 2: c_1 and c_2 differ by an *integer* and make a coefficient *indeterminate* when $c = c_1$. Substitute $c = c_1$ to obtain the complete solution.

Case 3: c_1 and c_2 $(c_1 < c_2)$ differ by an *integer* and make a coefficient *infinite* when $c = c_1$. Replace a_0 by $k(c - c_1)$. Two independent solutions then obtained by putting $c = c_1$ in the new series for y and for $\dfrac{\partial y}{\partial c}$.

Case 4: c_1 and c_2 are *equal*. Substitute $c = c_1$ in the series for y and for $\dfrac{\partial y}{\partial c}$. Make the substitution after differentiating. The second solution will consist of the product of the first solution and $\ln x$, together with a further series.

4. *Bessel's equation*

$x^2y_2 + xy_1 + (x^2 - v^2) = 0$ where v is a real constant.

Bessel functions: Express the two solutions obtained, in terms of gamma functions.

$$J_v(x) = \left(\frac{x}{2}\right)^v\left\{\frac{1}{\Gamma(v+1)} - \frac{x^2}{2^2(1!)\Gamma(v+2)} + \frac{x^4}{2^4(2!)\Gamma(v+3)}\cdots\right\}$$

This is the *Bessel function of the first kind of order v* — valid for v not a negative integer.

Also $J_{-v}(x) = \left(\frac{x}{2}\right)^v \left\{ \frac{1}{\Gamma(1-v)} - \frac{x^2}{2(1!)\Gamma(2-v)} + \frac{x^4}{2^2(2!)\Gamma(3-v)} \cdots \right\}$

provided v is not a positive integer.

Complete solution is therefore $y = A \times J_v(x) + B \times J_{-v}(x)$.

When $v = n$ *(an integer)* $\quad J_{-n}(x) = (-1)^n J_n(x)$

$$J_n(x) = \left(\frac{x}{2}\right)^n \left\{ \frac{1}{n!} - \frac{1}{(n+1)!}\left(\frac{x}{2}\right)^2 + \frac{1}{(2!)(n+2)!}\left(\frac{x}{2}\right)^4 - \frac{1}{(3!)(n+3)!}\left(\frac{x}{2}\right)^6 \cdots \right\}$$

In particular

$$J_0(x) = 1 - \frac{1}{(1!)^2}\left(\frac{x}{2}\right)^2 + \frac{1}{(2!)^2}\left(\frac{x}{2}\right)^4 - \frac{1}{(3!)^2}\left(\frac{x}{2}\right)^6 + \ldots$$

and

$$J_1(x) = \frac{x}{2}\left\{ 1 - \frac{1}{(1!)(2!)}\left(\frac{x}{2}\right)^2 + \frac{1}{(2!)(3!)}\left(\frac{x}{2}\right)^4 - \frac{1}{(3!)(4!)}\left(\frac{x}{2}\right)^6 + \ldots \right\}$$

5. *Legendre's equation*

$$(1 - x^2)y_2 - 2xy_1 + k(k+1)y = 0$$

where k is a real constant.

Solution by Frobenius gives

$$c = 0: \quad y = a_0\left\{ 1 - \frac{k(k+1)}{2!}x^2 + \frac{k(k-2)(k+1)(k+3)}{4!}x^4 \ldots \right\}$$

$$c = 1: \quad y = a_1\left\{ x - \frac{(k-1)(k+2)}{3!}x^3 + \frac{(k-1)(k-3)(k+2)(k+4)}{5!}x^5 \ldots \right\}$$

When k *is an integer*, one series terminates. The resulting polynomial in x, $P_n(x)$, is a *Legendre polynomial*, with a_0 or a_1 being chosen so that the polynomial has unit value when $x = 1$.

TEST EXERCISE V

1. If $y = e^{x^2+x}$, show that $y_2 = y_1(2x+1) + 2y$ and hence prove that $y_{n+2} = (2x+1)y_{n+1} + 2(n+1)y_n$.
2. Obtain a power series solution of the equation

$$(1+x^2)y_2 - 3xy_1 - 5y = 0$$

up to and including the term in x^6.

3. Determine a series solution for each of the following.

(a) $3xy_2 + 2y_1 + y = 0$

(b) $y_2 + x^2y = 0$

(c) $xy_2 + 3y_1 - y = 0$.

FURTHER PROBLEMS V

(a) *Use Leibnitz theorem for the following.*

1. If $y = x^3e^{4x}$, determine y_5.
2. Find the nth derivative of $y = x^3e^{-x}$ for $n > 3$.
3. If $y = x^3(2x+1)^2$, find y_4.
4. Find the 6th derivative of $y = x^4 \cos x$.
5. If $y = e^{-x} \sin x$, obtain an expression for y_4.
6. Determine y_3 when $y = x^4 \ln x$.
7. If $x^2y_2 + xy_1 + y = 0$, show that $x^2y_{n+2} + (2n+1)xy_{n+1} + (n^2+1)y_n = 0$.
8. If $y = (2x - \pi)^4 \sin\left(\dfrac{x}{2}\right)$, evaluate y_6 when $x = \pi/2$.
9. If $y = e^{-x} \cos x$, show that $y_4 + 4y = 0$.
10. Find the $(2n)$th derivative of (a) $y = x^2 \sinh x$

    (b) $y = x^3 \cosh x$.
11. If $y = (x^3 + 3x^2)e^{2x}$, determine an expression for y_6.
12. Find the nth derivative of $y = e^{-ax} \cos ax$ and hence determine y_3.

13. If $y = \dfrac{\sin x}{1 - x^2}$, show that

(a) $(1 - x^2)y_2 - 4xy_1 - (1 + x^2)y = 0$

(b) $y_{n+2} - (n^2 + 3n + 1)y_n - n(n - 1)y_{n-2} = 0$ at $x = 0$.

(b) *Use Leibnitz–Maclaurin method* to determine series solutions for the following.

14. $(1 + x^2)y_2 + xy_1 - 9y = 0$

15. $(x + 1)y_2 + (x - 1)y_1 - 2y = 0$

16. $(1 - x^2)y_2 - 7xy_1 - 9y = 0$

17. $(1 - x^2)y_2 - 2xy_1 + 2y = 0$

18. $xy_2 + y_1 + 2xy = 0$

(c) *Use the method of Frobenius* to obtain series solutions of the following.

19. $3xy_2 + y_1 - y = 0$

20. $y_2 + y = 0$

21. $y_2 - xy = 0$

22. $3xy_2 + 4y_1 + y = 0$

23. $y_2 - xy_1 + y = 0$

24. $xy_2 - 3y_1 + y = 0$

25. $xy_2 + y_1 - 3y = 0$.

Programme 6

Numerical Solution of Differential Equations

Prerequisites: Engineering Mathematics (fourth edition)
Programmes 14, 24, 25

Introduction

1 The range of differential equations that can be solved by straightforward analytical methods is relatively restricted. Even solution in series may not always be satisfactory, either because of the slow convergence of the resulting series or because of the involved manipulation in repeated stages of differentiation.

In such cases, where a differential equation and known boundary conditions are given, an approximate solution is often obtainable by the application of numerical methods, when the relevant particular solution is obtained as a set of function values for the range of values of the independent variable.

The solution of differential equations by numerical methods is a wide subject. The present programme introduces some of the simpler methods, which nevertheless are of practical use.

Taylor's Series

Let us start off by briefly revising the fundamentals of Maclaurin's and Taylor's series.

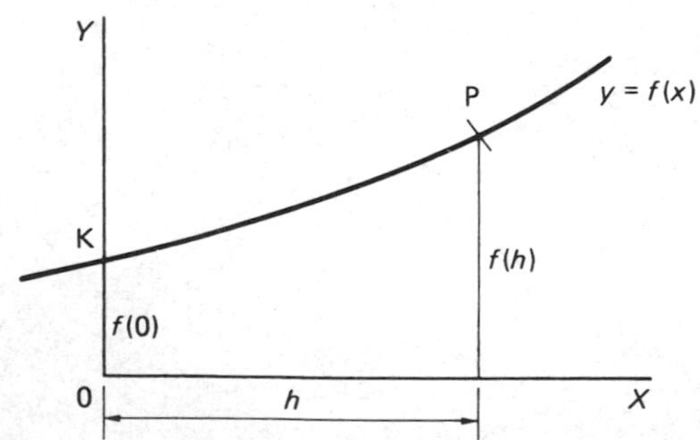

Maclaurin's series for $f(x)$ is

$$f(x) = f(0) + xf'(0) + \frac{x^2}{2!}f''(0) + \ldots + \frac{x^n}{n!}f^n(0) + \ldots \qquad (1)$$

and expresses the function $f(x)$ in terms of its successive differential coefficient at $x = 0$, i.e. at the point K.

Therefore, at P, $\quad f(h) = \ldots\ldots\ldots$

2

$$f(h) = f(0) = hf'(0) + \frac{h^2}{2!}f''(0) + \ldots + \frac{h^n}{n!}f^n(0) + \ldots \qquad (2)$$

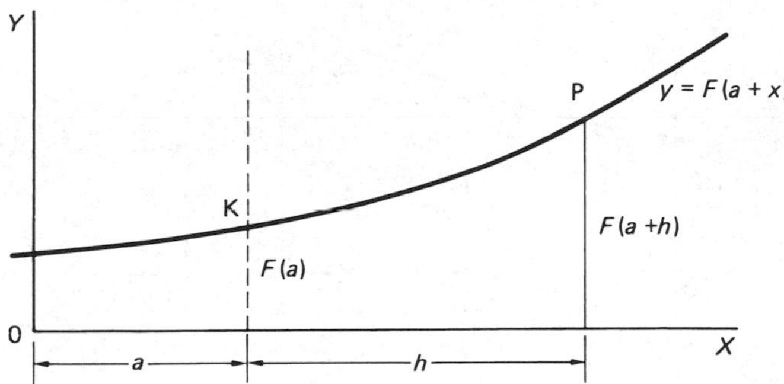

If the y-axis and origin are moved a units to the left, the equation of the same curve relative to the new axes, becomes $y = F(a + x)$ and the function value at K is $F(a)$.

At P, $F(a + h) = F(a) + hF'(a) + \frac{h^2}{2!}F''(a) + \ldots + \frac{h^n}{n!}F^n(a) + \ldots$

Writing $a = x$ in this result gives

$$f(x + h) = f(x) + hf'(x) + \frac{h^2}{2!}f''(x) + \ldots + \frac{h^n}{n!}f^n(x) + \ldots \qquad (3)$$

This is one common form of Taylor's series.

Make a note of it and then move on.

3

Function increment

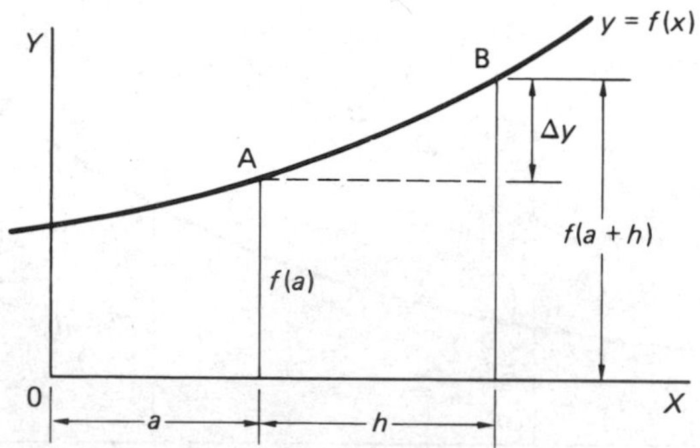

If we know the function value $f(a)$ at A, i.e. at $x = a$, we can apply Taylor's series to determine the function value at a neighbouring point B, i.e. at $x = a + h$.

$$f(a+h) = f(a) + hf'(a) + \frac{h^2}{2!}f''(a) + \frac{h^3}{3!}f'''(a) + \dots \qquad (4)$$

The *function increment* from A to B $= \Delta y = f(a+h) - f(a)$

i.e. $$f(a+h) = f(a) + \Delta y$$

where $\Delta y = hf'(a) + \frac{h^2}{2!}f''(a) + \frac{h^3}{3!}f'''(a) + \dots$

This entails evaluation of an infinite number of differential coefficients at $x = a$: in practice an approximation is accepted by restricting the number of terms that are used in the series .

This approximation of Taylor's series forms the basis of several numerical methods, some of which we shall now introduce.

On then to the next frame.

First-Order Differential Equations

Numerical solution of $\frac{dy}{dx} = f(x, y)$ with the initial condition that, at $x = x_0$, $y = y_0$. **4**

Euler's method

The simplest of the numerical methods for solving first-order differential equations is *Euler's method*, in which the Taylor's series

$$f(a+h) = f(a) + hf'(a) \quad + \frac{h^2}{2!}f''(a) + \frac{h^3}{3!}f'''(a) + \dots$$

is truncated after the second term to give

$$f(a+h) \approx f(a) + hf'(a) \qquad (5)$$

This is a severe approximation, but in practice the 'approximately equals' sign is replaced by the normal 'equals' sign, in the knowledge that the result we obtain will necessarily differ to some extent from the function value we seek. With this in mind, we write

$$\underline{f(a+h) = f(a) + hf'(a)}$$

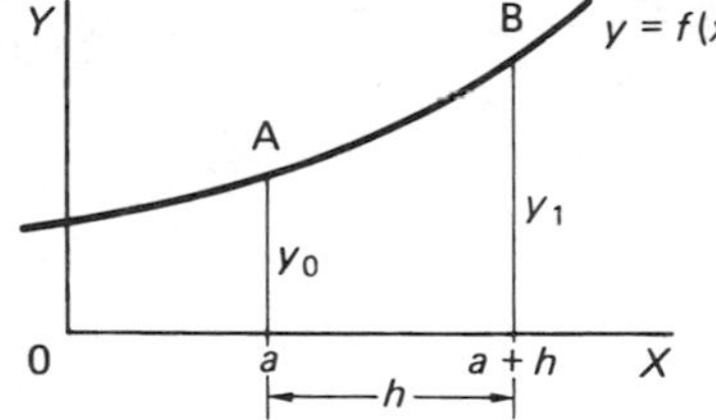

If h is the interval between two near ordinates and if we denote $f(a)$ by y_0, then the relationship

$$f(a+h) = f(a) + hf'(a)$$

becomes

$$\underline{y_1 = y_0 + h(y')_0} \qquad (6)$$

Hence, knowing y_0, h and $(y')_0$, we can compute y_1, an approximate value for the function value at B.

Make a note of result (6): we shall be using it quite a lot.

Then move on for an example.

5

Example 1 Given that $\frac{dy}{dx} = 2(1 + x) - y$ with the initial condition that at $x = 2$, $y = 5$, we can find an approximate value of y at $x = 2.2$, as follows.

We have $y' = 2(1 + x) - y$ with $x_0 = 2$, $y_0 = 5$

$$\therefore (y')_0 = \dots\dots\dots$$

6

$$\boxed{(y')_0 = 1}$$

We obtain this by referring x_0 and y_0 to the given equation

$$(y')_0 = 2(1 + x_0) - y_0 = 2(1 + 2) - 5 \quad \therefore (y')_0 = 1$$

So we have $x_0 = 2$; $y_0 = 5$; $(y')_0 = 1$; $x_1 = 2.2$; $h = 0.2$.

By Euler's relationship,

$$y_1 = y_0 + h(y')_0 \quad \therefore y_1 = \dots\dots\dots$$

7

$$\boxed{y_1 = 5.2}$$

for $y_1 = y_0 + h(y')_0 = 5 + (0.2)1 = 5.2$.

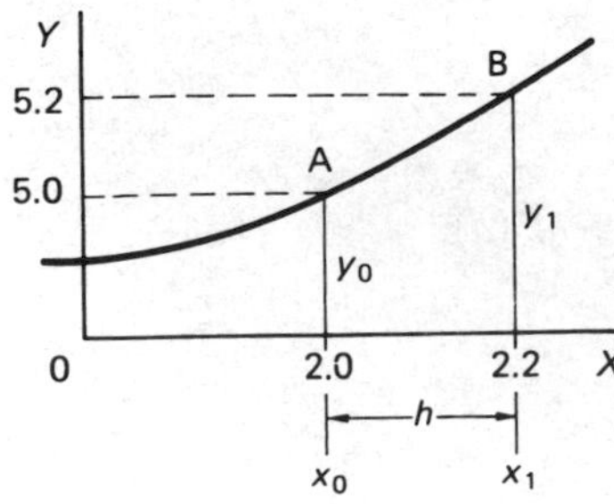

At B, $x_1 = 2.2$; $y_1 + 5.2$; and $(y')_1 = \dots\dots\dots$

8

$$\boxed{(y')_1 = 1.2}$$

$$(y')_1 = 2(1 + x_1) - y_1 = 2(1 + 2.2) - 5.2 = 1.2$$

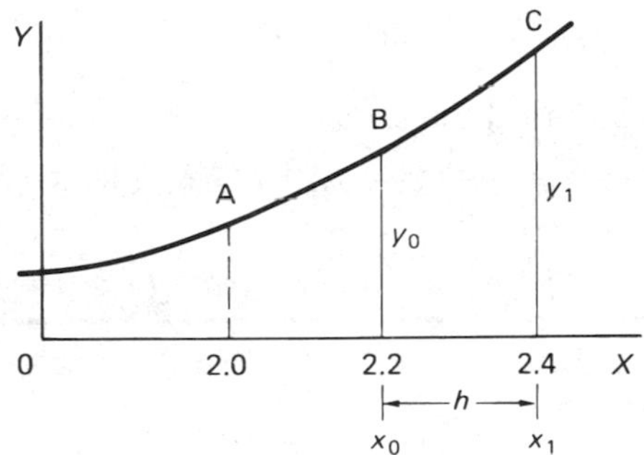

If we take the values of x, y and y' that we have just found for the point B and treat these as new starter values x_0, y_0, $(y')_0$, we can repeat the process and find values corresponding to the point C.

At B, $x_0 = 2.2$; $y_0 = 5.2$; $(y')_0 = 1.2$; $x_1 = 2.4$.

Then at C, $y_1 = \ldots\ldots\ldots$; $(y')_1 = \ldots\ldots\ldots$

9

$$\boxed{y_1 = 5.44; \quad (y')_1 = 1.36}$$

$y_1 = y_0 + h(y')_0 = 5.2 + (0.2)1.2 = 5.44$

$(y')_1 = 2(1 + x_1) - y_1 = 2(1 + 2.4) - 5.44 = 1.36$

So we could continue in a step-by-step method. At each stage, the determined values of x_1, y_1 and $(y')_1$ become the new starter values x_0, y_0 and $(y')_0$ for the next stage.

Our results so far can be tabulated thus

x_0	y_0	$(y')_0$	x_1	y_1	$(y')_1$
2.0	5.0	1.0	2.2	5.2	1.2
2.2	5.2	1.2	2.4	5.44	1.36
2.4	5.44	1.36			

Continue the table with a constant interval of $h = 0.2$. The third row can be completed to give

$x_1 = \ldots\ldots\ldots$; $y_1 = \ldots\ldots\ldots$; $(y')_1 = \ldots\ldots\ldots$

10

$x_1 = 2.6; \quad y_1 = 5.712; \quad (y')_1 = 1.488$

for
$$x_1 = x_0 + h = 2.4 + 0.2 = 2.6$$
$$y_1 = y_0 + h(y')_0 = 5.44 + (0.2)1.36 = 5.712$$
$$(y')_1 = 2(1 + x_1) - y_1 = 2(1 + 2.6) - 5.712 = 1.488$$

Now you can continue in the same way and complete the table for

$$x = 2.0,\ 2.2,\ 2.4,\ 2.6,\ 2.8,\ 3.0$$

Finish it off and compare results with the next frame.

11

Here is the result

x_0	y_0	$(y')_0$	x_1	y_1	$(y')_1$
2.0	5.0	1.0	2.2	5.2	1.2
2.2	5.2	1.2	2.4	5.44	1.36
2.4	5.44	1.36	2.6	5.712	1.488
2.6	5.712	1.488	2.8	6.009 6	1.590 4
2.8	6.009 6	1.590 4	3.0	6.327 68	1.672 32
3.0	6.327 68	1.672 32			

In practice, we do not, in fact, enter the values in the right-hand half of the table, but write them in directly as new starter values in the left-hand section of the table.

x_0	y_0	$(y')_0$
2.0	5.0	1.0
2.2	5.2	1.2
2.4	5.44	1.36
2.6	5.712	1.488
2.8	6.009 6	1.590 4
3.0	6.327 68	1.672 32

The particular solution is given by the values of y against x and a graph of the function can be drawn.

Draw the graph of the function carefully on graph paper.

12

Graph of the solution of $\frac{dy}{dx} = 2(1 + x) - y$ with $y = 5$ at $x = 2$.

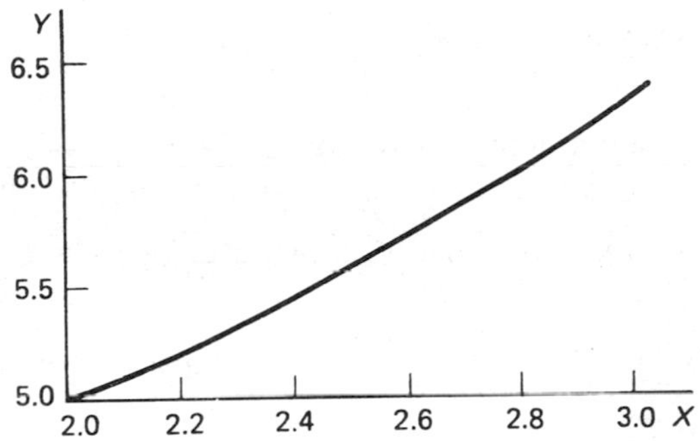

13

It is an advantage to plot the points step-by-step as the results are built up. In that way, one can check that there is a smooth progression and that no apparent errors in the calculations occur at any one stage.

In the example we have just worked, the actual function values for the range of x, $x = 2$ to $x = 3$, can be obtained by other means and are tabulated.

x	y (Euler)	y (actual)	Absolute error
2.0	5.0	5.0	0
2.2	5.2	5.218 731	0.018 731
2.4	5.44	5.470 320	0.030 320
2.6	5.712	5.748 812	0.036 812
2.8	6.009 6	6.049 329	0.039 729
3.0	6.327 68	6.367 879	0.040 199

The errors involved in the process are shown. These errors are due mainly to ....

14

the fact that Taylor's series was truncated after the second term.

Now another example.

Example 2 Obtain a numerical solution of the equation

$\dfrac{dy}{dx} = 1 + x - y$ with initial condition that $y = 2$ at $x = 1$, for the range

$x = 1.0(0.2)2.0$, i.e. from $x = 1.0$ to $x = 2.0$ at constant intervals of $x = 0.2$.

As starter values, we have $x_0 = 1$; $y_0 = 2$; $(y')_0 = \dots\dots\dots$

15

$$(y')_0 = 0$$

since $y' = 1 + x - y$ i.e. $(y')_0 = 1 + x_0 - y_0 = 1 + 1 - 2 = 0$

Euler's relationship is

$$y_1 = \dots\dots\dots$$

16

$$y_1 = y_0 + h(y')_0$$

First we prepare the table in the usual way

x_0	y_0	$(y')_0$
1.0	2.0	0
1.2		
1.4		
1.6		
1.8		
2.0		

Now, for each step, calculate y and y' and so complete the table.

Plot the points on a graph as you proceed. Complete the work.

17

Here is the finished table

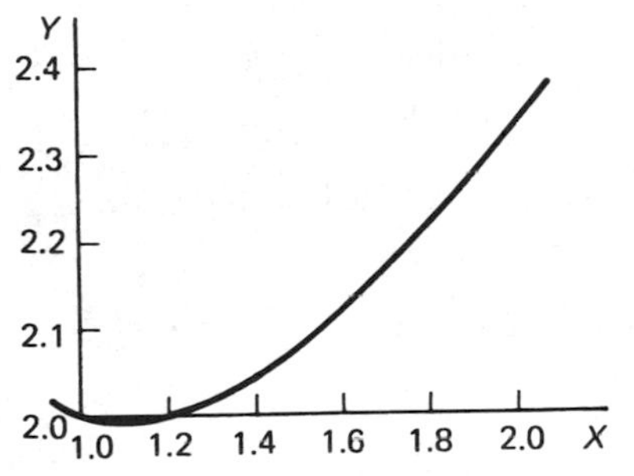

x_0	y_0	$(y')_0$
1.0	2.0	0
1.2	2.0	0.2
1.4	2.04	0.36
1.6	2.112	0.488
1.8	2.209 6	0.590 4
2.0	2.327 68	0.672 32

Comparison of the calculated and actual values of y shows the following

x	y (Eulcr)	y (actual)	Absolute error
1.0	2.0	2.0	0
1.2	2.0	2.018 731	0.018 731
1.4	2.04	2.070 320	0.030 320
1.6	2.112	2.148 812	0.036 812
1.8	2.209 6	2.249 329	0.039 729
2.0	2.327 68	2.367 879	0.040 199

Note once again that the errors involved are considerable.

So on to the next frame to see if we can improve matters.

18

Graphical interpretation of Euler's method

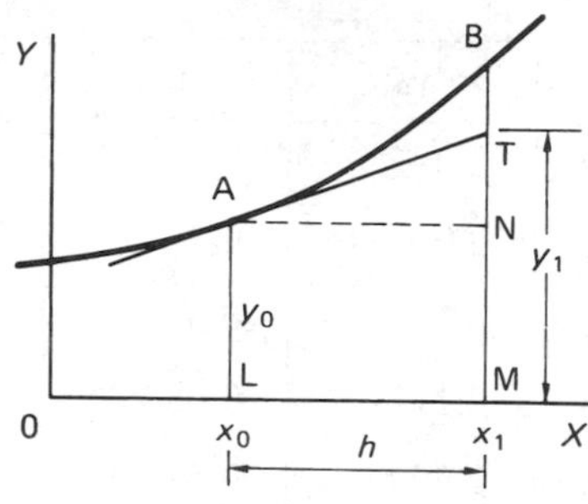

If AT is the tangent to the curve at A,

then $\dfrac{\text{NT}}{\text{AN}} = \left[\dfrac{\text{d}y}{\text{d}x}\right]_{x=x_0} = (y')_0$

$\dfrac{\text{NT}}{h} = (y')_0 \qquad \therefore \text{NT} = h(y')_0$

$\therefore$ At $x = x_1$, $\quad \text{MT} = y_0 + h(y')_0$

By Euler's relationship, $\quad y_1 = y_0 + h(y')_0$ i.e. MT.

The difference between the calculated value of y, i.e. MT, and the actual value of the function y, i.e. MB, at $x = x_1$, is indicated by TB. This error can be considerable, depending on the curvature of the graph and the size of the interval h. It is inherent to the method and corresponds to the truncation of the Taylor's series after the second term.

Euler's method, then

(a) is simple in procedure
(b) is lacking in accuracy, especially away from the starter values of the initial conditions
(c) is of use only for very small values of the interval h.

In spite of its practical limitations, it is the foundation of several more sophisticated methods and hence it is worthy of note.

Here is one more example to work on your own.

Example 3 Obtain the solution of $\dfrac{\text{d}y}{\text{d}x} = x + y$ with the initial condition that $y = 1$ at $x = 0$, for the range $x = 0(0.1)0.5$.

It will not take long. Compile the graph as the results are obtained.

The function values are given in the next frame.

Here are the results.

x_0	y_0	$(y')_0$
0	1.0	1.0
0.1	1.1	1.2
0.2	1.22	1.42
0.3	1.362	1.662
0.4	1.528 2	1.928 2
0.5	1.721 02	2.221 02

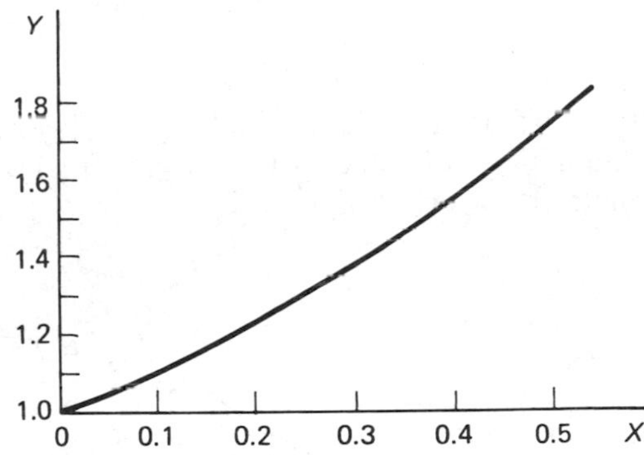

The actual function values are, in fact,

x	y
0	1.0
0.1	1.110 342
0.2	1.242 806
0.3	1.399 718
0.4	1.583 649
0.5	1.797 443

Plot these on your graph. Even with an h value as small as 0.1, the errors in the calculated values of y increase considerably as we move from the starter values, i.e. the initial conditions.

The percentage error at $x = 0.5$ is ....

20

4.252%

The Euler–Cauchy method—or the improved Euler method.

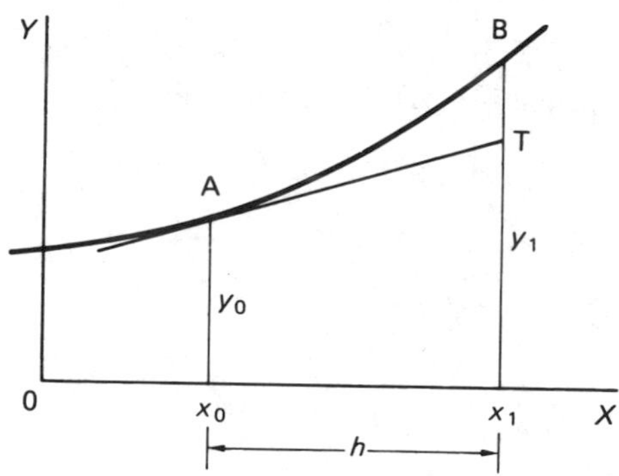

In Euler's method, we use the slope $(y')_0$ at $A(x_0, y_0)$ across the whole interval h to obtain an approximate value of y_1 at B. TB is the resulting error in the result.

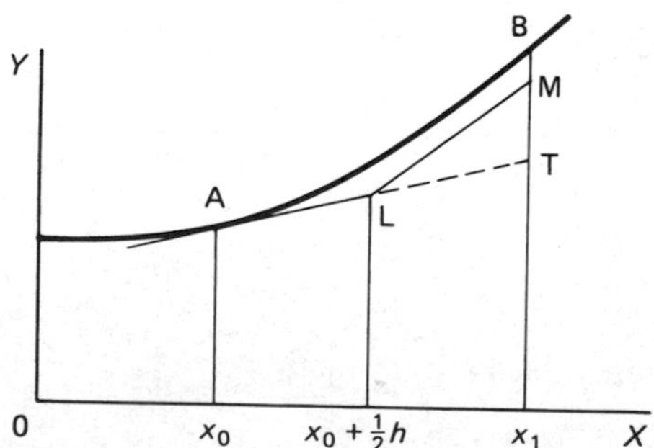

In the Euler–Cauchy method, we use the slope at $A(x_0, y_0)$ across half the interval and then continue with a line whose slope approximates to the slope of the curve at x_1.

Let $\bar{\bar{y}}_1$ be the y value of the point at T.

The error (MB) in the result is now considerably less than the error (TB) associated with the basic Euler method and the calculated results will accordingly be of greater accuracy.

21

Euler–Cauchy calculations

The steps in the Euler–Cauchy method are as follows.

1. We start with the given equation $y' = f(x, y)$ with the initial condition that at $x = x_0$, $y = y_0$. We have to determine function values for $x = x_0(h)x_n$.
2. From the equation and the initial condition we obtain $(y')_0 = f(x_0, y_0)$.
3. Knowing x_0, y_0, $(y')_0$ and h, we then evaluate
   (a) $x_1 = x_0 + h$
   (b) the auxiliary value of y, denoted by $\bar{\bar{y}}$ where $\bar{\bar{y}}_1 = y_0 + h(y')_0$. This is the same step as in Euler's method.
   (c) Then $y_1 = y_0 + \frac{1}{2}h\{(y')_0 + f(x_1, \bar{\bar{y}}_1)\}$
   Note that $f(x_1, y_1)$ is the right-hand side of the given equation with x and y replaced by the calculated values of x_1 and $\bar{\bar{y}}_1$.
   (d) Finally $(y')_1 = f(x_1, y_1)$.

We have thus evaluated x_1, y_1 and $(y')_1$.

The whole process is then repeated, the calculated values of x_1, y_1 and $(y')_1$ becoming the starter values x_0, y_0, $(y')_0$ for the next stage.

Make a note of the relationships above. We shall be using them quite often.

Then on to the next frame for an example of their use.

22

Example 1 Apply the Euler–Cauchy method to solve the equation $\frac{dy}{dx} = x + y$ with the initial condition that at $x = 0$, $y = 1$, for the range $x = 0(0.1)0.5$.

We proceed in a step-by-step routine as before.

We have the equation $y' = x + y$

initial condition $x_0 = 0$, $y_0 = 1$ $\therefore (y')_0 = x_0 + y_0 = 1$

Also $h = 0.1$ $x_1 = x_0 + h = 0.1$

(a) First we calculate the auxiliary y value, i.e. $\bar{\bar{y}}_1$, where

$\bar{\bar{y}}_1 = y_0 + h(y')_0$ (this is the same step as in Euler's method.)

$= \dots\dots\dots$

23

$$\boxed{\bar{\bar{y}}_1 = 1.1}$$

since $\bar{\bar{y}}_1 = y_0 + h(y')_0 = 1 + (0.1)1 = 1.1 \quad \therefore \bar{\bar{y}}_1 = 1.1$

(b) The second relationship is

$$y_1 = y_0 + \tfrac{1}{2}h\{(y')_0 + f(x_1, \bar{\bar{y}}_1)\}$$

$$= \dots\dots\dots$$

24

$$\boxed{y_1 = 1.11}$$

In this problem, $f(x, y) = x + y \quad \therefore f(x_1, \bar{\bar{y}}_1) = x_1 + \bar{\bar{y}}_1$

$$\therefore y_1 = 1 + \tfrac{1}{2}(0.1)\{1 + (0.1 + 1.1)\} = 1.11$$

So far, we have $x_1 = 0.1$; $y_1 = 1.11$;

$$\therefore (y')_1 = \dots\dots\dots$$

25

$$\boxed{(y')_1 = 1.21}$$

(c) $y' = x + y \quad \therefore (y')_1 = x_1 + y_1 = 0.1 + 1.11 = 1.21$

Collecting our results together, we now have

$$\underline{x_1 = 0.1; \quad y_1 = 1.11; \quad (y')_1 = 1.21}$$

Enter the initial values and these calculated values in a table with columns headed x, y, y'.

Then move on.

26

x	y	y'
0	1.0	1.0
0.1	1.11	1.21

We now take our last results as x_0, y_0 and $(y')_0$ for the next interval and repeat the process. This gives

$$x_1 = \dots\dots\dots; \quad \bar{\bar{y}}_1 = \dots\dots\dots$$

27

$$x_1 = 0.2; \quad \overline{\overline{y}}_1 = 1.231$$

for $\quad x_1 = x_0 + h = 0.1 + 0.1 = 0.2 \qquad\qquad x_1 = 0.2$

and $\quad \overline{\overline{y}}_1 = y_0 + h(y')_0 = 1.11 + (0.1)1.21 = 1.231 \qquad \overline{\overline{y}}_1 = 1.231$

Then $\quad y_1 = y_0 + \frac{1}{2}h\{(y')_0 + f(x_1, \overline{\overline{y}}_1)\}$

$\qquad = \dots\dots\dots$

28

$$y_1 = 1.242\,05$$

since $\quad y_1 = 1.11 + 0.05\{1.21 + (0.2 + 1.231)\} = 1.242\,05$

and $\quad (y')_1 = \dots\dots\dots$

29

$$(y')_1 = 1.442\,05$$

for $y' = x + y \qquad \therefore (y')_1 = x_1 + y_1 = 0.2 + 1.242\,05 = 1.442\,05$

We can now enter the next complete line in the table, which then becomes

. . . .

30

x	y	y'
0	1.0	1.0
0.1	1.11	1.21
0.2	1.242 05	1.442 05

Now we repeat the steps and obtain corresponding values for $x = 0.3$, 0.4 and 0.5.

Complete the table on your own and check with the next frame.

31

Here it is

x	y	y'
0	1.0	1.0
0.1	1.11	1.21
0.2	1.242 05	1.442 05
0.3	1.398 465	1.698 465
0.4	1.581 804	1.981 804
0.5	1.794 893	2.294 893

Actual y	Abs. error
1.0	0
1.110 342	0.000 342
1.242 806	0.000 756
1.399 718	0.001 253
1.583 649	0.001 845
1.797 443	0.002 550

If we compare the results obtained with the actual function values given in the right-hand table above, the errors are very much less than those obtained by the basic Euler method.

	Absolute errors	
x	Euler	Euler–Cauchy
0	0	0
0.1	0.010 342	0.000 342
0.2	0.022 806	0.000 756
0.3	0.037 718	0.001 253
0.4	0.055 449	0.001 845
0.5	0.076 423	0.002 550

With Euler's method, the percentage error at $x = 0.5$ was 4.252%.

With the Euler–Cauchy method, the percentage error at $x = 0.5$ is ....

32

$$\boxed{0.142\,\%}$$

Now for another example, but before we do so, complete the following without reference to your notes — if possible. In the Euler–Cauchy method, the relevant relationships are

$$x_1 = \dots\dots\dots$$
$$\bar{\bar{y}}_1 = \dots\dots\dots$$
$$y_1 = \dots\dots\dots$$
$$(y')_1 = \dots\dots\dots$$

Check with the next frame.

33

$$x_1 = x_0 + h$$
$$\bar{\bar{y}}_1 = y_0 + h(y')_0$$
$$y_1 = y_0 + \tfrac{1}{2}h\{(y')_0 + f(x_1, \bar{\bar{y}}_1)\}$$
$$(y')_1 = f(x_1,\ y_1)$$

Example 2 Determine a numerical solution of the equation $y' = 2(1 + x) - y$ with the initial condition that $y = 5$ at $x = 2$, for the range $x = 2.0(0.2)3.0$.

First of all we have

$$y' = 2(1 + x) - y \qquad x_0 = 2,\ \ y_0 = 5 \ \ \therefore\ (y')_0 = 1$$

Also $h = 0.2$ $\qquad x_1 = x_0 + h = 2.2$

Then $\quad \bar{\bar{y}}_1 = \dots\dots\dots$

34

$\boxed{\bar{\bar{y}}_1 = 5.2}$

for $\bar{\bar{y}}_1 = y_0 + h(y')_0 = 5 + (0.2)1 = 5.2$

Also $\quad y_1 = \ldots\ldots\ldots$

35

$\boxed{y_1 = 5.22}$

since
$$\begin{aligned} y_1 &= y_0 + \tfrac{1}{2}h\{(y')_0 + f(x_1, \bar{\bar{y}}_1)\} \\ &= 5 + 0.1\{1 + 2(1 + x_1) - \bar{\bar{y}}_1\} \\ &= 5 + 0.1\{1 + 2(1 + 2.2) - 5.2\} = 5.22 \end{aligned}$$

and $\quad (y')_1 = \ldots\ldots\ldots$

36

$\boxed{(y')_1 = 1.18}$

for $\quad (y')_1 = 2(1 + x_1) - y_1 = 2(1 + 2.2) - 5.22 = 1.18$

So we have
$$\begin{array}{lll} x_0 = 2 & y_0 = 5 & (y')_0 = 1 \\ x_1 = 2.2 & y_1 = 5.22 & (y')_1 = 1.18 \end{array}$$

Enter these as the first two lines in a suitable table and complete the next line.

37

x	y	y'
2.0	5.0	1.0
2.2	5.22	1.18
2.4	5.4724	1.3276

for $x_1 = x_0 + h = 2.2 + 0.2 = 2.4$ $\qquad x_1 = 2.4$

$\bar{\bar{y}}_1 = y_0 + h(y')_0 = 5.22 + (0.2)1.18 = 5.456$ $\qquad \bar{\bar{y}}_1 = 5.456$

$y_1 = y_0 + \frac{1}{2}h\{(y')_0 + f(x_1, \bar{\bar{y}}_1)\}$

$= 5.22 + 0.1\{1.18 + 2(1 + 2.4) - 5.456\} = 5.4724$ $\quad y_1 = 5.4724$

$(y')_1 = 2(1 + x_1) - y_1 = 2(1 + 2.4) - 5.4724 = 1.3276$ $\qquad (y')_1 = 1.3276$

Now you can carry on and complete the whole table up to $x = 3.0$.

38

x	y	y'
2.0	5.0	1.0
2.2	5.22	1.18
2.4	5.472 4	1.327 6
2.6	5.751 368	1.448 632
2.8	6.052 122	1.547 878
3.0	6.370 740	1.629 260

Each step-by-step stage is carried out in the same way.

Here is a further problem that you can easily complete on your own.

Example 3 Solve the equation $y' = y^2 + xy$ with initial condition that at $x = 1$, $y = 1$, for the range $x = 1.0(0.1)1.5$. Use the Euler–Cauchy method and work to six places of decimals.

We have $x_0 = 1$, $y_0 = 1$ and $h = 0.1$. Computation will be more convenient if we re-write the equation in the form $y' = y(y + x)$. Tabulate the results in the usual way and draw the graph of the function.

Complete the solution and check with the next frame.

39

x	y	y'
1.0	1.0	2.0
1.1	1.238	2.894 444
1.2	1.591 023	4.440 582
1.3	2.152 410	7.431 002
1.4	3.145 895	14.300 524
1.5	5.251 005	35.449 561

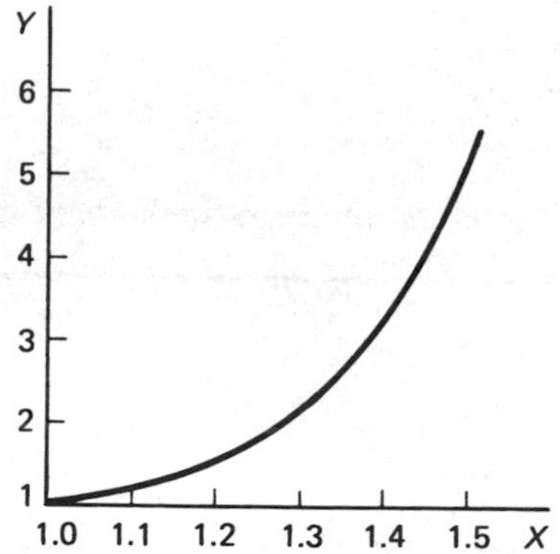

If you agree with these values, move straight on to the next frame. If not, check back through the previous examples that were worked in detail.

Note: Care and accuracy of working are essential in numerical work. Wherever possible, some form of checking should be employed. At this stage, it is sensible to plot the graph of the function as results are generated. Any deviation of points from the smooth curve then becomes apparent before the errors are involved in subsequent calculations.

Runge–Kutta method 40

The Runge–Kutta method for solving first-order differential equations is widely used and affords a high degree of accuracy. It is a further step-by-step process where a table of function values for a range of values of x is accumulated. Several intermediate calculations are required at each stage, but these are straightforward and present little difficulty.

In general terms, the method is as follows.

To solve $y' = f(x, y)$ with initial condition $y = y_0$ at $x = x_0$, for a range of values of $x = x_0(h)x_n$.
Starting as usual with $x = x_0$, $y = y_0$, $y' = (y')_0$ and h, we have

$$x_1 = x_0 + h$$

Finding y_1 requires four intermediate calculations

$$k_1 = hf(x_0, y_0) = h(y')_0$$
$$k_2 = hf(x_0 + \tfrac{1}{2}h,\ y_0 + \tfrac{1}{2}k_1)$$
$$k_3 = hf(x_0 + \tfrac{1}{2}h,\ y_0 + \tfrac{1}{2}k_2)$$
$$k_4 = hf(x_0 + h,\ y_0 + k_3)$$

The increment Δy_0 in the y-value from $x = x_0$ to $x = x_1$ is then

$$\Delta y_0 = \tfrac{1}{6}\{k_1 + 2k_2 + 2k_3 + k_4\}$$

and finally $$y_1 = y_0 + \Delta y_0.$$

We shall be using these repeatedly, so make a note of them for future reference. Then let us see an example.

Example 1 $y' = x + y$ with $y = 1$ at $x = 0$. Determine the function values of y for $x = 0(0.1)0.5$. 41

Our starting values are therefore

$$x_0 = 0, \quad y_0 = 1, \quad (y')_0 = 1, \quad h = 0.1$$

$$x_1 = x_0 + h = 0 + 0.1 = 0.1$$ $x_1 = 0.1$

$k_1 = h(y')_0 = (0.1)1 = 0.1$ $\therefore \underline{k_1 = 0.1}$

$k_2 = hf(x_0 + \frac{1}{2}h,\ y_0 + \frac{1}{2}k_1)$ i.e. we have the same function $f(x, y) = x + y$ with the x-value replaced by $x_0 + \frac{1}{2}h$, i.e. $0 + 0.05 = 0.05$
and the y-value replaced by $y_0 + \frac{1}{2}k_1$, i.e. $1 + 0.05 = 1.05$.

$$\therefore k_2 = \dots\dots\dots$$

42

$$\boxed{k_2 = 0.11}$$

since $k_2 = hf(x_0 + \frac{1}{2}h,\ y_0 + \frac{1}{2}k_1)$ $\qquad x_0 + \frac{1}{2}h = 0.05$

$= 0.1\{0.05 + 1.05\}$ $\qquad y_0 + \frac{1}{2}k_1 = 1.05$

$= 0.11$ $\qquad \therefore \underline{k_2 = 0.11}$

Similarly, $k_3 = hf(x_0 + \frac{1}{2}h,\ y_0 + \frac{1}{2}k_2)$ $x_0 + \frac{1}{2}h = 0.05;$ $y_0 + \frac{1}{2}k_2 = 1.055$

$= 0.1\{0.05 + 1.055\} = 0.1105$ $\qquad \underline{k_3 = 0.1105}$

$$k_4 = hf(x_0 + h,\ y_0 + k_3) = \dots\dots\dots$$

43

$$\boxed{k_4 = 0.121\ 05}$$

since $x_0 + h = 0 + 0.1 = 0.1;$ $\quad y_0 + k_3 = 1 + 0.1105 = 1.1105$

$\therefore k_4 = hf(x_0 + h,\ y_0 + k_3) = 0.1\{0.1 + 1.1105\} = 0.121\ 05$

So we have $k_1 = 0.1;$ $\quad k_2 = 0.11;$ $\quad k_3 = 0.110\ 5;$ $\quad k_4 = 0.121\ 05$

$$\therefore \Delta y_0 = \frac{1}{6}\{k_1 + 2k_2 + 2k_3 + k_4\} = \dots\dots\dots$$

and hence $y_1 = y_0 + \Delta y_0 = \dots\dots\dots$

44

$$\boxed{\Delta y_0 = 0.110\ 342; \quad y_1 = 1.110\ 342}$$

Finally, from the equation $(y')_1 = x_1 + y_1$

$$\therefore\ (y')_1 = 0.1 + 1.110\ 342 = 1.210\ 342$$

$$\therefore\ \underline{x_1 = 0.1}; \quad \underline{y_1 = 1.110\ 342}; \quad \underline{(y')_1 = 1.210\ 342}$$

Stage 2 We now repeat the whole process, taking the calculated values of x_1, y_1 and $(y')_1$ as the new x_0, y_0 and $(y')_0$.

$$y' = x + y \quad \therefore\ (y')_1 = x_1 + y_1$$

So now the new $x_0 = 0.1$; $y_0 = 1.110\ 342$; $(y')_0 = 1.210\ 342$; $h = 0.1$; $x_1 = 0.2$.

$$k_1 = h(y')_0 = (0.1)1.210\ 342 = 0.121\ 034 \qquad \therefore\ \underline{k_1 = 0.121\ 034}$$

$$k_2 = hf(x_0 + \tfrac{1}{2}h,\ y_0 + \tfrac{1}{2}k_1) \quad \therefore\ k_2 = \dots\dots\dots$$

45

$$\boxed{k_2 = 0.132\ 086}$$

since $\quad x_0 + \frac{1}{2}h = 0.1 + 0.05 = 0.15$

and $\quad y_0 + \frac{1}{2}k_1 = 1.110\ 342 + \frac{1}{2}(0.121\ 034) = 1.170\ 859$

$$\therefore\ k_2 = hf(x_0 + \tfrac{1}{2}h,\ y_0 + \tfrac{1}{2}k_1)$$
$$= 0.1\{0.15 + 1.170\ 859\} = 0.132\ 986 \quad \therefore\ \underline{k_2 = 0.132\ 086}$$

In the same way $k_3 = \dots\dots\dots$; $k_4 = \dots\dots\dots$

46

$$\boxed{k_3 = 0.132\ 639; \quad k_4 = 0.144\ 298}$$

$$k_3 = hf(x_0 + \tfrac{1}{2}h,\ y_0 + \tfrac{1}{2}k_2) = 0.1(0.15 + 1.176\ 385) = 0.132\ 639$$
$$k_4 = hf(x_0 + h,\ y_0 + k_3) = 0.1(0.2 + 1.242\ 981) = 0.144\ 298$$

Now you can finish it off.

$$\Delta y_0 = \dots\dots\dots; \qquad y_1 = \dots\dots\dots$$

47

$$\boxed{\Delta y_0 = 0.132\,464; \quad y_1 = 1.242\,806}$$

since $\Delta y_0 = \frac{1}{6}\{k_1 + 2k_2 + 2k_3 + 2k_4\}$ and $y_1 = y_0 + \Delta y_0$.

Finally, from the equation, $(y')_1 = \ldots\ldots\ldots$

48

$$\boxed{(y')_1 = 1.442\,806}$$

for $y' = x + y \quad \therefore (y')_1 = x_1 + y_1 = 0.2 + 1.242\,806 = 1.442\,806.$

Therefore, at this point we have

$$\underline{x_1 = 0.2}; \quad \underline{y_1 = 1.242\,806}; \quad \underline{(y')_1 = 1.442\,806}$$

Stage 3 Now we repeat the process again, using these values as the new starter values x_0, y_0 and $(y')_0$.

For this section $k_1 = \ldots\ldots\ldots; \quad k_2 = \ldots\ldots\ldots$
$k_3 = \ldots\ldots\ldots; \quad k_4 = \ldots\ldots\ldots$

49

$$\boxed{\begin{array}{ll} k_1 = 0.144\,281; & k_2 = 0.156\,495 \\ k_3 = 0.157\,105; & k_4 = 0.169\,991 \end{array}}$$

Check that you agree with these. Then go on and complete the stage for $x_1 = 0.3$.

$$\Delta y_0 = \ldots\ldots\ldots; \quad y_1 = \ldots\ldots\ldots; \quad (y')_1 = \ldots\ldots\ldots$$

50

$$\Delta y_0 = 0.156\,912; \quad y_1 = 1.399\,718; \quad (y')_1 = 1.699\,718;$$

for $\Delta y_0 = \frac{1}{6}\{k_1 + 2k_2 + 2k_3 + k_4\}; \qquad y_1 = y_0 + \Delta y_0$

$$(y')_1 = f(x_1, y_1) = x_1 + y_1$$

So, at the end of stage 3, we have

$$\underline{x_1 = 0.3}; \quad \underline{y_1 = 1.399\,718}; \quad \underline{(y')_1 = 1.699\,718}$$

Stage 4 Now work through stage 4 on your own, starting with

$$y' = x + y \qquad \therefore (y')_1 = x_1 + y_1$$

$$x_0 = 0.3; \quad y_0 = 1.399\,718; \quad (y')_0 = 1.699\,718; \quad h = 0.1; \quad x_1 = 0.4$$

Check the results with the next frame when you have completed the stage.

51

$$x_1 = 0.4; \quad y_1 = 1.583\,649; \quad (y')_1 = 1.983\,649$$

Intermediate working gives $k_1 = 0.169\,972 \qquad k_2 = 0.183\,470$

$k_3 = 0.184\,145 \qquad k_4 = 0.198\,386$

$\Delta y_0 = 0.183\,931 \qquad \therefore y_1 = 1.583\,649 \quad (y')_1 = 1.983\,649$

Stage 5 There now remains one more stage to obtain values for $x_1 = 0.5$ to complete the solution.

Starting with the last set of values as usual, complete the calculations to obtain $x_1 = 0.5; \quad y_1 = \ldots\ldots\ldots; \quad (y')_1 = \ldots\ldots\ldots$

52

$$x_1 = 0.5; \quad y_1 = 1.797\,442; \quad (y')_1 = 2.297\,442$$

As a check $\quad k_1 = 0.198\,365 \quad k_2 = 0.213\,283$

$k_3 = 0.214\,029 \quad k_4 = 0.229\,768$

$\Delta y_0 = 0.213\,793 \quad y_1 = y_0 + \Delta y_0 = 1.797\,442$

$(y')_1 = x_1 + y_1 \quad = 2.297\,442$

The final table of results is therefore as follows

x	y	y'	Actual y	Absolute error
0	1.0	1.0	1.0	0.000 000
0.1	1.110 342	1.210 342	1.110 342	0.000 000
0.2	1.242 806	1.442 806	1.242 806	0.000 000
0.3	1.399 718	1.699 718	1.399 718	0.000 000
0.4	1.583 649	1.983 649	1.583 649	0.000 000
0.5	1.797 442	2.297 442	1.797 443	0.000 001

Note the accuracy of the calculated function values.

53

We see from the table of results that, to six places of decimals, there is practically no error in the method. The computation involved in the procedure may be rather more tedious than in the previous methods considered, but the extra labour involved is rewarded by the higher degree of accuracy of the results.

We have obtained the solution to this same equation by the three methods we have covered. It might be useful to compare the errors in each case.

x	Absolute errors		
	Euler	Euler–Cauchy	Runge–Kutta
0	0	0	0.000 000
0.1	0.010 342	0.000 342	0.000 000
0.2	0.022 806	0.000 756	0.000 000
0.3	0.037 718	0.001 253	0.000 000
0.4	0.055 449	0.001 845	0.000 000
0.5	0.076 423	0.022 550	0.000 001

The supremacy of the Runge–Kutta method is self-evident and the method is popular where accuracy of results is all important. One disadvantage of Runge–Kutta, however, is that it does not readily lend itself to any self-checking procedures—hence the necessity of plotting the values determined to recognise any irregularity, should numerical errors occur.

Without reference to your notes, complete the expressions for

$$k_1 = \dots\dots\dots$$
$$k_2 = \dots\dots\dots$$
$$k_3 = \dots\dots\dots$$
$$k_4 = \dots\dots\dots$$
$$\Delta y_0 = \dots\dots\dots$$
$$y_1 = \dots\dots\dots$$

It speeds up the working if you can remember them.

54

$$\begin{aligned} k_1 &= h(y')_0 \\ k_2 &= hf(x_0 + \tfrac{1}{2}h,\ y_0 + \tfrac{1}{2}k_1) \\ k_3 &= hf(x_0 + \tfrac{1}{2}h,\ y_0 + \tfrac{1}{2}k_2) \\ k_4 &= hf(x_0 + h,\ y_0 + k_3) \\ \Delta y_0 &= \tfrac{1}{6}\{k_1 + 2k_2 + 2k_3 + k_4\} \\ y_1 &= y_0 + \Delta y_0 \end{aligned}$$

With those in mind, let us move on to a further example. Next frame.

55 *Example 2* Solve $y' = \sqrt{x^2 + y}$ for $x = 0(0.2)1.0$ given that at $x = 0$, $y = 0.8$.

From the equation and initial condition, we have

$$x_0 = 0;\quad y_0 = 0.8;\quad h = 0.2;\quad x_1 = 0.2;$$

$$(y')_0 = \{x_0^2 + y_0\}^{1/2} = 0.894\,427$$

Stage 1 $x_0 = 0$; $y_0 = 0.8$; $(y')_0 = 0.894\,427$; $h = 0.2$; $x_1 = 0.2$.

$$\therefore k_1 = \dots\dots\dots$$

56

$$\boxed{k_1 = 0.178\,885}$$

for $k_1 = h(y')_0 = (0.2)0.894\,427 = 0.178\,885$

Then $k_2 = \dots\dots\dots$

57

$$\boxed{k_2 = 0.189\,678}$$

for $x_0 + \frac{1}{2}h = 0.1$; $y_0 + \frac{1}{2}k_1 = 0.8 + 0.089\,443 = 0.889\,443$

$$\therefore k_2 = 0.2(0.1^2 + 0.889\,443)^{1/2} = 0.189\,678$$

Similarly, $k_3 = \dots\dots\dots$

58

$$\boxed{k_3 = 0.190\ 246}$$

for $k_3 = hf(x_0 + \frac{1}{2}h,\ y_0 + \frac{1}{2}k_2)$; $\quad y_0 + \frac{1}{2}k_2 = 0.8 + 0.094\ 839 = 0.894\ 839$
$\quad = 0.2(0.1^2 + 0.894\ 839)^{1/2} = 0.190\ 246$

and $\qquad k_4 = \dots\dots\dots$

59

$$\boxed{k_4 = 0.203\ 002}$$

for $k_4 = hf(x_0 + h,\ y_0 + k_3)$; $\quad x_0 + h = 0.2$; $\quad y_0 + k_3 = 0.990\ 246$
$\quad = 0.2(0.2^2 + 0.990\ 246)^{1/2} = 0.203\ 002.$

Collecting the results together

$$k_1 = 0.178\ 885 \qquad k_2 = 0.189\ 678$$
$$k_3 = 0.190\ 246 \qquad k_4 = 0.203\ 002$$

$$\therefore\ \Delta y_0 = \dots\dots\dots; \quad y_1 = \dots\dots\dots$$

60

$$\boxed{\begin{aligned} \Delta y_0 &= \tfrac{1}{6}\{k_1 + 2k_2 + 2k_3 + k_4\} = 0.190\ 289 \\ y_1 &= y_0 + \Delta y_0 = 0.8 + \Delta y_0 = 0.990\ 289 \end{aligned}}$$

$$(y')_1 = (x_1^2 + y_1)^{1/2} = (0.2^2 + 0.990\ 289)^{1/2} = 1.015\ 032$$

$$\therefore\ x_1 = 0.2; \quad y_1 = 0.190\ 289; \quad (y')_1 = 1.015\ 032$$

For stage 2 then

$$x_0 = 0.2; \quad y_0 = 0.190\ 289; \quad (y')_0 = 1.015\ 032; \quad h = 0.2; \quad x_1 = 0.4.$$

$$\therefore\ k_1 = \dots\dots\dots; \quad k_2 = \dots\dots\dots$$
$$k_3 = \dots\dots\dots; \quad k_4 = \dots\dots\dots$$

61

$k_1 = 0.203\,006$;	$k_2 = 0.217\,421$
$k_3 = 0.218\,083$;	$k_4 = 0.233\,955$

Then, working as before

$$\Delta y_0 = \dots\dots\dots; \quad y_1 = \dots\dots\dots; \quad (y')_1 = \dots\dots\dots$$

62

$$\boxed{\Delta y_0 = 0.217\,995; \quad y_1 = 1.208\,284; \quad (y')_1 = 1.169\,737}$$

$$\therefore\; x_1 = 0.4; \quad y_1 = 1.208\,284; \quad (y')_1 = 1.169\,737$$

For stage 3, we have

$$x_0 = 0.4; \quad y_0 = 1.208\,284; \quad (y')_0 = 1.169\,737.$$

Work on through the next stage and obtain

$$x_1 = \dots\dots\dots; \quad y_1 = \dots\dots\dots; \quad (y')_1 = \dots\dots\dots$$

63

$$x_1 = 0.6; \quad y_1 = 1.459\,816; \quad (y')_1 = 1.349\,006$$

So far, then, the table looks like this

x	y	y'
0	0.8	0.894 427
0.2	0.990 289	1.015 032
0.4	1.208 284	1.169 737
0.6	1.459 816	1.349 006
0.8		
1.0		

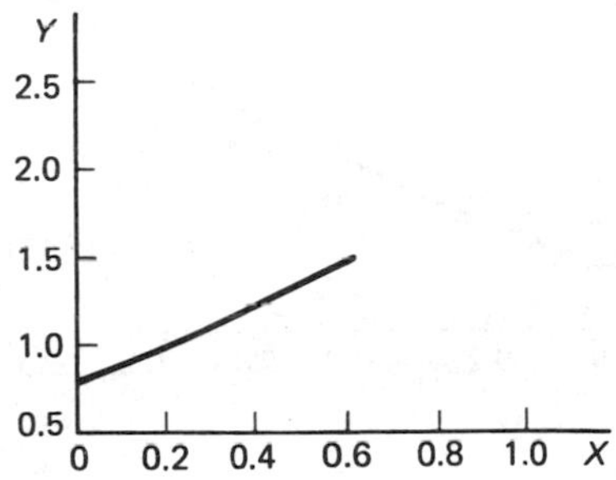

Now work through the final two stages: complete the table and the graph of the function. Finish it off.

64 Finally, we have

x	y	y'
0	0.8	0.894 427
0.2	0.990 289	1.015 032
0.4	1.208 284	1.169 737
0.6	1.459 816	1.349 006
0.8	1.749 040	1.545 652
1.0	2.078 899	1.754 679

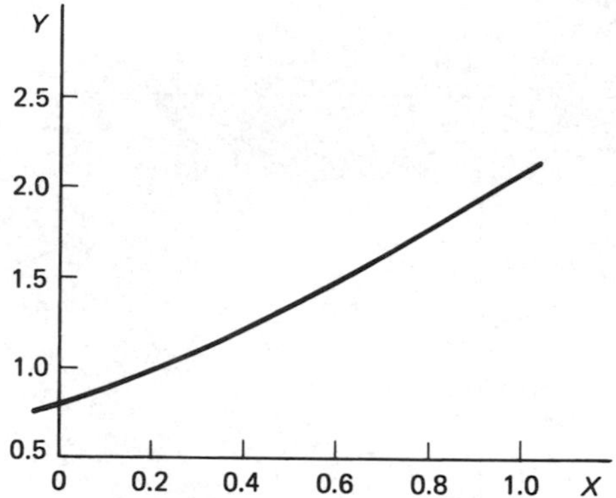

The method is the same every time.
If $y' = f(x, y)$ and $y = y_0$ at $x = x_0$, then y_1 at $x_1 = x_0 + h$ is calculated from

$$k_1 = \ldots\ldots\ldots$$
$$k_2 = \ldots\ldots\ldots$$
$$k_3 = \ldots\ldots\ldots$$
$$k_4 = \ldots\ldots\ldots$$
$$\Delta y_0 = \ldots\ldots\ldots$$
$$y_1 = \ldots\ldots\ldots$$
$$(y')_1 = \ldots\ldots\ldots$$

If you are in any doubt, check with your notes, if necessary.

65

$$\begin{aligned}
k_1 &= h(y')_0 \\
k_2 &= hf(x_0 + \tfrac{1}{2}h,\ y_0 + \tfrac{1}{2}k_1) \\
k_3 &= hf(x_0 + \tfrac{1}{2}h,\ y_0 + \tfrac{1}{2}k_2) \\
k_4 &= hf(x_0 + h,\ y_0 + k_3) \\
\Delta y_0 &= \tfrac{1}{6}\{k_1 + 2k_2 + 2k_3 + k_4\} \\
y_1 &= y_0 + \Delta y_0 \\
(y')_1 &= f(x_1,\ y_1)
\end{aligned}$$

That is it. Now we move on to the next frame where we make a new start and apply similar methods to the solution of second-order differential equations by numerical methods.

Second-Order Differential Equations

Method 1 The first method we will deal with is really an extension of the Euler method for the first-order equations and is a direct application of a truncated form of Taylor's series. We anticipate, therefore, that the method will be relatively easy, but the results will not be accurate to a high degree. 66

Taylor's series:

$$f(x+h) = f(x) + hf'(x) + \frac{h^2}{2!}f''(x) + \frac{h^3}{3!}f'''(x) + \dots$$

Differentiating term by term with respect to x, we obtain

$$f'(x+h) = f'(x) + hf''(x) + \frac{h^2}{2!}f'''(x) + \frac{h^3}{3!}f''''(x) + \dots$$

If we neglect terms in $f'''(x)$ and subsequent terms in each of these two series, we have the approximations

$$f(x+h) \approx \dots\dots\dots$$
$$f'(x+h) \approx \dots\dots\dots$$

67

$$f(x+h) \approx f(x) + hf'(x) + \frac{h^2}{2!}f''(x)$$
$$f'(x+h) \approx f'(x) + hf''(x)$$

Although these are approximations, in practice we tend to write them with the 'equals' sign. Therefore, at $x = a$, these become

....

and ....

68

$$f(a+h) = f(a) + hf'(a) + \frac{h^2}{2!}f''(a)$$
$$f'(a+h) = f'(a) + hf''(a)$$

and these, with the notation we have previously used, can be written

$$y_1 = y_0 + h(y')_0 + \frac{h^2}{2!}(y'')_0$$
$$(y')_1 = (y')_0 + h(y'')_0$$

Thus, if x_0, y_0, $(y')_0$ and $(y'')_0$ are known, we can find an approximate value of y_1 at $x_1 = x_0 + h$.

Make a note of these two relationships: then we can apply them.

69

Example 1 Solve the equation $\dfrac{d^2y}{dx^2} = x\dfrac{dy}{dx} + y$ for $x = 0(0.2)1.0$

given that at $x = 0$, $y = 1$ and $\dfrac{dy}{dx} = 0$.

From the equation and the initial conditions, we have

$$y'' = xy' + y$$

$x_0 = 0$; $y_0 = 1$; $(y')_0 = 0$; $h = 0.2$; $(y'')_0 =$ ....

70

$(y'')_0 = 1$ from the differential equation

We also have the two relationships

$$y_1 = y_0 + h(y')_0 + \frac{h^2}{2!}(y'')_0$$

and $$(y')_1 = (y')_0 + h(y'')_0$$

Stage 1 $x_0 = 0;\quad y_0 = 1;\quad (y')_0 = 0;\quad (y'')_0 = 1;\quad h = 0.2;\quad x_1 = 0.2.$

$$\therefore\ y_1 = 1 + (0.2)0 + \frac{0.2^2}{2!}(1) = 1.02 \quad \therefore\ y_1 = 1.02$$

$$(y')_1 = (y')_0 + h(y'')_0 = 0 + (0.2)(1) = 0.2 \quad \therefore\ (y')_1 = 0.2$$

$$\text{Now } y'' = xy' + y \quad \therefore\ (y'')_1 = x_1(y')_1 + y_1 = (0.2)(0.2) + 1.02$$

$$\therefore\ (y'')_1 = 1.06$$

$$\therefore\ x_1 = 0.2;\quad y_1 = 1.02;\quad (y')_1 = 0.2;\quad (y'')_1 = 1.06$$

These results become the starter values for the next stage.

Stage 2 $x_0 = 0.2;\quad y_0 = 1.02;\quad y'_0 = 0.2;\quad (y'')_0 = 1.06;\quad h = 0.2;$

$x_1 = 0.4.$

So $$y_1 = y_0 + h(y')_0 + \frac{h^2}{2!}(y'')_0$$

$$= 1.02 + (0.2)(0.2) + \frac{0.2^2}{2!}(1.06) = 1.081\,2 \qquad \therefore\ y_1 = 1.081\,2$$

Similarly, $$(y')_1 = (y')_0 + h(y'')_0 = \dots\dots\dots$$

and $$(y'')_1 = \dots\dots\dots$$

71

$$\boxed{(y')_1 = 0.412; \quad (y'')_1 = 1.246}$$

for $\quad (y'')_1 = x_1(y')_1 + y_1 = (0.4)(0.412) + 1.0812 = 1.246$

$$\therefore \underline{x_1 = 0.4; \;\; y_1 = 1.081\,2; \;\; (y')_1 = 0.412; \;\; (y'')_1 = 1.246}$$

Stage 3

$x_0 = 0.4$; $y_0 = 1.081\,2$; $(y')_0 = 0.412$; $(y'')_0 = 1.246$; $h = 0.2$; $x_1 = 0.6$.

Working as before $\quad y_1 = \ldots\ldots\ldots$; $\quad (y')_1 = \ldots\ldots\ldots$;
$(y'')_1 = \ldots\ldots\ldots$

72

$$\boxed{x_1 = 0.6; \quad y_1 = 1.188\,52; \quad (y')_1 = 0.661\,2; \quad (y'')_1 = 1.585\,24}$$

Now go ahead and complete stages 4 and 5 in the same way. Tabulate the results and complete the check graph of y against x as usual.

73

Here are the results

x	y	y'	y''
0	1.0	0	1.0
0.2	1.02	0.2	1.06
0.4	1.0812	0.412	1.246
0.6	1.188 52	0.661 2	1.585 24
0.8	1.352 465	0.978 248	2.135 063
1.0	1.590 816	1.405 261	2.996 077

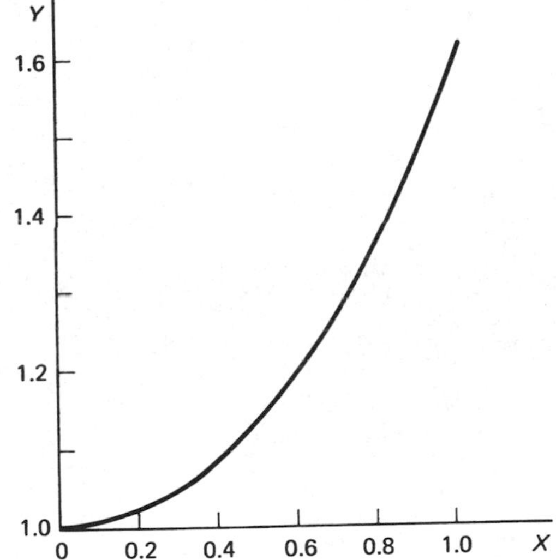

For interest, we can compare the function values obtained, with the real values.

x	y	Actual y	Error
0	1.0	1.0	0
0.2	1.02	1.020 201	0.000 201
0.4	1.081 2	1.083 287	0.002 087
0.6	1.188 52	1.197 217	0.008 697
0.8	1.352 465	1.377 128	0.024 663
1.0	1.590 816	1.648 721	0.057 905

As we might expect with this simple method, the error increases rapidly as we progress from the initial values. The main cause of the errors is

. . . .

74

the truncation of the Taylor's series on which the method is based.

A greater degree of accuracy can be obtained by using the Runge–Kutta method for second-order differential equations, which is an extension of the method we have already used for first-order equations. As before, more intermediate calculations are required, but the reliability of results reflects the extra work involved.

Runge–Kutta method for second-order differential equations

Starting with the given equation $y'' = f(x, y, y')$ and initial conditions that at $x = x_0$, $y = y_0$ and $y' = (y')_0$, we can obtain the value of y_1 at $x_1 = x_0 + h$ as follows.

(a) We evaluate

$$k_1 = \tfrac{1}{2}h^2 f\{x_0,\ y_0,\ (y')_0\} = \tfrac{1}{2}h^2(y'')_0$$

$$k_2 = \tfrac{1}{2}h^2 f\left\{x_0 + \tfrac{1}{2}h,\ y_0 + \tfrac{1}{2}h(y')_0 + \tfrac{1}{4}k_1,\ (y')_0 + \frac{k_1}{h}\right\}$$

$$k_3 = \tfrac{1}{2}h^2 f\left\{x_0 + \tfrac{1}{2}h,\ y_0 + \tfrac{1}{2}h(y')_0 + \tfrac{1}{4}k_1,\ (y')_0 + \frac{k_2}{h}\right\}$$

$$k_4 = \tfrac{1}{2}h^2 f\left\{x_0 + h,\ y_0 + h(y')_0 + k_3,\ (y')_0 + \frac{2k_3}{h}\right\}$$

(b) From these results, we then determine

$$P = \tfrac{1}{3}\{k_1 + k_2 + k_3\}$$

$$Q = \tfrac{1}{3}\{k_1 + 2k_2 + 2k_3 + k_4\}$$

(c) Finally, we have

$$x_1 = x_0 + h$$

$$y_1 = y_0 + h(y')_0 + P$$

$$(y')_1 = (y')_0 + \frac{Q}{h}$$

It is not as complicated as it looks at first sight. Copy down this list of relationships for reference when dealing with some examples that follow,

Then move on.

75

Note the following

1. Four evaluations for k are required to determine a single new point on the solution curve.
2. The method is self-starting in that no preliminary calculations are required. The equation and initial conditions are sufficient to provide the next point on the curve.
3. As with the Runge–Kutta method for first-order equations, the method contains no self-correcting element or indication of any error involved. Building up a graph of the calculated results is therefore highly advisable.

Example 1 Solve the equation $y'' = xy' + y$ for $x = 0(0.2)1.0$ given that at $x = 0$, $y = 1$ and $y' = 0$.

This is the same problem as the one we have just solved by the previous method. In due course, we can compare results.
From the initial conditions and the equation $(y'')_0 = x_0(y')_0 + y_0$

$$\therefore\ (y'')_0 = (0)(0) + 1 \qquad \therefore\ (y'')_0 = 1$$

Stage 1 $x_0 = 0$; $y_0 = 1$; $(y')_0 = 0$; $(y'')_0 = 1$; $h = 0.2$; $x_1 = 0.2$.

Now we have four calculations for k.

$$k_1 = \tfrac{1}{2}h^2 f\{x_0,\ y_0,\ (y')_0\} = \tfrac{1}{2}h^2\{x_0(y')_0 + y_0\} = \tfrac{1}{2}h^2(y'')_0$$

$$= \tfrac{1}{2}(0.2)^2(1) = 0.02 \qquad \therefore\ \underline{k_1 = 0.02}$$

$$k_2 = \tfrac{1}{2}h^2 f\left\{x_0 + \tfrac{1}{2}h,\ y_0 + \tfrac{1}{2}h(y')_0 + \tfrac{1}{4}k_1,\ (y')_0 + \frac{k_1}{h}\right\}$$

The function f is the right-hand side of the given equation with

x replaced by $x_0 + \tfrac{1}{2}h$, i.e. $0 + 0.1 = 0.1$

y replaced by $y_0 + \tfrac{1}{2}h(y')_0 + \tfrac{1}{4}k_1$, i.e. $1 + 0 + \tfrac{1}{4}(0.02) = 1.005$

and y' replaced by $(y')_0 + \dfrac{k_1}{h}$, i.e. $0 + \dfrac{0.02}{0.2} = 0.1$

$$\therefore\ k_2 = \ldots\ldots\ldots\ldots\ldots$$

76

$$\boxed{k_2 = 0.020\,3}$$

for $k_2 = \dfrac{0.04}{2}\{(0.1)(0.1) + 1.005\} = 0.020\,3 \qquad \underline{k_2 = 0.020\,3}$

Also $\quad k_3 = \frac{1}{2}h^2 f\left\{x_0 + \frac{1}{2}h,\ y_0 + \frac{1}{2}h(y')_0 + \frac{1}{4}k_1,\ (y')_0 + \dfrac{k_2}{h}\right\}$

Note that the values for the revised x and y terms are the same as for k_2, but that y' is replaced by $(y')_0 + \dfrac{k_2}{h} = 0 + \dfrac{0.020\,3}{0.2} = 0.101\,5$

$$\therefore k_3 = \ldots\ldots\ldots$$

77

$$\boxed{k_3 = 0.020\,303}$$

for $\quad k_3 = \dfrac{0.04}{2}\{(0.1)(0.101\,5) + 1.005\} = 0.020\,303 \qquad \underline{k_3 = 0.020\,303}$

Finally $\quad k_4 = \frac{1}{2}h^2 f\left\{x_0 + h,\ y_0 + h(y')_0 + k_3,\ (y')_0 + \dfrac{2k_3}{h}\right\}$

$$x_0 + h = 0 + 0.2 = 0.2$$

$$y_0 + h(y')_0 + k_3 = 1 + (0.2)(0) + 0.020\,303 = 1.020\,303$$

$$(y')_0 + \frac{2k_3}{h} = 0 + \frac{2(0.020\,303)}{0.2} = 0.203\,03$$

$$\therefore k_4 = \ldots\ldots\ldots\ldots\ldots$$

78

$$k_4 = 0.021\ 218$$

Collecting the results

$k_1 = 0.02$

$k_2 = 0.020\ 3$

$k_3 = 0.020\ 303$

$k_4 = 0.021\ 218$

Now $P = \frac{1}{3}(k_1 + k_2 + k_3)$

$= \dots\dots\dots$

and $Q = \frac{1}{3}(k_1 + 2k_2 + 2k_3 + k_4)$

$= \dots\dots\dots$

79

$$P = 0.020\ 201; \quad Q = 0.040\ 808$$

Then $x_1 = x_0 + h = \quad \dots\dots\dots$

$y_1 = y_0 + h(y')_0 + P = \quad \dots\dots\dots$

$(y')_1 = (y')_0 + \dfrac{Q}{h} = \quad \dots\dots\dots$

80

$$x_1 = 0.2; \quad y_1 = 1.020\ 201; \quad (y')_1 = 0.204\ 040$$

Finally $y'' = xy' + y$ $\qquad \therefore (y'')_1 = x_1(y')_1 + y_1$

$$(y'')_1 = \dots\dots\dots$$

81

$$(y'')_1 = 1.061\ 009$$

These results now complete the second line in the table.

x	y	y'	y''
0	1.0	0	1.0
0.2	1.020 201	0.204 040	1.061 009

These now become the starter values for the next stage.

Stage 2 $\quad x_0 = 0.2;\quad y_0 = 1.020\ 201;\quad (y')_0 = 0.204\ 040;$
$(y'')_0 = 1.061\ 009.$

$$h = 0.2 \qquad y'' = xy' + y \qquad x_1 = x_0 + h = 0.2 + 0.2 = 0.4$$

$$k_1 = \tfrac{1}{2}h^2 f\{x_0,\ y_0,\ (y')_0\} = \tfrac{1}{2}h^2 (y'')_0 = \tfrac{1}{2}(0.04)(1.061\ 009) = 0.021\ 220$$

$$\underline{k_1 = 0.021\ 220}$$

$$k_2 = \tfrac{1}{2}h_2 f\left\{x_0 + \tfrac{1}{2}h,\ y_0 + \tfrac{1}{2}h(y')_0 + \tfrac{1}{4}k_1,\ (y')_0 + \frac{k_1}{h}\right\}$$

$$x_0 + \tfrac{1}{2}h = 0.2 + 0.1 = 0.3$$

$$y_0 + \tfrac{1}{2}h(y')_0 + \tfrac{1}{4}k_1 = 1.020\ 201 + (0.1)0.204\ 040 + \tfrac{1}{4}(0.021\ 220)$$
$$= 1.045\ 910$$

$$(y')_0 + \frac{k_1}{h} = 0.204\ 040 + \frac{0.021\ 220}{0.2} = 0.310\ 140$$

$$\therefore\ k_2 = \frac{0.04}{2}\{(0.3)0.310\ 140 + 1.045\ 910\} = 0.022\ 779$$

$$\underline{k_2 = 0.022\ 779}$$

In the same way

$k_3 = \dots\dots\dots$
$k_4 = \dots\dots\dots$

82

$$\boxed{\begin{aligned} k_3 &= 0.022\,826 \\ k_4 &= 0.025\,135 \end{aligned}}$$

for $k_3 = \frac{1}{2}h^2 f\left\{x_0 + \frac{1}{2}h,\ y_0 + \frac{1}{2}h(y')_0 + \frac{1}{4}k_1,\ (y')_0 + \frac{k_2}{h}\right\}$

$$(y')_0 + \frac{k_2}{h} = 0.204\,040 + \frac{0.022\,779}{0.2} = 0.317\,935$$

$$\therefore\ k_3 = 0.02\{(0.3)0.317\,935 + 1.045\,910\} = \underline{0.022\,826}$$

and $k_4 = \frac{1}{2}h^2 f\left\{x_0 + h,\ y_0 + h(y')_0 + k_3,\ (y')_0 + \frac{2k_3}{h}\right\}$

$$\begin{aligned} x_0 + h &= 0.2 + 0.2 = 0.4 \\ y_0 + h(y')_0 + k_3 &= 1.020\,201 + (0.2)0.204\,040 + 0.022\,826 \\ &= 1.083\,835 \\ (y')_0 + \frac{2k_3}{h} &= 0.204\,040 + \frac{2(0.022\,826)}{0.2} = 0.432\,300 \end{aligned}$$

$$\therefore\ k_4 = 0.02\{(0.4)0.432\,300 + 1.083\,835\} = \underline{0.025\,135}$$

So

$$\begin{aligned} k_1 &= 0.021\,220 \\ k_2 &= 0.022\,779 \\ k_3 &= 0.022\,826 \\ k_4 &= 0.025\,135 \end{aligned}$$

$$\therefore\ x_1 = \dots\dots\dots;\quad y_1 = \dots\dots\dots;\quad (y')_1 = \dots\dots\dots;$$

$$(y'')_1 = \dots\dots\dots$$

83

$x_1 = 0.4; \quad y_1 = 1.083\,284; \quad (y')_1 = 0.433\,316; \quad (y'')_1 = 1.256\,610$

since $\quad P = \frac{1}{3}(k_1 + k_2 + k_3) = 0.022\,275$

and $\quad Q = \frac{1}{3}(k_1 + 2k_2 + 2k_3 + k_4) = 0.045\,855$

with $\quad x_1 = x_0 + h \qquad y_1 = y_0 + h(y')_0 + P$

$$(y')_1 = (y')_0 + \frac{Q}{h} \qquad (y'')_1 = x_1(y')_1 + y_1.$$

Now for stage 3. As usual, take the calculated values above as the new starter values. *Work right through it and then check your results with the next frame.*

84

$x_0 = 0.4; \quad y_0 = 1.083\,284; \quad (y')_0 = 0.433\,316; \quad (y'')_0 = 1.256\,610$

$k_1 = 0.025\,132$

$k_2 = 0.028\,248 \qquad P = 0.027\,261$

$k_3 = 0.028\,404 \qquad Q = 0.057\,003$

$k_4 = 0.032\,575$

$x_1 = 0.6; \quad y_1 = 1.197\,208; \quad (y')_1 = 0.718\,333; \quad (y'')_1 = 1.628\,208$

The table of results so far is as follows.

x	y	y'	y''
0	1.0	0	1.0
0.2	1.020 201	0.204 040	1.061 009
0.4	1.083 284	0.433 316	1.256 610
0.6	1.197 208	0.718 333	1.628 208

The process repeats. Complete stages 4 and 5 on your own and so complete the table for $x = 0(0.2)1.0$.

85

x	y	y'	y''
0.1	1.0	0	1.0
0.2	1.020 201	0.204 040	1.061 009
0.4	1.083 284	0.433 316	1.256 610
0.6	1.197 208	0.718 333	1.628 208
0.8	1.377 107	1.101 706	2.258 472
1.0	1.648 677	1.648 721	3.297 398

Comparing the calculated values of y with the real values, we have

x	y	Actual y	Absolute error
0	1.0	1.0	0
0.2	1.020 201	1.020 201	0.000 000
0.4	1.083 284	1.083 287	0.000 003
0.6	1.197 208	1.197 217	0.000 009
0.8	1.377 107	1.377 128	0.000 021
1.0	1.648 677	1.648 721	0.000 144

Notice the great increase in the accuracy of the results compared with those obtained by the previous method.

86

Now here is one for you to do entirely on your own. The method is exactly the same as before and there are no snags.

Example 2 Solve the equation $y'' = x - y^2$ for $x = 0(0.2)0.6$ given that at $x = 0$, $y = 0$ and $y' = 0$.

Tabulate the results as usual.

When you have finished, check the results with the next frame.

87

x	y	y'	y''
0	0	0	0
0.2	0.001 333	0.020 000	0.199 998
0.4	0.010 666	0.080 000	0.399 886
0.6	0.035 993	0.179 890	0.598 705

You can check the intermediate results from the details below.

x	k_1	k_2	k_3	k_4	P	Q
0	0	0.002 00	0.002 00	0.004 00	0.001 33	0.004 00
0.2	0.004 00	0.006 00	0.006 00	0.008 00	0.005 33	0.012 00
0.4	0.008 00	0.009 99	0.009 99	0.011 97	0.009 33	0.019 98

The required solution is therefore

x	0	0.2	0.4	0.6
y	0	0.001 333	0.010 666	0.035 993

And that is it. There are many other more sophisticated methods for the solution of differential equations by numerical methods. A detailed study could form a course in itself.

The methods we have used give an introduction to the processes and are practical in application.

The Runge–Kutta methods can readily be programmed for computer use.

The revision summary now follows as usual. Check down it carefully and refer back into the programme for any points that may need further brushing up. Then you will be ready for the Test Exercise.

REVISION SUMMARY

88

1. *Taylor's series*

$$f(a+h) = f(a) + hf'(a) + \frac{h}{2!}f''(a) + \frac{h}{3!}f'''(a) + \ldots$$

2. *Solution of first-order differential equations*

Equation $y' = f(x, y)$ with $y = y_0$ at $x = x_0$ for $x_0(h)x_n$.

(a) *Euler's method*

$$y_1 = y_0 + h(y')_0$$

(b) *Euler–Cauchy method*

$$\begin{aligned} x_1 &= x_0 + h \\ \bar{\bar{y}}_1 &= y_0 + h(y')_0 \\ y_1 &= y_0 + \tfrac{1}{2}h\{(y')_0 + f(x_1, \bar{\bar{y}}_1)\} \\ (y')_1 &= f(x_1,\ y_1). \end{aligned}$$

(c) *Runge–Kutta method*

$$\begin{aligned} x_1 &= x_0 + h \\ k_1 &= hf(x_0,\ y_0) = h(y')_0 \\ k_2 &= hf(x_0 + \tfrac{1}{2}h,\ y_0 + \tfrac{1}{2}k_1) \\ k_3 &= hf(x_0 + \tfrac{1}{2}h,\ y_0 + \tfrac{1}{2}k_2) \\ k_4 &= hf(x_0 + h,\ y_0 + k_3) \\ \Delta y_0 &= \tfrac{1}{6}(k_1 + 2k_2 + 2k_3 + k_4) \\ y_1 &= y_0 + \Delta y_0 \\ (y')_1 &= f(x_1,\ y_1). \end{aligned}$$

3. *Solution of second-order differential equations*

Equation $y'' = f(x, y, y')$ with $y = y_0$ and $y' = (y')_0$ at $x = x_0$ for $x = x_0(h)x_n$.

(a) *Euler's second-order method*

$$\begin{aligned} y_1 &= y_0 + h(y')_0 + \frac{h^2}{2!}(y'')_0 \\ (y')_1 &= (y')_0 + h(y'')_0 \end{aligned}$$

(b) *Runge–Kutta method*

$$x_1 = x_0 + h$$
$$k_1 = \tfrac{1}{2}h^2 f\{x_0, y_0, (y')_0\} = \tfrac{1}{2}h^2 (y'')_0$$
$$k_2 = \tfrac{1}{2}h^2 f\left\{x_0 + \tfrac{1}{2}h,\ y_0 + \tfrac{1}{2}h(y')_0 + \tfrac{1}{4}k_1,\ (y')_0 + \frac{k_1}{h}\right\}$$
$$k_3 = \tfrac{1}{2}h^2 f\left\{x_0 + \tfrac{1}{2}h,\ y_0 + \tfrac{1}{2}h(y')_0 + \tfrac{1}{4}k_1,\ (y')_0 + \frac{k_2}{h}\right\}$$
$$k_4 = \tfrac{1}{2}h^2 f\left\{x_0 + h,\ y_0 + h(y')_0 + k_3,\ (y')_0 + \frac{2k_3}{h}\right\}$$
$$P = \tfrac{1}{3}(k_1 + k_2 + k_3)$$
$$Q = \tfrac{1}{3}(k_1 + 2k_2 + 2k_3 + k_4)$$
$$y_1 = y_0 + h(y')_0 + P$$
$$(y')_1 = (y')_0 + \frac{Q}{h}$$
$$(y'')_1 = f\{x_1, y_1, (y')_1\}.$$

---

89 TEST EXERCISE VI

1. Apply Euler's method to solve the equation

$$\frac{dy}{dx} = 1 + xy \quad \text{for} \quad x = 0(0.1)0.5$$

given that at $x = 0$, $y = 1$.

2. The equation $\frac{dy}{dx} = x^2 - 2y$ is subject to the initial condition $y = 0$ at $x = 1$. Use the Euler–Cauchy method to obtain function values for $x = 1.0(0.2)2.0$.

3. Using the Runge–Kutta method, solve the equation

$$\frac{dy}{dx} = 1 + y - x \quad \text{for} \quad x = 0(0.1)0.5$$

given that $y = 1$ when $x = 0$.

4. Apply Euler's second-order method to solve the equation

$$y'' = y - x \quad \text{for} \quad x = 2.0(0.1)2.5$$

given that at $x = 2$, $y = 3$ and $y' = 0$.

5. Use the Runge–Kutta method to solve the equation

$$y'' = (y'/x) + y \quad \text{for} \quad x = 1.0(0.1)1.5$$

given the initial conditions that at $x = 1.0$, $y = 0$ and $y' = 1.0$.

FURTHER PROBLEMS VI

Solve the following differential equations by the methods indicated.

Euler's method

1. $y' = 2x - y$	$x = 0,\ y = 1$	$x = 0(0.2)1.0$
2. $y' = 2x + y^2$	$x = 0,\ y = 1.4$	$x = 0(0.1)0.5$

Euler–Cauchy method

3. $y' = 2 - y/x$	$x = 1,\ y = 2$	$x = 1.0(0.2)2.0$
4. $y' = x^2 - 2x + y$	$x = 0,\ y = 0.5$	$x = 0(0.1)0.5$
5. $y' = (y - x^2)^{\frac{1}{2}}$	$x = 0,\ y = 1$	$x = 0(0.1)0.5$
6. $y' = \dfrac{x + y}{xy}$	$x = 1,\ y = 1$	$x = 1.0(0.1)1.5$
7. $y' = y \sin x + \cos x$	$x = 0,\ y = 0$	$x = 0(0.1)0.5$

Runge Kutta method

8. $y' = 2x - y$	$x = 0,\ y = 1$	$x = 0(0.2)1.0$
9. $y' = x - y^2$	$x = 0,\ y = 1$	$x = 0(0.1)0.5$
10. $y' = y^2 - xy$	$x = 0,\ y = 0.4$	$x = 0(0.2)1.0$
11. $y' = \sqrt{2x + y}$	$x = 1,\ y = 2$	$x = 1.0(0.2)2.0$
12. $y' = 1 - x^3/y$	$x = 0,\ y = 1$	$x = 0(0.2)1.0$
13. $y' = \dfrac{y - x}{y + x}$	$x = 0,\ y = 1$	$x = 0(0.2)1.0$

Euler second-order method

14. $y'' = (x + 1)y' + y$	$x = 0,\ y = 1,\ y' = 1$	$x = 0(0.1)0.5$
15. $y'' = 2(xy' - 4y)$	$x = 0,\ y = 3,\ y' = 0$	$x = 0(0.1)0.5$

Runge–Kutta second-order method

16. $y'' = x - y - xy'$	$x = 0,\ y = 0,\ y' = 1$	$x = 0(0.2)1.0$
17. $y'' = (1 - x)y' - y$	$x = 0,\ y = 1,\ y' = 1$	$x = 0(0.2)1.0$
18. $y'' = 1 + x - y^2$	$x = 0,\ y = 2,\ y' = 1$	$x = 0(0.1)0.5$
19. $y'' = (x + 2)y - 2y'$	$x = 0,\ y = 1,\ y' = 0$	$x = 0(0.2)1.0$
20. $y'' = \dfrac{y - xy'}{x^2}$	$x = 1,\ y = 0,\ y' = 1$	$x = 1.0(0.2)2.0$

Programme 7

Laplace Transforms

PART 1

Prerequisites: Engineering Mathematics (fourth edition)
Programme 15

Introduction

1 The standard methods of solving second-order differential equations with constant coefficients, e.g. $a\dfrac{d^2y}{dx^2}+b\dfrac{dy}{dx}+cy=F(x)$, are either by substitution of an assumed solution or by using operator D methods. In either case, the general solution is first ascertained and the arbitrary constants evaluated by insertion of the initial conditions.

A much neater and less tedious method is by the use of *Laplace transforms*, in which the solution of the differential equation is obtained largely by algebraic processes. Furthermore, the initial conditions are involved from the early stages so that the determination of the particular solution is considerably shortened.

A further important advantage is that the method of Laplace transforms enables us to deal with situations where the function is discontinuous: classical methods necessarily require the function to be continuous.

The use of Laplace transforms is a powerful tool and has applications in numerous fields of technology. Let us see what it is all about.

Laplace transforms

The Laplace transform of a function $F(t)$ is denoted by $\mathscr{L}\{F(t)\}$ and is defined as the integral of $F(t)e^{-st}$ between the limits $t=0$ and $t=\infty$

i.e.
$$\mathscr{L}\{F(t)\}=\int_0^\infty F(t)e^{-st}dt$$

The constant parameter s is assumed to be positive and large enough to ensure that the product $F(t)e^{-st}$ converges to zero as $t\to\infty$, for most common functions $F(t)$.

In determining the transform of any function, you will appreciate that the limits are substituted for t, so that the result will be a function of s.

$$\therefore\ \mathscr{L}\{F(t)\}=\int_0^\infty F(t)e^{-st}dt=f(s) \qquad (1)$$

Make a note of this general definition: then we can apply it.

2

So we have $$\mathscr{L}\{F(t)\} = \int_0^\infty F(t)\mathrm{e}^{-st}\mathrm{d}t = f(s).$$

Example 1 To find the Laplace transform of $F(t) = a$ (constant).

$$\mathscr{L}\{a\} = \int_0^\infty a\mathrm{e}^{-st}\mathrm{d}t = a\left[\frac{\mathrm{e}^{-st}}{-s}\right]_0^\infty = -\frac{a}{s}[\mathrm{e}^{-st}]_0^\infty$$
$$= -\frac{a}{s}\{0-1\} = \frac{a}{s}.$$

$$\therefore\ \mathscr{L}\{a\} = \frac{a}{s} \qquad (2)$$

Example 2 To find the Laplace transform of $F(t) = \mathrm{e}^{at}$ (a constant). As with all cases, we multiply the function of t by e^{-st} and integrate between $t = 0$ and $t = \infty$.

$$\therefore\ \mathscr{L}\{\mathrm{e}^{at}\} = \int_0^\infty \mathrm{e}^{at}\mathrm{e}^{-st}\mathrm{d}t = \int_0^\infty \mathrm{e}^{-(s-a)t}\mathrm{d}t$$

$$= \ldots\ldots\ldots$$

Finish it off.

3

$$\boxed{\mathscr{L}\{\mathrm{e}^{at}\} = \frac{1}{s-a}}$$

for $$\mathscr{L}\{\mathrm{e}^{at}\} = \int_0^\infty \mathrm{e}^{at}\mathrm{e}^{-st}\mathrm{d}t = \int_0^\infty \mathrm{e}^{-(s-a)t}\mathrm{d}t = \left[\frac{\mathrm{e}^{-(s-a)t}}{-(s-a)}\right]_0^\infty$$
$$= -\frac{1}{s-a}\{0-1\} = \frac{1}{s-a}$$
$$\therefore\ \mathscr{L}\{\mathrm{e}^{at}\} = \frac{1}{s-a} \qquad (3)$$

So we already have two standard transforms

$$\mathscr{L}\{a\} = \frac{a}{s} \quad \text{and} \quad \mathscr{L}\{\mathrm{e}^{at}\} = \frac{1}{s-a}$$

$$\therefore\ \mathscr{L}\{4\} = \ldots\ldots\ldots; \qquad \mathscr{L}\{e^{4t}\} = \ldots\ldots\ldots$$

$$\mathscr{L}\{-5\} = \ldots\ldots\ldots; \qquad \mathscr{L}\{e^{-2t}\} = \ldots\ldots\ldots$$

4

$$\mathscr{L}\{4\} = \frac{4}{s}; \qquad \mathscr{L}\{e^{4t}\} = \frac{1}{s-4}$$

$$\mathscr{L}\{-5\} = -\frac{5}{s}; \qquad \mathscr{L}\{e^{-2t}\} = \frac{1}{s+2}$$

Note that, as we said earlier, the Laplace transform is always a function of s.

Now for some more examples

5

Example 3 To find the Laplace transform of $F(t) = \sin at$.

We could, of course, apply the definition and evaluate

$$\mathscr{L}\{\sin at\} = \int_0^\infty \sin at \cdot e^{-st} dt$$

using integration by parts.

However, it is much shorter if we use the fact that

$$e^{j\theta} = \cos\theta + j\sin\theta$$

so that $\sin\theta$ is the imaginary part of $e^{j\theta}$, written $\mathscr{I}(e^{j\theta})$

The function sin at can therefore be written $\mathscr{I}(e^{jat})$ so that

$$\mathscr{L}\{\sin at\} = \mathscr{L}\{\mathscr{I}(e^{jat})\} = \mathscr{I}\int_0^\infty e^{jat}e^{-st}dt = \mathscr{I}\int_0^\infty e^{-(s-ja)t}dt$$

$$= \mathscr{I}\left\{\left[\frac{e^{-(s-ja)t)}}{-(s-ja)}\right]_o^\infty\right\} = \mathscr{I}\left\{-\frac{1}{(s-ja)}[0-1]\right\}$$

$$= \mathscr{I}\left\{\frac{1}{s-ja}\right\}$$

We can rationalise the denominator by multiplying top and bottom by ....

6

$$\boxed{s + \mathrm{j}a}$$

$$\therefore \ \mathscr{L}\{\sin at\} = \mathscr{I}\left\{\frac{s + \mathrm{j}a}{s^2 + a^2}\right\} = \frac{a}{s^2 + a^2}$$

$$\therefore \ \mathscr{L}\{\sin at\} = \frac{a}{s^2 + a^2} \qquad (4)$$

We can use the same method to determine $\mathscr{L}\{\cos at\}$ since $\cos at$ is the real part of $\mathrm{e}^{\mathrm{j}at}$, written $\mathscr{R}(\mathrm{e}^{\mathrm{j}at})$.

Then $\mathscr{L}\{\cos at\} = \ldots\ldots\ldots$

7

$$\boxed{\mathscr{L}\{\cos at\} = \frac{s}{s^2 + a^2}} \qquad (5)$$

$$\text{for } \mathscr{L}\{\cos at\} = \mathscr{R}\left\{\frac{s + \mathrm{j}a}{s^2 + a^2}\right\} = \frac{s}{s^2 + a^2}$$

Recapping then $\mathscr{L}\{1\} = \ldots\ldots\ldots$; $\mathscr{L}\{\mathrm{e}^{3t}\} = \ldots\ldots\ldots$

$\mathscr{L}\{\sin 2t\} = \ldots\ldots\ldots$; $\mathscr{L}\{\cos 4t\} = \ldots\ldots\ldots$

8

$$\mathcal{L}\{1\} = \frac{1}{s}; \qquad \mathcal{L}\{e^{3t}\} = \frac{1}{s-3}$$

$$\mathcal{L}\{\sin 2t\} = \frac{2}{s^2+4}; \quad \mathcal{L}\{\cos 4t\} = \frac{s}{s^2+16}$$

Example 4 To find the transform of $F(t) = t^n$ where n is a positive integer.

By the definition $\quad \mathcal{L}\{t^n\} = \int_0^\infty t^n e^{-st} dt.$

Integrating by parts

$$\mathcal{L}\{t^n\} = \left[t^n\left(\frac{e^{-st}}{-s}\right)\right]_0^\infty + \frac{n}{s}\int_0^\infty e^{-st} t^{n-1} dt$$

$$= -\frac{1}{s}\left[t^n e^{-st}\right]_0^\infty + \frac{n}{s}\int_0^\infty t^{n-1} e^{-st} dt$$

We said earlier that in a product such as $t^n e^{-st}$ the numerical value of s is large enough to make the product converge to zero as $t \to \infty$

$$\therefore \left[t^n e^{-st}\right]_0^\infty = 0 - 0 = 0$$

$$\therefore \mathcal{L}\{t^n\} = \frac{n}{s}\int_0^\infty t^{n-1} e^{-st} dt \tag{6}$$

You will notice that $\int_0^\infty t^{n-1} e^{-st} dt$ is identical to $\int_0^\infty t^n e^{-st} dt$ except that n is replaced by $(n-1)$.

$$\therefore \text{If } I_n = \int_0^\infty t^n e^{-st} dt, \text{ then } I_{n-1} = \int_0^\infty t^{n-1} e^{-st} dt$$

and the result (6) becomes $\quad \underline{I_n = \frac{n}{s} \cdot I_{n-1}} \qquad (7)$

This is a reduction formula and, if we now replace n by $(n-1)$ we get

$$I_{n-1} = \ldots\ldots\ldots$$

9

$$I_{n-1} = \frac{n-1}{s} \cdot I_{n-2}$$

If we replace n by $(n-1)$ again in this last result, we have

$$I_{n-2} = \frac{n-2}{s} \cdot I_{n-3}$$

$$\text{So } I_n = \int_0^\infty t^n e^{-st} dt = \frac{n}{s} \cdot I_{n-1}$$

$$= \frac{n}{s} \cdot \frac{n-1}{s} \cdot I_{n-2}$$

$$= \frac{n}{s} \cdot \frac{n-1}{s} \cdot \frac{n-2}{s} \cdot I_{n-3} \quad \text{etc.}$$

$= \ldots\ldots\ldots$ (next line)

10

$$I_n = \frac{n}{s} \cdot \frac{n-1}{s} \cdot \frac{n-2}{s} \cdot \frac{n-3}{s} \cdot I_{n-4}$$

So finally, we have

$$I_n = \frac{n}{s} \cdot \frac{n-1}{s} \cdot \frac{n-2}{s} \cdot \frac{n-3}{s} \ldots \frac{n-(n-1)}{s} \cdot I_0$$

But $$I_0 = \mathscr{L}\{t^o\} = \mathscr{L}\{1\} = \frac{1}{s}$$

$$\therefore I_n = \frac{n(n-1)(n-2)(n-3)\ldots(3)(2)(1)}{s^{n+1}} = \frac{n!}{s^{n+1}}$$

$$\therefore \mathscr{L}\{t^n\} = \frac{n!}{s^{n+1}} \tag{8}$$

$$\therefore \mathscr{L}\{t\} = \frac{1}{s^2}; \quad \mathscr{L}\{t^2\} = \frac{2}{s^3}; \quad \mathscr{L}\{t^3\} = \frac{6}{s^4}$$

and, with $n = 0$, since $0! = 1$, the general result includes $\mathscr{L}\{1\} = \frac{1}{s}$ which we have already established.

Example 5 Laplace transforms of $F(t) = \sinh at$ and $F(t) = \cosh at$.

Starting from the exponential definitions of sinh at and cosh at, i.e.

$$\sinh at = \tfrac{1}{2}(e^{at} - e^{-at}) \quad \text{and} \quad \cosh at = \tfrac{1}{2}(e^{at} + e^{-at})$$

we proceed as follows.

(a) $F(t) = \sinh at$. $\quad \mathscr{L}\{\sinh at\} = \int_0^\infty \sinh at\, e^{-st} dt$

$$= \frac{1}{2}\int_0^\infty (e^{at} - e^{-at})e^{-st} dt$$

$$= \frac{1}{2}\int_0^\infty \{e^{-(s-a)t} - e^{-(s+a)t}\} dt = \dots\dots\dots$$

Complete it.

11

$$\boxed{\mathscr{L}\{\sinh at\} = \frac{a}{s^2 - a^2}}$$

for $\frac{1}{2}\int_0^\infty \{e^{-(s-a)t} - e^{-(s+a)t}\} dt = \frac{1}{2}\left[\frac{e^{-(s-a)t}}{-(s-a)} - \frac{e^{-(s+a)t}}{-(s+a)}\right]_0^\infty$

$$= \frac{1}{2}\left\{\frac{1}{s-a} - \frac{1}{s+a}\right\} = \frac{a}{s^2 - a^2}$$

$$\therefore \mathscr{L}\{\sinh at\} = \frac{a}{s^2 - a^2} \qquad (9)$$

(b) $F(t) = \cosh at$. Proceeding in the same way

$$\mathscr{L}\{\cosh at\} = \dots\dots\dots$$

$$\mathcal{L}\{\cosh at\} = \frac{s}{s^2 - a^2}$$

$$\mathcal{L}\{\cosh at\} = \frac{1}{2}\int_0^\infty (e^{at} + e^{-at})e^{-st}dt = \frac{1}{2}\int_0^\infty \{e^{-(s-a)t} + e^{-(s+a)t}\}dt$$

$$= \frac{1}{2}\left[\frac{e^{-(s-a)t}}{-(s-a)} + \frac{e^{-(s+a)t}}{-(s+a)}\right]_0^\infty = \frac{1}{2}\left\{\frac{1}{s-a} + \frac{1}{s+a}\right\}$$

$$= \frac{1}{2}\left\{\frac{2s}{s^2 - a^2}\right\} = \frac{s}{s^2 - a^2}$$

$$\therefore \mathcal{L}\{\cosh at\} = \frac{s}{s^2 - a^2} \qquad (10)$$

So we have amassed several standard results

$$\mathcal{L}\{a\} = \frac{a}{s}; \qquad \mathcal{L}\{e^{at}\} = \frac{1}{s-a}; \qquad \mathcal{L}\{t^n\} = \frac{n!}{s^{n+1}}$$

$$\mathcal{L}\{\sin at\} = \frac{a}{s^2 + a^2}; \qquad \mathcal{L}\{\cos at\} = \frac{s}{s^2 + a^2}$$

$$\mathcal{L}\{\sinh at\} = \frac{a}{s^2 - a^2}; \qquad \mathcal{L}\{\cosh at\} = \frac{s}{s^2 - a^2}$$

Make a note of this list if you have not already done so: it forms the basis of much that is to follow.

13

We can, of course, combine these transforms by adding or subtracting as necessary, but they must *not* be multiplied together to form the transform of a product.

Example 6

(a) $\mathscr{L}\{2\sin 3t + \cos 3t\} = 2\mathscr{L}\{\sin 3t\} + \mathscr{L}\{\cos 3t\}$

$$= 2 \cdot \frac{3}{s^2+9} + \frac{s}{s^2+9} = \underline{\frac{s+6}{s^2+9}}$$

(b) $\mathscr{L}\{4e^{2t} + 3\cosh 4t\} = 4\mathscr{L}\{e^{2t}\} + 3\mathscr{L}\{\cosh 4t\}$

$$= 4 \cdot \frac{1}{s-2} + 3 \cdot \frac{s}{s^2-16} = \frac{4}{s-2} + \frac{3s}{s^2-16}$$

$$= \underline{\frac{7s^2-6s-64}{(s-2)(s^2-16)}}$$

So 1. $\mathscr{L}\{2\sin 3t + 4\sinh 3t\} = \ldots\ldots\ldots$

2. $\mathscr{L}\{5e^{4t} + \cosh 2t\} = \ldots\ldots\ldots$

3. $\mathscr{L}\{t^3 + 2t^2 - 4t + 1\} = \ldots\ldots\ldots$

14

1. $\dfrac{18(s^2+3)}{s^4-81}$; 2. $\dfrac{6s^2-4s-20}{(s-4)(s^2-4)}$; 3. $\dfrac{1}{s^4}\{s^3-4s^2+4s+6\}$

The working is straightforward.

1. $\mathscr{L}\{2\sin 3t + 4\sinh 3t\} = 2\cdot\dfrac{3}{s^2+9} + 4\cdot\dfrac{3}{s^2-9}$

$$= \frac{6}{s^2+9} + \frac{12}{s^2-9} = \underline{\frac{18(s^2+3)}{s^4-81}}$$

2. $\mathscr{L}\{5e^{4t} + \cosh 2t\} = \dfrac{5}{s-4} + \dfrac{s}{s^2-4} = \underline{\dfrac{6s^2-4s-20}{(s-4)(s^2-4)}}$

3. $\mathscr{L}\{t^3 + 2t^2 - 4t + 1\} = \dfrac{3!}{s^4} + 2\cdot\dfrac{2!}{s^3} - 4\cdot\dfrac{1!}{s^2} + \dfrac{1}{s}$

$$\underline{= \frac{1}{s^4}\{s^3 - 4s^2 + 4s + 6\}}$$

We have been building up a list of standard transforms of simple functions. Before we leave this part of the work, there are three important and useful theorems which enable us to deal with rather more complicated expressions.

Theorem 1 The First Shift Theorem

We have seen that a transform of $F(t)$ is a function of s only, i.e.

$$\mathscr{L}\{F(t)\} = f(s)$$

The *first shift theorem* states that

if $\quad \mathscr{L}\{F(t)\} = f(s)$

then $\quad \mathscr{L}\{e^{-at}F(t)\} = f(s+a).$

The transform $\mathscr{L}\{e^{-at}F(t)\}$ is thus the same as $\mathscr{L}\{F(t)\}$ with s everywhere in the result replaced by $(s+a)$.

For example $\mathscr{L}\{\sin 2t\} = \dfrac{2}{s^2+4}$

then $\quad \mathscr{L}\{e^{-3t}\sin 2t\} = \dfrac{2}{(s+3)^2+4} = \underline{\dfrac{2}{s^2+6s+13}}$

Similarly, $\mathscr{L}\{t^2\} = \dfrac{2}{s^3} \quad \therefore\ \mathscr{L}\{t^2e^{4t}\} = \ldots\ldots\ldots$

15

$$\boxed{\frac{2}{(s-4)^3}}$$

for $\mathscr{L}\{t^2\} = \dfrac{2}{s^3}$. $\therefore\ \mathscr{L}\{t^2e^{4t}\}$ is the same with s replaced by $(s-4)$.

$$\therefore\ \mathscr{L}\{t^2e^{4t}\} = \frac{2}{(s-4)^3}$$

Here is a short exercise by way of practice.

Exercise Determine the following.

1. $\mathscr{L}\{e^{-2t}\cosh 3t\}$
2. $\mathscr{L}\{2e^{3t}\sin 3t\}$
3. $\mathscr{L}\{4t\,e^{-t}\}$
4. $\mathscr{L}\{e^{2t}\cos t\}$
5. $\mathscr{L}\{e^{3t}\sinh 2t\}$
6. $\mathscr{L}\{t^3e^{-4t}\}$

Complete all six and then check the results with the next frame.

16

Here they are.

1. $\mathscr{L}\{\cosh 3t\} = \dfrac{s}{s^2 - 9}$. $\quad \therefore\ \mathscr{L}\{e^{-2t}\cosh 3t\} = \dfrac{s+2}{(s+2)^2 - 9}$

$$= \underline{\frac{s+2}{s^2 + 4s - 5}}$$

2. $\mathscr{L}\{\sin 3t\} = \dfrac{3}{s^2 + 9}$. $\quad \therefore\ \mathscr{L}\{2e^{3t}\sin 3t\} = \dfrac{6}{(s-3)^2 + 9}$

$$= \underline{\frac{6}{s^2 - 6s + 18}}$$

3. $\mathscr{L}\{4t\} = 4 \cdot \dfrac{1}{s^2}$. $\quad \therefore\ \mathscr{L}\{4t\,e^{-t}\} = \underline{\dfrac{4}{(s+1)^2}}$

4. $\mathscr{L}\{\cos t\} = \dfrac{s}{s^2 + 1}$. $\quad \therefore\ \mathscr{L}\{e^{2t}\cos t\} = \dfrac{s-2}{(s-2)^2 + 1}$

$$= \underline{\frac{s-2}{s^2 - 4s + 5}}$$

5. $\mathscr{L}\{\sinh 2t\} = \dfrac{2}{s^2 - 4}$. $\quad \therefore\ \mathscr{L}\{e^{3t}\sinh 2t\} = \dfrac{2}{(s-3)^2 - 4}$

$$= \underline{\frac{2}{s^2 - 6s + 5}}$$

6. $\mathscr{L}\{t^3\} = \dfrac{3!}{s^4}$. $\quad \therefore\ \mathscr{L}\{t^3 e^{-4t}\} = \underline{\dfrac{6}{(s+4)^4}}$

Now let us deal with the next theorem.

17 Theorem 2 Multiplying by t

Theorem 2 states that, if $\mathscr{L}\{F(t)\} = f(s)$

then $$\mathscr{L}\{tF(t)\} = -\frac{\mathrm{d}}{\mathrm{d}s}\{f(s)\}.$$

For example, $\mathscr{L}\{\sin 2t\} = \dfrac{2}{s^2+4}$

$$\therefore \quad \mathscr{L}\{t\sin 2t\} = -\frac{\mathrm{d}}{\mathrm{d}s}\left(\frac{2}{s^2+4}\right) = \frac{4s}{(s^2+4)^2}$$

and similarly, $\mathscr{L}\{t\cosh 3t\} = \ldots\ldots\ldots$

18

$$\boxed{\frac{s^2+9}{(s^2-9)^2}}$$

for $\mathscr{L}\{t\cosh 3t\} = -\dfrac{\mathrm{d}}{\mathrm{d}s}\left(\dfrac{s}{s^2-9}\right) = -\dfrac{(s^2-9)-s(2s)}{(s^2-9)^2} = \dfrac{s^2+9}{(s^2-9)^2}$

We could, if necessary, take this a stage further and find $\mathscr{L}\{t^2\cosh 3t\}$

$$\mathscr{L}\{t^2\cosh 3t\} = \mathscr{L}\{t(t\cosh 3t)\} = -\frac{\mathrm{d}}{\mathrm{d}s}\left\{\frac{s^2+9}{(s^2-9)^2}\right\}$$

$$= \frac{2s(s^2+27)}{(s^2-9)^3}$$

Likewise, starting with $\mathscr{L}\{\sin 4t\} = \dfrac{4}{s^2+16}$

$\mathscr{L}\{t\sin 4t\} = \ldots\ldots\ldots$ and $\mathscr{L}\{t^2\sin 4t\} = \ldots\ldots\ldots$

19

$$\frac{8s}{(s^2+16)^2};\ \frac{8(3s^2-16)}{(s^2+16)^3}$$

applying $\mathscr{L}\{tF(t)\} = -\frac{\mathrm{d}}{\mathrm{d}s}\{f(s)\}$ in each case.

Theorem 2 obviously extends the range of functions that we can deal with.

So, in general, if $\mathscr{L}\{F(t)\} = f(s)$, then

$$\mathscr{L}\{t^n F(t)\} = (-1)^n \frac{\mathrm{d}^n}{\mathrm{d}s^n}\{f(s)\}$$

Make a note of this in your record book.

20

Theorem 3 Dividing by *t*

If $\mathscr{L}\{F(t)\} = f(s)$, then $\mathscr{L}\left\{\frac{F(t)}{t}\right\} = \int_s^\infty f(s)\,\mathrm{d}s$.

This rule is somewhat restricted in use, since it is applicable only if the limit of $\left\{\frac{F(t)}{t}\right\}$ as $t \to 0$, exists. In indeterminate cases, we use l'Hopital's rule to find out.

Example 1 Determine $\mathscr{L}\left\{\frac{\sin at}{t}\right\}$.

First we test $\lim_{t\to 0}\left\{\frac{\sin at}{t}\right\} = \frac{0}{0} = ?$

By l'Hopital's rule, we differentiate top and bottom separately and substitute $t = 0$ in the result to ascertain the limit of the new function.

$$\lim_{t\to 0}\left\{\frac{\sin at}{t}\right\} = \lim_{t\to 0}\left\{\frac{a\cos at}{1}\right\} = \frac{a}{1},$$ i.e. the limit exists.

The theorem can therefore be applied.

So $\mathscr{L}\{\sin at\} = \dfrac{a}{s^2 + a^2}$

$$\therefore \mathscr{L}\left\{\frac{\sin at}{t}\right\} = \int_s^\infty \frac{a}{s^2 + a^2}\,\mathrm{d}s = \left[\arctan\left(\frac{s}{a}\right)\right]_s^\infty$$

$$= \frac{\pi}{2} - \arctan\left(\frac{s}{a}\right) = \underline{\arctan\frac{a}{s}}$$

Example 2 Determine $\mathscr{L}\left\{\dfrac{1 - \cos 2t}{t}\right\}$.

First we test whether $\lim\limits_{t\to 0}\left\{\dfrac{1 - \cos 2t}{t}\right\}$ exists. Result?

21

the limit exists

$$\lim_{t\to 0}\left\{\frac{1 - \cos 2t}{t}\right\} = \frac{1 - 1}{0} = \frac{0}{0} = ? \;\therefore \text{ Apply l'Hopital's rule.}$$

$$\lim_{t\to 0}\left\{\frac{1 - \cos 2t}{t}\right\} = \lim_{t\to 0}\left\{\frac{2\sin 2t}{1}\right\} = \frac{0}{1} = 0. \;\therefore \text{ limit exists.}$$

$$\mathscr{L}\{1 - \cos 2t\} = \frac{1}{s} - \frac{s}{s^2 + 4}$$

Then, by Theorem 3,

$$\mathscr{L}\left\{\frac{1 - \cos 2t}{t}\right\} = \int_s^\infty \left\{\frac{1}{s} - \frac{s}{s^2 + 4}\right\}\mathrm{d}s$$

$$= \left[\ln s - \frac{1}{2}\ln(s^2 + 4)\right]_s^\infty = \frac{1}{2}\left[\ln\left(\frac{s^2}{s^2 + 4}\right)\right]_s^\infty$$

When $s \to \infty$, $\quad \ln\left(\dfrac{s^2}{s^2 + 4}\right) \to \ln 1 = 0.$

$$\therefore \mathscr{L}\left\{\frac{1 - \cos 2t}{t}\right\} = \ldots\ldots\ldots \quad \text{Complete it.}$$

22

$$\ln\sqrt{\frac{s^2+4}{s^2}}$$

since $\mathscr{L}\left\{\frac{1-\cos 2t}{t}\right\} = -\frac{1}{2}\ln\left(\frac{s^2}{s^2+4}\right) = \ln\left(\frac{s^2}{s^2+4}\right)^{-1/2}$

$$= \ln\sqrt{\frac{s^2+4}{s^2}}$$

Let us pause here a while and take stock, for we have met a number of results important in the future work.

1. *Standard transforms*

$F(t)$	$\mathscr{L}\{F(t)\} = f(s)$	
a	$\frac{a}{s}$	
e^{at}	$\frac{1}{s-a}$	
$\sin at$	$\frac{a}{s^2+a^2}$	
$\cos at$	$\frac{s}{s^2+a^2}$	
$\sinh at$	$\frac{a}{s^2-a^2}$	
$\cosh at$	$\frac{s}{s^2-a^2}$	
t^n	$\frac{n!}{s^{n+1}}$	(n a positive integer)

2. *Theorem 1* The first shift theorem
   If $\mathscr{L}\{F(t)\} = f(s)$, then $\mathscr{L}\{e^{-at}F(t)\} = f(s+a)$.
3. *Theorem 2* Multiplying by t
   If $\mathscr{L}\{F(t)\} = f(s)$, then $\mathscr{L}\{tF(t)\} = -\frac{d}{ds}\{f(s)\}$.
4. *Theorem 3* Dividing by t
   If $\mathscr{L}\{F(t)\} = f(s)$, then $\mathscr{L}\left\{\frac{F(t)}{t}\right\} = \int_s^\infty f(s)ds$.

provided $\lim_{t\to 0}\left\{\frac{F(t)}{t}\right\}$ exists.

Now let us work through a short revision exercise, so move on.

23 *Exercise* Determine the Laplace transforms of the following functions.

1. $\sin 3t$
2. $\cos 2t$
3. e^{4t}
4. $6t^2$
5. $\sinh 3t$
6. $t\cosh 4t$
7. $t^2 - 3t + 4$
8. $\dfrac{e^{3t} - 1}{t}$
9. $e^{3t}\cos 4t$
10. $t^2 \sin t$

Complete the whole set and then check results with the next frame.

24 Here are the results.

1. $\dfrac{3}{s^2+9}$
2. $\dfrac{s}{s^2+4}$
3. $\dfrac{1}{s-4}$
4. $\dfrac{12}{s^3}$
5. $\dfrac{3}{s^2-9}$
6. $\dfrac{s^2+16}{(s^2-16)^2}$
7. $\dfrac{1}{s^3}(4s^2-3s+2)$
8. $\ln\left(\dfrac{s}{s-3}\right)$
9. $\dfrac{s-3}{s^2-6s+25}$
10. $\dfrac{6s^2-2}{(s^2+1)^3}$

It is just a case of applying the standard transforms and the three theorems.

Now on to the next piece of work.

Inverse Transforms

Here we have the reverse process, i.e. given a Laplace transform, we have to find the function of t to which it belongs. 25

For example, we know that $\dfrac{a}{s^2+a^2}$ is the Laplace transform of $\sin at$, so we can now write $\mathscr{L}^{-1}\left\{\dfrac{a}{s^2+a^2}\right\} = \sin at$, the symbol $\mathscr{L}^{-1}$ indicating the inverse transform and *not* a reciprocal.

$$\therefore \text{(a) } \mathscr{L}^{-1}\left\{\frac{1}{s-2}\right\} = \dots\dots\dots; \quad \text{(c) } \mathscr{L}^{-1}\left\{\frac{4}{s}\right\} = \dots\dots\dots$$

$$\text{(b) } \mathscr{L}^{-1}\left\{\frac{2}{s^2+25}\right\} = \dots\dots\dots; \quad \text{(d) } \mathscr{L}^{-1}\left\{\frac{12}{s^2-9}\right\} = \dots\dots\dots$$

26

$$\text{(a) } \mathscr{L}^{-1}\left\{\frac{1}{s-2}\right\} = e^{2t}; \quad \text{(c) } \mathscr{L}^{-1}\left\{\frac{4}{s}\right\} = 4$$

$$\text{(b) } \mathscr{L}^{-1}\left\{\frac{s}{s^2+25}\right\} = \cos 5t; \quad \text{(d) } \mathscr{L}^{-1}\left\{\frac{12}{s^2-9}\right\} = 4\sinh 3t$$

Therefore, given a transform, we can write down the corresponding function in t, provided we can recognise it from our table of transforms.

But what about $\mathscr{L}^{-1}\left\{\dfrac{3s+1}{s^2-s-6}\right\}$? This certainly did not appear in our list of standard transforms.

In considering $\mathscr{L}^{-1}\left\{\dfrac{3s+1}{s^2-s-6}\right\}$, it happens that we can write $\dfrac{3s+1}{s^2-s-6}$ as the sum of two simpler functions $\dfrac{1}{s+2}+\dfrac{2}{s-3}$ which, of course, makes all the difference, since we can now proceed

$$\mathscr{L}^{-1}\left\{\frac{3s+1}{s^2-s-6}\right\} = \mathscr{L}^{-1}\left\{\frac{1}{s+2}+\frac{2}{s-3}\right\}$$

which we immediately recognise as ....

27

$$e^{-2t} + 2e^{3t}$$

The two simpler functions $\dfrac{1}{s-2}$ and $\dfrac{2}{s-3}$ are called the *partial fractions* of $\dfrac{3s+1}{s^2-s-6}$ and the ability to represent a complicated algebraic fraction in terms of its partial fractions is the key to much of this work. Let us take a closer look at the rules.

Rules of partial fractions

1. The numerator must be of lower degree than the denominator. If it is not, then we first divide out.
2. Factorise the denominator into its prime factors. These determine the shapes of the partial fractions.
3. A linear factor $(s+a)$ gives a partial fraction $\dfrac{A}{s+a}$ where A is a constant to be determined.
4. A repeated factor $(s+a)^2$ gives $\dfrac{A}{(s+a)} + \dfrac{B}{(s+a)^2}$.
5. Similarly $(s+a)^3$ gives $\dfrac{A}{(s+a)} + \dfrac{B}{(s+a)^2} + \dfrac{C}{(s+a)^3}$.
6. A quadratic factor (s^2+ps+q) gives $\dfrac{Ps+Q}{s^2+ps+q}$.
7. Repeated quadratic factors $(s^2+ps+q)^2$ gives

$$\frac{Ps+Q}{s^2+ps+q} + \frac{Rs+T}{(s^2+ps+q)^2}.$$

So $\dfrac{s-19}{(s+2)(s-5)}$ has partial fractions of the form ....

28

$$\boxed{\frac{A}{s+2}+\frac{B}{s-5}}$$

and $\dfrac{3s^2-4s+11}{(s+3)(s-2)^2}$ has partial fractions of the form ....

Be careful of the repeated factor.

29

$$\boxed{\frac{A}{s+3}+\frac{B}{s-2}+\frac{C}{(s-2)^2}}$$

Let us work through the various steps with an example.

Example 1 To determine $\mathscr{L}^{-1}\left\{\dfrac{5s+1}{s^2-s-12}\right\}$.

(a) First we check that the numerator is of lower degree than the denominator. In fact, this is so.

(b) Factorise the denominator $\dfrac{5s+1}{s^2-s-12}=\dfrac{5s+1}{(s-4)(s+3)}$

(c) Then the partial fractions are of the form ....

30

$$\boxed{\frac{A}{s-4}+\frac{B}{s+3}}$$

We therefore have an identity

$$\frac{5s+1}{s^2-s-12}\equiv\frac{A}{s-4}+\frac{B}{s+3}$$

which is true for any value of s we care to substitute—and our job now is to find the values of A and B.

If we multiply through by the denominator (s^2-s-12) we have

$$5s+1\equiv A(s+3)+B(s-4)$$

We now substitute convenient values for s

(i) Let $(s-4)=0$, i.e. $s=4$ $\therefore$ $21=A(7)+B(0)$ $\therefore$ $\underline{A=3}$

(ii) Let $(s+3)=0$, i.e. $s=-3$ and we get ....

31

$$\boxed{B = 2}$$

$$\therefore \frac{5s+1}{s^2-s-12} \equiv \frac{3}{s-4} + \frac{2}{s+3}$$

$$\therefore \mathscr{L}^{-1}\left\{\frac{5s+1}{s^2-s-12}\right\}$$

$$= \dots\dots\dots$$

32

$$\boxed{3e^{4t} + 2e^{-3t}}$$

Example 2 Determine $\mathscr{L}^{-1}\left\{\dfrac{9s-8}{s^2-2s}\right\}$.

Working as before, $F(t) = \dots\dots\dots$

33

$$\boxed{4 + 5e^{2t}}$$

for $\quad \mathscr{L}\{F(t)\} = \dfrac{9s-8}{s^2-2s}$.

(a) Numerator of first degree; denominator of second degree. Therefore rule satisfied.

(b) $\dfrac{9s-8}{s(s-2)} \equiv \dfrac{A}{s} + \dfrac{B}{s-2}$.

(c) Multiply by $s(s-2)$. $\quad \therefore 9s - 8 \equiv A(s-2) + B(s)$.

(d) Put $s = 0$. $\quad \therefore -8 = A(-2) + B(0) \quad \therefore \underline{A = 4}$

(e) Put $s - 2 = 0$, i.e. $s = 2$. $\quad \therefore 10 = A(0) + B(2) \quad \therefore B = 5$

$$\therefore F(t) = \mathscr{L}^{-1}\left\{\frac{4}{s} + \frac{5}{s-2}\right\} = \underline{4 + 5e^{2t}}$$

Example 3 Express $f(s) = \dfrac{s^2 - 15s + 41}{(s+2)(s-3)^2}$ in partial fractions and hence determine its inverse transform.

$\dfrac{s^2 - 15s + 41}{(s+2)(s-3)^2}$ has partial fractions of the form $\dots\dots\dots$

34

$$\boxed{\frac{A}{s+2}+\frac{B}{s-3}+\frac{C}{(s-3)^2}}$$

Now we multiply throughout by $(s+2)(s-3)^2$ and get

$$s^2 - 15s + 41 \equiv A(s-3)^2 + B(s+2)(s-3) + C(s+2)$$

Putting $(s-3) = 0$ and then $(s+2) = 0$ we obtain ....

35

$$\boxed{A = 3 \quad \text{and} \quad C = 1}$$

Now that we have run out of 'crafty' substitutions, we equate coefficients of the highest power of s on each side, i.e. the coefficients of s^2. This gives ....

36

$$\boxed{1 = A + B \quad \therefore 1 = 3 + B \quad \therefore \underline{B = -2}}$$

So $\dfrac{s^2 - 15s + 41}{(s+2)(s-3)^2} = \dfrac{3}{s+2} - \dfrac{2}{s-3} + \dfrac{1}{(s-3)^2}$

Now $\mathscr{L}^{-1}\left\{\dfrac{3}{s+2}\right\} = \ldots\ldots\ldots$ and $\mathscr{L}^{-1}\left\{\dfrac{2}{(s-3)}\right\} = \ldots\ldots\ldots$

37

$$\boxed{3e^{-2t} \quad \text{and} \quad 2e^{3t}}$$

But what about $\mathscr{L}^{-1}\left\{\dfrac{1}{(s-3)^2}\right\}$?

We remember that $\mathscr{L}^{-1}\left\{\dfrac{1}{s^2}\right\} = \ldots\ldots\ldots$

38

$$\boxed{t}$$

and that by Theorem 1, if $\mathscr{L}\{F(t)\} = f(s)$ then $\mathscr{L}\{e^{-at}F(t)\} = f(s+a)$.

$\therefore \dfrac{1}{(s-3)^2}$ is like $\dfrac{1}{s^2}$ with s replaced by $(s-3)$ i.e. $a = -3$

$$\therefore \mathscr{L}^{-1}\left\{\frac{1}{(s-3)^2}\right\} = te^{3t}$$

$$\therefore \mathscr{L}^{-1}\left\{\frac{s^2 - 15s + 41}{(s+2)(s-3)^2}\right\} = \underline{3e^{-2t} + 2e^{3t} + te^{3t}}$$

Example 4 Determine $\mathscr{L}^{-1}\left\{\dfrac{4s^2 - 5s + 6}{(s+1)(s^2+4)}\right\}$.

Notice that this time we have a quadratic factor in the denominator

$$\frac{4s^2 - 5s + 6}{(s+1)\,(s^2+4)} \equiv \frac{A}{s+1} + \frac{Bs + C}{s^2 + 4}$$

$$\therefore 4s^2 - 5s + 6 \equiv A(s^2 + 4) + (Bs + C)\,(s+1).$$

(i) Putting $(s+1) = 0$, i.e. $s = -1$, $\quad 15 = 5A \quad \therefore \underline{A = 3}$

(ii) Equate coefficients of highest power, i.e. s^2

$$4 = A + B \qquad \therefore 4 = 3 + B \qquad \therefore \underline{B = 1}$$

(iii) We now equate the lowest power on each side, i.e. the constant term

$$6 = 4A + C \qquad \therefore 6 = 12 + C \qquad \therefore \underline{C = -6}$$

Now you can finish it off. $\qquad F(t) = \ldots\ldots\ldots$

39

$$\boxed{F(t) = 3e^{-t} + \cos 2t - 3\sin 2t}$$

$$\text{for } \mathscr{L}\{F(t)\} = \frac{3}{s+1} + \frac{s}{s^2+4} - \frac{6}{s^2+4}$$

$$\therefore\ F(t) = 3e^{-t} + \cos 2t - 3\sin 2t$$

Poles and zeros

The Laplace transform in general has the form $f(s) = \dfrac{\phi(s)}{\theta(s)}$ where the degree of $\phi(s)$ is less than that of $\theta(s)$.

The denominator then factorises into

$$f(s) = \frac{\phi(s)}{(s-a)(s-b)(s-c)\dots}$$

Poles: The values a, b, c, ... that make the denominator zero and hence $f(s)$ infinite are called the *poles* of $f(s)$.
If there are no repeated factors, the poles are *simple poles*.
If there are repeated factors, the poles are *multiple poles*.

Zeros: Values of s that make the numerator $\phi(s)$ zero and hence $f(s)$ zero are called the *zeros* of $f(s)$.

For example

$\dfrac{s-5}{(s+2)(s-1)}$ has simple poles at $s = 1$ and $s = -2$, and a zero at $s = 5$.

$\dfrac{s+4}{(s+2)^2(2s+3)}$ has a simple pole at $s = -\frac{3}{2}$ and double poles at $s = -2$, and a zero at $s = -4$.

Similarly $\dfrac{s+5}{s(s-2)\,(s+3)\,(2s+1)}$ has ....

40

$$\boxed{\text{simple poles at } s = 0,\ 2,\ -3,\ -\tfrac{1}{2};\ \text{ zero at } s = -5}$$

Every entry in our table of standard transforms gives rise to a corresponding entry in a similar table of inverse transforms. Let us tabulate such a list.

41 Table of inverse transforms

$f(s)$	$F(t)$	
$\dfrac{a}{s}$	a	
$\dfrac{1}{s+a}$	e^{-at}	
$\dfrac{n!}{s^{n+1}}$	t^n	(n a positive integer)
$\dfrac{1}{s^n}$	$\dfrac{t^{n-1}}{(n-1)!}$	(n a positive integer)
$\dfrac{a}{s^2+a^2}$	$\sin at$	
$\dfrac{s}{s^2+a^2}$	$\cos at$	
$\dfrac{a}{s^2-a^2}$	$\sinh at$	
$\dfrac{s}{s^2-a^2}$	$\cosh at$	

Theorem 1 The first shift theorem can be stated as follows.

If $f(s)$ is the Laplace transform of $F(t)$ then $f(s+a)$ is the Laplace transform of $e^{-at}F(t)$.

Here is a short revision exercise.

Exercise

1. Find the inverse transforms of
   (a) $\dfrac{1}{2s-3}$; (b) $\dfrac{5}{(s-4)^3}$; (c) $\dfrac{3s+4}{s^2+9}$.

2. Express in partial fractions

   (a) $\dfrac{22s+16}{(s+1)(s-2)(s+3)}$; (b) $\dfrac{s^2-11s+6}{(s+1)(s-2)^2}$.

3. Determine
   (a) $\mathscr{L}^{-1}\left\{\dfrac{4s^2-17s-24}{s(s+3)(s-4)}\right\}$; (b) $\mathscr{L}^{-1}\left\{\dfrac{5s^2-4s-7}{(s-3)(s^2+4)}\right\}$.

1. (a) $\frac{1}{2}e^{3t/2}$; (b) $5t^2e^{4t}$; (c) $3\cos 3t + \frac{4}{3}\sin 3t$.

2. (a) $\frac{1}{s+1} + \frac{4}{s-2} - \frac{5}{s+3}$; (b) $\frac{2}{s+1} - \frac{1}{s-2} - \frac{4}{(s-2)^2}$.

3. (a) $2 + 3e^{-3t} - e^{4t}$; (b) $2e^{3t} + 3\cos 2t + \frac{5}{2}\sin 2t$.

Solution of Differential Equations by Laplace Transforms

42

To solve a differential equation by Laplace transforms, we go through four distinct stages.

(a) Re-write the equation in terms of Laplace transforms.
(b) Insert the given initial conditions.
(c) Rearrange the equation algebraically to give the transform of the solution.
(d) Determine the inverse transform to obtain the particular solution.

We have spent some time finding the transforms of a variety of functions of t and the inverse transforms of functions of s, i.e. we have largely covered steps (a) and (d) of the above list. However, to write a differential equation in Laplace transforms, we must obtain the transforms of the differential coefficients $\frac{dx}{dt}$ and $\frac{d^2x}{dt^2}$.

Transforms of derivatives

Let $F'(t)$ denote the first derivative of $F(t)$ with respect to t,
$F''(t)$ denote the second derivative of $F(t)$ with respect to t, etc.

Then $\mathscr{L}\{F'(t)\} = \int_0^\infty e^{-st}F'(t)dt$ by definition.

Integrating by parts

$$\mathscr{L}\{F'(t)\} = \left[e^{-st}F(t)\right]_0^\infty - \int_0^\infty F(t)\{-se^{-st}\}dt$$

When $t \to \infty$, $e^{-st}F(t) \to \dots\dots\dots$

43

$$\boxed{0}$$

since s is positive and large enough to ensure that e^{-st} decays faster than any possible growth of $F(t)$.

$$\therefore \underline{\mathscr{L}\{F'(t)\} = -F(0) + s\mathscr{L}F(t)}$$

Replacing $F(t)$ by $F'(t)$ gives

$$\mathscr{L}\{F''(t)\} = \dots\dots\dots$$

44

$$\boxed{\mathscr{L}\{F''(t)\} = s^2 f(s) - s\,F(0) - F'(0)}$$

for $\quad \mathscr{L}\{F'(t)\} = -F(0) + s\,\mathscr{L}\{F(t)\}$

then $\quad \mathscr{L}\{F''(t)\} = -F'(0) + s\,\mathscr{L}\{F'(t)\}$

$$= -F'(0) + s(-F(0) + s\,\mathscr{L}\{F(t)\})$$

Writing $\quad \mathscr{L}\{F(t)\} = f(s)$ as usual, we have

$$\mathscr{L}\{F(t)\} = f(s)$$
$$\mathscr{L}\{F'(t)\} = s \cdot f(s) - F(0)$$
$$\mathscr{L}\{F''(t)\} = s^2 \cdot f(s) - s\,F(0) - F'(0)$$

We can see a pattern emerging

$$\mathscr{L}\{F'''(t)\} = \dots\dots\dots$$

45

$$\mathscr{L}\{F'''(t)\} = s^3 \cdot f(s) - s^2F(0) - s\,F'(0) - F''(0)$$

Alternative notation We make the working neater by adopting the following notation.
Let $x = F(t)$ and at $t = 0$, we write

$$x = x_0 \qquad \text{i.e. } F(0) = x_0$$

$$\frac{dx}{dt} = x_1 \qquad \text{i.e. } F'(0) = x_1$$

$$\frac{d^2x}{dt^2} = x_2 \qquad \text{i.e. } F''(0) = x_2 \qquad \text{etc.}$$

$$\therefore \ \frac{d^nx}{dt^n} = x_n \qquad \text{i.e. } F^n(0) = x_n$$

Also we denote the Laplace transform of x by $\overline{x}$,
i.e. $\overline{x} = \mathscr{L}\{x\} = \mathscr{L}\{F(t)\} = f(s)$.
So, using the 'dot' notation for differential coefficients, the previous results can be written ....

46

$$\begin{aligned} \mathscr{L}\{x\} &= \overline{x} \\ \mathscr{L}\{\dot{x}\} &= s\overline{x} - x_0 \\ \mathscr{L}\{\ddot{x}\} &= s^2\overline{x} - sx_0 - x_1 \\ \mathscr{L}\{\dddot{x}\} &= s^3\overline{x} - s^2x_0 - sx_1 - x_2 \end{aligned}$$

In each case, the subscript indicates the order of the differential coefficient, i.e. $x_n =$ the value of $\dfrac{d^nx}{dt^n}$ at $t = 0$.

Notice the pattern of the results.

$$\mathscr{L}\{\ddddot{x}\} = \ ....$$

47

$$\mathscr{L}\{\dddot{x}\} = s^4\bar{x} - s^3x_0 - s^2x_1 - sx_2 - x_3$$

Now, at long last, we can start solving differential equations.

Solution of first-order differential equations

Example 1 Solve the equation $\dfrac{dx}{dt} - 2x = 4$ given that at $t = 0,\ x = 1$.
We go through the four stages.

(a) Re-write the equation in Laplace transforms, using the last notation

$$\mathscr{L}\{x\} = \bar{x}; \quad \mathscr{L}\{\dot{x}\} = \dots\dots\dots$$
$$\mathscr{L}\{4\} = \dots\dots\dots$$

48

$$\mathscr{L}\{\dot{x}\} = s\bar{x} - x_0; \quad \mathscr{L}\{4\} = \frac{4}{s}$$

Then the equation becomes $\qquad (s\bar{x} - x_0) - 2\bar{x} = \dfrac{4}{s}$

(b) Insert the initial condition that at $t = 0,\ x = 1$, i.e. $x_0 = 1$

$$\therefore\ s\bar{x} - 1 - 2\bar{x} = \frac{4}{s}$$

(c) Now we rearrange this to give an expression for $\bar{x}$

$$\bar{x} = \dots\dots\dots$$

49

$$\bar{x} = \frac{s+4}{s(s-2)}$$

(d) Finally, we take inverse transforms to obtain x.

$\dfrac{s+4}{s(s-2)}$ in partial fractions gives . . . .

50

$$\boxed{\frac{3}{s-2}-\frac{2}{s}}$$

for $\dfrac{s+4}{s(s-2)} \equiv \dfrac{A}{s} + \dfrac{B}{s-2}$ $\quad\therefore s+4 = A(s-2) + B(s)$

(i) Put $(s-2) = 0$, i.e. $s = 2$ $\quad\therefore 6 = B(2) \quad \therefore B = 3$

(ii) Put $s = 0$ $\quad\therefore 4 = A(-2) \quad \therefore A = -2$

$$\therefore \bar{x} = \frac{s+4}{s(s-2)} = \frac{3}{s-2} - \frac{2}{s}$$

Therefore, taking invcrse transforms

$$x - \mathcal{L}^{-1}\left\{\frac{s+4}{s(s-2)}\right\} = \mathcal{L}^{-1}\left\{\frac{3}{s-2} - \frac{2}{s}\right\} = \dots\dots\dots$$

51

$$\boxed{x = 3e^{2t} - 2}$$

Example 2 Solve the equation $\dfrac{dx}{dt} + 2x = 10e^{3t}$ given that at $t = 0$, $x = 6$.

(a) Convert the equations to Laplace transforms, i.e.

....

52

$$\boxed{(s\bar{x} - x_0) + 2\bar{x} = \frac{10}{s-3}}$$

(b) Insert the initial condition, $x_0 = 6$

$$\therefore s\bar{x} - 6 + 2\bar{x} = \frac{10}{s-3}$$

(c) Rearrange to obtain $\bar{x} = \dots\dots\dots$

53

$$\bar{x} = \frac{6s - 8}{(s+2)(s-3)}$$

(d) Taking inverse transforms to obtain x

$$x = \mathscr{L}^{-1}\left\{\frac{6s-8}{(s+2)(s-3)}\right\} = \dots\dots\dots$$

Complete the solution.

54

$$x = 4e^{-2t} + 2e^{3t}$$

for $$\frac{6s-8}{(s+2)(s-3)} \equiv \frac{A}{s+2} + \frac{B}{s-3}$$

$$\therefore\ 6s - 8 = A(s-3) + B(s+2)$$

(i) Put $(s-3) = 0$, i.e. $s = 3$ $\quad\therefore\quad 10 = B(5) \quad \therefore\ \underline{B = 2}$

(ii) Put $(s+2) = 0$, i.e. $s = -2$. $\quad\therefore\ -20 = A(-5) \quad \therefore\ \underline{A = 4}$

$$\therefore\ \bar{x} = \frac{6s-8}{(s+2)(s-3)} = \frac{4}{s+2} + \frac{2}{s-3}$$

$$\therefore\ x = \mathscr{L}^{-1}\left\{\frac{4}{s+2} + \frac{2}{s-3}\right\} = \underline{4e^{-2t} + 2e^{3t}}$$

Example 3 Solve the equation $\dfrac{dx}{dt} - 4x = 2e^{2t} + e^{4t}$, given that at $t = 0$, $x = 0$.

Work this through the four steps in the same way as before and complete it on your own.

$$x = \dots\dots\dots$$

55

$$x = e^{4t} - e^{2t} + te^{4t}$$

The working is quite standard.

$$\frac{dx}{dt} - 4x = 2e^{2t} + e^{4t}$$

(a) $$(s\bar{x} - x_0) - 4\bar{x} = \frac{2}{s-2} + \frac{1}{s-4}$$

(b) $$x_0 = 0 \quad \therefore s\bar{x} - 4\bar{x} = \frac{2}{s-2} + \frac{1}{s-4}$$

(c) $$\therefore \bar{x} = \frac{2}{(s-2)(s-4)} + \frac{1}{(s-4)^2}$$

(d) $$\frac{2}{(s-2)(s-4)} \equiv \frac{A}{s-2} + \frac{B}{s-4} \quad \therefore 2 = A(s-4) + B(s-2)$$

Putting $(s-2) = 0$, i.e. $s = 2$ $\quad \therefore 2 = A(-2) \quad \therefore \underline{A = -1}$
Putting $(s-4) = 0$, i.e. $s = 4$ $\quad \therefore 2 = B(2) \quad \therefore \underline{B = 1}$

$$\therefore \bar{x} = \frac{1}{s-4} - \frac{1}{s-2} + \frac{1}{(s-4)^2}$$

$$\therefore \underline{x = e^{4t} - e^{2t} + te^{4t}}$$

Now on to the next frame.

56

Solution of second-order differential equations

The method is, in effect, the same as before, going through the same four distinct stages.

Example 1 Solve the equation $\dfrac{d^2x}{dt^2} - 3\dfrac{dx}{dt} + 2x = 2e^{3t}$, given that at $t = 0$, $x = 5$ and $\dfrac{dx}{dt} = 7$.

(a) We re-write the equation in terms of its transforms, remembering that

$$\mathscr{L}\{x\} = \bar{x}$$
$$\mathscr{L}\{\dot{x}\} = s\bar{x} - x_0$$
$$\mathscr{L}\{\ddot{x}\} = s^2\bar{x} - sx_0 - x_1$$

The equation becomes ....

57

$$(s^2\bar{x} - sx_0 - x_1) - 3(s\bar{x} - x_0) + 2\bar{x} = \frac{2}{s-3}$$

(b) Insert the initial conditions. In this case $x_0 = 5$ and $x_1 = 7$

$$\therefore\ (s^2\bar{x} - 5s - 7) - 3(s\bar{x} - 5) + 2\bar{x} = \frac{2}{s-3}$$

(c) Rearrange to obtain $\bar{x} = \ldots\ldots\ldots$

58

$$\bar{x} = \frac{5s^2 - 23s + 26}{(s-1)(s-2)(s-3)}$$

for $s^2\bar{x} - 5s - 7 - 3s\bar{x} + 15 + 2\bar{x} = \dfrac{2}{s-3}$

$$(s^2 - 3s + 2)\bar{x} - 5s + 8 = \frac{2}{s-3}$$

$$(s-1)(s-2)\bar{x} = \frac{2}{s-3} + 5s - 8 = \frac{2 + 5s^2 - 23s + 24}{s-3}$$

$$\therefore\ \bar{x} = \frac{5s^2 - 23s + 26}{(s-1)(s-2)(s-3)}$$

(d) Now for partial fractions

$$\frac{5s^2 - 23s + 26}{(s-1)(s-2)(s-3)} = \frac{A}{s-1} + \frac{B}{s-2} + \frac{C}{s-3}$$

$$\therefore\ 5s^2 - 23s + 26 = A(s-2)(s-3) + B(s-1)(s-3) + C(s-1)(s-2)$$

So that $A = \ldots\ldots\ldots$; $B = \ldots\ldots\ldots$; $C = \ldots\ldots\ldots$

59

$$A = 4; \quad B = 0; \quad C = 1$$

$$\therefore \bar{x} = \frac{4}{s-1} + \frac{1}{s-3}$$

$\therefore x = \dots\dots\dots$

60

$$x = 4e^{t} + e^{3t}$$

As you see, the Laplace transform method is considerably shorter than the classical method which entails

(i) determination of the complementary function
(ii) determination of a particular integral
(iii) obtaining the general solution, before
(iv) arriving at the particular solution by substitution of the initial conditions in the general solution.

Here is another example.

Example 2 Solve $\frac{d^2x}{dt^2} - 4x = 24\cos 2t$, given that at $t = 0$, $x = 3$ and $\frac{dx}{dt} = 4$.

(a) In Laplace transforms $\dots\dots\dots$

61

$$(s^2\bar{x} - sx_0 - x_1) - 4\bar{x} = \frac{24s}{s^2+4}$$

(b) Insert initial conditions, i.e. $x_0 = 3$; $x_1 = 4$

$$s^2\bar{x} - 3s - 4 - 4\bar{x} = \frac{24s}{s^2+4}$$

$$\therefore (s^2 - 4)\bar{x} = 3s + 4 + \frac{24s}{s^2+4}$$

$$= \frac{3s^3 + 4s^2 + 36s + 16}{s^2+4}$$

(c)
$$\bar{x} = \frac{3s^3 + 4s^2 + 36s + 16}{(s^2+4)(s-2)(s+2)}$$

Expressed in partial fractions, this becomes

....

62

$$\frac{3s^3 + 4s^2 + 36s + 16}{(s^2+4)\,(s-2)\,(s+2)} \equiv \frac{As+B}{s^2+4} + \frac{C}{s-2} + \frac{D}{s+2}$$

$$\therefore \quad 3s^3 + 4s^2 + 36s + 16 \equiv (As+B)(s-2)(s+2) + C(s^2+4)(s+2) + D(s^2+4)(s-2)$$

Putting $(s-2) = 0$, i.e. $s = 2$, gives $C = 4$
Putting $(s+2) = 0$, i.e. $s = -2$, gives $D = 2$
Equating coefficients of s^3 and also the constant terms gives $A = -3$ and $B = 0$.

$$\therefore \bar{x} = \frac{3s^3 + 4s^2 + 36s + 16}{(s^2+4)(s-2)(s+2)} = \frac{4}{s-2} + \frac{2}{s+2} - \frac{3s}{s^2+4}$$

$$\therefore x = \ldots\ldots\ldots$$

63

$$x = 4e^{2t} + 2e^{-2t} - 3\cos 2t$$

We see the importance of partial fractions in this work. While we can always find A, B, C, etc. by the substitution method we have used, there are many cases when we can apply the 'cover up' method and write down the values of the constant coefficients almost on sight provided the denominator has non-repeated linear factors. If you do not already know the 'cover up' method, here it is.

'Cover up' rule

An example will explain it.

Example 1 We know that $f(s) = \dfrac{8s-6}{s(s-2)}$ has partial fractions of the form $\dfrac{A}{s} + \dfrac{B}{s-2}$. By the 'cover up' rule, the constant A, i.e. the coefficient of $\dfrac{1}{s}$, is found by temporarily covering up the factor s in the denominator of $f(s)$ and finding the limiting value of what remains when s (the factor covered up) tends to zero.

$$\therefore\ A = \text{coefficient of } \frac{1}{s} = \lim_{s\to 0}\left\{\frac{8s-6}{s-2}\right\} = 3 \qquad \therefore\ \underline{A = 3}$$

Similarly, B, the coefficient of $\dfrac{1}{s-2}$, is obtained by covering up the factor $(s-2)$ in the denominator of $f(s)$ and finding the limiting value of what remains when $(s-2) \to 0$, i.e. $s \to 2$.

$$\therefore\ B = \text{coefficient of } \frac{1}{s-2} = \lim_{s\to 0}\left\{\frac{8s-6}{s}\right\} = 5 \qquad \therefore\ \underline{B = 5}$$

$$\therefore\ \underline{\frac{8s-6}{s(s-2)} = \frac{3}{s} + \frac{5}{s-2}}$$

Another example.

64

Example 2 $f(s)=\dfrac{s+17}{(s-1)(s+2)(s-3)}\equiv\dfrac{A}{s-1}+\dfrac{B}{s+2}+\dfrac{C}{s-3}.$

A: cover up $(s-1)$ in $f(s)$ and find

$$\lim_{s\to 1}\left\{\frac{s+17}{(s+2)(s-3)}\right\}=\frac{18}{-6}\qquad \therefore \underline{A=-3}$$

Similarly

$B:\ldots\ldots\ldots \qquad \therefore B=\ldots\ldots\ldots$

$C:\ldots\ldots\ldots \qquad \therefore C=\ldots\ldots\ldots$

65

$$B=\lim_{s\to -2}\left\{\frac{s+17}{(s-1)(s-3)}\right\}=\frac{15}{(-3)(-5)}=1\qquad \therefore \underline{B=1}$$

$$C=\lim_{s\to 3}\left\{\frac{s+17}{(s-1)(s+2)}\right\}=\frac{20}{(2)(5)}=2\qquad \therefore \underline{C=2}$$

$$\therefore \bar{x}=\frac{1}{s+2}+\frac{2}{s-3}-\frac{3}{s-1}$$

$$\therefore \underline{x=e^{-2t}+2e^{3t}-3e^{t}}$$

Now let us solve another equation.

Example 3 Solve $\ddot{x}+5\dot{x}+6x=4t$, given that at $t=0$, $x=0$ and $\dot{x}=0$.

As usual we begin $(s^2\bar{x}-sx_0-x_1)+5(s\bar{x}-x_0)+6\bar{x}=\dfrac{4}{s^2}$

$x_0=0;\quad x_1=0 \qquad \therefore (s^2+5s+6)\bar{x}=\dfrac{4}{s^2}$

$$\therefore \bar{x}=\frac{4}{s^2(s+2)(s+3)}$$

The s^2 in the denominator can be awkward, so we introduce a useful trick and detach one factor s outside the main expression, thus

$$\bar{x}=\frac{1}{s}\left\{\frac{4}{s(s+2)(s+3)}\right\}=\frac{1}{s}\left\{\frac{A}{s}+\frac{B}{s+2}+\frac{C}{s+3}\right\}$$

Applying the cover up rule to the expressions within the brackets

$$\bar{x} = \frac{1}{s}\left\{\frac{4}{6}\cdot\frac{1}{s} - \frac{2}{s+2} + \frac{4}{3}\cdot\frac{1}{s+3}\right\}$$

Now we bring the external $\frac{1}{s}$ back into the fold

$$\bar{x} = \frac{2}{3}\cdot\frac{1}{s^2} - \frac{2}{s(s+2)} + \frac{4}{3}\cdot\frac{1}{s(s+3)}$$

and the second and third terms can be expressed in simple partial fractions so that

$$\bar{x} = \dots\dots\dots$$

66

$$\bar{x} = \frac{2}{3}\cdot\frac{1}{s^2} - \frac{1}{s} + \frac{1}{s+2} + \frac{4}{9}\cdot\frac{1}{s} - \frac{4}{9}\cdot\frac{1}{s+3}$$

which can now be simplified into

$$\bar{x} = \frac{2}{3}\cdot\frac{1}{s^2} - \frac{5}{9}\cdot\frac{1}{s} + \frac{1}{s+2} - \frac{4}{9}\cdot\frac{1}{s+3}$$

$$\therefore x = \dots\dots\dots$$

67

$$x = \frac{2}{3}t - \frac{5}{9} + e^{-2t} - \frac{4}{9}e^{-3t}$$

There are times when a quadratic coefficient of $\bar{x}$ cannot be expressed in simple linear factors. In that case, we merely complete the square converting the expression into $(s \pm k)^2 \pm a^2$. Let us see such an example.

Example 4 Solve $\ddot{x} - 2\dot{x} + 10x = e^{2t}$, given that at $t = 0$, $x = 0$ and $\dot{x} = 1$.

We find the expression for $\bar{x}$ as before.

$$\bar{x} = \ldots\ldots\ldots$$

68

$$\bar{x} = \frac{s-1}{(s-2)(s^2-2s+10)}$$

for $(s^2\bar{x} - sx_0 - x_1) - 2(s\bar{x} - x_0) + 10\bar{x} = \dfrac{1}{s-2}$

$x_0 = 0; x_1 = 1 \qquad \therefore s^2\bar{x} - 1 - 2s\bar{x} + 10\bar{x} = \dfrac{1}{s-2}$

$$\therefore (s^2 - 2s + 10)\bar{x} = 1 + \frac{1}{s-2} = \frac{s-1}{s-2}$$

$$\therefore \bar{x} = \frac{s-1}{(s-2)(s^2-2s+10)}$$

Expressing this in partial fractions

$\bar{x} = \ldots\ldots\ldots \qquad$ Evaluate the coefficients.

69

$$\bar{x} = \frac{1}{10}\left\{\frac{1}{s-2} - \frac{s-10}{s^2-2s+10}\right\}$$

for we have

$$\frac{s-1}{(s-2)(s^2-2s+10)} \equiv \frac{A}{(s-2)} + \frac{Bs+C}{s^2-2s+10}$$

$$\therefore\ s-1 = A(s^2-2s+10) + (s-2)(Bs+C)$$

Put $(s-2)=0$, i.e. $s=2$. $\quad \therefore\ 1 = A(4-4+10) \quad \therefore\ A = \frac{1}{10}$

$[s^2] \quad 0 = A + B \quad \therefore\ B = -\frac{1}{10}$

[CT] $\quad -1 = 10A - 2C \quad \therefore\ 2C = 2 \quad \therefore\ C = 1$

$$\therefore\ \bar{x} = \frac{1}{10}\left\{\frac{1}{s-2} - \frac{s-10}{s^2-2s+10}\right\}$$

Now we have to find the inverse transforms to obtain x. The first term $\dfrac{1}{s-2}$ is easy enough, but what of $\dfrac{s-10}{s^2-2s+10}$? The denominator will not factorise into simple linear factors; therefore we complete the square in the denominator and write it as

$$\frac{s-10}{s^2-2s+10} = \frac{s-10}{(s-1)^2+9}$$

and then we improve this still further and write it in the form $\dfrac{(s-1)-9}{(s-1)^2+9}$. We are quite happy with this, for $\dfrac{s-1}{(s-1)^2+9}$ is merely $\dfrac{s}{s^2+9}$ with s replaced by $(s-1)$, which indicates an extra factor e^t in the final function of t (Theorem 1).

So $\quad \bar{x} = \dfrac{1}{10}\left\{\dfrac{1}{s-2} - \dfrac{s-1}{(s-1)^2+9} + \dfrac{9}{(s-1)^2+9}\right\}$

$\therefore\ x = \ldots\ldots\ldots$

70

$$x = \frac{1}{10}\{e^{2t} - e^t \cos 3t + 3e^t \sin 3t\}$$

Before we leave this topic, the same general approach can be employed for solving simultaneous differential equations. *Let us see an example in the next frame.*

71

Simultaneous differential equations

Example 1 Solve the pair of simultaneous equations

$$\dot{y} - x = e^t$$
$$\dot{x} + y = e^{-t}$$

given that at $t = 0$, $x = 0$ and $y = 0$.

(a) We first express both equations in Laplace transforms.

$$(s\bar{y} - y_0) - \bar{x} = \frac{1}{s-1}$$
$$(s\bar{x} - x_0) + \bar{y} = \frac{1}{s+1}$$

(b) Then we insert the initial conditions, $x_0 = 0$ and $y_0 = 0$

$$\left.\begin{aligned} \therefore\ s\bar{y} - \bar{x} &= \frac{1}{s-1} \\ s\bar{x} + \bar{y} &= \frac{1}{s+1} \end{aligned}\right\} \qquad (1)$$

(c) We now solve these for $\bar{x}$ and $\bar{y}$ by the normal algebraic method. Eliminating $\bar{y}$ we have

$$s\bar{y} - \bar{x} = \frac{1}{s-1}$$

$$s\bar{y} + s^2\bar{x} = \frac{s}{s+1}$$

$$\therefore\ (s^2+1)\bar{x} = \frac{2}{s+1} - \frac{1}{s-1} = \frac{s^2 - 2s - 1}{(s+1)(s-1)}$$

$$\therefore\ \bar{x} = \frac{s^2 - 2s - 1}{(s-1)(s+1)(s^2+1)}$$

Representing this in partial fractions, gives ....

72

$$\bar{x} = -\frac{1}{2} \cdot \frac{1}{s-1} - \frac{1}{2} \cdot \frac{1}{s+1} + \frac{s}{s^2+1} + \frac{1}{s^2+1}$$

$$\text{for } \bar{x} = \frac{s^2 - 2s - 1}{(s-1)(s+1)(s^2+1)} \equiv \frac{A}{s-1} + \frac{B}{s+1} + \frac{Cs+D}{s^2+1}$$

$$\therefore s^2 - 2s - 1 = A(s+1)(s^2+1) + B(s-1)(s^2+1)$$
$$+(s-1)(s+1)(Cs+D)$$

Putting $s = 1$ and $s = -1$, gives $A = -\frac{1}{2}$ and $B = -\frac{1}{2}$.

Comparing coefficients of s^3 and the constant terms gives $C = 1$ and $D = 1$.

$$\therefore \bar{x} = \frac{1}{2} \cdot \frac{1}{s-1} - \frac{1}{2} \cdot \frac{1}{s+1} + \frac{s+1}{s^2+1}$$

$$\therefore x = \ldots\ldots\ldots$$

73

$$x = -\tfrac{1}{2}e^{t} - \tfrac{1}{2}e^{-t} + \cos t + \sin t$$

We now revert to equations (1) and eliminate $\bar{x}$ to obtain $\bar{y}$ and hence y, in the same way. Do this on your own.

$$y = \ldots\ldots\ldots$$

74

$$\boxed{y = \tfrac{1}{2}e^{t} + \tfrac{1}{2}e^{-t} - \cos t + \sin t}$$

Here is the working

$$\left.\begin{aligned} s^2\bar{y} - s\bar{x} &= \frac{s}{s-1} \\ \bar{y} + s\bar{x} &= \frac{1}{s+1} \end{aligned}\right\}$$

$$\therefore\ (s^2+1)\bar{y} = \frac{s}{s-1} + \frac{1}{s+1} = \frac{s^2+2s-1}{(s-1)(s+1)}$$

$$\therefore\ \bar{y} = \frac{s^2+2s-1}{(s-1)(s+1)(s^2+1)} \equiv \frac{A}{s-1} + \frac{B}{s+1} + \frac{Cs+D}{s^2+1}$$

$$\therefore\ s^2+2s-1 = A(s+1)(s^2+1) + B(s-1)(s^2+1)$$
$$+(s-1)(s+1)(Cs+D)$$

Putting $s = 1$ and $s = -1$ gives $A = \frac{1}{2}$ and $B = \frac{1}{2}$.

Equating coefficients of s^3 and the constant terms gives $C = -1$ and $D = 1$.

$$\therefore\ \bar{y} = \frac{1}{2}\cdot\frac{1}{s-1} + \frac{1}{2}\cdot\frac{1}{s+1} - \frac{s}{s^2+1} + \frac{1}{s^2+1}$$

$$\therefore\ \underline{y = \frac{1}{2}e^{t} + \frac{1}{2}e^{-t} - \cos t + \sin t}$$

So the results are

$$x = -\frac{1}{2}(e^{t} + e^{-t}) + \sin t + \cos t = \sin t + \cos t - \cosh t$$

$$y = \ \frac{1}{2}(e^{t} + e^{-t}) + \sin t - \cos t = \sin t - \cos t + \cosh t$$

$$\therefore\ \underline{x = \sin t + \cos t - \cosh t;\quad y = \sin t - \cos t + \cosh t}$$

Simultaneous equations are all solved in much the same way. Here is another.

Example 2 Solve the equations

$$2\dot{y} - 6y + 3x = 0$$
$$3\dot{x} - 3x - 2y = 0$$

given that at $t = 0$, $x = 1$ and $y = 3$.
Expressing these in Laplace transforms, we have

....

....

75

$$2(s\bar{y} - y_0) - 6\bar{y} + 3\bar{x} = 0$$
$$3(s\bar{x} - x_0) - 3\bar{x} - 2\bar{y} = 0$$

Then we insert the initial conditions and simplify, obtaining

....

....

76

$$3\bar{x} + (2s - 6)\bar{y} = 6 \qquad \text{(i)}$$
$$(3s - 3)\bar{x} - 2\bar{y} = 3 \qquad \text{(ii)}$$

(a) *To find* $\bar{x}$

(i) $\qquad 3\bar{x} + (2s - 6)\bar{y} = 6$

(ii) $\times(s - 3) \qquad (s - 3)(3s - 3)\bar{x} - (2s - 6)\bar{y} = 3(s - 3)$

Adding: $\qquad [(s - 3)(3s - 3) + 3]\bar{x} = 3s - 9 + 6$

$$\therefore\ (3s^2 - 12s + 12)\bar{x} = 3s - 3$$
$$(s^2 - 4s + 4)\bar{x} = s - 1$$

$$\therefore\ \bar{x} = \frac{s-1}{(s-2)^2} \equiv \frac{A}{s-2} + \frac{B}{(s-2)^2} = \frac{A(s-2)+B}{(s-2)^2}$$

$$\therefore\ s-1 = A(s-2)+B \qquad \text{giving} \quad A=1 \quad \text{and} \quad B=1$$

$$\therefore\ \bar{x} = \frac{1}{s-2} + \frac{1}{(s-2)^2} \qquad \therefore\ \underline{x = e^{2t} + te^{2t}}$$

(b) Going back to equations (i) and (ii), we can find y.

$$y = \dots\dots\dots$$

77

$$\boxed{y = \tfrac{1}{2}\{6e^{2t} + 3te^{2t}\}}$$

for, eliminating $\bar{x}$ we get

$$\bar{y} = \frac{6s-9}{2(s-2)^2} \equiv \frac{1}{2}\left\{\frac{A}{s-2} + \frac{B}{(s-2)^2}\right\} = \frac{1}{2}\left\{\frac{A(s-2)+B}{(s-2)^2}\right\}$$

$$\therefore\ 6s-9 = A(s-2)+B \qquad \therefore\ A=6; \quad B=3$$

$$\therefore\ \bar{y} = \frac{1}{2}\left\{\frac{6}{s-2} + \frac{3}{(s-2)^2}\right\} \qquad \therefore\ \underline{y = \tfrac{1}{2}\{6e^{2t} + 3te^{2t}\}}$$

Simultaneous second-order equations are solved in like manner.

---

78 *Example 3* If x and y are functions of t, solve the equations

$$\ddot{x} + 2x - y = 0$$
$$\ddot{y} + 2y - x = 0$$

given that at $t = 0$, $x_0 = 4$; $y_0 = 2$; $x_1 = 0$; $y_1 = 0$.

We start off as usual with $(s^2\bar{x} - sx_0 - x_1) + 2\bar{x} - \bar{y} = 0$

and $(s^2\bar{y} - sy_0 - y_1) + 2\bar{y} - \bar{x} = 0$

Inserting the initial conditions, we have

$$s^2\bar{x} - 4s + 2\bar{x} - \bar{y} = 0$$
$$s^2\bar{y} - 2s + 2\bar{y} - \bar{x} = 0$$

Simplifying these we can eliminate $\bar{y}$ to obtain $\bar{x}$ and hence x.

$$x = \dots\dots\dots$$

79

$$x = 3\cos t + \cos(\sqrt{3}t)$$

for
$$(s^2+2)\bar{x} - \bar{y} = 4s \qquad \text{(i)}$$
$$-\bar{x} + (s^2+2)\bar{y} = 2s \qquad \text{(ii)}$$

Eliminating $\bar{y}$ and simplifying gives

$$\bar{x} = \frac{4s^3+10s}{(s^2+1)(s^2+3)}$$

$$\therefore\ \bar{x} = \frac{4s^3+10s}{(s^2+1)(s^2+3)} \equiv \frac{As+B}{s^2+1} + \frac{Cs+D}{s^2+3}$$

$$\therefore\ 4s^3+10s = (s^2+3)(As+B) + (s^2+1)(Cs+D)$$

Equating coefficients of like powers of s

$[s^3]$ $\quad 4 = A + C \qquad \therefore\ A + C = 4$

$[\text{CT}]$ $\quad 0 = 3B + D \qquad \therefore\ 3B + D = 0$

Putting $s = 1$, $\quad 14 = 4A + 4B + 2C = 2D \quad \therefore\ 2A + 2B + C + D = 7$

Putting $s = -1$, $\quad -14 = -4A + 4B - 2C + 2D \quad \therefore\ 2A - 2B + C - D = 7$

Putting $C = 4 - A$ and $D = -3B$ in the last two leads to

$A = \dots\dots\dots$; $\quad B = \dots\dots\dots$; $\quad C = \dots\dots\dots$; $\quad D = \dots\dots\dots$

80

$$A = 3; \quad B = 0; \quad C = 1; \quad D = 0$$

$$\therefore\ \bar{x} = \frac{3s}{s^2+1} + \frac{s}{s^2+3}$$

$$\therefore\ x = \dots\dots\dots$$

81

$$\boxed{x = 3\cos t + \cos(\sqrt{3}t)}$$

To find y we could return to equations (i) and (ii) and repeat the process, eliminating $\bar{x}$ so as to obtain $\bar{y}$ and hence y.

But always keep an eye on the original equations, the first of which is

$$\ddot{x} + 2x - y = 0$$

Therefore, in this particular case, $y = \ddot{x} + 2x$.

So all we have to do is to differentiate x twice and substitute

$$x = 3\cos t + \cos(\sqrt{3}t)$$
$$\dot{x} = -3\sin t - \sqrt{3}\sin(\sqrt{3}t)$$
$$\ddot{x} = -3\cos t - 3\cos(\sqrt{3}t)$$

$$\therefore\ y = -3\cos t - 3\cos(\sqrt{3}t) + 6\cos t + 2\cos(\sqrt{3}t)$$

$$\therefore\ \underline{y = 3\cos t - \cos(\sqrt{3}t)}$$

which is a good deal quicker.

So, as we have seen, the method of solving differential equations by Laplace transforms follows a general routine.

(a) Express the equation in Laplace transforms
(b) Insert the initial conditions
(c) Simplify to obtain the transform of the solution
(d) Re-write the final transform in partial fractions
(e) Determine the inverse transforms

and, by now, you are fully aware of the importance of *partial fractions*!

That brings us to the end of this particular programme. We shall continue our study of Laplace transforms in the next programme. Meanwhile, be sure you are familiar with the items listed in the Revision Summary that follows: you will then have no difficulty with the Test Exercise.

82

REVISION SUMMARY

1. *Laplace transform* $\mathscr{L}\{F(t)\} = \int_0^\infty F(t)e^{-st}dt = f(s)$.
2. *Table of transforms*

$F(t)$	$\mathscr{L}\{F(t)\} = f(s)$
a	$\dfrac{a}{s}$
e^{at}	$\dfrac{1}{s-a}$
$\sin at$	$\dfrac{a}{s^2+a^2}$
$\cos at$	$\dfrac{s}{s^2+a^2}$
$\sinh at$	$\dfrac{a}{s^2-a^2}$
$\cosh at$	$\dfrac{s}{s^2-a^2}$
t^n	$\dfrac{n!}{s^{n+1}}$ (*n* a positive integer)

3. *Theorem 1* First shift theorem

If $\mathscr{L}\{F(t)\} = f(s)$, then $\mathscr{L}\{e^{-at}F(t)\} = f(s+a)$.

4. *Theorem 2* Multiplying by t

If $\mathscr{L}\{F(t)\} = f(s)$, then $\mathscr{L}\{tF(t)\} = -\dfrac{d}{ds}\{f(s)\}$.

5. *Theorem 3* Dividing by t

If $\mathscr{L}\{F(t)\} = f(s)$, then $\mathscr{L}\left\{\dfrac{F(t)}{t}\right\} = \int_s^\infty f(s)ds$

provided that $\lim\limits_{t\to 0}\left\{\dfrac{F(t)}{t}\right\}$ exists.

6. *Inverse transform*

$$\text{If } \mathcal{L}\{F(t)\} = f(s), \text{ then } \mathcal{L}^{-1}\{f(s)\} = F(t).$$

7. *Table of inverse transforms*

$f(s)$	$F(t)$	
$\dfrac{a}{s}$	a	
$\dfrac{1}{s+a}$	e^{-at}	
$\dfrac{n!}{s^{n+1}}$	t^n	(n a positive integer)
$\dfrac{1}{s^n}$	$\dfrac{t^{n-1}}{(n-1)!}$	
$\dfrac{a}{s^2+a^2}$	$\sin at$	
$\dfrac{s}{s^2+a^2}$	$\cos at$	
$\dfrac{a}{s^2-a^2}$	$\sinh at$	
$\dfrac{s}{s^2-a^2}$	$\cosh at$	

By the first shift theorem

If $f(s)$ is the Laplace transform of $F(t)$

then $f(s+a)$ is the Laplace transform of $e^{-at}F(t)$.

8. *Laplace transforms of differential coefficients*

$$\mathcal{L}\{x\} = \bar{x}$$

$$\mathcal{L}\left\{\frac{dx}{dt}\right\} = \mathcal{L}\{\dot{x}\} = s\bar{x} - x_0$$

$$\mathcal{L}\left\{\frac{d^2x}{dt^2}\right\} = \mathcal{L}\{\ddot{x}\} = s^2\bar{x} - sx_0 - x_1 \quad \text{etc.}$$

where $x_0 =$ value of x at $t = 0$

$x_1 =$ value of $\dfrac{dx}{dt}$ at $t = 0$, etc.

9. *Solution of differential equations*

(a) Re-write the equation in terms of Laplace transforms.

(b) Insert the given initial conditions.

(c) Rearrange the equation algebraically to give the transform of the solution.

(d) Express the transform in standard forms by partial fractions.

(e) Determine the inverse transforms to obtain the particular solution.

10. *Rules of partial fractions*

(a) The numerator must be of lower degree than the denominator. If not, divide out.

(b) Factorise the denominator into its prime factors.

(c) A linear factor $(s+a)$ gives a partial fraction $\dfrac{A}{s+a}$ where A is a constant to be determined.

(d) A repeated factor $(s+a)^2$ gives $\dfrac{A}{s+a}+\dfrac{B}{(s+a)^2}$.

(e) Similarly $(s+a)^3$ gives $\dfrac{A}{s+a}+\dfrac{B}{(s+a)^2}+\dfrac{C}{(s+a)^3}$.

(f) A quadratic factor (s^2+ps+q) gives $\dfrac{Ps+Q}{s^2+ps+q}$.

(g) A repeated quadratic factor $(s^2+ps+q)^2$ gives

$$\frac{Ps+Q}{s^2+ps+q}+\frac{Rs+T}{(s^2+ps+q)^2}.$$

11. *Poles and zeros*

$$\text{If } \mathcal{L}\{F(t)\}=f(s)=\frac{\phi(s)}{(s-a)(s-b)(s-c)\ldots}$$

(a) The values a, b, c, … that make the denominator zero and hence $f(s)$ infinite are called the *poles* of $f(s)$. If there are no repeated factors, the poles are *simple poles*. If there are repeated factors, the poles are *multiple poles*.

(b) The values of s that make the numerator $\phi(s)$ zero and hence $f(s)$ zero are called the *zeros* of $f(s)$.

83 TEST EXERCISE VII

1. Determine the Laplace transforms of the following functions.

(a) $3e^{-4t} - 5e^{4t}$ (b) $\sin 4t + \cos 4t$

(c) $t^3 + 2t^2 - t + 4$ (d) $e^{-2t} \cos 5t$

(e) $t \sin 3t$ (f) $\dfrac{e^{-t} - e^{-2t}}{t}$.

2. Determine the inverse transforms of the following.

(a) $\dfrac{s-5}{(s-3)(s-4)}$ (b) $\dfrac{s^2+3s-7}{(s-1)(s^2+2)}$

(c) $\dfrac{s^2-3s-4}{(s-3)(s-1)^2}$ (d) $\dfrac{2s^2-6s-1}{(s-3)(s^2-2s+5)}$.

3. Solve the following equations by Laplace transforms.

(a) $\dfrac{dx}{dt} + 3x = e^{-2t}$ given that $x = 2$ when $t = 0$

(b) $3\dot{x} - 6x = \sin 2t$ given that $x = 1$ when $t = 0$

(c) $\ddot{x} - 7\dot{x} + 12x = 2$ given that at $t = 0$, $x = 1$ and $\dot{x} = 5$

(d) $\ddot{x} - 2\dot{x} + x = te^t$ given that at $t = 0$, $x = 1$ and $\dot{x} = 0$.

4. Solve the following pair of simultaneous equations where x and y are functions of t.

$$\dot{x} + \dot{y} + x + 2y = e^{-3t}$$
$$\dot{x} + 3x + 5y = 5e^{-2t}$$

given that at $t = 0$, $x = 4$ and $y = -1$.

FURTHER PROBLEMS VII

1. Determine the Laplace transforms of the following functions.

(a) $e^{4t}\cos 2t$ (b) $t\sin 2t$ (c) $t^3 + 4t^2 + 5$

(d) $e^{3t}(t^2 + 4)$ (e) $t^2 \cos t$ (f) $\dfrac{\sinh 2t}{t}$.

2. Determine the inverse transforms of the following.

(a) $\dfrac{2s - 6}{(s-2)(s-4)}$ (b) $\dfrac{5s - 8}{s(s-4)}$

(c) $\dfrac{s^2 - 2s + 3}{(s-2)^3}$ (d) $\dfrac{2 - 11s}{(s-2)(s^2 + 2s + 2)}$

(e) $\dfrac{s}{(s^2+1)(s^2+4)}$ (f) $\dfrac{s-5}{s^2 + 4s + 20}$.

In Questions 3 to 11, solve the equations by Laplace transforms.

3. $\dot{x} - 4x = 8$ at $t = 0$, $x = 2$.

4. $3x - 4x = \sin 2t$ at $t = 0$, $x = \frac{1}{3}$.

5. $\ddot{x} - 2\dot{x} + x = 2(t + \sin t)$ at $t = 0$, $x = 6$, $\dot{x} = 5$.

6. $\ddot{x} - 6\dot{x} + 8x = e^{3t}$ at $t = 0$, $x = 0$, $\dot{x} = 2$.

7. $\ddot{x} + 9x = \cos 2t$ at $t = 0$, $x = 1$, $\dot{x} = 3$.

8. $\ddot{x} - 2\dot{x} + 5x = e^{2t}$ at $t = 0$, $x = 0$, $\dot{x} = 1$.

9. $\ddot{x} + 4\dot{x} + 4x = t^2 + e^{-2t}$ at $t = 0$, $x = \frac{1}{2}$, $\dot{x} = 0$.

10. $\ddot{x} + 8\dot{x} + 32x = 32\sin 4t$ at $t = 0$, $x = \dot{x} = 0$.

11. $\ddot{x} + 25x = 10(\cos 5t - 2\sin 5t)$ at $t = 0$, $x = 1$, $\dot{x} = 2$.

In Questions 12 to 17, solve the pairs of simultaneous equations by Laplace transforms.

12. $\left.\begin{aligned} \dot{y} + 3x &= e^{-2t} \\ \dot{x} - 3y &= e^{2t} \end{aligned}\right\}$ at $t = 0$, $x = y = 0$.

13. $\left.\begin{aligned} 4\dot{x} - 2\dot{y} + 10x - 5y &= 0 \\ \dot{y} - 18x + 15y &= 10 \end{aligned}\right\}$ at $t = 0$, $y = 4$, $x = 2$.

14. $\left.\begin{aligned} \dot{x} - 2\dot{y} - 3x + 6y &= 12 \\ 3\dot{y} + 5x + 2y &= 16 \end{aligned}\right\}$ at $t = 0$, $x = 12$, $y = 8$.

15. $$\left.\begin{aligned} 2\dot{x} + 3\dot{y} + 7x &= 14t + 7 \\ 5\dot{x} - 3\dot{y} + 4x + 6y &= 14t - 14 \end{aligned}\right\} \text{at } t = 0,\ x = y = 0.$$

16. $$\left.\begin{aligned} 2\dot{x} + 2x + 3\dot{y} + 6y &= 56e^{t} - 3e^{-t} \\ \dot{x} - 2x - \dot{y} - 3y &= -21e^{t} - 7e^{-t} \end{aligned}\right\} \text{at } t = 0,\ x = 8,\ y = 3.$$

17. $$\left.\begin{aligned} \ddot{x} - \ddot{y} + x - y &= 5e^{2t} \\ 2\dot{x} - \dot{y} + y &= 0 \end{aligned}\right\} t = 0,\ x = 1,\ y = 2,\ \dot{x} = 0.$$

18. Find an expression for x in terms of t, given that

$$\dot{y} - \dot{x} + 2x = 10\sin 2t$$
$$\dot{y} + 2y + x = 0$$

and when $t = 0, x = y = 0$.

19. If $\ddot{x} + 8x + 2y = 24\cos 4t$

and $\ddot{y} + 2x + 5y = 0$

and at $t = 0$, $x = y = 0$, $\dot{x} = 1$, $\dot{y} = 2$, determine an expression for y in terms of t.

20. Solve completely, the pair of simultaneous equations

$$5\ddot{x} + 12\ddot{y} + 6x = 0$$
$$5\ddot{x} + 16\ddot{y} + 6y = 0$$

given that, at $t = 0$, $x = \frac{7}{4}$, $y = 1$, $\dot{x} = 0$, $\dot{y} = 0$.

Programme 8

Laplace Transforms

PART 2

Introduction

1 In the previous programme, we dealt with the Laplace transforms of continuous functions of t. In practical applications, it is convenient to have a function which, in effect, 'switches on' or 'switches off' a given term at pre-described values of t. This we can do with the *Heaviside unit step function*.

Heaviside unit step function

Consider a function that maintains a zero value for all values of t up to $t = c$ and a unit value for $t = c$ and all values of $t \geq c$.

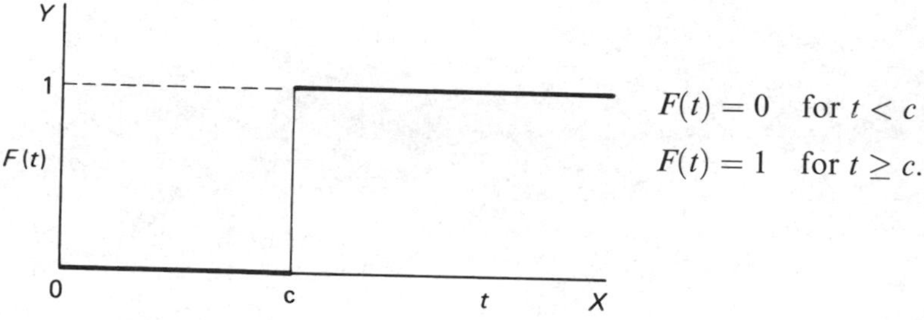

$$F(t) = 0 \quad \text{for } t < c$$
$$F(t) = 1 \quad \text{for } t \geq c.$$

This function is the *Heaviside unit step function* and is denoted by

$$F(t) = H(t - c)$$

where $H(t - c)$ is a single symbol in which the c indicates the value of t at which the function changes from a value 0 to a value 1.
Thus, the function

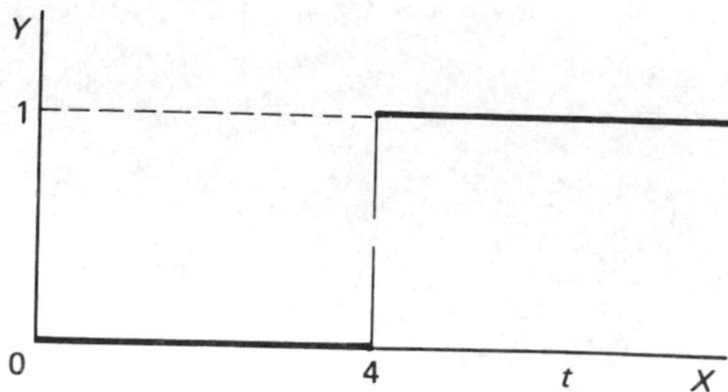

is denoted by $\quad F(t) = \dots\dots\dots$

2

$$F(t) = H(t-4)$$

Similarly, the graph of $F(t) = 2H(t-3)$ is

. . . .

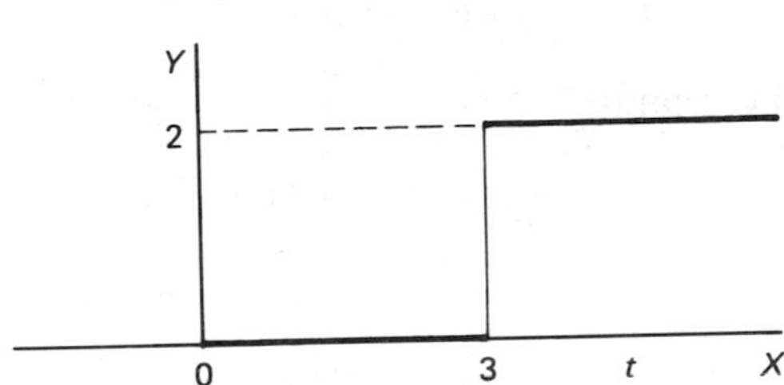

So $H(t-c)$ has just two values

for $t < c$, $H(t-c) = \dots\dots\dots$

for $t \geq c$, $H(t-c) = \dots\dots\dots$

4

$$t < c, \quad H(t-c) = 0; \qquad t \geq c, \quad H(t-c) = 1$$

Unit step at the origin

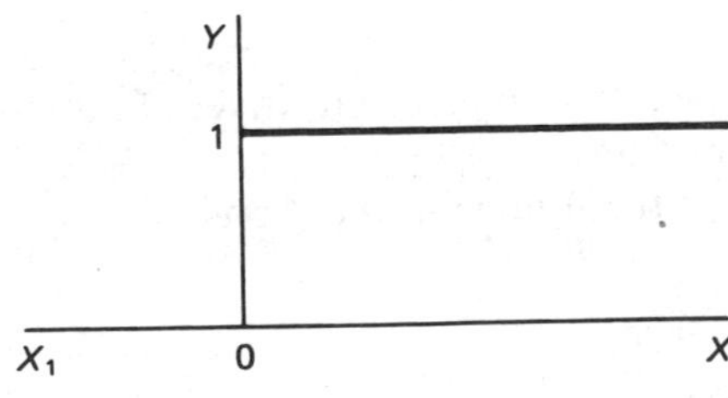

If the unit step occurs at the origin, then $c = 0$ and $F(t) = H(t-c)$ becomes

$$F(t) = H(t)$$

i.e.

$$H(t) = 0 \quad \text{for} \quad t < 0$$
$$H(t) = 1 \quad \text{for} \quad t \geq 0.$$

Effect of the unit step function

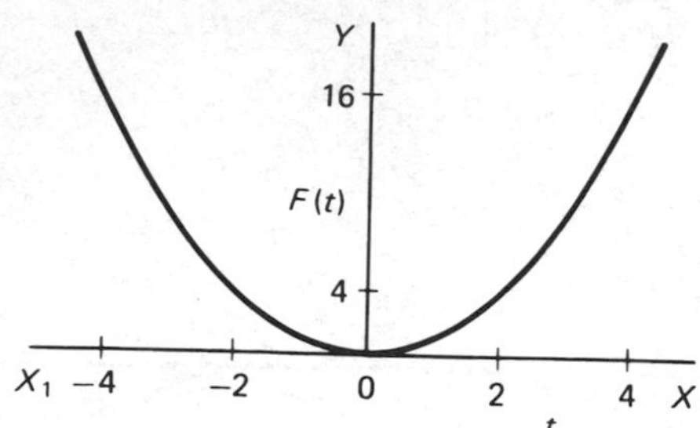

The graph of $F(t) = t^2$ is, of course, as shown.

Remembering the definition of $H(t - c)$, the graph of

$$F(t) = H(t - 2) \cdot t^2 \text{ is}$$

. . . .

5

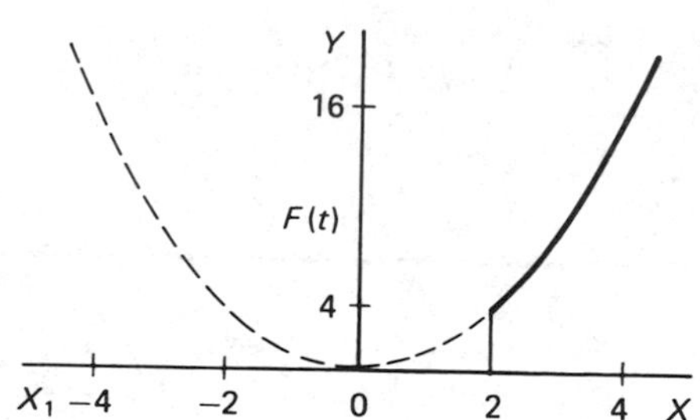

For $t < 2, \quad H(t - 2) = 0 \qquad \therefore H(t - 2) \cdot t^2 = 0 \cdot t^2 = 0$

$t \geq 2, \quad H(t - 2) = 1 \qquad \therefore H(t - 2) \cdot t^2 = 1 \cdot t^2 = t^2.$

So the function $H(t - 2)$ suppresses the function t^2 for all values of t up to $t = 2$ and 'switches on' the function t^2 at $t = 2$.

Now we can sketch the graphs of the following functions.

(a) $F(t) = \sin t \qquad \text{for } 0 < t < 2\pi$

(b) $F(t) = H(t - \pi/4) \cdot \sin t \qquad \text{for } 0 < t < 2\pi.$

These give . . . .

and . . . .

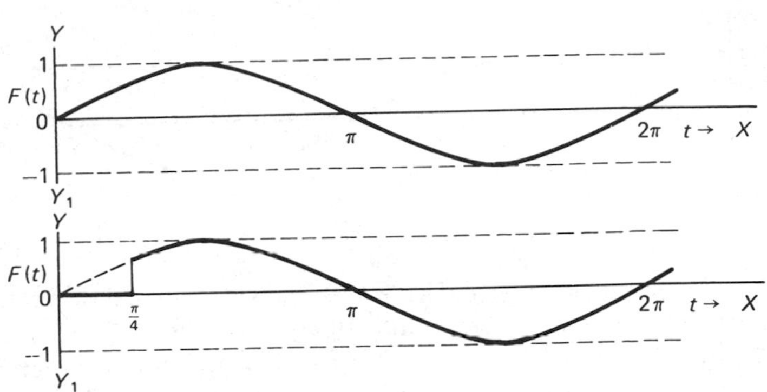

That is, the graph of $F(t) = H(t - \pi/4) \cdot \sin t$ is the graph of $F(t) = \sin t$ but suppressed for all values prior to $t = \pi/4$.

If we sketch the graph of $F(t) = \sin(t - \pi/4)$ we have

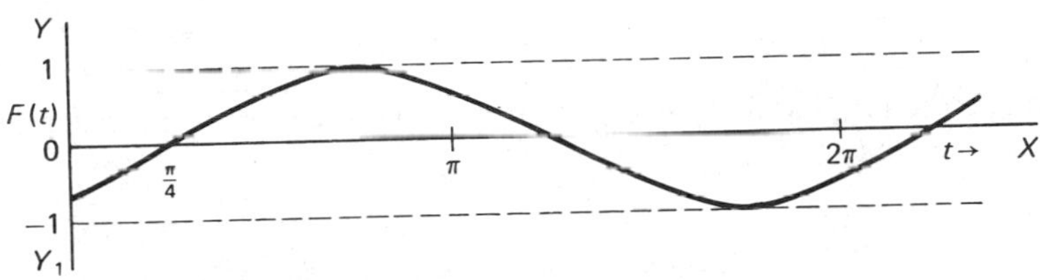

Since $H(t - c)$ has the effect of suppressing a function for $t < c$, then the graph of $F(t) = H(t - \pi/4) \cdot \sin(t - \pi/4)$ is

. . . .

7

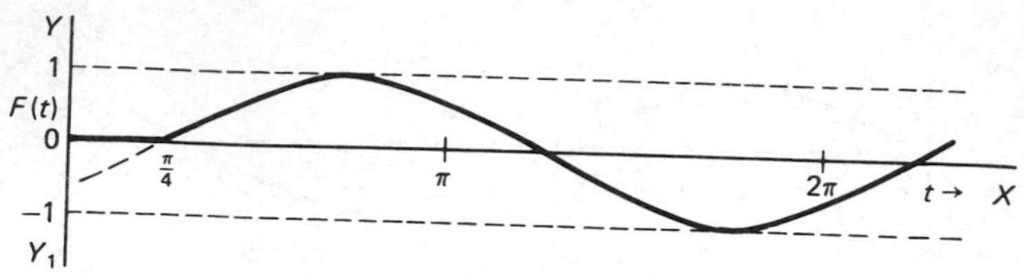

That is, the graph of $F(t) = H(t - \pi/4) \cdot \sin(t - \pi/4)$ is the graph of $F(t) = \sin t$ $(t > 0)$, shifted $\pi/4$ units along the t-axis.

In general, the graph of $F(t) = H(t - c) \cdot \sin(t - c)$ is the graph of $F(t) = \sin t$ $(t > 0)$, shifted along the t-axis through an interval of c units.

Similarly, for $t > 0$, sketch the graphs of

(a) $F(t) = e^{-t}$

(b) $F(t) = H(t - c) \cdot e^{-t}$

(c) $F(t) = H(t - c) \cdot e^{-(t-c)}$.

Arrange the graphs under each other to show the important differences.

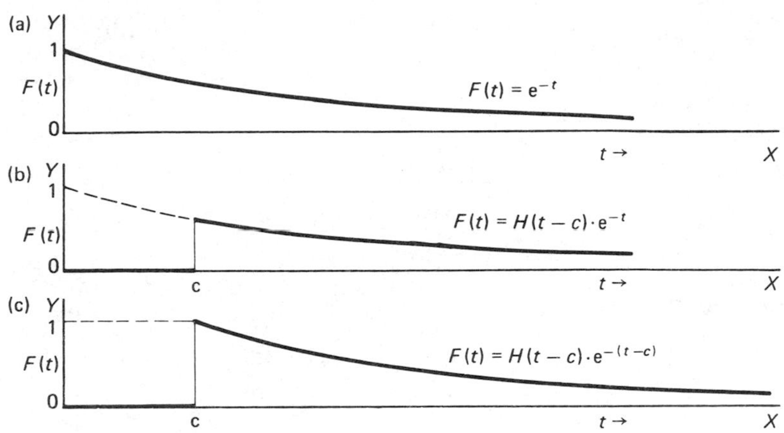

In (a), we have the graph of $F(t) = e^{-t}$
In (b), the same graph is suppressed prior to $t = c$
In (c), the graph of $F(t) = e^{-t}$ is shifted c units along the t-axis.

Laplace transform of $H(t - c)$

We simply apply the definition of a Laplace transform.

$$\mathscr{L}\{H(t-c) = \int_0^\infty e^{-st} \cdot H(t-c)dt$$

But
$$e^{-st} \cdot H(t-c) = 0 \quad \text{for} \quad 0 < t < c$$
$$= e^{-st} \quad \text{for} \quad t \geq c.$$

$$\therefore \mathscr{L}\{H(t-c)\} = \int_0^\infty e^{-st} \cdot H(t-c)dt = \int_c^\infty e^{-st}dt$$

$$= \left[\frac{e^{-st}}{-s}\right]_c^\infty = \frac{e^{-cs}}{s}$$

$$\therefore \mathscr{L}\{H(t-c)\} = \frac{e^{-cs}}{s}$$

Therefore, the Laplace transform of the unit step function at the origin is

$$\mathscr{L}\{H(t)\} = \ldots\ldots\ldots$$

9

$$\boxed{\frac{1}{s}}$$

since $c = 0$.

So, $$\mathscr{L}\{H(t-c)\} = \frac{e^{-cs}}{s}$$

and $$\mathscr{L}\{H(t)\} = \frac{1}{s}.$$

Also, from the definition of $H(t)$

$$\mathscr{L}(1) = \mathscr{L}\{1 \cdot \mathrm{H(t)}\};$$
$$\mathscr{L}(t) = \mathscr{L}\{t \cdot H(t)\};$$
$$\mathscr{L}\{f(t)\} = \mathscr{L}\{f(t) \cdot H(t)\}$$

Make a note of these results: we shall be using them.

10 As we have seen, the unit step function $H(t-c)$ is often combined with other functions of t, so we now consider the Laplace transform of $H(t-c) \cdot F(t-c)$.

Laplace transform of $H(t-c) \cdot F(t-c)$

By the definition of a Laplace transform

$$\mathscr{L}\{H(t-c) \cdot F(t-c)\} = \int_0^\infty e^{-st} \cdot H(t-c) \cdot F(t-c)\mathrm{d}t$$

As before, since $H(t-c) = 0 \quad \text{for} \quad 0 < t < c$

$$= 1 \quad \text{for} \quad t \geq c$$

$$\therefore \ \mathscr{L}\{H(t-c) \cdot F(t-c)\} = \int_c^\infty e^{-st} \cdot F(t-c)\mathrm{d}t$$

We now make the substitution $t - c = v \quad \therefore t = c + v \quad \therefore \mathrm{d}t \equiv \mathrm{d}v$

Also, for the limits, when $t = c$, $v = 0$ and when $t = \infty$, $v = \infty$.

$$\therefore \ \mathscr{L}\{H(t-c) \cdot F(t-c)\} = \int_0^\infty e^{-s(c+v)} \cdot F(v)\mathrm{d}v$$

$$= e^{-cs} \int_0^\infty e^{-sv} \cdot F(v)\mathrm{d}v$$

Now $\int_0^\infty e^{-sv} \cdot F(v)\mathrm{d}v$ has exactly the same value as $\int_0^\infty e^{-st} \cdot F(t)\mathrm{d}t$ which is, of course, the Laplace transform of $F(t)$.

$$\therefore \ \mathscr{L}\{H(t-c) \cdot F(t-c)\} = e^{-cs}\mathscr{L}\{F(t)\} = e^{-cs} \cdot f(s)$$

So $\underline{\mathscr{L}\{H(t-c) \cdot F(t-c)\} = e^{-cs} \cdot f(s) \qquad \text{where } f(s) = \mathscr{L}\{F(t)\}}$

Make a note of this important result.

11

$$\boxed{\mathscr{L}\{H(t-c)\cdot F(t-c)\} = \mathrm{e}^{-cs}\cdot f(s) \quad \text{where} \quad f(s) = \mathscr{L}\{F(t)\}}$$

So $\mathscr{L}\{H(t-4)\cdot(t-4)^2\} = \mathrm{e}^{-4s}\cdot f(s) \quad \text{where} \quad f(s) = \mathscr{L}\{t^2\}$

$$= \mathrm{e}^{-4s}\left(\frac{2!}{s^3}\right) = \underline{\frac{2\mathrm{e}^{-4s}}{s^3}}$$

Note that $f(s)$ is the transform of t^2 and *not* of $(t-4)^2$.

In the same way,

$$\mathscr{L}\{H(t-3)\cdot\sin(t-3)\} = \ldots\ldots\ldots$$

12

$$\boxed{\frac{\mathrm{e}^{-3s}}{s^2+1}}$$

for $\mathscr{L}\{H(t-3)\cdot\sin(t-3)\} = \mathrm{e}^{-3s}\cdot f(s) \quad \text{where} \quad f(s) = \mathscr{L}\{\sin t\}$

$$= \frac{1}{s^2+1}$$

$$\therefore \{H(t-3)\cdot\sin(t-3)\} = \underline{\mathrm{e}^{-3s}\left(\frac{1}{s^2+1}\right)}$$

So now do these in the same way.

(a) $\mathscr{L}\{H(t-2)\cdot(t-2)^3\} = \ldots\ldots\ldots$

(b) $\mathscr{L}\{H(t-1)\cdot\sin 3(t-1)\} = \ldots\ldots\ldots$

(c) $\mathscr{L}\{H(t-5)\cdot\mathrm{e}^{(t-5)}\} = \ldots\ldots\ldots$

(d) $\mathscr{L}\{H(t-\pi/2)\cdot\cos 2(t-\pi/2)\} = \ldots\ldots\ldots$

13 Here they are.

(a) $\mathscr{L}\{H(t-2)\cdot(t-2)^3\} = e^{-2s}\cdot f(s)$ where $f(s) = \mathscr{L}\{t^3\}$

$$= e^{-2s}\left(\frac{3!}{s^4}\right) = \underline{\frac{6e^{-2s}}{s^4}}$$

(b) $\mathscr{L}\{H(t-1)\cdot\sin 3(t-1)\} = e^{-s}\cdot f(s)$ where $f(s) = \mathscr{L}\{\sin 3t\}$

$$= e^{-s}\left(\frac{3}{s^2+9}\right) = \underline{\frac{3e^{-s}}{s^2+9}}$$

(c) $\mathscr{L}\{H(t-5)\cdot e^{(t-5)}\} = e^{-5s}\cdot f(s)$ where $f(s) = \mathscr{L}\{e^t\}$

$$= e^{-5s}\left(\frac{1}{s-1}\right) = \underline{\frac{e^{-5s}}{s-1}}$$

(d) $\mathscr{L}\{H(t-\pi/2)\cdot\cos 2(t-\pi/2)\} = e^{-\pi s/2}\cdot f(s)$ where

$$f(s) = \mathscr{L}\{\cos 2t\}$$

$$= e^{-\pi s/2}\left(\frac{s}{s^2+4}\right) = \underline{\frac{s\cdot e^{-\pi s/2}}{s^2+4}}$$

So $\mathscr{L}\{H(t-c)\cdot F(t-c)\} = e^{-cs}\cdot f(s)$ where $f(s) = \mathscr{L}\{F(t)\}$.

Written in reverse, this becomes

$$\underline{\text{If } f(s) = \mathscr{L}\{F(t)\}, \text{ then } e^{-cs}\cdot f(s) = \mathscr{L}\{H(t-c)\cdot F(t-c)\}}$$

where c is real and positive.

This is known as the *Second Shift Theorem. Make a note of it: then we will use it.*

14

$$\text{If } f(s) = \mathscr{L}\{F(t)\}, \text{ then } e^{-cs} \cdot f(s) = \mathscr{L}\{H(t-c) \cdot F(t-c)\}$$

This is useful in finding inverse transforms, as we shall now see.

Example 1 Find the function whose transform is $\dfrac{e^{-4s}}{s^2}$.

The numerator corresponds to e^{-cs} where $c = 4$ and therefore indicates $H(t-4)$.

Then $\dfrac{1}{s^2} = f(s) = \mathscr{L}\{t\} \quad \therefore F(t) = t$.

$$\therefore \mathscr{L}^{-1}\left\{\frac{e^{-4s}}{s^2}\right\} = H(t-4) \cdot (t-4)$$

Remember that in writing the final result, $F(t)$ is replaced by

....

15

$$F(t-c)$$

Example 2 Determine $\mathscr{L}^{-1}\left\{\dfrac{6e^{-2s}}{s^2+4}\right\}$.

The numerator contains e^{-2s} and therefore indicates ....

16

$$\boxed{H(t-2)}$$

The remainder of the transform, i.e. $\dfrac{6}{s^2+4}$, can be written as $3\left(\dfrac{2}{s^2+4}\right)$

$$\therefore \frac{6}{s^2+4} = f(s) = \mathscr{L}\{\,\ldots\ldots\ldots\}$$

17

$$\boxed{\mathscr{L}\{3\sin 2t\}}$$

$$\therefore \mathscr{L}^{-1}\left\{\frac{6\mathrm{e}^{-2s}}{s^2+4}\right\} = \ldots\ldots\ldots$$

18

$$\boxed{3H(t-2)\cdot\sin 2(t-2)}$$

for $\mathscr{L}^{-1}\left\{\dfrac{6\mathrm{e}^{-2s}}{s^2+4}\right\} = H(t-2)\cdot F(t-2)$ where $F(t) = \mathscr{L}^{-1}\left\{\dfrac{6}{s^2+4}\right\}$

$$= \underline{H(t-2)\cdot 3\sin 2(t-2)}$$

Example 3 Determine $\mathscr{L}^{-1}\left\{\dfrac{s\cdot\mathrm{e}^{-s}}{s^2+9}\right\}$.

This, in similar manner, is ………

19

$$\boxed{H(t-1)\cdot\cos 3(t-1)}$$

for the numerator contains e^{-s} which indicates $H(t-1)$.

Also $\dfrac{s}{s^2+9} = f(s) = \mathscr{L}\{\cos 3t\} \;\therefore\; F(t) = \cos 3t \;\therefore\; F(t-1) = \cos 3(t-1)$.

$$\therefore\; \mathscr{L}^{-1}\left\{\frac{s\cdot e^{-s}}{s^2+9}\right\} = H(t-1)\cdot\cos 3(t-1)$$

Remember that, having obtained $F(t)$, the result contains $F(t-c)$.

Here is a short exercise by way of practice.

Exercise Determine the inverse transforms of the following.

(a) $\dfrac{2e^{-5s}}{s^3}$ (d) $\dfrac{2s\cdot e^{-3s}}{s^2-16}$

(b) $\dfrac{3e^{-2s}}{s^2-1}$ (e) $\dfrac{5e^{-s}}{s}$

(c) $\dfrac{8c^{-4s}}{s^2+4}$ (f) $\dfrac{s\cdot e^{-s/2}}{s^2+2}$

Results—all very straightforward. 20

(a) $H(t-5)\cdot(t-5)^2$
(b) $3H(t-2)\cdot\sinh(t-2)$
(c) $4H(t-4)\cdot\sin 2(t-4)$
(d) $2H(t-3)\cdot\cosh 4(t-3)$
(e) $5H(t-1)$
(f) $H(t-1/2)\cdot\cos\sqrt{2}(t-1/2)$.

Before looking at a more interesting example, let us collect our results together as far as we have gone.

21

The main points are

(a)
$$\left.\begin{aligned} H(t-c) &= 0 \qquad 0 < t < c \\ &= 1 \qquad t \geq c \end{aligned}\right\} \tag{1}$$

(b)
$$\left.\begin{aligned} \mathscr{L}\{H(t-c)\} &= \frac{e^{-cs}}{s} \\ \mathscr{L}\{H(t)\} &= \frac{1}{s} \end{aligned}\right\} \tag{2}$$

(c) $\mathscr{L}\{H(t-c)\cdot F(t-c)\} = e^{-cs}\cdot f(s)$ where $f(s) = \mathscr{L}\{F(t)\}$ (3)

(d) If $f(s) = \mathscr{L}\{F(t)\}$, then $e^{-cs}\cdot f(s) = \mathscr{L}\{H(t-c)\cdot F(t-c)\}$ (4)

Now let us apply these to some further examples.

Example 1 Determine the function $F(t)$ for which

$$\mathscr{L}\{F(t)\} = \frac{3}{s} - \frac{4e^{-s}}{s^2} + \frac{5e^{-2s}}{s^2}.$$

We take each term in turn and find its inverse transform.

(i) $\mathscr{L}^{-1}\left\{\frac{3}{s}\right\} = 3\ \mathscr{L}^{-1}\left\{\frac{1}{s}\right\} = 3$ i.e. $3H(t)$

(ii) $\mathscr{L}^{-1}\left\{\frac{4e^{-s}}{s^2}\right\} = H(t-1)\cdot 4(t-1)$

(iii) $\mathscr{L}^{-1}\left\{\frac{5e^{-2s}}{s^2}\right\} = \ldots\ldots\ldots$

22

$$\boxed{H(t-2)\cdot 5(t-2)}$$

So we have $\mathscr{L}^{-1}\left\{\dfrac{3}{s}\right\} = 3H(t)$

$$\mathscr{L}^{-1}\left\{\frac{4e^{-s}}{s^2}\right\} = H(t-1)\cdot 4(t-1)$$

$$\mathscr{L}^{-1}\left\{\frac{5e^{-2s}}{s^2}\right\} = H(t-2)\cdot 5(t-2)$$

$$\therefore\ F(t) = 3H(t) - H(t-1)\cdot 4(t-1) + H(t-2)\cdot 5(t-2)$$

To sketch the graph of $F(t)$ we consider the values of the function within the three sections $0 < t < 1$, $1 < t < 2$, and $2 < t$.

Between $t = 0$ and $t = 1$, $F(t) = \dots\dots\dots$

23

$$\boxed{F(t) = 3}$$

since in this interval, $H(t) = 1$, but $H(t-1) = 0$ and $H(t-2) = 0$. In the same way, between $t = 1$ and $t = 2$, $F(t) = \dots\dots\dots$

24

$$\boxed{F(t) = 7 - 4t}$$

for, between $t = 1$ and $t = 2$, $H(t) = 1$, $H(t-1) = 1$, but $H(t-2) = 0$.

$$\therefore\ F(t) = 3 - 4(t-1) + 0 = 3 - 4t + 4 = 7 - 4t$$

Similarly, for $t > 2$, $F(t) = \dots\dots\dots$

25

$$\boxed{F(t) = t - 3}$$

since, for $t > 2$, $H(t) = 1$, $H(t-1) = 1$ and $H(t-2) = 1$

$$\therefore F(t) = 3 - 4(t-1) + 5(t-2)$$
$$= 3 - 4t + 4 + 5t - 10 = t - 3$$

So, collecting the results together, we have

for $0 < t < 1$, $F(t) = 3$

$1 < t < 2$, $F(t) = 7 - 4t$ $(t = 1,\ F(t) = 3;\ t = 2,\ F(t) = -1)$

$2 < t$, $F(t) = t - 3$ $(t = 2,\ F(t) = -1;\ t = 3,\ F(t) = 0)$

Using these facts we can sketch the graph of $F(t)$, which is

. . . .

26

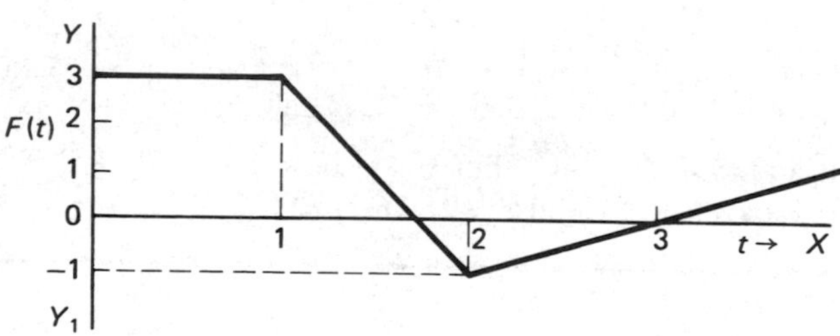

Here is another.

Example 2 Determine the function $F(t) = \mathscr{L}^{-1}\left\{\dfrac{2}{s} + \dfrac{3e^{-s}}{s^2} - \dfrac{3e^{-3s}}{s^2}\right\}$

and sketch the graph of $F(t)$.

First we express the inverse transform of each term in terms of the unit step function.

This gives . . . .

27

$$\mathscr{L}^{-1}\left\{\frac{2}{s}\right\} = 2H(t); \quad \mathscr{L}^{-1}\left\{\frac{3e^{-s}}{s^2}\right\} = H(t-1)\cdot 3(t-1)$$

$$\mathscr{L}^{-1}\left\{\frac{3e^{-3s}}{s^2}\right\} = H(t-3)\cdot 3(t-3)$$

$$\therefore\ F(t) = 2H(t) + H(t-1)\cdot 3(t-1) - H(t-3)\cdot 3(t-3)$$

So there are 'break points', i.e. changes of function, at $t = 1$ and $t = 3$, and we investigate $F(t)$ within the three intervals.

$0 < t < 1 \qquad F(t) = \dots\dots\dots$

$1 < t < 3 \qquad F(t) = \dots\dots\dots$

$3 < t \qquad F(t) = \dots\dots\dots$

28

$0 < t < 1,\ F(t) = 2; \quad 1 < t < 3,\ F(t) = 3t - 1; \quad 3 < t,\ F(t) = 8$

for with

$0 < t < 1, \quad H(t) = 1$, but $H(t-1) = H(t-3) = 0 \quad \therefore\ F(t) = 2$

$1 < t < 3, \quad H(t) = 1,\ H(t-1) = 1$, but $H(t-3) = 0$

$\therefore\ F(t) = 2 + 3(t-1) = 3t - 1 \quad \therefore\ F(t) = 3t - 1$

$3 < t, \quad H(t) = 1,\ H(t-1) = 1,\ H(t-3) = 1$

$\therefore\ F(t) = 2 + 3t - 3 - 3t + 9 \quad \therefore\ F(t) = 8$

Therefore, the graph of $F(t)$ is ………

29

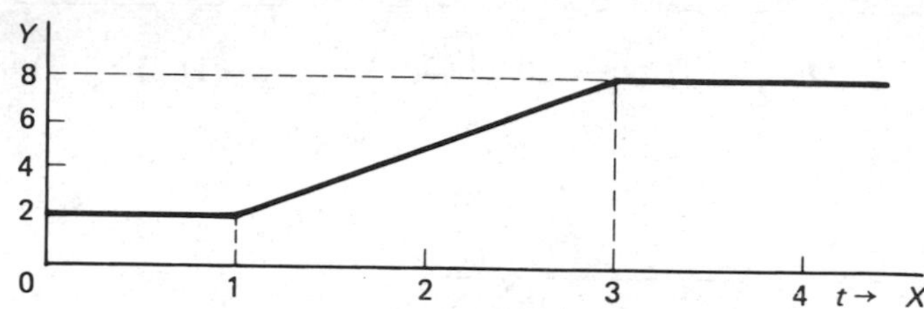

Between the 'break points', $F(t) = 3t - 1$ $\begin{cases} t = 1, \ F(t) = 2 \\ t = 3, \ F(t) = 8 \end{cases}$

Now turn on for the next example.

30

Example 3 If $F(t) = \mathscr{L}^{-1}\left\{\dfrac{(1 - e^{-2s})(1 + e^{-4s})}{s^2}\right\}$, determine $F(t)$ and sketch the graph of the function.

Although at first sight this looks more complicated, we simply multiply out the numerator and proceed as before.

$$F(t) = \mathscr{L}^{-1}\left\{\frac{1 - e^{-2s} + e^{-4s} - e^{-6s}}{s^2}\right\}$$

$$= \mathscr{L}^{-1}\left\{\frac{1}{s^2} - \frac{e^{-2s}}{s^2} + \frac{e^{-4s}}{s^2} - \frac{e^{-6s}}{s^2}\right\}$$

We now write down the inverse transform of each term in terms of the unit function, so that

$$F(t) = \dots\dots\dots$$

31

$$F(t) = H(t) \cdot t - H(t-2) \cdot (t-2) + H(t-4) \cdot (t-4) - H(t-6) \cdot (t-6)$$

and we can see there are 'break points' at $t = 2$, $t = 4$, $t = 6$.

$$\begin{array}{lll} \text{For } 0 < t < 2, & F(t) = t - 0 + 0 - 0 & F(t) = t \\ 2 < t < 4, & F(t) = t - (t-2) + 0 - 0 & F(t) = 2 \\ 4 < t < 6, & F(t) = t - (t-2) + (t-4) - 0 & F(t) = t - 2 \\ 6 < t, & F(t) = t - (t-2) + (t-4) - (t-6) & F(t) = 4. \end{array}$$

The second and fourth components are constant, but before sketching the graph of the function, we check the values of $F(t) = t$ and $F(t) = t - 2$ at the relevant break points.

$F(t) = t.$ At $t = 0$, $F(t) = 0$; at $t = 2$, $F(t) = 2$

$F(t) = t - 2.$ At $t = 4$, $F(t) = 2$; at $t = 6$, $F(t) = 4$.

So the graph of the function is ....

32

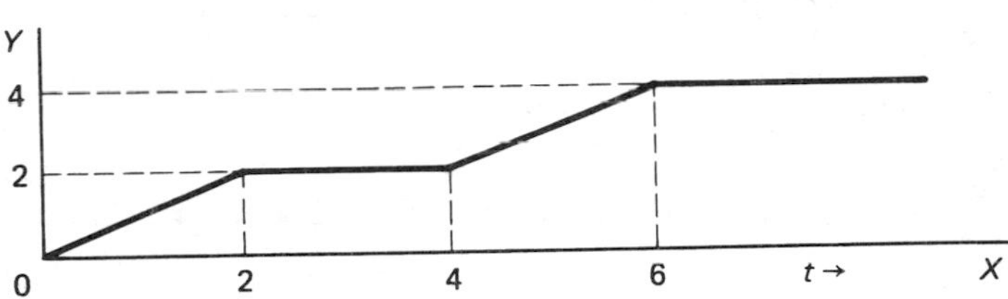

It is always wise to calculate the function values at break points, since discontinuities, or jumps, sometimes occur.

On to the next frame.

Now for one in reverse. 33

Example 4 A function $F(t)$ is defined by

$$\begin{aligned} F(t) &= 4 \quad \text{for} \quad 0 < t < 2 \\ &= 2t - 3 \quad \text{for} \quad 2 < t. \end{aligned}$$

Sketch the graph of the function and determine its Laplace transform.

We see that (i) for $t = 0$ to $t = 2$, $F(t) = 4$

and (ii) for $t > 2$, $F(t) = 2t - 3$ and at $t = 2$, $F(t) = 1$.

So the graph is ....

34

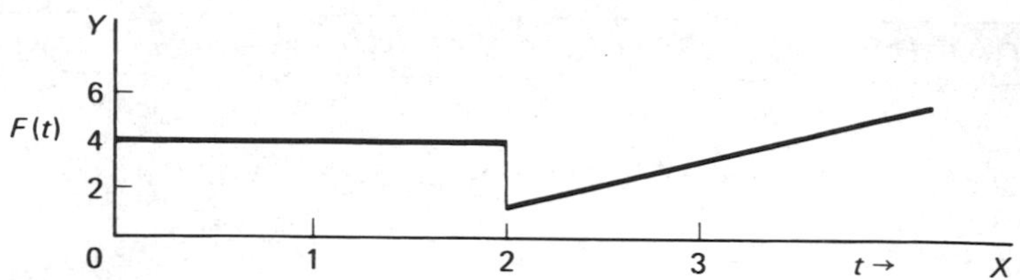

Notice the discontinuity at $t = 2$.

Expressing the function in unit step form

$$F(t) = 4H(t) - 4H(t-2) + H(t-2) \cdot (2t-3)$$

Note that the second term cancels $F(t) = 4$ at $t = 2$ and that the third term switches on $F(t) = 2t - 3$ at $t = 2$.

Before we can express this in Laplace transforms, $(2t - 3)$ in the third term must be written as a function of $(t - 2)$ to correspond to $H(t - 2)$. Therefore, we write $2t - 3$ as $2(t - 2) + 1$

Then $$\begin{aligned} F(t) &= 4H(t) - 4H(t-2) + H(t-2) \cdot \{2(t-2)+1\} \\ &= 4H(t) - 4H(t-2) + H(t-2) \cdot 2(t-2) + H(t-2) \\ &= 4H(t) - 3H(t-2) + H(t-2) \cdot 2(t-2) \end{aligned}$$

$\therefore \ \mathscr{L}\{F(t)\} = \ldots\ldots\ldots$

35

$$\mathscr{L}\{F(t)\} = \frac{4}{s} - \frac{3e^{-2s}}{s} + \frac{2e^{-2s}}{s^2}$$

Here is one for you to work through in much the same way.

Example 5 A function is defined by
$$\begin{aligned} F(t) &= 6 && 0 < t < 1 \\ &= 8 - 2t && 1 < t < 3 \\ &= 4 && 3 < t. \end{aligned}$$

Sketch the graph and find the Laplace transform of the function.

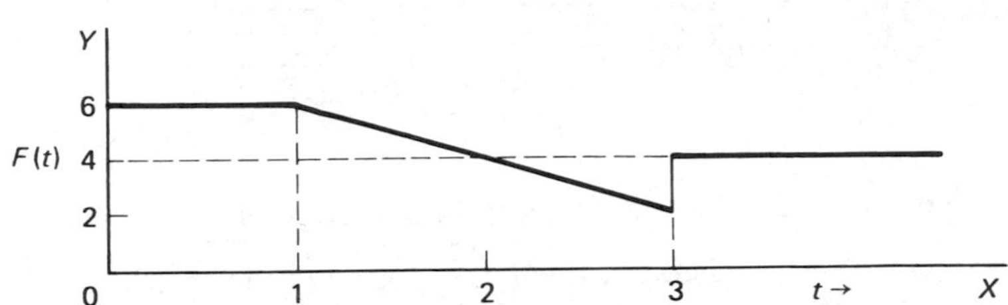

Expressing this in unit step form we have

$$F(t) = 6H(t) - 6H(t-1) + H(t-1)\cdot(8-2t)$$

$$-H(t-3)\cdot(8-2t) + H(t-3)\cdot 4$$

where the second term switches off the first function $F(t) = 6$ at $t = 1$ and the third term switches on the second function $F(t) = 8 - 2t$, which in turn is switched off by the fourth term at $t = 3$ and replaced by $F(t) = 4$ in the fifth term.

Before we can write down the transforms of the third and fourth terms, we must express $F(t) = 8 - 2t$ in terms of $(t-1)$ and $(t-3)$ respectively.

$$8 - 2t = 6 + 2 - 2t = 6 - 2(t-1)$$

$$8 - 2t = 2 + 6 - 2t = 2 - 2(t-3).$$

$$\begin{aligned}\therefore\ F(t) &= 6H(t) - 6H(t-1) + H(t-1)\cdot\{6 - 2(t-1)\}\\ &\quad - H(t-3)\cdot\{2 - 2(t-3)\} + 4H(t-3)\\ &= 6H(t) - 6H(t-1) + 6H(t-1)\\ &\quad - H(t-1)\cdot 2(t-1) - 2H(t-3)\\ &\quad + H(t-3)\cdot 2(t-3) + 4H(t-3)\end{aligned}$$

which simplifies finally to $F(t) = \dots\dots\dots$

37

$$\boxed{F(t) = 6H(t) - H(t-1)\cdot 2(t-1) + H(t-3)\cdot 2(t-3) + 2H(t-3)}$$

from which $\mathscr{L}\{F(t)\} = \dots\dots\dots$

38

$$\mathscr{L}\{F(t)\} = \frac{6}{s} - \frac{2e^{-s}}{s^2} + \frac{2e^{-3s}}{s^2} + \frac{2e^{-3s}}{s}$$

Note that, in building up the function in unit step form

(a) to 'switch on' a function $F(t)$ at $t = c$ we add the term $H(t-c) \cdot F(t-c)$

(b) to 'switch off' a function $F(t)$ at $t = c$, we subtract $H(t-c) \cdot F(t-c)$.

Now on to something rather different.

Laplace Transforms of Periodic Functions

39 In many technological problems, we are dealing with forms of mechanical vibrations or electrical oscillations and the necessity to express such periodic functions in Laplace transforms soon arises.

Let $F(t)$ be a periodic function of period ω, i.e. $F(t) = F(t + n\omega)$ for $n = 1, 2, 3, \ldots$

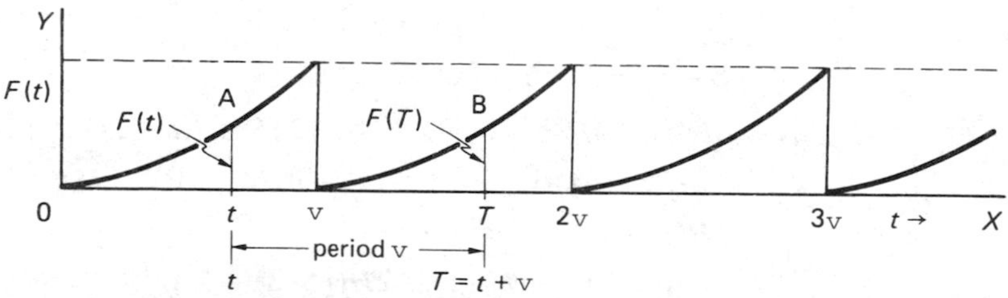

Let A and B be two corresponding points in successive cycles.

$$\therefore\ T = t + \omega \qquad \therefore\ t = T - \omega \qquad \text{Also, } F(T) = F(t)$$

By the definition of a Laplace transform,

$$\mathscr{L}\{F(t)\} = \int_0^\infty e^{-st} \cdot F(t)\,dt \qquad (s > 0) \qquad (1)$$

Now since $t = T - \omega$ $\quad \therefore\ dt \equiv dT$ and $F(T) = F(t)$

$\therefore$ when $t = 0$, $\quad T = \ldots\ldots\ldots$ and when $t = \infty$, $\quad T = \ldots\ldots\ldots$

40

$$t = 0,\ T = \omega; \quad t = \infty,\ T = \infty$$

In terms of T

$$\mathscr{L}\{F(t)\} = \int_{\omega}^{\infty} e^{-s(T-\omega)} \cdot F(T)dT$$

$$= \int_{\omega}^{\infty} e^{-sT} \cdot e^{\omega s} \cdot F(T)dT = e^{\omega s}\int_{\omega}^{\infty} e^{-sT} \cdot F(T)dT$$

$$\therefore\ e^{-\omega s}\mathscr{L}\{F(t)\} = \int_{\omega}^{\infty} e^{-sT} \cdot F(T)dT = \int_{\omega}^{\infty} e^{-st} \cdot F(t)dt \qquad (2)$$

since the value of the definite integral depends on the limits and not on the symbols chosen for the variable.

Subtracting result (2) from result (1), we have

....

41

$$(1 - e^{-\omega s})\mathscr{L}\{F(t)\} = \int_{0}^{\infty} e^{-st} \cdot F(t)dt - \int_{\omega}^{\infty} e^{-st} \cdot F(t)dt$$

$$\therefore\ (1 - e^{-\omega s})\mathscr{L}\{F(t)\} = \int_{0}^{\omega} e^{-st} \cdot F(t)dt$$

$$\therefore\ \mathscr{L}\{F(t)\} = \frac{1}{1 - e^{-\omega s}}\int_{0}^{\omega} e^{-st} \cdot F(t)dt$$

where ω is the period of the function $F(t)$.

Note that we integrate $e^{-st} \cdot F(t)$ over one cycle, i.e. from $t = 0$ to $t = \omega$ and not from $t = 0$ to $t = \infty$ as we did previously.

This is an important result. Make a note of it—then we will apply it.

Example 1 Find the Laplace transform of the function $F(t)$ defined by

$$\left.\begin{aligned} F(t) &= 3 \quad 0 < t < 2 \\ &= 0 \quad 2 < t < 4 \end{aligned}\right\} \quad F(t) = F(t+4).$$

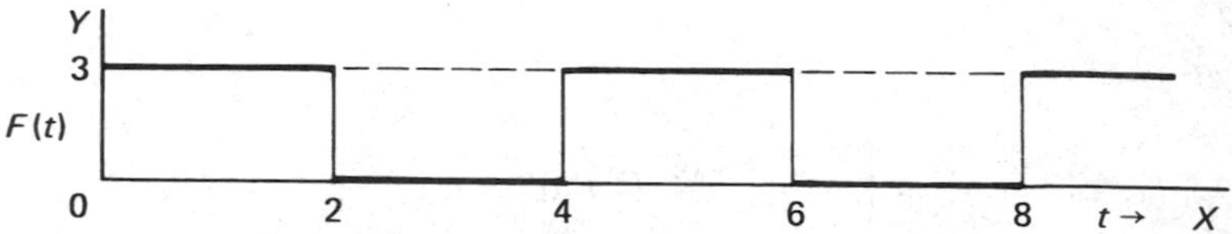

The expression for $\mathscr{L}\{F(t)\}$ is

.... (do not evaluate it yet)

42

$$\mathscr{L}\{F(t)\} = \frac{1}{1-\mathrm{e}^{-4s}} \int_0^4 \mathrm{e}^{-st} \cdot F(t)\mathrm{d}t$$

for the period $= 4$, i.e. $\omega = 4$.

The function $F(t) = 3$ for $0 < t < 2$ and $F(t) = 0$ for $2 < t < 4$.

$$\therefore \mathscr{L}\{F(t)\} = \frac{1}{1-\mathrm{e}^{-4s}} \int_0^2 \mathrm{e}^{-st} \cdot 3\,\mathrm{d}t = \ldots\ldots\ldots$$

43

$$\mathscr{L}\{F(t)\} = \frac{3}{s(1+\mathrm{e}^{-2s})}$$

for $\mathscr{L}\{F(t)\} = \dfrac{3}{1-\mathrm{e}^{-4s}} \left[\dfrac{\mathrm{e}^{-st}}{-s}\right]_0^2 = \dfrac{3}{1-\mathrm{e}^{-4s}} \left\{\left(\dfrac{\mathrm{e}^{-2s}}{-s}\right) - \left(\dfrac{1}{-s}\right)\right\}$

$$= \frac{3}{1-\mathrm{e}^{-4s}} \left\{\frac{1-\mathrm{e}^{-2s}}{s}\right\} = \underline{\frac{3}{s(1+\mathrm{e}^{-2s})}}$$

That is all there is to it. Now for another, so move on.

44

Example 2 Find the Laplace transform of the periodic function defined by

$$F(t) = t/2 \qquad 0 < t < 3$$
$$F(t) = F(t+3).$$

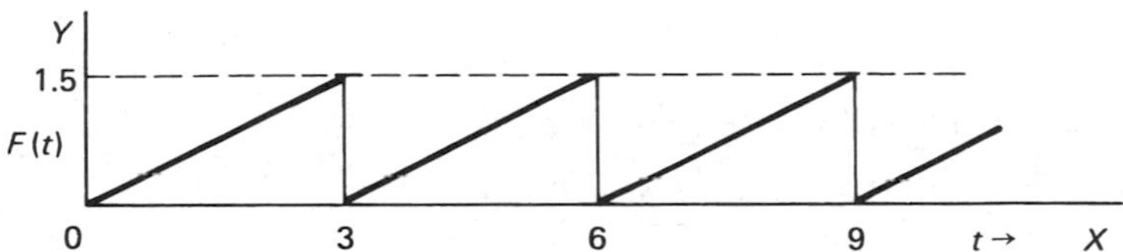

In this case, period $= 3$, i.e. $\omega = 3$.

$$\therefore \mathscr{L}\{F(t)\} = \frac{1}{1-e^{-\omega s}} \int_0^\omega e^{-st} \cdot F(t)\,dt$$

$$= \frac{1}{1-e^{-3s}} \int_0^3 e^{-st} \cdot \left(\frac{t}{2}\right) dt$$

$$\therefore 2(1-e^{-3s})\mathscr{L}\{F(t)\} = \int_0^3 t \cdot e^{-st}\,dt$$

Integrating by parts and simplifying the result gives

$$\mathscr{L}\{F(t)\} = \ldots\ldots\ldots$$

45

$$\boxed{\mathscr{L}\{F(t)\} = \frac{1}{2s^2}\left\{1 - \frac{3s}{e^{3s}-1}\right\}}$$

for $\quad 2(1-e^{-3s})\mathscr{L}\{F(t)\} = \int_0^3 te^{-st}\,dt$

$$= \left[t\left(\frac{e^{-st}}{-s}\right)\right]_0^3 + \frac{1}{s}\int_0^3 e^{-st}\,dt$$

$$= -\frac{3e^{-3s}}{s} + \frac{1}{s}\left[\frac{e^{-st}}{-s}\right]_0^3$$

$$= -\frac{3e^{-3s}}{s} - \frac{e^{-3s}}{s^2} + \frac{1}{s^2}$$

$$= \frac{1}{s^2}\{1 - 3se^{-3s} - e^{-3s}\}$$

$$\therefore\ \mathscr{L}\{F(t)\} = \frac{1}{2s^2}\left\{1 - \frac{3se^{-3s}}{1 - e^{-3s}}\right\} = \underline{\frac{1}{2s^2}\left\{1 - \frac{3s}{e^{3s} - 1}\right\}}$$

Example 3 Sketch the graph of the function

$$F(t) = e^t \qquad 0 < t < 5$$
$$F(t) = F(t + 5)$$

and determine its Laplace transform.

First we sketch the graph of $F(t)$, which is ....

46

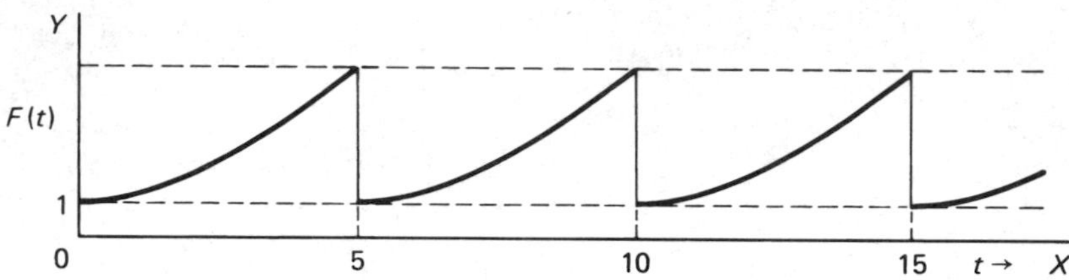

Clearly, period $= 5 \quad \therefore\ \omega = 5$

$$\mathscr{L}\{F(t)\} = \frac{1}{1 - e^{-\omega s}} \int_0^{\omega} e^{-st} \cdot F(t)dt \quad \text{gives}$$

$\mathscr{L}\{F(t)\} = \ldots\ldots\ldots$ Complete the working.

47

$$\mathcal{L}\{F(t)\} = \frac{1 - e^{-5(s-1)}}{(s-1)(1 - e^{-5s})}$$

For $$\mathcal{L}\{F(t)\} = \frac{1}{1 - e^{-5s}} \int_0^5 e^{-st} \cdot e^t \, dt$$

$$\therefore \; (1 - e^{-5s})\mathcal{L}\{F(t)\} = \int_0^5 e^{-(s-1)t} \, dt$$

$$= \left[\frac{e^{-(s-1)t}}{-(s-1)} \right]_0^5 = \frac{1}{s-1}\{1 - e^{-5(s-1)}\}$$

$$\therefore \; \underline{\mathcal{L}\{F(t)\} = \frac{1 - e^{-5(s-1)}}{(s-1)(1 - e^{-5s})}}$$

All very straightforward.

Example 4 Determine the Laplace transform of the half-wave rectifier output waveform defined by

$$\left. \begin{aligned} F(t) &= 8 \sin t \qquad 0 < t < \pi \\ F(t) &= 0 \qquad\quad \pi < t < 2\pi \end{aligned} \right\} \quad F(t) = F(t + 2\pi).$$

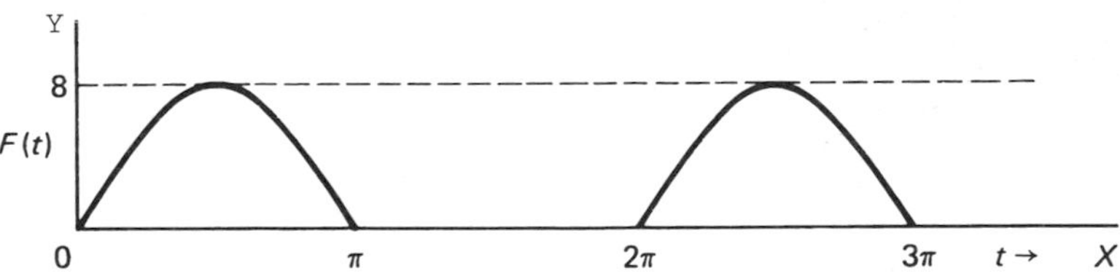

Here the period is 2π i.e. $\omega = 2\pi$

In general, for a periodic function of period ω,

$$\mathcal{L}\{F(t)\} = \ldots\ldots\ldots$$

48

$$\mathscr{L}\{F(t)\} = \frac{1}{1 - e^{-\omega s}} \int_0^\omega e^{-st} \cdot F(t) dt$$

So, for this example,

$$\mathscr{L}\{F(t)\} = \frac{1}{1 - e^{-2\pi s}} \int_0^{2\pi} e^{-st} \cdot F(t) dt$$

$$\therefore\ (1 - e^{-2\pi s})\mathscr{L}\{F(t)\} = \int_0^\pi e^{-st} \cdot 8 \sin t \, dt$$

Writing $\sin t$ as the imaginary part of e^{jt}, i.e. $\sin t \equiv \mathscr{I} e^{jt}$,

$$(1 - e^{-2\pi s})\mathscr{L}\{F(t)\} = 8\mathscr{I} \int_0^\pi e^{-st} \cdot e^{jt} dt$$

$$= 8\mathscr{I} \int_0^\pi e^{-(s-j)t} dt$$

and this you can finish off in the usual manner, giving

$$\mathscr{L}\{F(t)\} = \dots\dots\dots$$

49

$$\boxed{\mathscr{L}\{F(t)\} = \frac{8}{(s^2+1)(1-e^{-\pi s})}}$$

for $\quad (1-c^{-2\pi s})\mathscr{L}\{F(t)\} = 8 \cdot \mathscr{I}\displaystyle\int_0^{\pi} e^{-(s-j)t}\,dt$

$$= 8 \cdot \mathscr{I}\left[\frac{e^{-(s-j)t}}{-(s-j)}\right]_0^{\pi}$$

$$= \mathscr{I}\left\{\frac{-8}{s-j}\left[e^{-(s-j)\pi} - 1\right]\right\}$$

$$= 8 \cdot \mathscr{I}\left\{\frac{1}{s-j}\left[1 - e^{-s\pi}e^{j\pi}\right]\right\}$$

But $e^{j\pi} = \cos\pi + j\sin\pi = -1$.

$$\therefore\ (1-e^{-2\pi s})\mathscr{L}\{F(t)\} = 8 \cdot \mathscr{I}\left\{\frac{1}{s-j}(1+e^{-s\pi})\right\}$$

$$= 8 \cdot \mathscr{I}\left\{\frac{s+j}{s^2+1}(1+e^{-\pi s})\right\} = 8\left\{\frac{1+e^{-\pi s}}{s^2+1}\right\}$$

$$\therefore\ \mathscr{L}\{F(t)\} = \frac{1}{1-e^{-2\pi s}} \times 8\left\{\frac{1+e^{-\pi s}}{s^2+1}\right\}$$

$$= \underline{\frac{8}{(1-e^{-\pi s})(s^2+1)}}$$

Now let us consider the corresponding inverse transforms when periodic functions are involved.

50

Inverse transforms

Finding inverse transforms of functions of s which are transforms of periodic functions is not as straightforward as in earlier examples, for the transforms result from integration over one cycle and not from $t = 0$ to $t = \infty$. Hence we have no simple table of inverse transforms upon which to draw.

However, all difficulties can be surmounted and an example will show how we deal with this particular problem.

Example 1 Determine the inverse transform

$$\mathscr{L}^{-1}\left\{\frac{2+e^{-2s}-3e^{-s}}{s(1-e^{-2s})}\right\}.$$

The first thing we see is the factor $(1-e^{-2s})$ in the denominator, which suggests a periodic function of period 2 units, i.e. $\dfrac{1}{1-e^{-\omega s}}$ where $\omega = 2$.

The key to the solution is to write $(1-e^{-2s})$ in the denominator as $(1-e^{-2s})^{-1}$ in the numerator and to expand this as a binomial series. We remember that $(1-x)^{-1} = \ldots\ldots\ldots$

51

$$\boxed{(1-x)^{-1} = 1 + x + x^2 + x^3 + \ldots}$$

$$\begin{aligned}\therefore\ (1-e^{-2s})^{-1} &= 1 + (e^{-2s}) + (e^{-2s})^2 + (e^{-2s})^3 + \ldots\\ &= 1 + e^{-2s} + e^{-4s} + e^{-6s} + \ldots\end{aligned}$$

$$\begin{aligned}\therefore\ \mathscr{L}\{F(t)\} &= \frac{2+e^{-2s}-3e^{-s}}{s(1-e^{-2s})} = \frac{1}{s}(2+e^{-2s}-3e^{-s})(1-e^{-2s})^{-1}\\ &= \frac{1}{s}(2+e^{-2s}-3e^{-s})(1+e^{-2s}+e^{-4s}+e^{-6s}+e^{-8s}+\ldots)\end{aligned}$$

We now multiply the second series by each term of the first in turn and collect up like terms, giving

$$\mathscr{L}\{F(t)\} = \frac{1}{s}\left\{\begin{array}{lllllll} 2 & & +2e^{-2s} & & +2e^{-4s} & & +2e^{-6s}\ldots\\ & & +\ e^{-2s} & & +\ e^{-4s} & & +\ e^{-6s}\ldots\\ & -3e^{-s} & & -3e^{-3s} & & -3e^{-5s}- & \ldots \end{array}\right\}$$

$$= \ldots\ldots\ldots$$

52

$$\mathscr{L}\{F(t)\} = \frac{1}{s}\{2 - 3e^{-s} + 3e^{-2s} - 3e^{-3s} + 3e^{-4s} - 3e^{-5s} + \ldots\}$$

Each term is of the form $\dfrac{e^{-cs}}{s}$, so, expressing $F(t)$ in unit step form, we have

$$F(t) = \ldots\ldots\ldots$$

53

$$F(t) = 2H(t) - 3H(t-1) + 3H(t-2) - 3H(t-3) + 3H(t-4)\ldots$$

and from this we can sketch the waveform, which is therefore

....

54

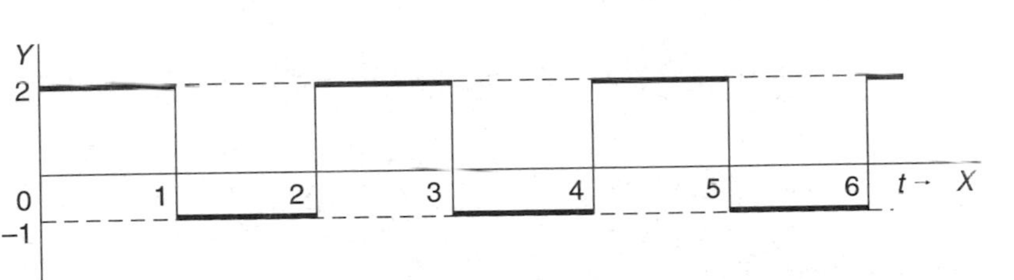

We can finally define this periodic function in analytical terms.

$$F(t) = \ldots\ldots\ldots$$

55

$$\left.\begin{aligned} F(t) &= 2 \qquad 0 < t < 1 \\ &= -1 \qquad 1 < t < 2 \end{aligned}\right\} F(t) = F(t+2)$$

The key to the whole process is thus to ....

56

express $(1 - e^{-\omega s})$ in the denominator as $(1 - e^{-\omega s})^{-1}$ in the numerator and to expand this as a binomial series.

We do this by making use of the basic series

$$(1 - x)^{-1} = \ldots\ldots\ldots$$

57

$$(1-x)^{-1} = 1 + x + x^2 + x^3 + x^4 + \dots$$

Example 2 Determine $\mathscr{L}^{-1}\left\{\dfrac{3(1-e^{-s})}{s(1-e^{-3s})}\right\}$ and sketch the resulting waveform of $F(t)$.

$$\mathscr{L}\{F(t)\} = \frac{3}{s}(1-e^{-s})(1-e^{-3s})^{-1}$$

$$= \dots\dots\dots \quad \text{(next step)}$$

58

$$\mathscr{L}\{F(t)\} = \frac{3}{s}(1-e^{-s})(1+e^{-3s}+e^{-6s}+e^{-9s}+\dots)$$

which multiplied out gives

$$\mathscr{L}\{F(t)\} = \frac{3}{s}(1-e^{-s}+e^{-3s}-e^{-4s}+e^{-6s}-e^{-7s}+\dots)$$

$$= \frac{3}{s} - \frac{3e^{-s}}{s} + \frac{3e^{-3s}}{s} - \frac{3e^{-4s}}{s} + \frac{3e^{-6s}}{s} - \dots$$

And in unit step form, this gives

$F(t) = \dots\dots\dots$

59

$$F(t) = 3H(t) - 3H(t-1) + 3H(t-3) - 3H(t-4) + \dots$$

The waveform is thus ………

60

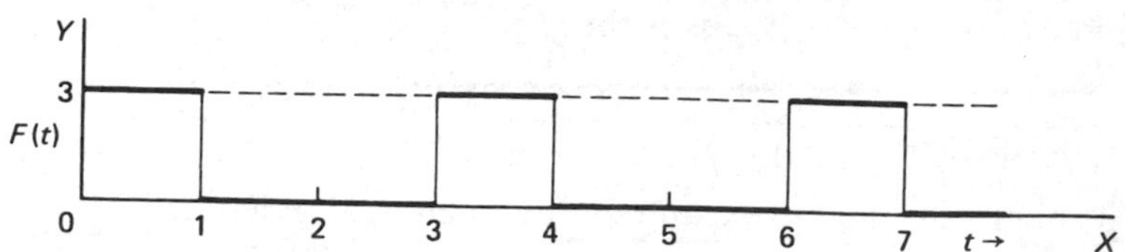

Written in analytical form this becomes

………

61

$$\left.\begin{array}{ll} F(t) = 3 & 0 < t < 1 \\ F(t) = 0 & 1 < t < 3 \end{array}\right\} F(t) = F(t+3)$$

And now, one more. They are all done in the same way.

Example 3 If $\mathscr{L}\{F(t)\} = \dfrac{1}{2s^2} - \dfrac{2e^{-4s}}{s(1-e^{-4s})}$, determine $F(t)$ and sketch the waveform.

$$\mathscr{L}\{F(t)\} = \frac{1}{2s^2} - \frac{2e^{-4s}}{s(1-e^{-4s})}$$

The first term is easy enough. In unit step form $\mathscr{L}^{-1}\left\{\dfrac{1}{2s^2}\right\} = \dfrac{t}{2} \cdot H(t)$

From the second term,

$$\frac{2e^{-4s}}{s(1-e^{-4s})} = \frac{2}{s}\{e^{-4s}(1-e^{-4s})^{-1}\}$$

$$= \frac{2}{s}\{e^{-4s}(1+e^{-4s}+e^{-8s}+e^{-12s}+\ldots)\}$$

$$= \frac{2e^{-4s}}{s} + \frac{2e^{-8s}}{s} + \frac{2e^{-12s}}{s} + \frac{2e^{-16s}}{s} + \ldots$$

$\therefore\ F(t) = \ldots\ldots\ldots$ (in unit step form)

62

$$F(t) = \frac{t}{2} \cdot H(t) - 2H(t-4) - 2H(t-8) - 2H(t-12) - \ldots$$

Now we have to draw the waveform. Consider the function terms up to each break point in turn.

$0 < t < 4 \qquad F(t) = \frac{t}{2} \qquad F(0) = 0; \quad F(4) = 2$

$4 < t < 8 \qquad F(t) = \frac{t}{2} - 2 \qquad F(4) = 0; \quad F(8) = 2$

$8 < t < 12 \qquad F(t) = \frac{t}{2} - 2 - 2 \quad F(8) = 0; \quad F(12) = 2$ etc.

So the waveform is ....

63

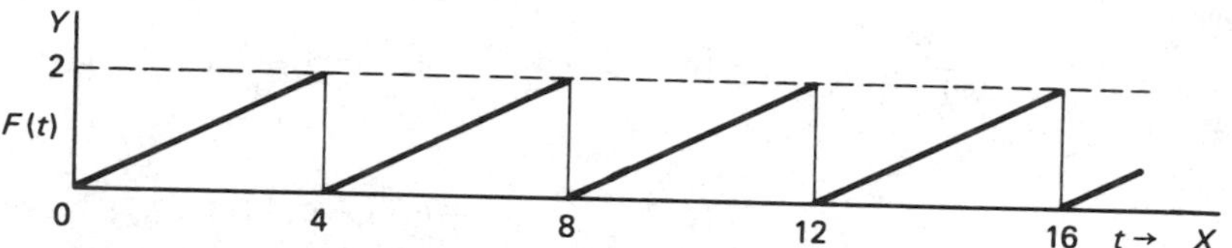

Expressed analytically, we finally have

$$F(t) = \frac{t}{2} \qquad 0 < t < 4, \qquad F(t) = F(t+4)$$

Now let us turn to something new.

The Dirac Delta Function—the impulse function

64

So far we have dealt with a number of standard Laplace transforms and then the Heaviside unit step function with some of its applications.

We now come to consider a function which is different from any of those we have used before, but which is important in numerous practical situations. This is the *Dirac delta function, or impulse function*, for it represents an extremely large force acting for a minutely small interval of time—the short, sharp shock—that may be a sudden mechanical blow or a shock voltage applied to an electrical circuit. This is clearly a special kind of function, so let us first consider it in some detail.

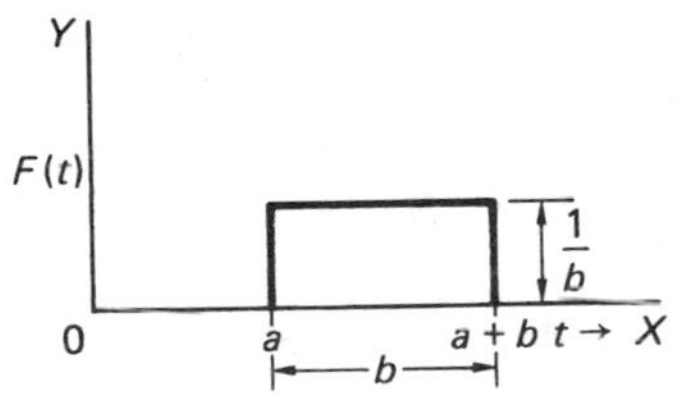

Consider a single rectangular pulse of width b and height $\frac{1}{b}$ occurring at $t = a$, as shown.

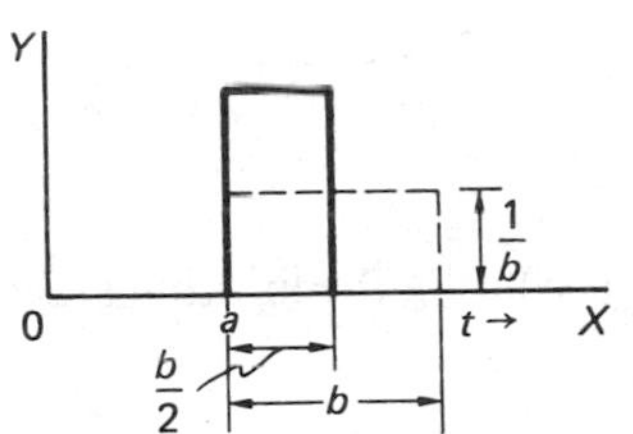

If we reduce the width of the pulse to $\frac{b}{2}$ and keep the area of the pulse unchanged, the height of the pulse will be ....

65

$$\boxed{\frac{2}{b}}$$

for the area of the pulse is 1 unit and, if the width is halved, the height must be doubled to maintain the same unit area.

If we continue the process of reducing still further the width of the pulse while retaining an area of unity, then as $b \rightarrow 0$, the height $\frac{1}{b} \rightarrow$ ....

66

∞

Thus we approach the final stage where the time interval is extremely small and the magnitude of the function extremely large.

The function represented by this limiting stage is known as the *unit impulse function*, or *Dirac delta function*, and is denoted by $\delta(t-a)$.

Graphically, we should require a rectangular pulse of zero width and infinite height and this is represented by a single vertical line surmounted by an arrow head.

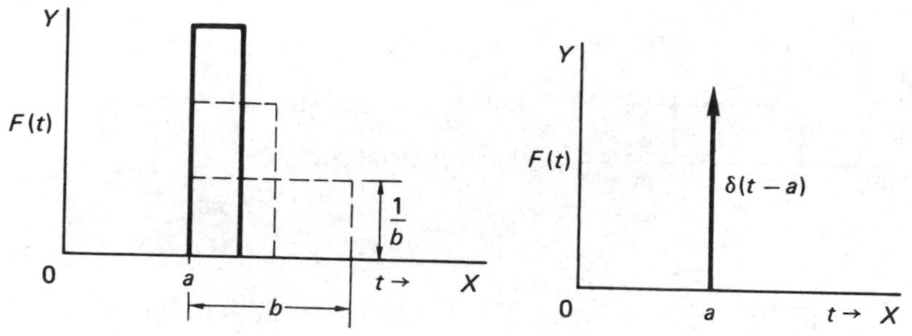

It is important to remember that at all stages, including the limiting stage, the area of the pulse is ....

67

1

We can express the impulse function analytically as follows.

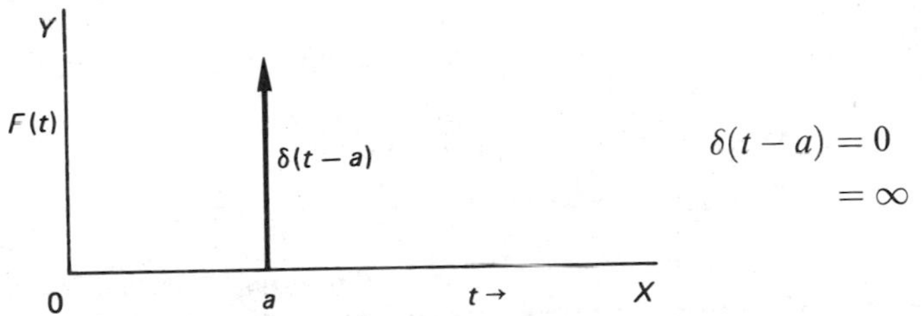

$$\delta(t-a) = 0 \qquad t \neq a$$
$$ = \infty \qquad t = a$$

for the function exists only at $t = a$, at which it is infinitely great.

Impulse function at the origin

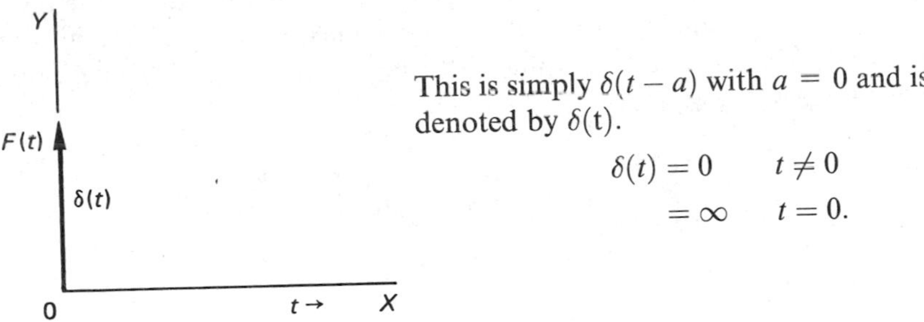

This is simply $\delta(t-a)$ with $a = 0$ and is denoted by $\delta(\mathrm{t})$.

$$\delta(t) = 0 \qquad t \neq 0$$
$$ = \infty \qquad t = 0.$$

Let us now consider one or two operations involving the impulse function, so on to the next frame.

68

Integration involving the impulse function

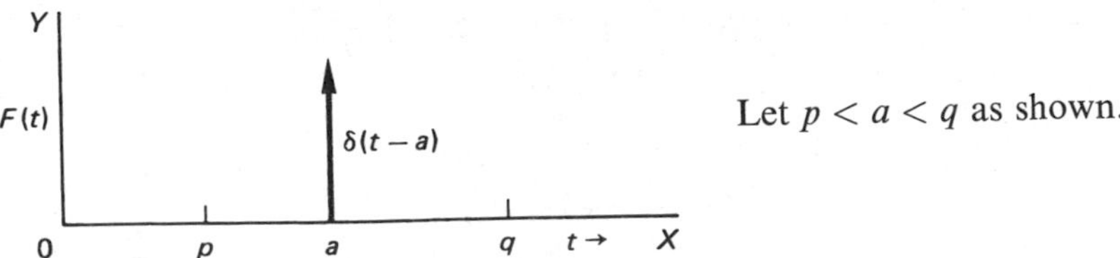

Let $p < a < q$ as shown.

Then, from the definition of $\delta(t-a)$

$$\int_p^q \delta(t-a)\,\mathrm{d}t = \ldots\ldots\ldots$$

69

$$\boxed{1}$$

for (i) $p < t < a$ $\qquad \delta(t - a) = 0$
(ii) $t = a$ $\qquad$ area of pulse $= 1$
(iii) $a < t < q$ $\qquad \delta(t - a) = 0$

Therefore, the total area between $t = p$ and $t = q$ is $0 + 1 + 0 = 1$

$$\therefore \int_p^q \delta(t - a)\mathrm{d}t = 1 \quad \text{provided } p < a < q \qquad (1)$$

Make a note and then move on.

70

Now let us consider $\int_p^q F(t) \cdot \delta(t - a)\,\mathrm{d}t$ where $F(t)$ is a given function of t and $p < a < q$.

Remembering the definition of $\delta(t - a)$, the product $F(t) \cdot \delta(t - a)$ is zero for all values of t within the range $p < t < q$, except at the one point $t = a$ and for a minute interval at $t = a$, $F(t)$ may be regarded as constant, i.e. $F(a)$. This constant $F(a)$ may then be taken outside the integral so that, for $p < t < q$,

$$\int_p^q F(t) \cdot \delta(t - a)\,\mathrm{d}t = F(a) \int_p^q \delta(t - a)\,\mathrm{d}t$$

$$= \dots\dots\dots$$

71

$$\boxed{F(a)}$$

since $\int_p^q \delta(t - a)\,\mathrm{d}t = 1$ provided $p < a < q$.

$$\therefore \int_p^q F(t) \cdot \delta(t - a)\,\mathrm{d}t = F(a) \quad \text{for } p < a < q \qquad (2)$$

This is also important, so make a note of it. Then on to the next frame.

72

So far, then, we have

(i) $\int_p^q \delta(t-a)\,dt = 1$

(ii) $\int_p^q F(t)\cdot\delta(t-a)\,dt = F(a)$

provided, in each case, that $p < a < q$.

Example 1 To evaluate $\int_1^3 (t^2+4)\cdot\delta(t-2)\,dt$.

The factor $\delta(t-2)$ shows that the impulse occurs at $t = 2$, i.e. $a = 2$.

$$F(t) = t^2 + 4 \quad \therefore\; F(a) = F(2) = 4 + 4 = 8$$

$$\therefore \int_1^3 (t^2+4)\cdot\delta(t-2)dt = F(2) = \underline{8}$$

Example 2 To evaluate $\int_0^\pi \cos 6t\cdot\delta(t-\pi/2)dt$.

$$\int_0^\pi \cos 6t\cdot\delta(t-\pi/2)dt = F(\pi/2) = \cos 3\pi = \underline{-1}$$

and in the same way

(a) $\int_0^6 5\cdot\delta(t-3)\,dt = \ldots\ldots\ldots$

(b) $\int_2^5 e^{-2t}\cdot\delta(t-4)\,dt = \ldots\ldots\ldots$

(c) $\int_0^\infty (3t^2-4t+5)\cdot\delta(t-2)\,dt = \ldots\ldots\ldots$

73

(a) $\int_0^6 5 \cdot \delta(t-3)\,dt = 5 \times 1 = \underline{5}$

(b) $\int_2^5 e^{-2t} \cdot \delta(t-4)\,dt = F(4) = [e^{-2t}]_{t=4} = \underline{e^{-8}}$

(c) $\int_0^\infty (3t^2 - 4t + 5) \cdot \delta(t-2)\,dt = 12 - 8 + 5 = \underline{9}$

Nothing could be easier. It all rests on the fact that, provided $p < a < q$,

$$\int_p^q F(t) \cdot \delta(t-a)\,dt = \ldots\ldots\ldots$$

74

$F(a)$

Now let us consider the Laplace transform of $\delta(t-a)$.

On then to the next frame.

75

Laplace transform of $\delta(t-a)$

We have already obtained that

$$\int_p^q F(t) \cdot \delta(t-a)\,dt = F(a) \qquad p < a < q$$

Therefore, if $p = 0$ and $q = \infty$

$$\int_0^\infty F(t) \cdot \delta(t-a)\,dt = F(a)$$

Hence, if $F(t) = e^{-st}$, this becomes

$$\int_0^\infty e^{-st} \cdot \delta(t-a)\,dt = \mathscr{L}\{\delta(t-a)\}$$

$$= \ldots\ldots\ldots$$

76

$$\boxed{e^{-as}}$$

i.e. the value of $F(t)$, i.e. e^{-st}, at $t = a$.

$$\underline{\mathcal{L}\{\delta(t-a)\} = e^{-as}} \qquad (3)$$

It follows from this that the Laplace transform of the impulse function at the origin is ....

77

$$\boxed{1}$$

since, for $a = 0$, $\mathcal{L}\{\delta(t-a)\} = \mathcal{L}\{\delta(t)\} = e^0 = 1$

$$\therefore \underline{\mathcal{L}\{\delta(t)\} = 1} \qquad (4)$$

Finally, let us deal with the more general case of $\mathcal{L}\{F(t) \cdot \delta(t-a)\}$.

We have $\mathcal{L}\{F(t) \cdot \delta(t-a)\} = \int_0^\infty e^{-st} \cdot F(t) \cdot \delta(t-a)\,dt$. Now the integrand $e^{-st} \cdot F(t) \cdot \delta(t-a) = 0$ for all values of t except at $t = a$ at which point $e^{-st} = e^{-as}$, and $F(t) = F(a)$.

$$\therefore \mathcal{L}\{F(t) \cdot \delta(t-a)\} = F(a) \cdot e^{-as} \int_0^\infty \delta(t-a)\,dt$$

$$= F(a) \cdot e^{-as}(1)$$

$$\therefore \underline{\mathcal{L}\{F(t) \cdot \delta(t-a)\} = F(a)e^{-as}} \qquad (5)$$

Another important result to note. Then let us deal with some examples.

78

We have $\mathscr{L}\{F(t) \cdot \delta(t-a)\} = F(a) \cdot e^{-as}$

Therefore

(a) $\mathscr{L}\{6 \cdot \delta(t-4)\}$ $a = 4$, $\therefore$ $\mathscr{L}\{6 \cdot \delta(t-4)\} = \underline{6e^{-4s}}$

(b) $\mathscr{L}\{t^3 \cdot \delta(t-2)\}$ $a = 2$ $\therefore$ $\mathscr{L}\{t^3 \cdot \delta(t-2)\} = \underline{8e^{-2s}}$

Similarly

(c) $\mathscr{L}\{\sin 3t \cdot \delta(t-\pi/2)\} = \dots\dots\dots$

79

$$\boxed{-e^{-\pi s/2}}$$

for $\mathscr{L}\{\sin 3t \cdot \delta(t-\pi/2)\} = [\sin 3t]_{t=\pi/2} \cdot e^{-\pi s/2} = \underline{-e^{-\pi s/2}}$

and

(d) $\mathscr{L}\{\cosh 2t \cdot \delta(t)\} = \dots\dots\dots$

80

$$\boxed{1}$$

for $\mathscr{L}\{\cosh 2t \cdot \delta(t)\} = [\cosh 2t]_{t=0} \cdot e^0 = \cosh 0 \cdot (1) = \underline{1}$

So our main conclusions so far are as follows.

(i) $\displaystyle\int_p^q \delta(t-a)\,dt = \dots\dots\dots$ provided $\dots\dots\dots$

(ii) $\displaystyle\int_p^q F(t) \cdot \delta(t-a)\,dt = \dots\dots\dots$ provided $\dots\dots\dots$

(iii) $\mathscr{L}\{\delta(t-a)\} = \dots\dots\dots$

(iv) $\mathscr{L}\{\delta(t)\} = \dots\dots\dots$

(v) $\mathscr{L}\{F(t) \cdot \delta(t-a)\} = \dots\dots\dots$

81

(i) $\int_p^q \delta(t-a)\mathrm{d}t = \underline{1}$ provided $p < a < q$

(ii) $\int_p^q F(t) \cdot \delta(t-a)\mathrm{d}t = \underline{F(a)}$ provided $p < a < q$

(iii) $\mathscr{L}\{\delta(t-a)\} = \underline{\mathrm{e}^{-as}}$

(iv) $\mathscr{L}\{\delta(t)\} = \underline{1}$

(v) $\mathscr{L}\{F(t) \cdot \delta(t-a)\} = \underline{F(a) \cdot \mathrm{e}^{-as}}$

Just check that you have noted this important list—the basis of all work on the Dirac delta function.

Now for one further example on this section.

Example Impulses of 1, 4, 7 units occur at $t = 1$, $t = 3$ and $t = 4$ respectively, in the directions shown.

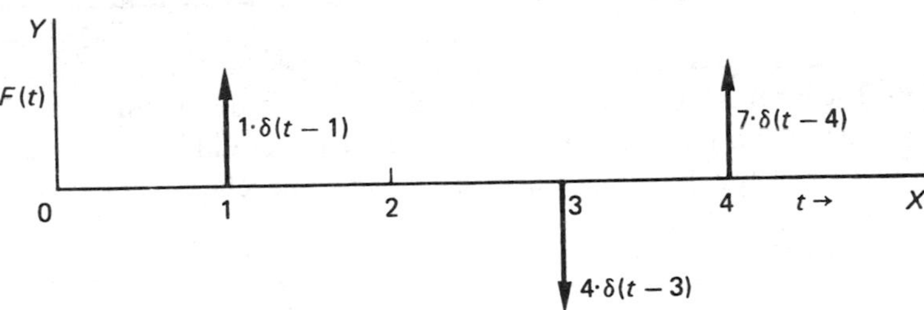

Write down an expression for $F(t)$ and determine its Laplace transform.

We have $F(t) = 1 \cdot \delta(t-1) - 4 \cdot \delta(t-3) + 7 \cdot \delta(t-4)$.

Then $\mathscr{L}\{F(t)\} = \dots\dots\dots$

82

$$\mathscr{L}\{F(t)\} = e^{-s} - 4e^{-3s} + 7e^{-4s}$$

and that is all there is to that.

One further consideration is interesting.

$$\text{By definition} \quad \int_0^t \delta(t-a)\mathrm{d}t = 0 \quad \text{for} \quad t < a$$
$$= 1 \quad \text{for} \quad t > a$$

which is very much like the definition of the Heaviside unit step function

$$H(t-a) = 0 \quad \text{for} \quad t < a$$
$$= 1 \quad \text{for} \quad t > a.$$

$$\therefore \int_0^t \delta(t-a)\mathrm{d}t = H(t-a) \quad \therefore \delta(t-a) = \frac{d}{\mathrm{d}t}\{H(t-a)\}$$

$$\text{and when } a = 0 \qquad \delta(t) = \frac{d}{\mathrm{d}t}\{H(t)\}$$

Can you suggest any reason for this relationship?

83

You will remember that we defined the Dirac delta function in reference to a rectangular pulse which, itself, could be expressed in terms of the unit step function, so the close relationship is not surprising after all.

Differential equations involving the impulse function

Example 1 A system has the equation of motion

$$\ddot{x} + 6\dot{x} + 8x = F(t)$$

where $F(t)$ is an impulse of 4 units applied at $t = 5$. At $t = 0$, $x = 0$ and $\dot{x} = 3$. Determine an expression for the displacement x in terms of t.

The impulse of 4 units is applied at $t = 5$. $\therefore F(t) = 4 \cdot \delta(t-5)$.

$$\therefore \ddot{x} + 6\dot{x} + 8x = 4 \cdot \delta(t-5) \quad \text{At } t = 0,\ x = 0,\ \dot{x} = 3.$$

Expressing the equation in Laplace transforms, we have

....

84

$$(s^2\bar{x} - sx_0 - x_1) + 6(s\bar{x} - x_0) + 8\bar{x} = 4e^{-5s}$$

Now $x_0 = 0$; $x_1 = 3$

$$\therefore s^2\bar{x} - 3 + 6s\bar{x} + 8\bar{x} = 4e^{-5s}$$

$$\therefore (s^2 + 6s + 8)\bar{x} = 3 + 4e^{-5s}$$

$$\therefore \bar{x} = (3 + 4e^{-5s})\frac{1}{(s+2)(s+4)}$$

Writing $\dfrac{1}{(s+2)(s+4)}$ in partial fractions, we get

$$\bar{x} = \ldots\ldots\ldots$$

85

$$x = (3 + 4e^{-5s})\left\{\frac{1}{2}\cdot\frac{1}{s+2} - \frac{1}{2}\cdot\frac{1}{s+4}\right\}$$

$$\therefore \bar{x} = \frac{3}{2}\left\{\frac{1}{s+2} - \frac{1}{s+4}\right\} + 2\left\{\frac{e^{-5s}}{s+2} - \frac{e^{-5s}}{s+4}\right\}$$

Taking inverse transforms

$$x = \frac{3}{2}\left\{e^{-2t} - e^{-4t}\right\} + 2\left\{e^{-2(t-5)}\cdot H(t-5) - e^{-4(t-5)}\cdot H(t-5)\right\}$$

$$= \frac{3}{2}\left\{e^{-2t} - e^{-4t}\right\} + 2\left\{e^{-2t}\cdot e^{10}\cdot H(t-5) - e^{-4t}\cdot e^{20}\cdot H(t-5)\right\}$$

which simplifies to $x = \ldots\ldots\ldots$

86

$$x = e^{-2t}\left\{\frac{3}{2} + 2e^{10} \cdot H(t-5)\right\} - e^{-4t}\left\{\frac{3}{2} + 2e^{20} \cdot H(t-5)\right\}$$

Example 2 Solve the equation $\ddot{x} + 4\dot{x} + 13x = 2 \cdot \delta(t)$ where, at $t = 0$, $x = 2$ and $\dot{x} = 0$.

$$\ddot{x} + 4\dot{x} + 13x = 2 \cdot \delta(t) \qquad x_0 = 2;\ x_1 = 0$$

Expressing in Laplace transforms, we have

....

87

$$(s^2\bar{x} - sx_0 - x_1) + 4(s\bar{x} - x_0) + 13\bar{x} = 2 \cdot (1)$$

Inserting the initial conditions and simplifying,

$$\bar{x} = \dots\dots\dots$$

88

$$\bar{x} = (2s + 10)\frac{1}{s^2 + 4s + 13}$$

Rearranging the denominator by completing the square, this can be written

$$\bar{x} = (2s + 10)\frac{1}{(s+2)^2 + 9}$$

$$\therefore\ x = \dots\dots\dots$$

89

$$\boxed{x = 2e^{-2t}\{\cos 3t + \sin 3t\}}$$

for $$\bar{x} = \frac{2(s+2)}{(s+2)^2+9} + \frac{6}{(s+2)^2+9}$$

$$\therefore x = 2e^{-2t}\cos 3t + 2e^{-2t}\sin 3t$$

$$\therefore \underline{x = 2e^{-2t}\{\cos 3t + \sin 3t\}}$$

Now for one further example for you to work through on your own.

So move on.

90

Example 3 The equation of motion of a system is

$$\ddot{x} + 5\dot{x} + 4x = F(t) \text{ where } F(t) = 3 \cdot \delta(t-2).$$

At $t = 0$, $x = 2$ and $\dot{x} = -2$. Determine an expression for the displacement x in terms of t.

We have $\ddot{x} + 5\dot{x} + 4x = 3 \cdot \delta(t-2)$ with $x_0 = 2$ and $x_1 = -2$.

As before, you can now express this in Laplace transforms, substitute the initial conditions, simplify to obtain an expression for x and finally take inverse transforms to determine the required expression for x.

Work right through it carefully. It is good revision and there are no snags.

$$x = \dots\dots\dots$$

91

$$x = e^{-t}\{2 + e^2 \cdot H(t-2)\} - e^8 \cdot e^{-4t} \cdot H(t-2)$$

Here is the working for you to check.

$$\ddot{x} + 5\dot{x} + 4x = 3 \cdot \delta(t-2) \quad \text{with} \quad x_0 = 2 \text{ and } x_1 = -2$$

$$(s^2\bar{x} - sx_0 - x_1) + 5(s\bar{x} - x_0) + 4\bar{x} = 3e^{-2s}$$

$$s^2\bar{x} - 2s + 2 + 5s\bar{x} - 10 + 4\bar{x} = 3e^{-2s}$$

$$(s^2 + 5s + 4)\bar{x} - 2s - 8 = 3e^{-2s}$$

$$\therefore (s+1)(s+4)\bar{x} = 2s + 8 + 3e^{-2s}$$

$$\therefore \bar{x} = \frac{2(s+4)}{(s+1)(s+4)} + e^{-2s} \cdot \frac{3}{(s+1)(s+4)}$$

$$= \frac{2}{s+1} + e^{-2s}\left\{\frac{1}{s+1} - \frac{1}{s+4}\right\}$$

$$\therefore \bar{x} = \frac{2}{s+1} + \frac{e^{-2s}}{s+1} - \frac{e^{-2s}}{s+4}$$

$$\therefore x = 2e^{-t} + H(t-2) \cdot e^{-(t-2)} - H(t-2) \cdot e^{-4(t-2)}$$

$$= 2e^{-t} + H(t-2) \cdot e^2 \cdot e^{-t} - H(t-2) \cdot e^8 \cdot e^{-4t}$$

$$\underline{x = e^{-t}\{2 + e^2 \cdot H(t-2)\} - e^8 \cdot e^{-4t} \cdot H(t-2)}$$

That completes this programme. The Heaviside unit step function, periodic functions and the Dirac delta function, or impulse function, have many practical applications, so check carefully down the Revision Summary before moving on to the Test Exercise that then follows.

92

REVISION SUMMARY

1. *Heaviside unit step function:* $H(t - c)$

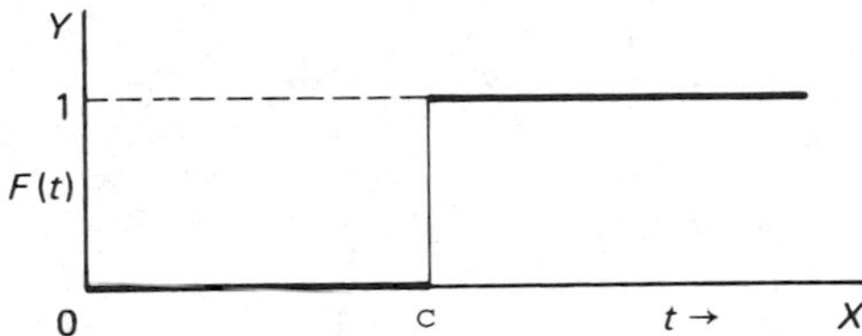

$$F(t) = 0 \qquad 0 < t < c$$
$$= 1 \qquad c < t.$$

2. *Suppression and shift*

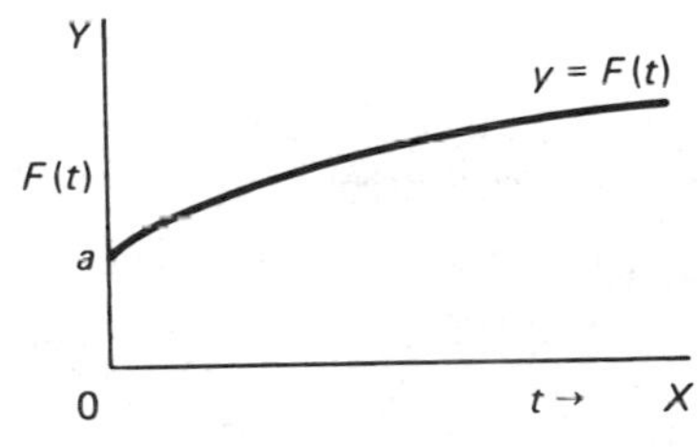

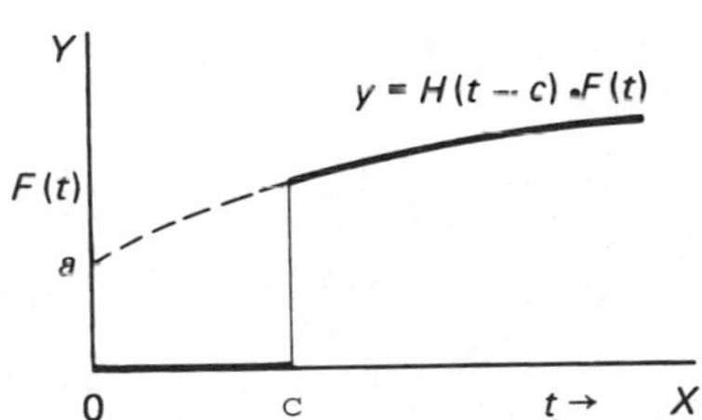

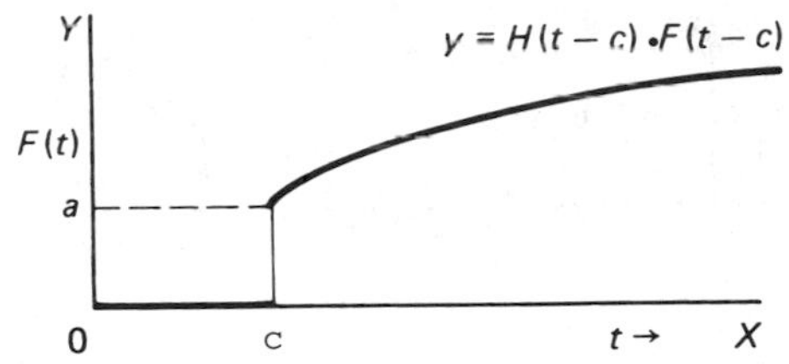

3. *Laplace transform of* $H(t - c)$

$$\mathcal{L}\{H(t-c)\} = \frac{e^{-cs}}{s}; \quad \mathcal{L}\{H(t)\} = \frac{1}{s}.$$

4. *Laplace transform of* $H(t - c) \cdot F(t - c)$

$$\mathcal{L}\{H(t-c) \cdot F(t-c)\} = e^{-cs} \cdot f(s) \quad \text{where} \quad f(s) = \mathcal{L}\{F(t)\}.$$

5. *Second shift theorem*

If $f(s) = \mathcal{L}\{F(t)\}$, then $e^{-cs} \cdot f(s) = \mathcal{L}\{H(t-c) \cdot F(t-c)\}$ where c is real and positive.

6. *Periodic functions*

$$F(t) = F(t + n\omega) \qquad n = 1, 2, 3, \ldots \qquad \text{Period} = \omega.$$

7. *Laplace transform of a periodic function* with period ω

$$\mathscr{L}\{F(t)\} = \frac{1}{1 - e^{-\omega s}} \int_0^{\omega} e^{-st} \cdot F(t)\, dt.$$

8. *Inverse transforms involving periodic functions*

$$\text{e.g.} \quad \mathscr{L}^{-1}\left\{\frac{1 + 2e^{-3s} - 3e^{-2s}}{s(1 - e^{-3s})}\right\}$$

Expand $(1 - e^{-3s})^{-1}$ as a binomial series, like

$$(1 - x)^{-1} = 1 + x + x^2 + x^3 + \ldots$$

Multiply out and take inverse transforms of each term in turn.

9. *Dirac delta function* or unit impulse function

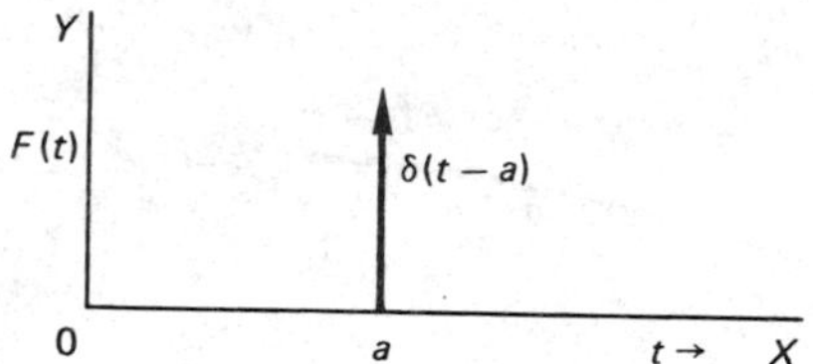

$$\delta(t - a) = 0 \qquad t \neq a$$
$$= \infty \qquad t = a.$$

10. *Delta function at the origin*

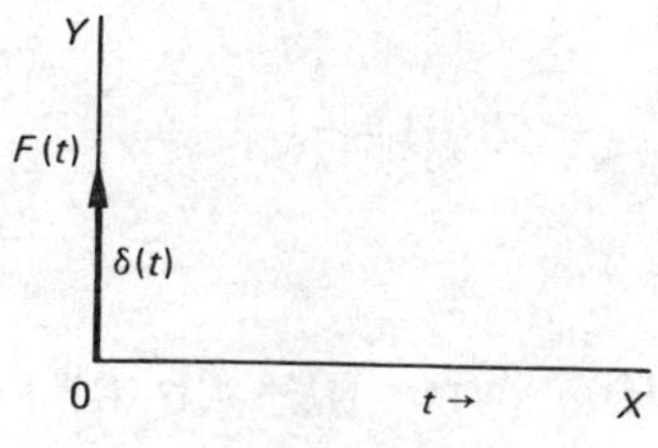

$$a = 0 \qquad \therefore \delta(t) = 0 \qquad t \neq 0$$
$$= \infty \qquad t = 0.$$

11. *Area of pulse = 1*

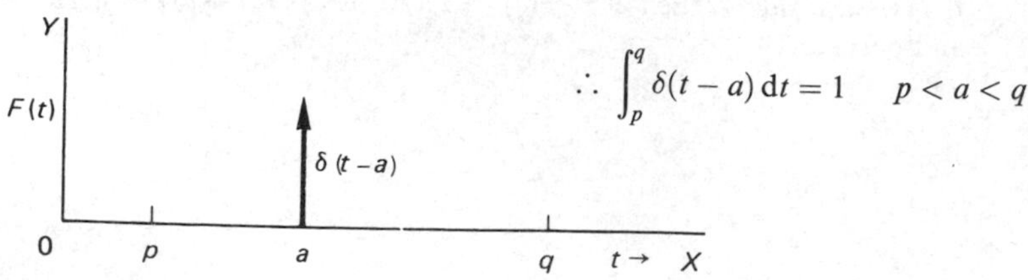

$$\therefore \int_p^q \delta(t - a)\, dt = 1 \qquad p < a < q$$

12. *Integration of the impulse function*

$$\int_p^q F(t) \cdot \delta(t-a)\,\mathrm{d}t = F(a) \qquad p < a < q$$

13. *Laplace transform of* $\delta(t-a)$

$$\mathscr{L}\{\delta(t-a)\} = \mathrm{e}^{-as}$$
$$\mathscr{L}\{\delta(t)\} = 1 \text{ since } a = 0$$
$$\mathscr{L}\{F(t) \cdot \delta(t-a)\} = F(a) \cdot \mathrm{e}^{-as}.$$

93 TEST EXERCISE VIII

1. In each of the following cases, sketch the graph of the function and find its Laplace transform.

(a) $F(t) = 3t \qquad 0 < t < 2$
$\quad = 6 \qquad 2 < t$

(b) $F(t) = e^{-2t} \qquad 0 < t < 3$
$\quad = 0 \qquad 3 < t$

(c) $F(t) = t^2 \qquad 0 < t < 2$
$\quad = 2 \qquad 2 < t < 3$
$\quad = 4 \qquad 3 < t$

(d) $F(t) = \sin 2t \qquad 0 < t < \pi$
$\quad = 0 \qquad \pi < t.$

2. Determine the function $F(t)$ whose transform $f(s)$ is

$$f(s) = \frac{1}{s}\left\{2 - 5e^{-s} + 8e^{-3s}\right\}.$$

Sketch the graph of the function between $t = 0$ and $t = 4$.

3. If $F(t) = \mathscr{L}^{-1}\left\{\dfrac{(1 + 3e^{-2s})(1 - e^{-3s})}{s^2}\right\}$, determine $F(t)$ and sketch the graph of the function.

4. Determine the function $F(t)$ for which

$$F(t) = \mathscr{L}^{-1}\left\{\frac{2(1 - e^{-s})}{s(1 - e^{-3s})}\right\}.$$

Sketch the waveform and express the function in analytical form.

5. Determine the Laplace transform of the periodic function shown.

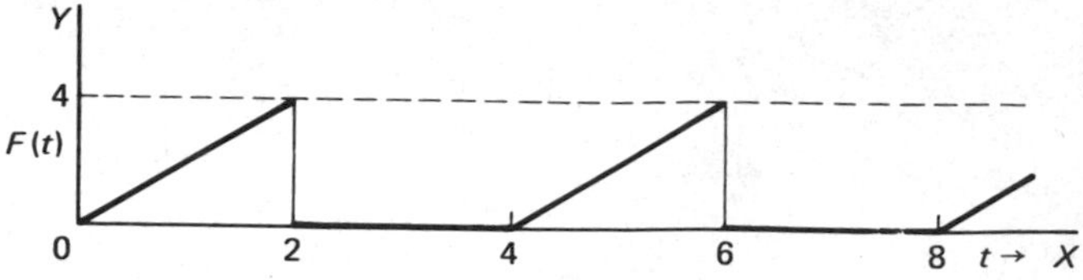

6. Evaluate (a) $\int_0^4 e^{-3t} \cdot \delta(t-2)\,dt$

(b) $\int_0^\infty \sin 3t \cdot \delta(t-\pi)\,dt$

(c) $\int_1^3 (2t^2+3) \cdot \delta(t-2)\,dt.$

7. Determine (a) $\mathscr{L}\{4 \cdot \delta(t-3)\}$, (b) $\mathscr{L}\{e^{-3t} \cdot \delta(t-2)\}$.

8. Sketch the graph of $F(t) = 3 \cdot \delta(t) + 4 \cdot \delta(t-2) - 3 \cdot \delta(t-4)$ and determine its Laplace transform.

9. Solve the equation $\ddot{x} + 6\dot{x} + 10x = 7 \cdot \delta(t)$ given that, at $t = 0$, $x = -1$ and $\dot{x} = 0$.

10. The equation of motion of a system is

$$\ddot{x} + 3\dot{x} + 2x = 3 \cdot \delta(t-4).$$

At $t = 0$, $x = 2$ and $\dot{x} = -4$. Determine an expression for the displacement x in terms of t.

FURTHER PROBLEMS VIII

1. If $\mathscr{L}\{F(t)\} = \frac{1}{s^2}\left\{3s + 2e^{-2s} - 2e^{-5s}\right\}$, determine $F(t)$.

2. If $F(t) = \mathscr{L}^{-1}\left\{\frac{(1-e^{-s})(1+e^{-2s})}{s^2}\right\}$, find $F(t)$ in terms of the unit step function.

3. A function $F(t)$ is defined by

$$F(t) = 4 \qquad 0 < t < 3$$
$$= 2t + 1 \qquad 3 < t.$$

Sketch the graph of the function and determine its Laplace transform.

4. Express in terms of the Heaviside unit step function

(a) $F(t) = t^2 \qquad 0 < t < 3$
$\quad = 5t \qquad 3 < t.$

(b) $F(t) = \cos t \qquad 0 < t < \pi$
$\quad = \cos 2t \qquad \pi < t < 2\pi$
$\quad = \cos 3t \qquad 2\pi < t.$

5. A function $F(t)$ is defined by

$$\begin{aligned} F(t) &= 0 && 0 < t < 2 \\ &= t+1 && 2 < t < 3 \\ &= 0 && 3 < t. \end{aligned}$$

Determine $\mathscr{L}\{F(t)\}$.

6. A function is defined by

$$\begin{aligned} F(t) &= t^2 && 0 < t < 2 \\ &= 4 && 2 < t < 5 \\ &= 0 && 5 < t. \end{aligned}$$

Determine (a) the function in terms of the unit step function

(b) the Laplace transform of $F(t)$.

7. If $\left.\begin{aligned} F(t) &= a\sin t && 0 < t < \pi \\ &= 0 && \pi < t < 2\pi \end{aligned}\right\} F(t) = F(t+2\pi)$,

prove that $\mathscr{L}\{F(t)\} = \dfrac{a}{(s^2+1)(1-e^{-\pi s})}$.

8. If $F(t) = a\sin t \quad 0 < t < \pi \quad F(t) = F(t+\pi)$, determine $\mathscr{L}\{F(t)\}$.

9. Find the Laplace transforms of the following periodic functions.

(a) $F(t) = t \quad 0 < t < \omega \quad F(t) = F(t+\omega)$

(b) $F(t) = e^t \quad 0 < t < 2\pi \quad F(t) = F(t+2\pi)$

(c) $\left.\begin{aligned} F(t) &= t && 0 < t < 1 \\ &= 0 && 1 < t < 2 \end{aligned}\right\} \quad F(t) = F(t+2)$

(d) $\left.\begin{aligned} F(t) &= t^2 && 0 < t < 2 \\ &= 4 && 2 < t < 3 \end{aligned}\right\} \quad F(t) = F(t+3)$.

10. A mass M is attached to a spring of stiffness $\omega^2 M$ and is set in motion at $t = 0$ by an impulsive force P. The equation of motion is

$$M\ddot{x} + M\omega^2 x = P \cdot \delta(t).$$

Obtain an expression for x in terms of t.

11. An impulsive voltage E is applied at $t = 0$ to a series circuit containing inductance L and capacitance C. Initially, the current and charge are zero. The current i at time t is given by

$$L\frac{di}{dt} + \frac{q}{C} = E \cdot \delta(t)$$

where q is the instantaneous value of the charge on the capacitor.
Since $i = \dfrac{dq}{dt}$, determine an expression for the current i in the circuit at time t.

12. A system has the equation of motion

$$\ddot{x} + 5\dot{x} + 6x = F(t)$$

where, at $t = 0$, $x = 0$ and $\dot{x} = 2$. If $F(t)$ is an impulse of 20 units applied at $t = 4$, determine an expression for x in terms of t.

Programme 9

Multiple Integrals

PART 1

Prerequisites: Engineering Mathematics (fourth edition)
Programme 23

Introduction

1

The introductory work on double and triple integrals was covered in detail in Programme 23 of the author's *Engineering Mathematics* (fourth edition) and another look at the main points before launching forth on the current development could well be worth while.

You will no doubt recognise the following.

1. *Double integrals*

$\int_{y_1}^{y_2}\int_{x_1}^{x_2} f(x,y)\,dx\,dy$ is a double integral and is evaluated from the inside outwards, i.e.

$$\boxed{\int_{y_1}^{y_2}\boxed{\int_{x_1}^{x_2} f(x,y)\,dx}^{①}\quad dy}^{②}$$

A double integral is sometimes expressed in the form

$$\int_{y_1}^{y_2} dy\quad \int_{x_1}^{x_2} f(x,y)\,dx$$

in which case, we evaluate from the right-hand end, i.e.

$$\int_{y_1}^{y_2} dy\ \boxed{\int_{x_1}^{x_2} f(x,y)\,dx}^{①}$$

then

$$\boxed{\int_{y_1}^{y_2}\boxed{\int_{x_1}^{x_2} f(x,y)\,dx}\ dy}^{②}$$

2. *Triple integrals*

Triple integrals follow the same procedure.

$\int_a^b\int_c^d\int_e^f f(x,y,z)\,dx\,dy\,dz$ is evaluated in the order

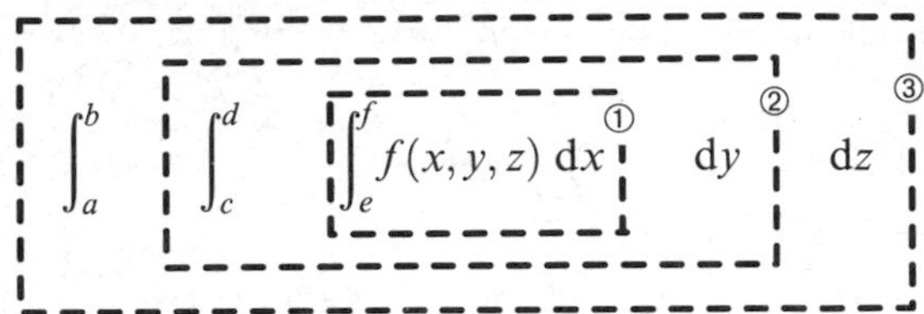

3. *Applications*

(a) *Areas of plane figures*

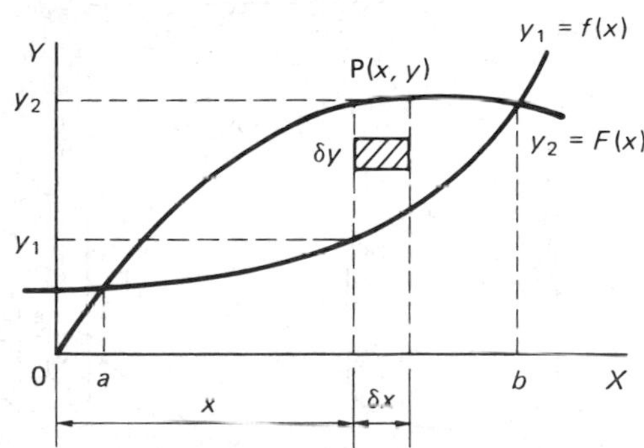

Area of element $\delta A = \delta x \delta y$

Area of strip $\approx \sum_{y=y_1}^{y=y_2} \delta x \delta y$

Area of all such strips $\approx \sum_{x=a}^{x=b} \left\{ \sum_{y=y_1}^{y=y_2} \delta x \delta y \right\}$

If $\delta x \to 0$ and $\delta y \to 0$, $\quad \underline{A = \int_a^b \int_{y_1}^{y_2} \mathrm{d}y \, \mathrm{d}x}$

(b) *Areas of plane figures bounded by a polar curve* $r = f(\theta)$ and radius vectors at $\theta = \theta_1$ and $\theta = \theta_2$.

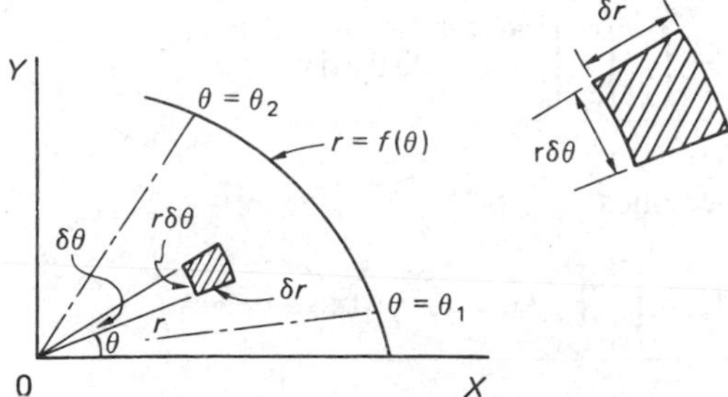

Small arc of circle of radius r, subtending angle $\delta\theta$ at centre.

$\therefore$ Arc $= r\,\delta\theta$

Area of element $\delta A \approx r\,\delta\theta\,\delta r$

Area of thin sector $\approx \sum_{r=0}^{r=f(\theta)} r\,\delta\theta\,\delta r$

$\therefore$ Total area of all such sectors $\approx \sum_{\theta=\theta_1}^{\theta=\theta_2} \left\{ \sum_{r=0}^{r=f(\theta)} r\,\delta r\,\delta\theta \right\}$

$\therefore$ If $\delta r \to 0$ and $\delta\theta \to 0$ $\quad \underline{A = \int_{\theta_1}^{\theta_2} \int_0^{r=f(\theta)} r\,\mathrm{d}r\,\mathrm{d}\theta}$

(c) *Volume of solids*

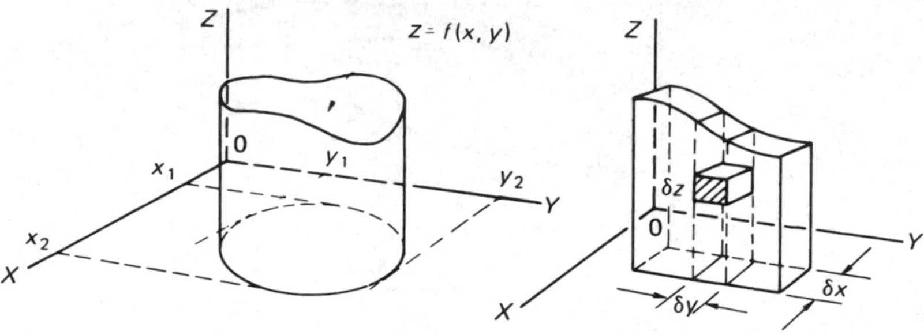

Volume of element $\delta V = \delta x\,\delta y\,\delta z$

Volume of column $\approx \sum_{z=0}^{z=f(x,\,y)} \delta x\,\delta y\,\delta z$

Volume of slice $\approx \sum_{y=y_1}^{y=y_2} \left\{ \sum_{z=0}^{z=f(x,\,y)} \delta x\,\delta y\,\delta z \right\}$

$\therefore$ Total volume $V \approx$ sum of all such slices

i.e. $$V \approx \sum_{x=x_1}^{x=x_2} \sum_{y=y_1}^{y=y_2} \sum_{z=0}^{z=f(x,\,y)} \delta x\,\delta y\,\delta z$$

Then, if $\delta x \to 0$, $\delta y \to 0$, $\delta z \to 0$,

$$\underline{V = \int_{x_1}^{x_2} \int_{y_1}^{y_2} \int_0^{z=f(x,\,y)} \mathrm{d}z\,\mathrm{d}y\,\mathrm{d}x}$$

If $z = f(x,\ y)$, this becomes

$$\underline{V = \int_{x_1}^{x_2} \int_{y_1}^{y_2} f(x,\ y)\,\mathrm{d}y\,\mathrm{d}x}$$

4. *Revision examples* As a means of 'warming up', let us work through one or two straightforward examples on the work of the previous year. **2**

Example 1 Find the area of the plane figure bounded by the curves $y_1 = (x-1)^2$ and $y_2 = 4-(x-3)^2$.

The first thing, as always, is to sketch the curves—each of which is a parabola—and to determine their points of intersection.

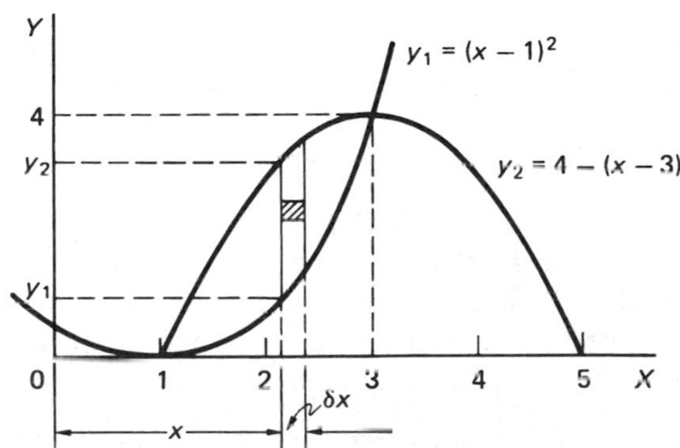

Points of intersection: $(x-1)^2 = 4-(x-3)^3$

$$x^2 - 2x + 1 = 4 - x^2 + 6x - 9 \quad \text{i.e.} \quad x^2 - 4x + 3 = 0$$

$$\therefore (x-1)(x-3) = 0 \qquad \therefore x = 1 \quad \text{or} \quad x = 3.$$

Now we have all the information to determine the required area, which is

....

3

$$A = 2\tfrac{2}{3} \text{ square units}$$

for $A = \displaystyle\int_{x=1}^{x=3}\int_{y_1}^{y_2} \mathrm{d}y\,\mathrm{d}x = \int_{x=1}^{x=3}\int_{y=(x-1)^2}^{y=4-(x-3)^2} \mathrm{d}y\,\mathrm{d}x$

$$= \int_1^3 \{4-(x-3)^2-(x-1)^2\}\,\mathrm{d}x = -2\int_1^3 (x^2-4x+3)\,\mathrm{d}x$$

$$= -2\left[\frac{x^3}{3} - 2x^2 + 3x\right]_1^3 = \underline{2\tfrac{2}{3} \text{ square units}}$$

Now for another.

Example 2 A rectangular plate is bounded by the x and y axes and the lines $x = 6$ and $y = 4$. The thickness t of the plate at any point is proportional to the square of the distance of the point from the origin. Determine the total volume of the plate.

First of all draw the figure and build up the appropriate double integral. Do not evaluate it yet. The expression is therefore

$$V = \ldots\ldots\ldots$$

4

$$V = \int_{x=0}^{x=6}\int_{y=0}^{y=4} k(x^2+y^2)\,\mathrm{d}y\,\mathrm{d}x$$

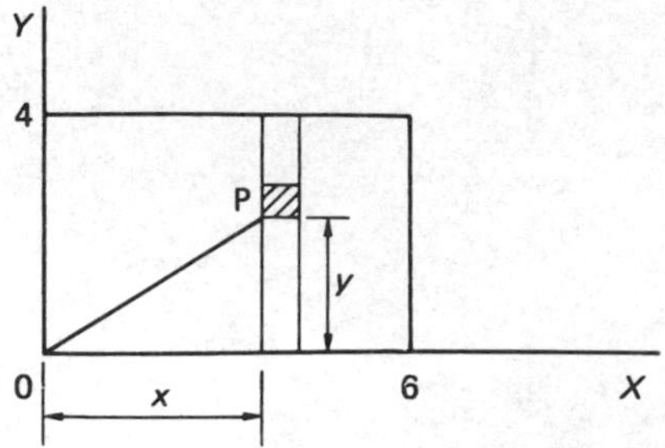

Thickness t of plate at P is

$$t = k\ \mathrm{OP}^2 = k(x^2+y^2)$$

Element of area $= \delta y\,\delta x$

$\therefore$ Element of volume at $P \approx k(x^2 + y^2)\,\delta y\,\delta x$

$$\therefore \text{ Total volume } V = \int_{x=0}^{x=6}\int_{y=0}^{y=4} k(x^2 + y^2)\,dy\,dx$$

Now we can evaluate the integral. We start from the inside with $\int_{y=2}^{y=4} k(x^2 + y^2)\,dy$, remembering that for this integral (volume of the strip) x is constant. This gives.........

5

$$k\left(4x^2 + \frac{64}{3}\right)$$

$$k\int_0^4 (x^2 + y^2)\,dy = k\left[x^2y + \frac{y^3}{3}\right]_{y=0}^{y=4} = k\left(4x^2 + \frac{64}{3}\right)$$

Then $\quad V = k\int_0^6 \left(4x^2 + \frac{64}{3}\right) dx = ...$

6

$$V = 416k \text{ cubic units}$$

That was easy enough. Now for one in polar coordinates.

Example 3 Express as a double integral the area enclosed by one loop of the curve $r = 3\cos 2\theta$ and evaluate the integral.

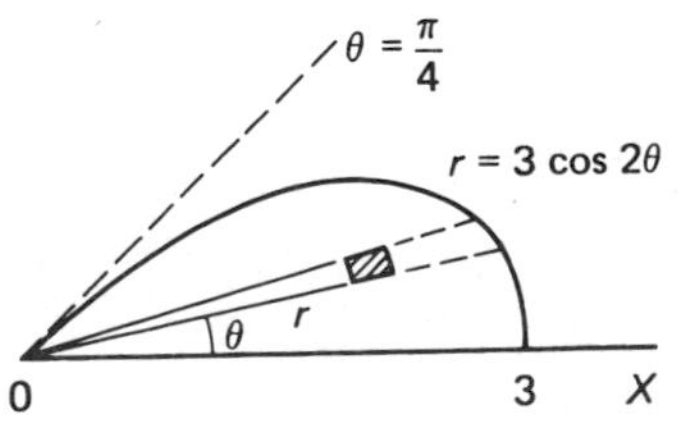

Consider the half loop shown

First set up the double integral which is

....

7

$$A = \int_{\theta=0}^{\theta=\pi/4} \int_{r=0}^{r=3\cos 2\theta} r \, \mathrm{d}r \, \mathrm{d}\theta$$

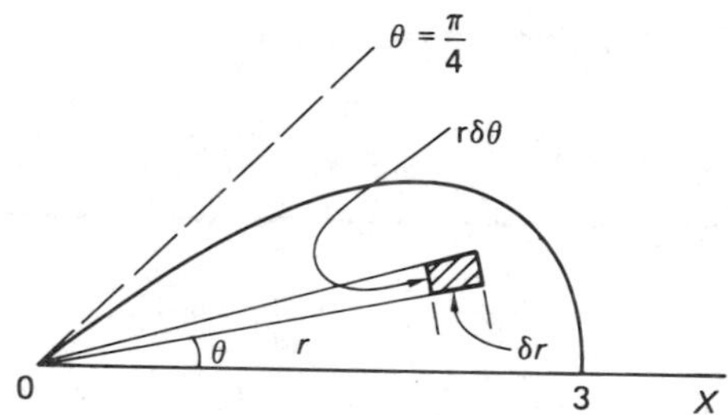

Area of element $= r\,\delta r\,\delta\theta$

$$\therefore \text{ Area of sector} \approx \sum_{r=0}^{r=3\cos 2\theta} r\,\delta r\,\delta\theta$$

$$\therefore \text{ Area of half loop} \approx \sum_{\theta=0}^{\theta=\pi/4} \sum_{r=0}^{r=3\cos 2\theta} r\,\delta r\,\delta\theta$$

If $\delta r \to 0$ and $\delta\theta \to 0$, $\quad A = \int_{\theta=0}^{\theta=\pi/4} \int_{r=0}^{r=3\cos 2\theta} r \, \mathrm{d}r \, \mathrm{d}\theta$

Now finish it off to find the area of the whole loop, which is

....

8

$$\frac{9\pi}{8} \text{ square units}$$

for $\quad A = \int_{\theta=0}^{\theta=\pi/4} \int_{r=0}^{r=3\cos 2\theta} r \, \mathrm{d}r \, \mathrm{d}\theta$

$$= \int_0^{\pi/4} \left[\frac{r^2}{2}\right]_0^{3\cos 2\theta} \mathrm{d}\theta = \frac{9}{2}\int_0^{\pi/4} \cos^2 2\theta \, \mathrm{d}\theta$$

$$= \frac{9}{4}\int_0^{\pi/4} (1 + \cos 4\theta)\, d\theta = \frac{9}{4}\left[\theta + \frac{\sin 4\theta}{4}\right]_0^{\pi/4} = \frac{9\pi}{16}$$

This is the area of a half loop. ∴ Required area $= \dfrac{9\pi}{8}$ square units

Now here is another.

Example 4 Find the volume of the solid bounded by the planes $z = 0$, $x = 1$, $x = 3$, $y = 1$, $y = 2$ and the surface $z = x^2 y^2$.

As always, we start off by sketching the figure. When you have done that, check the result with the next frame.

9

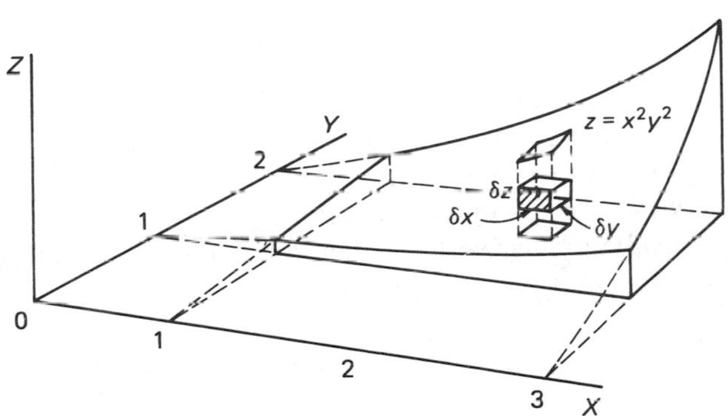

We now build up the integral which will give us the volume of the solid.

Element of volume $\delta V = \delta x\, \delta y\, \delta z$

Volume of column $\approx \displaystyle\sum_{z=0}^{z=x^2y^2} \delta x\, \delta y\, \delta z$

Volume of slice $\approx \displaystyle\sum_{y=1}^{y=2} \left\{ \sum_{z=0}^{z=x^2y^2} \delta x\, \delta y\, \delta z \right\}$

Volume of solid $\approx \displaystyle\sum_{x=1}^{x=3} \left\{ \sum_{y=1}^{y=2} \sum_{z=0}^{z=x^2y^2} \delta x\, \delta y\, \delta z \right\}$

When $\delta x \to 0,\ \delta y \to 0,\ \delta z \to 0,$ $\quad V = \displaystyle\int_{x=1}^{x=3} \int_{y=1}^{y=2} \int_{z=0}^{z=x^2y^2} dz\, dy\, dx$

Evaluating this, $\quad V = \ldots\ldots\ldots$

10

$$V = 20\tfrac{2}{9} \text{ cubic units}$$

for, starting with the innermost integral

$$V = \int_{x=1}^{x=3} \int_{y=1}^{y=2} \Big[z \Big]_0^{x^2y^2} \mathrm{d}y\,\mathrm{d}x = \int_1^3 \int_1^2 x^2y^2\,\mathrm{d}y\,\mathrm{d}x$$

$$= \int_1^3 \left[\frac{x^2\,y^3}{3} \right]_{y=1}^{y=2} \mathrm{d}x \qquad = \int_1^3 \frac{7x^2}{3}\,\mathrm{d}x = \underline{20\tfrac{2}{9}}$$

Now that we have revised the basics, let us move on to something rather different.

Differentials

11 It is convenient in various branches of the calculus to denote small increases in value of a variable by the use of *differentials*. The method is particularly useful in dealing with the effects of small finite increases and shortens the writing of calculus expressions.

We are already familiar with the diagram from which finite increases δy and δx in a function $y = f(x)$ are depicted.

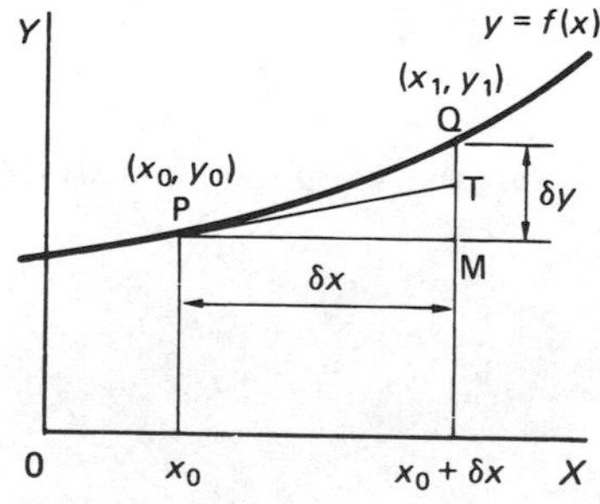

The increase in y from P to Q $= \mathrm{MQ} = \delta y = f(x_0 + \delta x) - f(x_0)$
If PT is the tangent at P, then
$\mathrm{MQ} = \mathrm{MT} + \mathrm{TQ}$.
Also $\dfrac{\mathrm{MT}}{\delta x} = f'(x_0) \quad \therefore \ \mathrm{MT} = f'(x_0)\delta x$

$\therefore \ \mathrm{MQ} = \delta y = f'(x_0) \cdot \delta x + \mathrm{TQ}$

and, if Q is close to P, then $\delta y \approx f'(x_0)\delta x$

We define the differentials $\mathrm{d}y$ and $\mathrm{d}x$ as finite quantities such that

$$\underline{\mathrm{d}y = f'(x_0)\,\mathrm{d}x}$$

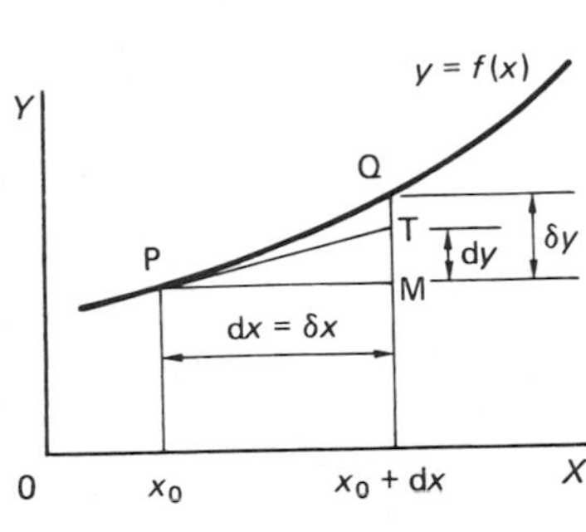

Note that the differentials $\mathrm{d}y$ and $\mathrm{d}x$ are finite quantities—not necessarily zero—and can therefore exist alone. They should not be confused with the symbols in $\dfrac{\mathrm{d}y}{\mathrm{d}x}$ which is the limiting value of $\dfrac{\delta y}{\delta x}$ as δx and $\delta y \to 0$. The differential $\mathrm{d}y$ is said to be the *principal part* of δy.

Note too that $\mathrm{d}x = \delta x$.

From the diagram, we can see that

δy is the increase in y as we move from P to Q along the curve.

$\mathrm{d}y$ is the increase in y as we move from P to T along the tangent.

As Q approaches P, the difference between δy and $\mathrm{d}y$ decreases to zero. The use of differentials simplifies the writing of many relationships and is based on the general statement $\underline{\mathrm{d}y = f'(x)\,\mathrm{d}x}$.

For example

(a) $y = x^5$ then $\mathrm{d}y = 5x^4\,\mathrm{d}x$
(b) $y = \sin 3x$ then $\mathrm{d}y = 3\cos 3x\,\mathrm{d}x$
(c) $y = \mathrm{e}^{4x}$ then $\mathrm{d}y = 4\,\mathrm{e}^{4x}\,\mathrm{d}x$
(d) $y = \cosh 2x$ then $\mathrm{d}y = 2\sinh 2x\,\mathrm{d}x$

Note that when the left-hand side is a differential $\mathrm{d}y$ the right-hand side must also contain a differential. Remember therefore to include the '$\mathrm{d}x$' on the right-hand side.

The product and quotient rules can also be expressed in differentials.

$$\frac{\mathrm{d}}{\mathrm{d}x}(uv) = u\frac{\mathrm{d}v}{\mathrm{d}x} + v\frac{\mathrm{d}u}{\mathrm{d}x} \quad \text{becomes} \quad \mathrm{d}(uv) = u\,\mathrm{d}v + v\,\mathrm{d}u$$

$$\frac{\mathrm{d}}{\mathrm{d}x}\left(\frac{u}{v}\right) = \frac{v\dfrac{\mathrm{d}u}{\mathrm{d}x} - u\dfrac{\mathrm{d}v}{\mathrm{d}x}}{v^2} \quad \text{becomes} \quad \mathrm{d}\left(\frac{u}{v}\right) = \frac{v\,\mathrm{d}u - u\,\mathrm{d}v}{v^2}$$

So, if $y = \mathrm{e}^{2x}\sin 4x$, $\mathrm{d}y = \dots\dots\dots$

and if $y = \dfrac{\cos 2t}{t^2}$ $\mathrm{d}y = \dots\dots\dots$

12

$y = e^{2x}\sin 4x,$	$dy = 2e^{2x}(2\cos 4x + \sin 4x)\,dx$
$y = \dfrac{\cos 2t}{t^2},$	$dy = -\dfrac{2}{t^3}\{t\sin 2t + \cos 2t\}\,dt$

That was easy enough. Let us now consider a function of two independent variables, $z = f(x, y)$.

If $z = f(x, y)$ then $z + \delta z = f(x + \delta x, y + \delta y)$

$$\therefore \delta z = f(x + \delta x,\ y + \delta y) - f(x,\ y)$$

Expanding δz in terms of δx and δy, gives

$\delta z = A\delta x + B\delta y$ + higher powers of δx and δy, where A and B are functions of x and y.

If y remains constant, i.e. $\delta y = 0$, then

$$\delta z = A\,\delta x + \text{higher powers of } \delta x \quad \therefore \frac{\delta z}{\delta x} \approx A$$

$$\therefore \text{If } \delta x \to 0, \text{ then } \underline{A = \frac{\partial z}{\partial x}}$$

Similarly, if x remains constant, i.e. $\delta x = 0$, then

$$\delta z = B\,\delta y + \text{ higher powers of } \delta y \quad \therefore \frac{\delta z}{\delta y} \approx B$$

$$\therefore \text{If } \delta y \to 0, \text{ then } \underline{B = \frac{\partial z}{\partial y}}$$

$$\therefore \delta z = \frac{\partial z}{\partial x}\delta x + \frac{\partial z}{\partial y}\delta y + \text{higher powers of small quantities}$$

$$\underline{\therefore \delta z = \frac{\partial z}{\partial x}\delta x + \frac{\partial z}{\partial y}\delta y}$$

In terms of differentials, this result can be written

$$\text{If } z = f(x,\ y), \text{ then } \underline{dz = \frac{\partial z}{\partial x}dx + \frac{\partial z}{\partial y}dy}$$

The result can be extended to functions of more than two independent variables.

$$\text{If } z = f(x, y, w), \quad \mathrm{d}z = \frac{\partial z}{\partial x}\mathrm{d}x + \frac{\partial z}{\partial y}\mathrm{d}y + \frac{\partial z}{\partial w}\mathrm{d}w$$

Make a note of these results in differential form as shown.

Exercise Determine the differential dz for each of the following functions.

1. $z = x^2 + y^2$
2. $z = x^3 \sin 2y$
3. $z = (2x - 1)\,\mathrm{e}^{3y}$
4. $z = x^2 + 2y^2 + 3w^2$
5. $z = x^3y^2w.$

Finish all five and then check the results.

13

1. $\mathrm{d}z = 2(x\,\mathrm{d}x + y\,\mathrm{d}y)$
2. $\mathrm{d}z = x^2(3\sin 2y\,\mathrm{d}x + 2x\cos 2y\,\mathrm{d}y)$
3. $\mathrm{d}z = \mathrm{e}^{3y}\{2\,\mathrm{d}x + (6x - 3)\mathrm{d}y\}$
4. $\mathrm{d}z = 2(x\,\mathrm{d}x + 2y\,\mathrm{d}y + 3w\,\mathrm{d}w)$
5. $\mathrm{d}z = x^2y(3yw\,\mathrm{d}x + 2xw\,\mathrm{d}y + xy\,\mathrm{d}w)$

Now move on.

14 Exact differential

We have just established that if $z = f(x, y)$

$$dz = \frac{\partial z}{\partial x}dx + \frac{\partial z}{\partial y}dy$$

We now work in reverse.

Any expression $dz = P\,dx + Q\,dy$, where P and Q are functions of x and y, is an *exact differential* if it can be integrated to determine z.

$$\therefore P = \frac{\partial z}{\partial x} \quad \text{and} \quad Q = \frac{\partial z}{\partial y}$$

Now $\dfrac{\partial P}{\partial y} = \dfrac{\partial^2 z}{\partial y\,\partial x}$ and $\dfrac{\partial Q}{\partial x} = \dfrac{\partial^2 z}{\partial x\,\partial y}$ and we know that

$$\frac{\partial^2 z}{\partial y\,\partial x} = \frac{\partial^2 z}{\partial x\,\partial y}.$$

Therefore, for dz to be an exact differential $\underline{\dfrac{\partial P}{\partial y} = \dfrac{\partial Q}{\partial x}}$ and this is the test we apply.

Example 1 $dz = (3x^2 + 4y^2)\,dx + 8xy\,dy$.

If we compare the right-hand side with $P\,dx + Q\,dy$, then

$$P = 3x^2 + 4y^2 \quad \therefore \frac{\partial P}{\partial y} = 8y$$

$$Q = 8xy \quad \therefore \frac{\partial Q}{\partial x} = 8y$$

$\dfrac{\partial P}{\partial y} = \dfrac{\partial Q}{\partial x}$ $\quad\therefore$ <u>dz is an exact differential</u>

Similarly, we can test this one.

Example 2 $dz = (1 + 8xy)\,dx + 5x^2\,dy$.

From this we find ....

dz is *not* an exact differential

for $dz = (1 + 8xy)\,dx + 5x^2\,dy$

$$\therefore P = 1 + 8xy \qquad \therefore \frac{\partial P}{\partial y} = 8x$$

$$Q = 5x^2 \qquad \therefore \frac{\partial Q}{\partial x} = 10x$$

$\dfrac{\partial P}{\partial y} \neq \dfrac{\partial Q}{\partial x}$ $\quad\therefore$ dz is not an exact differential.

Exercise Determine whether each of the following is an exact differential.

1. $dz = 4x^3y^3\,dx + 3x^4y^2\,dy$
2. $dz = (4x^3y + 2xy^3)\,dx + (x^4 + 3x^2y^2)\,dy$
3. $dz = (15y^2e^{3x} + 2xy^2)\,dx + (10ye^{3x} + x^2y)\,dy$
4. $dz = (3x^2e^{2y} - 2y^2e^{3x})\,dx + (2x^3e^{2y} - 2ye^{3x})\,dy$
5. $dz = (4y^3\cos 4x + 3x^2\cos 2y)\,dx + (3y^2\sin 4x - 2x^3\sin 2y)\,dy.$

16

1. Yes 2. Yes 3. No 4. No 5. Yes

We have just tested whether certain expressions are, in fact, exact differentials—and we said previously that, by definition, an exact differential can be integrated. But how exactly do we go about it? The following examples will show.

Integration of exact differentials

$$\mathrm{d}z = P\,\mathrm{d}x + Q\,\mathrm{d}y \quad \text{where} \quad P = \frac{\partial z}{\partial x} \quad \text{and} \quad Q = \frac{\partial z}{\partial y}$$

$$\therefore z = \int P\,\mathrm{d}x \quad \text{and also} \quad z = \int Q\,\mathrm{d}y$$

Example 1 $\mathrm{d}z = (2xy + 6x)\,\mathrm{d}x + (x^2 + 2y^3)\,\mathrm{d}y$.

$$P = \frac{\partial z}{\partial x} = 2xy + 6x \qquad \therefore z = \int (2xy + 6x)\,\mathrm{d}x$$

$\therefore z = x^2y + 3x^2 + f(y)$ where $f(y)$ is an arbitrary function of y only, and is akin to the constant of integration in a normal integral.

Also $Q = \dfrac{\partial z}{\partial y} = x^2 + 2y^3 \qquad \therefore z = \displaystyle\int (x^2 + 2y^3)\,\mathrm{d}y$

$$\therefore z = \ldots\ldots\ldots$$

17

$$z = x^2y + \frac{y^4}{2} + F(x) \text{ where } F(x) \text{ is an arbitrary function of } x \text{ only.}$$

So the two results tell us

$$z = x^2y + 3x^2 + f(y) \qquad \text{(i)}$$

and

$$z = x^2y + \frac{y^4}{2} + F(x) \qquad \text{(ii)}$$

For these two expressions to represent the same function, then

$f(y)$ in (i) must be $\frac{y^4}{2}$ already in (ii)

and $F(x)$ in (ii) must be $3x^2$ already in (i)

$$\therefore \; \underline{z = x^2y + 3x^2 + \frac{y^4}{2}}$$

Example 2 Integrate $\mathrm{d}z = (8\mathrm{e}^{4x} + 2xy^2)\,\mathrm{d}x + (4\cos 4y + 2x^2y)\,\mathrm{d}y$.

Argue through the working in just the same way, from which we obtain

$$z = \dots\dots\dots$$

18

$$z = 2e^{4x} + x^2y^2 + \sin 4y$$

Here it is. $\quad dz = (8e^{4x} + 2xy^2)\,dx + (4\cos 4y + 2x^2y)\,dy$

$$P = \frac{\partial z}{\partial x} = 8e^{4x} + 2xy^2 \qquad \therefore z = \int (8e^{4x} + 2xy^2)dx$$

$$\therefore z = 2e^{4x} + x^2y^2 + f(y) \qquad \text{(i)}$$

$$Q = \frac{\partial z}{\partial y} = 4\cos 4y + 2x^2y \quad \therefore z = \int (4\cos 4y + 2x^2y)\,dy$$

$$\therefore z = \sin 4y + x^2y^2 + F(x) \qquad \text{(ii)}$$

For (i) and (ii) to agree, $f(y) = \sin 4y$ and $F(x) = 2e^{4x}$

$$\therefore z = 2\,e^{4x} + x^2y^2 + \sin 4y$$

They are all done in the same way, so you will have no difficulty with the short exercise that follows. *On you go.*

Exercise

Integrate the following exact differentials to obtain the function z.

1. $dz = (6x^2 + 8xy^3)\,dx + (12x^2y^2 + 12y^3)\,dy$
2. $dz = (3x^2 + 2xy + y^2)\,dx + (x^2 + 2xy + 3y^2)\,dy$
3. $dz = 2(y+1)e^{2x}\,dx + (e^{2x} - 2y)\,dy$
4. $dz = (3y^2\cos 3x - 3\sin 3x)\,dx + (2y\sin 3x + 4)\,dy$
5. $dz = (\sinh y + y\sinh x)dx + (x\cosh y + \cosh x)\,dy.$

Finish all five before checking with the next frame.

19

1. $z = 2x^3 + 4x^2y^3 + 3y^4$
2. $z = x^3 + x^2y + xy^2 + y^3$
3. $z = e^{2x}(1+y) - y^2$
4. $z = y^2\sin 3x + \cos 3x + 4y$
5. $z = x\sinh y + y\cosh x.$

In the last one, of course, we find that the two expressions for z agree without any further addition of $f(y)$ or $F(x)$.

We shall be meeting exact differentials again later on, but for the moment let us deal with something different. On then to the next frame.

20

Area enclosed by a closed curve

One of the earliest applications of integration is finding the area of a plane figure bounded by the x-axis, the curve $y = f(x)$ and ordinates at $x = x_1$ and $x = x_2$.

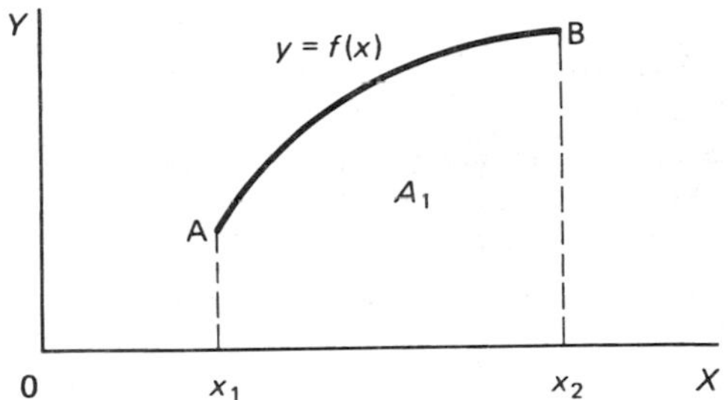

$$A_1 = \int_{x_1}^{x_2} y\,\mathrm{d}x = \int_{x_1}^{x_2} f(x)\,\mathrm{d}x$$

If points A and B are joined by another curve, $y = F(x)$

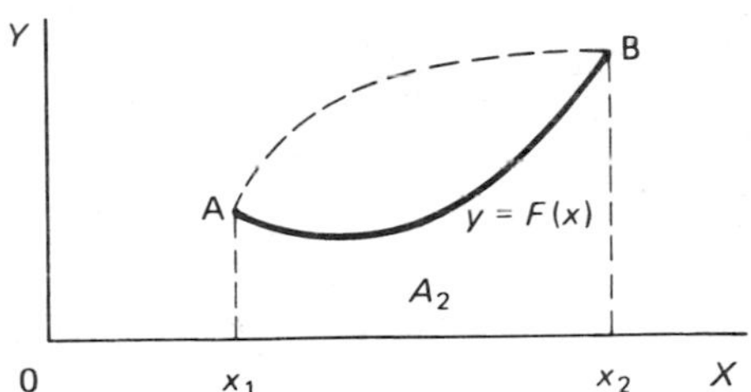

$$A_2 = \int_{x_1}^{x_2} F(x)\,\mathrm{d}x$$

Combining the two figures, we have

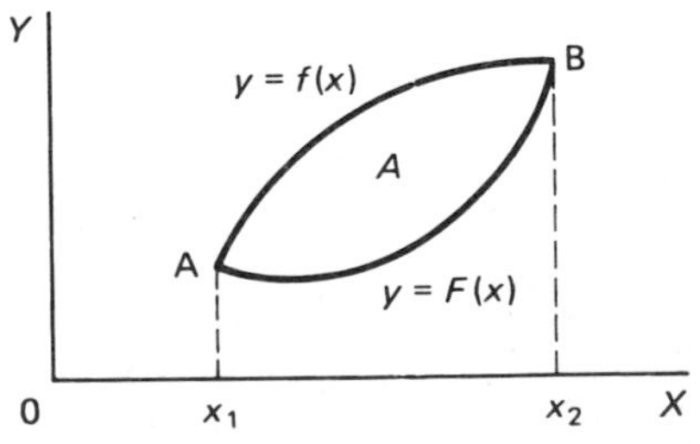

$$A = A_1 - A_2$$

$$\therefore\ A = \int_{x_1}^{x_2} f(x)\,\mathrm{d}x - \int_{x_1}^{x_2} F(x)\,\mathrm{d}x$$

It is convenient on occasions to arrange the limits so that the integration follows the path round the enclosed area in a regular order.

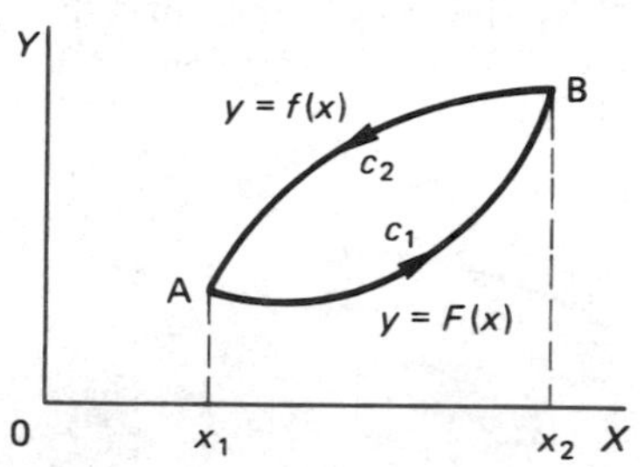

For example

$\int_{x_1}^{x_2} F(x)\,\mathrm{d}x$ gives A_2 as before, but integrating from B to A along c_2 with

$y=f(x)$, i.e. $\int_{x_2}^{x_1} f(x)\,\mathrm{d}x$, is the integral for A_1 with the sign changed, i.e.

$$\int_{x_2}^{x_1} f(x)\,\mathrm{d}x = -\int_{x_1}^{x_2} f(x)\,\mathrm{d}x$$

$\therefore$ The result $A = A_1 - A_2 = \int_{x_1}^{x_2} f(x)\,\mathrm{d}x - \int_{x_1}^{x_2} F(x)\,\mathrm{d}x$

becomes

$$A = \ldots\ldots\ldots$$

$$A = -\int_{x_1}^{x_2} F(x)\,dx - \int_{x_2}^{x_1} f(x)\,dx$$

i.e.
$$A = -\left\{\int_{x_1}^{x_2} F(x)\,dx + \int_{x_2}^{x_1} f(x)\,dx\right\}$$

If we proceed round the boundary in an *anticlockwise manner*, the enclosed area is kept on the *left-hand side* and the resulting area is considered *positive*. If we proceed round the boundary in a *clockwise manner*, the enclosed area remains on the *right-hand side* and the resulting area is *negative*.

The final result above can be written in the form

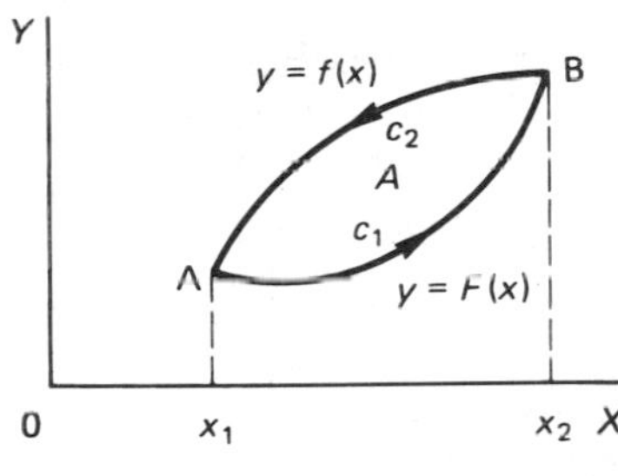

$$A = -\oint y\,dx$$

where the symbol $\oint$ indicates that the integral is to be evaluated round the closed boundary in the positive (i.e. anticlockwise) direction

$$\therefore\ A = -\oint y\,dx = -\left\{\int_{x_1}^{x_2} F(x)\,dx + \int_{x_2}^{x_1} f(x)\,dx\right\}$$

(along c_1) (along c_2)

Let us apply this result to a very simple case.

Example 1 Determine the area enclosed by the graphs of $y = x^3$ and $y = 4x$ for $x \geq 0$.

First we need to know the points of intersection. These are ....

22

$$x = 0 \quad \text{and} \quad x = 2$$

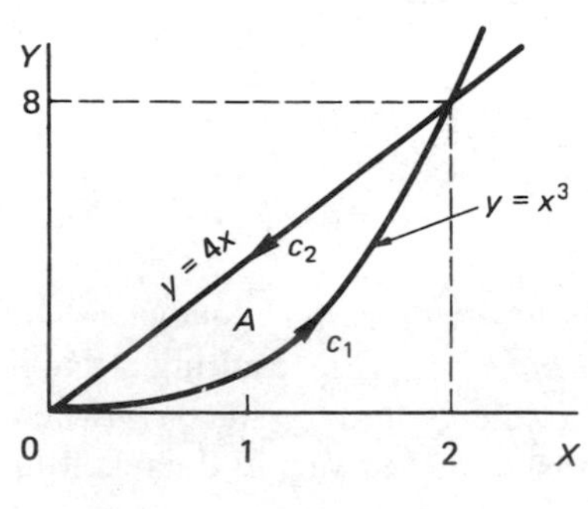

We integrate in an anticlockwise manner

c_1: $y = x^3$, limits $x = 0$ to $x = 2$

c_2: $y = 4x$, limits $x = 2$ to $x = 0$.

$$A = -\oint y\,dx = \ldots\ldots\ldots$$

23

$$A = 4 \text{ square units}$$

$$\text{for } A = -\oint y\,dx = -\left\{\int_0^2 x^3\,dx + \int_2^0 4x\,dx\right\}$$

$$= -\left\{\left[\frac{x^4}{4}\right]_0^2 + \left[2x^2\right]_2^0\right\} = \underline{4}$$

Another example.

Example 2 Find the area of the triangle with vertices (0, 0), (5, 3) and (2, 6).

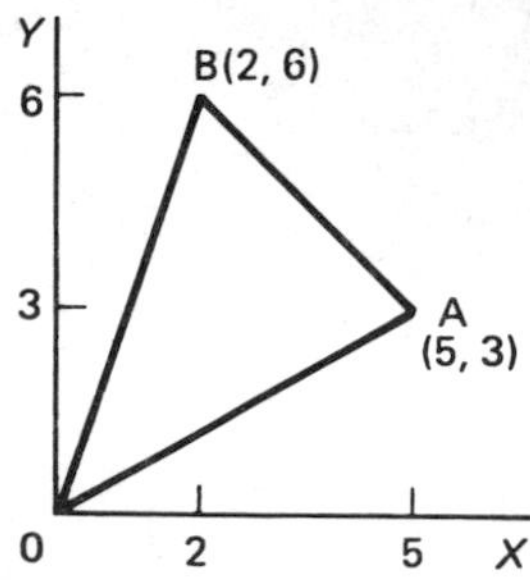

The equation of

OA is ....

BA is ....

OB is ....

24

OA is $y = \frac{3}{5}x$
BA is $y = 8 - x$
OB is $y = 3x$

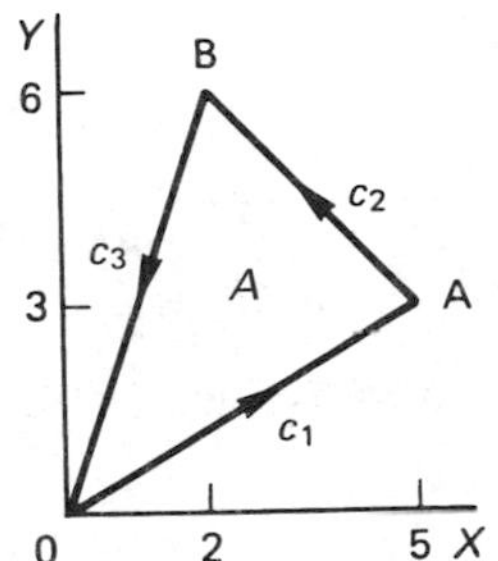

Then $A = -\oint y\,dx$

$= \dots\dots\dots$

Write down the component integrals with appropriate limits

25

$$A = -\oint y\,dx = -\left\{\int_0^5 \frac{3}{5}x\,dx + \int_5^2 (8 - x)\,dx + \int_2^0 3x\,dx\right\}$$

The limits chosen must progress the integration round the boundary of the figure in an *anticlockwise manner*. Finishing off the integration, we have

$A = \dots\dots\dots$

26

$A = 12$ square units

The actual integration is easy enough.

The work we have just done leads us on to consider ***line integrals****, so let us make a fresh start in the next frame.*

27

Line Integrals

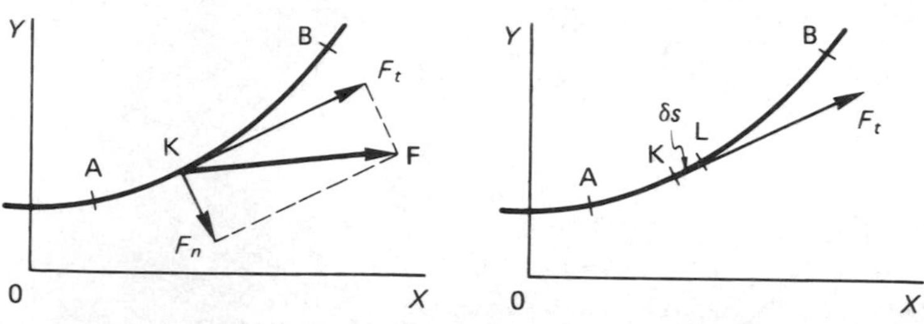

If a field exists in the xy-plane, producing a force F on a particle at K, then F can be resolved into two components

F_{t} along the tangent to the curve AB at K
F_{n} along the normal to the curve AB at K.

The work done in moving the particle through a small distance δs from K to L along the curve is then approximately $F_{\mathrm{t}}\,\delta s$. So the total work done in moving a particle along the curve from A to B is given by

. . . .

28

$$\lim_{\delta s \to 0} \sum F_{\mathrm{t}}\,\delta s = \int F_{\mathrm{t}}\,\mathrm{d}s \text{ from A to B}$$

This is normally written $\int_{\mathrm{AB}} F_{\mathrm{t}}\,\mathrm{d}s$ where A and B are the end points of the curve, or as $\int_{\mathrm{c}} F_{\mathrm{t}}\,\mathrm{d}s$ where the curve c connecting A and B is defined.

Such an integral thus formed, is called a *line integral* since integration is carried out along the path of the particular curve c joining A and B.

$$\therefore\ I = \int_{\mathrm{AB}} F_{\mathrm{t}}\,\mathrm{d}s = \int_{\mathrm{c}} F_{\mathrm{t}}\,\mathrm{d}s$$

where c is the curve $y = f(x)$ between $\mathrm{A}(x_1, y_1)$ and $\mathrm{B}(x_2, y_2)$.

There is in fact an alternative form of the integral which is often useful, so let us also consider that.

29

Alternative form of a line integral

It is often more convenient to integrate with respect to x or y than to take arc length as the variable.

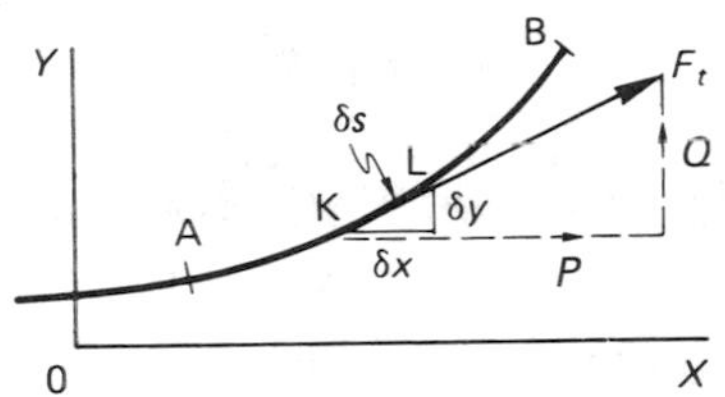

If F_t has a component

P in the x-direction
Q in the y-direction

then the work done from K to L can be stated as $P\,\delta x + Q\,\delta y$

$$\therefore \int_{AB} F_t\,ds = \int_{AB} (P\,dx + Q\,dy)$$

where P and Q are functions of x and y.

In general then, the line integral can be expressed as

$$I = \int_c F_t\,ds = \int_c (P\,dx + Q\,dy)$$

where c is the prescribed curve
and F, or P and Q, are functions of x and y.

Make a note of these results—then we will apply them to one or two examples.

30 *Example 1* Evaluate $\int_c (x+3y)dx$ from A(0, 1) to B(2, 5) along the curve $y = 1 + x^2$.

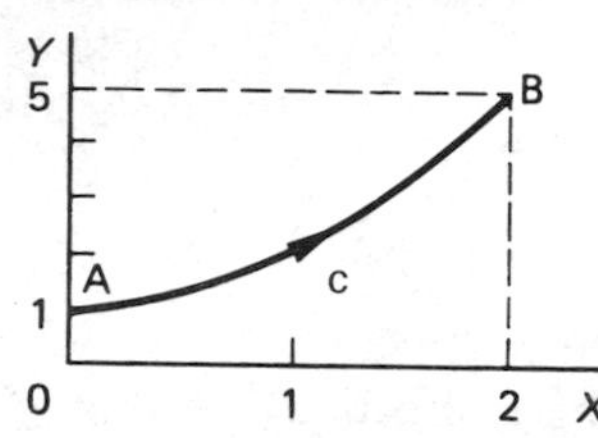

The line integral is of the form

$$\int_c (P\,dx + Q\,dy)$$

where, in this case, $Q = 0$ and c is the curve $y = 1 + x^2$.

It can be converted at once into an ordinary integral by substituting for y and applying the appropriate limits of x.

$$I = \int_c (P\,dx + Q\,dy) = \int_c (x+3y)\,dx = \int_0^2 (x+3+3x^2)\,dx$$

$$= \left[\frac{x^2}{2} + 3x + x^3\right]_0^2 = \underline{16}$$

Now for another, so move on.

31 *Example 2* Evaluate $I = \int_c (x^2+y)\,dx + (x-y^2)\,dy$ from A(0, 2) to B(3, 5) along the curve $y = 2 + x$.

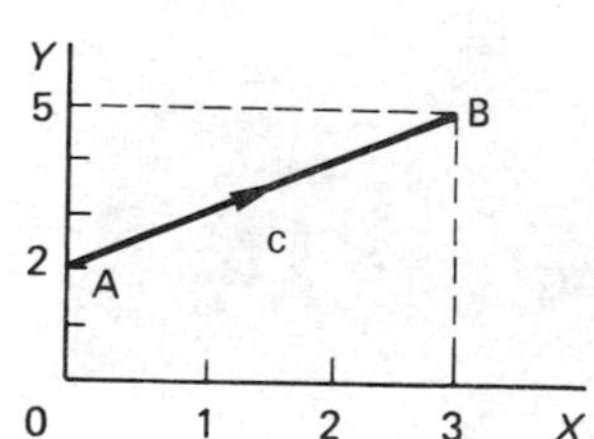

$$I = \int_c (P\,dx + Q\,dy)$$

$$P = x^2 + y = x^2 + 2 + x = x^2 + x + 2$$

$$Q = x - y^2 = x - (4 + 4x + x^2)$$

$$= -(x^2 + 3x + 4)$$

Also $y = 2 + x \quad \therefore\ dy = dx$ and the limits are $x = 0$ to $x = 3$.

$$\therefore\ I = \ldots\ldots\ldots$$

32

$$\boxed{I = -15}$$

for $I = \int_0^3 \{(x^2 + x + 2)\,\mathrm{d}x - (x^2 + 3x + 4)\,\mathrm{d}x\}$

$$\int_0^3 -(2x+2)\mathrm{d}x = \Big[x^2 - 2x\Big]_0^3 = -15$$

Here is another.

Example 3 Evaluate $I = \int_c \{(x^2 + 2y)\,\mathrm{d}x + xy\,\mathrm{d}y\}$ from O(0, 0) to B(1, 4) along the curve $y = 4x^2$.

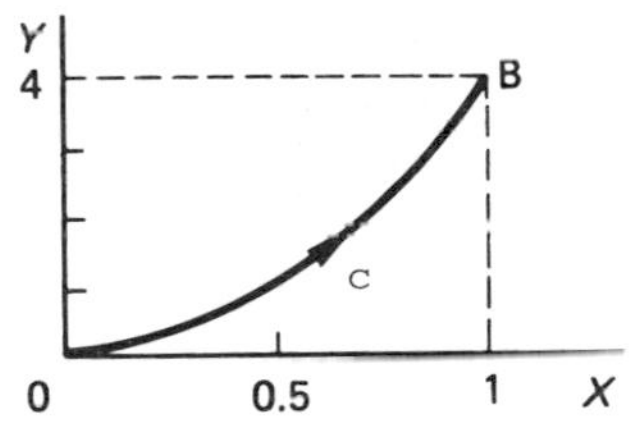

In this case, c is the curve $y = 4x^2$.

$$\therefore \mathrm{d}y = 8x\,\mathrm{d}x$$

Substitute for y in the integral and apply the limits.

Then $I = \ldots\ldots\ldots$

Finish it off: it is quite straightforward.

33

$$\boxed{I = 9.4}$$

for $I - \int_c \{(x^2 + 2y)\,\mathrm{d}x + xy\,\mathrm{d}y\}$ $\quad y = 4x^2 \quad \therefore \mathrm{d}y = 8x\,\mathrm{d}x$

Also $x^2 + 2y = x^2 + 8x^2 = 9x^2$; $\quad xy = 4x^3$

$$\therefore I = \int_0^1 \{9x^2\,\mathrm{d}x + 32x^4\,\mathrm{d}x\} = \int_0^1 (9x^2 + 32x^4)\,\mathrm{d}x = \underline{9.4}$$

They are all done in very much the same way.

Move on for Example 4.

34 *Example 4* Evaluate $I = \int_c \{(x^2 + 2y)\,dx + xy\,dy\}$ from O(0, 0) to A(1, 0) along line $y = 0$ and then from A(1, 0) to B(1, 4) along the line $x = 1$.

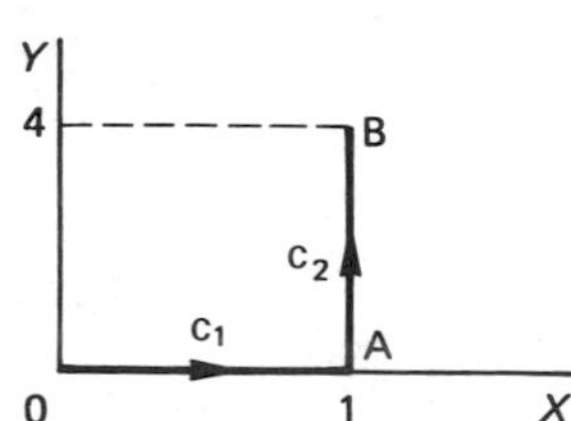

(i) OA: c_1 is the line $y = 0$ $\therefore dy = 0$. Substituting $y = 0$ and $dy = 0$ in the given integral gives

$$I_{OA} = \int_0^1 x^2\,dx = \left[\frac{x^3}{3}\right]_0^1 = \underline{\frac{1}{3}}$$

(ii) AB: Here c_2 is the line $x = 1$ $\quad \therefore dx = 0$

$$\therefore I_{AB} = \ldots\ldots\ldots$$

35

$$\boxed{I_{AB} = 8}$$

for $I_{AB} = \int_0^4 \{(1 + 2y)(0) + y\,dy\} = \int_0^4 y\,dy = \left[\frac{y^2}{2}\right]_0^4 = \underline{8}$

Then $I = I_{OA} + I_{AB} = \frac{1}{3} + 8 = 8\frac{1}{3}$ $\quad \underline{\therefore I = 8\frac{1}{3}}$

If we now look back to Examples 3 and 4 just completed, we find that we have evaluated the same integral between the same two end points, but ………

36

along different paths of integration

If we combine the two diagrams, we have

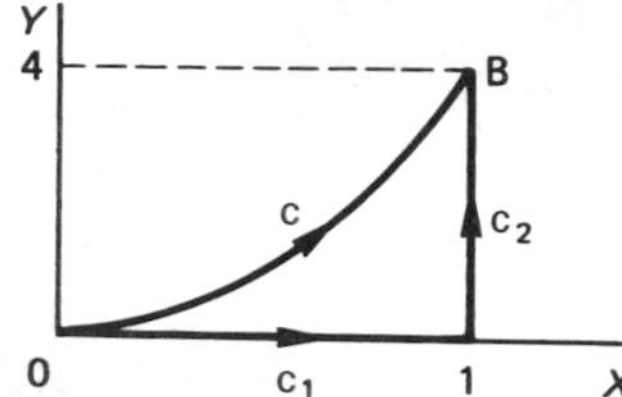

where c is the curve $y = 4x^2$ and $c_1 + c_2$ are the lines $y = 0$ and $x = 1$. The results obtained were

$\underline{I_c = 9\frac{2}{5}}$ and $\underline{I_{c_1 + c_2} = 8\frac{1}{3}}$

Notice therefore that integration along two distinct paths joining the same two end points does not necessarily give the same results.

37

Let us pause here a moment and list the main properties of line integrals.

Properties of line integrals

1. $\int_c F ds = \int_c \{P\,dx + Q\,dy\}$

2. $\int_{AB} F ds = -\int_{BA} F ds$ and $\int_{AB} \{P\,dx + Q\,dy\} = -\int_{BA} \{P\,dx + Q\,dy\}$

   i.e. the sign of a line integral is reversed when the direction of the integration along the path is reversed.

3. (a) For a path of integration parallel to the y-axis, i.e. $x = k$,

   $$dx = 0. \quad \therefore \int_c P\,dx = 0 \quad \therefore I_c = \int_c Q\,dy.$$

   (b) For a path of integration parallel to the x-axis, i.e. $y = k$,

   $$dy = 0. \quad \therefore \int_c Q\,dy = 0 \quad \therefore I_c = \int_c P\,dx.$$

4. If the path of integration c joining A to B is divided into two parts AK and KB, then $I_c = I_{AB} = I_{AK} + I_{KB}$.

5. If the path of integration c is not single valued for part of its extent, the path is divided into two sections.

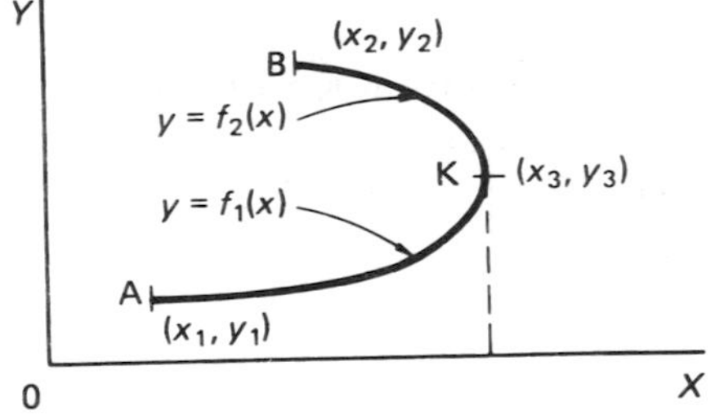

$y = f_1(x)$ from A to K

$y = f_2(x)$ from K to B.

6. In all cases, the actual path of integration involved must be continuous and single-valued—or dealt with as in item 5 above.

Make a note of this list for future reference and revision.

38 *Example* Evaluate $I = \int_c (x + y)\,dx$ from A(0, 1) to B(0, −1) along the semi-circle $x^2 + y^2 = 1$ for $x \geq 0$.

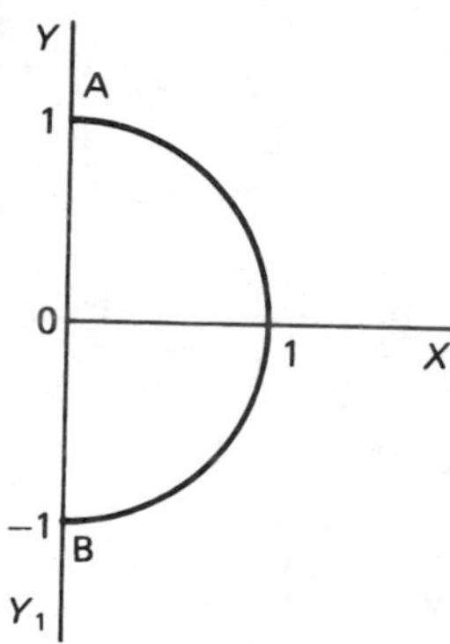

The first thing we notice is that

. . . .

39

the path of integration c is *not* single-valued

For any value of x, $y = \pm\sqrt{1 - x^2}$. Therefore, we divide c into two parts

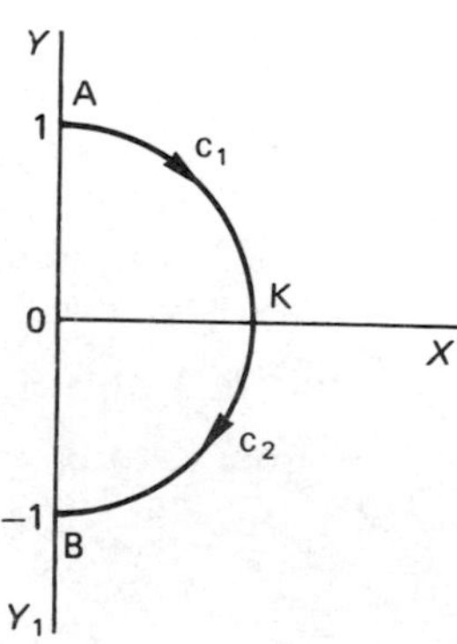

(i) $y = \sqrt{1 - x^2}$ from A to K

(ii) $y = -\sqrt{1 - x^2}$ from K to B.

As usual, $I = \int_c (P\,dx + Q\,dy)$ and in this particular case, Q = ...

40

$$\boxed{Q = 0}$$

$$\therefore\ I = \int_c P\,\mathrm{d}x = \int_0^1 (x + \sqrt{(1 - x^2)}\,\mathrm{d}x + \int_1^0 (x - \sqrt{1 - x^2})\,\mathrm{d}x$$

$$= \int_0^1 (x + \sqrt{1 - x^2} - x + \sqrt{1 - x^2})\,\mathrm{d}x = 2\int_0^1 \sqrt{1 - x^2}\,\mathrm{d}x$$

Now substitute $x = \sin\theta$ and finish it off.

$$I = \ldots\ldots\ldots$$

41

$$\boxed{I = \frac{\pi}{2}}$$

for $I = 2\int_0^1 \sqrt{1 - x^2}\,\mathrm{d}x \qquad x = \sin\theta \qquad \therefore\ \mathrm{d}x = \cos\theta\,\mathrm{d}\theta$

$$\sqrt{1 - x^2} = \cos\theta$$

Limits: $x = 0,\quad \theta = 0;\quad x = 1,\quad \theta = \dfrac{\pi}{2}$

$$\therefore\ I = 2\int_0^{\pi/2} \cos^2\theta\,\mathrm{d}\theta = \int_0^{\pi/2} (1 + \cos 2\theta)\,\mathrm{d}\theta = \left[\theta + \frac{\sin 2\theta}{2}\right]_0^{\pi/2} = \underline{\frac{\pi}{2}}$$

Now let us extend this line of development a stage further.

42 Regions enclosed by closed curves

A region is said to be *simply connected* if a path joining A and B can be deformed to coincide with any other line joining A and B without going outside the region.

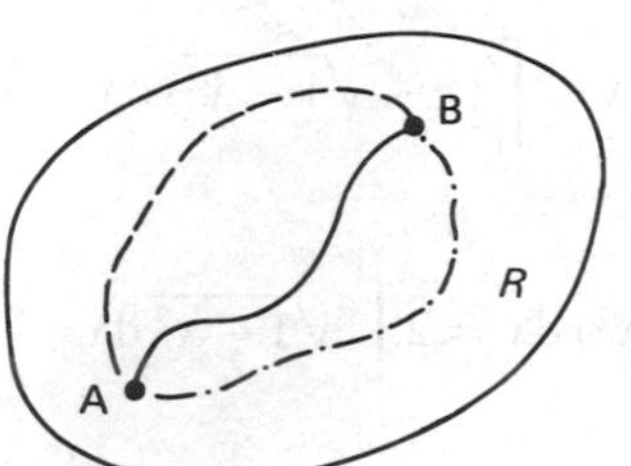

Another definition is that a region is simply connected if any closed path in the region can be contracted to a single point without leaving the region.

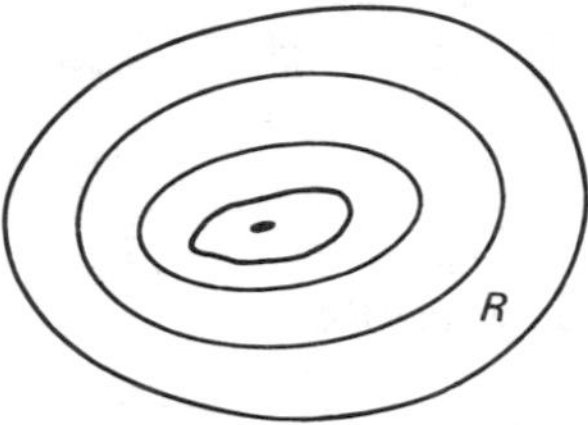

Clearly, this would not be satisfied in the case where the region R contains one or more 'holes'.

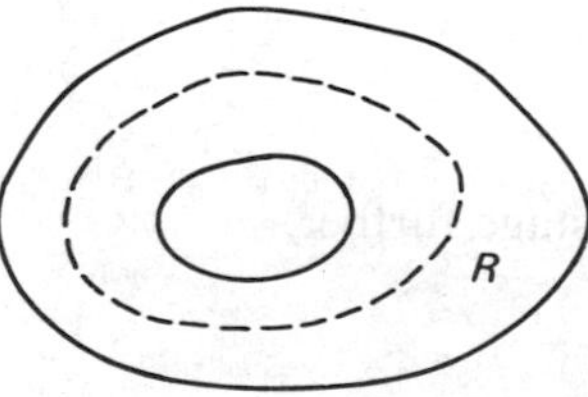

The closed curves involved in problems in this programme all relate to simply connected regions, so no difficulties will arise.

43

Line integrals round a closed curve

We have already introduced the symbol $\oint$ to indicate that an integral is to be evaluated round a closed curve in the positive (anticlockwise) direction.

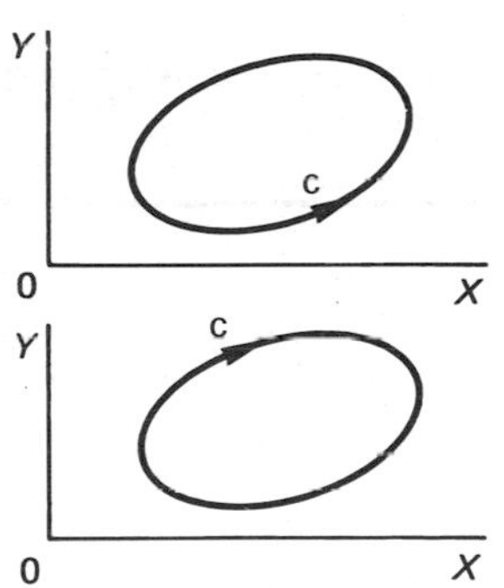

Positive direction (anticlockwise) line integral denoted by $\oint$.

Negative direction (clockwise) line integral denoted by $-\oint$.

With a closed curve, the path c cannot be single-valued. Therefore, we divide the path into two or more parts and treat each separately as a single-valued curve.

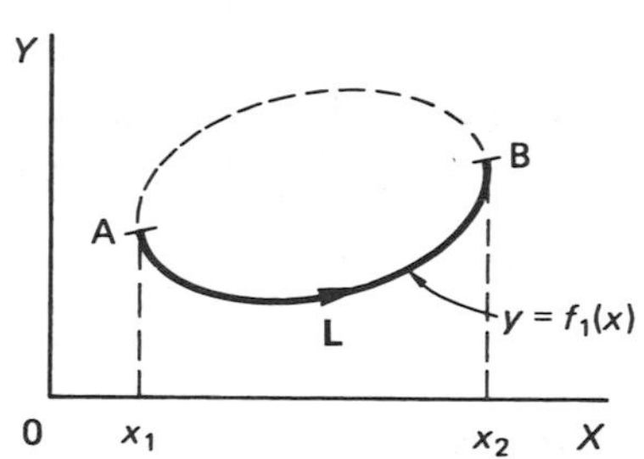

(i) Use $y = f_1(x)$ for ALB

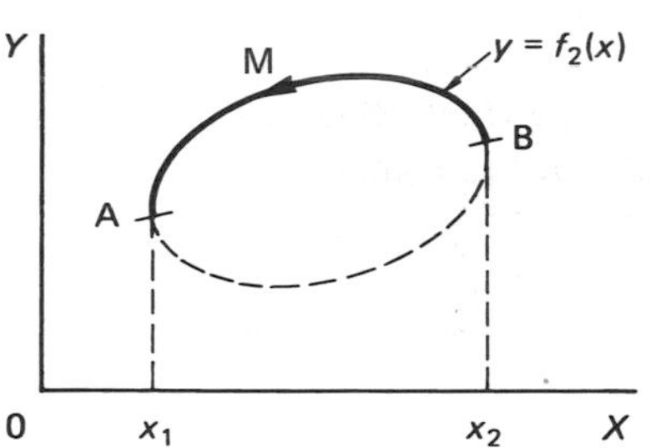

(ii) Use $y = f_2(x)$ for BMA.

Unless specially required otherwise, we always proceed round the closed curve in an ....

44

anticlockwise direction.

Example 1 Evaluate the line integral $I = \oint_c (x^2\,dx - 2xy\,dy)$ where c comprises the three sides of the triangle joining O(0, 0), A(1, 0) and B(0, 1).

First draw the diagram and mark in c_1, c_2 and c_3, the proposed directions of integration. Do just that.

45

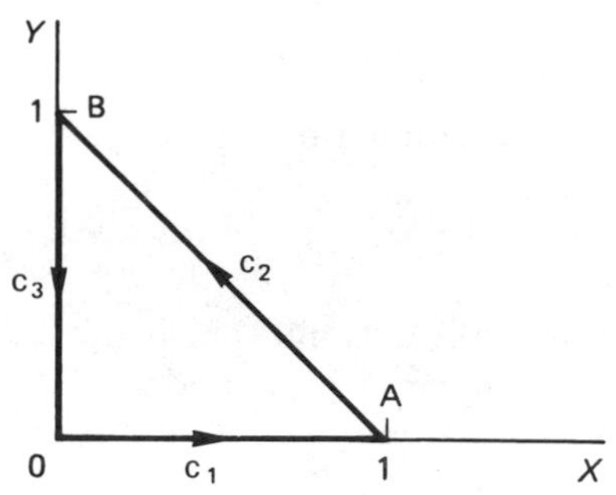

The three sections of the path of integration must be arranged in an anticlockwise manner round the figure. Now we deal with each part separately.

(a) OA: c_1 is the line $y = 0 \quad \therefore\ dy = 0$.

Then $I = \oint (x^2\,dx - 2xy\,dy)$ for this part becomes

$$I_1 = \int_0^1 x^2\,dx = \left[\frac{x^3}{3}\right]_0^1 = \frac{1}{3} \qquad \therefore\ I_1 = \frac{1}{3}$$

(b) AB: c_2 is the line $y = 1 - x \quad \therefore\ dy = -dx$

$I_2 = \ldots\ldots\ldots$ (evaluate it)

46

$$\boxed{I_2 = -\tfrac{2}{3}}$$

for c_2 is the line $y = 1 - x$ $\quad \therefore\ \mathrm{d}y = -\mathrm{d}x.$

$$I_2 = \int_1^0 \{x^2\,\mathrm{d}x + 2x(1-x)\,\mathrm{d}x\} = \int_1^0 (x^2 + 2x - 2x^2)\,\mathrm{d}x$$

$$= \int_1^0 (2x - x^2)\mathrm{d}x = \left[x^2 - \frac{x^3}{3}\right]_1^0 = -\frac{2}{3} \qquad \therefore\ I_2 = -\frac{2}{3}$$

Note that anticlockwise progression is obtained by arranging the limits in the appropriate order.

Now we have to determine I_3 for BO.

(c) BO: $\quad c_3$ is the line $x = 0$

$$I_3 = \ldots\ldots\ldots$$

47

$$\boxed{I_3 = 0}$$

For c_3, $\quad x = 0 \quad \therefore\ dx = 0 \quad \therefore\ I_3 = \int 0\,dy = 0 \qquad \underline{\therefore\ I_3 = 0}$

Finally, $I = I_1 + I_2 + I_3 = \frac{1}{3} - \frac{2}{3} + 0 = -\frac{1}{3} \qquad \underline{\therefore\ I = -\frac{1}{3}}$

Let us work through another example.

Example 2 Evaluate $\oint_c y\,dx$ when c is the circle $x^2 + y^2 = 4$.

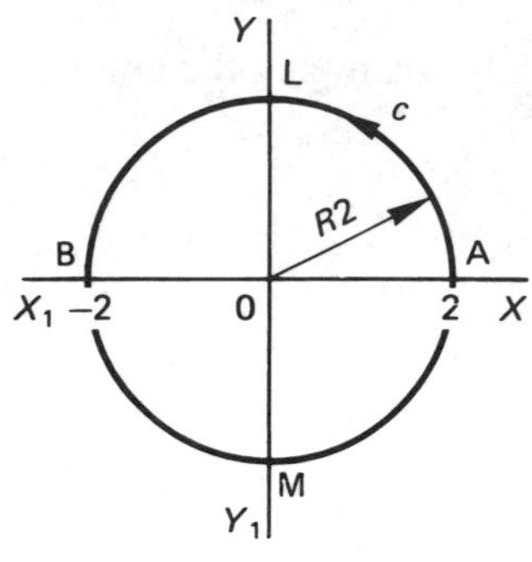

$x^2 + y^2 = 4 \qquad \therefore\ y = \pm\sqrt{4 - x^2}$

y is thus not single-valued. Therefore use $y = \sqrt{4 - x^2}$ for ALB between $x = 2$ and $x = -2$ and $y = -\sqrt{4 - x^2}$ for BMA between $x = -2$ and $x = 2$.

$$\begin{aligned}
\therefore\ I &= \int_2^{-2} \sqrt{4 - x^2}\,dx + \int_{-2}^{2} \{-\sqrt{4 - x^2}\}\,dx \\
&= 2\int_2^{-2} \sqrt{4 - x^2}\,dx \;=\; -2\int_{-2}^{2} \sqrt{4 - x^2}\,dx \\
&= -4\int_0^2 \sqrt{4 - x^2}\,dx.
\end{aligned}$$

To evaluate this integral, substitute $x = 2\sin\theta$ and finish it off.

$$I = \ldots\ldots\ldots$$

48

$$I = -4\pi$$

for $x = 2\sin\theta \qquad \therefore \mathrm{d}x = 2\cos\theta\,\mathrm{d}\theta \qquad \therefore \sqrt{4-x^2} = 2\cos\theta$

limits: $x = 0,\ \ \theta = 0;\ \ x = 2,\ \ \theta = \dfrac{\pi}{2}$

$$\therefore I = -4\int_0^{\pi/2} 2\cos\theta\, 2\cos\theta\,\mathrm{d}\theta = -16\int_0^{\pi/2} \cos^2\theta\,\mathrm{d}\theta$$

$$= -8\int_0^{\pi/2} (1+\cos 2\theta)\,\mathrm{d}\theta = -8\left[\theta + \frac{\sin\ 2\theta}{2}\right]_0^{\pi/2} = \underline{-4\pi}$$

Now for one more

Example 3 Evaluate $I = \oint_c \{xy\,\mathrm{d}x + (1+y^2)\,\mathrm{d}y\}$ where c is the boundary of the rectangle joining A(1, 0), B(3, 0), C(3, 2), D(1, 2).

First draw the diagram and insert $c_1,\ c_2,\ c_3,\ c_4$.

That gives ....

49

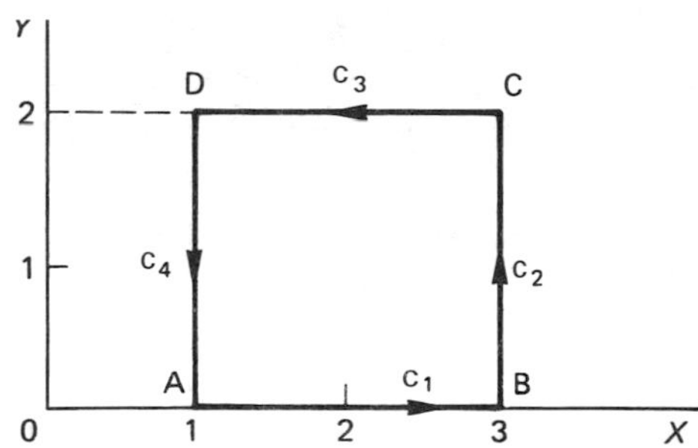

Now evaluate I_1 for AB; I_2 for BC; I_3 for CD; I_4 for DA; and finally I.

Complete the working and then check with the next frame.

50

$$I_1 = 0; \quad I_2 = 4\tfrac{2}{3}; \quad I_3 = -8; \quad I_4 = -4\tfrac{2}{3}; \quad I = -8$$

Here is the complete working.

$$I = \oint_c \{xy\,dx + (1 + y^2)\,dy\}$$

(a) AB: c_1 is $y = 0$ $\quad \therefore dy = 0 \quad \therefore \underline{I_1 = 0}$

(b) BC: c_2 is $x = 3$ $\quad \therefore dx = 0$

$$\therefore I_2 = \int_0^2 (1 + y^2)dy = \left[y + \frac{y^3}{3}\right]_0^2 = 4\tfrac{2}{3} \qquad \therefore \underline{I_2 = 4\tfrac{2}{3}}$$

(c) CD: c_3 is $y = 2$ $\quad \therefore dy = 0$

$$\therefore I_3 = \int_3^1 2x\,dx = \left[x^2\right]_3^1 = -8 \qquad \therefore \underline{I_3 = -8}$$

(d) DA: c_4 is $x = 1$ $\quad \therefore dx = 0$

$$\therefore I_4 = \int_2^0 (1 + y^2)\,dy = \left[y + \frac{y^3}{3}\right]_2^0 = -4\tfrac{2}{3} \qquad \therefore \underline{I_4 = -4\tfrac{2}{3}}$$

Finally $I = I_1 + I_2 + I_3 + I_4$

$$= 0 + 4\tfrac{2}{3} - 8 - 4\tfrac{2}{3} = -8 \qquad \therefore \underline{I = -8}$$

Remember that, unless we are directed otherwise, we always proceed round the closed boundary in an anticlockwise manner.

On now to the next piece of work.

Line integral with respect to arc length 51

We have already established that

$$I=\int_{AB} F_t\,ds=\int_{AB}\{P\,dx+Q\,dy\}$$

where F_t denoted the tangential force along the curve c at the sample point K(x, y).

The same kind of integral can, of course, relate to any function $f(x, y)$ which is a function of the position of a point on the stated curve, so that $I=\int_c f(x, y)\,ds$.

This can readily be converted into an integral in terms of x:

$$I=\int_c f(x, y)\,ds=\int_c f(x, y)\frac{ds}{dx}\,dx \qquad \text{where } \frac{ds}{dx}=\sqrt{1+\left(\frac{dy}{dx}\right)^2}$$

$$\therefore \int_c f(x, y)\,ds=\int_{x_1}^{x_2} f(x, y)\sqrt{1+\left(\frac{dy}{dx}\right)^2}\,dx \qquad (1)$$

Example Evaluate $I=\int_c (4x+3xy)\,ds$ where c is the straight line joining O(0, 0) to A(1, 2).

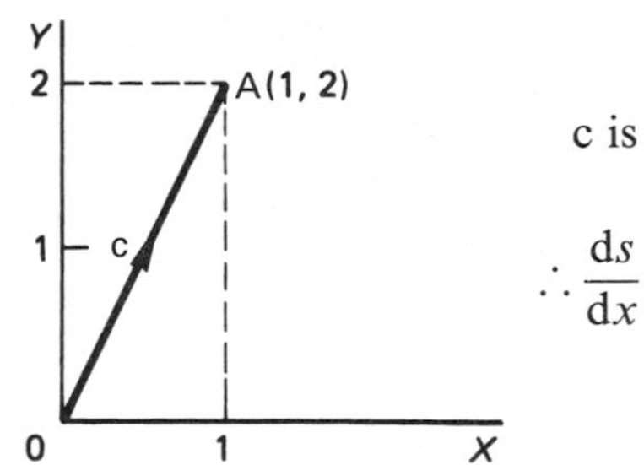

c is the line $y=2x \quad \therefore \frac{dy}{dx}=2$

$$\therefore \frac{ds}{dx}=\sqrt{1+\left(\frac{dy}{dx}\right)^2}=\sqrt{5}$$

$$\therefore I=\int_{x=0}^{x=1}(4x+3xy)\,ds=\int_0^1(4x+3xy)(\sqrt{5})\,dx. \quad \text{But } y=2x$$

$$\therefore I=\ldots\ldots\ldots$$

52

$$\boxed{I = 4\sqrt{5}}$$

for $I = \int_0^1 (4x + 6x^2)(\sqrt{5})\,dx = 2\sqrt{5}\int_0^1 (2x + 3x^2)\,dx = 4\sqrt{5}.$

Parametric equations

When x and y are expressed in parametric form, e.g. $x = f(t), y = g(t)$,

then $\dfrac{ds}{dt} = \sqrt{\left(\dfrac{dx}{dt}\right)^2 + \left(\dfrac{dy}{dt}\right)^2}$ $\quad\therefore\ ds = \sqrt{\left(\dfrac{dx}{dt}\right)^2 + \left(\dfrac{dy}{dt}\right)^2}\,dt$

and result (1) above becomes

$$I = \int_c f(x, y)\,ds = \int_{t_1}^{t_2} f(x, y)\sqrt{\left(\frac{dx}{dt}\right)^2 + \left(\frac{dy}{dt}\right)^2}\,dt \qquad (2)$$

Make a note of results (1) *and* (2) *for future use.*

53 *Example* Evaluate $I = \oint_c 4xy\,ds$ where c is defined as the curve $x = \sin t,\ y = \cos t$ between $t = 0$ and $t = \dfrac{\pi}{4}$.

We have $x = \sin t \qquad \therefore\ \dfrac{dx}{dt} = \cos t$

$y = \cos t \qquad \therefore\ \dfrac{dy}{dt} = -\sin t$

$$\therefore\ \frac{ds}{dt} = \dots\dots\dots$$

54

$$\boxed{\frac{ds}{dt} = 1}$$

for $\displaystyle \frac{ds}{dt} = \sqrt{\left(\frac{dx}{dt}\right)^2 + \left(\frac{dy}{dt}\right)^2} = \sqrt{\cos^2 t + \sin^2 t} = 1$

$$\therefore I = \int_{t_1}^{t_2} f(x, y)\sqrt{\left(\frac{dx}{dt}\right)^2 + \left(\frac{dy}{dt}\right)^2}\, dt = \int_0^{\pi/4} 4 \sin t \cos t \, dt$$

$$= 2\int_0^{\pi/4} \sin 2t \, dt = -2\left[\frac{\cos 2t}{2}\right]_0^{\pi/4} = 1 \qquad \underline{\therefore I = 1}$$

Dependence of the line integral on the path of integration

We saw earlier in the programme that integration along two separate paths joining the same two end points does not necessarily give identical results.

With this in mind, let us investigate the following problem.

Example Evaluate $I = \oint_c \{3x^2y^2\, dx + 2x^3y\, dy\}$ between O(0, 0) and A(2, 4)

(a) along c_1 i.e. $y = x^2$

(b) along c_2 i.e. $y = 2x$

(c) along c_3 i.e. $x = 0$ from (0, 0) to (0, 4) and $y = 4$ from (0, 4) to (2, 4).

Let us concentrate on section (a).

First we draw the figure and insert relevant information.

This gives ....

55

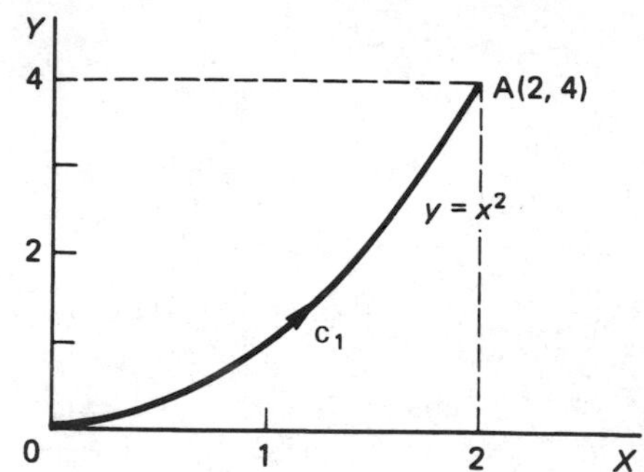

(a) $I = \int_c \{3x^2y^2\,dx + 2x^3y\,dy\}$

The path c_1 is $y = x^2$ $\quad \therefore dy = 2x\,dx$

$$\therefore I_1 = \int_0^2 \{3x^2x^4\,dx + 2x^3x^2 2x\,dx\} = \int_0^2 (3x^6 + 4x^4)\,dx$$

$$= \left[x^7\right]_0^2 = 128 \qquad \therefore \underline{I_1 = 128}$$

(b) In (b), the path of integration changes to c_2, i.e. $y = 2x$

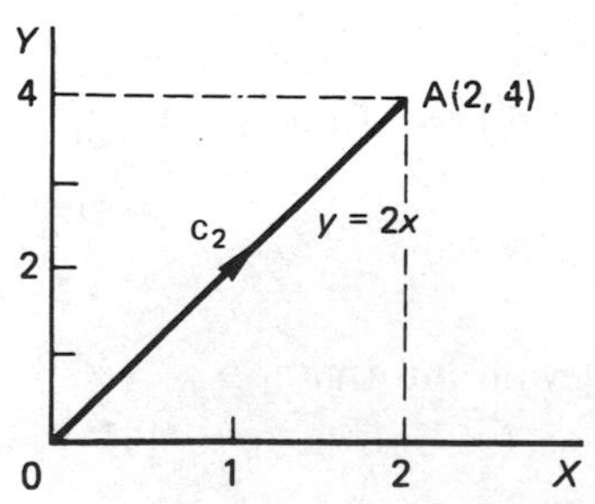

So, in this case,

$I_2 = \dots\dots\dots$

56

$$I_2 = 128$$

for with c_2, $\quad y = 2x \quad \therefore\ dy = 2\,dx$

$$\therefore I_2 = \int_0^2 \{3x^2\,4x^2\,dx + 2x^3\,2x2\,dx\} = \int_0^2 20x^4\,dx$$

$$= 4\left[x^5\right]_0^2 = 128 \qquad \therefore I_2 = 128$$

(c) In the third case, the path c_3 is split

$x = 0$ from (0, 0) to (0, 4)
$y = 4$ from (0, 4) to (2, 4)

Sketch the diagram and determine I_3.

$I_3 = \ldots\ldots\ldots$

57

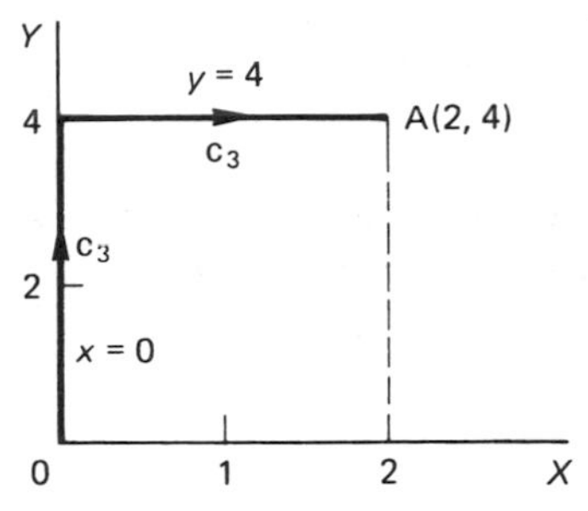

$$I_3 = 128$$

from (0, 0) to (0, 4) $\quad x = 0 \quad \therefore\ dx = 0 \quad \therefore\ I_{3a} = 0$

from (0, 4) to (2, 4) $\quad y = 4 \quad \therefore\ dy = 0 \quad \therefore\ I_{3b} = 48\int_0^2 x^2 dx = 128$

$$\therefore I_3 = 128$$

On to the next frame.

58 In the example we have just worked through, we took three different paths and in each case, the line integral produced the same result. It appears, therefore, that in this case, the value of the integral is independent of the path of integration taken.

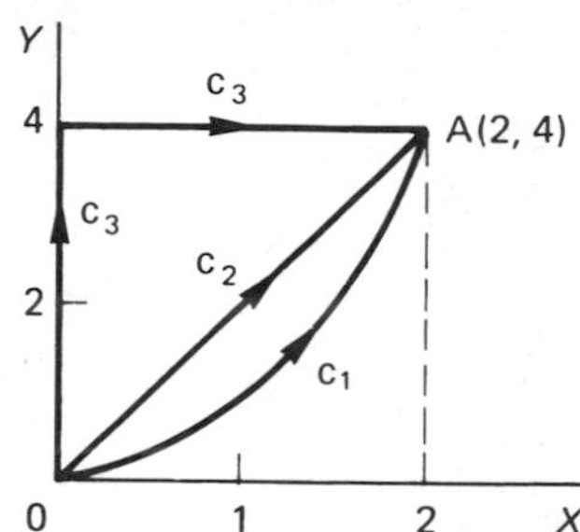

How then does this integral perhaps differ from those of previous cases?

Let us investigate.

---

59 We have been dealing with $I = \int_c \{3x^2y^2\,dx + 2x^3y\,dy\}$

On reflection, we see that the integrand $3x^2y^2\,dx + 2x^3y\,dy$ is of the form $P\,dx + Q\,dy$ which we have met before and that it is, in fact, an *exact differential* of the function $z = x^3y^2$, for

$$\frac{\partial z}{\partial x} = 3x^2y^2 \quad \text{and} \quad \frac{\partial z}{\partial y} = 2x^3y$$

This always happens. If the integrand of the given integral is seen to be an *exact differential*, then the value of the line integral is *independent of the path taken and depends only on the coordinates of the two end points.*

Make a note of this. It is important.

60

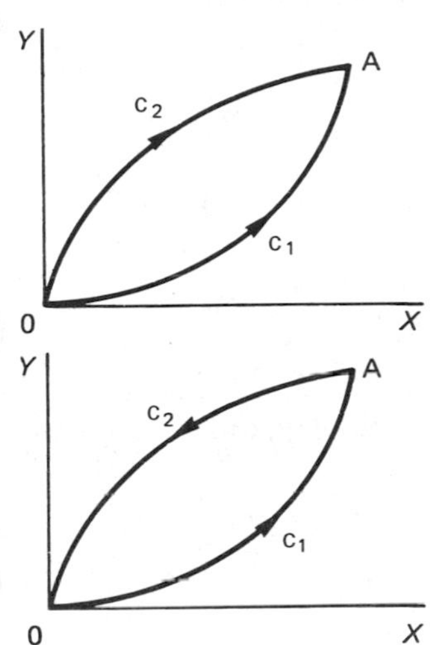

If $I = \int_c \{P\,dx + Q\,dy\}$ and $(P\,dx + Q\,dy)$ is an exact differential, then

$$I_{c_1} = I_{c_2}$$

If we reverse the direction of c_2, then

$$I_{c_1} = -I_{c_2}$$

i.e. $\quad I_{c_1} + I_{c_2} = 0$

Hence, *the integration taken round a closed curve is zero, provided* $(P\,dx + Q\,dy)$ *is an exact differential.*

$\therefore$ If $(P\,dx + Q\,dy)$ is an exact differential, $\oint (P\,dx + Q\,dy) = 0$

61

Example 1 Evaluate $I = \int_c \{3y\,dx + (3x + 2y)\,dy\}$ from A(1, 2) to B(3, 5).

No path is given, so the integrand is doubtless an exact differential of some function $z = f(x, y)$. In fact $\frac{\partial P}{\partial y} = 3 = \frac{\partial Q}{\partial x}$.

We have already dealt with the integration of exact differentials, so there is no difficulty. Compare with $I = \int_c \{P\,dx + Q\,dy\}$.

$$P = \frac{\partial z}{\partial x} = 3y \qquad \therefore z = \int 3y\,dx = 3xy + f(y) \qquad \text{(i)}$$

$$Q = \frac{\partial z}{\partial y} = 3x + 2y \qquad \therefore z = \int (3x + 2y)\,dy = 3xy + y^2 + F(x) \qquad \text{(ii)}$$

For (i) and (ii) to agree

$f(y) = \ldots\ldots\ldots$ and $F(x) = \ldots\ldots\ldots$

62

$$f(y) = y^2; \quad F(x) = 0$$

Hence $z = 3xy + y^2$

$$\therefore I = \int_c \{3y\,dx + (3x + 2y)\,dy\} = \int_{(1,2)}^{(3,5)} d(3xy + y^2) = \Big[3xy + y^2\Big]_{(1,2)}^{(3,5)}$$

$$= (45 + 25) - (6 + 4) = \underline{60}$$

Example 2 Evaluate $I = \int_c \{(x^2 + ye^x)\,dx + (e^x + y)\,dy\}$ between A(0, 1) and B(1, 2).

As before, compare with $\int_c \{P\,dx + Q\,dy\}$.

$$P = \frac{\partial z}{\partial x} = x^2 + ye^x \qquad \therefore z = \ldots\ldots\ldots$$

$$Q = \frac{\partial z}{\partial y} = e^x + y \qquad \therefore z = \ldots\ldots\ldots$$

Continue the working and complete the evaluation.

When you have finished, check the result with the next frame.

63

$$z = \frac{x^3}{3} + ye^x + f(y)$$

$$z = ye^x + \frac{y^2}{2} + F(x)$$

For these expressions to agree, $f(y) = \dfrac{y^2}{2}$; $F(x) = \dfrac{x^3}{3}$

Then $I = \left[\dfrac{x^3}{3} + ye^x + \dfrac{y^2}{2}\right]_{(0,1)}^{(1,2)} = \underline{\dfrac{5}{6} + 2e}$

So the main points are that, if $(P\,dx + Q\,dy)$ is an exact differential

(a) $I = \int_c (P\,dx + Q\,dy)$ is independent of the path of integration

(b) $I = \oint_c (P\,dx + Q\,dy)$ is zero.

On to the next frame.

64

Exact differentials in three independent variables

A line integral in space naturally involves three independent variables, but the method is very much like that for two independent variables.

$\mathrm{d}z = P\mathrm{d}x + Q\mathrm{d}y + R\,\mathrm{d}w$ is an exact differential of $z = f(x, y, w)$

if
$$\frac{\partial P}{\partial y} = \frac{\partial Q}{\partial x}; \quad \frac{\partial P}{\partial w} = \frac{\partial R}{\partial x}; \quad \frac{\partial R}{\partial y} = \frac{\partial Q}{\partial w}$$

If the test is successful, then

(a) $\int_c (P\,\mathrm{d}x + Q\,\mathrm{d}y + R\,\mathrm{d}w)$ is independent of the path of integration.

(b) $\oint_c (P\,\mathrm{d}x + Q\,\mathrm{d}y + R\,\mathrm{d}w)$ is zero.

Example Verify that

$\mathrm{d}z = (3x^2yw + 6x)\mathrm{d}x + (x^3w - 8y)\mathrm{d}y + (x^3y + 1)\mathrm{d}w$ is an exact differential and hence evaluate $\int_c \mathrm{d}z$ from A(1, 2, 4) to B(2, 1, 3).

First check that dz is an exact differential by finding the partial derivatives above, when $P = 3x^2yw + 6x$; $Q = x^3w - 8y$; and $R = x^3y + 1$.

We have ....

65

$$\frac{\partial P}{\partial y} = 3x^2w; \quad \frac{\partial Q}{\partial x} = 3x^2w \quad \therefore \frac{\partial P}{\partial y} = \frac{\partial Q}{\partial x}$$

$$\frac{\partial P}{\partial w} = 3x^2y; \quad \frac{\partial R}{\partial x} = 3x^2y \quad \therefore \frac{\partial P}{\partial w} = \frac{\partial R}{\partial x}$$

$$\frac{\partial R}{\partial y} = x^3; \quad \frac{\partial Q}{\partial w} = x^3 \quad \therefore \frac{\partial R}{\partial y} = \frac{\partial Q}{\partial w}$$

$\therefore \mathrm{d}z$ is an exact differential

Now to find z. $\quad P = \frac{\partial z}{\partial x}; \quad Q = \frac{\partial z}{\partial y}; \quad R = \frac{\partial z}{\partial w}$

$$\therefore \quad \frac{\partial z}{\partial x} = 3x^2yw + 6x \qquad \therefore z = \int (3x^2yw + 6x)\mathrm{d}x$$

$$= x^3yw + 3x^2 + f(y, w)$$

$$\frac{\partial z}{\partial y} = x^3w - 8y \qquad \therefore z = \int (x^3w - 8y)\,\mathrm{d}y$$

$$= x^3wy - 4y^2 + F(x, w)$$

$$\frac{\partial z}{\partial w} = x^3y + 1 \qquad \therefore z = \int (x^3y + 1)\,\mathrm{d}w$$

$$= x^3yw + w + g(x, y)$$

For these three expressions for z to agree

$f(y, w) = \ldots\ldots\ldots; \; F(x, w) = \ldots\ldots\ldots; \; g(x, y) = \ldots\ldots\ldots$

66

$$f(y, w) = -4y^2; \quad F(x, w) = w; \quad g(x, y) = 3x^2$$

$$\therefore z = x^3yw + 3x^2 - 4y^2 + w$$

$$\therefore I = \left[x^3yw + 3x^2 - 4y^2 + w\right]_{(1,2,4)}^{(2,1,3)}$$

$$= \ldots\ldots\ldots$$

67

$$I = 36$$

for $I = \left[x^3yw + 3x^2 - 4y^2 + w\right]_{(1,2,4)}^{(2,1,3)}$

$$= (24 + 12 - 4 + 3) - (8 + 3 - 16 + 4) = \underline{36}$$

The extension to line integrals in space is thus quite straightforward.

Finally, we have a theorem that can be very helpful on occasions and which links up with the work we have been doing.

It is important, so let us start a new section.

Green's Theorem

68 Let P and Q be two functions of x and y that are finite and continuous inside and on the boundary c of a region R in the xy-plane.

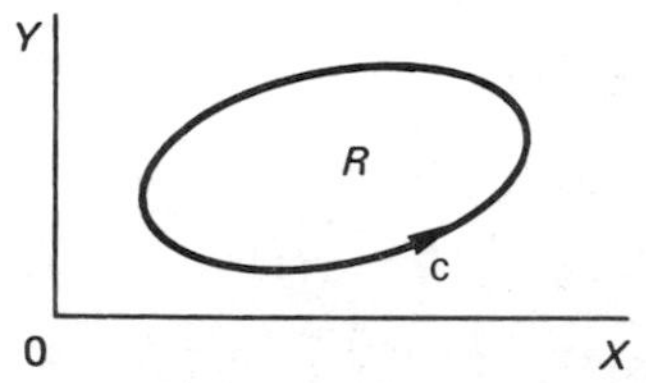

If the first partial derivatives are continuous within the region and on the boundary, then Green's theorem states that

$$\int_R\int\left(\frac{\partial P}{\partial y}-\frac{\partial Q}{\partial x}\right)\mathrm{d}x\,\mathrm{d}y=-\oint_c(P\,\mathrm{d}x+Q\,\mathrm{d}y)$$

That is, a double integral over the plane region R can be transformed into a line integral over the boundary c of the region—and the action is reversible.

Let us see how it works.

Example 1 Evaluate $I=\oint_c\{(2x-y)\,\mathrm{d}x+(2y+x)\,\mathrm{d}y\}$ around the boundary c of the ellipse $x^2+9y^2=16$.

The integral is of the form $I=\oint_c\{P\,\mathrm{d}x+Q\,\mathrm{d}y\}$ where

$$P=2x-y\quad\therefore\quad\frac{\partial P}{\partial y}=-1\ \text{ and }\ Q=2y+x\quad\therefore\quad\frac{\partial Q}{\partial x}=1.$$

$$\therefore I=-\int_R\int\left(\frac{\partial P}{\partial y}-\frac{\partial Q}{\partial x}\right)\mathrm{d}x\,\mathrm{d}y=-\int_R\int(-1-1)\,\mathrm{d}x\,\mathrm{d}y$$

$$=2\int_R\int\mathrm{d}x\,\mathrm{d}y$$

But $\int_R\int\mathrm{d}x\,\mathrm{d}y$ over any closed region gives ....

69

the area of the figure

In this case, then, $I = 2A$ where A is the area of the ellipse

$$x^2 + 9y^2 = 16 \text{ i.e. } \frac{x^2}{16} + \frac{9y^2}{16} = 1 \quad \therefore a = 4; b = \frac{4}{3} \quad \therefore A = \frac{16\pi}{3}$$

$$\therefore I = 2A = \frac{32\pi}{3}$$

To demonstrate the advantage of Green's theorem, let us work through the next example (a) by the previous method, and (b) by applying Green's theorem.

Example 2 Evaluate $I = \oint_c \{(2x + y)\,dx + (3x - 2y)\,dy\}$ taken in anticlockwise manner round the triangle with vertices at O(0, 0), A(1, 0), B(1, 2).

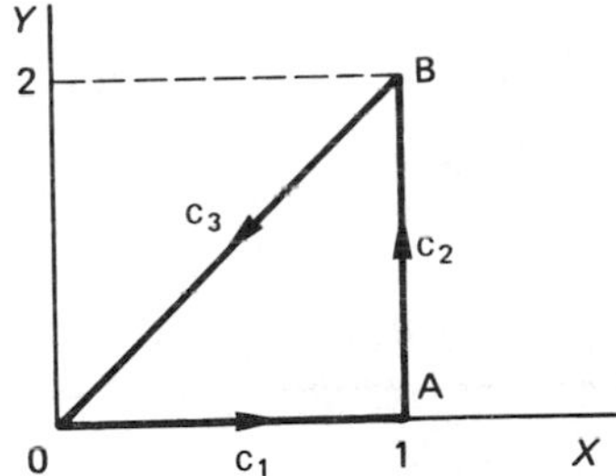

$$I = \oint_c \{(2x + y)\,dx + (3x - 2y)\,dy\}$$

(a) *By the previous method*

There are clearly three stages with c_1, c_2, c_3. Work through the complete evaluation to determine the value of I. It will be good revision.

When you have finished, check the result with the solution in the next frame.

70

$$I = 2$$

(a) (i) c_1 is $y = 0 \quad \therefore \; \mathrm{d}y = 0$

$$\therefore I_1 = \int_0^1 2x\,\mathrm{d}x = \left[x^2\right]_0^1 = 1 \qquad \therefore I_1 = 1$$

(ii) c_2 is $x = 1 \quad \therefore \; \mathrm{d}x = 0$

$$\therefore I_2 = \int_0^2 (3 - 2y)\,\mathrm{d}y = \left[3y - y^2\right]_0^2 = 2 \qquad \therefore I_2 = 2$$

(iii) c_3 is $y = 2x \quad \therefore \; \mathrm{d}y = 2\,\mathrm{d}x$

$$\therefore I_3 = \int_1^0 \{4x\,\mathrm{d}x + (3x - 4x)2\,\mathrm{d}x\}$$

$$= \int_1^0 2x\,\mathrm{d}x = \left[x^2\right]_1^0 = -1 \qquad \therefore I_3 = -1$$

$$I = I_1 + I_2 + I_3 = 1 + 2 + (-1) = 2 \qquad \therefore \underline{I = 2}$$

Now we will do the same problem by applying Green's theorem, so move on.

71

(b) *By Green's theorem*

$$I = \oint_c \{(2x + y)\,\mathrm{d}x + (3x - 2y)\,\mathrm{d}y\}$$

$$P = 2x + y \quad \therefore \; \frac{\partial P}{\partial y} = 1; \quad Q = 3x - 2y \quad \therefore \; \frac{\partial Q}{\partial x} = 3$$

$$I = -\int_R\int\left(\frac{\partial P}{\partial y} - \frac{\partial Q}{\partial x}\right)\mathrm{d}x\,\mathrm{d}y$$

Finish it off. $I = \ldots\ldots\ldots$

72

$$\boxed{I = 2}$$

$$\text{for } I = -\int_R\int (1-3)\,\mathrm{d}x\,\mathrm{d}y = 2\int_R\int \mathrm{d}x\,\mathrm{d}y = 2A$$

$$= 2 \times \text{ the area of the triangle} = 2 \times 1 = 2 \quad \therefore \underline{I = 2}$$

Application of Green's theorem is not always the quickest method. It is useful, however, to have both methods available. If you have not already done so, make a note of Green's theorem.

$$\underline{\int_R\int\left(\frac{\partial P}{\partial y} - \frac{\partial Q}{\partial x}\right)\mathrm{d}x\,\mathrm{d}y = -\oint_c (P\,\mathrm{d}x + Q\,\mathrm{d}y)}$$

73

Note Green's theorem can, in fact, be applied to a region that is not simply connected by arranging a link between outer and inner boundaries, provided the path of integration is such that the region is kept on the left-hand side.

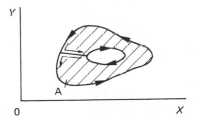

For the examples in this programme, the regions concerned are all simply connected.

Example 3 Evaluate the line integral $I = \oint_c \{xy\,\mathrm{d}x + (2x - y)\,\mathrm{d}y\}$ round the region bounded by the curves $y = x^2$ and $x = y^2$ by the use of Green's theorem.

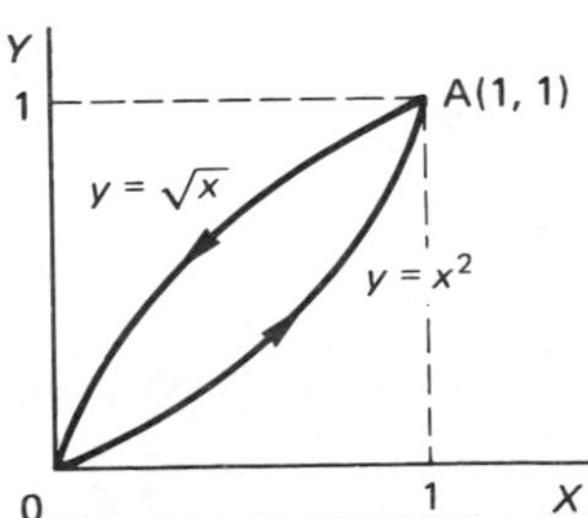

Points of intersection are O(0, 0) and A(1, 1). P and Q are known, so there is no difficulty.

Complete the working. $I = \ldots\ldots\ldots$

74

$$I = \frac{31}{60}$$

Here is the working.

$$I = \oint_c \{xy\,dx + (2x - y)\,dy\}$$

$$\oint_c \{P\,dx + Q\,dy\} = -\int_R\int\left(\frac{\partial P}{\partial y} - \frac{\partial Q}{\partial x}\right) dx\,dy$$

$$P = xy \quad \therefore \quad \frac{\partial P}{\partial y} = x; \qquad Q = 2x - y \quad \therefore \quad \frac{\partial Q}{\partial x} = 2$$

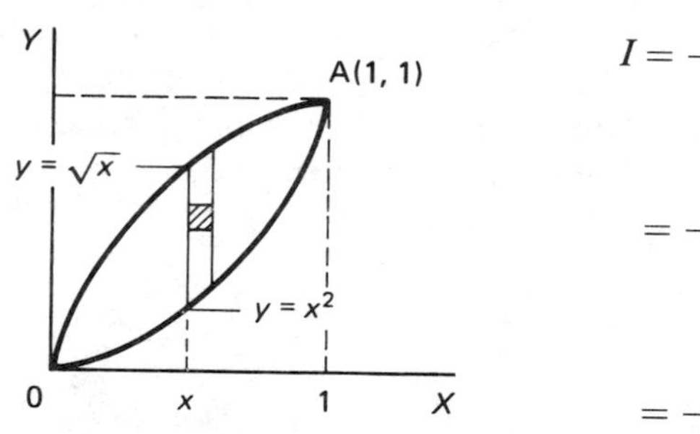

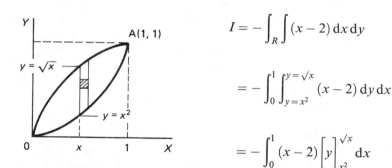

$$I = -\int_R\int (x-2)\,dx\,dy$$

$$= -\int_0^1\int_{y=x^2}^{y=\sqrt{x}} (x-2)\,dy\,dx$$

$$= -\int_0^1 (x-2)\Big[y\Big]_{x^2}^{\sqrt{x}}\,dx$$

$$\therefore I = -\int_0^1 (x-2)(\sqrt{x} - x^2)\,dx$$

$$= -\int_0^1 (x^{3/2} - x^3 - 2x^{1/2} + 2x^2)\,dx$$

$$= -\left[\frac{2}{5}x^{5/2} - \frac{1}{4}x^4 - \frac{4}{3}x^{3/2} + \frac{2}{3}x^3\right]_0^1 = \underline{\frac{31}{60}}$$

Before we finally leave this section of the work, there is one more result to note.

In the special case when $P = y$ and $Q = -x$

$$\frac{\partial P}{\partial y} = 1 \quad \text{and} \quad \frac{\partial Q}{\partial x} = -1$$

Green's theorem then states

$$\int_R\int \{1 - (-1)\}\,\mathrm{d}x\,\mathrm{d}y = -\oint_c (P\mathrm{d}x + Q\,\mathrm{d}y)$$

i.e.

$$2\int_R\int \mathrm{d}x\,\mathrm{d}y = -\oint_c (y\,\mathrm{d}x - x\,\mathrm{d}y)$$

$$= \oint_c (x\,\mathrm{d}y - y\,\mathrm{d}x)$$

Therefore, the area of the closed region

$$\underline{A = \int_R\int \mathrm{d}x\,\mathrm{d}y = \frac{1}{2}\oint_c (x\,\mathrm{d}y - y\,\mathrm{d}x)}$$

Note this result in your record book. Then let us see an example.

Example 1 Determine the area of the figure enclosed by $y = 3x^2$ **75**
and $y = 6x$.

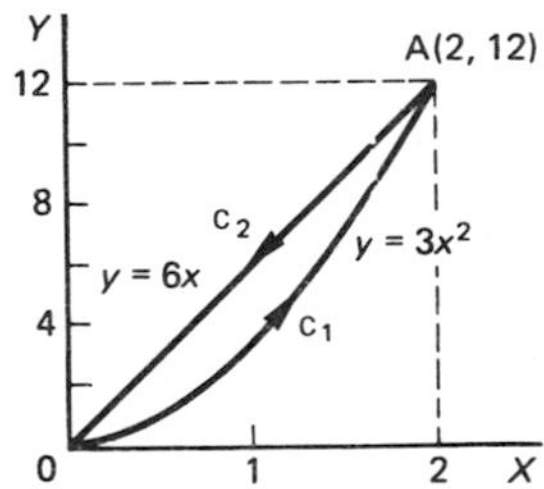

Points of intersection:

$$3x^2 = 6x \qquad \therefore\ x = 0 \text{ or } 2$$

$$\text{Area } A = \tfrac{1}{2}\oint_c (x\,\mathrm{d}y - y\,\mathrm{d}x)$$

We evaluate the integral in two parts, i.e. OA along c_1
and AO along c_2

$$2A = \int_{c_1 \text{ (along OA)}} (x\,\mathrm{d}y - y\,\mathrm{d}x) + \int_{c_2 \text{ (along AO)}} (x\,\mathrm{d}y - y\,\mathrm{d}x) = I_1 + I_2$$

I_1 : c_1 is $y = 3x^2$ $\therefore \mathrm{d}y = 6x\,\mathrm{d}x$

$$\therefore I_1 = \int_0^2 (6x^2\,\mathrm{d}x - 3x^2\,\mathrm{d}x) = \int_0^2 3x^2\,\mathrm{d}x = \Big[x^3\Big]_0^2 = 8$$

$\therefore \underline{I_1 = 8}$

Similarly, $I_2 = \ldots\ldots\ldots$

76

$$\boxed{I_2 = 0}$$

for c_2 is $y = 6x$ $\therefore \mathrm{d}y = 6\,\mathrm{d}x$

$$\therefore I_2 = \int_2^0 (6x\,\mathrm{d}x - 6x\,\mathrm{d}x) = 0 \quad \therefore \underline{I_2 = 0}$$

$$\therefore I = I_1 + I_2 = 8 + 0 = 8 \quad \therefore \underline{A = 4 \text{ square units}}$$

Finally, here is one for you to do entirely on your own.

Example 2 Determine the area bounded by the curves $y = 2x^3$, $y = x^3 + 1$ and the axis $x = 0$ for $x \geq 0$.

Complete the working and see if you agree with the working in the next frame.

77

Here it is.

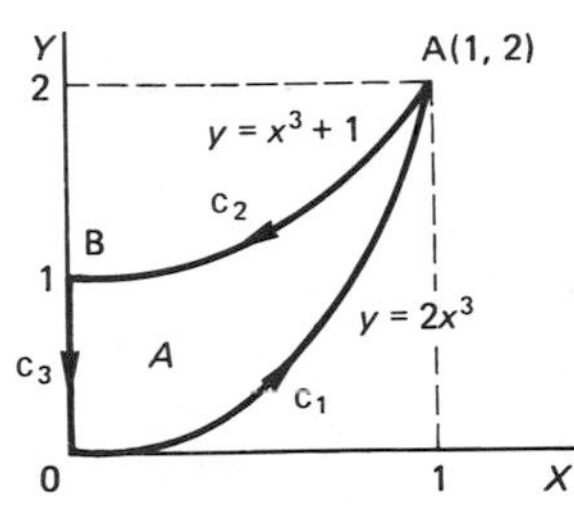

$y = 2x^3$; $y = x^3 + 1$; $x = 0$

Point of intersection

$$2x^3 = x^3 + 1 \quad \therefore x^3 = 1 \quad \therefore x = 1$$

Area $\quad A = \frac{1}{2}\oint_c (x\,dy - y\,dx)$

$$\therefore 2A = \oint_c (x\,dy - y\,dx)$$

(a) OA: $\quad c_1$ is $y = 2x^3 \quad \therefore dy = 6x^2\,dx$

$$\therefore I_1 = \int_{c_1} (x\,dy - y\,dx) = \int_0^1 (6x^3\,dx - 2x^3\,dx)$$

$$= \int_0^1 4x^3\,dx = \Big[x^4\Big]_0^1 = 1 \qquad \therefore \underline{I_1 = 1}$$

(b) AB: $\quad c_2$ is $y = x^3 + 1 \quad \therefore dy = 3x^2\,dx$

$$\therefore I_2 = \int_1^0 \{3x^3\,dx - (x^3 + 1)\,dx\} = \int_1^0 (2x^3 - 1)\,dx$$

$$= \left[\frac{x^4}{2} - x\right]_1^0 = -(\tfrac{1}{2} - 1) = \tfrac{1}{2} \qquad \therefore \underline{I_2 = \tfrac{1}{2}}$$

(c) BO: $\quad c_3$ is $x = 0 \quad \therefore dx = 0$

$$I_3 = \int_{y=1}^{y=0} (x\,dy - y\,dx) = 0 \qquad \therefore \underline{I_3 = 0}$$

$$\therefore 2A = I = I_1 + I_2 + I_3 = 1 + \tfrac{1}{2} + 0 = 1\tfrac{1}{2}$$

$$\therefore \underline{A = \tfrac{3}{4} \text{ square units}}$$

And that brings this programme to an end. We have covered some important topics, so check down the Revision Summary that follows and revise any part of the text if necessary, before working through the Test Exercise.

78 REVISION SUMMARY

1. *Differentials* dy and dx

(a)

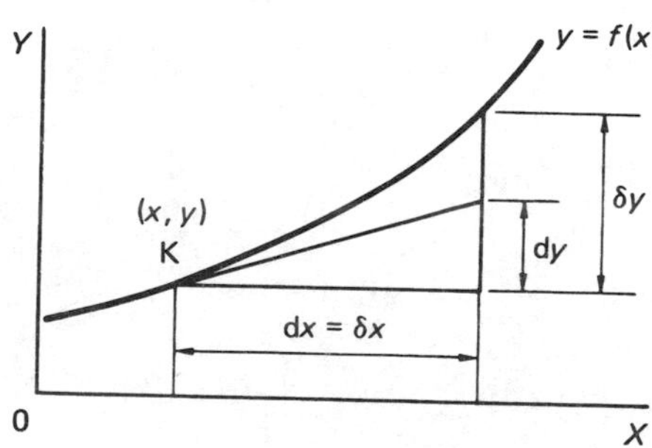

$$dy = f'(x)dx$$

(b) If $z = f(x, y)$, $\quad dz = \dfrac{\partial z}{\partial x} dx + \dfrac{\partial z}{\partial y} dy$

If $z = f(x, y, w)$, $\quad dz = \dfrac{\partial z}{\partial x} dx + \dfrac{\partial z}{\partial y} dy + \dfrac{\partial z}{\partial w} dw.$

(c) $dz = P\,dx + Q\,dy$, where P and Q are functions of x and y, is an *exact differential* if $\dfrac{\partial P}{\partial y} = \dfrac{\partial Q}{\partial x}$.

2. *Line integrals* – definition

$$I = \int_c f(x, y)\,ds = \int_c (P\,dx + Q\,dy)$$

3. *Properties of line integrals*

(a) Sign of line integral is reversed when the direction of integration along the path is reversed.

(b) Path of integration parallel to y-axis, $dx = 0$ $\quad \therefore I_c = \int_c Q\,dy$.

Path of integration parallel to x-axis, $dy = 0$ $\quad \therefore I_c = \int_c P\,dx$.

(c) Path of integration must be continuous and single-valued.

4. *Line of integral round a closed curve* $\oint$

Positive direction $\circlearrowleft$ anticlockwise

Negative direction $\circlearrowright$ clockwise, i.e. $\oint_{\circlearrowright} = -\oint_{\circlearrowleft}$.

5. *Line integral related to arc length*

$$I = \int_{AB} F\,ds = \int_{AB} (P\,dx + Q\,dy)$$

$$= \int_{x_1}^{x_2} f(x, y)\sqrt{1 + \left(\frac{dy}{dx}\right)^2}\,dx.$$

With parametric equations, x and y in terms of t,

$$I = \int_c f(x, y)\,ds = \int_{t_1}^{t_2} f(x, y)\sqrt{\left(\frac{dx}{dt}\right)^2 + \left(\frac{dy}{dt}\right)^2}\,dt$$

6. *Dependence of line integral on path of integration*

In general, the value of the line integral depends on the particular path of integration.

7. *Exact differential*

If $P\,dx + Q\,dy$ is an exact differential

(a) $\frac{\partial P}{\partial y} = \frac{\partial Q}{\partial x}$

(b) $I = \int_c (P\,dx + Q\,dy)$ is independent of the path of integration

(c) $I = \oint_c (P\,dx + Q\,dy)$ is zero.

8. *Exact differentials in three variables*

If $P\,dx + Q\,dy + R\,dw$ is an exact differential

(a) $\frac{\partial P}{\partial y} = \frac{\partial Q}{\partial x}$; $\frac{\partial P}{\partial w} = \frac{\partial R}{\partial x}$; $\frac{\partial R}{\partial y} = \frac{\partial Q}{\partial w}$

(b) $\int_c (P\,dx + Q\,dy + R\,dw)$ is independent of the path of integration

(c) $\oint_c (P\,dx + Q\,dy + R\,dw)$ is zero.

9. *Green's theorem*

$$\oint_c (P\,dx + Q\,dy) = -\int_R\int \left\{\frac{\partial P}{\partial y} - \frac{\partial Q}{\partial x}\right\} dx\,dy$$

and, for a simple closed curve,

$$\oint_c (x\,dy - y\,dx) = 2\int_R\int dx\,dy = 2A$$

where A is the area of the enclosed figure.

79 TEST EXERCISE IX

1. Determine the differential dz of each of the following

(a) $z = x^4 \cos 3y$; (b) $z = e^{2y} \sin 4x$; (c) $z = x^2yw^3$.

2. Determine which of the following are exact differentials and integrate where appropriate to determine z.

(a) $dz = (3x^2y^4 + 8x)\,dx + (4x^3y^3 - 15y^2)\,dy$

(b) $dz = (2x \cos 4y - 6 \sin 3x)dx - 4(x^2 \sin 4y - 2y)\,dy$

(c) $dz = 3e^{3x}(1 - y)\,dx + (e^{3x} + 3y^2)\,dy$.

3. Calculate the area of the triangle with vertices at O(0, 0), A(4, 2) and B(1, 5).

4. Evaluate the following

(a) $I = \int_c\{(x^2 - 3y)\,dx + xy^2\,dy\}$ from A(1, 2) to B(2, 8) along the curve $y = 2x^2$.

(b) $I = \int_c(2x + y)\,dx$ from A(0, 1) to B(0, −1) along the semi-circle $x^2 + y^2 = 1$ for $x \geq 0$.

(c) $I = \oint_c\{(1 + xy)\,dx + (1 + x^2)\,dy\}$ where c is the boundary of the rectangle joining A(1, 0), B(4, 0), C(4, 3), D(1, 3).

(d) $I = \int_c 2xy\,ds$ where c is defined by the parametric equations $x = 4\cos\theta$, $y = 4\sin\theta$ between $\theta = 0$ and $\theta = \dfrac{\pi}{3}$.

(e) $I = \int_c\{(8xy + y^3)\,dx + (4x^2 + 3xy^2)\,dy\}$ from A(1, 3) to B(2, 1).

(f) $I = \oint_c\{(3x + y)\,dx + (y - 2x)\,dy\}$ round the boundary of the ellipse $x^2 + 4y^2 = 36$.

5. Apply Green's theorem to determine the area of the plane figure bounded by the curves $y = x^3$ and $y = \sqrt{x}$.

FURTHER PROBLEMS IX

1. Show that $I = \int_c \{xy^2w^2\,dx + x^2yw^2\,dy + x^2y^2w\,dx\}$ is independent of the path of integration c and evaluate the integral from A(1, 3, 2) to B(2, 4, 1).

2. Determine whether $dz = 3x^2(x^2 + y^2)\,dx + 2y(x^3 + y^4)\,dy$ is an exact differential. If so, determine z and hence evaluate $\int_c dz$ from A(1, 2) to B(2, 1).

3. Evaluate the line integral $I = \oint_c \left\{ \dfrac{x\,dy - y\,dx}{x^2 + y^2 + 4} \right\}$ where c is the boundary of the segment formed by the arc of the circle $x^2 + y^2 = 4$ and the chord $y = 2 - x$ for $x \geq 0$.

4. Show that

   $I = \int_c \{(3x^2 \sin y + 2\sin 2x + y^3)\,dx + (x^3 \cos y + 3xy^2)\,dy\}$

   is independent of the path of integration and evaluate it from A(0, 0) to $B\left(\frac{\pi}{2}, \pi\right)$.

5. Evaluate the integral $I = \int_c xy\,ds$ where c is defined by the parametric equations $x = \cos^3 t$, $y = \sin^3 t$ from $t = 0$ to $t = \frac{\pi}{2}$.

6. Verify that $dz = \dfrac{x\,dx}{x^2 - y^2} - \dfrac{y\,dy}{x^2 - y^2}$ for $x^2 > y^2$ is an exact differential and evaluate $z = f(x, y)$ from A(3, 1) to B(5, 3).

7. The parametric equations of a circle, centre (1, 0) and radius 1, can be expressed as $x = 2\cos^2\theta$, $y = 2\cos\theta\sin\theta$.

   Evaluate $I = \int_c \{(x + y)\,dx + x^2\,dy\}$ along the semi-circle for which $y \geq 0$ from O(0, 0) to A(2, 0).

8. Evaluate $\oint_c \{x^3y^2\,dx + x^2y\,dy\}$ where c is the boundary of the region enclosed by the curve $y = 1 - x^2$, $x = 0$ and $y = 0$ in the first quadrant.

9. Use Green's theorem to evaluate

   $$I = \oint_c \{(4x + y)\,dx + (3x - 2y)\,dy\}$$

   where c is the boundary of the trapezium with vertices A(0, 1), B(5, 1), C(3, 3), D(1, 3).

10. Evaluate $I = \int_c \{(3x^2y^2 + 2\cos 2x - 2xy)\,dx + (2x^3y + 8y - x^2)\,dy\}$

    (a) along the curve $y = x^2 - x$ from A(0, 0) to B(2, 2)

    (b) round the boundary of the quadrilateral joining the points (1, 0), (3, 1), (2, 3), (0, 3)

Programme 10

Multiple Integrals

PART 2

Surface Integrals

1

Let us start off with an example with which we are already familiar.

Example 1 A solid is enclosed by the planes $z = 0$, $y = 1$, $y = 2$, $x = 0$, $x = 3$ and the surface $z = x + y^2$. We have to determine the volume of the solid so formed.

First take some care in sketching the figure, which is

....

2

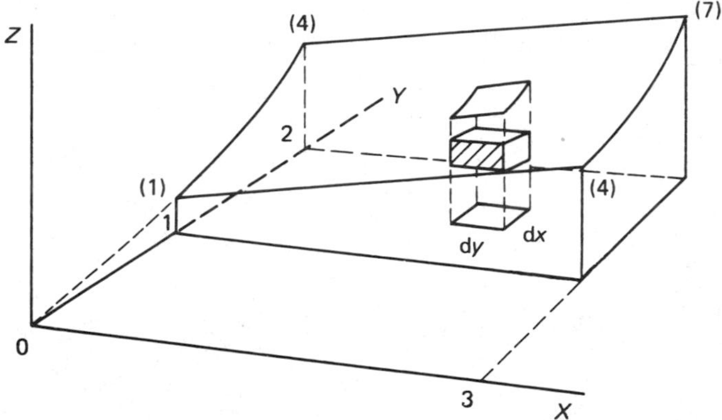

In the plane $y = 1$, $z = x + 1$, i.e. a straight line joining (0, 1, 1) and (3, 1, 4)

In the plane $y = 2$, $z = x + 4$, i.e. a straight line joining (0, 2, 4) and (3, 2, 7)

In the plane $x = 0$, $z = y^2$, i.e. a parabola joining (0, 1, 1) and (0, 2, 4)

In the plane $x = 3$, $z = 3 + y^2$, i.e. a parabola joining (3, 1, 4) and (3, 2, 7).

Consideration like this helps us to visualise the problem and the time involved is well spent.

Now we can proceed.

The element of volume $\delta v = \delta x \, \delta y \, \delta z$

Then the total volume $V = \int\int\int \mathrm{d}x \, \mathrm{d}y \, \mathrm{d}z$ between appropriate limits

in each case.

We could have said that the element of area on the $z = 0$ plane

$$\delta a = \delta y \, \delta x$$

and that the volume of the column $\delta v_{\mathrm{c}} = z \, \delta a = z \, \delta x \, \delta y$
Then, since $z = x + y^2$, this becomes $\delta v_{\mathrm{c}} = (x + y^2) \, \delta x \, \delta y$

Summing in the usual way then gives

$$V = \int z \, \mathrm{d}a = \int_R \int (x + y^2) \, \mathrm{d}x \, \mathrm{d}y$$

where R is the region bounded in the xy-plane.

Now we insert the appropriate limits and complete the integration

$$V = \ldots\ldots\ldots$$

3

$$V = 11.5 \text{ cubic units}$$

for $V = \int_{y=1}^{y=2} \int_{x=0}^{x=3} (x + y^2)\, dx\, dy = \int_1^2 \left[\frac{x^2}{2} + xy^2\right]_{x=0}^{x=3} dy$

$$= \int_1^2 \left(\frac{9}{2} + 3y^2\right) dy = \left[\frac{9}{2}y + y^3\right]_1^2 = 11.5$$

$$\therefore \underline{V = 11.5 \text{ cubic units}}$$

Although we have found a volume, this is, in fact, an example of a *surface integral* since the expression for z was a function of position in the xy-plane within the closed region

i.e. $$I = \int_R \int f(x, y)\, da = \int_R \int f(x, y)\, dy\, dx$$

In this particular case, R is the region in the xy-plane bounded by $x = 0$, $x = 3$, $y = 1$, $y = 2$.

Example 2 A triangular thin plate has the dimensions shown and a variable density ρ where $\rho = 1 + x + xy$.

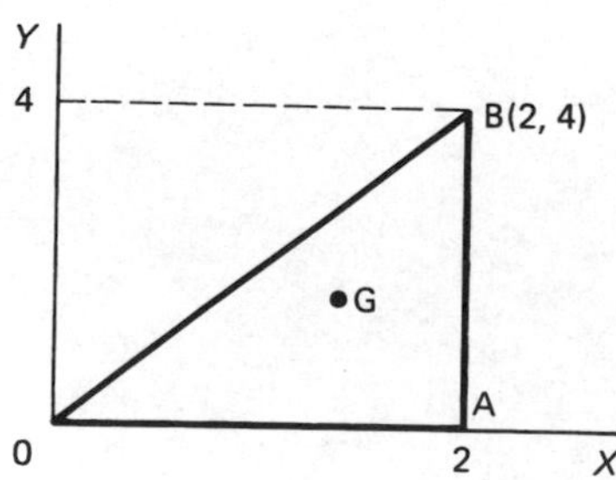

We have to determine

(a) the mass of the plate

(b) the position of its centre of gravity G.

(a) Consider an element of area at the point P (x, y) in the plate

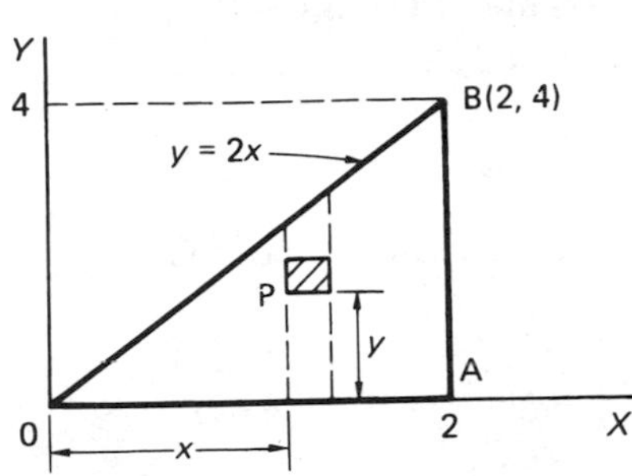

$$\delta a = \delta x\, \delta y$$

The mass δm of the element is then

$$\delta m = \rho\, \delta x\, \delta y$$

$$\therefore \text{ Total mass } M = \int_R \int \mathrm{d}m = \int_R \int \rho \, \mathrm{d}x \, \mathrm{d}y$$

Now we insert the limits and complete the integration, remembering that $\rho = (1 + x + xy)$

$$M = \dots\dots\dots$$

4

$$\boxed{M = 17\frac{1}{3}}$$

for we have

$$M = \int_R \int \rho \, \mathrm{d}x \, \mathrm{d}y = \int_{x=0}^{x=2} \int_{y=0}^{y=2x} (1 + x + xy) \, \mathrm{d}y \, \mathrm{d}x$$

$$= \int_0^2 \left[y + xy + \frac{xy^2}{2} \right]_{y=0}^{y=2x} \mathrm{d}x$$

$$= \int_0^2 \{2x + 2x^2 + 2x^3\} \mathrm{d}x$$

$$= \left[x^2 + \frac{2x^3}{3} + \frac{x^4}{2} \right]_0^2 = \underline{17\frac{1}{3}}$$

(b) To find the position of the centre of gravity, we need to know

$$\dots\dots\dots$$

5

the sum of the moments of mass about OY and OX.

(i) To find $\overline{x}$, we take moments about OY.

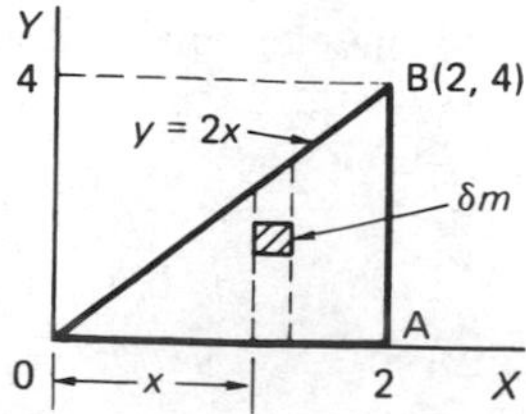

Moment of mass of element about OY

$$= x\,\delta m$$
$$= x(1 + x + xy)\,\delta x\,\delta y$$

$$\therefore \text{ Sum of first moments } = \int_R\int (x + x^2 + x^2y)\,\mathrm{d}x\,\mathrm{d}y$$
$$= \ldots\ldots\ldots$$

6

$$26\frac{2}{15}$$

$$\text{for sum of first moments } = \int_{x=0}^{x=2}\int_{y=0}^{y=2x} (x + x^2 + x^2y)\,\mathrm{d}y\,\mathrm{d}x$$
$$= \int_0^2 \left[xy + x^2y + \frac{x^2y^2}{2}\right]_{y=0}^{y=2x} \mathrm{d}x$$
$$= \int_0^2 \{2x^2 + 2x^3 + 2x^4\}\mathrm{d}x = 2\int_0^2 (x^2 + x^3 + x^4)\,\mathrm{d}x$$
$$= 2\left[\frac{x^3}{3} + \frac{x^4}{4} + \frac{x^5}{5}\right]_0^2 = \underline{26\frac{2}{15}}$$

Now $M\overline{x} =$ sum of moments $\quad\therefore\ \overline{x} = \ldots\ldots\ldots$

7

$$\overline{x} = 1.508$$

We found previously that $M = 17\frac{1}{3}\quad\therefore\ \left(17\frac{1}{3}\right)\overline{x} = 26\frac{2}{15}$
which gives $\overline{x} = 1\frac{33}{65} = \underline{1.508}$

(ii) To find $\overline{y}$ we proceed in just the same way, this time taking moments about OX. Work right through it on your own.

$$\overline{y} = \ldots\ldots\ldots$$

8

$$\boxed{\bar{y} = 1.754}$$

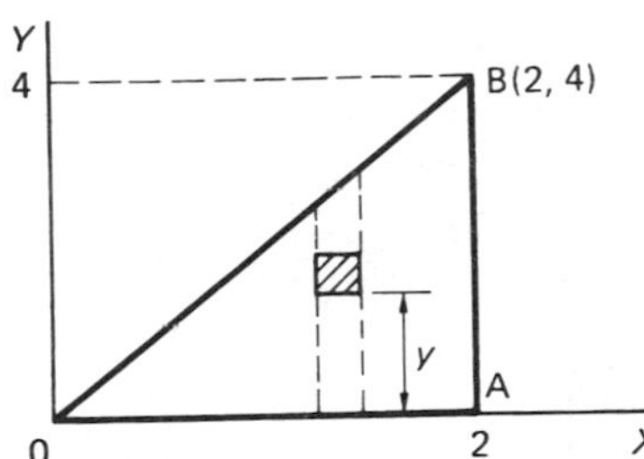

Moment of element of mass δm about OX
$= y\,\delta m = y(1 + x + xy)\,\delta x\,\delta y$

$\therefore$ Sum of first moments about OX $= \int_R\int (y + xy + xy^2)\,dx\,dy$

$$= \int_{x=0}^{x=2}\int_{y=0}^{y=2x} (y + xy + xy^2)\,dy\,dx = \int_0^2 \left[\frac{y^2}{2} + \frac{xy^2}{2} + \frac{xy^3}{3}\right]_{y=0}^{y=2x} dx$$

$$= \int_0^2 \left\{2x^2 + 2x^3 + \frac{8x^4}{3}\right\} dx$$

$$= \left[\frac{2x^3}{3} + \frac{x^4}{2} + \frac{8x^5}{15}\right]_0^2 = \underline{30\frac{2}{5}}$$

$$\therefore M\bar{y} = 30\frac{2}{5} \qquad \therefore \bar{y} = 30\frac{2}{5}\Big/17\frac{1}{3} = \underline{1.754}$$

So we finally have

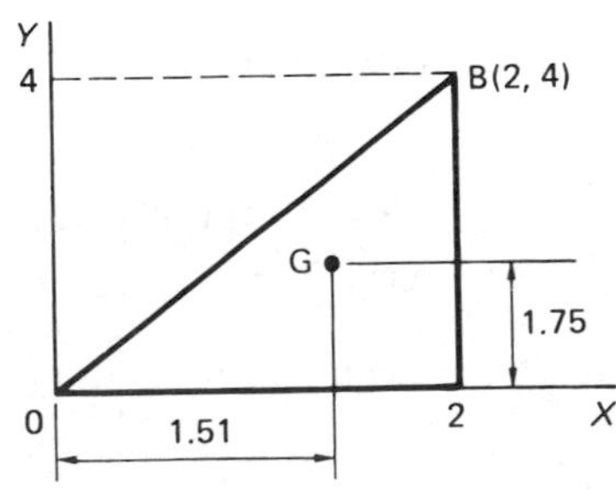

Note that this again referred to a plane figure in the xy-plane.

Now let us move on to something slightly different.

9

Surface in space

When the surface is not restricted to the xy-plane, matters become rather

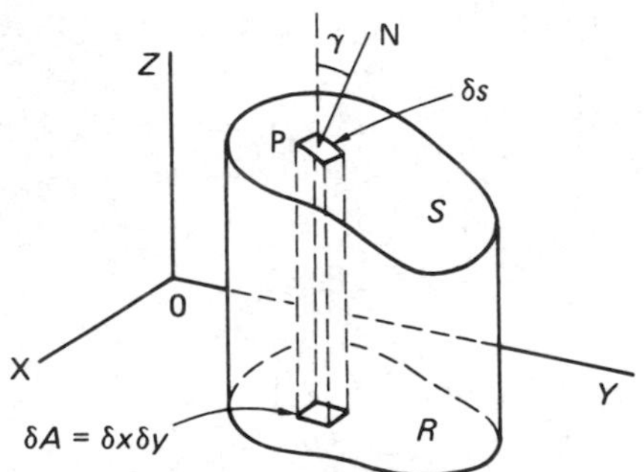

more involved, but also more interesting.

If S is a two-sided surface in space and R is its projection on the xy-plane, then the equation of S is of the form $z = f(x, y)$ where f is a single-valued function and continuous throughout R.

Let δA denote an element of R and δS the corresponding element of area of S at the point $P(x, y, z)$ in S.

Let also $\phi(x, y, z)$ be a function of position on S (e.g. potential) and let γ denote the angle between the outward normal PN to the surface at P and the positive z-axis.

Then $\delta A \approx \delta S \cos\gamma$ i.e. $\delta S \approx \dfrac{\delta A}{\cos\gamma} = \delta A \sec\gamma$ and

$\Sigma\phi(x, y, z)\delta S$ is the total value of $\phi(x, y, z)$ taken over the surface S.

As $\delta S \to 0$, this sum becomes the integral

$$\underline{I = \int_S \phi(x, y, z)\,\mathrm{d}S}$$

and, since $\delta S \approx \delta A \sec\gamma$, the result can be written

$$\underline{I = \int_R\int \phi(x, y, z)\sec\gamma\,\mathrm{d}x\,\mathrm{d}y} \qquad \left(\gamma < \frac{\pi}{2}\right)$$

With limits inserted for x and y, the integral seems straightforward, except for the factor $\sec\gamma$, which naturally varies over the surface S.

We can, in fact, show that $\sec\gamma = \sqrt{1+\left(\dfrac{\partial z}{\partial x}\right)^2+\left(\dfrac{\partial z}{\partial y}\right)^2}$

Therefore, the *surface integral* of $\phi(x, y, z)$ over the surface S is given by

$$\text{(a)}\quad I=\int_S \phi(x,y,z)\,\mathrm{d}S \qquad (1)$$

or

$$\text{(b)}\quad I=\int_R\int \phi(x,y,z)\sqrt{1+\left(\frac{\partial z}{\partial x}\right)^2+\left(\frac{\partial z}{\partial y}\right)^2}\,\mathrm{d}x\,\mathrm{d}y \qquad (2)$$

where $z=f(x,\ y)$

Note that, when $\phi(x,y,z)=1$, then $I=\int_S \mathrm{d}S$ gives the area of the surface S.

$$\therefore\ S=\int_S \mathrm{d}S=\int_R\int\sqrt{1+\left(\frac{\partial z}{\partial x}\right)^2+\left(\frac{\partial z}{\partial y}\right)^2}\,\mathrm{d}x\,\mathrm{d}y \qquad (3)$$

Make a note of these three important results.
Then we will apply them to a few examples.

10

Example 1 Find the area of the surface $z=\sqrt{x^2+y^2}$ over the region bounded by $x^2+y^2=1$.

$$S=\int_R\int\sqrt{1+\left(\frac{\partial z}{\partial x}\right)^2+\left(\frac{\partial z}{\partial y}\right)^2}\,\mathrm{d}x\,\mathrm{d}y$$

So we now find $\dfrac{\partial z}{\partial x}$ and $\dfrac{\partial z}{\partial y}$ and determine $\sqrt{1+\left(\dfrac{\partial z}{\partial x}\right)^2+\left(\dfrac{\partial z}{\partial y}\right)^2}$

which is ....

11

$$\boxed{\sqrt{2}}$$

since $z = (x^2 + y^2)^{1/2}$ $\quad \therefore \dfrac{\partial z}{\partial x} = \dfrac{1}{2}(x^2 + y^2)^{-1/2}2x = \dfrac{x}{\sqrt{x^2 + y^2}}$

$$\frac{\partial z}{\partial y} = \frac{1}{2}(x^2 + y^2)^{-1/2}2y = \frac{y}{\sqrt{x^2 + y^2}}$$

$$\therefore 1 + \left(\frac{\partial z}{\partial x}\right)^2 + \left(\frac{\partial z}{\partial y}\right)^2 = 1 + \frac{x^2 + y^2}{x^2 + y^2} = 2$$

$$\therefore \sqrt{1 + \left(\frac{\partial z}{\partial x}\right)^2 + \left(\frac{\partial z}{\partial y}\right)^2} = \sqrt{2}$$

$$\therefore S = \sqrt{2}\int_R\int \mathrm{d}x\,\mathrm{d}y = \sqrt{2} \times \ldots\ldots\ldots$$

12

the area of the region R

But R is bounded by $x^2 + y^2 = 1$, i.e. a circle, centre the origin and radius 1. $\therefore$ area $= \pi$

$$\therefore S = \sqrt{2}\int_R\int dx\,dy = \underline{\sqrt{2}\pi}$$

Example 2 Find the area of the surface S of the paraboloid $z = x^2 + y^2$ cut off by the cone $z = 2\sqrt{x^2 + y^2}$.

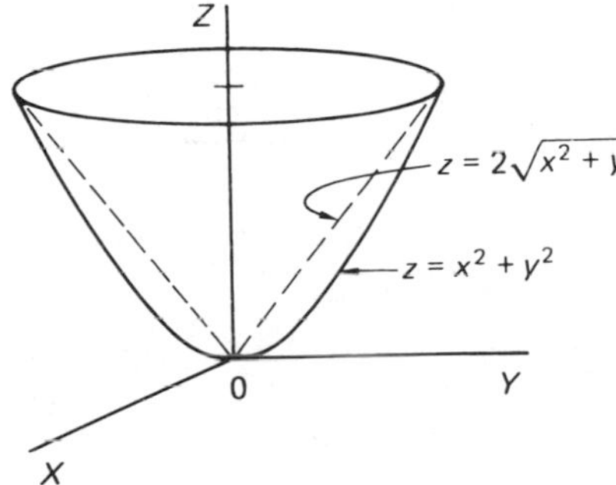

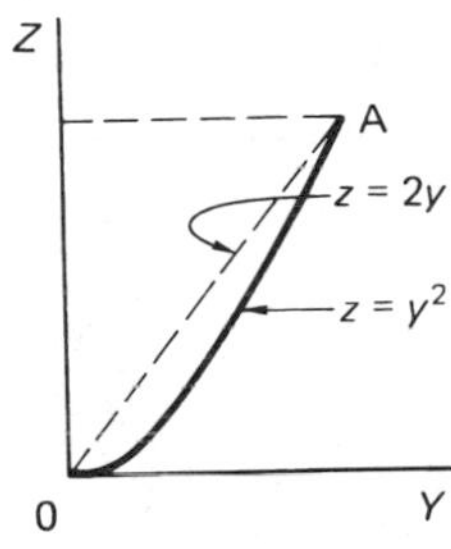

We can find the point of intersection A by considering the yz-plane, i.e. put $x = 0$.

Coordinates of A are ....

13

A(2, 4)

Therefore the projection of the surface S on the xy-plane is

....

14

the circle $x^2 + y^2 = 4$

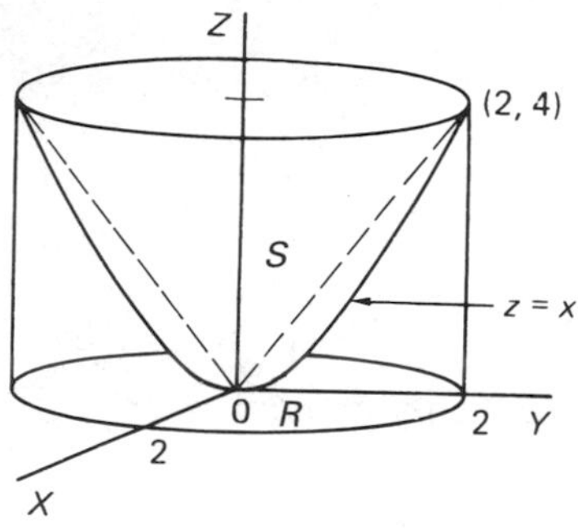

$$S = \int_R\int \sqrt{1 + \left(\frac{\partial z}{\partial x}\right)^2 + \left(\frac{\partial z}{\partial y}\right)^2}\,dx\,dy$$

For this we use the equation of the surface S. The information from the projection R on the xy-plane will later provide the limits of the two stages of integration.

For the time being, then, $S = \ldots\ldots\ldots$

15

$$S = \int_R\int \sqrt{1 + 4x^2 + 4y^2}\,dx\,dy$$

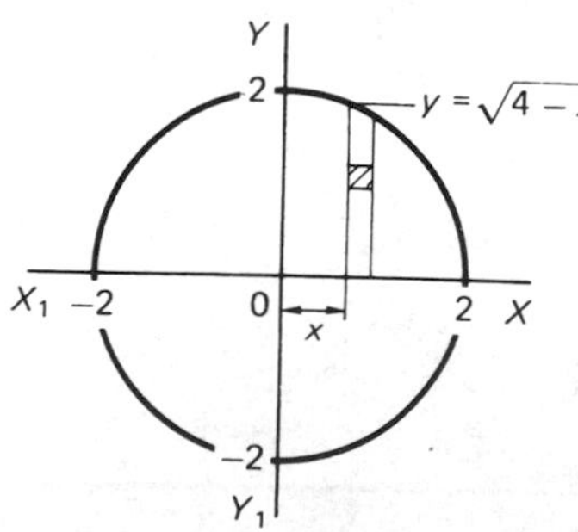

Using cartesian coordinates, we could integrate with respect to y from $y = 0$ to $y = \sqrt{4 - x^2}$ and then with respect to x from $x = 0$ to $x = 2$. Finally, we should multiply by four to cover all four quadrants.

i.e. $$S = 4\int_{x=0}^{x=2}\int_{y=0}^{y=\sqrt{4-x^2}} \sqrt{1 + 4x^2 + 4y^2}\,dy\,dx$$

But how do we carry out the actual integration?
It becomes a lot easier if we use polar coordinates.

The same integral in polar coordinates is $\ldots\ldots\ldots$

16

$$S = \int_{\theta=0}^{\theta=2\pi} \int_{r=0}^{r=2} \sqrt{1 + 4r^2}\; r\,\mathrm{d}r\,\mathrm{d}\theta$$

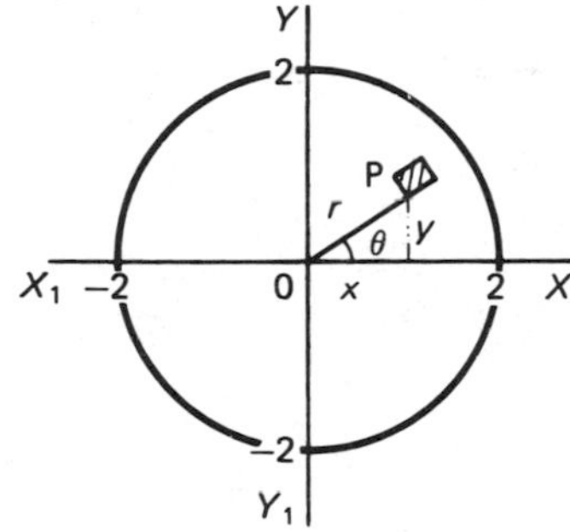

$x = r\cos\theta; \qquad y = r\sin\theta$

$x^2 + y^2 = r^2 \qquad \mathrm{d}x\,\mathrm{d}y = r\,\mathrm{d}r\,\mathrm{d}\theta$

$$S = \int_{\theta=0}^{\theta=2\pi} \int_{r=0}^{r=2} \sqrt{1 + 4r^2}\; r\,\mathrm{d}r\,\mathrm{d}\theta$$

$\therefore S = \ldots\ldots\ldots$

Finish it off.

17

$$S = 36.18 \text{ square units}$$

for $S = \displaystyle\int_{\theta=0}^{\theta=2\pi} \int_{r=0}^{r=2} (1 + 4r^2)^{1/2} r\,\mathrm{d}r\,\mathrm{d}\theta = \int_0^{2\pi} \left[\frac{1}{12}(1 + 4r^2)^{3/2}\right]_0^2 \mathrm{d}\theta$

$$= \frac{1}{12}\int_0^{2\pi} \{17^{3/2} - 1\}\,\mathrm{d}\theta = 5.7577\Big[\theta\Big]_0^{2\pi} = \underline{36.18}$$

Now on to Example 3.

18

Example 3 To determine the moment of inertia of a thin spherical shell of radius a about a diameter as axis. The mass per unit area of shell is ρ.

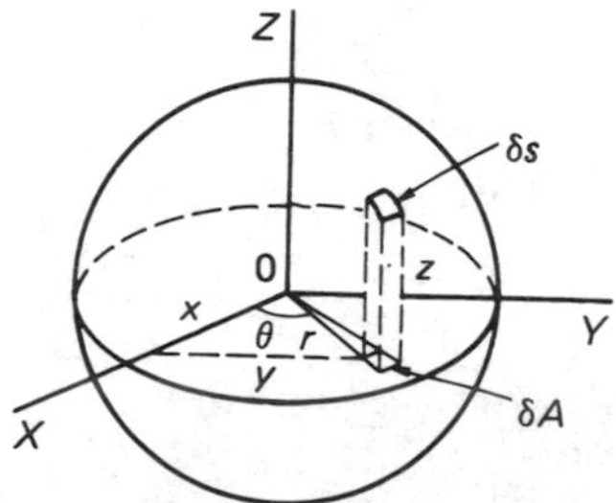

Equation of sphere

$$x^2 + y^2 + z^2 = a^2$$

Mass of element $= m = \rho\,\delta S$

$$I \approx \Sigma m r^2 \approx \Sigma \rho\,\delta S r^2$$

Let us deal with the upper hemisphere

$$\therefore I_H = \int_S \rho r^2\,dS$$

$$= \int_R \int \rho r^2 \sqrt{1 + \left(\frac{\partial z}{\partial x}\right)^2 + \left(\frac{\partial z}{\partial y}\right)^2}\,dx\,dy$$

Now determine the partial derivatives and simplify the integral as far as possible in cartesian coordinates.

$$I_H = \dots\dots\dots$$

19

$$I_H = \int_R \int \rho r^2 \frac{a}{\sqrt{a^2 - x^2 - y^2}}\,dx\,dy$$

In this particular example, R is, of course, the region bounded by the circle $x^2 + y^2 = a^2$ in the xy-plane.

Converting to polar coordinates

$$x = r\cos\theta; \qquad y = r\sin\theta; \qquad dx\,dy = r\,dr\,d\theta$$

the integral becomes $I_H = \dots\dots\dots$

20

$$I_{\mathrm{H}} = \rho a \int_{\theta=0}^{\theta=2\pi} \int_{r=0}^{r=a} \frac{r^3}{\sqrt{a^2 - r^2}} \, \mathrm{d}r \, \mathrm{d}\theta$$

for $x^2 + y^2 = r^2$: limits of r: $r = 0$ to $r = a$
limits of θ: $\theta = 0$ to $\theta = 2\pi$

$$I_{\mathrm{H}} = \int_R \int \rho r^2 \frac{a}{\sqrt{a^2 - r^2}} r \, \mathrm{d}r \, \mathrm{d}\theta = \rho a \int_{\theta=0}^{\theta=2\pi} \int_{r=0}^{r=a} \frac{r^3}{\sqrt{a^2 - r^2}} \, \mathrm{d}r \, \mathrm{d}\theta$$

First we have to evaluate

$$I_r = \int_0^a \frac{r^3}{\sqrt{a^2 - r^2}} \, \mathrm{d}r$$

If we substitute $r = a \sin\phi$, $\mathrm{d}r = a\cos\phi \, \mathrm{d}\phi$ and the integral becomes

$$I_r = \dots\dots\dots$$

21

$$I_r = \int_0^{\pi/2} a^3 \sin^3\phi \, \mathrm{d}\phi$$

for we have

$$I_r = \int_0^a \frac{r^3}{\sqrt{a^2 - r^2}} \, \mathrm{d}r = \int_0^{\pi/2} \frac{a^3 \sin^3\phi}{a\cos\phi} a\cos\phi \, \mathrm{d}\phi$$

$$= \int_0^{\pi/2} a^3 \sin^3\phi \, \mathrm{d}\phi$$

$r = a\sin\phi \quad \therefore$ when $r = 0, \ \phi = 0$
$r = a, \ \phi = \pi/2$

$$\therefore I_r = \int_0^{\pi/2} a^3 \sin^3\phi \, \mathrm{d}\phi$$

Then, using Wallis' integral $\displaystyle\int_0^{\pi/2} \sin^n\phi \, \mathrm{d}\phi = \frac{2 \times 4 \times (n-1)}{1 \times 3 \dots n}$ for n odd

$$\therefore I_r = a^3 \times \frac{2}{3} = \frac{2a^3}{3}$$

Now, to complete I_{H}, we have $I_{\mathrm{H}} = \rho a \displaystyle\int_0^{2\pi} \frac{2a^3}{3} \, \mathrm{d}\theta$

$$= \dots\dots\dots$$

22

$$I_{\text{H}} = \frac{4\pi\rho a^4}{3}$$

since $I_{\text{H}} = \rho a \int_0^{2\pi} \frac{2a^3}{3} \, \mathrm{d}\theta = \frac{2a^4\rho}{3} \left[\theta \right]_0^{2\pi} = \frac{4\pi a^4 \rho}{3}$

Therefore, the moment of inertia for the complete spherical shell is

$$I_s = \frac{8\pi a^4 \rho}{3}$$

The total mass of the shell $M = 4\pi a^2 \rho \quad \therefore\ I = \frac{2Ma^2}{3}$

Now let us turn our attention towards *Volume integrals* and in preparation review systems of space coordinates.

23

Space coordinate systems

1. *Cartesian coordinates* (x, y, z)—referred to three coordinate axes OX, OY, OZ at right angles to each other. These are arranged in a *right-handed* manner, i.e. turning from OX to OY gives a right-handed screw action in the positive direction of OZ.

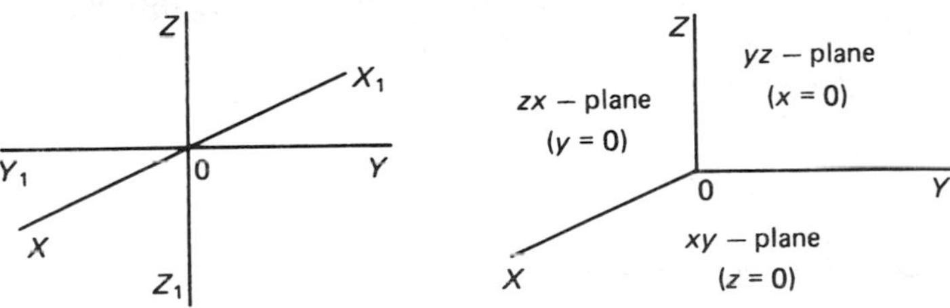

The three coordinate planes, $x = 0$, $y = 0$, $z = 0$, divide the space into eight sections called *octants*. The section containing $x \geq 0$, $y \geq 0$, $z \geq 0$ is called the *first octant*.

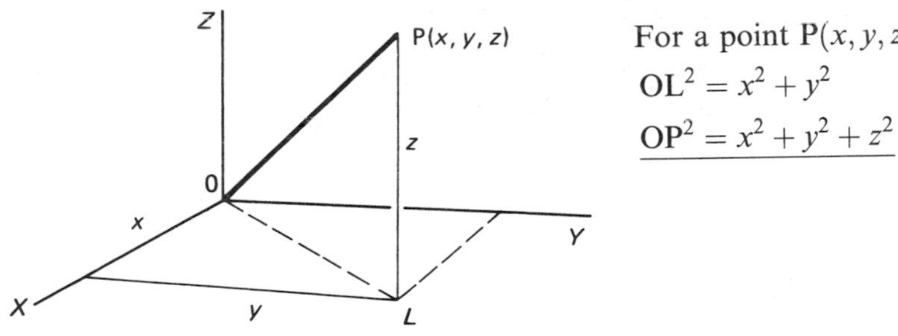

For a point $P(x, y, z)$

$$OL^2 = x^2 + y^2$$

$$OP^2 = x^2 + y^2 + z^2$$

We are all familiar with this system of coordinates.

24

2. *Cylindrical coordinates* (r, θ, z) are useful where an axis of symmetry occurs.

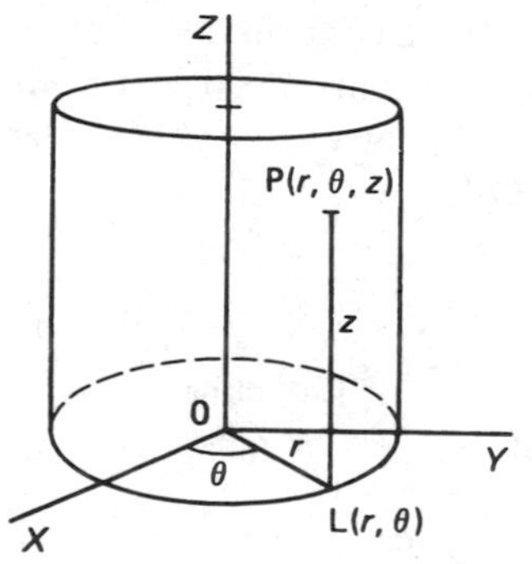

Any point P is considered as having a position on a cylinder. If L is the projection of P on the xy-plane, then (r, θ) are the usual polar coordinates of L. The cylindrical coordinates of P then merely require the addition of the z-coordinate.

$$r \geq 0$$

Relationship between cartesian and cylindrical coordinates

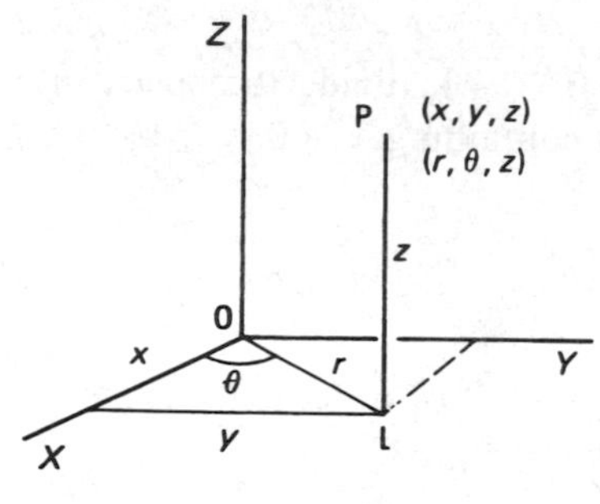

If we consider a combined figure, we can easily relate the two systems.

Expressing each of the following in terms of the alternative system,

$x = \ldots\ldots\ldots$ $r = \ldots\ldots\ldots$

$y = \ldots\ldots\ldots$ $\theta = \ldots\ldots\ldots$

$z = \ldots\ldots\ldots$ $z = \ldots\ldots\ldots$

25

$$x = r\cos\theta; \quad r = \sqrt{x^2 + y^2}$$

$$y = r\sin\theta; \quad \theta = \arctan(y/x)$$

$$z = z; \quad z = z$$

So, in cylindrical coordinates, the surface defined by

(i) $r = 5$ is ....

(ii) $\theta = \pi/6$ is ....

(iii) $z = 4$ is ....

26

(i) $r = 5$ is a right cylinder, radius 5, with OZ as axis.

(ii) $\theta = \pi/6$ is a plane through OZ, making an angle $\pi/6$ with OX.

(iii) $z = 4$ is a plane parallel to the xy-plane, cutting OZ at 4 units above the origin.

So position P(2, 3, 4) in cartesian coordinates

= .... in cylindrical coordinates

and position Q(2.5, $\pi/3$, 6) in cylindrical coordinates

= .... in cartesian coordinates.

27

$P(2, 3, 4) = (\sqrt{13}, 0.983, 4)$ in cylindrical coordinates
$Q(2.5, \pi/3, 6) = (1.25, 2.165, 6)$ in cartesian coordinates.

3. *Spherical coordinates* (r, θ, ϕ) are appropriate where a centre of symmetry occurs. The position of a point is considered as being a point on a sphere.

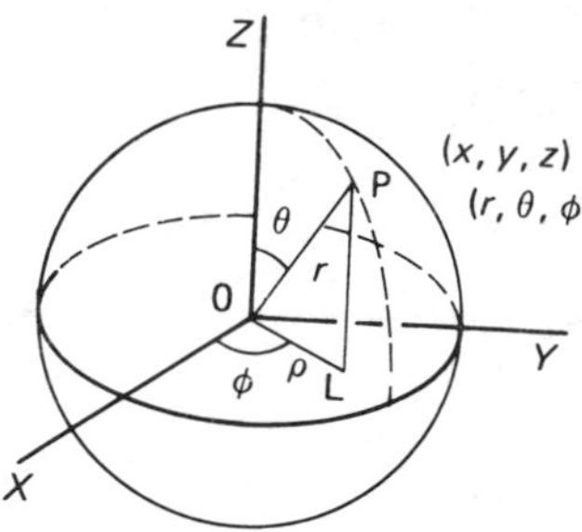

r is the distance of P from the origin and is always taken as positive.

L is the projection of P on the xy-plane;
θ is the angle between OP and the positive OZ axis;
ϕ is the angle between OL and the OX axis.

Note that (i) ϕ may be regarded as the longitude of P from OX
(ii) θ may be regarded as the complement of the latitude of P.

Relationship between cartesian and spherical coordinates

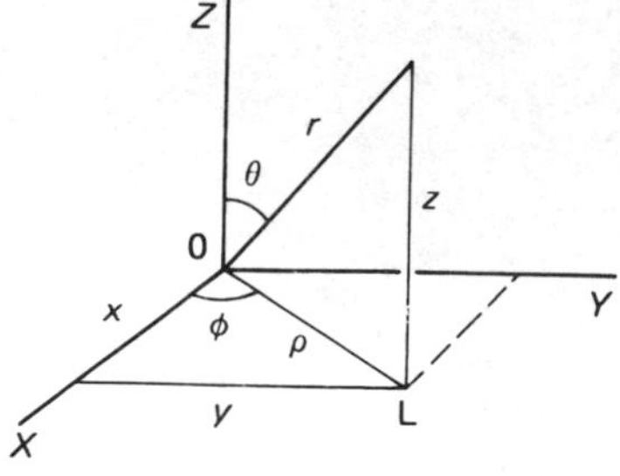

The combined figure shows the connection between the two systems, so

$x = \ldots\ldots\ldots$ $r = \ldots\ldots\ldots$

$y = \ldots\ldots\ldots$ $\theta = \ldots\ldots\ldots$

$z = \ldots\ldots\ldots$ $\phi = \ldots\ldots\ldots$

28

$$\begin{array}{ll} x = r\sin\theta\cos\phi & r = \sqrt{x^2+y^2+z^2} \\ y = r\sin\theta\sin\phi & \theta = \arccos(z/r) \\ z = r\cos\theta & \phi = \arctan(y/x) \end{array}$$

For the spherical coordinates of any point in space

$$r \geq 0; \quad 0 \leq \theta \leq \pi; \quad 0 \leq \phi \leq 2\pi$$

So, converting cartesian coordinates (2, 3, 4) to spherical coordinates gives ....

29

$$P(r, \theta, \phi) = (5.385,\ 0.734,\ 0.983)$$

for $x = 2,\ y = 3,\ z = 4$

$$\therefore\ r = \sqrt{x^2+y^2+z^2} = \sqrt{4+9+16} = \sqrt{29} = 5.385$$

$$\theta = \arccos(z/r) = \arccos(4/\sqrt{29}) = 0.734$$

$$\phi = \arctan(y/x) = \arctan 1.5 = 0.983$$

And, in reverse, spherical coordinates $(5, \pi/4, \pi/3)$ transform into cartesian coordinates ....

30

$$P(x, y, z) = (1.768,\ 3.061,\ 3.536)$$

for $x = r\sin\theta\cos\phi = 5\sin\frac{\pi}{4}\cos\frac{\pi}{3} = 5(0.707)(0.5) = 1.768$

$$y = r\sin\theta\sin\phi = 5\sin\frac{\pi}{4}\sin\frac{\pi}{3} = 5(0.707)(0.866) = 3.061$$

$$z = r\cos\theta = 5\cos\frac{\pi}{4} = 5(0.707) = 3.536.$$

One of the main uses of cylindrical and spherical coordinates occurs in integrals dealing with volumes of solids. In preparation for this, let us consider the next important section of the work.

So move on.

31 *Element of volume in space* in the three coordinate systems

1. *Cartesian coordinates*

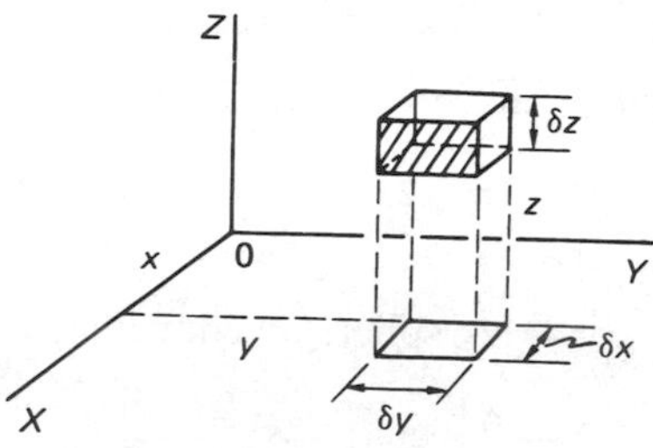

We have already used this many times.

$$\underline{\delta v = \delta x\, \delta y\, \delta z}$$

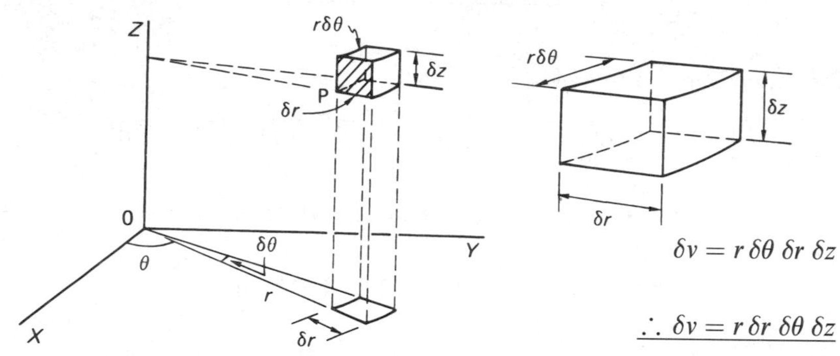

$$\delta v = r\,\delta\theta\,\delta r\,\delta z$$

$$\underline{\therefore\ \delta v = r\,\delta r\,\delta\theta\,\delta z}$$

2. *Cylindrical coordinates*

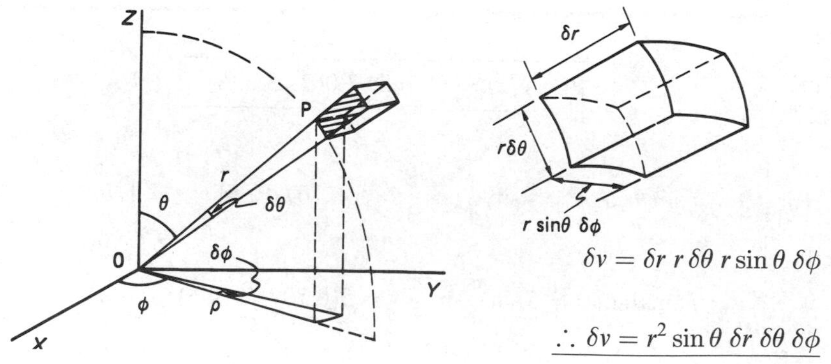

$$\delta v = \delta r\; r\,\delta\theta\; r\sin\theta\,\delta\phi$$

$$\underline{\therefore\ \delta v = r^2\sin\theta\,\delta r\,\delta\theta\,\delta\phi}$$

3. *Spherical coordinates*

It is important to make a note of these results, since they are required when we change the variables in various types of integrals. We shall meet them again before long, so be sure of them now.

Volume Integrals

32

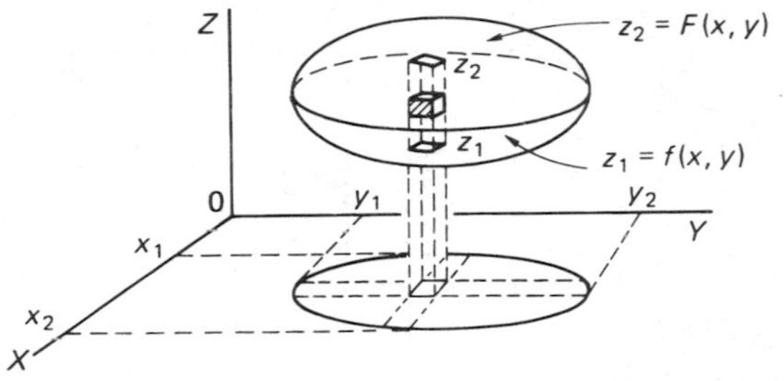

A solid is enclosed by a lower surface $z_1 = f(x, y)$ and an upper surface $z_2 = F(x, y)$.

Then, in general, using cartesian coordinates, the element of volume is $\delta v = \delta x\, \delta y\, \delta z$.

The approximate value of the total volume V is then found

(a) by summing δv from $z = z_1$ to $z = z_2$ to obtain the volume of the column
(b) by summing all such columns from $y = y_1$ to $y = y_2$ to obtain the volume of the slice
(c) by summing all such slices from $x = x_1$ to $x = x_2$ to obtain the total volume V.

Then, when $\delta x \to 0$, $\delta y \to 0$, $\delta z \to 0$, the summation becomes an integral

$$V = \int_{x=x_1}^{x=x_2} \int_{y=y_1}^{y=y_2} \int_{z=z_1}^{z=z_2} dz\, dy\, dx$$

Example 1 Find the volume of the solid bounded by the planes $z = 0$, $x = 0$, $y = 0$, $x^2 + y^2 = 4$ and $z = 6 - xy$ for $x \geq 0$, $y \geq 0$, $z \geq 0$.

First sketch the figure, so that we can see what we are doing. Take time over it.

33

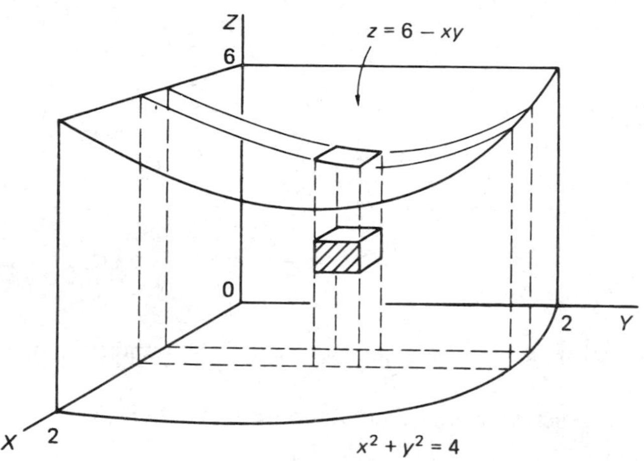

$$\delta v = \delta x\, \delta y\, \delta z$$

Volume of column $\approx \displaystyle\sum_{z=0}^{z=6-xy} \delta x\, \delta y\, \delta z$

Volume of slice $\approx \displaystyle\sum_{y=0}^{\sqrt{4-x^2}} \left\{ \sum_{z=0}^{6-xy} \delta x\, \delta y\, \delta z \right\}$

Total volume $\approx \displaystyle\sum_{x=0}^{2} \sum_{y=0}^{\sqrt{4-x^2}} \sum_{z=0}^{6-xy} \delta x\, \delta y\, \delta z$

If $\delta x \to 0$, $\delta y \to 0$, $\delta z \to 0$, then

$$V = \int_0^2 \int_0^{\sqrt{4-x^2}} \int_0^{6-xy} \mathrm{d}z\, \mathrm{d}y\, \mathrm{d}x$$

Starting with the innermost integral

$$\int_0^{6-xy} \mathrm{d}z = \Big[z \Big]_0^{6-xy} = 6 - xy$$

Then $\displaystyle\int_0^{\sqrt{4-x^2}} (6 - xy)\, \mathrm{d}y = \dots\dots\dots$

34

$$6\sqrt{4-x^2} - \frac{x}{2}(4-x^2)$$

for $\displaystyle\int_0^{\sqrt{4-x^2}} (6-xy)\,\mathrm{d}y = \left[6y - \frac{xy^2}{2}\right]_{y=0}^{y=\sqrt{4-x^2}} = 6\sqrt{4-x^2} - \frac{x}{2}(4-x^2).$

Then finally $V = \displaystyle\int_0^2 \left\{6(4-x^2)^{1/2} - 2x + \frac{x^3}{2}\right\} \mathrm{d}x.$

Now we are faced with $\displaystyle\int (4-x^2)^{1/2}\,\mathrm{d}x$. You may remember that this is a

standard form $\displaystyle\int \sqrt{a^2-x^2}\,\mathrm{d}x = \tfrac{1}{2}\left\{x\sqrt{a^2-x^2} + a^2 \arcsin\frac{x}{a}\right\}.$

If not, to evaluate $\displaystyle\int_0^2 \sqrt{4-x^2}\,\mathrm{d}x$, put $x = 2\sin\theta$ and proceed from there.

Finish off the main integral, so that we have

$$V = \ldots\ldots\ldots$$

35

$$\boxed{V = 6\pi - 2 \approx 16.8 \text{ cubic units}}$$

for we had $V = \int_0^2 \left\{ 6(4 - x^2)^{1/2} - 2x + \frac{x^3}{2} \right\} \mathrm{d}x$

$$= 3\left[x\sqrt{4 - x^2} + 4 \arcsin \frac{x}{2} \right]_0^2 - \left[x^2 - \frac{x^4}{8} \right]_0^2$$

$$= 3\{4 \arcsin 1 - 4 \arcsin 0\} - 4 + 2$$

$$= 3\{2\pi\} - 2 = 6\pi - 2 \approx \underline{16.8}$$

Alternative method

We could, of course, have used cylindrical coordinates in this problem.

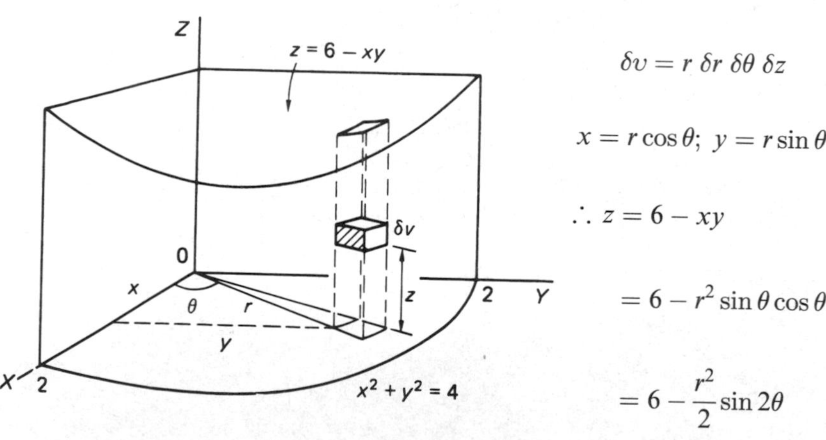

$\delta v = r\,\delta r\,\delta\theta\,\delta z$

$x = r\cos\theta;\ y = r\sin\theta$

$\therefore z = 6 - xy$

$= 6 - r^2 \sin\theta\cos\theta$

$= 6 - \frac{r^2}{2}\sin 2\theta$

$$\therefore V = \int_{r=0}^{2} \int_{\theta=0}^{\pi/2} \int_{z=0}^{6-(r^2/2)\sin 2\theta} r\,\mathrm{d}r\,\mathrm{d}\theta\,\mathrm{d}z$$

$$= \int_{\theta=0}^{\pi/2} \int_{r=0}^{2} \int_{z=0}^{6-(r^2/2)\sin 2\theta} \mathrm{d}z\,r\,\mathrm{d}r\,\mathrm{d}\theta = \dots\dots\dots$$

Finish it.

36

$$V = 6\pi - 2 \text{ (as before)}$$

$$V = \int_{\theta=0}^{\pi/2} \int_{r=0}^{2} \left(6 - \frac{r^2}{2}\sin 2\theta\right) r \, dr \, d\theta$$

$$= \int_{\theta=0}^{\pi/2} \int_{r=0}^{2} \left(6r - \frac{r^3}{2}\sin 2\theta\right) dr \, d\theta$$

$$= \int_0^{\pi/2} \left[3r^2 - \frac{r^4}{8}\sin 2\theta\right]_{r=0}^{r=2} d\theta = \int_0^{\pi/2} (12 - 2\sin 2\theta) \, d\theta$$

$$= \left[12\theta + \cos 2\theta\right]_0^{\pi/2} = (6\pi - 1) - 1 \qquad \therefore \underline{V = 6\pi - 2}$$

In this case, the use of cylindrical coordinates facilitates the evaluation.

Let us consider another example.

37

Example 2 To find the moment of inertia and radius of gyration of a thick hollow sphere about a diameter as axis. Outer radius $= a$; inner radius $= b$; density of material $= c$.

It is convenient to deal with one-eighth of the sphere in the first octant.

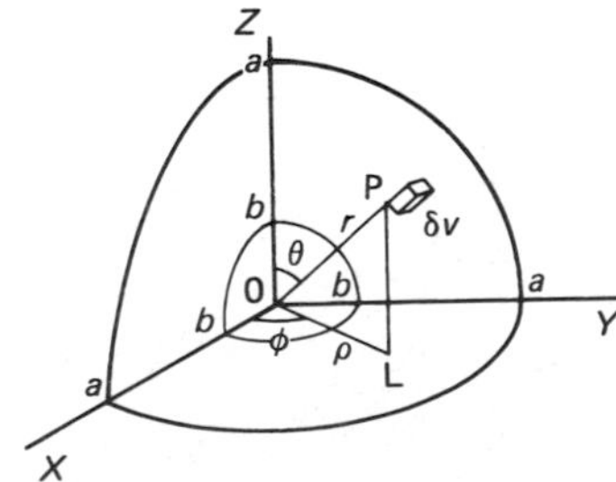

$\therefore$ Total mass of the solid $M_1 = \frac{1}{8}M$

$$M_1 = \frac{1}{8} \cdot \frac{4}{3}\pi(a^3 - b^3)c = \frac{\pi}{6}(a^3 - b^3)c$$

Using spherical coordinates, the element of volume $\delta v = \ldots\ldots\ldots$

38

$$\delta v = r^2 \sin\theta \, \delta r \, \delta\theta \, \delta\phi$$

Also the element of mass $m = c\delta v$

Second moment of mass of the element about OZ

$$= m\rho^2 = m(r \sin\theta)^2$$
$$= c\, r^2 \sin\theta \, \delta r \, \delta\theta \, \delta\phi \, r^2 \sin^2\theta$$
$$= c\, r^4 \sin^3\theta \, \delta r \, \delta\theta \, \delta\phi$$

$\therefore$ Total second moment for the solid

$$I_1 \approx \sum_{\phi=0}^{\pi/2} \sum_{\theta=0}^{\pi/2} \sum_{r=b}^{a} c\, r^4 \, \delta r \, \sin^3\theta \, \delta\theta \, \delta\phi$$

Then, as usual, if $\delta r \to 0$, $\delta\theta \to 0$, $\delta\phi \to 0$, we finally obtain

$$I_1 = \int_{\phi=0}^{\pi/2} \int_{\theta=0}^{\pi/2} \int_{r=b}^{a} c\, r^4 \, \mathrm{d}r \, \sin^3\theta \, \mathrm{d}\theta \, \mathrm{d}\phi$$

which you can evaluate without any difficulty and obtain

$$I_1 = \ldots\ldots\ldots$$

39

$$\boxed{I_1 = \frac{\pi}{15}(a^5 - b^5)c}$$

$$\text{for } I_1 = \int_0^{\pi/2}\int_0^{\pi/2}\left[c\frac{r^5}{5}\right]_b^a \sin^3\theta\,d\theta\,d\phi$$

$$= \int_0^{\pi/2}\int_0^{\pi/2}\frac{c}{5}(a^5 - b^5)\sin^3\theta\,d\theta\,d\phi$$

$$= \frac{c}{5}(a^5 - b^5)\int_0^{\pi/2}\int_0^{\pi/2}(1 - \cos^2\theta)\sin\theta\,d\theta\,d\phi$$

$$= \frac{c}{5}(a^5 - b^5)\int_0^{\pi/2}\left[-\cos\theta + \frac{\cos^3\theta}{3}\right]_0^{\pi/2}d\phi$$

$$= \frac{c}{5}(a^5 - b^5)\int_0^{\pi/2}\left(1 - \frac{1}{3}\right)d\phi$$

$$= \frac{2c}{15}(a^5 - b^5)\Big[\phi\Big]_0^{\pi/2} = \underline{\frac{c\pi}{15}(a^5 - b^5)}$$

Therefore, the moment of inertia for the whole sphere I is

$$I = 8I_1 \qquad \text{i.e.} \quad \underline{I = \frac{8\pi}{15}(a^5 - b^5)c}$$

Radius of gyration (k) $\quad Mk^2 = I$

$$\therefore k = \ldots\ldots\ldots$$

40

$$k = \sqrt{\frac{2}{5}\left(\frac{a^5 - b^5}{a^3 - b^3}\right)}$$

We had already calculated the total mass $M = \dfrac{4\pi}{3}(a^3 - b^3)c$ and since $I = \dfrac{8\pi}{15}(a^5 - b^5)c$ then

$$\frac{4\pi}{3}(a^3 - b^3)ck^2 = \frac{8\pi}{15}(a^5 - b^5)c$$

$$\therefore k^2 = \frac{2}{5}\left(\frac{a^5 - b^5}{a^3 - b^3}\right) \qquad \therefore k = \sqrt{\frac{2}{5}\left(\frac{a^5 - b^5}{a^3 - b^3}\right)}$$

We have set the working out in considerable detail, since spherical coordinates may be a new topic. Many of the statements can be streamlined when one is familiar with the system.

Now move on for another example.

41

Example 3 Find the total mass of a solid sphere of radius a, enclosed by the surface $x^2 + y^2 + z^2 = a^2$ and having variable density c where $c = 1 + r|z|$ and r is the distance of any point from the origin.

This is a case where spherical coordinates can clearly be used with advantage.

In the element of volume, the three dimensions are

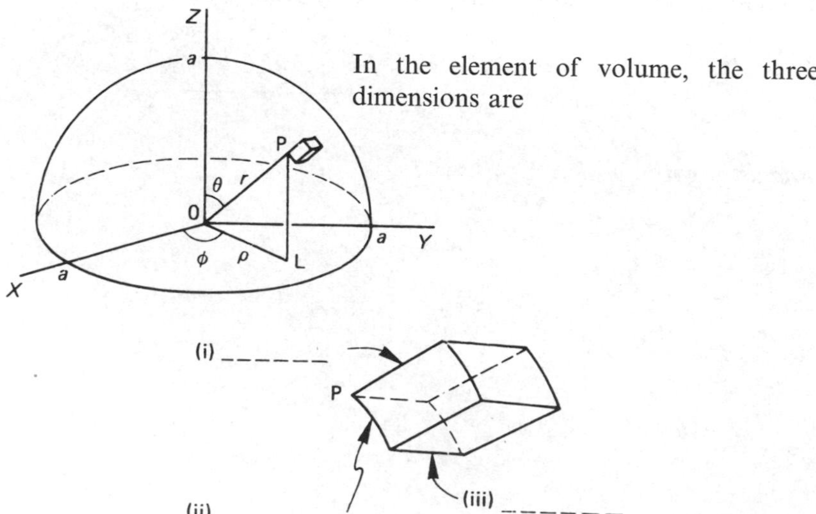

42

$$\text{(i) } \delta r; \quad \text{(ii) } r\,\delta\theta; \quad \text{(iii) } \rho\delta\phi = r\sin\theta\,\delta\phi$$

so that $\qquad \delta v = \ldots\ldots\ldots$

43

$$\delta v = r^2 \sin\theta\,\delta r\,\delta\theta\,\delta\phi$$

Then the mass of the element $= c\,\delta v = (1 + r|z|)\,\delta v$

and $\qquad z = r\cos\theta$

$$\therefore\ m = c\,\delta v = (1 + r^2\cos\theta)\,r^2\sin\theta\,\delta r\,\delta\theta\,\delta\phi$$

If we consider the *upper hemisphere* only, the integral for the total mass M_1 is $\qquad M_1 = \ldots\ldots\ldots$

Write out the integral and insert the limits.

44

$$M_1 = \int_{\phi=0}^{\phi=2\pi}\int_{\theta=0}^{\theta=\frac{\pi}{2}}\int_{r=0}^{r=a}(1 + r^2\cos\theta)r^2\sin\theta\,\mathrm{d}r\,\mathrm{d}\theta\,\mathrm{d}\phi$$

$$\text{i.e. } M_1 = \int_{\phi=0}^{2\pi}\int_{\theta=0}^{\pi/2}\int_{r=0}^{a}\{r^2\sin\theta\,\mathrm{d}r\,\mathrm{d}\theta\,\mathrm{d}\phi + r^4\sin\theta\cos\theta\,\mathrm{d}r\,\mathrm{d}\theta\,d\phi\}$$

$$= \qquad I_1 \qquad + \qquad I_2$$

$$I_1 = \int_0^{2\pi}\int_0^{\pi/2}\int_0^a r^2\sin\theta\,\mathrm{d}r\,\mathrm{d}\theta\,\mathrm{d}\phi \text{ gives } \ldots\ldots\ldots$$

Do *not* work it out. You can doubtless recognise what the result would represent.

45

The volume of the hemisphere

for the integral is simply the summation of elements of volume throughout the region of the hemisphere.

Thus, without more ado, $\quad I_1 = \frac{2}{3}\pi a^3.$

Now for I_2.

$$I_2 = \int_0^{2\pi}\int_0^{\pi/2}\int_0^a r^4 \sin\theta\cos\theta \,\mathrm{d}r\,\mathrm{d}\theta\,\mathrm{d}\phi$$

$$= \dots\dots\dots \quad \text{Evaluate the triple integral.}$$

46

$$I_2 = \frac{\pi a^5}{5}$$

for $I_2 = \int_0^{2\pi}\int_0^{\pi/2} \frac{a^5}{5}\sin\theta\cos\theta\,\mathrm{d}\theta\,\mathrm{d}\phi = \frac{a^5}{5}\int_0^{2\pi}\left[\frac{\sin^2\theta}{2}\right]_0^{\pi/2}\mathrm{d}\phi$

$$= \frac{a^5}{10}\int_0^{2\pi} 1\mathrm{d}\phi = \frac{a^5}{10}\Big[\phi\Big]_0^{2\pi} = \frac{\pi a^5}{5} \qquad \therefore\ I_2 = \frac{\pi a^5}{5}$$

So now finish it off. For the complete sphere

$$M = \dots\dots\dots$$

47

$$M = \frac{2\pi a^3}{15}(10 + 3a^2)$$

for $M_1 = I_1 + I_2 = \frac{2}{3}\pi a^3 + \frac{\pi a^5}{5} = \frac{\pi a^3}{15}(10 + 3a^2)$

Then, for the whole sphere, $M = 2M_1 = \underline{\frac{2\pi a^3}{15}(10 + 3a^2)}$

Each problem, then, is tackled in much the same way.

(a) Draw a careful sketch diagram, inserting all relevant information.
(b) Decide on the most appropriate coordinate system to use.
(c) Build up the multiple integral and insert correct limits.
(d) Evaluate the integral.

And now we can apply the general guide lines to a final problem.

Example 4 Determine the volume of the solid bounded by the planes $x = 0$, $y = 0$, $z = x$, $z = 2$ and $y = 4 - x^2$ in the first quadrant.

First we sketch the diagram.

48

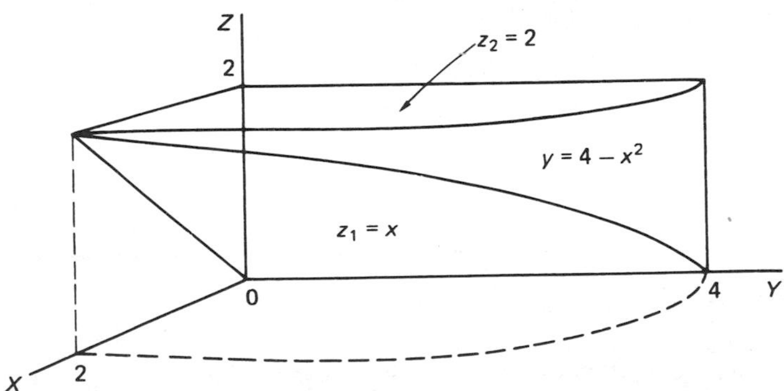

There is no axis of symmetry and no spherical centre. We shall therefore use .... coordinates.

49

cartesian

So off you go on your own. There are no snags.

$V = \dots\dots\dots$

50

$V = 6\frac{2}{3}$ cubic units

Here is the complete solution.

$$V \approx \sum_{x=0}^{2}\sum_{y=0}^{4-x^2}\sum_{z=x}^{2} \delta x\, \delta y\, \delta z$$

$$\therefore\ V = \int_{x=0}^{2}\int_{y=0}^{4-x^2}\int_{z=x}^{2} \mathrm{d}z\, \mathrm{d}y\, \mathrm{d}x = \int_0^2\int_0^{4-x^2} (2-x)\, \mathrm{d}y\, \mathrm{d}x$$

$$= \int_0^2 \Big[2y - xy\Big]_{y=0}^{4-x^2} \mathrm{d}x \qquad = \int_0^2 \{8 - 2x^2 - 4x + x^3\}\mathrm{d}x$$

$$= \left[8x - \frac{2x^3}{3} - 2x^2 + \frac{x^4}{4}\right]_0^2 = 6\frac{2}{3}$$

And that is it. *Now we move to the next section of work.*

51 **Change of Variables** in multiple integrals

In cartesian coordinates, we use the variables (x, y, z); in cylindrical coordinates, we use the variables (r, θ, z); in spherical coordinates, we use the variables (r, θ, ϕ); and we have established relationships connecting these systems of variables, permitting us to transfer from one system to another. These relationships, you will remember, were obtained geometrically in frames 23 to 30 of this programme.

There are occasions, however, when it is expedient to make other transformations beside those we have used and it is worth looking at the problem in a rather more general manner.

This we will now do.

52

First, however, let us revise a result from an earlier programme on determinants to find the area of the triangle ABC.

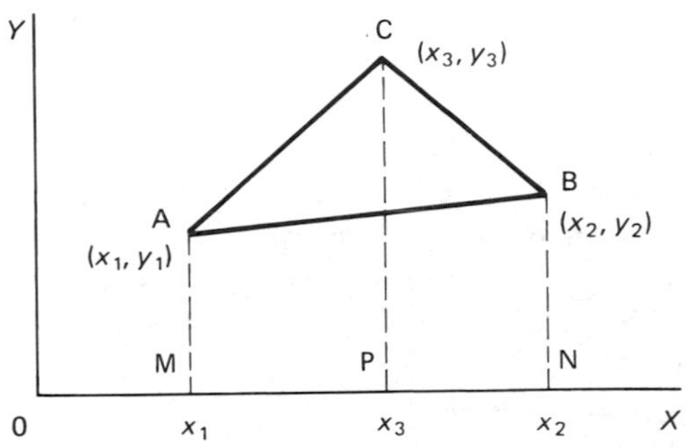

If we arrange the vertices A(x_1, y_1) B(x_2, y_2) C(x_3, y_3) in an anticlockwise manner then

area triangle ABC = trapezium AMPC + trapezium CPNB – trapezium AMNB

$$= \tfrac{1}{2}\{(x_3 - x_1)(y_1 + y_3) + (x_2 - x_3)(y_2 + y_3) - (x_2 - x_1)(y_1 + y_2)\}$$

$$= \tfrac{1}{2}\{x_3y_1 - x_1y_1 + x_3y_3 - x_1y_3 + x_2y_2 + x_2y_3 - x_3y_2 - x_3y_3 - x_2y_1 - x_2y_2 + x_1y_1 + x_1y_2\}$$

$$= \tfrac{1}{2}\{(x_2y_3 - x_3y_2) + (x_3y_1 - x_1y_3) + (x_1y_2 - x_2y_1)\}$$

$$= \tfrac{1}{2}\begin{vmatrix} 1 & 1 & 1 \\ x_1 & x_2 & x_3 \\ y_1 & y_2 & y_3 \end{vmatrix}$$

The determinant is positive if the points A, B, C are taken in an anticlockwise manner.

We shall need to use this result in a short while, so keep it in mind.

On to the next frame.

Curvilinear Coordinates

53

Consider the double integral $\int_R\int \phi(x,y)\mathrm{d}x\,\mathrm{d}y$. If we introduce two new independent variables, u and v, with $x = f(u,v)$ and $y = g(u,v)$, where $f(u,v)$ and $g(u,v)$ are two single-valued functions, then every point $\mathrm{P}(x,y)$ within the region R in the xy-plane maps on to a unique point $\mathrm{P}_1(u,v)$ in the corresponding region R_1 in the uv-plane.

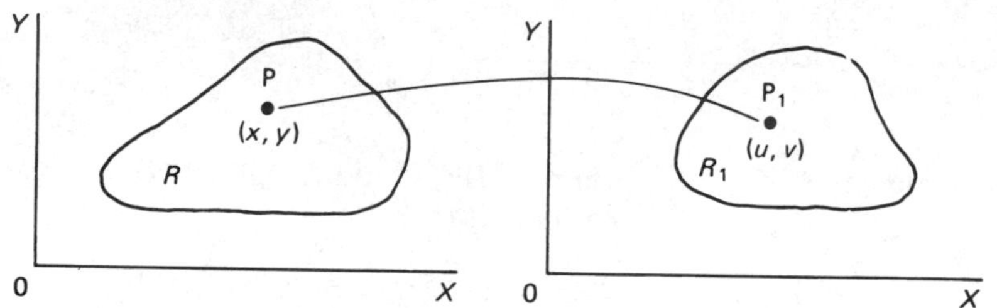

Let us see where this leads us, so on to the next frame.

54

Now let us consider an element of area in the xy-plane and the corresponding element in the uv-plane.

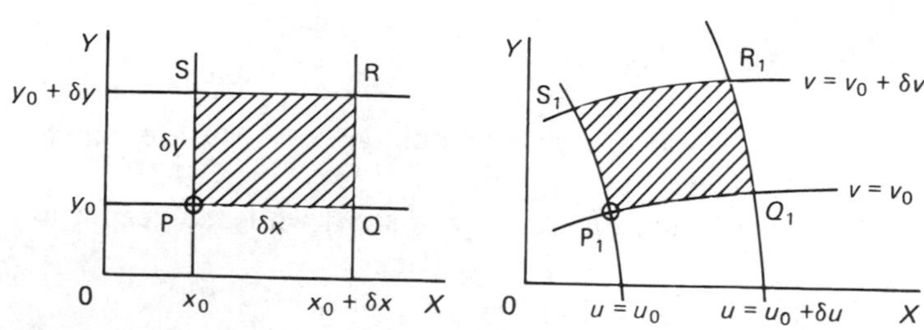

Points P, Q, R, S in the xy-plane map into P_1, Q_1, R_1, S_1 in the uv-plane. Let the point $\mathrm{P}(x_0, y_0)$ have the value $\mathrm{P}_1(u_0, v_0)$ in the new variables. If the transformation equations are $x = f(u,v)$ and $y = g(u,v)$, these can be treated as a pair of simultaneous equations and solved for u and v, giving

$$u = F(x,y) \quad \text{and} \quad v = G(x,y)$$

Then $u = F(x,y)$ will be a family of curves depending on the particular constant value given to u in each case.

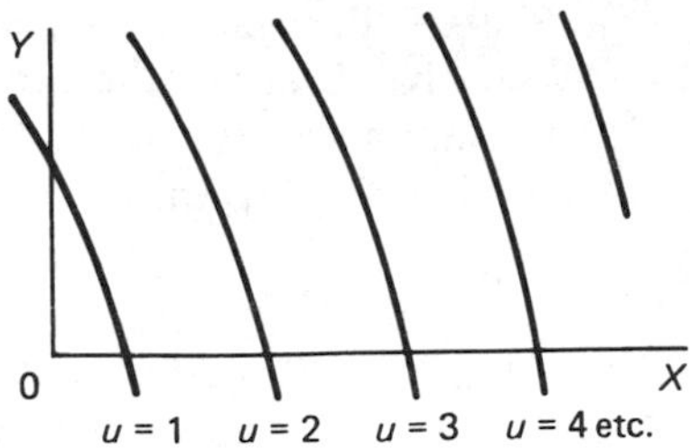

Curves $u = F(x, y)$ for different constant values of u.

Similarly, $v = G(x, y)$ will be a family of curves depending on the particular constant value assigned to v in each case.

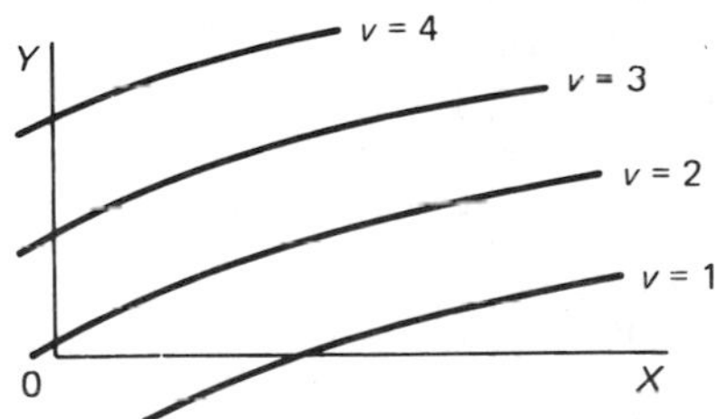

Curves $v = G(x, y)$ for different constant values of v.

These two sets of curves will therefore cover the region R and form a network, and to any point $P(x_0, y_0)$ there will be a pair of curves $u = u_0$ (constant) and $v = v_0$ (constant) that intersect at that point.

The u and v values relating to any particular point are known as its *curvilinear coordinates* and $x = f(u, v)$ and $y = g(u, v)$ are the *transformation equations* between the two systems.

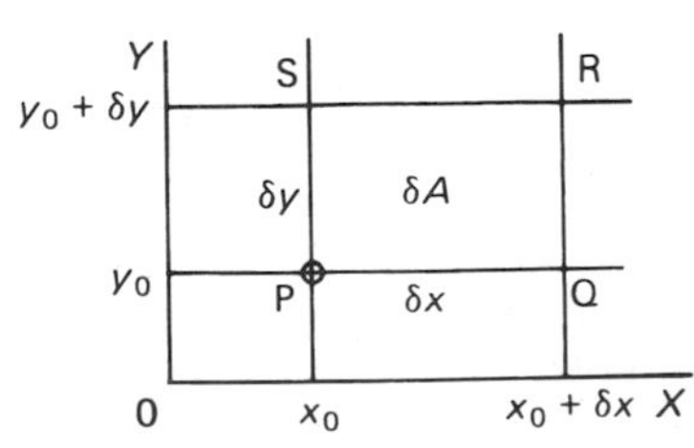

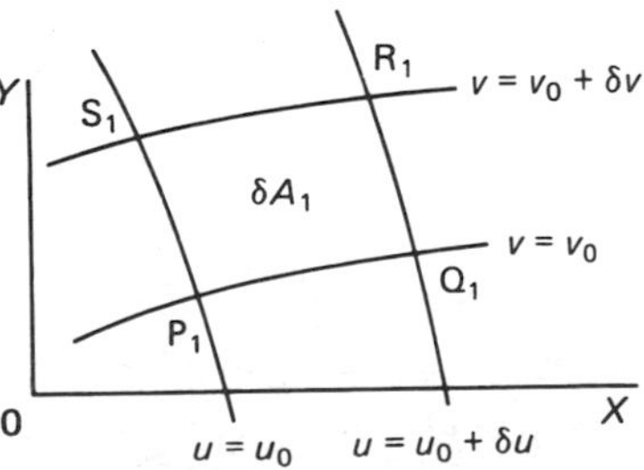

In the cartesian coordinates (x, y) system, the element of area $\delta A = \delta x \delta y$ and is the area bounded by the lines $x = x_0$, $x = x_0 + \delta x$, $y = y_0$, and $y = y_0 + \delta y$.

In the new system of *curvilinear coordinates* (u, v) the element of area δA_1 can be taken as that of the figure P_1, Q_1, R_1, S_1, i.e. the area bounded by the curves $u = u_0$, $u = u_0 + \delta u$, $v = v_0$ and $v = v_0 + \delta v$.

Since δA_1 is small, $P_1Q_1R_1S_1$ may be regarded as a rectangle

i.e. $$\delta A_1 \approx 2 \times \text{area of triangle } P_1Q_1S_1$$

and this is where we make use of the result previously revised that the area of a triangle ABC with vertices (x_1, y_1), (x_2, y_2), (x_3, y_3) can be expressed in determinant form as

Area = ...

55

$$\text{Area} = \frac{1}{2}\begin{vmatrix} 1 & 1 & 1 \\ x_1 & x_2 & x_3 \\ y_1 & y_2 & y_3 \end{vmatrix}$$

Before we can apply this, we must find the cartesian coordinates of P_1, Q_1 and S_1 in the bottom diagram on page 489.
If $x = f(u, v)$, then a small increase δx in x is given by

$\delta x =$ ....

56

$$\delta x = \frac{\partial f}{\partial u}\delta u + \frac{\partial f}{\partial v}\delta v$$

i.e. $$\delta x = \frac{\partial x}{\partial u}\delta u + \frac{\partial x}{\partial v}\delta v$$

and, for $y = g(u, v)$

$\delta y =$ ....

57

$$\delta y = \frac{\partial y}{\partial u}\delta u + \frac{\partial y}{\partial v}\delta v$$

Now P_1 coincides with $P(x, y)$, therefore

(a) P_1 is the point (x, y)
(b) Q_1 corresponds to Q, i.e. small changes from P.

$$\delta x = \frac{\partial x}{\partial u}\delta u + \frac{\partial x}{\partial v}\delta v \quad \text{and} \quad \delta y = \frac{\partial y}{\partial u}\delta u + \frac{\partial y}{\partial v}\delta v$$

But along P_1Q_1 v is constant. $\therefore \delta v = 0$.

$$\therefore \delta x = \frac{\partial x}{\partial u}\delta u \quad \text{and} \quad \delta y = \frac{\partial y}{\partial u}\delta u$$

i.e. Q is the point $\left(x + \frac{\partial x}{\partial u}\delta u,\ y + \frac{\partial y}{\partial u}\delta u\right)$.

(c) Similarly for S_1, since u is constant along P_1S_1 $\delta u = 0$ and

$\therefore$ S_1 is the point $\left(x + \frac{\partial x}{\partial v}\delta v,\ y + \frac{\partial y}{\partial v}\delta v\right)$

So the cartesian coordinates of P_1, Q_1, S_1 are

$$P_1(x, y); \qquad Q_1\left(x + \frac{\partial x}{\partial u}\delta u,\ y + \frac{\partial y}{\partial u}\delta u\right); \qquad S_1\left(x + \frac{\partial x}{\partial v}\delta v,\ y + \frac{\partial y}{\partial v}\delta v\right)$$

$\therefore$ The determinant for the area $P_1Q_1S_1$ is ....

58

$$\text{Area} = \frac{1}{2}\begin{vmatrix} 1 & 1 & 1 \\ x & x + \dfrac{\partial x}{\partial u}\delta u & x + \dfrac{\partial x}{\partial v}\delta v \\ y & y + \dfrac{\partial y}{\partial u}\delta u & y + \dfrac{\partial y}{\partial v}\delta v \end{vmatrix}$$

Subtracting column 1 from columns 2 and 3 gives

$$\text{Area} = \frac{1}{2}\begin{vmatrix} 1 & 0 & 0 \\ x & \dfrac{\partial x}{\partial u}\delta u & \dfrac{\partial x}{\partial v}\delta v \\ y & \dfrac{\partial y}{\partial u}\delta u & \dfrac{\partial y}{\partial v}\delta v \end{vmatrix}$$

which simplifies immediately to

. . . .

59

$$\text{Area} = \frac{1}{2}\begin{vmatrix} \dfrac{\partial x}{\partial u}\delta u & \dfrac{\partial x}{\partial v}\delta v \\ \dfrac{\partial y}{\partial u}\delta u & \dfrac{\partial y}{\partial v}\delta v \end{vmatrix}$$

Then, taking out the factor δu from the first column and the factor δv from the second column, this becomes

Area = ...

60

$$\frac{1}{2}\begin{vmatrix} \dfrac{\partial x}{\partial u} & \dfrac{\partial x}{\partial v} \\ \dfrac{\partial y}{\partial u} & \dfrac{\partial y}{\partial v} \end{vmatrix} \delta u\,\delta v$$

The area of the approximate rectangle is twice the area of the triangle.

$$\therefore \text{ Area of rectangle} = \delta A = \begin{vmatrix} \dfrac{\partial x}{\partial u} & \dfrac{\partial x}{\partial v} \\ \dfrac{\partial y}{\partial u} & \dfrac{\partial y}{\partial v} \end{vmatrix} \delta u\,\delta v$$

Expressing this in differentials

$$\mathrm{d}A = \begin{vmatrix} \dfrac{\partial x}{\partial u} & \dfrac{\partial x}{\partial v} \\ \dfrac{\partial y}{\partial u} & \dfrac{\partial y}{\partial v} \end{vmatrix} \mathrm{d}u\,\mathrm{d}v$$

and, for convenience, this is often written

$$\mathrm{d}A = \frac{\partial(x, y)}{\partial(u, v)}\,\mathrm{d}u\,\mathrm{d}v$$

$\dfrac{\partial(x, y)}{\partial(u, v)}$ is called the *Jacobian of the transformation* from the cartesian coordinates (x, y) to the curvilinear coordinates (u, v).

$$\therefore J(u, v) = \frac{\partial(x, y)}{\partial(u, v)} = \begin{vmatrix} \dfrac{\partial x}{\partial u} & \dfrac{\partial x}{\partial v} \\ \dfrac{\partial y}{\partial u} & \dfrac{\partial y}{\partial v} \end{vmatrix}$$

So, if the transformation equations are

$$x = u(u + v) \qquad \text{and} \qquad y = uv^2$$

$$J(u, v) = \ldots\ldots\ldots$$

61

$$\boxed{J(u, v) = uv(4u + v)}$$

for $\quad \dfrac{\partial x}{\partial u} = 2u + v \qquad \dfrac{\partial x}{\partial v} = u$

$$\frac{\partial y}{\partial u} = v^2 \qquad \frac{\partial y}{\partial v} = 2uv$$

$$\therefore J(u, v) = \begin{vmatrix} 2u + v & u \\ v^2 & 2uv \end{vmatrix} = 4u^2v + 2uv^2 - uv^2$$

$$= 4u^2v + uv^2 = \underline{uv(4u + v)}$$

There is one further point to note in this piece of work, so move on.

62 *Note* In the transformation, it is possible for the order of the points P_1, Q_1, R_1, S_1 to be reversed for that of P, Q, R, S with the result that δA may give a negative result when the determinant is evaluated. To ensure a positive element of area, the result is finally written

$$dA = \left|\frac{\partial(x, y)}{\partial(u, v)}\right| du\,dv$$

where the 'modulus' lines indicate the numerical value of the relevant determinant.

Therefore, to re-write the integral $\int_R \int F(x, y)\,dx\,dy$ in terms of the new variables, u and v, where $x = f(u, v)$ and $y = g(u, v)$, we substitute for x and y in $F(x, y)$ and replace $dx\,dy$ with $\left|\dfrac{\partial(x, y)}{\partial(u, v)}\right| du\,dv$.

The integral then becomes

$$\underline{\int_R \int F\{f(u, v), g(u, v)\} \left|\frac{\partial(x, y)}{\partial(u, v)}\right| du\,dv}$$

Make a note of this result.

63

Example 1 Express $I = \int_R\int xy^2 \, dx \, dy$ in polar coordinates, making the substitutions $x = r\cos\theta$, $y = r\sin\theta$.

$$\frac{\partial x}{\partial r} = \cos\theta \qquad \frac{\partial x}{\partial \theta} = -r\sin\theta$$

$$\frac{\partial y}{\partial r} = \sin\theta \qquad \frac{\partial y}{\partial \theta} = r\cos\theta$$

$$\therefore\ J(r, \theta) = \ldots\ldots\ldots$$

64

$$\boxed{J(r, \theta) = r}$$

$$J(r, \theta) = \begin{vmatrix} \cos\theta & -r\sin\theta \\ \sin\theta & r\cos\theta \end{vmatrix} = r\cos^2\theta + r\sin^2\theta = r$$

Then $I = \int_R\int xy^2 \, dx \, dy$ becomes ....

65

$$\boxed{I = \int_R\int r^3 \sin^2\theta \cos\theta \, r \, dr \, d\theta}$$

since $xy^2 = r\cos\theta \, r^2 \sin^2\theta = r^3 \sin^2\theta \cos\theta$

$$\left|\frac{\partial(x, y)}{\partial(r, \theta)}\right| dr \, d\theta = r \, dr \, d\theta$$

$$\therefore\ I = \int_R\int r^3 \sin^2\theta \cos\theta \, r \, dr \, d\theta = \underline{\int_R\int r^4 \sin^2\theta \cos\theta \, dr \, d\theta}$$

Now this one.

Example 2 Express $I = \int_R\int (x^2 + y^2) dx \, dy$ in terms of u and v, given that $x = u^2 - v^2$ and $y = 2uv$.

First of all, the expression for $\dfrac{\partial(x, y)}{\partial(u, v)}$ gives ......

66

$$\boxed{4(u^2 + v^2)}$$

for $\quad x = u^2 - v^2 \quad \therefore \dfrac{\partial x}{\partial u} = 2u \qquad \dfrac{\partial x}{\partial v} = -2v$

$\qquad y = 2uv \qquad \therefore \dfrac{\partial y}{\partial u} = 2v \qquad \dfrac{\partial y}{\partial v} = 2u$

$$\therefore \frac{\partial(x, y)}{\partial(u, v)} = \begin{vmatrix} \dfrac{\partial x}{\partial u} & \dfrac{\partial y}{\partial u} \\ \dfrac{\partial x}{\partial v} & \dfrac{\partial y}{\partial v} \end{vmatrix} = \begin{vmatrix} 2u & 2v \\ -2v & 2u \end{vmatrix} = \underline{4(u^2 + v^2)}$$

Also $\quad x^2 + y^2 = (u^2 - v^2)^2 + (2uv)^2 = u^4 - 2u^2v^2 + v^4 + 4u^2v^2$

$$= u^4 + 2u^2v^2 + v^4 = (u^2 + v^2)^2$$

Then $I = \int\int_R (x^2 + y^2)\,dx\,dy$ becomes $I = \dots\dots\dots$

67

$$\boxed{I = 4\int_R\int (u^2 + v^2)^3 du\,dv}$$

One more.

Example 3 By substituting $x = 2uv$ and $y = u(1 - v)$ where $u > 0$ and $v > 0$, express the integral $I = \displaystyle\int_R\int x^2 y\,dx\,dy$ in terms of u and v.

Complete it: there are no snags. $\qquad I = \dots\dots\dots$

68

$$I = 8 \int_R \int u^4 v^2 (1 - v)\, du\, dv$$

Working : $x = 2uv \quad \therefore \dfrac{\partial x}{\partial u} = 2v \quad \dfrac{\partial x}{\partial v} = 2u$

$$y = u - uv \quad \frac{\partial y}{\partial u} = 1 - v \quad \frac{\partial y}{\partial v} = -u$$

$$\therefore J(u, v) = \frac{\partial(x, y)}{\partial(u, v)} = \begin{vmatrix} \dfrac{\partial x}{\partial u} & \dfrac{\partial y}{\partial u} \\ \dfrac{\partial x}{\partial v} & \dfrac{\partial y}{\partial v} \end{vmatrix} = \begin{vmatrix} 2v & 1 - v \\ 2u & -u \end{vmatrix}$$

$$= 2u \begin{vmatrix} v & 1 - v \\ 1 & -1 \end{vmatrix} = 2u \begin{vmatrix} v & 1 \\ 1 & 0 \end{vmatrix} = -2u$$

$$\therefore \left| \frac{\partial(x, y)}{\partial(u, v)} \right| = 2u$$

$$x^2 y = 4u^2 v^2 (u - uv) = 4u^3 v^2 (1 - v)$$

$$\therefore I = \int_R \int 4u^3 v^2 (1 - v)\, 2u\, du\, dv$$

$$\underline{I = 8 \int_R \int u^4 v^2 (1 - v)\, du\, dv}$$

Transformation in three dimensions

If we extend the previous results to convert variables (x, y, z) to (u, v, w), we proceed in just the same way.

If $\quad x = f(u, v, w); \qquad y = g(u, v, w); \qquad z = h(u, v, w)$

Then $$J(u, v, w) = \frac{\partial(x, y, z)}{\partial(u, v, w)} = \begin{vmatrix} \dfrac{\partial x}{\partial u} & \dfrac{\partial y}{\partial u} & \dfrac{\partial z}{\partial u} \\ \dfrac{\partial x}{\partial v} & \dfrac{\partial y}{\partial v} & \dfrac{\partial z}{\partial v} \\ \dfrac{\partial x}{\partial w} & \dfrac{\partial y}{\partial w} & \dfrac{\partial z}{\partial w} \end{vmatrix}$$

and the element of volume $\mathrm{d}V = \mathrm{d}x\,\mathrm{d}y\,\mathrm{d}z$ becomes

$$\mathrm{d}V = |J(u, v, w)|\,\mathrm{d}u\,\mathrm{d}v\,\mathrm{d}w$$

Also $\iiint F(x, y, z)\,\mathrm{d}x\,\mathrm{d}y\,\mathrm{d}z$ is transformed into

$$\iiint G(u, v, w)\left|\frac{\partial(x, y, z)}{\partial(u, v, w)}\right|\mathrm{d}u\,\mathrm{d}v\,\mathrm{d}w$$

Now for an example, so move on.

69

Example 4 To transform a triple integral $I = \iiint F(x, y, z)\,\mathrm{d}x\,\mathrm{d}y\,\mathrm{d}z$ in cartesian coordinates to spherical coordinates by the transformation equations

$$x = r\sin\theta\cos\phi$$
$$y = r\sin\theta\sin\phi$$
$$z = r\cos\theta.$$

First we need the partial derivatives, from which to build up the Jacobian.

These are ....

70

$$\frac{\partial x}{\partial r} = \sin\theta\cos\phi \qquad \frac{\partial y}{\partial r} = \sin\theta\sin\phi \qquad \frac{\partial z}{\partial r} = \cos\theta$$
$$\frac{\partial x}{\partial \theta} = r\cos\theta\cos\phi \qquad \frac{\partial y}{\partial \theta} = r\cos\theta\sin\phi \qquad \frac{\partial z}{\partial \theta} = -r\sin\theta$$
$$\frac{\partial x}{\partial \phi} = -r\sin\theta\sin\phi \qquad \frac{\partial y}{\partial \phi} = r\sin\theta\cos\phi \qquad \frac{\partial z}{\partial \phi} = 0$$

$$\therefore\ J(r,\theta,\phi) = \begin{vmatrix} \sin\theta\cos\phi & \sin\theta\sin\phi & \cos\theta \\ r\cos\theta\cos\phi & r\cos\theta\sin\phi & -r\sin\theta \\ -r\sin\theta\sin\phi & r\sin\theta\cos\phi & 0 \end{vmatrix}$$

$$= \cos\theta \begin{vmatrix} r\cos\theta\cos\phi & r\cos\theta\sin\phi \\ -r\sin\theta\sin\phi & r\sin\theta\cos\phi \end{vmatrix} + r\sin\theta \begin{vmatrix} \sin\theta\cos\phi & \sin\theta\sin\phi \\ -r\sin\theta\sin\phi & r\sin\theta\cos\phi \end{vmatrix}$$

$$= \ldots\ldots\ldots$$

71

$$\boxed{r^2\sin\theta}$$

for $\quad J(r,\theta,\phi) = r^2\cos^2\theta\sin\theta \begin{vmatrix} \cos\phi & \sin\phi \\ -\sin\phi & \cos\phi \end{vmatrix}$

$$+ r^2\sin^3\theta \begin{vmatrix} \cos\phi & \sin\phi \\ -\sin\phi & \cos\phi \end{vmatrix}$$

$$= (r^2\sin^3\theta + r^2\sin\theta\cos^2\theta) \begin{vmatrix} \cos\phi & \sin\phi \\ -\sin\phi & \cos\phi \end{vmatrix}$$

$$= r^2\sin\theta(\sin^2\theta + \cos^2\theta)(\cos^2\phi + \sin^2\phi) = \underline{r^2\sin\theta}$$

$$\therefore\ I = \underline{\iiint G(u, v, w)\, r^2\sin\theta\, dr\, d\theta\, d\phi}$$

which agrees, of course, with the result we had previously obtained by a geometric consideration.

And that is about it. Check carefully down the Revision Summary that now follows, before working through the Test Exercise.

72 REVISION SUMMARY

1. *Surface integrals*

$$I = \int_R f(x, y)\, da = \int_R \int f(x, y)\, dy\, dx$$

2. *Surface in space*

$$I = \int_S \phi(x, y, z)\, dS = \int_R \int \phi(x, y, z) \sec \gamma\, dx\, dy \qquad (\gamma < \pi/2)$$

$$= \int_R \int \phi(x, y, z) \sqrt{1 + \left(\frac{\partial z}{\partial x}\right)^2 + \left(\frac{\partial z}{\partial y}\right)^2}\, dx\, dy$$

3. *Space coordinate systems*

(a) *Cartesian coordinates* (x, y, z)

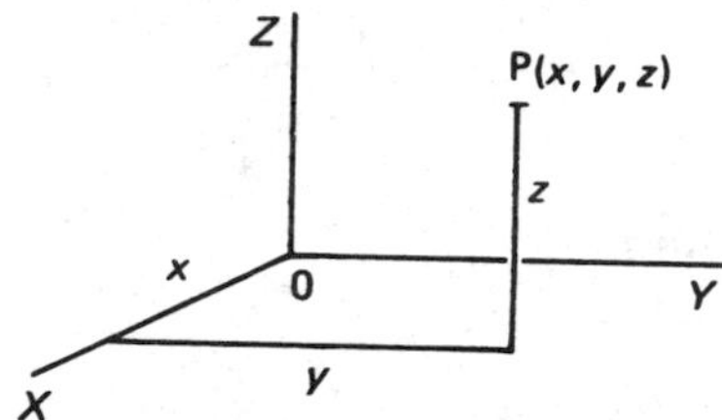

First octant: $x \geq 0$; $y \geq 0$; $z \geq 0$

(b) *Cylindrical coordinates* (r, θ, z) $\qquad r \geq 0$

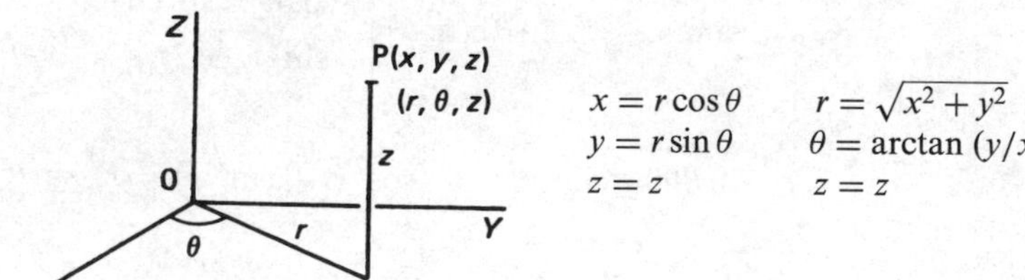

$x = r\cos\theta \qquad r = \sqrt{x^2 + y^2}$

$y = r\sin\theta \qquad \theta = \arctan(y/x)$

$z = z \qquad z = z$

(c) *Spherical coordinates* (r, θ, ϕ) $\qquad r \geq 0$

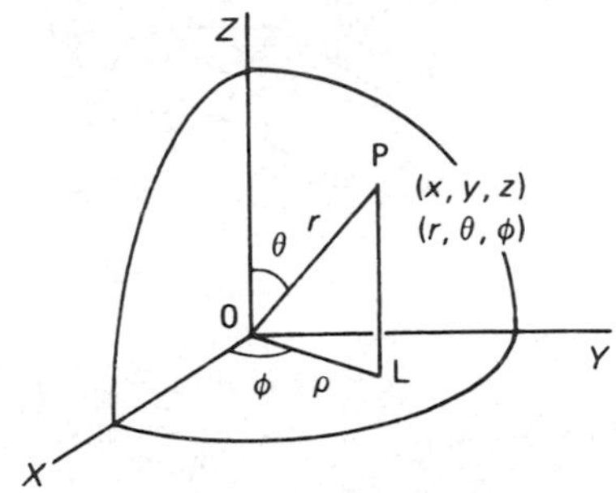

$$x = r\sin\theta\cos\phi \qquad r = \sqrt{x^2 + y^2 + z^2}$$
$$y = r\sin\theta\sin\phi \qquad \theta = \arccos(z/r)$$
$$z = r\cos\theta \qquad \phi = \arctan(y/x)$$

4. *Elements of volume*

(a) *Cartesian coordinates*

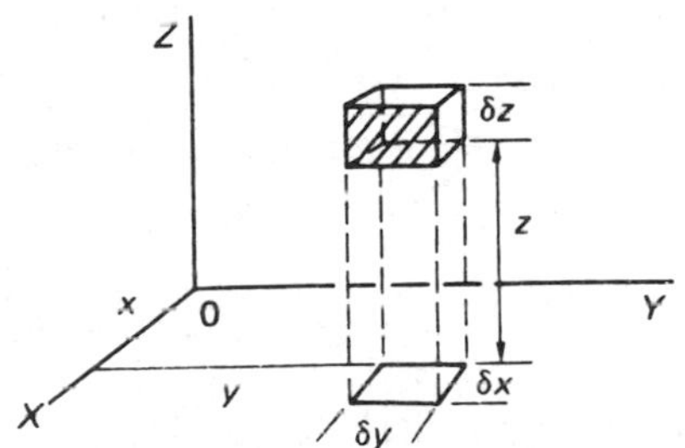

$$\delta v = \delta x\, \delta y\, \delta z$$

(b) *Cylindrical coordinates* $\qquad r \geq 0$

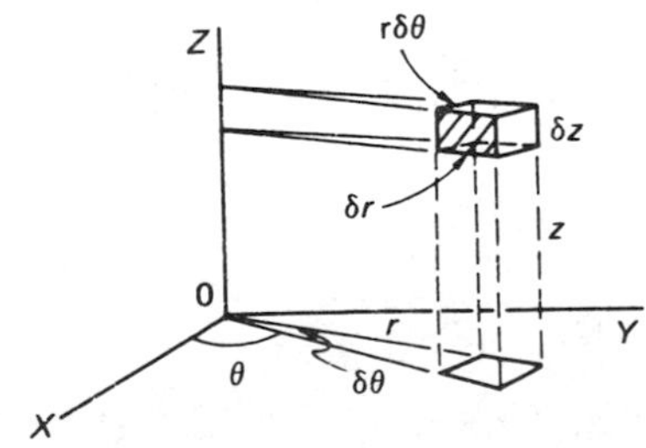

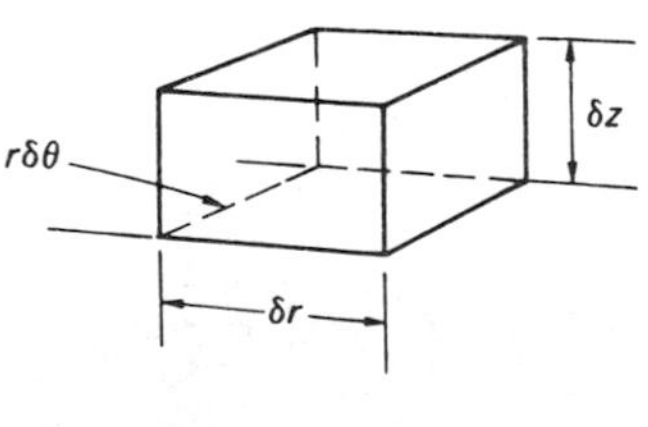

$$\delta v = r\, \delta r\, \delta\theta\, \delta z$$

(c) *Spherical coordinates*

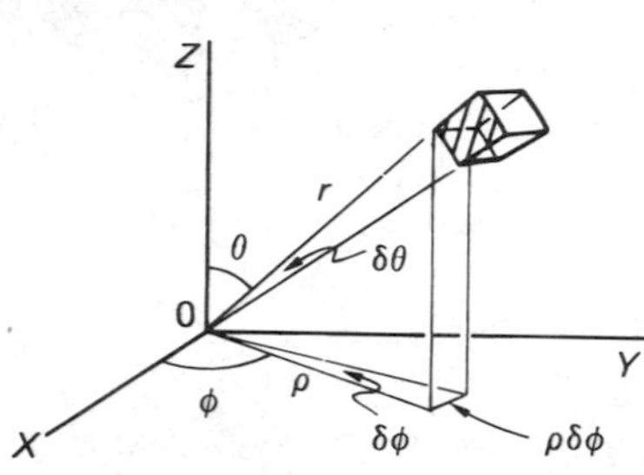

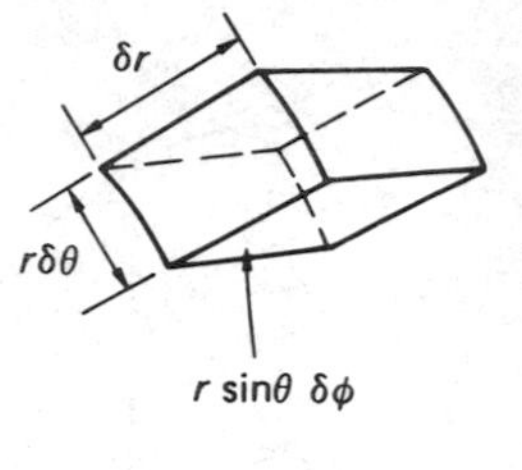

$\delta v = r^2 \sin\theta\, \delta r\, \delta\theta\, \delta\phi$

5. *Volume integrals*

$$V = \iiint \mathrm{d}z\, \mathrm{d}y\, \mathrm{d}x$$

$$I = \iiint f(x, y, z)\, \mathrm{d}z\, \mathrm{d}y\, \mathrm{d}x$$

6. *Change of variables* in multiple integrals

(a) *Double integrals* $x = f(u, v); \quad y = g(u, v)$

$$\mathrm{d}A = \left|\frac{\partial(x, y)}{\partial(u, v)}\right| \mathrm{d}u\, \mathrm{d}v; \qquad J(u, v) = \frac{\partial(x, y)}{\partial(u, v)} = \begin{vmatrix} \frac{\partial x}{\partial u} & \frac{\partial y}{\partial u} \\ \frac{\partial x}{\partial v} & \frac{\partial y}{\partial v} \end{vmatrix}$$

$$I = \int_R (x, y)\, \mathrm{d}x\, \mathrm{d}y = \int_R \int F\{f(u, v), g(u, v)\} \left|\frac{\partial(x, y)}{\partial(u, v)}\right| \mathrm{d}u\, \mathrm{d}v$$

(b) *Triple integrals* $x = f(u, v, w); \ y = g(u, v, w); \ z = h(u, v, w)$

$$J(u, v, w) = \frac{\partial(x, y, z)}{\partial(u, v, w)} = \begin{vmatrix} \frac{\partial x}{\partial u} & \frac{\partial y}{\partial u} & \frac{\partial z}{\partial u} \\ \frac{\partial x}{\partial v} & \frac{\partial y}{\partial v} & \frac{\partial z}{\partial v} \\ \frac{\partial x}{\partial w} & \frac{\partial y}{\partial w} & \frac{\partial z}{\partial w} \end{vmatrix}$$

Then $I = \iiint F(x, y, z)\, \mathrm{d}x\, \mathrm{d}y\, \mathrm{d}z$

$$= \iiint G(u, v, w) \left|\frac{\partial(x, y, z)}{\partial(u, v, w)}\right| \mathrm{d}u\, \mathrm{d}v\, \mathrm{d}w$$

73

TEST EXERCISE X

1. Determine the area of the surface $z = \sqrt{x^2 + y^2}$ over the region bounded by $x^2 + y^2 = 4$.
2. Evaluate the surface integral $I = \int_S \phi \, dS$ where $\phi = \dfrac{1}{\sqrt{x^2 + y^2}}$ over the surface of the sphere $x^2 + y^2 + z^2 = a^2$ in the first octant.
3. (a) Transform the cartesian coordinates
   (i) (4, 2, 3) to cylindrical coordinates (r, θ, z)
   (ii) (3, 1, 5) to spherical coordinates (r, θ, ϕ).
   (b) Express in cartesian coordinates (x, y, z)
   (i) the cylindrical coordinates $(5, \pi/4, 3)$
   (ii) the spherical coordinates $(4, \pi/6, 2)$.
4. Determine the volume of the solid bounded by the plane $z = 0$ and the surfaces $x^2 + y^2 = 4$ and $z = x^2 + y^2 + 1$.
5. Determine the total mass of a solid hemisphere, bounded by the plane $z = 0$ and the surface $x^2 + y^2 + z^2 = a^2 \quad (z \geq 0)$ if the density at any point is given by $\rho = 1 - z \quad (z < a)$.
6. (a) Express the integral $I = \int_R \int (x - y) \, dx \, dy$ in terms of u and v, where $x = u(1 + v)$ and $y = u - v$.
   (b) Express the triple integral $I = \iiint \left(\dfrac{x+z}{y}\right) dx \, dy \, dz$ in terms of u, v, w using the transformation equations

$$x = u + v + w; \quad y = v^2 w; \quad z = u - w.$$

FURTHER PROBLEMS X

1. Evaluate the surface integral $I = \int_S (x^2 + y^2) \, dS$ over the surface of the cone $z^2 = 4(x^2 + y^2)$ between $z = 0$ and $z = 4$.
2. Find the position of the centre of gravity of that part of a thin spherical shell $x^2 + y^2 + z^2 = a^2$ which exists in the first octant.
3. Determine the surface area of the plane $6x + 3y + 4z = 60$ cut off by $x = 0, y = 0, x = 5, y = 8$.
4. Find the surface area of the plane $3x + 2y + 3z = 12$ cut off by the planes $x = 0$, $y = 0$, and the cylinder $x^2 + y^2 = 16$ for $x \geq 0, y \geq 0$.

5. Determine the area of the paraboloid $z = 2(x^2 + y^2)$ cut off by the cone $z = \sqrt{x^2 + y^2}$.

6. Find the area of the cone $z^2 = 4(x^2 + y^2)$ which is inside the paraboloid $z = 2(x^2 + y^2)$.

7. Cylinders $x^2 + y^2 = a^2$ and $x^2 + z^2 = a^2$ intersect. Determine the total external surface area of the common portion.

8. Determine the surface area of the sphere $x^2 + y^2 + z^2 = a^2$ cut off by the cylinder $x^2 + y^2 = ax$.

9. A cylinder of radius b, with the z-axis as its axis of symmetry, is removed from a sphere of radius a, $a > b$, with centre at the origin. Calculate the total curved surface area of the ring so formed, including the inner cylindrical surface.

10. Find the volume enclosed by the cylinder $x^2 + y^2 = 9$ and the planes $z = 0$ and $z = 5 - x$.

11. Determine the volume of the solid bounded by the surfaces $y = x^2$, $x = y^2$, $z = 2$ and $x + y + z = 4$.

12. Find the volume of the solid bounded by the plane $z = 0$, the cylinder $x^2 + y^2 = a^2$ and the surface $z = x^2 + y^2$.

13. A solid is bounded by the planes $x = 0$, $y = 0$, $z = 2$, $z = x$ and the surface $x^2 + y^2 = 4$. Determine the volume of the solid.

14. Find the position of the centre of gravity of the part of the solid sphere $x^2 + y^2 + z^2 = a^2$ in the first octant.

15. A solid is bounded by the cone $z = 2\sqrt{x^2 + y^2}$, $z \geq 0$, and the sphere $x^2 + y^2 + (z - a)^2 = 2a^2$. Determine the volume of the solid so formed.

16. Determine the volume enclosed by the ellipsoid $\dfrac{x^2}{a^2} + \dfrac{y^2}{b^2} + \dfrac{z^2}{c^2} = 1$.

17. Find the volume of the solid in the first octant bounded by the planes $x = 0$, $y = 0$, $z = 0$, $z = x + y$ and the surface $x^2 + y^2 = a^2$.

18. Express the integral $\displaystyle\iint (x^2 + y^2)\, dx\, dy$ in terms of u and v, using the transformations $u = x + y$, $v = x - y$.

19. Determine an expression for the element of volume $dx\, dy\, dz$ in terms of u, v, w using the transformations $x = u(1 - v)$, $y = uv$, $z = uvw$.

20. A solid sphere of radius a has variable density c at any point (x, y, z) given by $c = k(a - z)$ where k is a constant. Determine the position of the centre of gravity of the sphere.

21. Calculate $\displaystyle\iint x^2 y^2\, dx\, dy$ over the triangular region in the xy-plane with vertices $(0, 0)$, $(1, 1)$, $(1, 2)$.

22. Evaluate the integral $I = \int_0^2 \int_{\sqrt{y(2-y)}}^{\sqrt{4-y^2}} \frac{y}{x^2+y^2}\, dx\, dy$ by transforming to polar coordinates.

23. Evaluate $I = \int_0^1 \int_0^y \frac{xy^2}{\sqrt{x^2+y^2}}\, dx\, dy$.

24. Find the volume bounded by the cylinder $x^2+y^2=a^2$, the plane $z=0$ and the surface $z=x^2+y^2$. Convert to polar coordinates and show that $V = \frac{\pi a^4}{2}$.

25. By changing the order of integration in the integral

$$I = \int_0^a \int_x^a \frac{y^2\, dy\, dx}{\sqrt{x^2+y^2}}$$

show that $I = \frac{1}{3}a^3 \ln(1+\sqrt{2})$.

Programme 11

Matrix Algebra

Prerequisites: Engineering Mathematics (fourth edition)
Programmes 4, 5

Introduction

1 The work of this programme extends the material covered in Programme 5 of the author's *Engineering Mathematics* (fourth edition) which dealt with the basic principles of matrices and their applications, mainly to the solution of sets of equations. The current development assumes a knowledge of this earlier work and a review of that previous programme would provide worthwhile revision before embarking on the new topics of study.

Rank of a matrix

A knowledge of the rank of a matrix is important when dealing with the uniqueness of solutions of sets of simultaneous linear equations and can save a lot of unnecessary work.

The *rank* of a matrix can be described as the order of the largest non-zero determinant that can be formed from the elements of the given matrix.

A matrix **A** is said to have rank r if it contains at least one square sub-matrix of r rows with a non-zero determinant, while all square sub-matrices of $(r + 1)$ rows, or more, have zero determinants.

For example, the matrix $\mathbf{A} = \begin{bmatrix} 4 & 2 \\ 1 & 5 \end{bmatrix}$ is of rank 2, since

$\begin{vmatrix} 4 & 2 \\ 1 & 5 \end{vmatrix} = 20 - 2 = 18$, i.e. not zero.

However, $\mathbf{B} = \begin{bmatrix} 6 & 3 \\ 8 & 4 \end{bmatrix}$ gives $\begin{vmatrix} 6 & 3 \\ 8 & 4 \end{vmatrix} = 24 - 24 = 0$. Therefore, **B** is *not* of rank 2. It is, of course, of rank 1, since the sub-matrices [6], [3], [8], [4] are not zero.

Note. As a special case, if all the elements of a matrix **A** are zero, then **A** is said to be of *rank zero*.

Example 1 To determine the rank of $\mathbf{A} = \begin{bmatrix} 1 & 2 & 8 \\ 4 & 7 & 6 \\ 9 & 5 & 3 \end{bmatrix}$

we test the highest-order determinant, i.e. third order.

$$\begin{vmatrix} 1 & 2 & 8 \\ 4 & 7 & 6 \\ 9 & 5 & 3 \end{vmatrix} = \dots\dots\dots$$

2

$$\boxed{-269}$$

$$\begin{vmatrix} 1 & 2 & 8 \\ 4 & 7 & 6 \\ 9 & 5 & 3 \end{vmatrix} = 1(21-30) - 2(12-54) + 8(20-63) = \underline{-269}$$

This is not zero. Therefore the rank of matrix **A** is

. . . .

3

$$\boxed{3}$$

Example 2

Similarly, with $\mathbf{A} = \begin{bmatrix} 3 & 4 & 5 \\ 1 & 2 & 3 \\ 4 & 5 & 6 \end{bmatrix}$ we try, first of all, the full third-order determinant.

$$|\mathbf{A}| = \begin{vmatrix} 3 & 4 & 5 \\ 1 & 2 & 3 \\ 4 & 5 & 6 \end{vmatrix} = \ldots\ldots\ldots$$

4

$$\boxed{0}$$

for $|\mathbf{A}| = 3(12-15) - 4(6-12) + 5(5-8)$

$$= -9 + 24 - 15 = 0$$

Therefore, we can say that the rank of **A** is . . . .

5

not 3

We now try sub-matrices of order 2.

$$\begin{vmatrix} 3 & 4 \\ 1 & 2 \end{vmatrix} = 6 - 4 = 2 \neq 0. \qquad \therefore \underline{\text{Rank of } \mathbf{A} \text{ is } 2}$$

We could equally well have tested $\begin{vmatrix} 4 & 5 \\ 2 & 3 \end{vmatrix}$; $\begin{vmatrix} 1 & 2 \\ 4 & 5 \end{vmatrix}$; $\begin{vmatrix} 2 & 3 \\ 5 & 6 \end{vmatrix}$; etc

Example 3 Determine the rank of the matrix $\mathbf{A} = \begin{bmatrix} 3 & 9 & 2 \\ 1 & 5 & 6 \\ 2 & 7 & 4 \end{bmatrix}$.

As before, we start with the largest determinant, i.e. of third order in this case. If $|\mathbf{A}| = 0$, then we test lower-order determinants.

So the rank of **A** is ....

6

2

for we have $\begin{vmatrix} 3 & 9 & 2 \\ 1 & 5 & 6 \\ 2 & 7 & 4 \end{vmatrix} = 3(20 - 42) - 9(4 - 12) + 2(7 - 10) = 0.$

$$\therefore \underline{\mathbf{A} \text{ is not of rank } 3}$$

But $\begin{vmatrix} 3 & 9 \\ 1 & 5 \end{vmatrix} = 15 - 9 = 6 \qquad \therefore \underline{\mathbf{A} \text{ is of rank } 2}$

With a third-order determinant $|\mathbf{A}| = 0$, it may be necessary to test all possible second-order minors to find one that is not zero.

The tests for rank can be applied in like manner to rectangular matrices, by considering the largest square sub-matrix formed by the elements. If a rectangular matrix is of order $(n \times m)$ then the rank cannot be greater than ....

7

n or m, whichever is the least in value

Example 4 If we consider $\mathbf{B} = \begin{bmatrix} 2 & 2 & 3 & 1 \\ 0 & 8 & 2 & 4 \\ 1 & 7 & 3 & 2 \end{bmatrix}$ the largest square sub-matrix cannot be greater than order 3.

We try $\begin{vmatrix} 2 & 2 & 3 \\ 0 & 8 & 2 \\ 1 & 7 & 3 \end{vmatrix} = \dots\dots\dots$

8

0

for $\begin{vmatrix} 2 & 2 & 3 \\ 0 & 8 & 2 \\ 1 & 7 & 3 \end{vmatrix} = 2(24 - 14) - 2(0 - 2) + 3(0 - 8) = 0.$

But we must also try

$$\begin{vmatrix} 2 & 3 & 1 \\ 8 & 2 & 4 \\ 7 & 3 & 2 \end{vmatrix} = \dots\dots\dots$$

9

30

for $\begin{vmatrix} 2 & 3 & 1 \\ 8 & 2 & 4 \\ 7 & 3 & 2 \end{vmatrix} = 2(4 - 12) - 3(16 - 28) + 1(24 - 14) = 30$ i.e. not zero.

$\therefore$ <u>**B** is of rank 3</u>

It is just a case of applying the tests, starting with the largest square sub-matrices that can be formed by the elements.

Singular and non-singular matrices

If **A** is a square matrix of order n and if the rank of $\mathbf{A} = n$, then $|\mathbf{A}| \neq 0$ and **A** is said to be *non-singular*.
If, however, $|\mathbf{A}| = 0$, then **A** is said to be *singular*.

We have already seen that

$$\mathbf{A} = \begin{bmatrix} 1 & 2 & 8 \\ 4 & 7 & 6 \\ 9 & 5 & 3 \end{bmatrix} \text{ is of rank 3}$$

and

$$\mathbf{B} = \begin{bmatrix} 3 & 9 & 2 \\ 1 & 5 & 6 \\ 2 & 7 & 4 \end{bmatrix} \text{ is of rank 2}$$

$$\therefore \ \mathbf{A} \text{ is} \ldots\ldots\ldots \qquad \text{and} \quad \mathbf{B} \text{ is} \ldots\ldots\ldots$$

10

A is non-singular; **B** is singular

for, since **A** is of rank 3, then $|\mathbf{A}| \neq 0$
and since **B** is of rank 2, then $|\mathbf{B}| = 0$.

Exercise Determine whether each of the following is singular or non-singular.

1. $\mathbf{A} = \begin{bmatrix} 4 & 5 \\ 2 & 3 \end{bmatrix}$

2. $\mathbf{B} = \begin{bmatrix} 3 & -4 \\ -6 & 8 \end{bmatrix}$

3. $\mathbf{C} = \begin{bmatrix} 4 & 1 & -2 \\ 1 & 7 & 3 \\ 5 & 8 & 1 \end{bmatrix}$

4. $\mathbf{D} = \begin{bmatrix} 3 & 2 & 4 \\ 5 & 1 & 6 \\ 2 & 0 & 3 \end{bmatrix}$

1. non-singular	2. singular
3. singular	4. non-singular

Straightforward evaluation of the relevant determinants gives

1. $|\mathbf{A}| = 2$ 2. $|\mathbf{B}| = 0$

3. $|\mathbf{C}| = 0$ 4. $|\mathbf{D}| = -5.$

Consistency of equations

In solving sets of simultaneous equations, we can express the equations in matrix form. For example

$$a_{11}x_1 + a_{12}x_2 + a_{13}x_3 = b_1$$
$$a_{21}x_1 + a_{22}x_2 + a_{23}x_3 = b_2$$
$$a_{31}x_1 + a_{32}x_2 + a_{33}x_3 = b_3$$

can be written in the form

$$\begin{bmatrix} a_{11} & a_{12} & a_{13} \\ a_{21} & a_{22} & a_{23} \\ a_{31} & a_{32} & a_{33} \end{bmatrix} \begin{bmatrix} x_1 \\ x_2 \\ x_3 \end{bmatrix} = \begin{bmatrix} b_1 \\ b_2 \\ b_3 \end{bmatrix}$$

i.e. $$\mathbf{Ax} = \mathbf{b}$$

The set of three equations is said to be *consistent* if solutions for x_1, x_2, x_3 exist and *inconsistent* if no such solutions can be found.

In practice, we can solve the equations by operating on the *augmented coefficient matrix*, i.e. we write the constant terms as a fourth column of the coefficient matrix to form $\mathbf{A}_b$.

$$\mathbf{A}_b = \begin{bmatrix} a_{11} & a_{12} & a_{13} & b_1 \\ a_{21} & a_{22} & a_{23} & b_2 \\ a_{31} & a_{32} & a_{33} & b_3 \end{bmatrix}$$

which, of course, is a (3×4) matrix.

The general test for consistency is then:

A set of n simultaneous equations in n unknowns is consistent if the rank of the coefficient matrix $\mathbf{A}$ is equal to the rank of the augmented matrix $\mathbf{A}_b$.

If the rank of $\mathbf{A}$ is less than the rank of $\mathbf{A}_b$, then the equations are inconsistent and have no solution.

Make a note of this test. It can save time in working.

12 *Example*

$$\text{If } \begin{bmatrix} 1 & 3 \\ 2 & 6 \end{bmatrix}\begin{bmatrix} x_1 \\ x_2 \end{bmatrix} = \begin{bmatrix} 4 \\ 5 \end{bmatrix}$$

then $\mathbf{A} = \begin{bmatrix} 1 & 3 \\ 2 & 6 \end{bmatrix}$ and $\mathbf{A}_b = \begin{bmatrix} 1 & 3 & 4 \\ 2 & 6 & 5 \end{bmatrix}$

Rank of $\mathbf{A}$: $\begin{vmatrix} 1 & 3 \\ 2 & 6 \end{vmatrix} = 6 - 6 = 0 \quad \therefore$ rank of $\mathbf{A} = 1$

Rank of $\mathbf{A}_b$: $\begin{vmatrix} 1 & 3 \\ 2 & 6 \end{vmatrix} = 0$ as before

but $\begin{vmatrix} 3 & 4 \\ 6 & 5 \end{vmatrix} = 15 - 24 = -9 \quad \therefore$ rank of $\mathbf{A}_b = 2$

In this case, rank of $\mathbf{A}$ < rank of $\mathbf{A}_b$

$\therefore$ ....

13

no solution exists

Remember that, for consistency,

rank of $\mathbf{A} =$ ....

14

rank of $\mathbf{A}_b$

Uniqueness of solutions

1. With a set of n equations in n unknowns, the equations are consistent if the coefficient matrix $\mathbf{A}$ and the augmented matrix $\mathbf{A}_b$ are each of rank n. There is then a *unique* solution for the n equations.
2. If the rank of $\mathbf{A}$ and that of $\mathbf{A}_b$ is m, where $m < n$, then the matrix $\mathbf{A}$ is singular, i.e. $|\mathbf{A}| = 0$, and there will be an *infinite number* of solutions for the equations.
3. As we have already seen, if the rank of $\mathbf{A}$ < the rank of $\mathbf{A}_b$, then *no solution* exists.

Copy these up in your record; they are important.

15

Writing the results in a slightly different way:

With a set of n equations in n unknowns, checking the rank of the coefficient matrix $\mathbf{A}$ and that of the augmented matrix $\mathbf{A}_b$, enables us to see whether

(a) a unique solution exists

$$\text{rank } \mathbf{A} = \text{rank } \mathbf{A}_b = n$$

(b) an infinite number of solutions exist

$$\text{rank } \mathbf{A} = \text{rank } \mathbf{A}_b = m < n$$

(c) no solution exists

$$\text{rank } \mathbf{A} < \text{rank } \mathbf{A}_b$$

Example $$\begin{bmatrix} -4 & 5 \\ -8 & 10 \end{bmatrix} \begin{bmatrix} x_1 \\ x_2 \end{bmatrix} = \begin{bmatrix} -3 \\ -6 \end{bmatrix}$$

Finding the rank of $\mathbf{A}$ and of $\mathbf{A}_b$ leads us to the conclusion that

. . . .

16

there is an infinite number of solutions

$$\text{for}\quad \mathbf{A} = \begin{bmatrix} -4 & 5 \\ -8 & 10 \end{bmatrix} \quad \text{and}\quad \mathbf{A}_b = \begin{bmatrix} -4 & 5 & -3 \\ -8 & 10 & -6 \end{bmatrix}$$

$$\text{Rank of } \mathbf{A}: \begin{vmatrix} -4 & 5 \\ -8 & 10 \end{vmatrix} = -40 + 40 = 0 \quad \therefore \ \underline{\text{Rank of } \mathbf{A} = 1}$$

$$\text{Rank of } \mathbf{A}_b: \begin{vmatrix} -4 & 5 \\ -8 & 10 \end{vmatrix} = 0; \begin{vmatrix} 5 & -3 \\ 10 & -6 \end{vmatrix} = 0; \begin{vmatrix} -4 & -3 \\ -8 & -6 \end{vmatrix} = 0$$

$$\therefore \ \underline{\text{Rank of } \mathbf{A}_b = 1}$$

$$\therefore \ \text{Rank of } \mathbf{A} = \text{rank of } \mathbf{A}_b = \mathbf{1}$$

But there are two equations in two unknowns, i.e. $n = 2$

$\therefore$ Rank of $\mathbf{A}$ = rank of $\mathbf{A}_b = 1 < n$ $\therefore$ <u>Infinite number of solutions.</u>

You will recall that, for a unique solution of n equations in n unknowns

. . . .

17

rank $\mathbf{A}$ = rank $\mathbf{A}_b = n$

Now for one or two examples. In each of the following cases, apply the rank tests to determine the nature of the solutions. Do not solve the sets of equations.

Example 1

$$\begin{bmatrix} 1 & 2 & -1 \\ 3 & 4 & 2 \\ 1 & 4 & 3 \end{bmatrix} \begin{bmatrix} x_1 \\ x_2 \\ x_3 \end{bmatrix} = \begin{bmatrix} 1 \\ -2 \\ 3 \end{bmatrix}$$

$$\mathbf{A} = \begin{bmatrix} 1 & 2 & -1 \\ 3 & 4 & 2 \\ 1 & 4 & 3 \end{bmatrix} \quad \text{and} \quad \mathbf{A}_b = \begin{bmatrix} 1 & 2 & -1 & 1 \\ 3 & 4 & 2 & -2 \\ 1 & 4 & 3 & 3 \end{bmatrix}$$

Finish it off and we find that . . . .

18

a unique solution exists

for $n = 3$; rank of $\mathbf{A} = 3$; rank of $\mathbf{A}_b = 3$.
$\therefore$ rank of $\mathbf{A}$ = rank of $\mathbf{A}_b = 3 = n$ $\therefore$ Solution unique

And this one.

Example 2

$$\begin{bmatrix} 2 & -1 & 7 \\ 4 & 2 & 2 \\ 3 & 1 & 3 \end{bmatrix} \begin{bmatrix} x_1 \\ x_2 \\ x_3 \end{bmatrix} = \begin{bmatrix} 2 \\ 5 \\ 1 \end{bmatrix}$$

This time we find that ....

19

no solution is possible

for $\mathbf{A} = \begin{bmatrix} 2 & -1 & 7 \\ 4 & 2 & 2 \\ 3 & 1 & 3 \end{bmatrix}$; $\mathbf{A}_b = \begin{bmatrix} 2 & -1 & 7 & 2 \\ 4 & 2 & 2 & 5 \\ 3 & 1 & 3 & 1 \end{bmatrix}$

$n = 3$; rank of $\mathbf{A} = 2$; rank of $\mathbf{A}_b = 3$

$\therefore$ rank of $\mathbf{A}$ < rank of $\mathbf{A}_b$ $\therefore$ No solution exists

and finally

Example 3

$$\begin{bmatrix} 1 & 2 & -3 \\ 1 & 3 & 4 \\ 2 & 5 & 1 \end{bmatrix} \begin{bmatrix} x_1 \\ x_2 \\ x_3 \end{bmatrix} = \begin{bmatrix} 1 \\ 2 \\ 3 \end{bmatrix}$$

In this case, we find that ....

20

infinite number of solutions possible

$$\mathbf{A} = \begin{bmatrix} 1 & 2 & -3 \\ 1 & 3 & 4 \\ 2 & 5 & 1 \end{bmatrix}; \quad \mathbf{A}_b = \begin{bmatrix} 1 & 2 & -3 & 1 \\ 1 & 3 & 4 & 2 \\ 2 & 5 & 1 & 3 \end{bmatrix}$$

Rank of **A**: $\begin{vmatrix} 1 & 2 & -3 \\ 1 & 3 & 4 \\ 2 & 5 & 1 \end{vmatrix} = 1(3-20) - 2(1-8) - 3(5-6) = 0$

$$\begin{vmatrix} 1 & 2 \\ 1 & 3 \end{vmatrix} = 3 - 2 = 1 \qquad \therefore \ \underline{\text{Rank of A} = 2}$$

Rank of $\mathbf{A}_b$: $\begin{vmatrix} 1 & 2 & -3 \\ 1 & 3 & 4 \\ 2 & 5 & 1 \end{vmatrix} = 0$ as before

$$\begin{vmatrix} 2 & -3 & 1 \\ 3 & 4 & 2 \\ 5 & 1 & 3 \end{vmatrix} = 2(12-2) + 3(9-10) + 1(3-20) = 0$$

Similarly $\begin{vmatrix} 1 & 2 & 1 \\ 1 & 3 & 2 \\ 2 & 5 & 3 \end{vmatrix}$ and $\begin{vmatrix} 1 & -3 & 1 \\ 1 & 4 & 2 \\ 2 & 1 & 3 \end{vmatrix}$ are zero

$$\begin{vmatrix} 1 & 2 \\ 1 & 3 \end{vmatrix} = 3 - 2 = 1, \text{ i.e. } \neq 0 \qquad \therefore \ \underline{\text{Rank of } \mathbf{A}_b = 2}$$

$\therefore$ Rank of **A** = rank of $\mathbf{A}_b = 2 < n$ (i.e. 3).

$\therefore$ Infinite number of solutions

Now let us move on to a new section of the work.

Solution of Sets of Equations

21

1. *Inverse method*

Let us work through an example by way of explanation.

Example 1

To solve
$$\begin{aligned}3x_1 + 2x_2 - \ \ x_3 &= \ \ 4\\ 2x_1 - \ \ x_2 + 2x_3 &= 10\\ x_1 - 3x_2 - 4x_3 &= \ \ 5.\end{aligned}$$

We first write this in matrix form, which is

....

22

$$\begin{bmatrix}3 & 2 & -1\\ 2 & -1 & 2\\ 1 & -3 & -4\end{bmatrix}\begin{bmatrix}x_1\\ x_2\\ x_3\end{bmatrix} = \begin{bmatrix}4\\ 10\\ 5\end{bmatrix}$$

Then if $\mathbf{A} = \begin{bmatrix}3 & 2 & -1\\ 2 & -1 & 2\\ 1 & -3 & -4\end{bmatrix}$ then $\begin{bmatrix}x_1\\ x_2\\ x_3\end{bmatrix} = \mathbf{A}^{-1}\begin{bmatrix}4\\ 10\\ 5\end{bmatrix}$

where $\mathbf{A}^{-1}$ is the *inverse* of $\mathbf{A}$.

To find $\mathbf{A}^{-1}$

(a) Form the determinant of $\mathbf{A}$ and evaluate it.

$$|\mathbf{A}| = \begin{vmatrix}3 & 2 & -1\\ 2 & -1 & 2\\ 1 & -3 & -4\end{vmatrix} = 3(4+6) - 2(-8-2) - 1(-6+1) = \underline{55}$$

(b) Form a new matrix $\mathbf{C}$ consisting of the cofactors of the elements in $\mathbf{A}$.

The cofactor of any one element is its minor together with its 'place sign'

i.e. $\mathbf{C} = \begin{bmatrix}A_{11} & A_{12} & A_{13}\\ A_{21} & A_{22} & A_{23}\\ A_{31} & A_{32} & A_{33}\end{bmatrix}$ where A_{11} is the cofactor of a_{11} in $\mathbf{A}$.

$$A_{11} = \begin{vmatrix} -1 & 2 \\ -3 & -4 \end{vmatrix} = 10; \qquad A_{12} = -\begin{vmatrix} 2 & 2 \\ 1 & -4 \end{vmatrix} = 10;$$

$$A_{13} = \begin{vmatrix} 2 & -1 \\ 1 & -3 \end{vmatrix} = -5$$

$$A_{21} = -\begin{vmatrix} 2 & -1 \\ -3 & -4 \end{vmatrix} = 11; \qquad A_{22} = \begin{vmatrix} 3 & -1 \\ 1 & -4 \end{vmatrix} = -11;$$

$$A_{23} = -\begin{vmatrix} 3 & 2 \\ 1 & -3 \end{vmatrix} = 11$$

$A_{31} = \ldots\ldots\ldots$; $\qquad A_{32} = \ldots\ldots\ldots$; $\qquad A_{33} = \ldots\ldots\ldots$;

23

$$A_{31} = \begin{vmatrix} 2 & -1 \\ -1 & 2 \end{vmatrix} = 3; \; A_{32} = -\begin{vmatrix} 3 & -1 \\ 2 & 2 \end{vmatrix} = -8; \; A_{33} = \begin{vmatrix} 3 & 2 \\ 2 & -1 \end{vmatrix} = -7$$

$$\text{So} \quad \mathbf{C} = \begin{bmatrix} 10 & 10 & -5 \\ 11 & -11 & 11 \\ 3 & -8 & -7 \end{bmatrix}$$

We now write the transpose of **C**, i.e. $\mathbf{C}^{\mathrm{T}}$ in which we write rows as columns and columns as rows.

$$\mathbf{C}^{\mathrm{T}} = \ldots\ldots\ldots$$

24

$$\mathbf{C}^{\mathrm{T}} = \begin{bmatrix} 10 & 11 & 3 \\ 10 & -11 & -8 \\ -5 & 11 & -7 \end{bmatrix}$$

This is called the *adjoint* of the original matrix **A**

$$\text{i.e.} \quad \text{adj } \mathbf{A} = \mathbf{C}^{\mathrm{T}}$$

Then the inverse of **A**, i.e. $\mathbf{A}^{-1}$ is given by

$$\mathbf{A}^{-1} = \frac{1}{|\mathbf{A}|} \times \mathbf{C}^{\mathrm{T}} = \frac{1}{55} \begin{bmatrix} 10 & 11 & 3 \\ 10 & -11 & -8 \\ -5 & 11 & -7 \end{bmatrix}$$

So $\begin{bmatrix} x_1 \\ x_2 \\ x_3 \end{bmatrix} = \mathbf{A}^{-1} \begin{bmatrix} 4 \\ 10 \\ 5 \end{bmatrix}$ becomes

$$\begin{bmatrix} x_1 \\ x_2 \\ x_3 \end{bmatrix} = \frac{1}{55} \begin{bmatrix} 10 & 11 & 3 \\ 10 & -11 & -8 \\ -5 & 11 & -7 \end{bmatrix} \begin{bmatrix} 4 \\ 10 \\ 5 \end{bmatrix}$$

$$= \ldots\ldots\ldots$$

25

$$x_1 = 3; \quad x_2 = -2; \quad x_3 = 1$$

since $\begin{bmatrix} x_1 \\ x_2 \\ x_3 \end{bmatrix} = \dfrac{1}{55} \begin{bmatrix} 10 & 11 & 3 \\ 10 & -11 & -8 \\ -5 & 11 & -7 \end{bmatrix} \begin{bmatrix} 4 \\ 10 \\ 5 \end{bmatrix}$

$$= \frac{1}{55} \begin{bmatrix} 40 & +110 & +15 \\ 40 & -110 & -40 \\ -20 & +110 & -35 \end{bmatrix} = \begin{bmatrix} 3 \\ -2 \\ 1 \end{bmatrix}$$

$$\therefore x_1 = 3; \quad x_2 = -2; \quad x_3 = 1$$

The method is the same every time.

To solve $\quad \mathbf{Ax} = \mathbf{b} \qquad \mathbf{x} = \mathbf{A}^{-1}\mathbf{b}$

To find $\mathbf{A}^{-1}$
(i) evaluate $|\mathbf{A}|$
(ii) form $\mathbf{C}$, the matrix of cofactors of $\mathbf{A}$
(iii) write $\mathbf{C}^{\mathrm{T}}$, the transpose of $\mathbf{C}$
(iv) then $\mathbf{A}^{-1} = \dfrac{1}{|\mathbf{A}|} \times \mathbf{C}^{\mathrm{T}}$.

Now apply the method to

Example 2

$$4x_1 + 5x_2 + x_3 = 2$$
$$x_1 - 2x_2 - 3x_3 = 7$$
$$3x_1 - x_2 - 2x_3 = 1.$$

$x_1 = \ldots\ldots\ldots; \qquad x_2 = \ldots\ldots\ldots; \qquad x_3 = \ldots\ldots\ldots$

26

$$\boxed{x_1 = -2; \quad x_2 = 3; \quad x_3 = -5}$$

Here is the complete working.

$$\mathbf{A} = \begin{bmatrix} 4 & 5 & 1 \\ 1 & -2 & -3 \\ 3 & -1 & -2 \end{bmatrix} \quad \therefore \ |\mathbf{A}| = \begin{vmatrix} 4 & 5 & 1 \\ 1 & -2 & -3 \\ 3 & -1 & -2 \end{vmatrix} = \underline{-26}$$

$$\mathbf{C} = \begin{bmatrix} A_{11} & A_{12} & A_{13} \\ A_{21} & A_{22} & A_{23} \\ A_{31} & A_{32} & A_{33} \end{bmatrix}$$

$$A_{11} = \begin{vmatrix} -2 & -3 \\ -1 & -2 \end{vmatrix} = 1 \qquad A_{12} = -\begin{vmatrix} 1 & -3 \\ 3 & -2 \end{vmatrix} = -7 \quad A_{13} = \begin{vmatrix} 1 & -2 \\ 3 & -1 \end{vmatrix} = 5$$

$$A_{21} = -\begin{vmatrix} 5 & 1 \\ -1 & -2 \end{vmatrix} = 9 \quad A_{22} = \begin{vmatrix} 4 & 1 \\ 3 & -2 \end{vmatrix} = -11 \quad A_{23} = -\begin{vmatrix} 4 & 5 \\ 3 & -1 \end{vmatrix} = 19$$

$$A_{31} = \begin{vmatrix} 5 & 1 \\ -2 & -3 \end{vmatrix} = -13 \quad A_{32} = -\begin{vmatrix} 4 & 1 \\ 1 & -3 \end{vmatrix} = 13 \quad A_{33} = \begin{vmatrix} 4 & 5 \\ 1 & -2 \end{vmatrix} = -13$$

$$\therefore \ \mathbf{C} = \begin{bmatrix} 1 & -7 & 5 \\ 9 & -11 & 19 \\ -13 & 13 & -13 \end{bmatrix} \quad \therefore \ \mathbf{C}^{\mathrm{T}} = \begin{bmatrix} 1 & 9 & -13 \\ -7 & -11 & 13 \\ 5 & 19 & -13 \end{bmatrix}$$

$$\mathbf{A}^{-1} = \frac{1}{|\mathbf{A}|} \times \mathbf{C}^{\mathrm{T}} = -\frac{1}{26} \begin{bmatrix} 1 & 9 & -13 \\ -7 & -11 & 13 \\ 5 & 19 & -13 \end{bmatrix}$$

$$\begin{bmatrix} x_1 \\ x_2 \\ x_3 \end{bmatrix} = \mathbf{A}^{-1}\begin{bmatrix} 2 \\ 7 \\ 1 \end{bmatrix} = -\frac{1}{26}\begin{bmatrix} 1 & 9 & -13 \\ -7 & -11 & 13 \\ 5 & 19 & -13 \end{bmatrix}\begin{bmatrix} 2 \\ 7 \\ 1 \end{bmatrix}$$

$$= -\frac{1}{26}\begin{bmatrix} 2 & +63 & -13 \\ -14 & -77 & +13 \\ 10 & +133 & -13 \end{bmatrix}$$

$$= -\frac{1}{26}\begin{bmatrix} 52 \\ -78 \\ 130 \end{bmatrix} = -\begin{bmatrix} 2 \\ 3 \\ 5 \end{bmatrix}$$

$$\therefore\ x_1 = -2; \quad x_2 = -3; \quad x_3 = -5$$

With a set of four equations with four unknowns, the method becomes somewhat tedious as there are then sixteen cofactors to be evaluated and each one is a third-order determinant! There are, however, other methods that can be applied—so let us see method 2.

27

2. *Row transformation method*

Elementary row transformations that can be applied are as follows

(a) Interchange any two rows.
(b) Multiply (or divide) every element in a row by a non-zero scalar (constant) k.
(c) Add to (or subtract from) all the elements of any row k times the corresponding elements of any other row.

Equivalent matrices
Two matrices, **A** and **B**, are said to be equivalent if **B** can be obtained from **A** by a sequence of elementary transformations.

Solutions of equations

The method is best described by working through a typical example.

Example 1 Solve

$$\begin{aligned} 2x_1 + x_2 + x_3 &= 5 \\ x_1 + 3x_2 + 2x_3 &= 1 \\ 3x_1 - 2x_2 - 4x_3 &= -4. \end{aligned}$$

This can be written $\begin{bmatrix} 2 & 1 & 1 \\ 1 & 3 & 2 \\ 3 & -2 & -4 \end{bmatrix} \begin{bmatrix} x_1 \\ x_2 \\ x_3 \end{bmatrix} = \begin{bmatrix} 5 \\ 1 \\ -4 \end{bmatrix}$

and for convenience we introduce the unit matrix

$$\begin{bmatrix} 2 & 1 & 1 \\ 1 & 3 & 2 \\ 3 & -2 & -4 \end{bmatrix} \begin{bmatrix} x_1 \\ x_2 \\ x_3 \end{bmatrix} = \begin{bmatrix} 1 & 0 & 0 \\ 0 & 1 & 0 \\ 0 & 0 & 1 \end{bmatrix} \begin{bmatrix} 5 \\ 1 \\ -4 \end{bmatrix}$$

where $\begin{bmatrix} 1 & 0 & 0 \\ 0 & 1 & 0 \\ 0 & 0 & 1 \end{bmatrix}$ may be regarded as the coefficient of $\begin{bmatrix} 5 \\ 1 \\ -4 \end{bmatrix}$

We then form the combined coefficient matrix

$$\begin{bmatrix} 2 & 1 & 1 & 1 & 0 & 0 \\ 1 & 3 & 2 & 0 & 1 & 0 \\ 3 & -2 & -4 & 0 & 0 & 1 \end{bmatrix}$$

and work on this matrix from now on.

On then to the next frame.

28

The rest of the working is mainly concerned with applying row transformations to convert the left-hand half of the matrix to a unit matrix, eventually obtaining

$$\begin{bmatrix} 1 & 0 & 0 & a & b & c \\ 0 & 1 & 0 & d & e & f \\ 0 & 0 & 1 & g & h & i \end{bmatrix}$$

with $a, b, c, \ldots g, h, i$ being evaluated in the process.

The following notation will be helpful to denote the transformation used:

(1) ~ (2) denotes 'interchange rows 1 and 2'

(3) – 2(1) denotes 'subtract twice row 1 from row 3', etc.

So off we go.

$$(1) \sim (2) \begin{bmatrix} 1 & 3 & 2 & 0 & 1 & 0 \\ 2 & 1 & 1 & 1 & 0 & 0 \\ 3 & -2 & -4 & 0 & 0 & 1 \end{bmatrix} \qquad \begin{matrix} \\ (2)-2(1) \\ (3)-3(1) \end{matrix} \begin{bmatrix} 1 & 3 & 2 & 0 & 1 \\ 0 & -5 & -3 & 1 & -2 \\ 0 & -11 & -10 & 0 & -3 \end{bmatrix}$$

$$\begin{matrix} \\ \\ (3)-2(2) \end{matrix} \begin{bmatrix} 1 & 3 & 2 & 0 & 1 & 0 \\ 0 & -5 & -3 & 1 & -2 & 0 \\ 0 & -1 & -4 & -2 & 1 & 1 \end{bmatrix} \; -(2) \sim -(3) \begin{bmatrix} 1 & 3 & 2 & 0 & 1 \\ 0 & 1 & 4 & 2 & -1 \\ 0 & 5 & 3 & -1 & 2 \end{bmatrix}$$

$$\begin{matrix} \\ \\ (3)-5(2) \end{matrix} \begin{bmatrix} 1 & 3 & 2 & 0 & 1 & 0 \\ 0 & 1 & 4 & 2 & -1 & -1 \\ 0 & 0 & -17 & -11 & 7 & 5 \end{bmatrix}$$

$$\begin{matrix} (1)-3(2) \\ \\ (3)\div(-17) \end{matrix} \begin{bmatrix} 1 & 0 & -10 & -6 & 4 & 3 \\ 0 & 1 & 4 & 2 & -1 & -1 \\ 0 & 0 & 1 & 11/17 & -7/17 & -5/17 \end{bmatrix}$$

$$\begin{matrix} (1)+10(3) \\ (2)-4(3) \\ \\ \end{matrix} \begin{bmatrix} 1 & 0 & 0 & 8/17 & -2/17 & 1/17 \\ 0 & 1 & 0 & -10/17 & 11/17 & 3/17 \\ 0 & 0 & 1 & 11/17 & -7/17 & -5/17 \end{bmatrix}$$

We now have

$$\begin{bmatrix} 1 & 0 & 0 \\ 0 & 1 & 0 \\ 0 & 0 & 1 \end{bmatrix} \begin{bmatrix} x_1 \\ x_2 \\ x_3 \end{bmatrix} = \frac{1}{17} \begin{bmatrix} 8 & -2 & 1 \\ -10 & 11 & 3 \\ 11 & -7 & -5 \end{bmatrix} \begin{bmatrix} 5 \\ 1 \\ -4 \end{bmatrix}$$

$\therefore x_1 = \ldots\ldots\ldots; \qquad x_2 = \ldots\ldots\ldots; \qquad x_3 = \ldots\ldots\ldots$

29

$$x_1 = 2; \qquad x_2 = -3; \qquad x_3 = 4$$

$$\begin{bmatrix} x_1 \\ x_2 \\ x_3 \end{bmatrix} = \frac{1}{17}\begin{bmatrix} 40 & -2 & -4 \\ -50 & +11 & -12 \\ 55 & -7 & +20 \end{bmatrix} = \frac{1}{17}\begin{bmatrix} 34 \\ -51 \\ 68 \end{bmatrix} = \begin{bmatrix} 2 \\ -3 \\ 4 \end{bmatrix}$$

$$\underline{x_1 = 2; \qquad x_2 = -3; \qquad x_3 = 4}$$

Of course, there is no set pattern of how to carry out the row transformations. It depends on one's ingenuity and every case is different. Here is a further example.

Example 2

$$2x_1 - x_2 - 3x_3 = 1$$
$$x_1 + 2x_2 + x_3 = 3$$
$$2x_1 - 2x_2 - 5x_3 = 2.$$

First write the set of equations in matrix form—with the unit matrix included. This gives ....

30

$$\begin{bmatrix} 2 & -1 & -3 \\ 1 & 2 & 1 \\ 2 & -2 & -5 \end{bmatrix}\begin{bmatrix} x_1 \\ x_2 \\ x_3 \end{bmatrix} \begin{bmatrix} 1 & 0 & 0 \\ 0 & 1 & 0 \\ 0 & 0 & 1 \end{bmatrix}\begin{bmatrix} 1 \\ 3 \\ 2 \end{bmatrix}$$

The combined coefficient matrix is now ....

31

$$\begin{bmatrix} 2 & -1 & -3 & 1 & 0 & 0 \\ 1 & 2 & 1 & 0 & 1 & 0 \\ 2 & -2 & -5 & 0 & 0 & 1 \end{bmatrix}$$

If we start off by interchanging the top two rows, we obtain a 1 at the beginning of the top row which is a help.

$$(1) \sim (2) \begin{bmatrix} 1 & 2 & 1 & 0 & 1 & 0 \\ 2 & -1 & -3 & 1 & 0 & 0 \\ 2 & -2 & -5 & 0 & 0 & 1 \end{bmatrix}$$

Now, if we subtract $2 \times$ row 1 from row 2

and $2 \times$ row 1 from row 3, we get

. . . .

32

$$\begin{bmatrix} 1 & 2 & 1 & 0 & 1 & 0 \\ 0 & -5 & -5 & 1 & -2 & 0 \\ 0 & -6 & -7 & 0 & -2 & 1 \end{bmatrix}$$

Continuing with the same line of reasoning, we then have

$$(2)-(3) \begin{bmatrix} 1 & 2 & 1 & 0 & 1 & 0 \\ 0 & 1 & 2 & 1 & 0 & -1 \\ 0 & -6 & -7 & 0 & -2 & 1 \end{bmatrix}$$

$$(3)+6(2) \begin{bmatrix} 1 & 2 & 1 & 0 & 1 & 0 \\ 0 & 1 & 2 & 1 & 0 & -1 \\ 0 & 0 & 5 & 6 & -2 & -5 \end{bmatrix}$$

$$\begin{matrix} (1)-2(2) \\ \\ (3) \div 5 \end{matrix} \begin{bmatrix} 1 & 0 & -3 & -2 & 1 & 2 \\ 0 & 1 & 2 & 1 & 0 & -1 \\ 0 & 0 & 1 & \frac{6}{5} & -\frac{2}{5} & -1 \end{bmatrix}$$

Notice the three diagonal 1's appearing at the left-hand end.

What do you suggest we should do now?

33

Add three times row 3 to row 1
and subtract twice row 3 from row 2.

Right. That gives

$$\begin{array}{l} (1)+3(3) \\ \\ (2)-3(3) \\ \\ \\ \end{array}\begin{bmatrix} 1 & 0 & 0 & \frac{8}{5} & -\frac{1}{5} & -1 \\ & & & & & \\ 0 & 1 & 0 & -\frac{7}{5} & \frac{4}{5} & 1 \\ & & & & & \\ 0 & 0 & 1 & \frac{6}{5} & -\frac{2}{5} & -1 \end{bmatrix}$$

$$\therefore \begin{bmatrix} 1 & 0 & 0 \\ 0 & 1 & 0 \\ 0 & 0 & 1 \end{bmatrix}\begin{bmatrix} x_1 \\ x_2 \\ x_3 \end{bmatrix} = \frac{1}{5}\begin{bmatrix} 8 & -1 & -5 \\ -7 & 4 & 5 \\ 6 & -2 & -5 \end{bmatrix}\begin{bmatrix} 1 \\ 3 \\ 2 \end{bmatrix}$$

Now you can finish it off. $x_1 = \ldots\ldots\ldots$; $x_2 = \ldots\ldots\ldots$; $x_3 = \ldots\ldots\ldots$

34

$x_1 = -1$; $x_2 = 3$; $x_3 = -2$

$$\text{for} \quad \begin{bmatrix} x_1 \\ x_2 \\ x_3 \end{bmatrix} = \frac{1}{5}\begin{bmatrix} 8 & -3 & -10 \\ -7 & +12 & +10 \\ 6 & -6 & -10 \end{bmatrix} = \frac{1}{5}\begin{bmatrix} -5 \\ 15 \\ -10 \end{bmatrix} = \begin{bmatrix} -1 \\ 3 \\ -2 \end{bmatrix}$$

Let us now look at a somewhat similar method with rather fewer steps involved.

So move on.

35 3. *Gaussian elimination method*

Once again we will demonstrate the method by a typical example.

Example 1

$$\begin{aligned} 2x_1 - 3x_2 + 2x_3 &= 9 \\ 3x_1 + 2x_2 - x_3 &= 4 \\ x_1 - 4x_2 + 2x_3 &= 6. \end{aligned}$$

We start off as usual

$$\begin{bmatrix} 2 & -3 & 2 \\ 3 & 2 & -1 \\ 1 & -4 & 2 \end{bmatrix} \begin{bmatrix} x_1 \\ x_2 \\ x_3 \end{bmatrix} = \begin{bmatrix} 9 \\ 4 \\ 6 \end{bmatrix}$$

We then form the augmented coefficient matrix by including the constants as an extra column in the matrix

$$\left[\begin{array}{ccc|c} 2 & -3 & 2 & 9 \\ 3 & 2 & -1 & 4 \\ 1 & -4 & 2 & 6 \end{array}\right]$$

Now we operate on the rows to convert the first three columns into an upper triangular matrix

$$(1) \sim (3) \begin{bmatrix} 1 & -4 & 2 & 6 \\ 3 & 2 & -1 & 4 \\ 2 & -3 & 2 & 9 \end{bmatrix} \quad (2) \sim (3) \begin{bmatrix} 1 & -4 & 2 & 6 \\ 2 & -3 & 2 & 9 \\ 3 & 2 & -1 & 4 \end{bmatrix}$$

$$\begin{matrix} \\ (2) - 2(1) \\ (3) - 3(1) \end{matrix} \begin{bmatrix} 1 & -4 & 2 & 6 \\ 0 & 5 & -2 & -3 \\ 0 & 14 & -7 & -14 \end{bmatrix} \quad \begin{matrix} \\ (2) \div 5 \\ (3) \div 7 \end{matrix} \begin{bmatrix} 1 & -4 & 2 & 6 \\ 0 & 1 & -\frac{2}{5} & -\frac{3}{5} \\ 0 & 2 & -1 & -2 \end{bmatrix}$$

$$\begin{matrix} \\ \\ (3) - 2(2) \end{matrix} \begin{bmatrix} 1 & -4 & 2 & 6 \\ 0 & 1 & -\frac{2}{5} & -\frac{3}{5} \\ 0 & 0 & -\frac{1}{5} & -\frac{4}{5} \end{bmatrix} \quad \begin{matrix} \\ \\ (3) \times (-5) \end{matrix} \begin{bmatrix} 1 & -4 & 2 & 6 \\ 0 & 1 & -\frac{2}{5} & -\frac{3}{5} \\ 0 & 0 & 1 & 4 \end{bmatrix}$$

The first three columns now form an upper triangular matrix which has been our purpose. If we now detach the fourth column back to its original position on the right-hand side of the matrix equation, we have

. . . .

36

$$\begin{bmatrix} 1 & -4 & 2 \\ 0 & 1 & -\frac{2}{5} \\ 0 & 0 & 1 \end{bmatrix} \begin{bmatrix} x_1 \\ x_2 \\ x_3 \end{bmatrix} = \begin{bmatrix} 6 \\ -\frac{3}{5} \\ 4 \end{bmatrix}$$

Expanding from the bottom row, working upwards

$$x_3 = 4 \qquad \therefore \underline{x_3 = 4}$$

$$x_2 - \tfrac{2}{5}x_3 = -\tfrac{3}{5} \qquad \therefore x_2 = -\tfrac{3}{5} + \tfrac{8}{5} = 1 \qquad \therefore \underline{x_2 = 1}$$

$$x_1 - 4x_2 + 2x_3 = 6 \qquad \therefore x_1 - 4 + 8 = 6 \qquad \therefore \underline{x_1 = 2}$$

$$\underline{\therefore x_1 = 2; \quad x_2 = 1; \quad x_3 = 4}$$

It is a very useful method and entails fewer tedious steps, and can be used to solve four equations in four unknowns in the same way.

Example 2

$$\begin{aligned} x_1 + 3x_2 - 2x_3 + x_4 &= -1 \\ 2x_1 - 2x_2 + x_3 - 2x_4 &= 1 \\ x_1 + x_2 - 3x_3 + x_4 &= 6 \\ 3x_1 - x_2 + 2x_3 - x_4 &= 3. \end{aligned}$$

First we write this in matrix form and compile the augmented matrix which is ....

37

$$\left[\begin{array}{cccc|c} 1 & 3 & -2 & 1 & -1 \\ 2 & -2 & 1 & -2 & 1 \\ 1 & 1 & -3 & 1 & 6 \\ 3 & -1 & 2 & -1 & 3 \end{array}\right]$$

Next we operate on rows to convert the left-hand side to an upper triangular matrix. There is no set way of doing this. Use any trickery to save yourself unnecessary work.

So now you can go ahead and complete the transformations and obtain

$x_1 =$ ....; $x_2 =$ ....; $x_3 =$ ....; $x_4 =$ ....

38

$$x_1 = 2; \quad x_2 = -3; \quad x_3 = -1; \quad x_4 = 4$$

Here is one way. You may well have taken quite a different route.

$$\left[\begin{array}{cccc|c} 1 & 3 & -2 & 1 & -1 \\ 2 & -2 & 1 & -2 & 1 \\ 1 & 1 & -3 & 1 & 6 \\ 3 & -1 & 2 & -1 & 3 \end{array}\right]$$

(2) – 2(1)
(3) – (1)
(4) – [(1) + (2)]

$$\left[\begin{array}{cccc|c} 1 & 3 & -2 & 1 & -1 \\ 0 & -8 & 5 & -4 & 3 \\ 0 & -2 & -1 & 0 & 7 \\ 0 & -2 & 3 & 0 & 3 \end{array}\right]$$

(2) – 4(4)
(3) – (4)

$$\left[\begin{array}{cccc|c} 1 & 3 & -2 & 1 & -1 \\ 0 & 0 & -7 & -4 & -9 \\ 0 & 0 & -4 & 0 & 4 \\ 0 & -2 & 3 & 0 & 3 \end{array}\right]$$

(2) ~ (4)
(3) ÷ 4

$$\left[\begin{array}{cccc|c} 1 & 3 & -2 & 1 & -1 \\ 0 & -2 & 3 & 0 & 3 \\ 0 & 0 & -1 & 0 & 1 \\ 0 & 0 & -7 & -4 & -9 \end{array}\right]$$

(4) – 7(3)

$$\left[\begin{array}{cccc|c} 1 & 3 & -2 & 1 & -1 \\ 0 & -2 & 3 & 0 & 3 \\ 0 & 0 & -1 & 0 & 1 \\ 0 & 0 & 0 & -4 & -16 \end{array}\right]$$

Returning the right-hand column to its original position

$$\begin{bmatrix} 1 & 3 & -2 & 1 \\ 0 & -2 & 3 & 0 \\ 0 & 0 & -1 & 0 \\ 0 & 0 & 0 & -4 \end{bmatrix} \begin{bmatrix} x_1 \\ x_2 \\ x_3 \\ x_4 \end{bmatrix} = \begin{bmatrix} -1 \\ 3 \\ 1 \\ -16 \end{bmatrix}$$

Expanding from the bottom row, we have

$$\begin{array}{rrrr} & -4x_4 = -16 & \therefore x_4 = 4 & \therefore \underline{x_4 = 4} \\ & -x_3 = 1 & \therefore x_3 = -1 & \therefore \underline{x_3 = -1} \\ -2x_2 + 3x_3 = 3 & \therefore -2x_2 = 6 & \therefore x_2 = -3 & \therefore \underline{x_2 = -3} \\ x_1 + 3x_2 - 2x_3 + x_4 = -1 & \therefore x_1 - 9 + 2 + 4 = -1 & & \therefore \underline{x_1 = 2} \end{array}$$

$$\therefore \underline{x_1 = 2; \quad x_2 = -3; \quad x_3 = -1; \quad x_4 = 4}$$

39

We still have a further method for solving sets of simultaneous equations.

4. *Triangular decomposition method*

A square matrix **A** can usually be written as a product of a lower-triangular matrix **L** and an upper-triangular matrix **U**, where **A** = **LU**.

For example, if $\mathbf{A} = \begin{bmatrix} 1 & 2 & 3 \\ 3 & 5 & 8 \\ 4 & 9 & 10 \end{bmatrix}$, **A** can be expressed as

$$\mathbf{A} = \mathbf{LU} = \underset{(\mathbf{L})}{\begin{bmatrix} l_{11} & 0 & 0 \\ l_{21} & l_{22} & 0 \\ l_{31} & l_{32} & l_{33} \end{bmatrix}} \underset{(\mathbf{U})}{\begin{bmatrix} u_{11} & u_{12} & u_{13} \\ 0 & u_{22} & u_{23} \\ 0 & 0 & u_{33} \end{bmatrix}}$$

$$= \begin{bmatrix} l_{11}u_{11} & l_{11}u_{12} & l_{11}u_{13} \\ l_{21}u_{11} & l_{21}u_{12} + l_{22}u_{22} & l_{21}u_{13} + l_{22}u_{23} \\ l_{31}u_{11} & l_{31}u_{12} + l_{32}u_{22} & l_{31}u_{13} + l_{32}u_{23} + l_{33}u_{33} \end{bmatrix}$$

Note that, in **L** and **U**, elements occur in the major diagonal in each case. These are related in the product and whatever values we choose to put for $u_{11}, u_{22}, u_{33} \dots$ then the corresponding values of $l_{11}, l_{22}, l_{33} \dots$ will be determined – and vice versa.
For convenience, we put $u_{11} = u_{22} = u_{33} \dots = 1$

$$\text{Then } \mathbf{A} = \mathbf{LU} = \begin{bmatrix} l_{11} & l_{11}u_{12} & l_{11}u_{13} \\ l_{21} & l_{21}u_{12} + l_{22} & l_{21}u_{13} + l_{22}u_{23} \\ l_{31} & l_{31}u_{12} + l_{32} & l_{31}u_{13} + l_{32}u_{23} + l_{33} \end{bmatrix}$$

$$\text{In our example, } \mathbf{A} = \begin{bmatrix} 1 & 2 & 3 \\ 3 & 5 & 8 \\ 4 & 9 & 10 \end{bmatrix}$$

$\therefore l_{11} = 1;\quad l_{11}u_{12} = 2 \quad \therefore u_{12} = 2;\quad l_{11}u_{13} = 3 \quad \therefore u_{13} = 3;$

$l_{21} = 3;\quad$ Similarly $l_{22} = \dots\dots\dots;\quad u_{23} = \dots\dots\dots;$

$l_{31} = 4;\quad l_{32} = \dots\dots\dots;\quad l_{33} = \dots\dots\dots;$

40

$$l_{22} = -1;\quad u_{23} = 1;\quad l_{32} = 1;\quad l_{33} = -3$$

Now we substitute all these values back into the upper and lower triangular matrices and obtain

$$\mathbf{A} = \mathbf{LU} = \dots\dots\dots$$

41

$$\mathbf{A} = \mathbf{LU} = \begin{bmatrix} 1 & 0 & 0 \\ 3 & -1 & 0 \\ 4 & 1 & -3 \end{bmatrix} \begin{bmatrix} 1 & 2 & 3 \\ 0 & 1 & 1 \\ 0 & 0 & 1 \end{bmatrix}$$

We have thus expressed the given matrix **A** as the product of lower and upper triangular matriccs. Let us now see how we use them.

Example 1

$$\begin{aligned} x_1 + 2x_2 + 3x_3 &= 16 \\ 3x_1 + 5x_2 + 8x_3 &= 43 \\ 4x_1 + 9x_2 + 10x_3 &= 57. \end{aligned}$$

$$\text{i.e. } \begin{bmatrix} 1 & 2 & 3 \\ 3 & 5 & 8 \\ 4 & 9 & 10 \end{bmatrix} \begin{bmatrix} x_1 \\ x_2 \\ x_3 \end{bmatrix} = \begin{bmatrix} 16 \\ 43 \\ 57 \end{bmatrix} \text{ i.e. } \mathbf{Ax} = \mathbf{b}.$$

We have seen above that **A** can be written as **LU** where

$$\mathbf{A} = \mathbf{LU} = \begin{bmatrix} 1 & 0 & 0 \\ 3 & -1 & 0 \\ 4 & 1 & -3 \end{bmatrix} \begin{bmatrix} 1 & 2 & 3 \\ 0 & 1 & 1 \\ 0 & 0 & 1 \end{bmatrix}$$

To solve $\mathbf{Ax} = \mathbf{b}$, we have $\mathbf{LUx} = \mathbf{b}$ i.e. $\mathbf{L}(\mathbf{Ux}) = \mathbf{b}$
Putting $\mathbf{Ux} = \mathbf{y}$, we solve $\mathbf{Ly} = \mathbf{b}$ to obtain $\mathbf{y}$
and then $\mathbf{Ux} = \mathbf{y}$ to obtain $\mathbf{x}$.

(a) Solving $\mathbf{Ly} = \mathbf{b}$

$$\begin{bmatrix} 1 & 0 & 0 \\ 3 & -1 & 0 \\ 4 & 1 & -3 \end{bmatrix} \begin{bmatrix} y_1 \\ y_2 \\ y_3 \end{bmatrix} = \begin{bmatrix} 16 \\ 43 \\ 57 \end{bmatrix}$$

Expanding from the top $y_1 = 16$; $3y_1 - y_2 = 43$ $\therefore y_2 = 5$;
and $4y_1 + y_2 - 3y_3 = 57$ $\therefore 64 + 5 - 3y_3 = 57$ $\therefore y_3 = 4$

$$\therefore \begin{bmatrix} y_1 \\ y_2 \\ y_3 \end{bmatrix} = \begin{bmatrix} 16 \\ 5 \\ 4 \end{bmatrix}$$

(b) Solving $\mathbf{Ux} = \mathbf{y}$

$$\begin{bmatrix} 1 & 2 & 3 \\ 0 & 1 & 1 \\ 0 & 0 & 1 \end{bmatrix} \begin{bmatrix} x_1 \\ x_2 \\ x_3 \end{bmatrix} = \begin{bmatrix} 16 \\ 5 \\ 4 \end{bmatrix}$$

Expanding from the bottom, we then have

$x_1 = \ldots\ldots\ldots$; $x_2 = \ldots\ldots\ldots$; $x_3 = \ldots\ldots\ldots$

42

$$x_1 = 2; \quad x_2 = 1; \quad x_3 = 4$$

Note:

1. If $l_{ii} = 0$, then either decomposition is not possible, or, if **A** is singular, i.e. $|\mathbf{A}| = 0$, there is an infinite number of possible decompositions.
2. Instead of putting $u_{11} = u_{22} = u_{33} \ldots = 1$, we could have used the alternative substitution $l_{11} = l_{22} = l_{33} \ldots = 1$ and obtained values of u_{11}, u_{22}, u_{33} ... etc. The working is as before.
3. One advantage of employing **LU** decomposition over gaussian elimination is in the solution of a sequence of problems in which the same coefficient matrix occurs.

Now for another example.

43

Example 2

$$x_1 + 3x_2 + 2x_3 = 19$$
$$2x_1 + x_2 + x_3 = 13$$
$$4x_1 + 2x_2 + 3x_3 = 31.$$

$$\therefore \begin{bmatrix} 1 & 3 & 2 \\ 2 & 1 & 1 \\ 4 & 2 & 3 \end{bmatrix} \begin{bmatrix} x_1 \\ x_2 \\ x_3 \end{bmatrix} = \begin{bmatrix} 19 \\ 13 \\ 31 \end{bmatrix} \text{ i.e. } \mathbf{Ax} = \mathbf{b}$$

$$\mathbf{A} = \mathbf{LU} = \begin{bmatrix} l_{11} & 0 & 0 \\ l_{21} & l_{22} & 0 \\ l_{31} & l_{32} & l_{33} \end{bmatrix} \begin{bmatrix} 1 & u_{12} & u_{13} \\ 0 & 1 & u_{23} \\ 0 & 0 & 1 \end{bmatrix}$$

$$= \begin{bmatrix} l_{11} & l_{11}u_{12} & l_{11}u_{13} \\ l_{21} & l_{21}u_{12} + l_{22} & l_{21}u_{13} + l_{22}u_{23} \\ l_{31} & l_{31}u_{12} + l_{32} & l_{31}u_{13} + l_{32}u_{23} + l_{33} \end{bmatrix}$$

$$= \begin{bmatrix} 1 & 3 & 2 \\ 2 & 1 & 1 \\ 4 & 2 & 3 \end{bmatrix}$$

Now we have to find the values of the various elements. The usual order of doing this is shown by the diagram.

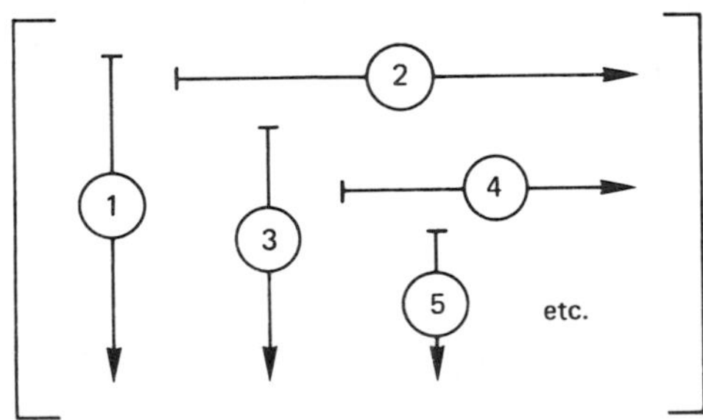

That is, first we can write down values for l_{11}, l_{21}, l_{31} from the left-hand column; then follow this by finding u_{12}, u_{13} from the top row; and proceed for the others.

So, completing the two triangular matrices, we have

$$\mathbf{A} = \mathbf{LU} = \ldots\ldots\ldots$$

44

$$\mathbf{A} = \mathbf{LU} = \begin{bmatrix} 1 & 0 & 0 \\ 2 & -5 & 0 \\ 4 & -10 & 1 \end{bmatrix} \begin{bmatrix} 1 & 3 & 2 \\ 0 & 1 & \frac{3}{5} \\ 0 & 0 & 1 \end{bmatrix}$$

As we stated before: $\mathbf{Ax} = \mathbf{b}$; $\mathbf{L}(\mathbf{Ux}) = \mathbf{b}$. Put $\mathbf{Ux} = \mathbf{y}$
then (a) solve $\mathbf{Ly} = \mathbf{b}$ to obtain $\mathbf{y}$
and (b) solve $\mathbf{Ux} = \mathbf{y}$ to obtain $\mathbf{x}$.

Solving $\mathbf{Ly} = \mathbf{b}$, gives $\begin{bmatrix} y_1 \\ y_2 \\ y_3 \end{bmatrix} = \begin{bmatrix} \dots \\ \dots \\ \dots \end{bmatrix}$

45

$$\begin{bmatrix} y_1 \\ y_2 \\ y_3 \end{bmatrix} = \begin{bmatrix} 19 \\ 5 \\ 5 \end{bmatrix}$$

for $\begin{bmatrix} 1 & 0 & 0 \\ 2 & -5 & 0 \\ 4 & -10 & 1 \end{bmatrix} \begin{bmatrix} y_1 \\ y_2 \\ y_3 \end{bmatrix} = \begin{bmatrix} 19 \\ 13 \\ 31 \end{bmatrix}$. Expanding from the top gives

$$y_1 = 19; \qquad y_2 = 5; \qquad y_3 = 5.$$

(b) Now solve $\mathbf{Ux} = \mathbf{y}$ from which $x_1 = \dots\dots\dots$; $x_2 = \dots\dots\dots$; $x_3 = \dots\dots\dots$

46

$$\boxed{x_1 = 3; \quad x_2 = 2; \quad x_3 = 5}$$

for we have $\quad \mathbf{Ux} = \mathbf{y}$

i.e.
$$\begin{bmatrix} 1 & 3 & 2 \\ 0 & 1 & \frac{3}{5} \\ 0 & 0 & 1 \end{bmatrix} \begin{bmatrix} x_1 \\ x_2 \\ x_3 \end{bmatrix} = \begin{bmatrix} 19 \\ 5 \\ 5 \end{bmatrix}$$

Expanding from the bottom $\quad x_3 = 5;\ x_2 + \frac{3}{5}x_3 = 5 \quad \therefore x_2 = 2$

and $x_1 + 3x_2 + 2x_3 = 19 \quad \therefore x_1 + 6 + 10 = 19 \quad \therefore x_1 = 3$

$$\therefore \underline{x_1 = 3; \quad x_2 = 2; \quad x_3 = 5}$$

We can of course apply the same method to a set of four equations.

Example 3
$$\begin{aligned} x_1 + 2x_2 - x_3 + 3x_4 &= 9 \\ 2x_1 - x_2 + 3x_3 + 2x_4 &= 23 \\ 3x_1 + 3x_2 + x_3 + x_4 &= 5 \\ 4x_1 + 5x_2 - 2x_3 + 2x_4 &= -2. \end{aligned}$$

i.e.
$$\begin{bmatrix} 1 & 2 & -1 & 3 \\ 2 & -1 & 3 & 2 \\ 3 & 3 & 1 & 1 \\ 4 & 5 & -2 & 2 \end{bmatrix} \begin{bmatrix} x_1 \\ x_2 \\ x_3 \\ x_4 \end{bmatrix} = \begin{bmatrix} 9 \\ 23 \\ 5 \\ -2 \end{bmatrix} \quad \text{i.e. } \mathbf{Ax} = \mathbf{b}$$

$$\mathbf{A} = \mathbf{LU} = \begin{bmatrix} l_{11} & 0 & 0 & 0 \\ l_{21} & l_{22} & 0 & 0 \\ l_{31} & l_{32} & l_{33} & 0 \\ l_{41} & l_{42} & l_{43} & l_{44} \end{bmatrix} \begin{bmatrix} 1 & u_{12} & u_{13} & u_{14} \\ 0 & 1 & u_{23} & u_{24} \\ 0 & 0 & 1 & u_{34} \\ 0 & 0 & 0 & 1 \end{bmatrix} = \begin{bmatrix} 1 & 2 & -1 & 3 \\ 2 & -1 & 3 & 2 \\ 3 & 3 & 1 & 1 \\ 4 & 5 & -2 & 2 \end{bmatrix}$$

$$\mathbf{A} = \begin{bmatrix} l_{11} & l_{11}u_{12} & l_{11}u_{13} & l_{11}u_{14} \\ l_{21} & l_{21}u_{12} + l_{22} & l_{21}u_{13} + l_{22}u_{23} & l_{21}u_{14} + l_{22}u_{24} \\ l_{31} & l_{31}u_{12} + l_{32} & l_{31}u_{13} + l_{32}u_{23} + l_{33} & l_{31}u_{14} + l_{32}u_{24} + l_{33}u_{34} \\ l_{41} & l_{41}u_{12} + l_{42} & l_{41}u_{13} + l_{42}u_{23} + l_{43} & l_{41}u_{14} + l_{42}u_{24} + l_{43}u_{34} + l_{44} \end{bmatrix}$$

Now we have to find the values of the individual elements. It is easy enough if we follow the order indicated in the diagram earlier. So the two triangular matrices are

$$\mathbf{A} = \mathbf{LU} = [\dots\dots\dots]\ [\dots\dots\dots]$$

47

$$\mathbf{A}=\mathbf{LU}=\begin{bmatrix}1&0&0&0\\2&-5&0&0\\3&-3&1&0\\4&-3&-1&-\frac{66}{5}\end{bmatrix}\begin{bmatrix}1&2&-1&3\\0&1&-1&\frac{4}{5}\\0&0&1&-\frac{28}{5}\\0&0&0&1\end{bmatrix}$$

As usual $\mathbf{Ax}=\mathbf{b}$; $\mathbf{L}(\mathbf{Ux})=\mathbf{b}$. Put $\mathbf{Ux}=\mathbf{y}$ $\therefore\ \mathbf{Ly}=\mathbf{b}$

(a) Solving $\mathbf{Ly}=\mathbf{b}$

$$\begin{bmatrix}1&0&0&0\\2&-5&0&0\\3&-3&1&0\\4&-3&-1&-\frac{66}{5}\end{bmatrix}\begin{bmatrix}y_1\\y_2\\y_3\\y_4\end{bmatrix}=\begin{bmatrix}9\\23\\5\\-2cr\end{bmatrix}$$

$$\therefore\ \begin{bmatrix}y_1\\y_2\\y_3\\y_4\end{bmatrix}=\begin{bmatrix}\dots\\\dots\\\dots\\\dots\end{bmatrix}$$

48

$$\begin{bmatrix}y_1\\y_2\\y_3\\y_4\end{bmatrix}=\begin{bmatrix}9\\-1\\-25\\5\end{bmatrix}$$

(b) Solving $\mathbf{Ux}=\mathbf{y}$

$$\begin{bmatrix}1&2&-1&3\\0&1&-1&\frac{4}{5}\\0&0&1&-\frac{28}{5}\\0&0&0&1\end{bmatrix}\begin{bmatrix}x_1\\x_2\\x_3\\x_4\end{bmatrix}=\begin{bmatrix}9\\-1\\-25\\5\end{bmatrix}$$

which finally gives $x_1=\dots\dots$; $x_2=\dots\dots$; $x_3=\dots\dots$; $x_4=\dots\dots$

49

$x_1 = 1; \quad x_2 = -2; \quad x_3 = 3; \quad x_4 = 5$

Now let us proceed to something rather different, so move on to the next frame for a new start.

Eigenvalues and Eigenvectors

50

Matrices commonly appear in technological problems, for example those involving coupled oscillations and vibrations, and give rise to equations of the form

$$\mathbf{Ax} = \lambda\mathbf{x}$$

where $\mathbf{A} = [a_{ij}]$ is a square matrix, $\mathbf{x}$ is a column matrix $[x_i]$ and λ is a scalar quantity, i.e. a number.

For non-trivial solutions, i.e. for $x \neq 0$, the values of λ are called the *eigenvalues*, *characteristic values*, or *latent roots* of the matrix $\mathbf{A}$ and the corresponding solutions of the given equations $\mathbf{Ax} = \lambda\mathbf{x}$ are called the *eigenvectors*, or *characteristic vectors* of $\mathbf{A}$.

The set of equations

$$\begin{bmatrix} a_{11} & a_{12} & \dots & a_{1n} \\ a_{21} & a_{22} & \dots & a_{2n} \\ \cdot & \cdot & & \cdot \\ \cdot & \cdot & & \cdot \\ \cdot & \cdot & & \cdot \\ a_{n1} & a_{n2} & \dots & a_{nn} \end{bmatrix} \begin{bmatrix} x_1 \\ x_2 \\ \cdot \\ \cdot \\ \cdot \\ x_n \end{bmatrix} = \lambda \begin{bmatrix} x_1 \\ x_2 \\ \cdot \\ \cdot \\ \cdot \\ x_n \end{bmatrix}$$

then simplifies to

$$\begin{bmatrix} (a_{11}-\lambda) & a_{12} & \dots & a_{1n} \\ a_{21} & (a_{22}-\lambda) & \dots & a_{2n} \\ \cdot & \cdot & & \cdot \\ \cdot & \cdot & & \cdot \\ \cdot & \cdot & & \cdot \\ a_{n1} & a_{n2} & & (a_{nn}-\lambda) \end{bmatrix} \begin{bmatrix} x_1 \\ x_2 \\ \cdot \\ \cdot \\ \cdot \\ x_n \end{bmatrix} = \begin{bmatrix} 0 \\ 0 \\ \cdot \\ \cdot \\ \cdot \\ 0 \end{bmatrix}$$

That is, $\mathbf{Ax} = \lambda\mathbf{x}$ becomes $\mathbf{Ax} - \lambda\mathbf{x} = \mathbf{0}$

i.e. $\underline{(\mathbf{A} - \lambda\mathbf{I})\mathbf{x} = \mathbf{0}}$

the unit matrix $\mathbf{I}$ being introduced since we can subtract only a matrix from another matrix.

For this set of homogeneous linear equations (right-hand side constant terms all zero) to have non-trivial solutions

$$|\mathbf{A} - \lambda\mathbf{I}| \text{ must be zero}$$

This is called the *characteristic determinant* of $\mathbf{A}$ and $|\mathbf{A} - \lambda\mathbf{I}| = 0$ is the *characteristic equation*, the solution of which gives the values of λ, i.e. the eigenvalues of $\mathbf{A}$.

---

51 *Example 1* Find the eigenvalues and corresponding eigenvectors of $\mathbf{Ax} = \lambda\mathbf{x}$ where $\mathbf{A} = \begin{bmatrix} 2 & 3 \\ 4 & 1 \end{bmatrix}$.

The characteristic equation is $|\mathbf{A} - \lambda\mathbf{I}| = 0$

i.e. $\begin{vmatrix} 2-\lambda & 3 \\ 4 & 1-\lambda \end{vmatrix} = 0$, which, when expanded, gives

$$\lambda_1 = \dots\dots\dots \quad \text{and} \quad \lambda_2 = \dots\dots\dots$$

52

$$\boxed{\lambda_1 = -2 \quad \text{and} \quad \lambda_2 = 5}$$

for $(2-\lambda)(1-\lambda) - 12 = 0 \quad \therefore 2 - 3\lambda + \lambda^2 - 12 = 0$

$\lambda^2 - 3\lambda - 10 = 0 \quad (\lambda - 5)(\lambda + 2) = 0 \qquad \therefore \underline{\lambda = -2 \text{ or } 5}$

Now we substitute each value of λ in turn in the equation

$$(\mathbf{A} - \lambda\mathbf{I})\mathbf{x} = 0$$

With $\lambda = -2$

$$\left\{\begin{bmatrix} 2 & 3 \\ 4 & 1 \end{bmatrix} - (-2)\begin{bmatrix} 1 & 0 \\ 0 & 1 \end{bmatrix}\right\}\begin{bmatrix} x_1 \\ x_2 \end{bmatrix} = \begin{bmatrix} 0 \\ 0 \end{bmatrix}$$

$$\left\{\begin{bmatrix} 2 & 3 \\ 4 & 1 \end{bmatrix} + \begin{bmatrix} 2 & 0 \\ 0 & 2 \end{bmatrix}\right\}\begin{bmatrix} x_1 \\ x_2 \end{bmatrix} = \begin{bmatrix} 0 \\ 0 \end{bmatrix}$$

$$\begin{bmatrix} 4 & 3 \\ 4 & 3 \end{bmatrix}\begin{bmatrix} x_1 \\ x_2 \end{bmatrix} = \begin{bmatrix} 0 \\ 0 \end{bmatrix}$$

Multiplying out the left-hand side, we get

. . . .

53

$$\boxed{4x_1 + 3x_2 = 0}$$

from which we get $x_2 = -\frac{4}{3}x_1$ i.e. not specific values for x_1 and x_2, but a relationship between them. Whatever value we assign to x_1 we obtain a corresponding value of x_2.

$$\mathbf{x}_1 = \begin{bmatrix} x_1 \\ x_2 \end{bmatrix} = \begin{bmatrix} 3 \\ -4 \end{bmatrix} \text{ or } \begin{bmatrix} 6 \\ -8 \end{bmatrix} \text{ or } \begin{bmatrix} 9 \\ -12 \end{bmatrix}, \quad \text{etc.}$$

i.e. $\mathbf{x}_1 = \alpha\begin{bmatrix} 3 \\ -4 \end{bmatrix}$ where α is a constant multiplifer.

The simplest result, with $\alpha = 1$, is the one normally quoted.

$$\therefore \text{ for } \lambda_1 = -2, \quad \mathbf{x}_1 = \begin{bmatrix} 3 \\ -4 \end{bmatrix}$$

Similarly, for $\lambda_2 = 5$, the corresponding eigenvector is ....

54

$$\boxed{\mathbf{x}_2 = \begin{bmatrix} 1 \\ 1 \end{bmatrix}}$$

for, with $\lambda_2 = 5$, $(\mathbf{A} - \lambda\mathbf{I})\mathbf{x} = \mathbf{0}$ becomes

$$\left\{\begin{bmatrix} 2 & 3 \\ 4 & 1 \end{bmatrix} - 5\begin{bmatrix} 1 & 0 \\ 0 & 1 \end{bmatrix}\right\}\begin{bmatrix} x_1 \\ x_2 \end{bmatrix} = \begin{bmatrix} 0 \\ 0 \end{bmatrix}$$

$$\left\{\begin{bmatrix} 2 & 3 \\ 4 & 1 \end{bmatrix} - \begin{bmatrix} 5 & 0 \\ 0 & 5 \end{bmatrix}\right\}\begin{bmatrix} x_1 \\ x_2 \end{bmatrix} = \begin{bmatrix} 0 \\ 0 \end{bmatrix}$$

$$\begin{bmatrix} -3 & 3 \\ 4 & -4 \end{bmatrix}\begin{bmatrix} x_1 \\ x_2 \end{bmatrix} = \begin{bmatrix} 0 \\ 0 \end{bmatrix}$$

$\therefore\ -3x_1 + 3x_2 = 0$ i.e. $x_2 = x_1$

$\therefore$ with $\lambda_2 = 5$, the corresponding eigenvector is $\mathbf{x}_2 = \beta\begin{bmatrix} 1 \\ 1 \end{bmatrix}$

Again, taking $\beta = 1$, for $\underline{\lambda_2 = 5, \quad \mathbf{x}_2 = \begin{bmatrix} 1 \\ 1 \end{bmatrix}}$

So the required eigenvectors are

$$\mathbf{x}_1 = \begin{bmatrix} 3 \\ -4 \end{bmatrix} \text{ corresponding to } \lambda_1 = -2$$

$$\mathbf{x}_2 = \begin{bmatrix} 1 \\ 1 \end{bmatrix} \text{ corresponding to } \lambda_2 = 5.$$

Example 2 Determine the eigenvalues and corresponding eigenvectors of

$$\mathbf{A}\mathbf{x} = \lambda\mathbf{x} \quad \text{where} \quad \mathbf{A} = \begin{bmatrix} 3 & 10 \\ 2 & 4 \end{bmatrix}.$$

The characteristic equation is $|\mathbf{A} - \lambda\mathbf{I}| = 0$, which in this case, can be written as ....

55

$$\begin{vmatrix} 3-\lambda & 10 \\ 2 & 4-\lambda \end{vmatrix} = 0$$

Expanding the determinant and solving the equation, gives

$$\lambda_1 = \dots\dots\dots; \qquad \lambda_2 = \dots\dots\dots$$

56

$$\lambda_1 = -1; \quad \lambda_2 = 8$$

for the equation is $(3-\lambda)(4-\lambda) - 20 = 0 \quad \therefore \lambda^2 - 7\lambda - 8 = 0$

$$\therefore (\lambda+1)(\lambda-8) = 0 \quad \therefore \underline{\lambda = -1 \text{ or } 8}$$

(a) With $\lambda_1 = -1$, we solve $(\mathbf{A} - \lambda\mathbf{I})\mathbf{x} = \mathbf{0}$ to obtain an eigenvector, which is ………

57

$$\mathbf{x}_1 = \begin{bmatrix} 5 \\ -2 \end{bmatrix}$$

$$\mathbf{A} = \begin{bmatrix} 3 & 10 \\ 2 & 4 \end{bmatrix} \therefore \left\{ \begin{bmatrix} 3 & 10 \\ 2 & 4 \end{bmatrix} - (-1)\begin{bmatrix} 1 & 0 \\ 0 & 1 \end{bmatrix} \right\} \begin{bmatrix} x_1 \\ x_2 \end{bmatrix} = \begin{bmatrix} 0 \\ 0 \end{bmatrix}$$

$$\left\{ \begin{bmatrix} 3 & 10 \\ 2 & 4 \end{bmatrix} + \begin{bmatrix} 1 & 0 \\ 0 & 1 \end{bmatrix} \right\} \begin{bmatrix} x_1 \\ x_2 \end{bmatrix} = \begin{bmatrix} 0 \\ 0 \end{bmatrix}$$

$$\begin{bmatrix} 4 & 10 \\ 2 & 5 \end{bmatrix} \begin{bmatrix} x_1 \\ x_2 \end{bmatrix} = \begin{bmatrix} 0 \\ 0 \end{bmatrix}$$

$$\therefore 4x_1 + 10x_2 = 0 \quad \therefore x_2 = -\frac{2}{5}x_1 \quad \mathbf{x}_1 = \alpha \begin{bmatrix} 5 \\ -2 \end{bmatrix}$$

$$\therefore \text{with } \alpha = 1 \quad \underline{\lambda_1 = -1 \quad \text{and} \quad \mathbf{x}_1 = \begin{bmatrix} 5 \\ -2 \end{bmatrix}}$$

(b) In the same way the corresponding eigenvector $\mathbf{x}_2$ for $\lambda_2 = 8$ is

………

58

$$\boxed{\mathbf{x}_2 = \begin{bmatrix} 2 \\ 1 \end{bmatrix}}$$

working :
$$\left\{\begin{bmatrix} 3 & 10 \\ 2 & 4 \end{bmatrix} - 8\begin{bmatrix} 1 & 0 \\ 0 & 1 \end{bmatrix}\right\}\begin{bmatrix} x_1 \\ x_2 \end{bmatrix} = \begin{bmatrix} 0 \\ 0 \end{bmatrix}$$

$$\left\{\begin{bmatrix} 3 & 10 \\ 2 & 4 \end{bmatrix} - \begin{bmatrix} 8 & 0 \\ 0 & 8 \end{bmatrix}\right\}\begin{bmatrix} x_1 \\ x_2 \end{bmatrix} = \begin{bmatrix} 0 \\ 0 \end{bmatrix}$$

$$\begin{bmatrix} -5 & 10 \\ 2 & -4 \end{bmatrix}\begin{bmatrix} x_1 \\ x_2 \end{bmatrix} = \begin{bmatrix} 0 \\ 0 \end{bmatrix}$$

$$\therefore\ -5x_1 + 10x_2 = 0 \quad \therefore\ x_2 = \frac{1}{2}x_1 \quad \mathbf{x}_2 = \beta\begin{bmatrix} 2 \\ 1 \end{bmatrix}$$

$$\therefore \text{ with } \beta = 1, \quad \underline{\lambda_2 = 8 \quad \text{and} \quad \mathbf{x}_2 = \begin{bmatrix} 2 \\ 1 \end{bmatrix}}$$

The same basic method can similarly be applied to third-order sets of equations.

Example 3 Determine the eigenvalues and eigenvectors of $\mathbf{Ax} = \lambda\mathbf{x}$ where $\mathbf{A} = \begin{bmatrix} 1 & 0 & 4 \\ 0 & 2 & 0 \\ 3 & 1 & -3 \end{bmatrix}$.

As before, we have $(\mathbf{A} - \lambda\mathbf{I})\mathbf{x} = \mathbf{0}$ with characteristic equation $|\mathbf{A} - \lambda\mathbf{I}| = 0$.

i.e.
$$\begin{vmatrix} 1-\lambda & 0 & 4 \\ 0 & 2-\lambda & 0 \\ 3 & 1 & -3-\lambda \end{vmatrix} = 0$$

Expanding this we have

$$\lambda_1 = \dots\dots\dots; \quad \lambda_2 = \dots\dots\dots; \quad \lambda_3 = \dots\dots\dots$$

59

$$\boxed{\lambda_1 = 2; \quad \lambda_2 = 3; \quad \lambda_3 = -5}$$

for $(1-\lambda)\{(2-\lambda)(-3-\lambda) - 0\} + 4\{0 - 3(2-\lambda)\} = 0$

$$(1-\lambda)(2-\lambda)(-3-\lambda) - 12(2-\lambda) = 0$$

$$\therefore\ (2-\lambda)\{(1-\lambda)(-3-\lambda) - 12\} = 0$$

$$\therefore\ \lambda = 2 \quad \text{or} \quad \lambda^2 + 2\lambda - 15 = 0 \quad \therefore\ (\lambda-3)(\lambda+5) = 0$$

$$\therefore\ \underline{\lambda = 2,\ 3,\ \text{or}\ -5}$$

(a) With $\lambda_1 = 2$, $\quad (\mathbf{A} - \lambda\mathbf{I})\mathbf{x} = \mathbf{0}$ becomes

$$\left\{\begin{bmatrix} 1 & 0 & 4 \\ 0 & 2 & 0 \\ 3 & 1 & -3 \end{bmatrix} - 2\begin{bmatrix} 1 & 0 & 0 \\ 0 & 1 & 0 \\ 0 & 0 & 1 \end{bmatrix}\right\}\begin{bmatrix} x_1 \\ x_2 \\ x_3 \end{bmatrix} = \begin{bmatrix} 0 \\ 0 \\ 0 \end{bmatrix}$$

$$\left\{\begin{bmatrix} 1 & 0 & 4 \\ 0 & 2 & 0 \\ 3 & 1 & -3 \end{bmatrix} - \begin{bmatrix} 2 & 0 & 0 \\ 0 & 2 & 0 \\ 0 & 0 & 2 \end{bmatrix}\right\}\begin{bmatrix} x_1 \\ x_2 \\ x_3 \end{bmatrix} = \begin{bmatrix} 0 \\ 0 \\ 0 \end{bmatrix}$$

$$\therefore\ \begin{bmatrix} -1 & 0 & 4 \\ 0 & 0 & 0 \\ 3 & 1 & -5 \end{bmatrix}\begin{bmatrix} x_1 \\ x_2 \\ x_3 \end{bmatrix} = \begin{bmatrix} 0 \\ 0 \\ 0 \end{bmatrix}$$

from which a corresponding eigenvector $\mathbf{x}_1$ is ....

60

$$\mathbf{x}_1 = \begin{bmatrix} 4 \\ -7 \\ 1 \end{bmatrix}$$

for we have $-x_1 + 4x_3 = 0$ $\therefore\ x_3 = \frac{1}{4}x_1$

$3x_1 + x_2 - 5x_3 = 0 \quad \therefore\ 3x_1 + x_2 - \frac{5}{4}x_1 = 0 \quad \therefore\ x_2 = -\frac{7}{4}x_1$

$\therefore\ x_1,\ x_2,\ x_3$ are in the ratio $1 : -\frac{7}{4} : \frac{1}{4}$ i.e. $4 : -7 : 1$ $\quad \therefore\ \mathbf{x}_1 = \begin{bmatrix} 4 \\ -7 \\ 1 \end{bmatrix}$

(b) Similarly for $\lambda_2 = 3,\ (\mathbf{A} - \lambda\mathbf{I})\mathbf{x} = \mathbf{0}$

$$\left\{\begin{bmatrix} 1 & 0 & 4 \\ 0 & 2 & 0 \\ 3 & 1 & -3 \end{bmatrix} - 3\begin{bmatrix} 1 & 0 & 0 \\ 0 & 1 & 0 \\ 0 & 0 & 1 \end{bmatrix}\right\}\begin{bmatrix} x_1 \\ x_2 \\ x_3 \end{bmatrix} = \begin{bmatrix} 0 \\ 0 \\ 0 \end{bmatrix}$$

from which a corresponding eigenvector is

$\mathbf{x}_2 = \dots\dots\dots$

61

$$\mathbf{x}_2 = \begin{bmatrix} 2 \\ 0 \\ 1 \end{bmatrix}$$

for $$\left\{\begin{bmatrix} 1 & 0 & 4 \\ 0 & 2 & 0 \\ 3 & 1 & -3 \end{bmatrix} - \begin{bmatrix} 3 & 0 & 0 \\ 0 & 3 & 0 \\ 0 & 0 & 3 \end{bmatrix}\right\}\begin{bmatrix} x_1 \\ x_2 \\ x_3 \end{bmatrix} = \begin{bmatrix} 0 \\ 0 \\ 0 \end{bmatrix}$$

$$\begin{bmatrix} -2 & 0 & 4 \\ 0 & -1 & 0 \\ 3 & 1 & -6 \end{bmatrix}\begin{bmatrix} x_1 \\ x_2 \\ x_3 \end{bmatrix} = \begin{bmatrix} 0 \\ 0 \\ 0 \end{bmatrix}$$

$\therefore\ -2x_1 + 4x_3 = 0 \qquad \therefore\ x_3 = \frac{1}{2}x_1$

Also $\quad -x_2 = 0 \qquad \therefore\ x_2 = 0 \qquad \therefore\ \mathbf{x}_2 = \begin{bmatrix} 2 \\ 0 \\ 1 \end{bmatrix}$

(c) All that now remains is $\lambda_3 = -5$. A corresponding eigenvector $\mathbf{x}_3$ is

$\mathbf{x}_3 = \dots\dots\dots$

Finish it on your own. Method just the same as before.

62

$$\mathbf{x_3} = \begin{bmatrix} 2 \\ 0 \\ -3 \end{bmatrix}$$

Check the working.

$$\mathbf{A} = \begin{bmatrix} 1 & 0 & 4 \\ 0 & 2 & 0 \\ 3 & 1 & -3 \end{bmatrix} \quad \text{and} \quad \lambda_3 = -5 \quad \text{with} \quad (\mathbf{A} - \lambda\mathbf{I})\mathbf{x} = \mathbf{0}.$$

$$\left\{ \begin{bmatrix} 1 & 0 & 4 \\ 0 & 2 & 0 \\ 3 & 1 & -3 \end{bmatrix} + 5 \begin{bmatrix} 1 & 0 & 0 \\ 0 & 1 & 0 \\ 0 & 0 & 1 \end{bmatrix} \right\} \begin{bmatrix} x_1 \\ x_2 \\ x_3 \end{bmatrix} = \begin{bmatrix} 0 \\ 0 \\ 0 \end{bmatrix}$$

$$\begin{bmatrix} 6 & 0 & 4 \\ 0 & 7 & 0 \\ 3 & 1 & 2 \end{bmatrix} \begin{bmatrix} x_1 \\ x_2 \\ x_3 \end{bmatrix} = \begin{bmatrix} 0 \\ 0 \\ 0 \end{bmatrix}$$

$\therefore 6x_1 + 4x_3 - 0 \qquad \therefore x_3 = -\frac{3}{2}x_1$

$7x_2 = 0 \qquad \therefore x_2 = 0 \qquad \therefore \mathbf{x_3} = \begin{bmatrix} 2 \\ 0 \\ -3 \end{bmatrix}$

Collecting the results together, we finally have

$$\lambda_1 = 2,\ \mathbf{x_1} = \begin{bmatrix} 4 \\ -7 \\ 1 \end{bmatrix};\quad \lambda_2 = 3,\ \mathbf{x_2} = \begin{bmatrix} 2 \\ 0 \\ 1 \end{bmatrix};\quad \lambda_3 = -5,\ \mathbf{x_3} = \begin{bmatrix} 2 \\ 0 \\ -3 \end{bmatrix}$$

Now for something new.

Matrix Transformation

63

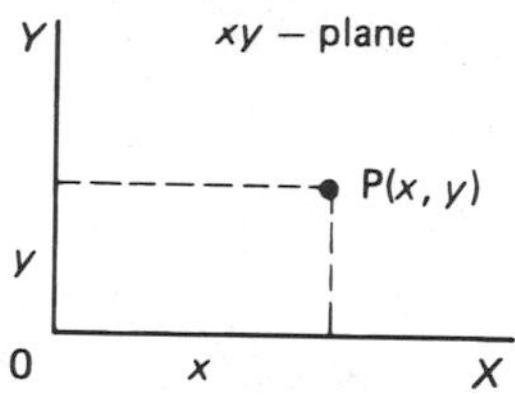

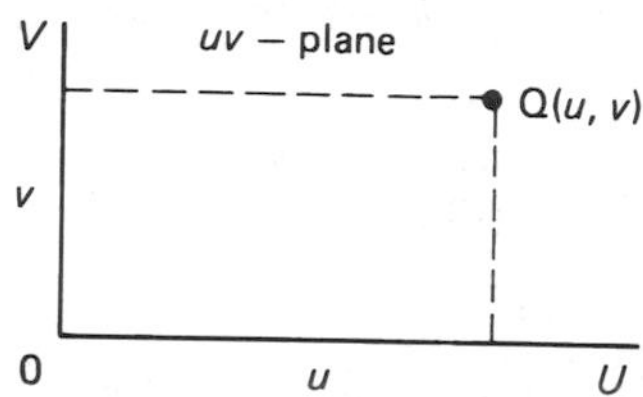

If a point Q(u, v) in the uv-plane corresponds to a point P(x, y) in the xy-plane, then there is a relationship between the two sets of coordinates.

If $u = ax$ and $v = by$

then we can combine these in matrix form

$$\begin{bmatrix} u \\ v \end{bmatrix} = \begin{bmatrix} a & 0 \\ 0 & b \end{bmatrix} \begin{bmatrix} x \\ y \end{bmatrix}$$

The matrix $\begin{bmatrix} a & 0 \\ 0 & b \end{bmatrix}$ then provides the transformation between the vector $\begin{bmatrix} x \\ y \end{bmatrix}$ in one set of coordinates and the vector $\begin{bmatrix} u \\ v \end{bmatrix}$ in the other set of coordinates.

Similarly, if we solve the two equations for x and y, we have

$$x = \frac{1}{a}u \quad \text{and} \quad y = \frac{1}{b}v$$

$$\therefore \begin{bmatrix} x \\ y \end{bmatrix} = \begin{bmatrix} 1/a & 0 \\ 0 & 1/b \end{bmatrix} \begin{bmatrix} u \\ v \end{bmatrix}$$

which allows us to transform back from the uv-plane coordinates to the xy-plane coordinates.

Now for some examples

Example 1 If $\mathbf{X} = \begin{bmatrix} x \\ y \end{bmatrix} = \begin{bmatrix} 2 \\ 1 \end{bmatrix}$ with the transformation $\mathbf{T} = \begin{bmatrix} -2 & 0 \\ 2 & 1 \end{bmatrix}$ determine $\mathbf{U} = \begin{bmatrix} u \\ v \end{bmatrix} = \mathbf{TX}$ and show the positions on the xy and uv planes.

In this case,

$$\begin{bmatrix} u \\ v \end{bmatrix} = \begin{bmatrix} -2 & 0 \\ 2 & 1 \end{bmatrix}\begin{bmatrix} 2 \\ 1 \end{bmatrix} = \begin{bmatrix} -4 \\ 5 \end{bmatrix}$$

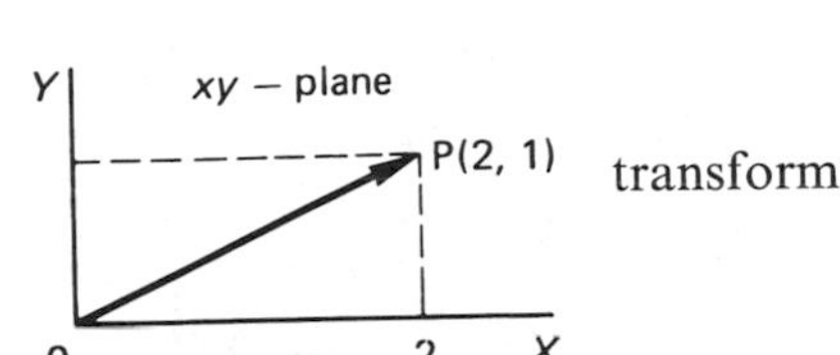

transforms into

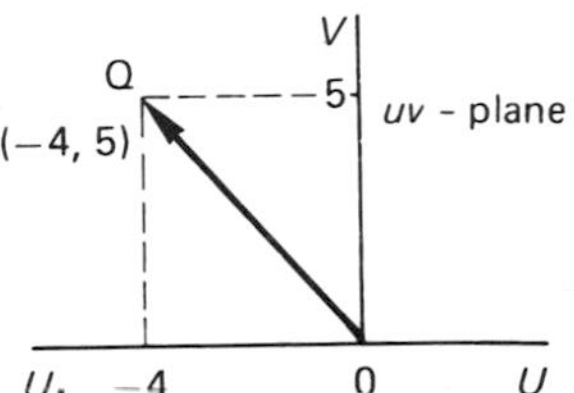

If **T** is non-singular and $\mathbf{U} = \mathbf{TX}$ then $\mathbf{X} = \mathbf{T}^{-1}\mathbf{U}$ and since

$\mathbf{T} = \begin{bmatrix} -2 & 0 \\ 2 & 1 \end{bmatrix}$ then $\mathbf{T}^{-1} = \ldots\ldots\ldots$

64

$$\boxed{\mathbf{T}^{-1} = \begin{bmatrix} -1/2 & 0 \\ 1 & 1 \end{bmatrix}}$$

There are several ways of finding the inverse of a matrix. One method is as follows. $\mathbf{T} = \begin{bmatrix} -2 & 0 \\ 2 & 1 \end{bmatrix}$

$$\therefore \left[\begin{array}{cc|cc} -2 & 0 & 1 & 0 \\ 2 & 1 & 0 & 1 \end{array}\right] = \left[\begin{array}{cc|cc} -2 & 0 & 1 & 0 \\ 0 & 1 & 1 & 1 \end{array}\right] = \left[\begin{array}{cc|cc} 1 & 0 & -1/2 & 0 \\ 0 & 1 & 1 & 1 \end{array}\right]$$

$$\therefore \mathbf{T}^{-1} = \begin{bmatrix} -1/2 & 0 \\ 1 & 1 \end{bmatrix}$$

So we have $\mathbf{U} = \mathbf{TX}$ $\therefore \mathbf{X} = \mathbf{T}^{-1}\mathbf{U}$

$$\therefore \begin{bmatrix} x \\ y \end{bmatrix} = \begin{bmatrix} -1/2 & 0 \\ 1 & 1 \end{bmatrix}\begin{bmatrix} u \\ v \end{bmatrix}$$

Hence a vector $\begin{bmatrix} 1 \\ 4 \end{bmatrix}$ in the *uv*-plane transforms into $\begin{bmatrix} x \\ y \end{bmatrix}$ in the *xy*-plane where $\begin{bmatrix} x \\ y \end{bmatrix} = \ldots\ldots\ldots$

65

$$\begin{bmatrix} x \\ y \end{bmatrix} = \begin{bmatrix} -1/2 \\ 5 \end{bmatrix}$$

$$\begin{bmatrix} x \\ y \end{bmatrix} = \begin{bmatrix} -1/2 & 0 \\ 1 & 1 \end{bmatrix} \begin{bmatrix} 1 \\ 4 \end{bmatrix} = \begin{bmatrix} -1/2 \\ 5 \end{bmatrix}$$

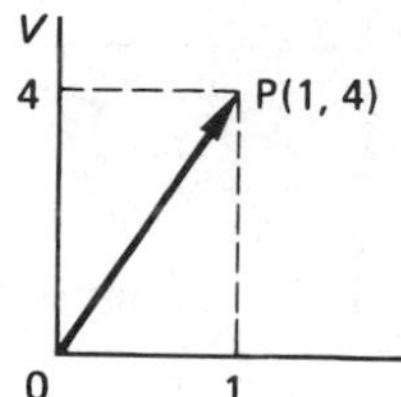

transforms into

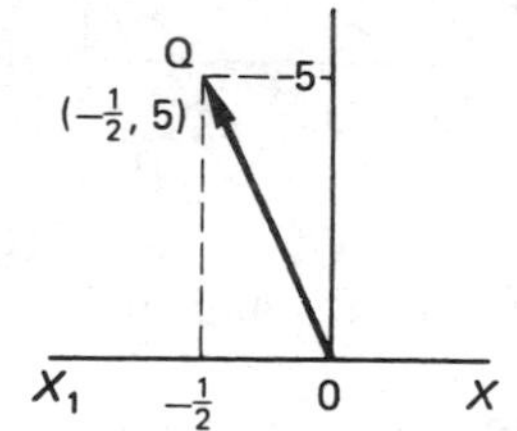

Rotation of axes

A more interesting case occurs with a degree of rotation between the two sets of coordinate axes.

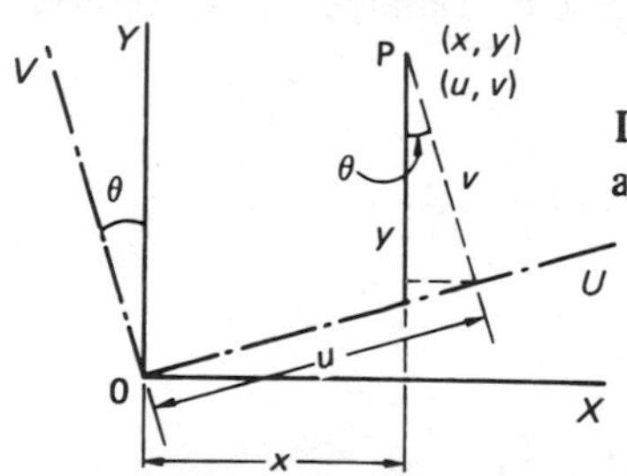

Let P be the point (x, y) in the xy-plane and the point (u, v) in the uv-plane.

Let θ be the angle of rotation between the two systems. From the diagram we can see that

$$\left.\begin{aligned} x &= u\cos\theta - v\sin\theta \\ y &= u\sin\theta + v\cos\theta \end{aligned}\right\} \qquad (1)$$

In matrix form, this becomes $\begin{bmatrix} x \\ y \end{bmatrix} = \begin{bmatrix} \cos\theta & -\sin\theta \\ \sin\theta & \cos\theta \end{bmatrix} \begin{bmatrix} u \\ v \end{bmatrix}$

which enables us to transform from the uv-plane coordinates to the corresponding xy-plane coordinates.

Make a note of this and then move on.

66

If we solve equations (1) for u and v, we have

$$x\sin\theta = u\sin\theta\cos\theta - v\sin^2\theta$$

$$y\cos\theta = u\sin\theta\cos\theta + v\cos^2\theta$$

$$\therefore\ y\cos\theta - x\sin\theta = v(\cos^2\theta + \sin^2\theta) = v$$

Also
$$x\cos\theta = u\cos^2\theta - v\sin\theta\cos\theta$$

$$y\sin\theta = u\sin^2\theta + v\sin\theta\cos\theta$$

$$\therefore\ x\cos\theta + y\sin\theta = u(\cos^2\theta + \sin^2\theta) = u$$

So
$$u = x\cos\theta + y\sin\theta$$

$$v = -x\sin\theta + y\cos\theta$$

and written in matrix form, this is ....

67

$$\begin{bmatrix} u \\ v \end{bmatrix} = \begin{bmatrix} \cos\theta & \sin\theta \\ -\sin\theta & \cos\theta \end{bmatrix}\begin{bmatrix} x \\ y \end{bmatrix}$$

So we have

$$\begin{bmatrix} x \\ y \end{bmatrix} = \begin{bmatrix} \cos\theta & -\sin\theta \\ \sin\theta & \cos\theta \end{bmatrix}\begin{bmatrix} u \\ v \end{bmatrix}$$

and
$$\begin{bmatrix} u \\ v \end{bmatrix} = \begin{bmatrix} \cos\theta & \sin\theta \\ -\sin\theta & \cos\theta \end{bmatrix}\begin{bmatrix} x \\ y \end{bmatrix}$$

i.e.
$$\mathbf{X} = \mathbf{TU} \quad \text{and} \quad \mathbf{U} = \mathbf{T}^{-1}\mathbf{X}$$

where $\mathbf{T}$ is the matrix of transformation and the equations provide linear transformation between the two sets of coordinates.

Example If the uv-plane axes rotate through 30° in an anticlockwise manner from the xy-plane axes, determine the (u, v) coordinates of a point whose (x, y) coordinates are $x = 2$, $y = 3$ in the xy-plane.

This is a straightforward application of the results above.

So
$$\begin{bmatrix} u \\ v \end{bmatrix} = \ldots\ldots\ldots$$

68

$$\begin{bmatrix} u \\ v \end{bmatrix} = \begin{bmatrix} \sqrt{3} + 3/2 \\ -1 + 3\sqrt{3}/2 \end{bmatrix} = \begin{bmatrix} 3.23 \\ 1.60 \end{bmatrix}$$

for $\begin{bmatrix} u \\ v \end{bmatrix} = \begin{bmatrix} \cos\theta & \sin\theta \\ -\sin\theta & \cos\theta \end{bmatrix} \begin{bmatrix} 2 \\ 3 \end{bmatrix}$ $\quad \cos\theta = \sqrt{3}/2$, $\sin\theta = 1/2$

$$= \begin{bmatrix} \sqrt{3}/2 & 1/2 \\ -1/2 & \sqrt{3}/2 \end{bmatrix} \begin{bmatrix} 2 \\ 3 \end{bmatrix}$$

$$= \begin{bmatrix} \sqrt{3} + 3/2 \\ -1 + 3\sqrt{3}/2 \end{bmatrix} = \begin{bmatrix} 3.23 \\ 1.60 \end{bmatrix}$$

Now we will consider one further operation with matrices, so on to a fresh frame.

69

Diagonalisation of a matrix

Modal matrix We have already discussed the eigenvalues and eigenvectors of a matrix **A** of order n. If these n eigenvectors are arranged as columns of a square matrix, the *modal matrix* of **A**, denoted by **M**, is formed

i.e. $$\mathbf{M} = [\mathbf{x}_1, \mathbf{x}_2, \mathbf{x}_3, \ldots \mathbf{x}_n].$$

For example, we have seen earlier that if

$$\mathbf{A} = \begin{bmatrix} 1 & 0 & 4 \\ 0 & 2 & 0 \\ 3 & 1 & -3 \end{bmatrix} \quad \text{then } \lambda_1 = 2,\ \lambda_2 = 3,\ \lambda_3 = -5$$

and the corresponding eigenvectors are $\mathbf{x}_1 = \begin{bmatrix} 4 \\ -7 \\ 1 \end{bmatrix}$, $\mathbf{x}_2 = \begin{bmatrix} 2 \\ 0 \\ 1 \end{bmatrix}$,

$$\mathbf{x}_3 = \begin{bmatrix} 2 \\ 0 \\ -3 \end{bmatrix}$$

Then the modal matrix $\mathbf{M} = \begin{bmatrix} 4 & 2 & 2 \\ -7 & 0 & 0 \\ 1 & 1 & -3 \end{bmatrix}$

Spectral matrix Also, we define the *spectral matrix* of **A**, i.e. **S**, as a diagonal matrix with the eigenvalues only on the main diagonal

i.e. $$\mathbf{S} = \begin{bmatrix} \lambda_1 & 0 & 0 & \ldots & 0 \\ 0 & \lambda_2 & 0 & \ldots & 0 \\ \cdot & \cdot & \cdot & & \cdot \\ \cdot & \cdot & \cdot & & \cdot \\ \cdot & \cdot & \cdot & & \cdot \\ 0 & 0 & 0 & \ldots & \lambda_n \end{bmatrix}$$

So, in the example above, $\mathbf{S} = \ldots\ldots\ldots$

70

$$\mathbf{S} = \begin{bmatrix} 2 & 0 & 0 \\ 0 & 3 & 0 \\ 0 & 0 & -5 \end{bmatrix}$$

Note that the eigenvalues of **S** and **A** are the same.

So, if $\mathbf{A} = \begin{bmatrix} 5 & -6 & 1 \\ 1 & 1 & 0 \\ 3 & 0 & 1 \end{bmatrix}$ has eigenvalues $\lambda = 1,\ 2,\ 4$ and corresponding

eigenvectors $\begin{bmatrix} 0 \\ 1 \\ 6 \end{bmatrix}, \begin{bmatrix} 1 \\ 1 \\ 3 \end{bmatrix}, \begin{bmatrix} 3 \\ 1 \\ 3 \end{bmatrix}$

then $\mathbf{M} = \dots\dots\dots$ and $\mathbf{S} = \dots\dots\dots$

71

$$\mathbf{M} = \begin{bmatrix} 0 & 1 & 3 \\ 1 & 1 & 1 \\ 6 & 3 & 3 \end{bmatrix}; \quad \mathbf{S} = \begin{bmatrix} 1 & 0 & 0 \\ 0 & 2 & 0 \\ 0 & 0 & 4 \end{bmatrix}$$

Now how are these connected? Let us investigate.

The eigenvectors $\mathbf{x}$ arranged in the modal matrix satisfy the original equation

$$\mathbf{Ax} = \lambda\mathbf{x}$$

Also $\quad \mathbf{M} = [\mathbf{x}_1 \quad \mathbf{x}_2 \quad \ldots \quad \mathbf{x}_n]$

Then $\quad \mathbf{AM} = \mathbf{A}[\mathbf{x}_1 \quad \mathbf{x}_2 \quad \ldots \quad \mathbf{x}_n]$

$$= [\mathbf{Ax}_1 \quad \mathbf{Ax}_2 \quad \ldots \quad \mathbf{Ax}_n]$$

$$= [\lambda_1\mathbf{x}_1 \quad \lambda_2\mathbf{x}_2 \quad \ldots \quad \lambda_n\mathbf{x}_n] \quad \text{since } \mathbf{Ax} = \lambda\mathbf{x}$$

Now $\quad \mathbf{S} = \begin{bmatrix} \lambda_1 & 0 & \ldots & 0 \\ 0 & \lambda_2 & \ldots & 0 \\ \cdot & \cdot & & \cdot \\ \cdot & \cdot & & \cdot \\ \cdot & \cdot & & \cdot \\ 0 & 0 & \ldots & \lambda_n \end{bmatrix} \quad \therefore [\lambda_1\mathbf{x}_1 \quad \lambda_2\mathbf{x}_2 \quad \ldots \quad \lambda_n\mathbf{x}_n] = \mathbf{MS}$

$$\therefore \mathbf{AM} = \mathbf{MS}$$

If we now pre-multiply both sides by $\mathbf{M}^{-1}$ we have

$$\mathbf{M}^{-1}\mathbf{AM} = \mathbf{M}^{-1}\mathbf{MS} \qquad \text{But } \mathbf{M}^{-1}\mathbf{M} = \mathbf{I}$$

$$\therefore \underline{\mathbf{M}^{-1}\mathbf{AM} = \mathbf{S}}$$

Make a note of this result. Then we will consider an example.

72 *Example 1* From the results of a previous example, if

$$\mathbf{A} = \begin{bmatrix} 1 & 0 & 4 \\ 0 & 2 & 0 \\ 3 & 1 & -3 \end{bmatrix} \quad \text{then } \lambda_1 = 2, \quad \lambda_2 = 3, \quad \lambda_3 = -5 \quad \text{and}$$

$$\mathbf{x}_1 = \begin{bmatrix} 4 \\ -7 \\ 1 \end{bmatrix}, \quad \mathbf{x}_2 = \begin{bmatrix} 2 \\ 0 \\ 1 \end{bmatrix}, \quad \mathbf{x}_3 = \begin{bmatrix} 2 \\ 0 \\ -3 \end{bmatrix}.$$

$$\text{Also } \mathbf{M} = \begin{bmatrix} 4 & 2 & 2 \\ -7 & 0 & 0 \\ 1 & 1 & -3 \end{bmatrix}.$$

We can find $\mathbf{M}^{-1}$ by any of the methods we have established previously.

$$\mathbf{M}^{-1} = \ldots\ldots\ldots$$

73

$$\mathbf{M}^{-1} = \begin{bmatrix} 0 & -1/7 & 0 \\ 3/8 & 1/4 & 1/4 \\ 1/8 & 1/28 & -1/4 \end{bmatrix}$$

Here is one way of determining the inverse. You may have done it by another.

$$\left[\begin{array}{rrr|rrr} 4 & 2 & 2 & 1 & 0 & 0 \\ -7 & 0 & 0 & 0 & 1 & 0 \\ 1 & 1 & -3 & 0 & 0 & 1 \end{array}\right] = \left[\begin{array}{rrr|rrr} 7 & 0 & 0 & 0 & -1 & 0 \\ 1 & 1 & -3 & 0 & 0 & 1 \\ 4 & 2 & 2 & 1 & 0 & 0 \end{array}\right]$$

$$\left[\begin{array}{rrr|rrr} 1 & 0 & 0 & 0 & -1/7 & 0 \\ 0 & 1 & -3 & 0 & 1/7 & 1 \\ 0 & 2 & 2 & 1 & 4/7 & 0 \end{array}\right] = \left[\begin{array}{rrr|rrr} 1 & 0 & 0 & 0 & -1/7 & 0 \\ 0 & 1 & -3 & 0 & 1/7 & 1 \\ 0 & 0 & 8 & 1 & 2/7 & -2 \end{array}\right]$$

$$\left[\begin{array}{rrr|rrr} 1 & 0 & 0 & 0 & -1/7 & 0 \\ 0 & 1 & -3 & 0 & 1/7 & 1 \\ 0 & 0 & 1 & 1/8 & 1/28 & -1/4 \end{array}\right]$$

$$= \left[\begin{array}{rrr|rrr} 1 & 0 & 0 & 0 & -1/7 & 0 \\ 0 & 1 & 0 & 3/8 & 7/28 & 1/4 \\ 0 & 0 & 1 & 1/8 & 1/28 & -1/4 \end{array}\right]$$

$$\therefore \mathbf{M}^{-1} = \begin{bmatrix} 0 & -1/7 & 0 \\ 3/8 & 1/4 & 1/4 \\ 1/8 & 1/28 & -1/4 \end{bmatrix}$$

So now $\mathbf{A} = \begin{bmatrix} 1 & 0 & 4 \\ 0 & 2 & 0 \\ 3 & 1 & -3 \end{bmatrix}$ and $\mathbf{M} = \begin{bmatrix} 4 & 2 & 2 \\ -7 & 0 & 0 \\ 1 & 1 & -3 \end{bmatrix}$

$$\therefore \mathbf{AM} = \begin{bmatrix} 1 & 0 & 4 \\ 0 & 2 & 0 \\ 3 & 1 & -3 \end{bmatrix} \begin{bmatrix} 4 & 2 & 2 \\ -7 & 0 & 0 \\ 1 & 1 & -3 \end{bmatrix} = \begin{bmatrix} 8 & 6 & -10 \\ -14 & 0 & 0 \\ 2 & 3 & 15 \end{bmatrix}$$

Then $\mathbf{M}^{-1}\mathbf{AM} = \begin{bmatrix} 0 & -1/7 & 0 \\ 3/8 & 1/4 & 1/4 \\ 1/8 & 1/28 & -1/4 \end{bmatrix} \begin{bmatrix} 8 & 6 & -10 \\ -14 & 0 & 0 \\ 2 & 3 & 15 \end{bmatrix}$

$= \ldots\ldots\ldots$

74

$$\mathbf{M}^{-1}\mathbf{AM} = \begin{bmatrix} 2 & 0 & 0 \\ 0 & 3 & 0 \\ 0 & 0 & -5 \end{bmatrix}$$

So we have transformed the original matrix **A** into a diagonal matrix and notice that the elements on the main diagonal are, in fact, the eigenvalues of **A**

$$\text{i.e. } \underline{\mathbf{M}^{-1}\mathbf{AM} = \mathbf{S}}$$

Therefore, let us list a few relevant facts

1. $\mathbf{M}^{-1}\mathbf{AM}$ transforms the square matrix **A** into a diagonal matrix **S**.
2. A square matrix **A** of order n can be so transformed if the matrix has n independent eigenvectors.
3. A matrix **A** always has n linearly independent eigenvectors if it has n distinct eigenvalues or if it is a symmetric matrix.
4. If the matrix has repeated eigenvalues, it may or may not have n linearly independent eigenvectors.

Now here is one straightforward example with which to finish.

Example 2

If $\mathbf{A} = \begin{bmatrix} -6 & 5 \\ 4 & 2 \end{bmatrix}$, $\quad \mathbf{M} = \ldots\ldots\ldots; \quad \mathbf{M}^{-1} = \ldots\ldots\ldots;$

and hence $\quad \mathbf{M}^{-1}\mathbf{AM} = \ldots\ldots\ldots$

Work through it entirely on your own:

(i) Determine the eigenvalues and corresponding eigenvectors.

(ii) Hence form the matrix **M**.

(iii) Determine $\mathbf{M}^{-1}$, the inverse of **M**.

(iv) Finally form the matrix products **AM** and $\mathbf{M}^{-1}(\mathbf{AM})$.

75

$$\mathbf{M} = \begin{bmatrix} 1 & 5 \\ 2 & -2 \end{bmatrix}; \quad \mathbf{M}^{-1} = \begin{bmatrix} 1/6 & 5/12 \\ 1/6 & -1/12 \end{bmatrix}; \quad \mathbf{M}^{-1}\mathbf{A}\mathbf{M} = \begin{bmatrix} 4 & 0 \\ 0 & -8 \end{bmatrix}$$

Here is the working. See whether you agree.

$$\mathbf{A} = \begin{bmatrix} -6 & 5 \\ 4 & 2 \end{bmatrix} \quad \therefore \begin{vmatrix} -6-\lambda & 5 \\ 4 & 2-\lambda \end{vmatrix} = 0$$

$$(-6-\lambda)(2-\lambda) - 20 = 0 \qquad \therefore \lambda^2 + 4\lambda - 32 = 0$$

$$(\lambda - 4)(\lambda + 8) = 0 \qquad \therefore \lambda = 4 \text{ or } -8$$

(a) $\lambda_1 = 4$
$$\left\{\begin{bmatrix} -6 & 5 \\ 4 & 2 \end{bmatrix} - \begin{bmatrix} 4 & 0 \\ 0 & 4 \end{bmatrix}\right\}\begin{bmatrix} x_1 \\ x_2 \end{bmatrix} = \begin{bmatrix} 0 \\ 0 \end{bmatrix}$$

$$\begin{bmatrix} -10 & 5 \\ 4 & -2 \end{bmatrix}\begin{bmatrix} x_1 \\ x_2 \end{bmatrix} = \begin{bmatrix} 0 \\ 0 \end{bmatrix}$$

$$\therefore -10x_1 + 5x_2 = 0 \quad \therefore x_2 = 2x_1 \qquad \mathbf{x}_1 = \begin{bmatrix} 1 \\ 2 \end{bmatrix}$$

(b) $\lambda_2 = -8$
$$\left\{\begin{bmatrix} -6 & 5 \\ 4 & 2 \end{bmatrix} + \begin{bmatrix} 8 & 0 \\ 0 & 8 \end{bmatrix}\right\}\begin{bmatrix} x_1 \\ x_2 \end{bmatrix} = \begin{bmatrix} 0 \\ 0 \end{bmatrix}$$

$$\begin{bmatrix} 2 & 5 \\ 4 & 10 \end{bmatrix}\begin{bmatrix} x_1 \\ x_2 \end{bmatrix} = \begin{bmatrix} 0 \\ 0 \end{bmatrix}$$

$$\therefore 2x_1 + 5x_2 = 0 \quad \therefore x_2 = -\tfrac{2}{5}x_1 \quad \therefore \mathbf{x}_2 = \begin{bmatrix} 5 \\ -2 \end{bmatrix}$$

$$\therefore \underline{\mathbf{M} = \begin{bmatrix} 1 & 5 \\ 2 & -2 \end{bmatrix}}$$

To find $\mathbf{M}^{-1}$ $\left[\begin{array}{cc|cc} 1 & 5 & 1 & 0 \\ 2 & -2 & 0 & 1 \end{array}\right]$

Operating on rows, we have

$$\left[\begin{array}{cc|cc} 0 & 5 & 1 & 0 \\ 0 & -12 & -2 & 1 \end{array}\right] = \left[\begin{array}{cc|cc} 1 & 5 & 1 & 0 \\ 0 & 1 & 1/6 & -1/12 \end{array}\right]$$

$$= \left[\begin{array}{cc|cc} 1 & 0 & 1/6 & 5/12 \\ 0 & 1 & 1/6 & -1/12 \end{array}\right]$$

$$\therefore \mathbf{M}^{-1} = \begin{bmatrix} 1/6 & 5/12 \\ 1/6 & -1/12 \end{bmatrix}$$

$$\therefore \mathbf{AM} = \begin{bmatrix} -6 & 5 \\ 4 & 2 \end{bmatrix} \begin{bmatrix} 1 & 5 \\ 2 & -2 \end{bmatrix} = \begin{bmatrix} 4 & -40 \\ 8 & 16 \end{bmatrix}$$

$$\therefore \mathbf{M}^{-1}\mathbf{AM} = \begin{bmatrix} 1/6 & 5/12 \\ 1/6 & -1/12 \end{bmatrix} \begin{bmatrix} 4 & -40 \\ 8 & 16 \end{bmatrix} = \begin{bmatrix} 4 & 0 \\ 0 & -8 \end{bmatrix}$$

$$\therefore \mathbf{M}^{-1}\mathbf{AM} = \begin{bmatrix} 4 & 0 \\ 0 & -8 \end{bmatrix}$$

So there it is. All that now remains is to check down the Revision Summary that follows—then you will have no trouble with the Test Exercise.

76

REVISION SUMMARY

1. *Rank of a matrix*—order of the largest non-zero determinant that can be formed from the elements of the matrix.

2. *Singular* square matrix: $|\mathbf{A}| = 0$

   Non-singular square matrix $|\mathbf{A}| \neq 0$.

3. *Consistency* of a set of n equations in n unknowns with coefficient matrix $\mathbf{A}$ and augmented matrix $\mathbf{A}_b$.

   (a) Consistent if rank of $\mathbf{A}$ = rank of $\mathbf{A}_b$

   (b) Inconsistent if rank of $\mathbf{A} <$ rank of $\mathbf{A}_b$.

4. *Uniqueness of solutions*—n equations with n unknowns.

   (a) rank of $\mathbf{A}$ = rank of $\mathbf{A}_b = n$ *unique solutions*

   (b) rank of $\mathbf{A}$ = rank of $\mathbf{A}_b = m < n$ *infinite number of solutions*

   (c) rank of $\mathbf{A} <$ rank of $\mathbf{A}_b$ *no solution*

5. *Solution of sets of equations*

   (a) *Inverse matrix method* $\mathbf{Ax} = \mathbf{b}$ $\mathbf{x} = \mathbf{A}^{-1}\mathbf{b}$

   To find $\mathbf{A}^{-1}$ (i) evaluate $|\mathbf{A}|$

   (ii) form $\mathbf{C}$, the matrix of cofactors of $\mathbf{A}$

   (iii) write $\mathbf{C}^T$, the transpose of $\mathbf{A}$

   (iv) $\mathbf{A}^{-1} = \frac{1}{|\mathbf{A}|} \times \mathbf{C}^T$.

   (b) *Row transformation* method $\mathbf{Ax} = \mathbf{b}$; $\mathbf{Ax} = \mathbf{Ib}$

   (i) form the combined coefficient matrix $[\mathbf{A} \mid \mathbf{I}]$

   (ii) row transformations to convert to $[\mathbf{I} \mid \mathbf{A}^{-1}]$

   (iii) then solve $\mathbf{x} = \mathbf{A}^{-1}\mathbf{b}$.

   (c) *Gauss elimination* method $\mathbf{Ax} = \mathbf{b}$

   (i) form augmented matrix $[\mathbf{A} \mid \mathbf{b}]$

   (ii) operate on rows to convert to $[\mathbf{U} \mid \mathbf{b}']$ where $\mathbf{U}$ is the upper-triangular matrix.

   (iii) expand from bottom row to obtain $\mathbf{x}$.

   (d) *Triangular decomposition* method $\mathbf{Ax} = \mathbf{b}$.

   Write $\mathbf{A}$ as the product of upper and lower triangular matrices.

   $\mathbf{A} = \mathbf{LU}$, $\mathbf{L}(\mathbf{Ux}) = \mathbf{b}$. Put $\mathbf{Ux} = \mathbf{y}$ $\therefore \mathbf{Ly} = \mathbf{b}$

   (i) Solve $\mathbf{Ly} = \mathbf{b}$ to obtain $\mathbf{y}$

   (ii) Solve $\mathbf{Ux} = \mathbf{y}$ to obtain $\mathbf{x}$.

6. *Eigenvalues and eigenvectors* $\mathbf{Ax} = \lambda\mathbf{x}$

Sets of equations of form $\mathbf{Ax} = \lambda\mathbf{x}$, where $\mathbf{A}$ = coefficient matrix, $\mathbf{x}$ = column matrix, λ = scalar quantity.

Equations become $(\mathbf{A} - \lambda\mathbf{I})\mathbf{x} = \mathbf{0}$.

For non-trivial solutions, $|\mathbf{A} - \lambda\mathbf{I}| = 0$ is the *characteristic equation* and gives values of λ i.e. the *eigenvalues.*

Substitution of each eigenvalue gives a corresponding *eigenvector.*

7. *Matrix transformation*

(a) $\mathbf{U} = \mathbf{TX}$, where $\mathbf{T}$ is a transformation matrix, transforms a vector in the xy-plane to a corresponding vector in the uv-plane. Similarly, $\mathbf{X} = \mathbf{T}^{-1}\mathbf{U}$ converts a vector in the uv-plane to a corresponding vector in the xy-plane.

(b) *Rotation of axes*

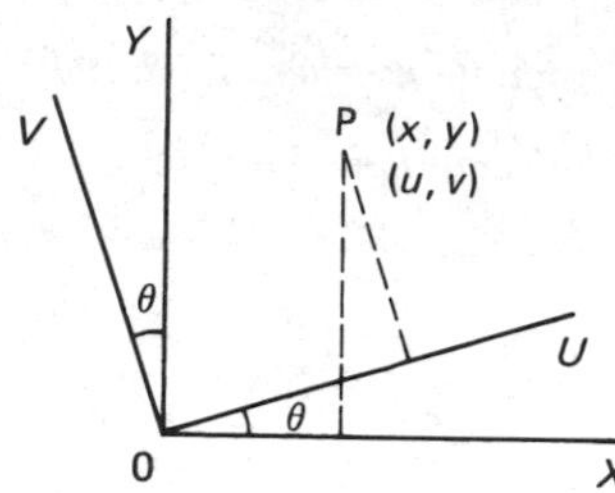

$$\begin{bmatrix} u \\ v \end{bmatrix} = \begin{bmatrix} \cos\theta & \sin\theta \\ -\sin\theta & \cos\theta \end{bmatrix} \begin{bmatrix} x \\ y \end{bmatrix}$$

$$\begin{bmatrix} x \\ y \end{bmatrix} = \begin{bmatrix} \cos\theta & -\sin\theta \\ \sin\theta & \cos\theta \end{bmatrix} \begin{bmatrix} u \\ v \end{bmatrix}$$

(c) *Diagonalisation of a matrix*

Modal matrix of $\mathbf{A}$, $\mathbf{M} = [\mathbf{x}_1, \mathbf{x}_2, \dots \mathbf{x}_n]$

where $x_1, x_2, \dots x_n$ are eigenvectors of $\mathbf{A}$.

Then $\mathbf{M}^{-1}\mathbf{AM} = \mathbf{S}$ where $\mathbf{S}$ is the *spectral matrix* of $\mathbf{A}$

where $$\mathbf{S} = \begin{bmatrix} \lambda_1 & 0 & \dots & 0 \\ 0 & \lambda_2 & \dots & 0 \\ \cdot & \cdot & & \cdot \\ \cdot & \cdot & & \cdot \\ \cdot & \cdot & & \cdot \\ 0 & 0 & \dots & \lambda_n \end{bmatrix}$$

$\lambda_1, \lambda_2, \dots \lambda_n$ are the eigenvalues of $\mathbf{A}$.

77

TEST EXERCISE XI

1. Determine the rank of $\mathbf{A}$ and of $\mathbf{A}_b$ for the following sets of equations and hence determine the nature of the solutions. Do *not* solve the equations.

(a) $x_1 + 3x_2 - 2x_3 = 6$
$4x_1 + 5x_2 + 2x_3 = 3$
$x_1 + 3x_2 + 4x_3 = 7$

(b) $x_1 + 2x_2 - 4x_3 = 3$
$x_1 + 2x_2 + 3x_3 = -4$
$2x_1 + 4x_2 + x_3 = -3.$

2. If $\mathbf{Ax} = \mathbf{b}$ where $\mathbf{A} = \begin{bmatrix} 2 & 3 & -2 \\ 3 & 5 & -4 \\ 1 & 2 & -3 \end{bmatrix}$ and $\mathbf{b} = \begin{bmatrix} 4 \\ 10 \\ 9 \end{bmatrix}$ determine $\mathbf{A}^{-1}$ and hence solve the set of equations.

3. Given that $3x_1 + 2x_2 + x_3 = 1$
$x_1 - x_2 + 3x_3 = 5$
$2x_1 + 5x_2 - 2x_3 = 0$
apply the method of row transformation to obtain the value of $x_1,\ x_2,\ x_3$.

4. By the method of gaussian elimination, solve the equations $\mathbf{Ax} = \mathbf{b}$, where $\mathbf{A} = \begin{bmatrix} 1 & -2 & -4 \\ 2 & 1 & -3 \\ 1 & 3 & 2 \end{bmatrix}$ and $\mathbf{b} = \begin{bmatrix} -3 \\ 4 \\ 5 \end{bmatrix}$.

5. If $\mathbf{Ax} = \mathbf{b}$ where $\mathbf{A} = \begin{bmatrix} 1 & -2 & 1 \\ 3 & 1 & -2 \\ 5 & 3 & 3 \end{bmatrix}$ and $\mathbf{b} = \begin{bmatrix} 7 \\ -3 \\ 5 \end{bmatrix}$ express $\mathbf{A}$ as the product $\mathbf{A} = \mathbf{LU}$ where $\mathbf{L}$ and $\mathbf{U}$ are lower and upper triangular matrices and hence determine the values of $x_1,\ x_2,\ x_3$.

6. Determine the eigenvalues and corresponding eigenvectors of $\mathbf{Ax} = \lambda\mathbf{x}$ where $\mathbf{A} = \begin{bmatrix} 1 & 3 & 0 \\ 1 & 2 & 1 \\ -2 & 1 & -1 \end{bmatrix}$.

7. (a) Determine the vector in the *uv*-plane formed by $\mathbf{U} = \mathbf{TX}$, where the transformation matrix is $\mathbf{T} = \begin{bmatrix} -2 & 1 \\ 3 & 4 \end{bmatrix}$ and $\mathbf{X} = \begin{bmatrix} 3 \\ -2 \end{bmatrix}$ is a vector in the *xy*-plane.

(b) The coordinate axes in the *xy*-plane and in the *uv*-plane have the same origin O, but OU is inclined to OX at an angle of 60° in an anticlockwise manner. Transform a vector $\mathbf{X} = \begin{bmatrix} 4 \\ 6 \end{bmatrix}$ in the *xy*-plane into the corresponding vector in the *uv*-plane.

8. If $\mathbf{x}_1$ and $\mathbf{x}_2$ are eigenvectors of $\mathbf{Ax} = \lambda\mathbf{x}$ where $\mathbf{A} = \begin{bmatrix} 3 & 2 \\ 4 & 1 \end{bmatrix}$ determine (i) $\mathbf{M} = [\mathbf{x}_1 \mathbf{x}_2]$

(ii) $\mathbf{M}^{-1}$

(iii) $\mathbf{M}^{-1}\mathbf{AM}$.

FURTHER PROBLEMS XI

1. If $\mathbf{Ax} = \mathbf{b}$ where $\mathbf{A} = \begin{bmatrix} 5 & 2 & 3 \\ 3 & -2 & -2 \\ 4 & 3 & 1 \end{bmatrix}$ and $\mathbf{b} = \begin{bmatrix} 6 \\ 5 \\ -5 \end{bmatrix}$ determine $\mathbf{A}^{-1}$ and hence solve the set of equations.

2. Apply the method of row transformation to solve the following sets of equations.

(a)
$$\begin{aligned} x_1 - 3x_2 - 2x_3 &= 8 \\ 2x_1 + 2x_2 + x_3 &= 4 \\ 3x_1 - 4x_2 + 2x_3 &= -3 \end{aligned}$$

(b)
$$\begin{aligned} x_1 - 3x_2 + 2x_3 &= 8 \\ 2x_1 - x_2 + x_3 &= 9 \\ 3x_1 + 2x_2 + 3x_3 &= 5. \end{aligned}$$

3. Solve the following sets of equations by gaussian elimination.

(a)
$$\begin{aligned} x_1 - 2x_2 - x_3 + 3x_4 &= 4 \\ 2x_1 + x_2 + x_3 - 4x_4 &= 3 \\ 3x_1 - x_2 - 2x_3 + 2x_4 &= 6 \\ x_1 + 3x_2 - x_3 + x_4 &= 8. \end{aligned}$$

(b)
$$\begin{aligned} 2x_1 + 3x_2 - 2x_3 + 2x_4 &= 2 \\ 4x_1 + 2x_2 - 3x_3 - x_4 &= 6 \\ x_1 - x_2 + 4x_3 - 2x_4 &= 7 \\ 3x_1 + 2x_2 + x_3 - x_4 &= 5. \end{aligned}$$

(c)
$$\begin{aligned} x_1 + 2x_2 + 5x_3 + x_4 &= 4 \\ 3x_1 - 4x_2 + 3x_3 - 2x_4 &= 7 \\ 4x_1 + 3x_2 + 2x_3 - x_4 &= 1 \\ x_1 - 2x_2 - 4x_3 - x_4 &= 2. \end{aligned}$$

4. Using the method of triangular decomposition, solve the following sets of equations.

(a)
$$\begin{bmatrix} 1 & 4 & -1 \\ 4 & 2 & 3 \\ 7 & -3 & 2 \end{bmatrix}\begin{bmatrix} x_1 \\ x_2 \\ x_3 \end{bmatrix} = \begin{bmatrix} -2 \\ -1 \\ -18 \end{bmatrix}$$

(b)
$$\begin{bmatrix} 1 & -2 & 3 \\ 2 & 1 & -5 \\ 6 & -3 & 2 \end{bmatrix}\begin{bmatrix} x_1 \\ x_2 \\ x_3 \end{bmatrix} = \begin{bmatrix} -2 \\ 17 \\ 22 \end{bmatrix}$$

(c)
$$\begin{bmatrix} 1 & -2 & 3 & -1 \\ 3 & 1 & -3 & 2 \\ 5 & 3 & 2 & 3 \\ 2 & -4 & -2 & 4 \end{bmatrix}\begin{bmatrix} x_1 \\ x_2 \\ x_3 \\ x_4 \end{bmatrix} = \begin{bmatrix} -3 \\ 14 \\ 21 \\ -10 \end{bmatrix}$$

5. If $\mathbf{Ax} = \lambda\mathbf{x}$, determine the eigenvalues and corresponding eigenvectors in each of the following cases.

(a) $\mathbf{A} = \begin{bmatrix} 4 & 3 \\ 2 & 5 \end{bmatrix}$ (b) $\mathbf{A} = \begin{bmatrix} 2 & -5 \\ 1 & -4 \end{bmatrix}$

(c) $\mathbf{A} = \begin{bmatrix} -6 & 5 \\ 4 & 2 \end{bmatrix}$ (d) $\mathbf{A} = \begin{bmatrix} -5 & 9 \\ 1 & 3 \end{bmatrix}$

(e) $\mathbf{A} = \begin{bmatrix} 2 & 7 & 0 \\ 1 & 3 & 1 \\ 5 & 0 & 8 \end{bmatrix}$ (f) $\mathbf{A} = \begin{bmatrix} 5 & -6 & 1 \\ 1 & 1 & 0 \\ 3 & 0 & 1 \end{bmatrix}$

(g) $\mathbf{A} = \begin{bmatrix} -3 & 0 & 6 \\ 4 & 5 & 3 \\ 1 & 2 & 1 \end{bmatrix}$ (h) $\mathbf{A} = \begin{bmatrix} 4 & 10 & -8 \\ 1 & 2 & 1 \\ -1 & 2 & 3 \end{bmatrix}$

6. If $\mathbf{A} = \begin{bmatrix} 1 & 3 & 0 \\ 3 & 10 & -3 \\ 0 & -3 & 9 \end{bmatrix}$, determine the three eigenvalues λ_1, λ_2, λ_3 of $\mathbf{A}$

and verify that, if $\mathbf{M} = \begin{bmatrix} -9 & 1 & 1 \\ 3 & 2 & 4 \\ 1 & 3 & -3 \end{bmatrix}$ then $\mathbf{M}^{-1}\mathbf{AM} = \mathbf{S}$, where $\mathbf{S}$ is a

diagonal matrix with elements λ_1, λ_2, λ_3.

7. Invert the matrix $\mathbf{A} = \begin{bmatrix} 8 & 10 & 7 \\ 5 & 9 & 4 \\ 9 & 11 & 8 \end{bmatrix}$ and hence solve the equations

$$\begin{aligned} 8I_1 + 10I_2 + 7I_3 &= 0 \\ 5I_1 + 9I_2 + 4I_3 &= -9 \\ 9I_1 + 11I_2 + 8I_3 &= 1. \end{aligned}$$

8. If $\mathbf{A} = \begin{bmatrix} 1 & 2 & 3 \\ 4 & 6 & 7 \\ 5 & 8 & 9 \end{bmatrix}$ and $\mathbf{B} = \begin{bmatrix} -2 & 6 & -4 \\ -1 & -6 & 5 \\ 2 & 2 & -2 \end{bmatrix}$, verify that $\mathbf{AB} = k\mathbf{I}$ where $\mathbf{I}$ is a unit matrix and k is a constant. Hence solve the equations

$$\begin{aligned} x_1 + 2x_2 + 3x_3 &= 2 \\ 4x_1 + 6x_2 + 7x_3 &= 2 \\ 5x_1 + 8x_2 + 9x_3 &= 3. \end{aligned}$$

Programme 12

Vector Analysis

PART 1

Prerequisites: Engineering Mathematics (fourth edition)
Programme 6

Introduction

1

The initial work on Vectors was covered in detail in Programme 6 of *Engineering Mathematics* and, if you are in any doubt, spend some time reviewing that section of the work before proceeding further.

The current programmes on Vector Analysis build on these early foundations, so, for quick reference, the essential results of the previous work are summarised in the following list.

Summary of prerequisites

1. A *scalar* quantity has magnitude only; a *vector* quantity has both magnitude and direction.

2. The axes of reference, OX, OY, OZ, form a right-handed set. The symbols **i**, **j**, **k** denote *unit vectors* in the directions OX, OY, OZ, respectively.

   If $\overline{\mathrm{OP}} = \mathbf{r} = a_x\mathbf{i} + a_y\mathbf{j} + a_z\mathbf{k}$ then $\mathrm{OP} = |\mathbf{r}| = \sqrt{a_x^2 + a_y^2 + a_z^2}$

3. The *direction cosines* [*l*, *m*, *n*] are the cosines of the angles between the vector **r** and the axes OX, OY, OZ, respectively. For any vector $\mathbf{r} = a_x\mathbf{i} + a_y\mathbf{j} + a_z\mathbf{k}$

   $$l = \frac{a_x}{|\mathbf{r}|}; \quad m = \frac{a_y}{|\mathbf{r}|}; \quad n = \frac{a_z}{|\mathbf{r}|}$$

   and $$l^2 + m^2 + n^2 = 1.$$

4. *Scalar product* ('dot product')
   $\mathbf{A} \cdot \mathbf{B} = AB\cos\theta$ where θ is the angle between **A** and **B**.
   If $\mathbf{A} = a_x\mathbf{i} + a_y\mathbf{j} + a_z\mathbf{k}$ and $\mathbf{B} = b_x\mathbf{i} + b_y\mathbf{j} + b_z\mathbf{k}$ then

   $$\mathbf{A} \cdot \mathbf{B} = a_xb_x + a_yb_y + a_zb_z$$

5. *Vector product* ('cross product')

   $\mathbf{A} \times \mathbf{B} = AB\sin\theta$ in a direction perpendicular to **A** and **B**

   so that $\mathbf{A}, \mathbf{B}, (\mathbf{A} \times \mathbf{B})$ form a right-handed set.

   Therefore $|\mathbf{A} \times \mathbf{B}| = AB\sin\theta$

   Also $$\mathbf{A} \times \mathbf{B} = \begin{vmatrix} \mathbf{i} & \mathbf{j} & \mathbf{k} \\ a_x & a_y & a_z \\ b_x & b_y & b_z \end{vmatrix}$$

6. *Angle between two vectors*

$$\cos\theta = l_1 l_2 + m_1 m_2 + n_1 n_2$$

where l_1, m_1, n_1 and l_2, m_2, n_2 are the direction cosines of vectors $\mathbf{r}_1$ and $\mathbf{r}_2$ respectively.
For perpendicular vectors $\quad l_1 l_2 + m_1 m_2 + n_1 n_2 = 0$
For parallel vectors $\quad l_1 l_2 + m_1 m_2 + n_1 n_2 = 1.$

One or two examples will no doubt help to recall the main points.

Example 1 (Direction cosines)
If $\mathbf{i}$, $\mathbf{j}$, $\mathbf{k}$ are unit vectors in the directions OX, OY, OZ, respectively, then any position vector $\overline{\text{OP}}$ (= $\mathbf{r}$) can be represented in the form

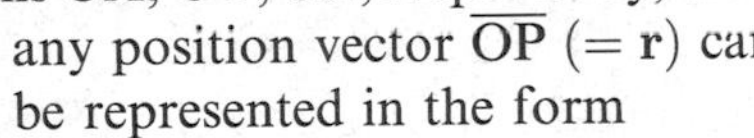

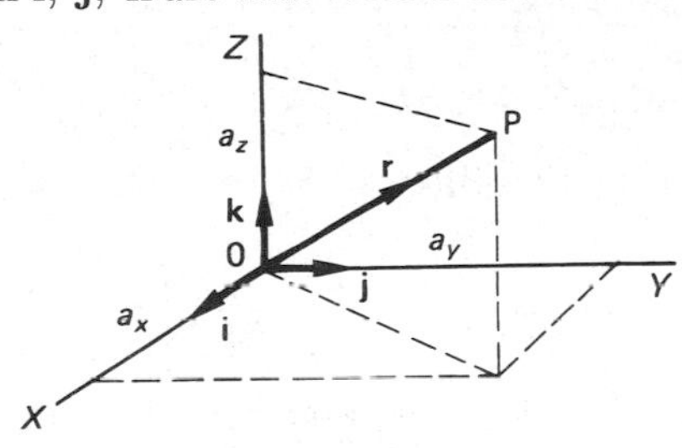

$$\overline{\text{OP}} = \mathbf{r} = a_x\mathbf{i} + a_y\mathbf{j} + a_z\mathbf{k}.$$

Then $|\mathbf{r}| = \ldots\ldots\ldots$

2

$$|\mathbf{r}| = \sqrt{a_x^2 + a_y^2 + a_z^2}$$

The direction of OP is denoted by stating the direction cosines of the angles made by OP and the three coordinate axes.

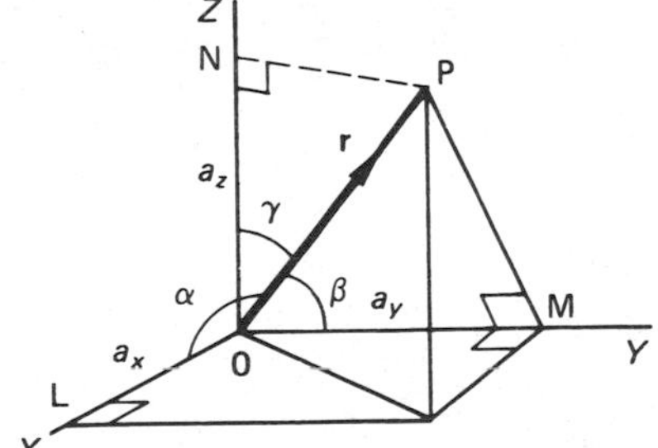

$$l = \cos\alpha = \frac{\text{OL}}{\text{OP}} = \frac{a_x}{|\mathbf{r}|}$$

$$m = \cos\beta = \frac{\text{OM}}{\text{OP}} = \frac{a_y}{|\mathbf{r}|}$$

$$n = \cos\gamma = \frac{\text{ON}}{\text{OP}} = \frac{a_z}{|\mathbf{r}|}$$

$$\therefore l, \quad m, \quad n = \cos\alpha, \quad \cos\beta, \quad \cos\gamma$$

So, if P is the point (3, 2, 6), then

$$|\mathbf{r}| = \ldots\ldots\ldots;$$

$$l = \ldots\ldots\ldots; \quad m = \ldots\ldots\ldots; \quad n = \ldots\ldots\ldots$$

3

$$|\mathbf{r}| = 7;$$
$$l = 0.429; \quad m = 0.286; \quad n = 0.857$$

since $(|\mathbf{r}|)^2 = 9 + 4 + 36 = 49 \qquad \therefore |\mathbf{r}| = 7$

$$l = \cos\alpha = \tfrac{3}{7} = 0.4286$$
$$m = \cos\beta = \tfrac{2}{7} = 0.2857$$
$$n = \cos\gamma = \tfrac{6}{7} = 0.8571.$$

Example 2 (Angle between two vectors)

If the direction cosines of **A** are l_1, m_1, n_1 and those of **B** are l_2, m_2, n_2, then the angle between the vectors is given by

$$\cos\theta = l_1 l_2 + m_1 m_2 + n_1 n_2. \qquad (1)$$

If $\mathbf{A} = 2\mathbf{i} + 3\mathbf{j} + 4\mathbf{k}$ and $\mathbf{B} = \mathbf{i} - 2\mathbf{j} + 3\mathbf{k}$, we can find the direction cosines of each and hence θ which is ....

4

$$\theta = 66° \; 36'$$

For **A**: $|\mathbf{r}_1| = \sqrt{4 + 9 + 16} = \sqrt{29}$

$$\therefore l_1 = \frac{2}{\sqrt{29}}; \quad m_1 = \frac{3}{\sqrt{29}}; \quad n = \frac{4}{\sqrt{29}}$$

For **B**: $|\mathbf{r}_2| = \sqrt{1 + 4 + 9} = \sqrt{14}$

$$\therefore l_2 = \frac{1}{\sqrt{14}}; \quad m_2 = \frac{-2}{\sqrt{14}}; \quad n_2 = \frac{3}{\sqrt{14}}$$

Then $\cos\theta = \dfrac{1}{\sqrt{14 \times 29}}\{2 - 6 + 12\} = 0.3970$

$$\therefore \underline{\theta = 66° \; 36'}$$

Let us now look at the question of scalar and vector products.

On to the next frame.

5

Example 3 (Scalar product)

If **A** and **B** are two vectors, the scalar product of **A** and **B** is defined as

$$\mathbf{A}\cdot\mathbf{B} = AB\cos\theta \tag{2}$$

where θ is the angle between the two vectors.

If we consider the *scalar products of the unit vectors* **i**, **j**, **k**, which are mutually perpendicular, then

$$\mathbf{i}\cdot\mathbf{j} = (1)(1)\cos 90^\circ = 0 \qquad \therefore \mathbf{i}\cdot\mathbf{j} = \mathbf{j}\cdot\mathbf{k} = \mathbf{k}\cdot\mathbf{i} = 0$$

and $\quad \mathbf{i}\cdot\mathbf{i} = (1)(1)\cos 0^\circ = 1 \qquad \therefore \mathbf{i}\cdot\mathbf{i} = \mathbf{j}\cdot\mathbf{j} = \mathbf{k}\cdot\mathbf{k} = 1.$

In general, if $\mathbf{A} = a_x\mathbf{i} + a_y\mathbf{j} + a_z\mathbf{k}$ and $\mathbf{B} = b_x\mathbf{i} + b_y\mathbf{j} + b_z\mathbf{k}$ then $\mathbf{A}\cdot\mathbf{B} = a_xb_x + a_yb_y + a_zb_z$ which is, of course, a scalar quantity.

So, if $\mathbf{A} = 2\mathbf{i} - 3\mathbf{j} + 4\mathbf{k}$ and $\mathbf{B} = \mathbf{i} + 2\mathbf{j} + 5\mathbf{k}$, then

$$\mathbf{A}\cdot\mathbf{B} = \ldots\ldots\ldots$$

6

$$\boxed{\mathbf{A}\cdot\mathbf{B} = 2 - 6 + 20 = 16}$$

Also, since $\mathbf{A}.\mathbf{B} = AB\cos\theta$, we can determine the angle θ between the vectors. In this case $\theta = \ldots\ldots\ldots$

7

$$\boxed{\theta = 57^\circ\ 9'}$$

$\mathbf{A} = 2\mathbf{i} - 3\mathbf{j} + 4\mathbf{k} \qquad \therefore A = |\mathbf{A}| = \sqrt{4 + 9 + 16} = \sqrt{29}$

$\mathbf{B} = \mathbf{i} + 2\mathbf{j} + 5\mathbf{k} \qquad \therefore B = |\mathbf{B}| = \sqrt{1 + 4 + 25} = \sqrt{30}$

We have already found that $\mathbf{A}\cdot\mathbf{B} = 16$ and $\mathbf{A}\cdot\mathbf{B} = AB\cos\theta$

$$\therefore 16 = \sqrt{29}\sqrt{30}\cos\theta \quad \therefore \cos\theta = 0.5425 \quad \therefore \underline{\theta = 57^\circ\ 9'}$$

So, the *scalar product* of $\mathbf{A} = a_x\mathbf{i} + a_y\mathbf{j} + a_z\mathbf{k}$ and $\mathbf{B} = b_x\mathbf{i} + b_y\mathbf{j} + b_z\mathbf{k}$

is $\quad \mathbf{A}\cdot\mathbf{B} = a_xb_x + a_yb_y + a_zb_z$

and $\quad \mathbf{A}\cdot\mathbf{B} = AB\cos\theta$ where θ is the angle between

the vectors. It can also be shown that

(i) $\mathbf{A}\cdot\mathbf{B} = \mathbf{B}\cdot\mathbf{A}$

and (ii) $\mathbf{A}\cdot(\mathbf{B} + \mathbf{C}) = \mathbf{A}\cdot\mathbf{B} + \mathbf{A}\cdot\mathbf{C}$

Make a note of these results.

8

Example 4 (Vector product)

If $\mathbf{A} = a_x\mathbf{i} + a_y\mathbf{j} + a_z\mathbf{k}$ and $\mathbf{B} = b_x\mathbf{i} + b_y\mathbf{j} + b_z\mathbf{k}$ the vector product $\mathbf{A} \times \mathbf{B}$ is defined as $\mathbf{A} \times \mathbf{B} = AB\sin\theta$ in the direction perpendicular to $\mathbf{A}$ and $\mathbf{B}$ such that $\mathbf{A}$, $\mathbf{B}$ and $(\mathbf{A} \times \mathbf{B})$ form a right-handed set.

We can write this as

$$\mathbf{A} \times \mathbf{B} = (AB\sin\theta)\mathbf{n} \tag{3}$$

where $\mathbf{n}$ is defined as a unit vector in the positive normal direction to the plane of $\mathbf{A}$ and $\mathbf{B}$, i.e. forming a right-handed set.

Also
$$\mathbf{A} \times \mathbf{B} = \begin{vmatrix} \mathbf{i} & \mathbf{j} & \mathbf{k} \\ a_x & a_y & a_z \\ b_x & b_y & b_z \end{vmatrix} \tag{4}$$

If we consider the *vector products of the unit vectors*, $\mathbf{i}$, $\mathbf{j}$, $\mathbf{k}$, then

$\mathbf{i} \times \mathbf{j} = (1)(1)\sin 90° = 1 \qquad \therefore\ \mathbf{i} \times \mathbf{j} = \mathbf{j} \times \mathbf{k} = \mathbf{k} \times \mathbf{i} = 1$

Note that $\mathbf{j} \times \mathbf{i} = -(\mathbf{i} \times \mathbf{j}) = -1 \quad \therefore\ \mathbf{j} \times \mathbf{i} = \mathbf{k} \times \mathbf{j} = \mathbf{i} \times \mathbf{k} = -1$

Also $\mathbf{i} \times \mathbf{i} = (1)(1)\sin 0° = 0 \qquad \therefore\ \mathbf{i} \times \mathbf{i} = \mathbf{j} \times \mathbf{j} = \mathbf{k} \times \mathbf{k} = 0.$

It can also be shown that

(i) $\mathbf{A} \times (\mathbf{B} + \mathbf{C}) = \mathbf{A} \times \mathbf{B} + \mathbf{A} \times \mathbf{C}$

and (ii) $\mathbf{A} \times \mathbf{B} = -(\mathbf{B} \times \mathbf{A})$ (5)

Make a note of these results (3), (4) and (5).

Then, if $\mathbf{A} = 3\mathbf{i} - 2\mathbf{j} + 4\mathbf{k}$ and $\mathbf{B} = 2\mathbf{i} - 3\mathbf{j} - 2\mathbf{k}$

$$\mathbf{A} \times \mathbf{B} = \dots\dots\dots$$

9

$$\boxed{\mathbf{A} \times \mathbf{B} = 16\mathbf{i} + 14\mathbf{j} - 5\mathbf{k}}$$

We simply evaluate the determinant

$$\mathbf{A} \times \mathbf{B} = \begin{vmatrix} \mathbf{i} & \mathbf{j} & \mathbf{k} \\ 3 & -2 & 4 \\ 2 & -3 & -2 \end{vmatrix}$$

$$= \mathbf{i}(4 + 12) - \mathbf{j}(-6 - 8) + \mathbf{k}(-9 + 4) = \underline{16\mathbf{i} + 14\mathbf{j} - 5\mathbf{k}}$$

Move on to the next frame.

10

We have seen therefore that

the scalar product of two vectors is a scalar
but that the vector product of two vectors is a vector.

We know also that $|\mathbf{A} \times \mathbf{B}| = AB\sin\theta$
Therefore, the angle between the vectors **A** and **B** given in Example 4 is

$$\theta = \ldots\ldots\ldots$$

11

$$\boxed{\theta = 79^\circ\ 40'}$$

for $\mathbf{A} = 3\mathbf{i} - 2\mathbf{j} + 4\mathbf{k}$; $\mathbf{B} = 2\mathbf{i} - 3\mathbf{j} - 2\mathbf{k}$; and $\mathbf{A} \times \mathbf{B} = 16\mathbf{i} + 14\mathbf{j} - 5\mathbf{k}$

$$\therefore\ |\mathbf{A} \times \mathbf{B}| = \sqrt{16^2 + 14^2 + 5^2} = \sqrt{477} = 21.84$$

$$A = |\mathbf{A}| = \sqrt{3^2 + 2^2 + 4^2} = \sqrt{29} = 5.385$$

$$B = |\mathbf{B}| = \sqrt{2^2 + 3^2 + 2^2} = \sqrt{17} = 4.123$$

$$\therefore\ 21.84 = (5.385)(4.123)\sin\theta$$

$$\therefore\ \sin\theta = 0.9838 \quad \therefore\ \underline{\theta = 79^\circ\ 40'}$$

So, to recapitulate:

If $\mathbf{A} = a_x\mathbf{i} + a_y\mathbf{j} + a_z\mathbf{k}$ and $\mathbf{B} = b_x\mathbf{i} + b_y\mathbf{j} + b_z\mathbf{k}$ and θ is the angle between them

(i) Scalar product $= \mathbf{A} \cdot \mathbf{B} = a_x b_x + a_y b_y + a_z b_z$
$= AB\cos\theta$

(ii) Vector product $= \mathbf{A} \times \mathbf{B} = \begin{vmatrix} \mathbf{i} & \mathbf{j} & \mathbf{k} \\ a_x & a_y & a_z \\ b_x & b_y & b_z \end{vmatrix}$

and $\quad |\mathbf{A} \times \mathbf{B}| = AB\sin\theta.$

Make a note of these fundamental results: we shall certainly need them. Then, in the next frame, we can set off on some new work.

12

Triple products

We now deal with the various products that we form with three vectors.

Scalar triple product of three vectors.

If $\mathbf{A}, \mathbf{B}, \mathbf{C}$ are three vectors, the scalar formed by the product $\mathbf{A} \cdot (\mathbf{B} \times \mathbf{C})$ is called the scalar triple product.

If $\mathbf{A} = a_x\mathbf{i} + a_y\mathbf{j} + a_z\mathbf{k}$; $\mathbf{B} = b_x\mathbf{i} + b_y\mathbf{j} + b_z\mathbf{k}$; $\mathbf{C} = c_x\mathbf{i} + c_y\mathbf{j} + c_z\mathbf{k}$;

then $$\mathbf{B} \times \mathbf{C} = \begin{vmatrix} \mathbf{i} & \mathbf{j} & \mathbf{k} \\ b_x & b_y & b_z \\ c_x & c_y & c_z \end{vmatrix}$$

$$\therefore \mathbf{A} \cdot (\mathbf{B} \times \mathbf{C}) = (a_x\mathbf{i} + a_y\mathbf{j} + a_z\mathbf{k}) \cdot \begin{vmatrix} \mathbf{i} & \mathbf{j} & \mathbf{k} \\ b_x & b_y & b_z \\ c_x & c_y & c_z \end{vmatrix}$$

Multiplying the top row by the external bracket and remembering that

$$\mathbf{i} \cdot \mathbf{j} = \mathbf{j} \cdot \mathbf{k} = \mathbf{k} \cdot \mathbf{i} = 0 \quad \text{and} \quad \mathbf{i} \cdot \mathbf{i} = \mathbf{j} \cdot \mathbf{j} = \mathbf{k} \cdot \mathbf{k} = 1$$

we have $$\mathbf{A} \cdot (\mathbf{B} \times \mathbf{C}) = \begin{vmatrix} a_x & a_y & a_z \\ b_x & b_y & b_z \\ c_x & c_y & c_z \end{vmatrix} \qquad (6)$$

Example If $\mathbf{A} = 2\mathbf{i} - 3\mathbf{j} + 4\mathbf{k}$; $\mathbf{B} = \mathbf{i} - 2\mathbf{j} - 3\mathbf{k}$; $\mathbf{C} = 2\mathbf{i} + \mathbf{j} + 2\mathbf{k}$;

then $$\mathbf{A} \cdot (\mathbf{B} \times \mathbf{C}) = \begin{vmatrix} 2 & -3 & 4 \\ 1 & -2 & -3 \\ 2 & 1 & 2 \end{vmatrix}$$

$$= \ldots\ldots\ldots$$

13

$$\boxed{\mathbf{A} \cdot (\mathbf{B} \times \mathbf{C}) = 42}$$

for $$\mathbf{A} \cdot (\mathbf{B} \times \mathbf{C}) = \begin{vmatrix} 2 & -3 & 4 \\ 1 & -2 & -3 \\ 2 & 1 & 2 \end{vmatrix}$$

$$= 2(-4 + 3) + 3(2 + 6) + 4(1 + 4) = \underline{42}$$

As simple as that.

Properties of scalar triple products

(a) $$\mathbf{B}\cdot(\mathbf{C}\times\mathbf{A}) = \begin{vmatrix} b_x & b_y & b_z \\ c_x & c_y & c_z \\ a_x & a_y & a_z \end{vmatrix} = -\begin{vmatrix} a_x & a_y & a_z \\ c_x & c_y & c_z \\ b_x & b_y & b_z \end{vmatrix}$$

since interchanging two rows in a determinant reverses the sign. If we now interchange rows 2 and 3 and again change the sign, we have

$$\mathbf{B}\cdot(\mathbf{C}\times\mathbf{A}) = \begin{vmatrix} a_x & a_y & a_z \\ b_x & b_y & b_z \\ c_x & c_y & c_z \end{vmatrix} = \mathbf{A}\cdot(\mathbf{B}\times\mathbf{C})$$

$$\therefore\ \mathbf{A}\cdot(\mathbf{B}\times\mathbf{C}) = \mathbf{B}\cdot(\mathbf{C}\times\mathbf{A}) = \mathbf{C}\cdot(\mathbf{A}\times\mathbf{B}) \qquad (7)$$

i.e. the scalar triple product is unchanged by a cyclic change of the vectors involved.

(b) $$\mathbf{B}\cdot(\mathbf{A}\times\mathbf{C}) = \begin{vmatrix} b_x & b_y & b_z \\ a_x & a_y & a_z \\ c_x & c_y & c_z \end{vmatrix} = -\begin{vmatrix} a_x & a_y & a_z \\ b_x & b_y & b_z \\ c_x & c_y & c_z \end{vmatrix}$$

$$\therefore\ \mathbf{B}\cdot(\mathbf{A}\times\mathbf{C}) = -\mathbf{A}\cdot(\mathbf{B}\times\mathbf{C}) \qquad (8)$$

i.e. a change of vectors not in cyclic order, changes the sign of the scalar triple product.

(c) $\mathbf{A}\cdot(\mathbf{B}\times\mathbf{A}) = \begin{vmatrix} a_x & a_y & a_z \\ b_x & b_y & b_z \\ a_x & a_y & a_z \end{vmatrix} = 0$ since two rows are identical.

$$\therefore\ \mathbf{A}\cdot(\mathbf{B}\times\mathbf{A}) = \mathbf{B}\cdot(\mathbf{C}\times\mathbf{B}) = \mathbf{C}\cdot(\mathbf{A}\times\mathbf{C}) = 0 \qquad (9)$$

Example If $\mathbf{A} = \mathbf{i} + 2\mathbf{j} + 3\mathbf{k}$; $\mathbf{B} = 2\mathbf{i} - 3\mathbf{j} + \mathbf{k}$; $\mathbf{C} = 3\mathbf{i} + \mathbf{j} - 2\mathbf{k}$

$\mathbf{A}\cdot(\mathbf{B}\times\mathbf{C}) = \ldots\ldots\ldots$ $\qquad$ $\mathbf{C}\cdot(\mathbf{B}\times\mathbf{A}) = \ldots\ldots\ldots$

15

$$\boxed{\mathbf{A}\cdot(\mathbf{B}\times\mathbf{C}) = 52; \quad \mathbf{C}\cdot(\mathbf{A}\times\mathbf{B}) = -52}$$

$$\text{for } \mathbf{A}\cdot(\mathbf{B}\times\mathbf{C}) = \begin{vmatrix} 1 & 2 & 3 \\ 2 & -3 & 1 \\ 3 & 1 & -2 \end{vmatrix} = 1(6-1) - 2(-4-3) + 3(2+9) = \underline{52}$$

$\mathbf{C}\cdot(\mathbf{B}\times\mathbf{A})$ is not a cyclic change from the above. Therefore

$$\mathbf{C}\cdot(\mathbf{B}\times\mathbf{A}) = -\mathbf{A}\cdot(\mathbf{B}\times\mathbf{C}) = \underline{-52}$$

Coplanar vectors

The scalar triple product provides a test to show whether three given vectors lie in the same plane.

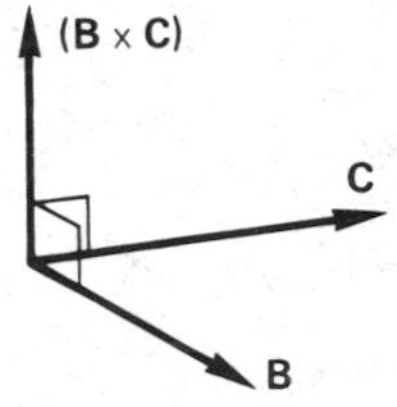

By definition, $(\mathbf{B}\times\mathbf{C})$ is a vector product of $\mathbf{B}$ and $\mathbf{C}$, of magnitude $|\mathbf{B}||\mathbf{C}|$ acting in a direction perpendicular to the plane of $\mathbf{B}$ and $\mathbf{C}$ and forming a right-handed set.

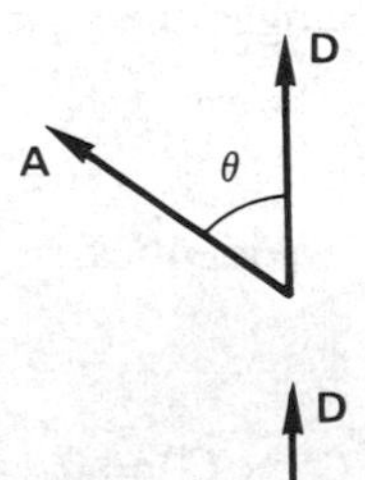

The scalar product of two vectors $\mathbf{A}$ and $\mathbf{D}$ is $\mathbf{A}\cdot\mathbf{D}$ where
$\mathbf{A}\cdot\mathbf{D} = |\mathbf{A}||\mathbf{D}|\cos\theta$, θ being the angle between $\mathbf{A}$ and $\mathbf{D}$.

If $\theta = \dfrac{\pi}{2}$, $\mathbf{A}\cdot\mathbf{D} = |\mathbf{A}||\mathbf{D}|\cos\dfrac{\pi}{2} = 0$

$$\therefore\ \mathbf{A}\cdot\mathbf{D} = 0$$

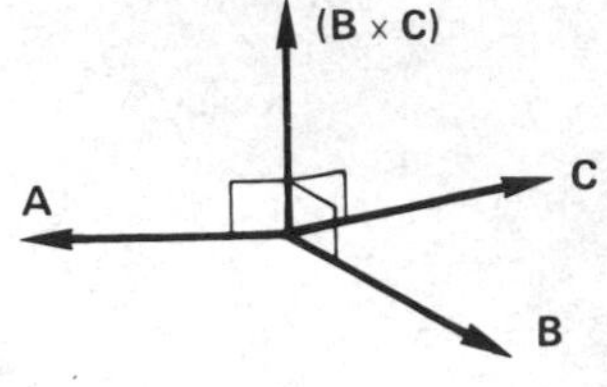

Therefore, if $\mathbf{D} = (\mathbf{B}\times\mathbf{C})$, combining the two results above, we have this conclusion:
If $\mathbf{A}$ is a third vector perpendicular to $(\mathbf{B}\times\mathbf{C})$, then
(i) $\mathbf{A}, \mathbf{B}$ and $\mathbf{C}$ are coplanar
(ii) $\mathbf{A}\cdot(\mathbf{B}\times\mathbf{C}) = 0$.

Therefore, three vectors $\underline{\mathbf{A},\ \mathbf{B},\ \mathbf{C}\text{ are coplanar if }\mathbf{A}\cdot(\mathbf{B}\times\mathbf{C}) = 0.}$

Example 1 Show that $\mathbf{A} = \mathbf{i} + 2\mathbf{j} - 3\mathbf{k}$; $\mathbf{B} = 2\mathbf{i} - \mathbf{j} + 2\mathbf{k}$; and $\mathbf{C} = 3\mathbf{i} + \mathbf{j} - \mathbf{k}$ are coplanar.

We just evaluate $\mathbf{A} \cdot (\mathbf{B} \times \mathbf{C}) = \dots\dots\dots$ and apply the test.

16

$$\boxed{\mathbf{A} \cdot (\mathbf{B} \times \mathbf{C}) = 0}$$

for $\mathbf{A} \cdot (\mathbf{B} \times \mathbf{C}) = \begin{vmatrix} 1 & 2 & -3 \\ 2 & -1 & 2 \\ 3 & 1 & -1 \end{vmatrix} = 1(1-2) - 2(-2-6) - 3(2+3) = 0.$

Therefore $\underline{\mathbf{A}, \mathbf{B}, \mathbf{C} \text{ are coplanar.}}$

Example 2 If $\mathbf{A} = 2\mathbf{i} - \mathbf{j} + 3\mathbf{k}$; $\mathbf{B} = 3\mathbf{i} + 2\mathbf{j} + \mathbf{k}$; $\mathbf{C} = \mathbf{i} + p\mathbf{j} + 4\mathbf{k}$ are coplanar, find the value of p.

The method is clear enough. We merely set up and evaluate the determinant and solve the equation $\mathbf{A} \cdot (\mathbf{B} \times \mathbf{C}) = 0$.

$$p = \dots\dots\dots$$

17

$$\boxed{p = -3}$$

since $\mathbf{A} \cdot (\mathbf{B} \times \mathbf{C}) = 0 \quad \therefore \begin{vmatrix} 2 & -1 & 3 \\ 3 & 2 & 1 \\ 1 & p & 4 \end{vmatrix} = 0$

$\therefore 2(8-p) + 1(12-1) + 3(3p-2) = 0 \quad \therefore 7p = -21 \quad \therefore \underline{p = -3}$

One more.

Example 3 Determine whether the three vectors $\mathbf{A} = 3\mathbf{i} + 2\mathbf{j} - \mathbf{k}$; $\mathbf{B} = 2\mathbf{i} - \mathbf{j} + 3\mathbf{k}$; $\mathbf{C} = \mathbf{i} - 2\mathbf{j} + 2\mathbf{k}$ are coplanar.

Work through it on your own. The result shows that

$\dots\dots\dots\dots\dots\dots$

18

$$\boxed{\mathbf{A},\ \mathbf{B},\ \mathbf{C} \text{ are not coplanar}}$$

In this case $\mathbf{A}\cdot(\mathbf{B}\times\mathbf{C}) = \begin{vmatrix} 3 & 2 & -1 \\ 2 & -1 & 3 \\ 1 & -2 & 2 \end{vmatrix} = 13$

$\therefore\ \mathbf{A}\cdot(\mathbf{B}\times\mathbf{C}) \neq 0 \qquad \therefore$ <u>$\mathbf{A}$, $\mathbf{B}$, $\mathbf{C}$ are not coplanar.</u>

Now on to something different.

19 Vector triple products of three vectors

If **A**, **B** and **C** are three vectors, then

$$\left.\begin{matrix} \mathbf{A}\times(\mathbf{B}\times\mathbf{C}) \\ \text{and} \quad (\mathbf{A}\times\mathbf{B})\times\mathbf{C} \end{matrix}\right\}\text{are called the vector triple products.} \qquad (10)$$

Consider $\mathbf{A}\times(\mathbf{B}\times\mathbf{C})$ where $\mathbf{A} = a_x\mathbf{i} + a_y\mathbf{j} + a_z\mathbf{k}$; $\mathbf{B} = b_x\mathbf{i} + b_y\mathbf{i} + \mathbf{b}_z\mathbf{k}$; and $\mathbf{C} = c_x\mathbf{i} + c_y\mathbf{j} + c_z\mathbf{k}$.

Then $(\mathbf{B}\times\mathbf{C})$ is a vector perpendicular to the plane of **B** and **C** and $\mathbf{A}\times(\mathbf{B}\times\mathbf{C})$ is a vector perpendicular to the plane containing **A** and $(\mathbf{B}\times\mathbf{C})$, i.e. coplanar with **B** and **C**.

Now

$$(\mathbf{B}\times\mathbf{C}) = \begin{vmatrix} \mathbf{i} & \mathbf{j} & \mathbf{k} \\ b_x & b_y & b_z \\ c_x & c_y & c_z \end{vmatrix} = \mathbf{i}\begin{vmatrix} b_y & b_z \\ c_y & c_z \end{vmatrix} - \mathbf{j}\begin{vmatrix} b_x & b_z \\ c_x & c_z \end{vmatrix} + \mathbf{k}\begin{vmatrix} b_x & b_y \\ c_x & c_y \end{vmatrix}$$

Then $$\mathbf{A}\times(\mathbf{B}\times\mathbf{C}) = \begin{vmatrix} \mathbf{i} & \mathbf{j} & \mathbf{k} \\ a_x & a_y & a_z \\ \begin{vmatrix} b_y & b_z \\ c_y & c_z \end{vmatrix} & -\begin{vmatrix} b_x & b_z \\ c_x & c_z \end{vmatrix} & \begin{vmatrix} b_x & b_y \\ c_x & c_y \end{vmatrix} \end{vmatrix}$$

$$= \begin{vmatrix} \mathbf{i} & \mathbf{j} & \mathbf{k} \\ a_x & a_y & a_z \\ \begin{vmatrix} b_y & b_z \\ c_y & c_z \end{vmatrix} & \begin{vmatrix} b_z & b_x \\ c_z & c_x \end{vmatrix} & \begin{vmatrix} b_x & b_y \\ c_x & c_y \end{vmatrix} \end{vmatrix}$$

In symbolic form, further expansion of the determinant becomes somewhat tedious. However a numerical example will clarify the method.

So make a note of the definition (10) above and then go on to the next fra

---

Example 1 If $\mathbf{A} = 2\mathbf{i} - 3\mathbf{j} + \mathbf{k}$; $\mathbf{B} = \mathbf{i} + 2\mathbf{j} - \mathbf{k}$; $\mathbf{C} = 3\mathbf{i} + \mathbf{j} + 3\mathbf{k}$; determine the vector triple product $\mathbf{A} \times (\mathbf{B} \times \mathbf{C})$. **20**

We start off with $\mathbf{B} \times \mathbf{C} = \ldots\ldots\ldots$

21

$$\boxed{\mathbf{B} \times \mathbf{C} = 7\mathbf{i} - 6\mathbf{j} - 5\mathbf{k}}$$

for $\mathbf{B} \times \mathbf{C} = \begin{vmatrix} \mathbf{i} & \mathbf{j} & \mathbf{k} \\ 1 & 2 & -1 \\ 3 & 1 & 3 \end{vmatrix} = \mathbf{i}(6+1) - \mathbf{j}(3+3) + \mathbf{k}(1-6)$
$= 7\mathbf{i} - 6\mathbf{j} - 5\mathbf{k}$

Then $\mathbf{A} \times (\mathbf{B} \times \mathbf{C}) = \ldots\ldots\ldots$

22

$$\boxed{\mathbf{A} \times (\mathbf{B} \times \mathbf{C}) = 21\mathbf{i} + 17\mathbf{j} + 9\mathbf{k}}$$

for $\mathbf{A} \times (\mathbf{B} \times \mathbf{C}) = \begin{vmatrix} \mathbf{i} & \mathbf{j} & \mathbf{k} \\ 2 & -3 & 1 \\ 7 & -6 & 5 \end{vmatrix}$

$$= \mathbf{i}(15+6) - \mathbf{j}(-10-7) + \mathbf{k}(-12+21)$$
$$= \underline{21\mathbf{i} + 17\mathbf{j} + 9\mathbf{k}}$$

That is fundamental enough. There is, however, an even easier way of determining a vector triple product. It can be proved that

$$\mathbf{A} \times (\mathbf{B} \times \mathbf{C}) = (\mathbf{A} \cdot \mathbf{C})\mathbf{B} - \mathbf{A} \cdot \mathbf{B})\mathbf{C} \qquad (11)$$

and $$(\mathbf{A} \times \mathbf{B}) \times \mathbf{C} = (\mathbf{C} \cdot \mathbf{A})\mathbf{B} - (\mathbf{C} \cdot \mathbf{B})\mathbf{A}$$

The proof of this is given in the Appendix. For the moment, make a careful note of the expressions: then we will apply the method to the example we have just completed.

23 $\mathbf{A} = 2\mathbf{i} - 3\mathbf{j} + \mathbf{k}$; $\mathbf{B} = \mathbf{i} + 2\mathbf{j} - \mathbf{k}$; $\mathbf{C} = 3\mathbf{i} + \mathbf{j} + 3\mathbf{k}$ and we have

$$\begin{aligned}\mathbf{A} \times (\mathbf{B} \times \mathbf{C}) &= (\mathbf{A} \cdot \mathbf{C})\mathbf{B} - (\mathbf{A} \cdot \mathbf{B})\mathbf{C} \\ &= (6 - 3 + 3)(\mathbf{i} + 2\mathbf{j} - \mathbf{k}) - (2 - 6 - 1)(3\mathbf{i} + \mathbf{j} + 3\mathbf{k}) \\ &= 6\,(\mathbf{i} + 2\mathbf{j} - \mathbf{k}) + 5(3\mathbf{i} + \mathbf{j} + 3\mathbf{k}) = \underline{21\mathbf{i} + 17\mathbf{j} + 9\mathbf{k}}\end{aligned}$$

which is, of course, the result we achieved before.

Here is another.

Example 2 If $\mathbf{A} = 3\mathbf{i} + 2\mathbf{j} - 2\mathbf{k}$; $\mathbf{B} = 4\mathbf{i} - \mathbf{j} + 3\mathbf{k}$; $\mathbf{C} = 2\mathbf{i} - 3\mathbf{j} + \mathbf{k}$ determine $(\mathbf{A} \times \mathbf{B}) \times \mathbf{C}$ using the relationship

$(\mathbf{A} \times \mathbf{B}) \times \mathbf{C} = (\mathbf{C} \cdot \mathbf{A})\mathbf{B} - (\mathbf{C} \cdot \mathbf{B})\mathbf{A}$.

$$(\mathbf{A} \times \mathbf{B}) \times \mathbf{C} = \ldots\ldots\ldots$$

24

$$\boxed{-50\mathbf{i} - 26\mathbf{j} + 22\mathbf{k}}$$

$$\begin{aligned}\text{for } (\mathbf{A} \times \mathbf{B}) \times \mathbf{C} &= (\mathbf{C} \cdot \mathbf{A})\mathbf{B} - (\mathbf{C} \cdot \mathbf{B})\mathbf{A} \\ &= (6 - 6 - 2)(4\mathbf{i} - \mathbf{j} + 3\mathbf{k}) - (8 + 3 + 3)(3\mathbf{i} + 2\mathbf{j} - 2\mathbf{k}) \\ &= -2(4\mathbf{i} - \mathbf{j} + 3\mathbf{k}) - 14(3\mathbf{i} + 2\mathbf{j} - 2\mathbf{k}) \\ &= \underline{-50\mathbf{i} - 26\mathbf{j} + 22\mathbf{k}}\end{aligned}$$

Now one more.

Example 3 If $\mathbf{A} = \mathbf{i} + 3\mathbf{j} + 2\mathbf{k}$; $\mathbf{B} = 2\mathbf{i} + 5\mathbf{j} - \mathbf{k}$; $\mathbf{C} = \mathbf{i} + 2\mathbf{j} + 3\mathbf{k}$

$$\mathbf{A} \times (\mathbf{B} \times \mathbf{C}) = \ldots\ldots\ldots$$

$$(\mathbf{A} \times \mathbf{B}) \times \mathbf{C} = \ldots\ldots\ldots$$

Finish them both.

25

$$\boxed{\begin{aligned} \mathbf{A} \times (\mathbf{B} \times \mathbf{C}) &= 11\mathbf{i} + 35\mathbf{j} - 58\mathbf{k} \\ (\mathbf{A} \times \mathbf{B}) \times \mathbf{C} &= 17\mathbf{i} + 38\mathbf{j} - 31\mathbf{k} \end{aligned}}$$

for
$$\begin{aligned} \mathbf{A} \times (\mathbf{B} \times \mathbf{C}) &= (\mathbf{A} \cdot \mathbf{C})\mathbf{B} - (\mathbf{A} \cdot \mathbf{B})\mathbf{C} \\ &= (1 + 6 + 6)(2\mathbf{i} + 5\mathbf{j} - \mathbf{k}) - (2 + 15 - 2)(\mathbf{i} + 2\mathbf{j} + 3\mathbf{k}) \\ &= 13(2\mathbf{i} + 5\mathbf{j} - \mathbf{k}) - 15(\mathbf{i} + 2\mathbf{j} + 3\mathbf{k}) \\ &= \underline{11\mathbf{i} + 35\mathbf{j} - 58\mathbf{k}} \end{aligned}$$

and
$$\begin{aligned} (\mathbf{A} \times \mathbf{B}) \times \mathbf{C} &= (\mathbf{C} \cdot \mathbf{A})\mathbf{B} - (\mathbf{C} \cdot \mathbf{B})\mathbf{A} \\ &= (1 + 6 + 6)(2\mathbf{i} + 5\mathbf{j} - \mathbf{k}) - (2 + 10 - 3)(\mathbf{i} + 3\mathbf{j} + 2\mathbf{k}) \\ &= 13(2\mathbf{i} + 5\mathbf{j} - \mathbf{k}) - 9(\mathbf{i} + 3\mathbf{j} + 2\mathbf{k}) = \underline{17\mathbf{i} + 38\mathbf{j} - 31\mathbf{k}} \end{aligned}$$

As these two results clearly show

$$\mathbf{A} \times (\mathbf{B} \times \mathbf{C}) \neq (\mathbf{A} \times \mathbf{B}) \times \mathbf{C} \qquad \textit{so beware!}$$

Before we proceed, note the following concerning the unit vectors.

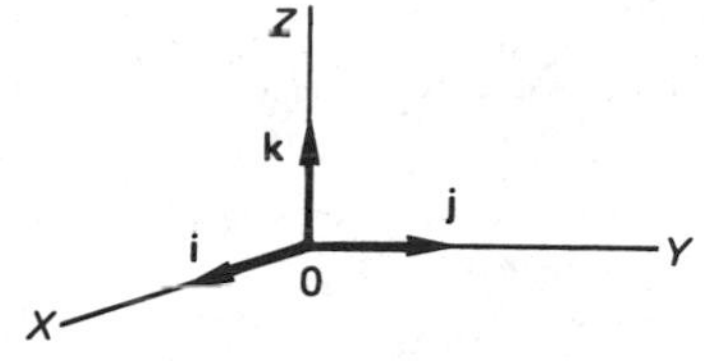

(i) $(\mathbf{i} \times \mathbf{j}) = \mathbf{k}$

$\therefore\ \mathbf{i} \times (\mathbf{i} \times \mathbf{j}) = \mathbf{i} \times \mathbf{k} = -\mathbf{j}$

$\therefore\ \underline{\mathbf{i} \times (\mathbf{i} \times \mathbf{j}) = -\mathbf{j}}$

(ii) $(\mathbf{i} \times \mathbf{i}) \times \mathbf{j} = (0) \times \mathbf{j} = 0$

$\therefore\ \underline{(\mathbf{i} \times \mathbf{i}) \times \mathbf{j} = 0}$

and once again, we see that $\underline{\mathbf{i} \times (\mathbf{i} \times \mathbf{j}) \neq (\mathbf{i} \times \mathbf{i}) \times \mathbf{j}}$

On to the next.

26

Finally, by way of revision:

Example 4 If $\mathbf{A} = 5\mathbf{i} - 2\mathbf{j} + 3\mathbf{k}$; $\mathbf{B} = 3\mathbf{i} + \mathbf{j} - 2\mathbf{k}$; $\mathbf{C} = \mathbf{i} - 3\mathbf{j} + 4\mathbf{k}$; determine

(a) the scalar triple product $\mathbf{A} \cdot (\mathbf{B} \times \mathbf{C})$
(b) the vector triple products (i) $\mathbf{A} \times (\mathbf{B} \times \mathbf{C})$
(ii) $(\mathbf{A} \times \mathbf{B}) \times \mathbf{C}$.

Finish all these and then check with the next frame.

27

$$
\boxed{\begin{array}{l}
\text{(a) } \mathbf{A}\cdot(\mathbf{B}\times\mathbf{C}) = -12 \\
\text{(b) } \mathbf{A}\times(\mathbf{B}\times\mathbf{C}) = 62\mathbf{i}+44\mathbf{j}-74\mathbf{k} \\
\quad\ (\mathbf{A}\times\mathbf{B})\times\mathbf{C} = 109\mathbf{i}+7\mathbf{j}-22\mathbf{k}
\end{array}}
$$

Here is the working.

(a)

$$\mathbf{A}\cdot(\mathbf{B}\times\mathbf{C}) = \begin{vmatrix} 5 & -2 & 3 \\ 3 & 1 & -2 \\ 1 & -3 & 4 \end{vmatrix}$$

$$= 5(4-6) + 2(12+2) + 3(-9-1) = \underline{-12}$$

(b) (i) $\mathbf{A}\times(\mathbf{B}\times\mathbf{C}) = (\mathbf{A}\cdot\mathbf{C})\mathbf{B} - (\mathbf{A}\cdot\mathbf{B})\mathbf{C}$

$$= (5+6+12)(3\mathbf{i}+\mathbf{j}-2\mathbf{k}) - (15-2-6)(\mathbf{i}-3\mathbf{j}+4\mathbf{k})$$

$$= 23(3\mathbf{i}+\mathbf{j}-2\mathbf{k}) - 7(\mathbf{i}-3\mathbf{j}+4\mathbf{k}) = \underline{62\mathbf{i}+44\mathbf{j}-74\mathbf{k}}$$

(ii) $(\mathbf{A}\times\mathbf{B})\times\mathbf{C} = (\mathbf{C}\cdot\mathbf{A})\mathbf{B} - (\mathbf{C}\cdot\mathbf{B})\mathbf{A}$

$$= 23(3\mathbf{i}+\mathbf{j}-2\mathbf{k}) - (-8)(5\mathbf{i}-2\mathbf{j}+3\mathbf{k})$$

$$= \underline{109\mathbf{i}+7\mathbf{j}-22\mathbf{k}}$$

Let us now move to the next topic.

28

Differentiation of vectors

In many practical problems, we often deal with vectors that change with time, e.g. velocity, acceleration, etc. If a vector **A** depends on a scalar variable t, then **A** can be represented as $\mathbf{A}(t)$ and **A** is then said to be a function of t.

If $\mathbf{A} = a_x\mathbf{i} + a_y\mathbf{j} + a_z\mathbf{k}$ then a_x, a_y, a_z will also be dependent on the parameter t.

i.e. $$\mathbf{A}(t) = a_x(t)\mathbf{i} + a_y(t)\mathbf{j} + a_z(t)\mathbf{k}$$

Differentiating with respect to t gives ....

29

$$\frac{\mathrm{d}}{\mathrm{d}t}\{\mathbf{A}(t)\} = \mathbf{i}\frac{\mathrm{d}}{\mathrm{d}t}\{a_x(t)\} + \mathbf{j}\frac{\mathrm{d}}{\mathrm{d}t}\{a_y(t)\} + \mathbf{k}\frac{\mathrm{d}}{\mathrm{d}t}\{a_z(t)\}$$

In short $\dfrac{\mathrm{d}\mathbf{A}}{\mathrm{d}t} = \mathbf{i}\dfrac{\mathrm{d}a_x}{\mathrm{d}t} + \mathbf{j}\dfrac{\mathrm{d}a_y}{\mathrm{d}t} + \mathbf{k}\dfrac{\mathrm{d}a_z}{\mathrm{d}t}$.

The independent scalar variable is not, of course, restricted to t. In general, if u is the parameter, then

$$\frac{\mathrm{d}\mathbf{A}}{\mathrm{d}u} = \ldots\ldots\ldots$$

30

$$\frac{d\mathbf{A}}{du} = \mathbf{i}\frac{da_x}{du} + \mathbf{j}\frac{da_y}{du} + \mathbf{k}\frac{da_z}{du}$$

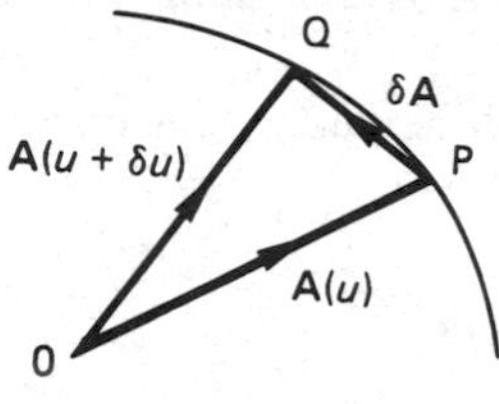

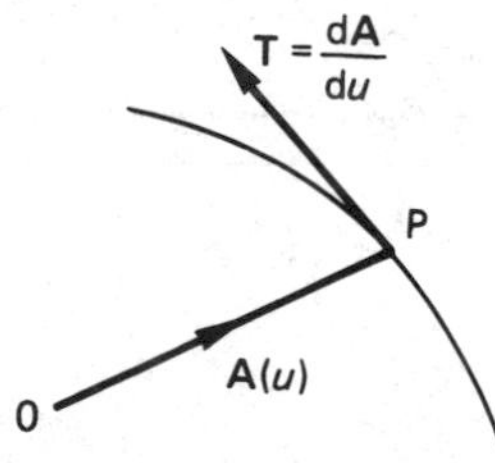

If a position vector $\overline{OP}$ moves to $\overline{OQ}$ when u becomes $u + \delta u$, then as $\delta u \to 0$, the direction of the chord $\overline{PQ}$ becomes that of the tangent to the curve at **P**, i.e. the direction of $\dfrac{d\mathbf{A}}{du}$ is along the tangent to the locus of **P**.

Example 1 If $\mathbf{A} = (3u^2 + 4)\mathbf{i} + (2u - 5)\mathbf{j} + 4u^3\mathbf{k}$, then

$$\frac{d\mathbf{A}}{du} = \ldots\ldots\ldots\ldots\ldots\ldots\ldots$$

31

$$\frac{d\mathbf{A}}{du} = 6u\mathbf{i} + 2\mathbf{j} + 12u^2\mathbf{k}$$

If we differentiate this again, we get $\dfrac{d^2\mathbf{A}}{du^2} = 6\mathbf{i} + 24u\mathbf{k}$

When $u = 2$, $\dfrac{d\mathbf{A}}{du} = 12\mathbf{i} + 2\mathbf{j} + 48\mathbf{k}$ and $\dfrac{d^2\mathbf{A}}{du^2} = 6\mathbf{i} + 48\mathbf{k}$

Then

$$\left|\frac{d\mathbf{A}}{du}\right| = \ldots\ldots\ldots \quad \text{and} \quad \left|\frac{d^2\mathbf{A}}{du^2}\right| = \ldots\ldots\ldots$$

32

$$\left|\frac{d\mathbf{A}}{du}\right| = 49.52; \qquad \left|\frac{d^2\mathbf{A}}{du^2}\right| = 48.37$$

for $\left|\frac{d\mathbf{A}}{du}\right| = \{12^2 + 2^2 + 48^2\}^{1/2} = \{2452\}^{1/2} = \underline{49.52}$

and $\left|\frac{d^2\mathbf{A}}{du^2}\right| = \{6^2 + 48^2\}^{1/2} = \{2340\}^{1/2} = \underline{48.37}$

Example 2 If $\mathbf{F} = \mathbf{i}\sin 2t + \mathbf{j}e^{3t} + \mathbf{k}(t^3 - 4t)$, then when $t = 1$

$\frac{d\mathbf{F}}{dt} = \dots\dots\dots; \quad \frac{d^2\mathbf{F}}{dt^2} = \dots\dots\dots$

33

$$\frac{d\mathbf{F}}{dt} = 2\cos 2\mathbf{i} + 3e^3\mathbf{j} - \mathbf{k}$$
$$\frac{d^2\mathbf{F}}{dt^2} = -4\sin 2\mathbf{i} + 9e^3\mathbf{j} + 6\mathbf{k}$$

From these, we could if required find the magnitudes of $\frac{d\mathbf{F}}{dt}$ and $\frac{d^2\mathbf{F}}{dt^2}$.

$$\left|\frac{d\mathbf{F}}{dt}\right| = \dots\dots \left|\frac{d^2\mathbf{F}}{dt^2}\right| = \dots\dots\dots$$

34

$$\left|\frac{d\mathbf{F}}{dt}\right| = 60.27; \qquad \left|\frac{d^2\mathbf{F}}{dt^2}\right| = 180.9$$

for $\left|\frac{d\mathbf{F}}{dt}\right| = \{(2\cos 2)^2 + 9e^6 + 1\}^{1/2}$

$= \{0.6927 + 3631 + 1\}^{1/2} = \underline{60.27}$

and $\left|\frac{d^2\mathbf{F}}{dt^2}\right| = \{(-4\sin 2)^2 + 81e^6 + 36\}^{1/2}$

$= \{13.23 + 32\,678 + 36\}^{1/2} = \underline{180.9}$

One more example.

Example 3 If $\mathbf{A} = (u + 3)\mathbf{i} - (2 + u^2)\mathbf{j} + 2u^3\mathbf{k}$, determine

(a) $\frac{d\mathbf{A}}{du}$ (b) $\frac{d^2\mathbf{A}}{du^2}$ (c) $\left|\frac{d\mathbf{A}}{du}\right|$ (d) $\left|\frac{d^2\mathbf{A}}{du^2}\right|$ at $u = 3$.

Work through all sections and then check with the next frame.

35 Here is the working. $\mathbf{A} = (u+3)\mathbf{i} - (2+u^2)\mathbf{j} + 2u^3\mathbf{k}$

(a) $\dfrac{d\mathbf{A}}{du} = \mathbf{i} - 2u\mathbf{j} + 6u^2\mathbf{k}$ At $u = 3$, $\underline{\dfrac{d\mathbf{A}}{du} = \mathbf{i} - 6\mathbf{j} + 54\mathbf{k}}$

(b) $\dfrac{d^2\mathbf{A}}{du^2} = -2\mathbf{j} + 12u\mathbf{k}$ At $u = 3$, $\underline{\dfrac{d^2\mathbf{A}}{du^2} = -2\mathbf{j} + 36\mathbf{k}}$

(c) $\left|\dfrac{d\mathbf{A}}{du}\right| = \{1 + 36 + 2916\}^{1/2} = (2953)^{1/2} = \underline{54.34}$

(d) $\left|\dfrac{d^2\mathbf{A}}{du^2}\right| = \{4 + 1296\}^{1/2} = (1300)^{1/2} = \underline{36.06}$

The next example is of a rather different kind, so move on.

---

36 *Example 4* A particle moves in space so that at time t its position is stated as $x = 2t + 3$, $y = t^2 + 3t$, $z = t^3 + 2t^2$. We are required to find the components of its velocity and acceleration in the direction of the vector $2\mathbf{i} + 3\mathbf{j} + 4\mathbf{k}$ when $t = 1$.

First we can write the position as a vector $\mathbf{r}$

$$\mathbf{r} = (2t+3)\mathbf{i} + (t^2 + 3t)\mathbf{j} + (t^3 + 2t^2)\mathbf{k}$$

Then, at $t = 1$

$\dfrac{d\mathbf{r}}{dt} = \ldots\ldots\ldots;\quad \dfrac{d^2\mathbf{r}}{dt^2} = \ldots\ldots\ldots$

37

$$\boxed{\dfrac{d\mathbf{r}}{dt} = 2\mathbf{i} + 5\mathbf{j} + 7\mathbf{k};\qquad \dfrac{d^2\mathbf{r}}{dt^2} = 2\mathbf{j} + 10\mathbf{k}}$$

for $\dfrac{d\mathbf{r}}{dt} = 2\mathbf{i} + (2t+3)\mathbf{j} + (3t^2 + 4t)\mathbf{k}$

$$\therefore \text{At } t = 1, \quad \underline{\dfrac{d\mathbf{r}}{dt} = 2\mathbf{i} + 5\mathbf{j} + 7\mathbf{k}}$$

and $\dfrac{d^2\mathbf{r}}{dt^2} = 2\mathbf{j} + (6t + 4)\mathbf{k}$

$$\therefore \text{At } t = 1, \quad \underline{\dfrac{d^2\mathbf{r}}{dt^2} = 2\mathbf{j} + 10\mathbf{k}}$$

Now, a unit vector parallel to $2\mathbf{i} + 3\mathbf{j} + 4\mathbf{k}$ is ………

38

$$\boxed{\frac{2\mathbf{i}+3\mathbf{j}+4\mathbf{k}}{\sqrt{4+9+16}}=\frac{1}{\sqrt{29}}(2\mathbf{i}+3\mathbf{j}+4\mathbf{k})}$$

Denote this unit vector by **I**. Then the component of $\frac{d\mathbf{r}}{dt}$ in the direction of **I**

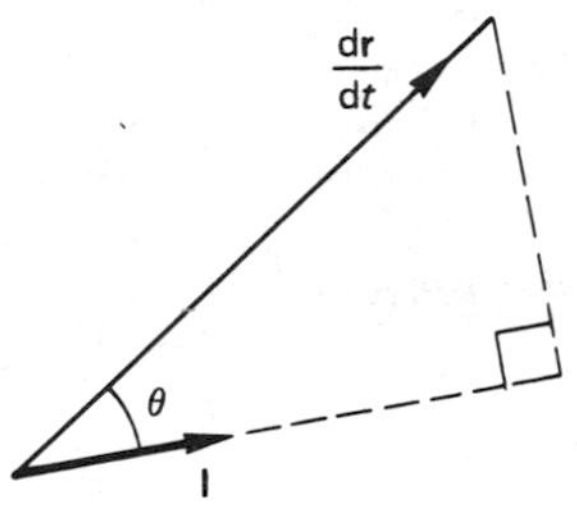

$$= \frac{d\mathbf{r}}{dt}\cos\theta = \frac{d\mathbf{r}}{dt}\cdot\mathbf{I}$$

$$= \frac{1}{\sqrt{29}}(2\mathbf{i}+5\mathbf{j}+7\mathbf{k})\cdot(2\mathbf{i}+3\mathbf{j}+4\mathbf{k})$$

$= \ldots\ldots\ldots$

39

$$\boxed{8.73}$$

since $\frac{1}{\sqrt{29}}(2\mathbf{i}+5\mathbf{j}+7\mathbf{k})\cdot(2\mathbf{i}+3\mathbf{j}+4\mathbf{k}) = \frac{1}{\sqrt{29}}(4+15+28)$

$$= \frac{47}{\sqrt{29}} = \underline{8.73}$$

Similarly, the component of $\frac{d^2\mathbf{r}}{dt^2}$ in the direction of **I** is

$\ldots\ldots\ldots$

40

8.54

for $$\frac{d^2\mathbf{r}}{dt^2}\cos\theta = \frac{d^2\mathbf{r}}{dt^2}\cdot\mathbf{I} = \frac{1}{\sqrt{29}}(2\mathbf{j}+10\mathbf{k})\cdot(2\mathbf{i}+3\mathbf{j}+4\mathbf{k})$$
$$= \frac{1}{\sqrt{29}}(6+40) = \frac{46}{\sqrt{29}} = \underline{8.54}$$

Differentiation of sums and products of vectors

If $\mathbf{A} = \mathbf{A}(u)$ and $\mathbf{B} = \mathbf{B}(u)$, then

(a) $\dfrac{d}{du}\{c\mathbf{A}\} = c\dfrac{d\mathbf{A}}{du}$

(b) $\dfrac{d}{du}\{\mathbf{A}+\mathbf{B}\} = \dfrac{d\mathbf{A}}{du}+\dfrac{d\mathbf{B}}{du}$

(c) $\dfrac{d}{du}\{\mathbf{A}\cdot\mathbf{B}\} = \mathbf{A}\cdot\dfrac{d\mathbf{B}}{du}+\dfrac{d\mathbf{A}}{du}\cdot\mathbf{B}$

(d) $\dfrac{d}{du}\{\mathbf{A}\times\mathbf{B}\} = \mathbf{A}\times\dfrac{d\mathbf{B}}{du}+\dfrac{d\mathbf{A}}{du}\times\mathbf{B}.$

These are very much like the normal rules of differentiation.

However, if $\mathbf{A}(u) = a_x\mathbf{i}+a_y\mathbf{j}+a_z\mathbf{k}$ $\quad$ (a_i const.)

$\mathbf{A}(u)\cdot\mathbf{A}(u) = a_x^2+a_y^2+a_z^2 = |\mathbf{A}|^2 = A^2$ i.e. constant.

Also $$\frac{d}{du}\{\mathbf{A}(u)\cdot\mathbf{A}(u)\} = \mathbf{A}(u)\cdot\frac{d}{du}\{\mathbf{A}(u)\}+\mathbf{A}(u)\cdot\frac{d}{du}\{\mathbf{A}(u)\}$$
$$= 2\mathbf{A}(u)\cdot\frac{d}{du}\cdot\{\mathbf{A}(u)\} = \frac{d}{du}\{\mathbf{A}^2\} = 0$$

Assuming that $\mathbf{A}(u) \neq 0$, then since $\mathbf{A}(u)\cdot\dfrac{d}{du}\{\mathbf{A}(u)\} = 0$ it follows that

$\mathbf{A}(u)$ and $\dfrac{d}{du}\{\mathbf{A}(u)\}$ are perpendicular vectors because

....

41

$$\mathbf{A}(u) \cdot \frac{\mathrm{d}}{\mathrm{d}u}\{\mathbf{A}(u)\} = |\mathbf{A}(u)| \left|\frac{\mathrm{d}}{\mathrm{d}u}\{\mathbf{A}(u)\}\right| \cos\theta = 0$$

$$\therefore \cos\theta = 0 \qquad \therefore \theta = \frac{\pi}{2}$$

Now let us deal with unit tangent vectors.

Unit tangent vectors

We have already established in frame 30 of this programme that if $\overline{\mathrm{OP}}$ is a position vector $\mathbf{A}(u)$ in space, then the direction of the vector denoting $\frac{\mathrm{d}}{\mathrm{d}u}\{\mathbf{A}(u)\}$ is ....

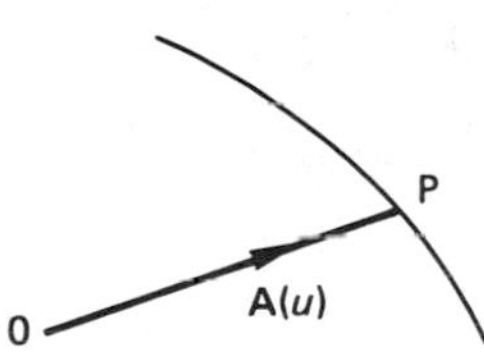

42

parallel to the tangent to the curve at **P**

Then the unit tangent vector **T** at **P** can be found from

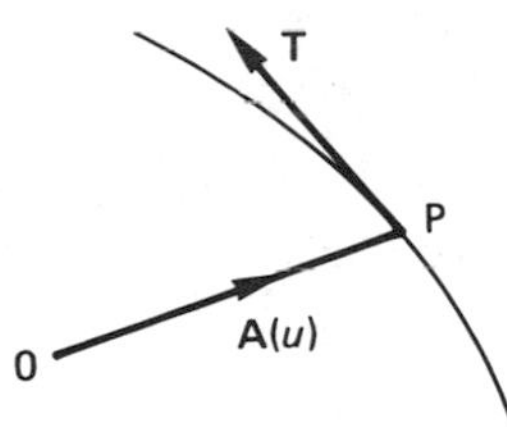

$$\mathbf{T} = \frac{\frac{\mathrm{d}}{\mathrm{d}u}\{\mathbf{A}(u)\}}{\left|\frac{\mathrm{d}}{\mathrm{d}u}\{\mathbf{A}(u)\}\right|}$$

In simpler notation, this becomes:

If $\mathbf{r} = a_x\mathbf{i} + a_y\mathbf{j} + a_z\mathbf{k}$ then the unit tangent vector **T** is given by

$$\mathbf{T} = \frac{\mathrm{d}\mathbf{r}/\mathrm{d}u}{|\mathrm{d}\mathbf{r}/\mathrm{d}u|}$$

Example 1 Determine the unit tangent vector at the point (2, 4, 7) for the curve with parametric equations $x = 2u$; $y = u^2 + 3$; $z = 2u^2 + 5$.

First we see that the point (2, 4, 7) corresponds to $u = 1$.
The vector equation of the curve is

$$\mathbf{r} = a_x\mathbf{i} + a_y\mathbf{j} + a_z\mathbf{k} = 2u\mathbf{i} + (u^2 + 3)\mathbf{j} + (2u^2 + 5)\mathbf{k}$$

$$\therefore \frac{\mathrm{d}\mathbf{r}}{\mathrm{d}u} = \ldots\ldots\ldots$$

43

$$\frac{d\mathbf{r}}{du} = 2\mathbf{i} + 2u\mathbf{j} + 4u\mathbf{k}$$

and at $u = 1$, $\dfrac{d\mathbf{r}}{du} = 2\mathbf{i} + 2\mathbf{j} + 4\mathbf{k}$

Hence $\left|\dfrac{d\mathbf{r}}{du}\right| = \dots\dots\dots$ and $\mathbf{T} = \dots\dots\dots$

44

$$\left|\frac{d\mathbf{r}}{du}\right| = 2\sqrt{6}; \quad \mathbf{T} = \frac{1}{\sqrt{6}}\{\mathbf{i} + \mathbf{j} + 2\mathbf{k}\}$$

for $\left|\dfrac{d\mathbf{r}}{du}\right| = \{4 + 4 + 16\}^{1/2} = 24^{1/2} = \underline{2\sqrt{6}}$

$$\mathbf{T} = \frac{\dfrac{d\mathbf{r}}{du}}{\left|\dfrac{d\mathbf{r}}{du}\right|} = \frac{2\mathbf{i} + 2\mathbf{j} + 4\mathbf{k}}{2\sqrt{6}} = \underline{\frac{1}{\sqrt{6}}\{\mathbf{i} + \mathbf{j} + 2\mathbf{k}\}}$$

Let us do another.

Example 2 Find the unit tangent vector at the point $(2, 0, \pi)$ for the curve with parametric equations $x = 2\sin\theta$; $y = 3\cos\theta$; $z = 2\theta$.

We see that the point $(2, 0, \pi)$ corresponds to $\theta = \pi/2$.
Writing the curve in vector from $\mathbf{r} = \dots\dots\dots$

45

$$\mathbf{r} = 2\sin\theta\,\mathbf{i} + 3\cos\theta\,\mathbf{j} + 2\theta\,\mathbf{k}$$

Then, at $\theta = \pi/2$, $\dfrac{d\mathbf{r}}{d\theta} = \dots\dots\dots$

$\left|\dfrac{d\mathbf{r}}{d\theta}\right| = \dots\dots\dots$

$\mathbf{T} = \dots\dots\dots$ Finish it off.

46

$$\frac{d\mathbf{r}}{d\theta} = -3\mathbf{j} + 2\mathbf{k}; \quad \left|\frac{d\mathbf{r}}{d\theta}\right| = \sqrt{13}$$

$$\mathbf{T} = \frac{1}{\sqrt{13}}(-3\mathbf{j} + 2\mathbf{k})$$

And now

Example 3 Determine the unit tangent vector for the curve

$$x = 3t; \quad y = 2t^2; \quad z = t^2 + t$$

at the point (6, 8, 6).

On your own. $\mathbf{T} = \ldots\ldots\ldots$

47

$$\mathbf{T} = \frac{1}{\sqrt{98}}(3\mathbf{i} + 8\mathbf{j} + 5\mathbf{k})$$

The point (6, 8, 6) corresponds to $t = 2$

$$\mathbf{r} = 3t\mathbf{i} + 2t^2\mathbf{j} + (t^2 + t)\mathbf{k}$$

$$\therefore \frac{d\mathbf{r}}{dt} = 3\mathbf{i} + 4t\mathbf{j} + (2t + 1)\mathbf{k}$$

At $t = 2$, $\mathbf{r} = 6\mathbf{i} + 8\mathbf{j} + 6\mathbf{k}$ and $\dfrac{d\mathbf{r}}{dt} = 3\mathbf{i} + 8\mathbf{j} + 5\mathbf{k}$

$$\therefore \left|\frac{d\mathbf{r}}{dt}\right| = (9 + 64 + 25)^{1/2} = \sqrt{98}$$

$$\therefore \mathbf{T} = \frac{d\mathbf{r}/dt}{|d\mathbf{r}/dt|} = \underline{\frac{1}{\sqrt{98}}(3\mathbf{i} + 8\mathbf{j} + 5\mathbf{k})}$$

Partial Differentiation of Vectors

48 If a vector **F** is a function of two independent variables u and v, then the rules of differentiation follow the usual pattern.

If $\mathbf{F} = x\mathbf{i} + y\mathbf{j} + z\mathbf{k}$ then x, y, z will also be functions of u and v.

Then
$$\frac{\partial \mathbf{F}}{\partial u} = \frac{\partial x}{\partial u}\mathbf{i} + \frac{\partial y}{\partial u}\mathbf{j} + \frac{\partial z}{\partial u}\mathbf{k}$$

$$\frac{\partial \mathbf{F}}{\partial v} = \frac{\partial x}{\partial v}\mathbf{i} + \frac{\partial y}{\partial v}\mathbf{j} + \frac{\partial z}{\partial v}\mathbf{k}$$

$$\frac{\partial^2 \mathbf{F}}{\partial u^2} = \frac{\partial^2 x}{\partial u^2}\mathbf{i} + \frac{\partial^2 y}{\partial u^2}\mathbf{j} + \frac{\partial^2 z}{\partial u^2}\mathbf{k}$$

$$\frac{\partial^2 \mathbf{F}}{\partial v^2} = \frac{\partial^2 x}{\partial v^2}\mathbf{i} + \frac{\partial^2 y}{\partial v^2}\mathbf{j} + \frac{\partial^2 z}{\partial v^2}\mathbf{k}$$

$$\frac{\partial^2 \mathbf{F}}{\partial u \partial v} = \frac{\partial^2 x}{\partial u \partial v}\mathbf{i} + \frac{\partial^2 y}{\partial u \partial v}\mathbf{j} + \frac{\partial^2 z}{\partial u \partial v}\mathbf{k}$$

and for small finite changes du and dv in u and v, we have

$$\mathrm{d}\mathbf{F} = \frac{\partial \mathbf{F}}{\partial u}\mathrm{d}u + \frac{\partial \mathbf{F}}{\partial v}\mathrm{d}v$$

Example If $\mathbf{F} = 2uv\mathbf{i} + (u^2 - 2v)\mathbf{j} + (u + v^2)\mathbf{k}$

$$\frac{\partial \mathbf{F}}{\partial u} = \ldots\ldots\ldots; \quad \frac{\partial \mathbf{F}}{\partial v} = \ldots\ldots\ldots; \quad \frac{\partial^2 \mathbf{F}}{\partial u^2} = \ldots\ldots\ldots; \quad \frac{\partial^2 \mathbf{F}}{\partial u \partial v} = \ldots\ldots\ldots$$

49

$$\frac{\partial \mathbf{F}}{\partial u} = 2v\mathbf{i} + 2u\mathbf{j} + \mathbf{k}; \quad \frac{\partial \mathbf{F}}{\partial v} = 2u\mathbf{i} - 2\mathbf{j} + 2v\mathbf{k}$$

$$\frac{\partial^2 \mathbf{F}}{\partial u^2} = 2\mathbf{j}; \quad \frac{\partial^2 \mathbf{F}}{\partial u \partial v} = 2\mathbf{i}.$$

This is straightforward enough.

Integration of vector functions

The process is the reverse of that for differentiation. If a vector $\mathbf{F} = x\mathbf{i} + y\mathbf{j} + z\mathbf{k}$ where $\mathbf{F}$, x, y, z are expressed as functions of u, then

$$\int_a^b \mathbf{F}\,\mathrm{d}u = \mathbf{i}\int_a^b x\,\mathrm{d}u + \mathbf{j}\int_a^b y\,\mathrm{d}u + \mathbf{k}\int_a^b z\,\mathrm{d}u.$$

Example 1 If $\mathbf{F} = (3t^2 + 4t)\mathbf{i} + (2t - 5)\mathbf{j} + 4t^3\mathbf{k}$, then

$$\int_1^3 \mathbf{F}\,\mathrm{d}t = \mathbf{i}\int_1^3 (3t^2 + 4t)\,\mathrm{d}t + \mathbf{j}\int_1^3 (2t - 5)\,\mathrm{d}t + \mathbf{k}\int_1^3 4t^3\,\mathrm{d}t$$

$$= \ldots\ldots\ldots$$

50

$$42\mathbf{i} - 2\mathbf{j} + 80\mathbf{k}$$

for $$\int_1^3 \mathbf{F}\,\mathrm{d}t = \Big[\mathbf{i}(t^3 + 2t^2) + \mathbf{j}(t^2 - 5t) + \mathbf{k}t^4\Big]_1^3$$

$$= (45\mathbf{i} - 6\mathbf{j} + 81\mathbf{k}) - (3\mathbf{i} - 4\mathbf{j} + \mathbf{k}) = \underline{42\mathbf{i} - 2\mathbf{j} + 80\mathbf{k}}$$

Here is a slightly different one.

Example 2 If $\mathbf{F} = 3u\mathbf{i} + u^2\mathbf{j} + (u + 2)\mathbf{k}$

and $\mathbf{V} = 2u\mathbf{i} - 3u\mathbf{j} + (u - 2)\mathbf{k}$

evaluate $$\int_0^2 (\mathbf{F} \times \mathbf{V})\mathrm{d}u.$$

First we must determine $\mathbf{F} \times \mathbf{V}$ in terms of u.

$$\mathbf{F} \times \mathbf{V} = \ldots\ldots\ldots$$

51

$$\mathbf{F} \times \mathbf{V} = (u^3 + u^2 + 6u)\mathbf{i} - (u^2 - 10u)\mathbf{j} - (2u^3 + 9u^2)\mathbf{k}$$

for $\mathbf{F} \times \mathbf{V} = \begin{vmatrix} \mathbf{i} & \mathbf{j} & \mathbf{k} \\ 3u & u^2 & (u+2) \\ 2u & -3u & (u-2) \end{vmatrix}$

which gives the result above.

Then $\displaystyle\int_0^2 (\mathbf{F} \times \mathbf{V})\, du = \ldots\ldots\ldots$

52

$$\tfrac{4}{3}\{14\mathbf{i} + 13\mathbf{j} - 24\mathbf{k}\}$$

$$\int (\mathbf{F} \times \mathbf{V}) du = \left(\frac{u^4}{4} + \frac{u^3}{3} + 3u^2\right)\mathbf{i} - \left(\frac{u^3}{3} - 5u^2\right)\mathbf{j} - \left(\frac{u^4}{2} + 3u^3\right)\mathbf{k}$$

$$\therefore \int_0^2 (\mathbf{F} \times \mathbf{V}) du = (4 + \tfrac{8}{3} + 12)\mathbf{i} - (\tfrac{8}{3} - 20)\mathbf{j} - (8 + 24)\mathbf{k}$$

$$= \underline{\tfrac{4}{3}\{14\mathbf{i} + 13\mathbf{j} - 24\mathbf{k}\}}$$

Example 3 If $\mathbf{F} = \mathbf{A} \times (\mathbf{B} \times \mathbf{C})$ where

$$\mathbf{A} = 3t^2\mathbf{i} + (2t - 3)\mathbf{j} + 4t\mathbf{k}$$
$$\mathbf{B} = 2\mathbf{i} + 4t\mathbf{j} + 3(1 - t)\mathbf{k}$$
$$\mathbf{C} = 2t\mathbf{i} - 3t^2\mathbf{j} - 2t\mathbf{k}$$

determine $\displaystyle\int_0^1 \mathbf{F}\, dt$.

First we need to find $\mathbf{A} \times (\mathbf{B} \times \mathbf{C})$. The simplest way to do this is to use the relationship

$$\mathbf{A} \times (\mathbf{B} \times \mathbf{C}) = \ldots\ldots\ldots$$

53

$$\boxed{\mathbf{A} \times (\mathbf{B} \times \mathbf{C}) = (\mathbf{A} \cdot \mathbf{C})\mathbf{B} - (\mathbf{A} \cdot \mathbf{B})\mathbf{C}}$$

So $\qquad \mathbf{A} \cdot \mathbf{C} = \ldots\ldots\ldots$

and $\qquad \mathbf{A} \cdot \mathbf{B} = \ldots\ldots\ldots$

54

$$\mathbf{A} \cdot \mathbf{C} = 6t^3 - 6t^3 + 9t^2 - 8t^2 = \underline{t^2}$$
$$\mathbf{A} \cdot \mathbf{B} = 6t^2 + 8t^2 - 12t + 12t - 12t^2 = \underline{2t^2}$$

Then $\mathbf{F} = \mathbf{A} \times (\mathbf{B} \times \mathbf{C}) = t^2\{2\mathbf{i} + 4t\mathbf{j} + 3(1-t)\mathbf{k}\} - 2t^2\{2t\mathbf{i} - 3t^2\mathbf{j} - 2t\mathbf{k}\}$

$$\therefore \int_0^1 \mathbf{F}\,dt = \ldots\ldots\ldots$$

Finish off the simplification and complete the integration.

55

$$\boxed{\tfrac{1}{60}\{-20\mathbf{i} + 132\mathbf{j} + 75\mathbf{k}\}}$$

for $\mathbf{F} = \mathbf{A} \times (\mathbf{B} \times \mathbf{C}) = (2t^2 - 4t^3)\mathbf{i} + (4t^3 + 6t^4)\mathbf{j} + (3t^2 + t^3)\mathbf{k}$

Integration with respect to t then gives the result stated above.

Now let us move on to the next stage of our development.

Scalar and vector fields

56

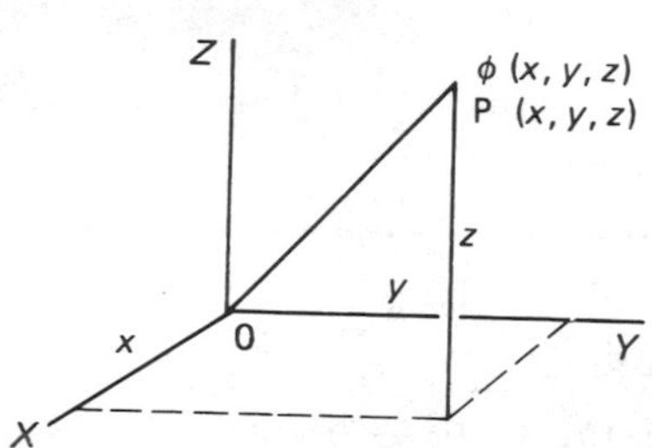

If every point P(x, y, z) of a region R of space has associated with it a scalar quantity $\phi(x, y, z)$, then $\phi(x, y, z)$ is a *scalar function* and a *scalar field* is said to exist in the region R.

Examples of scalar fields are temperature, potential, etc.

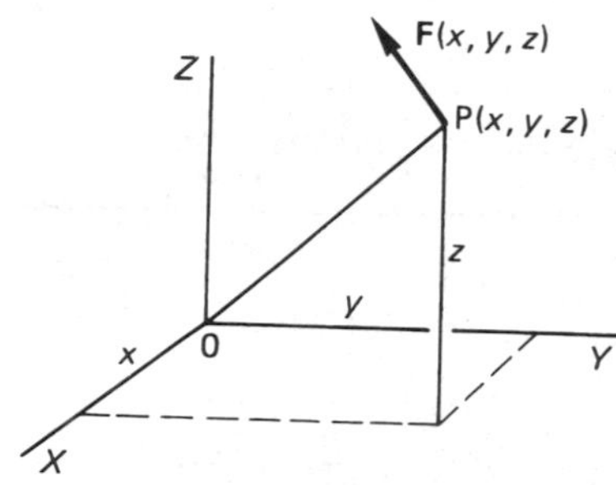

Similarly, if every point P(x, y, z) of a region R has associated with it a vector quantity $\mathbf{F}(x, y, z)$, then $\mathbf{F}(x, y, z)$ is a *vector function* and a *vector field* is said to exist in the region R.

Examples of vector fields are force, velocity, acceleration, etc. $\mathbf{F}(x, y, z)$ can be defined in terms of its components parallel to the coordinate axes, OX, OY, OZ.

That is, $$\mathbf{F}(x, y, z) = F_x\mathbf{i} + F_y\mathbf{j} + F_z\mathbf{k}.$$

Note these important definitions: we shall be making good use of them as we proceed.

57

Grad (gradient of a scalar function)

If a scalar function $\phi(x, y, z)$ is continuously differentiable with respect to its variables x, y, z, throughout the region, then the *gradient* of ϕ, written *grad* ϕ, is defined as the vector

$$\text{grad } \phi = \frac{\partial \phi}{\partial x}\mathbf{i} + \frac{\partial \phi}{\partial y}\mathbf{j} + \frac{\partial \phi}{\partial z}\mathbf{k} \qquad (12)$$

Note that, while ϕ is a scalar function, grad ϕ is a vector function. For example, if ϕ depends upon the position of P and is defined by $\phi = 2x^2yz^3$, then

$$\text{grad } \phi = 4xyz^3\mathbf{i} + 2x^2z^3\mathbf{j} + 6x^2yz^2\mathbf{k}$$

Notation The expression (12) above can be written

$$\text{grad } \phi = \left\{\mathbf{i}\frac{\partial}{\partial x} + \mathbf{j}\frac{\partial}{\partial y} + \mathbf{k}\frac{\partial}{\partial z}\right\}\phi$$

where $\left(\mathbf{i}\frac{\partial}{\partial x} + \mathbf{j}\frac{\partial}{\partial y} + \mathbf{k}\frac{\partial}{\partial z}\right)$ is called a *vector differential operator* and is denoted by the symbol ∇ (pronounced 'del' or sometimes 'nabla')

i.e.
$$\nabla \equiv \left(\mathbf{i}\frac{\partial}{\partial x} + \mathbf{j}\frac{\partial}{\partial y} + \mathbf{k}\frac{\partial}{\partial z}\right)$$

Beware! ∇ cannot exist alone: it is an operator and must operate on a stated scalar function $\phi(x, y, z)$.

If $\mathbf{F}$ is a vector function, $\nabla\mathbf{F}$ has no meaning.

So we have:

$$\begin{aligned}\nabla\phi = \text{grad } \phi &= \left(\mathbf{i}\frac{\partial}{\partial x} + \mathbf{j}\frac{\partial}{\partial y} + \mathbf{k}\frac{\partial}{\partial z}\right)\phi \\ &= \mathbf{i}\frac{\partial \phi}{\partial x} + \mathbf{j}\frac{\partial \phi}{\partial y} + \mathbf{k}\frac{\partial \phi}{\partial z}\end{aligned} \qquad (13)$$

Make a note of this definition and then let us see how to use it.

58

Example 1 If $\phi = x^2yz^3 + xy^2z^2$, determine grad ϕ at the point P(1, 3, 2).

By the definition, grad $\phi = \nabla\phi = \dfrac{\partial\phi}{\partial x}\mathbf{i} + \dfrac{\partial\phi}{\partial y}\mathbf{j} + \dfrac{\partial\phi}{\partial z}\mathbf{k}$.

All we have to do then is to find the partial derivatives at $x = 1,\ y = 3,\ z = 2$ and insert their values.

$$\therefore\ \nabla\phi = \ldots\ldots\ldots$$

59

$$\boxed{4(21\mathbf{i} + 8\mathbf{j} + 18\mathbf{k})}$$

since $\quad \phi = x^2yz^3 + xy^2z^2 \quad \therefore\ \dfrac{\partial\phi}{\partial x} = 2xyz^3 + y^2z^2;$

$$\frac{\partial\phi}{\partial y} = x^2z^3 + 2xyz^2; \qquad \frac{\partial\phi}{\partial z} = 3x^2yz^2 + 2xy^2z;$$

Then, at $(1, 3, 2)$ $\quad \dfrac{\partial\phi}{\partial x} = 48 + 36 \quad \therefore\ \dfrac{\partial\phi}{\partial x} = 84$

$$\frac{\partial\phi}{\partial y} = 8 + 24 \quad \therefore\ \frac{\partial\phi}{\partial y} = 32$$

$$\frac{\partial\phi}{\partial z} = 36 + 36 \quad \therefore\ \frac{\partial\phi}{\partial z} = 72$$

$$\therefore\ \text{grad}\,\phi = \nabla\phi = 84\mathbf{i} + 32\mathbf{j} + 72\mathbf{k} = \underline{4(21\mathbf{i} + 8\mathbf{j} + 18\mathbf{k})}$$

Example 2 If $\mathbf{A} = x^2z\mathbf{i} + xy\mathbf{j} + y^2z\mathbf{k}$

and $\mathbf{B} = yz^2\mathbf{i} + xz\mathbf{j} + x^2z\mathbf{k}$

determine an expression for grad $(\mathbf{A}\cdot\mathbf{B})$.

This we can soon do since we know that $\mathbf{A}\cdot\mathbf{B}$ is a scalar function of x, y and z.

First then, $\quad \mathbf{A}\cdot\mathbf{B} = \ldots\ldots\ldots$

60

$$\boxed{\mathbf{A}\cdot\mathbf{B} = x^2yz^3 + x^2yz + x^2y^2z^2}$$

Then $\quad \nabla(\mathbf{A}\cdot\mathbf{B}) = \dots\dots\dots$

61

$$\boxed{2xyz(z^2+1+yz)\mathbf{i} + x^2z(z^2+1+2yz)\mathbf{j} + x^2y(3z^2+1+2yz)\mathbf{k}}$$

for if $\phi = \mathbf{A}\cdot\mathbf{B} = (x^2z\mathbf{i} + xy\mathbf{j} + y^2z\mathbf{k})\cdot(yz^2\mathbf{i} + xz\mathbf{j} + x^2z\mathbf{k})$

$$\phi = \mathbf{A}\cdot\mathbf{B} = x^2yz^3 + x^2yz + x^2y^2z^2$$

$$\frac{\partial\phi}{\partial x} = 2xyz^3 + 2xyz + 2xy^2z^2 = 2xyz(z^2+1+yz)$$

$$\frac{\partial\phi}{\partial y} = x^2z^3 + x^2z + 2x^2yz^2 = x^2z(z^2+1+2yz)$$

$$\frac{\partial\phi}{\partial z} = 3x^2yz^2 + x^2y + 2x^2y^2z = x^2y(3z^2+1+2yz)$$

$$\therefore\ \nabla(\mathbf{A}\cdot\mathbf{B}) = 2xyz(z^2+1+yz)\mathbf{i} + x^2z(z^2+1+2yz)\mathbf{j} + x^2y(3z^2+1+2yz)\mathbf{k}$$

Now let us obtain another useful relationship.

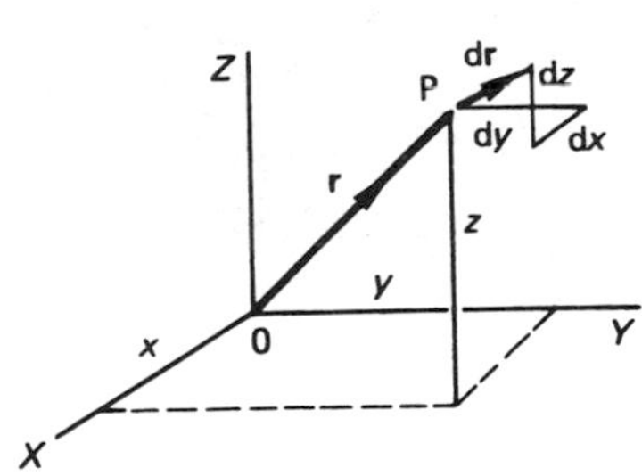

If $\overline{\text{OP}}$ is a position vector $\mathbf{r}$ where $\mathbf{r} = x\mathbf{i} + y\mathbf{j} + z\mathbf{k}$ and $\mathrm{d}\mathbf{r}$ is a small displacement corresponding to changes $\mathrm{d}x$, $\mathrm{d}y$, $\mathrm{d}z$ in x, y, z respectively, then

$$\mathrm{d}\mathbf{r} = \mathrm{d}x\mathbf{i} + \mathrm{d}y\mathbf{j} + \mathrm{d}z\mathbf{k}$$

If $\phi(x, y, z)$ is a scalar function at P, we know that

$$\text{grad}\,\phi = \nabla\phi = \frac{\partial\phi}{\partial x}\mathbf{i} + \frac{\partial\phi}{\partial y}\mathbf{j} + \frac{\partial\phi}{\partial z}\mathbf{k}$$

Then $\quad \text{grad}\,\phi\cdot\mathrm{d}\mathbf{r} = \dots\dots\dots$

62

$$\boxed{\text{grad}\,\phi\cdot\mathbf{dr}=\frac{\partial\phi}{\partial x}\mathrm{d}x+\frac{\partial\phi}{\partial y}\mathrm{d}y+\frac{\partial\phi}{\partial z}\mathrm{d}z}$$

for $\text{grad}\,\phi\cdot\mathbf{dr}=\left(\frac{\partial\phi}{\partial x}\mathbf{i}+\frac{\partial\phi}{\partial y}\mathbf{j}+\frac{\partial\phi}{\partial z}\mathbf{k}\right)\cdot(\mathrm{d}x\,\mathbf{i}+\mathrm{d}y\,\mathbf{j}+\mathrm{d}z\,\mathbf{k})$

$$=\frac{\partial\phi}{\partial x}\mathrm{d}x+\frac{\partial\phi}{\partial y}\mathrm{d}y+\frac{\partial\phi}{\partial z}\mathrm{d}z$$

$=$ the total differential $\mathrm{d}\phi$ of ϕ

$$\therefore\ \mathrm{d}\phi=\mathbf{dr}\cdot\text{grad}\,\phi \qquad (14)$$

This will certainly be useful, so make a note of it.

63

Directional derivatives

We have just established that

$$d\phi = d\mathbf{r} \cdot \text{grad}\,\phi$$

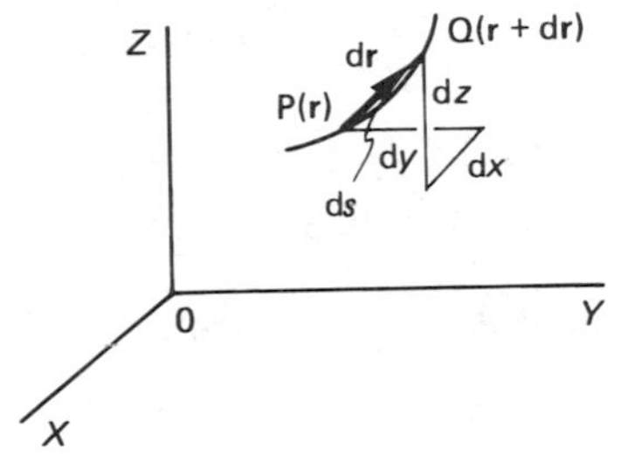

If ds is the small element of arc between $P(\mathbf{r})$ and $Q(\mathbf{r}+d\mathbf{r})$ then $ds = |d\mathbf{r}|$

$$\frac{d\mathbf{r}}{ds} = \frac{d\mathbf{r}}{|d\mathbf{r}|}$$

and $\dfrac{d\mathbf{r}}{ds}$ is thus a unit vector in the direction of $d\mathbf{r}$.

$$\therefore \frac{d\phi}{ds} = \frac{d\mathbf{r}}{ds} \cdot \text{grad}\,\phi$$

If we denote this unit vector by $\hat{\mathbf{a}}$, i.e. $\dfrac{d\mathbf{r}}{ds} = \hat{\mathbf{a}}$, the result becomes

$$\underline{\frac{d\phi}{ds} = \hat{\mathbf{a}} \cdot \text{grad}\,\phi}$$

$\dfrac{d\phi}{ds}$ is thus the projection of grad ϕ on the unit vector $\hat{\mathbf{a}}$ and is called the *directional derivative* of ϕ in the direction of $\hat{\mathbf{a}}$. It gives the rate of change of ϕ with distance measured in the direction of $\hat{\mathbf{a}}$ and $\dfrac{d\phi}{ds} = \hat{\mathbf{a}} \cdot \text{grad}\,\phi$ will

be a maximum when $\hat{\mathbf{a}}$ and grad ϕ have the same direction, since then

$$\hat{\mathbf{a}} \cdot \text{grad}\,\phi = |\,\hat{\mathbf{a}}\,||\,\text{grad}\,\phi\,|\cos\theta \text{ and } \theta \text{ will be zero.}$$

Thus the direction of grad ϕ gives the direction in which the maximum rate of change of ϕ occurs.

Example 1 Find the directional derivative of the function $\phi = x^2z + 2xy^2 + yz^2$ at the point (1, 2, –1) in the direction of the vector $\mathbf{A} = 2\mathbf{i} + 3\mathbf{j} - 4\mathbf{k}$.

We start off with $\phi = x^2z + 2xy^2 + yz^2$

$$\therefore \nabla\phi = \dots\dots\dots$$

64

$$\nabla\phi = (2xz + 2y^2)\mathbf{i} + (4xy + z^2)\mathbf{j} + (x^2 + 2yz)\mathbf{k}$$

Since $\dfrac{\partial\phi}{\partial x} = 2xz + 2y^2$; $\quad \dfrac{\partial\phi}{\partial y} = 4xy + z^2$; $\quad \dfrac{\partial\phi}{\partial z} = x^2 + 2yz$

then, at (1, 2, –1)

$$\nabla\phi = (-2 + 8)\mathbf{i} + (8 + 1)\mathbf{j} + (1 - 4)\mathbf{k} = \underline{6\mathbf{i} + 9\mathbf{j} - 3\mathbf{k}}$$

Next we have to find the unit vector $\hat{\mathbf{a}}$ where $\mathbf{A} = 2\mathbf{i} + 3\mathbf{j} - 4\mathbf{k}$

$$\hat{\mathbf{a}} = \dots\dots\dots$$

65

$$\hat{\mathbf{a}} = \frac{1}{\sqrt{29}}(2\mathbf{i} + 3\mathbf{j} - 4\mathbf{k})$$

for $\mathbf{A} = 2\mathbf{i} + 3\mathbf{j} - 4\mathbf{k} \quad \therefore \; |\,\mathbf{A}\,| = \sqrt{4 + 9 + 16} = \sqrt{29}$

$$\hat{\mathbf{a}} = \frac{\mathbf{A}}{|\,\mathbf{A}\,|} = \frac{1}{\sqrt{29}}(2\mathbf{i} + 3\mathbf{j} - 4\mathbf{k})$$

So we have $\nabla\phi = 6\mathbf{i} + 9\mathbf{j} - 3\mathbf{k}$ and $\hat{\mathbf{a}} = \dfrac{1}{\sqrt{29}}(2\mathbf{i} + 3\mathbf{j} - 4\mathbf{k})$

$$\therefore \; \frac{\mathrm{d}\phi}{\mathrm{d}s} = \hat{\mathbf{a}} \cdot \nabla\phi = \dots\dots\dots$$

66

$$\boxed{\frac{d\phi}{ds} = \frac{51}{\sqrt{29}} = 9.47}$$

since $\dfrac{d\phi}{ds} = \hat{\mathbf{a}} \cdot \nabla\phi = \dfrac{1}{\sqrt{29}}(2\mathbf{i} + 3\mathbf{j} - 4\mathbf{k}) \cdot (6\mathbf{i} + 9\mathbf{j} - 3\mathbf{k})$

$$= \frac{1}{\sqrt{29}}(12 + 27 + 12) = \frac{51}{\sqrt{29}} = \underline{9.47}$$

That is all there is to it.

(i) From the given scalar function ϕ, determine $\nabla\phi$.
(ii) Find the unit vector $\hat{\mathbf{a}}$ in the direction of the given vector $\mathbf{A}$.
(iii) Then $\dfrac{d\phi}{ds} = \hat{\mathbf{a}} \cdot \nabla\phi$.

Example 2 Find the directional derivative of $\phi = x^2y + y^2z + z^2x$ at the point (1, –1, 2) in the direction of the vector $\mathbf{A} = 4\mathbf{i} + 2\mathbf{j} - 5\mathbf{k}$.

Same as before. *Work through it and check the result with the next frame.*

67

$$\boxed{\frac{d\phi}{ds} = \frac{-23}{3\sqrt{5}} = -3.43}$$

for $\phi = x^2y + y^2z + z^2x$

$\therefore\ \nabla\phi = (2xy + z^2)\mathbf{i} + (x^2 + 2yz)\mathbf{j} + (y^2 + 2zx)\mathbf{k}$

$\therefore$ At (1, –1, 2), $\qquad \nabla\phi = 2\mathbf{i} - 3\mathbf{j} + 5\mathbf{k}$

$\mathbf{A} = 4\mathbf{i} + 2\mathbf{j} - 5\mathbf{k} \quad \therefore\ |\mathbf{A}| = \sqrt{16 + 4 + 25} = \sqrt{45} = 3\sqrt{5}$

$$\therefore\ \hat{\mathbf{a}} = \frac{1}{3\sqrt{5}}(4\mathbf{i} + 2\mathbf{j} - 5\mathbf{k})$$

$$\therefore\ \frac{d\phi}{ds} = \hat{\mathbf{a}} \cdot \nabla\phi = \frac{1}{3\sqrt{5}}(4\mathbf{i} + 2\mathbf{j} - 5\mathbf{k}) \cdot (2\mathbf{i} - 3\mathbf{j} + 5\mathbf{k})$$

$$= \frac{1}{3\sqrt{5}}(8 - 6 - 25) = \frac{-23}{3\sqrt{5}} = \underline{-3.43}$$

Example 3 Find the direction from the point (1, 1, 0) which gives the greatest rate of increase of the function $\phi = (x + 3y)^2 + (2y - z)^2$.

This appears to be different, but it rests on the fact that the greatest rate of increase of ϕ with respect to distance is in

....

68

the direction of $\nabla\phi$

All we need then is to find the vector $\nabla\phi$, which is

....

69

$$\nabla\phi = 4(2\mathbf{i} + 8\mathbf{j} - \mathbf{k})$$

for $\phi = (x + 3y)^2 + (2y - z)^2$

$$\therefore \frac{\partial\phi}{\partial x} = 2(x + 3y); \quad \frac{\partial\phi}{\partial y} = 6(x + 3y) + 4(2y - z); \quad \frac{\partial\phi}{\partial z} = -2(2y - z)$$

$$\therefore \text{At } (1, 1, 0), \quad \frac{\partial\phi}{\partial x} = 8; \quad \frac{\partial\phi}{\partial y} = 32; \quad \frac{\partial\phi}{\partial z} = -4$$

$$\therefore \nabla\phi = 8\mathbf{i} + 32\mathbf{j} - 4\mathbf{k} = 4(2\mathbf{i} + 8\mathbf{j} - \mathbf{k})$$

$\therefore$ greatest rate of increase occurs in direction $2\mathbf{i} + 8\mathbf{j} - \mathbf{k}$

So on we go.

70

Unit normal vectors

If $\phi(x, y, z)$ = constant, this relationship represents a surface in space, depending on the value ascribed to the constant.

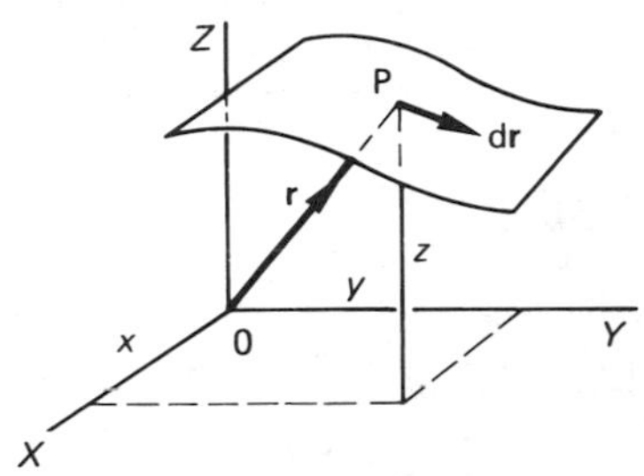

If **dr** is a displacement in this surface, then $d\phi = 0$ since ϕ is constant over the surface.

Therefore our previous relationship $\mathbf{dr} \cdot \text{grad } \phi = d\phi$ becomes

$$\mathbf{dr} \cdot \text{grad } \phi = 0$$

for all such small displacements **dr** in the surface.

But $\quad \mathbf{dr} \cdot \text{grad } \phi = |\,\mathbf{dr}\,||\,\text{grad } \phi\,| \cos\theta = 0.$

$\therefore\ \theta = \dfrac{\pi}{2}$ $\therefore$ grad ϕ is perpendicular to **dr**, i.e. grad ϕ is a vector perpendicular to the surface at P, in the direction of maximum rate of change of ϕ. The magnitude of that maximum rate of change is given by $|\,\text{grad } \phi\,|$.

The unit vector **N** in the direction of grad ϕ is called the *unit normal vector* at P.

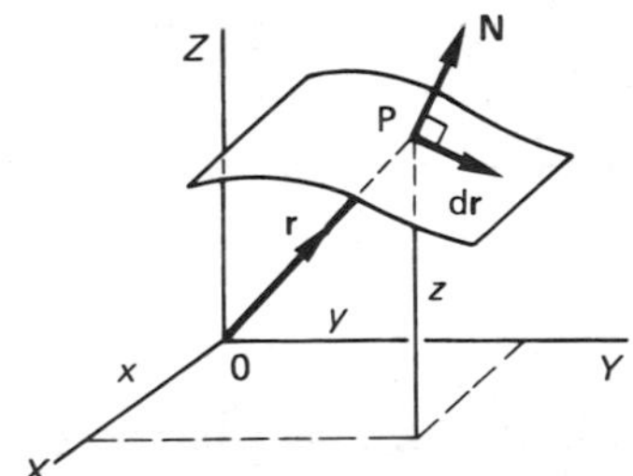

$\therefore$ Unit normal vector

$$\mathbf{N} = \frac{\nabla\phi}{|\,\nabla\phi\,|} \qquad (15)$$

Example 1 Find the unit normal vector to the surface $x^3y + 4xz^2 + xy^2z + 2 = 0$ at the point (1, 3, –1).

Vector normal = $\nabla\phi$ = ...

71

$$\nabla\phi = (3x^2y + 4z^2 + y^2z)\mathbf{i} + (x^3 + 2xyz)\mathbf{j} + (8xz + xy^2)\mathbf{k}$$

Then, at (1, 3, –1), $\nabla\phi = 4\mathbf{i} - 5\mathbf{j} + \mathbf{k}$

and the unit normal at (1, 3, –1) is ....

72

$$\frac{1}{\sqrt{42}}(4\mathbf{i} - 5\mathbf{j} + \mathbf{k})$$

since $|\nabla\phi| = \sqrt{16 + 25 + 1} = \sqrt{42}$

and $\mathbf{N} = \dfrac{\nabla\phi}{|\nabla\phi|} = \dfrac{1}{\sqrt{42}}(4\mathbf{i} - 5\mathbf{j} + \mathbf{k})$

One more.

Example 2 Determine the unit normal to the surface

$xyz + x^2y - 5yz - 5 = 0$ at the point (3, 1, 2).

All very straightforward. Complete it.

73

$$\text{Unit normal} = \mathbf{N} = \frac{1}{\sqrt{93}}(8\mathbf{i} + 5\mathbf{j} - 2\mathbf{k})$$

for $\phi = xyz + x^2y - 5yz - 5$

$\therefore\ \nabla\phi = (yz + 2xy)\mathbf{i} + (xz + x^2 - 5z)\mathbf{j} + (xy - 5y)\mathbf{k}$

At (3, 1, 2), $\nabla\phi = 8\mathbf{i} + 5\mathbf{j} - 2\mathbf{k}$; $|\nabla\phi| = \sqrt{64 + 25 + 4} = \sqrt{93}$

$\therefore$ Unit normal $= \mathbf{N} = \dfrac{\nabla\phi}{|\nabla\phi|} = \dfrac{1}{\sqrt{93}}(8\mathbf{i} + 5\mathbf{j} - 2\mathbf{k})$

Collecting our results so far, we have, for $\phi(x, y, z)$ a scalar function

(a) grad $\phi = \nabla\phi = \dfrac{\partial\phi}{\partial x}\mathbf{i} + \dfrac{\partial\phi}{\partial y}\mathbf{j} + \dfrac{\partial\phi}{\partial z}\mathbf{k}$

(b) $\mathrm{d}\phi = \mathrm{d}\mathbf{r} \cdot \text{grad}\,\phi$ where $\mathrm{d}\phi = \dfrac{\partial\phi}{\partial x}\mathrm{d}x + \dfrac{\partial\phi}{\partial y}\mathrm{d}y + \dfrac{\partial\phi}{\partial z}\mathrm{d}z$

(c) directional derivative $\dfrac{\mathrm{d}\phi}{\mathrm{d}s} = \hat{\mathbf{a}} \cdot \text{grad}\,\phi$

(d) unit normal vector $\mathbf{N} = \dfrac{\nabla\phi}{|\nabla\phi|}$.

Copy out this brief summary for future reference. It will help.

74

Grad of sums and products of scalars

(a) $\nabla(A+B) = \mathbf{i}\left\{\frac{\partial}{\partial x}(A+B)\right\} + \mathbf{j}\left\{\frac{\partial}{\partial y}(A+B)\right\} + \mathbf{k}\left\{\frac{\partial}{\partial z}(A+B)\right\}$

$$= \left\{\frac{\partial A}{\partial x}\mathbf{i} + \frac{\partial A}{\partial y}\mathbf{j} + \frac{\partial A}{\partial z}\mathbf{k}\right\} + \left\{\frac{\partial B}{\partial x}\mathbf{i} + \frac{\partial B}{\partial y}\mathbf{j} + \frac{\partial B}{\partial z}\mathbf{k}\right\}$$

$$\therefore\ \underline{\nabla(A+B) = \nabla A + \nabla B}$$

(b) $\nabla(AB) = \mathbf{i}\left\{\frac{\partial}{\partial x}(AB)\right\} + \mathbf{j}\left\{\frac{\partial}{\partial y}(AB)\right\} + \mathbf{k}\left\{\frac{\partial}{\partial z}(AB)\right\}$

$$= \mathbf{i}\left\{A\frac{\partial B}{\partial x} + B\frac{\partial A}{\partial x}\right\} + \mathbf{j}\left\{A\frac{\partial B}{\partial y} + B\frac{\partial A}{\partial y}\right\} + \mathbf{k}\left\{A\frac{\partial B}{\partial z} + B\frac{\partial A}{\partial z}\right\}$$

$$= \left\{A\frac{\partial B}{\partial x}\mathbf{i} + A\frac{\partial B}{\partial y}\mathbf{j} + A\frac{\partial B}{\partial z}\mathbf{k}\right\} + \left\{B\frac{\partial A}{\partial x}\mathbf{i} + B\frac{\partial A}{\partial y}\mathbf{j} + B\frac{\partial A}{\partial z}\mathbf{k}\right\}$$

$$= A\left\{\frac{\partial B}{\partial x}\mathbf{i} + \frac{\partial B}{\partial y}\mathbf{j} + \frac{\partial B}{\partial z}\mathbf{k}\right\} + B\left\{\frac{\partial A}{\partial x}\mathbf{i} + \frac{\partial A}{\partial y}\mathbf{j} + \frac{\partial A}{\partial z}\mathbf{k}\right\}$$

$$\therefore\ \underline{\nabla(AB) = A(\nabla B) + B(\nabla A)}$$

Remember that in these results A and B are scalars. The operator ∇ acting on a vector.........

75

has no meaning

Example 1 If $A = x^2yz + xz^2$ and $B = xy^2z - z^3$ evaluate $\nabla(AB)$ at the point (2, 1, 3).

We know that $\quad \nabla(AB) = A(\nabla B) + B(\nabla A)$

At (2, 1, 3), $\quad \nabla B = ...; \quad \nabla A = ...$

76

$$\nabla B = 3\mathbf{i} + 12\mathbf{j} - 25\mathbf{k}; \qquad \nabla A = 21\mathbf{i} + 12\mathbf{j} + 16\mathbf{k}$$

$$\nabla B = \frac{\partial B}{\partial x}\mathbf{i} + \frac{\partial B}{\partial y}\mathbf{j} + \frac{\partial B}{\partial z}\mathbf{k} = y^2 z\mathbf{i} + 2xyz\mathbf{j} + (xy^2 - 3z^2)\mathbf{k}$$

$$= \underline{3\mathbf{i} + 12\mathbf{j} - 25\mathbf{k}} \quad \text{at } (2, 1, 3)$$

$$\nabla A = \frac{\partial A}{\partial x}\mathbf{i} + \frac{\partial A}{\partial y}\mathbf{j} + \frac{\partial A}{\partial z}\mathbf{k} = (2xyz + z^2)\mathbf{i} + x^2 z\mathbf{j} + (x^2 y + 2xz)\mathbf{k}$$

$$= \underline{21\mathbf{i} + 12\mathbf{j} + 16\mathbf{k}} \quad \text{at } (2, 1, 3)$$

Now $\nabla(AB) = A(\nabla B) + B(\nabla A) = \ldots\ldots\ldots$

Finish it.

77

$$\nabla(AB) = 3(-117\mathbf{i} + 36\mathbf{j} - 362\mathbf{k})$$

$$\nabla(AB) = A(\nabla B) + B(\nabla A)$$

$A = x^2 yz + xz^2 \qquad \therefore$ at $(2, 1, 3)$, $\qquad A = 12 + 18 = 30$

$B = xy^2 z - z^3 \qquad \therefore$ at $(2, 1, 3)$, $\qquad B = 6 - 27 = -21$

$$\therefore \nabla(AB) = 30(3\mathbf{i} + 12\mathbf{j} - 25\mathbf{k}) - 21(21\mathbf{i} + 12\mathbf{j} + 16\mathbf{k})$$

$$= -351\mathbf{i} + 108\mathbf{j} - 1086\mathbf{k} = \underline{3(-117\mathbf{i} + 36\mathbf{j} - 362\mathbf{k})}$$

So add these to the list of results.

$$\nabla(A + B) = \nabla A + \nabla B$$
$$\nabla(AB) = A(\nabla B) + B(\nabla A)$$

where A and B are scalars.

Now on to the next page.

78

Div (divergence of a vector function)

The operator $\nabla\cdot$ (notice the 'dot'; it makes all the difference) can be applied to a vector function $\mathbf{A}(x, y, z)$ to give the *divergence* of **A**, written in short as *div* **A**.

$$\text{If } \mathbf{A} = a_x\mathbf{i} + a_y\mathbf{j} + a_z\mathbf{k}$$

$$\text{div } \mathbf{A} = \nabla \cdot \mathbf{A} = \left(\mathbf{i}\frac{\partial}{\partial x} + \mathbf{j}\frac{\partial}{\partial y} + \mathbf{k}\frac{\partial}{\partial z}\right) \cdot \left(a_x\mathbf{i} + a_y\mathbf{j} + a_z\mathbf{k}\right)$$

$$\therefore \text{div } \mathbf{A} = \nabla \cdot \mathbf{A} = \frac{\partial a_x}{\partial x} + \frac{\partial a_y}{\partial y} + \frac{\partial a_z}{\partial z}$$

Note that
(a) the grad operator ∇ acts on a scalar and gives a vector
(b) the div operation $\nabla\cdot$ acts on a vector and gives a scalar.

Example 1 If $\mathbf{A} = x^2y\mathbf{i} - xyz\mathbf{j} + yz^2\mathbf{k}$ then

$\text{div } \mathbf{A} = \nabla \cdot \mathbf{A} = \ldots\ldots\ldots$

79

$$\boxed{\text{div } \mathbf{A} = \nabla \cdot \mathbf{A} = 2xy - xz + 2yz}$$

We simply take the appropriate partial derivatives of the coefficients of **i**, **j** and **k**. It could hardly be easier.

Example 2 If $\mathbf{A} = 2x^2y\mathbf{i} - 2(xy^2 + y^3z)\mathbf{j} + 3y^2z^2\mathbf{k}$ determine $\nabla \cdot \mathbf{A}$ i.e. div **A**.

Complete it. $\nabla \cdot \mathbf{A} = \ldots\ldots\ldots$

80

$$\boxed{\nabla \cdot \mathbf{A} = 0}$$

for $\mathbf{A} = 2x^2y\mathbf{i} - 2(xy^2 + y^3z)\mathbf{j} + 3y^2z^2\mathbf{k}$

$$\begin{aligned}\nabla \cdot \mathbf{A} &= \frac{\partial a_x}{\partial x} + \frac{\partial a_y}{\partial y} + \frac{\partial a_z}{\partial z} \\ &= 4xy - 2(2xy + 3y^2z) + 6y^2z \\ &= 4xy - 4xy - 6y^2z + 6y^2z = \underline{0}\end{aligned}$$

Such a vector $\mathbf{A}$ for which $\nabla \cdot \mathbf{A} = 0$ at all points, i.e. for all values of x, y, z, is called a *solenoid vector*. It is rather a special case.

Curl (curl of a vector function)

The *curl operator* denoted by $\nabla\times$, acts on a vector and gives another vector as a result.

If $\mathbf{A} = a_x\mathbf{i} + a_y\mathbf{j} + a_z\mathbf{k}$, then curl $\mathbf{A} = \nabla \times \mathbf{A}$.

i.e. $$\text{curl } \mathbf{A} = \nabla \times \mathbf{A} = \left(\mathbf{i}\frac{\partial}{\partial x} + \mathbf{j}\frac{\partial}{\partial y} + \mathbf{k}\frac{\partial}{\partial z}\right) \times (a_x\mathbf{i} + a_y\mathbf{j} + a_z\mathbf{k})$$

$$= \begin{vmatrix} \mathbf{i} & \mathbf{j} & \mathbf{k} \\ \dfrac{\partial}{\partial x} & \dfrac{\partial}{\partial y} & \dfrac{\partial}{\partial z} \\ a_x & a_y & a_z \end{vmatrix}$$

$$\therefore\ \underline{\nabla \times \mathbf{A} = \mathbf{i}\left(\frac{\partial a_z}{\partial y} - \frac{\partial a_y}{\partial z}\right) + \mathbf{j}\left(\frac{\partial a_x}{\partial z} - \frac{\partial a_z}{\partial x}\right) + \mathbf{k}\left(\frac{\partial a_y}{\partial x} - \frac{\partial a_x}{\partial y}\right)}$$

Curl $\mathbf{A}$ is thus a vector function. *It is best remembered in its determinant form, so make a note of it.*

Then on for an example.

81

Example 1 If $\mathbf{A} = (y^4 - x^2z^2)\mathbf{i} + (x^2 + y^2)\mathbf{j} - x^2yz\mathbf{k}$, determine curl **A** at the point (1, 3, –2).

$$\text{curl } \mathbf{A} = \nabla \times \mathbf{A} = \begin{vmatrix} \mathbf{i} & \mathbf{j} & \mathbf{k} \\ \dfrac{\partial}{\partial x} & \dfrac{\partial}{\partial y} & \dfrac{\partial}{\partial z} \\ y^4 - x^2z^2 & x^2 + y^2 & -x^2yz \end{vmatrix}$$

Now we expand the determinant

$$\nabla \times \mathbf{A} = \mathbf{i}\left\{\frac{\partial}{\partial y}(-x^2yz) - \frac{\partial}{\partial z}(x^2 + y^2)\right\} - \mathbf{j}\left\{\frac{\partial}{\partial x}(-x^2yz) - \frac{\partial}{\partial z}(y^4 - x^2z^2)\right\}$$
$$+\mathbf{k}\left\{\frac{\partial}{\partial x}(x^2 + y^2) - \frac{\partial}{\partial y}(y^4 - x^2z^2)\right\}$$

All that now remains is to obtain the partial derivatives and substitute the values of x, y, z.

$$\therefore \nabla \times \mathbf{A} = \ldots\ldots\ldots$$

82

$2\mathbf{i} - 8\mathbf{j} - 106\mathbf{k}$

$$\nabla \times \mathbf{A} = \mathbf{i}\{-x^2z\} - \mathbf{j}\{-2xyz + 2x^2z\} + \mathbf{k}\{2x - 4y^3\}.$$

$$\therefore \text{ At } (1, 3, -2), \quad \nabla \times \mathbf{A} = \mathbf{i}(2) - \mathbf{j}(12 - 4) + \mathbf{k}(2 - 108)$$
$$= \underline{2\mathbf{i} - 8\mathbf{j} - 106\mathbf{k}}$$

Example 2 Determine curl **F** at the point (2, 0, 3) given that $\mathbf{F} = ze^{2xy}\mathbf{i} + 2xz\cos y\mathbf{j} + (x + 2y)\mathbf{k}$.

In determinant form, curl $\mathbf{F} = \nabla \times \mathbf{F} = \ldots\ldots\ldots$

83

$$\begin{vmatrix} \mathbf{i} & \mathbf{j} & \mathbf{k} \\ \dfrac{\partial}{\partial x} & \dfrac{\partial}{\partial y} & \dfrac{\partial}{\partial z} \\ ze^{2xy} & 2xz\cos y & x + 2y \end{vmatrix}$$

Now expand the determinant and substitute the values of x, y and z, finally obtaining curl $\mathbf{F} = \ldots\ldots\ldots$

84

$$\boxed{\text{curl } \mathbf{F} = \nabla \times \mathbf{F} = -2(\mathbf{i} + 3\mathbf{k})}$$

$$\nabla \times \mathbf{F} = \mathbf{i}\{2 - 2x\cos y\} - \mathbf{j}\{1 - e^{2xy}\} + \mathbf{k}\{2z\cos y - 2xze^{2xy}\}$$

$\therefore$ At (2, 0, 3) $\quad \nabla \times \mathbf{F} = \mathbf{i}(2-4) - \mathbf{j}(1-1) + \mathbf{k}(6-12)$

$$= -2\mathbf{i} - 6\mathbf{k} = \underline{-2(\mathbf{i} + 3\mathbf{k})}$$

Every one is done in the same way.

Summary of grad, div and curl

(a) *Grad* operator ∇ acts on a *scalar* field to give a *vector* field
(b) *Div* operator $\nabla\cdot$ acts on a *vector* field to give a *scalar* field
(c) *Curl* operator $\nabla\times$ acts on a *vector* field to give a *vector* field.
(d) With a scalar *function* $\phi(x, y, z)$

$$\text{grad}\,\phi = \nabla\phi = \frac{\partial\phi}{\partial x}\mathbf{i} + \frac{\partial\phi}{\partial y}\mathbf{j} + \frac{\partial\phi}{\partial z}\mathbf{k}$$

(e) With a *vector function* $\mathbf{A} = a_x\mathbf{i} + a_y\mathbf{j} + a_z\mathbf{k}$

(i) $\text{div}\,\mathbf{A} = \nabla\cdot\mathbf{A} = \dfrac{\partial a_x}{\partial x} + \dfrac{\partial a_y}{\partial y} + \dfrac{\partial a_z}{\partial z}$

(ii) $\text{curl}\,\mathbf{A} = \nabla\times\mathbf{A} = \begin{vmatrix} \mathbf{i} & \mathbf{j} & \mathbf{k} \\ \dfrac{\partial}{\partial x} & \dfrac{\partial}{\partial y} & \dfrac{\partial}{\partial z} \\ a_x & a_y & a_z \end{vmatrix}$

Check through that list, just to make sure. We shall need them all.

85 By way of revision, here is one further example.

Example 3 If $\phi = x^2y^2 + x^3yz - yz^2$
and $\mathbf{F} = xy^2\mathbf{i} - 2yz\mathbf{j} + xyz\mathbf{k}$

determine for the point P(1, –1, 2), (a) $\nabla\phi$, (b) unit normal, (c) $\nabla\cdot\mathbf{F}$, (d) $\nabla\times\mathbf{F}$.

Complete all four parts and then check the results with the next frame.

86

Here is the working in full. $\phi = x^2y^2 + x^3yz - yz^2$

(a) $\nabla\phi = \dfrac{\partial\phi}{\partial x}\mathbf{i} + \dfrac{\partial\phi}{\partial y}\mathbf{j} + \dfrac{\partial\phi}{\partial z}\mathbf{k}$

$= (2xy^2 + 3x^2yz)\mathbf{i} + (2x^2y + x^3z - z^2)\mathbf{j} + (x^3y - 2yz)\mathbf{k}$

$\therefore$ At $(1, -1, 2)$ $\quad \underline{\nabla\phi = -4\mathbf{i} - 4\mathbf{j} + 3\mathbf{k}}$

(b) $\mathbf{N} = \dfrac{\nabla\phi}{|\nabla\phi|} \qquad |\nabla\phi| = \sqrt{16 + 16 + 9} = \sqrt{41}$

$$\underline{\therefore \mathbf{N} = \frac{-1}{\sqrt{41}}(4\mathbf{i} + 4\mathbf{j} - 3\mathbf{k})}$$

(c) $\mathbf{F} = xy^2\mathbf{i} - 2yz\mathbf{j} + xyz\mathbf{k} \qquad \nabla \cdot \mathbf{F} = \dfrac{\partial a_x}{\partial x} + \dfrac{\partial a_y}{\partial y} + \dfrac{\partial a_z}{\partial z}$

$$\therefore \nabla \cdot \mathbf{F} = y^2 - 2z + xy$$

$\therefore$ At $(1, -1, 2)$, $\qquad \nabla \cdot \mathbf{F} = 1 - 4 - 1 - -4 \qquad \therefore \underline{\nabla \cdot \mathbf{F} = -4}$

(d)
$$\nabla \times \mathbf{F} = \begin{vmatrix} \mathbf{i} & \mathbf{j} & \mathbf{k} \\ \dfrac{\partial}{\partial x} & \dfrac{\partial}{\partial y} & \dfrac{\partial}{\partial z} \\ xy^2 & -2yz & xyz \end{vmatrix}$$

$$\therefore \nabla \times \mathbf{F} = \mathbf{i}(xz + 2y) - \mathbf{j}(yz - 0) + \mathbf{k}(0 - 2xy)$$
$$= (xz + 2y)\mathbf{i} - yz\mathbf{j} - 2xy\mathbf{k}$$

$\therefore$ At $(1, -1, 2)$, $\quad \nabla \times \mathbf{F} = 2\mathbf{j} + 2\mathbf{k} \qquad \therefore \underline{\nabla \times \mathbf{F} = 2(\mathbf{j} + \mathbf{k})}$

Now let us combine some of these operations.

87

Multiple operations

We can combine the operators grad, div and curl in multiple operations, as in the examples that follow.

Example 1 If $\mathbf{A} = x^2y\mathbf{i} + yz^3\mathbf{j} - zx^3\mathbf{k}$

$$\text{then div } \mathbf{A} = \nabla \cdot \mathbf{A} = \left(\frac{\partial}{\partial x}\mathbf{i} + \frac{\partial}{\partial y}\mathbf{j} + \frac{\partial}{\partial z}\mathbf{k}\right) \cdot (x^2y\mathbf{i} + yz^3\mathbf{j} - zx^3\mathbf{k})$$

$$= 2xy + z^3 + x^3 = \phi \quad \text{say}$$

$$\text{Then} \qquad \text{grad (div } \mathbf{A}) = \nabla(\nabla \cdot \mathbf{A}) = \frac{\partial \phi}{\partial x}\mathbf{i} + \frac{\partial \phi}{\partial y}\mathbf{j} + \frac{\partial \phi}{\partial z}\mathbf{k}$$

$$= (2y + 3x^2)\mathbf{i} + (2x)\mathbf{j} + (3z^2)\mathbf{k}$$

$$\text{i.e.} \qquad \text{grad div } \mathbf{A} = \nabla(\nabla \cdot \mathbf{A}) = \underline{(2y + 3x^2)\mathbf{i} + 2x\mathbf{j} + 3z^2\mathbf{k}}$$

Move on for the next example.

88

Example 2 If $\phi = xyz - 2y^2z + x^2z^2$ determine div grad ϕ at the point (2, 4, 1).

First find grad ϕ and then the div of the result.

At (2, 4, 1), div grad $\phi = \nabla \cdot (\nabla \phi) = \dots\dots\dots$

89

$$\boxed{\text{div grad } \phi = 6}$$

We have $\qquad \phi - xyz - 2y^2z + x^2z^2$

$$\text{grad } \phi = \nabla \phi = \frac{\partial \phi}{\partial x}\mathbf{i} + \frac{\partial \phi}{\partial y}\mathbf{j} + \frac{\partial \phi}{\partial z}\mathbf{k}$$

$$= (yz + 2xz^2)\mathbf{i} + (xz - 4yz)\mathbf{j} + (xy - 2y^2 + 2x^2z)\mathbf{k}$$

$$\therefore \text{ div grad } \phi = \nabla \cdot (\nabla \phi) = 2z^2 - 4z + 2x^2$$

$$\therefore \text{ At } (2, 4, 1), \qquad \text{div grad } \phi = \nabla \cdot (\nabla \phi) = 2 - 4 + 8 = \underline{6}$$

Example 3 If $\mathbf{F} = x^2yz\mathbf{i} + xyz^2\mathbf{j} + y^2z\mathbf{k}$ determine curl curl $\mathbf{F}$ at the point (2, 1, 1).

Determine an expression for curl $\mathbf{F}$ in the usual way, which will be a vector, and then the curl of the result. Finally substitute values.

curl curl $\mathbf{F} = \dots\dots\dots$

90

$$\boxed{\text{curl curl } \mathbf{F} = \nabla \times (\nabla \times \mathbf{F}) = \mathbf{i} + 2\mathbf{j} + 6\mathbf{k}}$$

for $\quad \text{curl } \mathbf{F} = \begin{vmatrix} \mathbf{i} & \mathbf{j} & \mathbf{k} \\ \dfrac{\partial}{\partial x} & \dfrac{\partial}{\partial y} & \dfrac{\partial}{\partial z} \\ x^2yz & xyz^2 & y^2z \end{vmatrix}$

$$= (2yz - 2xyz)\mathbf{i} + x^2y\mathbf{j} + (yz^2 - x^2z)\mathbf{k}$$

Then $\quad \text{curl curl } \mathbf{F} = \begin{vmatrix} \mathbf{i} & \mathbf{j} & \mathbf{k} \\ \dfrac{\partial}{\partial x} & \dfrac{\partial}{\partial y} & \dfrac{\partial}{\partial z} \\ 2yz - 2xyz & x^2y & yz^2 - x^2z \end{vmatrix}$

$$= z^2\mathbf{i} - (-2xz - 2y + 2xy)\mathbf{j} + (2xy - 2z + 2xz)\,\mathbf{k}$$

$\therefore$ At (2, 1, 1), $\quad \text{curl curl } \mathbf{F} = \nabla \times (\nabla \times \mathbf{F}) = \underline{\mathbf{i} + 2\mathbf{j} + 6\mathbf{k}}$

91

Remember that grad, div and curl are operators and that they must act on a scalar or vector as appropriate. They cannot exist alone and must be followed by a function.

One or two interesting general results appear.

(a) *Curl grad* ϕ where ϕ is a scalar

$$\text{grad}\,\phi = \frac{\partial\phi}{\partial x}\mathbf{i} + \frac{\partial\phi}{\partial y}\mathbf{j} + \frac{\partial\phi}{\partial z}\mathbf{k}$$

$$\therefore\ \text{curl grad}\,\phi = \begin{vmatrix} \mathbf{i} & \mathbf{j} & \mathbf{k} \\ \dfrac{\partial}{\partial x} & \dfrac{\partial}{\partial y} & \dfrac{\partial}{\partial z} \\ \dfrac{\partial\phi}{\partial x} & \dfrac{\partial\phi}{\partial y} & \dfrac{\partial\phi}{\partial z} \end{vmatrix}$$

$$= \mathbf{i}\left\{\frac{\partial^2\phi}{\partial y\partial z} - \frac{\partial^2\phi}{\partial z\partial y}\right\} - \mathbf{j}\left\{\frac{\partial^2\phi}{\partial z\partial x} - \frac{\partial^2\phi}{\partial x\partial z}\right\}$$

$$+ \mathbf{k}\left\{\frac{\partial^2\phi}{\partial x\partial y} - \frac{\partial^2\phi}{\partial y\partial x}\right\} = 0$$

$$\therefore\ \underline{\text{curl grad}\,\phi = \nabla \times (\nabla\phi) = 0}$$

(b) *Div curl* **A** where **A** is a vector. $\mathbf{A} = a_x\mathbf{i} + a_y\mathbf{j} + a_z\mathbf{k}$

$$\text{curl}\,\mathbf{A} = \nabla \times \mathbf{A} = \begin{vmatrix} \mathbf{i} & \mathbf{j} & \mathbf{k} \\ \dfrac{\partial}{\partial x} & \dfrac{\partial}{\partial y} & \dfrac{\partial}{\partial z} \\ a_x & a_y & a_z \end{vmatrix}$$

$$= \mathbf{i}\left(\frac{\partial a_z}{\partial y} - \frac{\partial a_y}{\partial z}\right) - \mathbf{j}\left(\frac{\partial a_z}{\partial x} - \frac{\partial a_x}{\partial z}\right) + \mathbf{k}\left(\frac{\partial a_y}{\partial x} - \frac{\partial a_x}{\partial y}\right)$$

Then $\text{div curl}\,\mathbf{A} = \nabla \cdot (\nabla \times \mathbf{A}) = \left(\mathbf{i}\dfrac{\partial}{\partial x} + \mathbf{j}\dfrac{\partial}{\partial y} + \mathbf{k}\dfrac{\partial}{\partial z}\right) \cdot (\nabla \times \mathbf{A})$

$$= \frac{\partial^2 a_z}{\partial x\partial y} - \frac{\partial^2 a_y}{\partial z\partial x} - \frac{\partial^2 a_z}{\partial x\partial y} + \frac{\partial^2 a_x}{\partial y\partial z} + \frac{\partial^2 a_y}{\partial z\partial x} - \frac{\partial^2 a_x}{\partial y\partial z}$$

$$= 0$$

$$\therefore\ \underline{\text{div curl}\,\mathbf{A} = \nabla \cdot (\nabla \times \mathbf{A}) = \mathbf{0}}$$

(c) Div *grad* ϕ where ϕ is a scalar

$$\text{grad}\ \phi = \frac{\partial\phi}{\partial x}\mathbf{i} + \frac{\partial\phi}{\partial y}\mathbf{j} + \frac{\partial\phi}{\partial z}\mathbf{k}$$

Then div grad $\phi = \nabla \cdot (\nabla\phi)$

$$= \left(\mathbf{i}\frac{\partial}{\partial x} + \mathbf{j}\frac{\partial}{\partial y} + \mathbf{k}\frac{\partial}{\partial z}\right) \cdot \left(\frac{\partial\phi}{\partial x}\mathbf{i} + \frac{\partial\phi}{\partial y}\mathbf{j} + \frac{\partial\phi}{\partial z}\mathbf{k}\right)$$

$$= \frac{\partial^2\phi}{\partial x^2} + \frac{\partial^2\phi}{\partial y^2} + \frac{\partial^2\phi}{\partial z^2}$$

$$\therefore \text{div grad}\ \phi = \nabla \cdot (\nabla\phi) = \frac{\partial^2\phi}{\partial x^2} + \frac{\partial^2\phi}{\partial y^2} + \frac{\partial^2\phi}{\partial z^2}$$

This result is sometimes denoted by $\nabla^2\phi$.

So these general results are

(a) curl grad $\phi = \nabla \times (\nabla\phi) = 0$

(b) div curl $\mathbf{A} = \nabla \cdot (\nabla \times \mathbf{A}) = 0$

(c) div grad $\phi = \nabla \cdot (\nabla\phi) = \dfrac{\partial^2\phi}{\partial x^2} + \dfrac{\partial^2\phi}{\partial y^2} + \dfrac{\partial^2\phi}{\partial z^2}$.

That brings us to the end of this particular programme. We have covered quite a lot of new material, so check carefully through the Revision Summary that follows: then you can deal with the Test Exercise.

92 REVISION SUMMARY

If $\mathbf{A} = a_x\mathbf{i} + a_y\mathbf{j} + a_z\mathbf{k}$; $\mathbf{B} = b_x\mathbf{i} + b_y\mathbf{j} + b_z\mathbf{k}$; $\mathbf{C} = c_x\mathbf{i} + c_y\mathbf{j} + c_z\mathbf{k}$; then we have the following relationships.

1. *Scalar product* (dot product) $\mathbf{A} \cdot \mathbf{B} = AB\cos\theta$

$$\mathbf{A} \cdot \mathbf{B} = \mathbf{B} \cdot \mathbf{A} \quad \text{and} \quad \mathbf{A} \cdot (\mathbf{B} + \mathbf{C}) = \mathbf{A} \cdot \mathbf{B} + \mathbf{A} \cdot \mathbf{C}$$

2. *Vector product* (cross product) $\mathbf{A} \times \mathbf{B} = (AB\sin\theta)\mathbf{N}$

$\mathbf{N} =$ unit normal vector where **A, B, N** form a right-handed set.

$$\mathbf{A} \times \mathbf{B} = \begin{vmatrix} \mathbf{i} & \mathbf{j} & \mathbf{k} \\ a_x & a_y & a_z \\ b_x & b_y & b_z \end{vmatrix}$$

$$\mathbf{A} \times \mathbf{B} = -(\mathbf{B} \times \mathbf{A}) \quad \text{and} \quad \mathbf{A} \times (\mathbf{B} + \mathbf{C}) = \mathbf{A} \times \mathbf{B} + \mathbf{A} \times \mathbf{C}$$

3. *Unit vectors*

(a) $\mathbf{i} \cdot \mathbf{i} = \mathbf{j} \cdot \mathbf{j} = \mathbf{k} \cdot \mathbf{k} = 1$
$\mathbf{i} \cdot \mathbf{j} = \mathbf{j} \cdot \mathbf{k} = \mathbf{k} \cdot \mathbf{i} = 0.$

(b) $\mathbf{i} \times \mathbf{i} = \mathbf{j} \times \mathbf{j} = \mathbf{k} \times \mathbf{k} = 0$
$\mathbf{i} \times \mathbf{j} = \mathbf{j} \times \mathbf{k} = \mathbf{k} \times \mathbf{i} = 1.$

4. *Scalar triple product* $\mathbf{A} \cdot (\mathbf{B} \times \mathbf{C})$

$$\mathbf{A} \cdot (\mathbf{B} \times \mathbf{C}) = \begin{vmatrix} a_x & a_y & a_z \\ b_x & b_y & b_z \\ c_x & c_y & c_z \end{vmatrix}$$

$$\mathbf{A} \cdot (\mathbf{B} \times \mathbf{C}) = \mathbf{B} \cdot (\mathbf{C} \times \mathbf{A}) = \mathbf{C} \cdot (\mathbf{A} \times \mathbf{B})$$

Unchanged by cyclic change of vectors.

Sign reversed by non-cyclic change of vectors.

5. *Coplanar vectors* $\mathbf{A} \cdot (\mathbf{B} \times \mathbf{C}) = 0.$

6. *Vector triple product* $\mathbf{A} \times (\mathbf{B} \times \mathbf{C})$ and $(\mathbf{A} \times \mathbf{B}) \times \mathbf{C}$

$$\mathbf{A} \times (\mathbf{B} \times \mathbf{C}) = (\mathbf{A} \cdot \mathbf{C})\mathbf{B} - (\mathbf{A} \cdot \mathbf{B})\mathbf{C}$$

and

$$(\mathbf{A} \times \mathbf{B}) \times \mathbf{C} = (\mathbf{C} \cdot \mathbf{A})\mathbf{B} - (\mathbf{C} \cdot \mathbf{B})\mathbf{A}.$$

7. *Differentiation of vectors*

If $\mathbf{A}$, a_x, a_y, a_z are functions of u,

$$\frac{d\mathbf{A}}{du} = \frac{da_x}{du}\mathbf{i} + \frac{da_y}{du}\mathbf{j} + \frac{da_z}{du}\mathbf{k}$$

8. *Unit tangent vector* $\mathbf{T}$

$$\mathbf{T} = \frac{\dfrac{d\mathbf{A}}{du}}{\left|\dfrac{d\mathbf{A}}{du}\right|}$$

9. *Integration of vectors*

$$\int_a^b \mathbf{A}\, du = \mathbf{i}\int_a^b a_x\, du + \mathbf{j}\int_a^b a_y\, du + \mathbf{k}\int_a^b a_z\, du$$

10. *Grad* (gradient of a scalar function ϕ)

$$\text{grad}\ \phi = \nabla\phi = \frac{\partial\phi}{\partial x}\mathbf{i} + \frac{\partial\phi}{\partial y}\mathbf{j} + \frac{\partial\phi}{\partial z}\mathbf{k}$$

$$\text{'del'} = \text{operator}\ \nabla = \left(\mathbf{i}\frac{\partial}{\partial x} + \mathbf{j}\frac{\partial}{\partial y} + \mathbf{k}\frac{\partial}{\partial z}\right)$$

(a) *Directional derivative* $\dfrac{d\phi}{ds} = \hat{\mathbf{a}} \cdot \text{grad}\ \phi = \hat{\mathbf{a}} \cdot \nabla\phi$ where $\hat{\mathbf{a}}$ is a unit vector in

a stated direction. Grad ϕ gives the direction for maximum rate of change of ϕ .

(b) *Unit normal vector* $\mathbf{N}$ to surface $\phi(x, y, z) = \text{constant}$.

$$\mathbf{N} = \frac{\nabla\phi}{|\nabla\phi|}$$

11. *Div* (divergence of a vector function $\mathbf{A}$)

$$\text{div}\ \mathbf{A} = \nabla \cdot \mathbf{A} = \frac{\partial a_x}{\partial x} + \frac{\partial a_y}{\partial y} + \frac{\partial a_z}{\partial z}$$

If $\nabla \cdot \mathbf{A} = 0$ for all points, $\mathbf{A}$ is a solenoid vector.

12. *Curl* (curl of a vector function $\mathbf{A}$)

$$\text{curl}\ \mathbf{A} = \nabla \times \mathbf{A} = \begin{vmatrix} \mathbf{i} & \mathbf{j} & \mathbf{k} \\ \dfrac{\partial}{\partial x} & \dfrac{\partial}{\partial y} & \dfrac{\partial}{\partial z} \\ a_x & a_y & a_z \end{vmatrix}$$

13. *Operators*

grad (∇) acts on a *scalar* and gives a *vector*

div ($\nabla\cdot$) acts on a *vector* and gives a *scalar*

curl ($\nabla\times$) acts on a *vector* and gives a *vector*.

14. *Multiple operations*

(a) curl grad $\phi = \nabla \times (\nabla\phi) = 0$

(b) div curl $\mathbf{A} = \nabla \cdot (\nabla \times \mathbf{A}) = 0$

(c) div grad $\phi = \nabla \cdot (\nabla\phi) = \dfrac{\partial^2\phi}{\partial x^2} + \dfrac{\partial^2\phi}{\partial y^2} + \dfrac{\partial^2\phi}{\partial z^2}$.

93

TEST EXERCISE XII

1. Find (a) the scalar product and (b) the vector product of the vectors $\mathbf{A} = 3\mathbf{i} - 2\mathbf{j} + 4\mathbf{k}$ and $\mathbf{B} = \mathbf{i} + 5\mathbf{j} - 2\mathbf{k}$.

2. If $\mathbf{A} = 2\mathbf{i} + 3\mathbf{j} - 5\mathbf{k}$; $\mathbf{B} = 3\mathbf{i} + \mathbf{j} + 2\mathbf{k}$; $\mathbf{C} = \mathbf{i} - \mathbf{j} + 3\mathbf{k}$ determine

   (a) the scalar triple product $\mathbf{A} \cdot (\mathbf{B} \times \mathbf{C})$

   (b) the vector triple product $\mathbf{A} \times (\mathbf{B} \times \mathbf{C})$.

3. Determine whether the three vectors $\mathbf{A} = 2\mathbf{i} + 3\mathbf{j} + \mathbf{k}$;

   $\mathbf{B} = \mathbf{i} - 2\mathbf{j} + 2\mathbf{k}$; $\mathbf{C} = 3\mathbf{i} + \mathbf{j} + 3\mathbf{k}$ are coplanar.

4. If $\mathbf{A} = (u^2 + 5)\mathbf{i} - (u^2 + 3)\mathbf{j} + 2u^3\mathbf{k}$, determine

   (a) $\dfrac{d\mathbf{A}}{du}$; (b) $\dfrac{d^2\mathbf{A}}{du^2}$; (c) $\left|\dfrac{d\mathbf{A}}{du}\right|$; all at $u = 2$.

5. Determine the unit tangent vector at the point (2, 4, 3) for the curve with parametric equations

$$x = 2u^2; \quad y = u + 3; \quad z = 4u^2 - u.$$

6. If $\mathbf{F} = 2\mathbf{i} + 4u\mathbf{j} + u^2\mathbf{k}$ and $\mathbf{G} = u^2\mathbf{i} - 2u\mathbf{j} + 4\mathbf{k}$ determine

$$\int_0^2 (\mathbf{F} \times \mathbf{G})du.$$

7. Find the directional derivative of the function $\phi = x^2y - 2xz^2 + y^2z$ at the point (1, 3, 2) in the direction of the vector $\mathbf{A} = 3\mathbf{i} + 2\mathbf{j} - \mathbf{k}$.

8. Find the unit normal to the surface $\phi = 2x^3z + x^2y^2 + xyz + 4 = 0$ at the point (2, 1, 0).

9. If $\mathbf{A} = x^2y\mathbf{i} + (xy + yz)\mathbf{j} + xz^2\mathbf{k}$; $\mathbf{B} = yz\mathbf{i} - 3xz\mathbf{j} + 2xy\mathbf{k}$; and $\phi = 3x^2y + xyz - 4y^2z^2 - 3$ determine, at the point (1, 2, 1),

   (a) $\nabla\phi$; (b) $\nabla \cdot \mathbf{A}$; (c) $\nabla \times \mathbf{B}$; (d) grad div $\mathbf{A}$; (e) curl curl $\mathbf{A}$.

FURTHER PROBLEMS XII

1. If $\mathbf{A} = 2\mathbf{i} + 3\mathbf{j} - 4\mathbf{k}$; $\mathbf{B} = 3\mathbf{i} + 5\mathbf{j} + 2\mathbf{k}$; $\mathbf{C} = \mathbf{i} - 2\mathbf{j} + 3\mathbf{k}$; determine $\mathbf{A} \cdot (\mathbf{B} \times \mathbf{C})$.

2. If $\mathbf{A} = 2\mathbf{i} + \mathbf{j} - 3\mathbf{k}$; $\mathbf{B} = \mathbf{i} - 2\mathbf{j} + 2\mathbf{k}$; $\mathbf{C} = 3\mathbf{i} + 2\mathbf{j} - \mathbf{k}$; find $\mathbf{A} \times (\mathbf{B} \times \mathbf{C})$.

3. If $\mathbf{A} = \mathbf{i} - 2\mathbf{j} + 3\mathbf{k}$; $\mathbf{B} = 2\mathbf{i} + \mathbf{j} - 2\mathbf{k}$; $\mathbf{C} = 3\mathbf{i} + 2\mathbf{j} + \mathbf{k}$; find

   (a) $\mathbf{A} \times (\mathbf{B} \times \mathbf{C})$, (b) $(\mathbf{A} \times \mathbf{B}) \times \mathbf{C}$.

4. If $\mathbf{F} = x^2\mathbf{i} + (3x + 2)\mathbf{j} + \sin x\mathbf{k}$, find (a) $\dfrac{d\mathbf{F}}{dx}$; (b) $\dfrac{d^2\mathbf{F}}{dx^2}$; (c) $\left|\dfrac{d\mathbf{F}}{dx}\right|$; (d) $\dfrac{d}{dx}(\mathbf{F} \cdot \mathbf{F})$ at $x = 1$.

5. If $\mathbf{F} = u\mathbf{i} + (1 - u)\mathbf{j} + 3u\mathbf{k}$ and $\mathbf{G} = 2\mathbf{i} - (1 + u)\mathbf{j} - u^2\mathbf{k}$, determine (a) $\dfrac{d}{du}(\mathbf{F} \cdot \mathbf{G})$; (b) $\dfrac{d}{du}(\mathbf{F} \times \mathbf{G})$; (c) $\dfrac{d}{du}(\mathbf{F} + \mathbf{G})$.

6. Find the unit normal to the surface $4x^2y^2 - 3xz^2 - 2y^2z + 4 = 0$ at the point $(2, -1, -2)$.

7. Find the unit normal to the surface $2xy^2 + y^2z + x^2z - 11 = 0$ at the point $(-2, 1, 3)$.

8. Determine the unit vector normal to the surface

   $xz^2 + 3xy - 2yz^2 + 1 = 0$ at the point $(1, -2, -1)$.

9. Find the unit normal to the surface $x^2y - 2yz^2 + y^2z = 3$ at the point $(2, -3, 1)$.

10. Determine the directional derivative of $\phi = xe^y + yz^2 + xyz$ at the point $(2, 0, 3)$ in the direction of $\mathbf{A} = 3\mathbf{i} - 2\mathbf{j} + \mathbf{k}$.

11. Find the directional derivative of $\phi = (x + 2y + z)^2 - (x - y - z)^2$ at the point $(2, 1, -1)$ in the direction of $\mathbf{A} = \mathbf{i} - 4\mathbf{j} + 2\mathbf{k}$.

12. Find the scalar triple product of

    (a) $\mathbf{A} = \mathbf{i} + 2\mathbf{j} - 3\mathbf{k}$; $\mathbf{B} = 2\mathbf{i} - \mathbf{j} + 4\mathbf{k}$; $\mathbf{C} = 3\mathbf{i} + \mathbf{j} - 2\mathbf{k}$.

    (b) $\mathbf{A} = 2\mathbf{i} - 3\mathbf{j} + \mathbf{k}$; $\mathbf{B} = 3\mathbf{i} + \mathbf{j} + 2\mathbf{k}$; $\mathbf{C} = \mathbf{i} + 4\mathbf{j} - 2\mathbf{k}$.

    (c) $\mathbf{A} = -2\mathbf{i} + 3\mathbf{j} - 2\mathbf{k}$; $\mathbf{B} = 3\mathbf{i} - \mathbf{j} + 3\mathbf{k}$; $\mathbf{C} = 2\mathbf{i} - 5\mathbf{j} + \mathbf{k}$.

13. Find the vector triple products $\mathbf{A} \times (\mathbf{B} \times \mathbf{C})$ of the following.

    (a) $\mathbf{A} = 3\mathbf{i} + \mathbf{j} - 2\mathbf{k}$; $\mathbf{B} = 2\mathbf{i} + 4\mathbf{j} + 3\mathbf{k}$; $\mathbf{C} = \mathbf{i} - 2\mathbf{j} + \mathbf{k}$.

    (b) $\mathbf{A} = 2\mathbf{i} - \mathbf{j} + 3\mathbf{k}$; $\mathbf{B} = \mathbf{i} + 4\mathbf{j} - 5\mathbf{k}$; $\mathbf{C} = 3\mathbf{i} - 2\mathbf{j} + \mathbf{k}$.

    (c) $\mathbf{A} = 4\mathbf{i} + 2\mathbf{j} - 3\mathbf{k}$; $\mathbf{B} = 2\mathbf{i} - 3\mathbf{j} + 2\mathbf{k}$; $\mathbf{C} = 3\mathbf{i} - 3\mathbf{j} + \mathbf{k}$.

14. If $\mathbf{F} = 4t^3\mathbf{i} - 2t^2\mathbf{j} + 4t\mathbf{k}$, determine when $t = 1$ (a) $\dfrac{d\mathbf{F}}{dt}$; (b) $\dfrac{d^2\mathbf{F}}{dt^2}$; (c) $\dfrac{d}{dt}(\mathbf{F} \cdot \mathbf{F})$.

15. If $\phi = x^2 \sin z + ze^y$ find, at the point $(1, 3, 2)$, the values of (a) grad ϕ and (b) $|\text{grad}\ \phi|$.

16. Given that $\phi = xy^2 + yz^2 - x^2$ find the derivative of ϕ with respect to distance at the point (1, 2, –1), measured parallel to the vector $2\mathbf{i} - 3\mathbf{j} + 4\mathbf{k}$.

17. Find unit vectors normal to the surfaces $x^2 + y^2 - z^2 + 3 = 0$ and $xy - yz + zx - 10 = 0$ at the point (3, 2, 4) and hence find the angle between the two surfaces at that point.

18. If $\mathbf{r} = (t^2 + 3t)\mathbf{i} - 2\sin 3t\mathbf{j} + 3e^{2t}\mathbf{k}$ determine (a) $\dfrac{d\mathbf{r}}{dt}$; (b) $\dfrac{d^2\mathbf{r}}{dt^2}$; (c) the value of $\left|\dfrac{d^2\mathbf{r}}{dt^2}\right|$ at $t = 0$.

19. (a) Show that curl $(-y\mathbf{i} + x\mathbf{j})$ is a constant vector.

(b) Show that the vector field $(yz\mathbf{i} + zx\mathbf{j} + xy\mathbf{k})$ has zero divergence and zero curl.

20. If $\mathbf{A} = 2xz^2\mathbf{i} - xz\mathbf{j} + (y + z)\mathbf{k}$, find curl curl $\mathbf{A}$.

21. Determine grad ϕ where $\phi = x^2\cos(2yz - 0.5)$ and obtain its value at the point (1, 3, 1).

22. Determine the value of p such that the three vectors **A**, **B**, **C** are coplanar when $\mathbf{A} = 2\mathbf{i} + \mathbf{j} + 4\mathbf{k}$; $\mathbf{B} = 3\mathbf{i} + 2\mathbf{j} + p\mathbf{k}$; $\mathbf{C} = \mathbf{i} + 4\mathbf{j} + 2\mathbf{k}$.

23. If $\mathbf{A} = p\mathbf{i} - 6\mathbf{j} - 3\mathbf{k}$; $\mathbf{B} = 4\mathbf{i} + 3\mathbf{j} - \mathbf{k}$; $\mathbf{C} = \mathbf{i} - 3\mathbf{j} + 2\mathbf{k}$:

(a) find the values of p for which (i) **A** and **B** are perpendicular to each other, (ii) **A**, **B** and **C** are coplanar

(b) determine a unit vector perpendicular to both **A** and **B** when $p = 2$.

Programme 13

Vector Analysis

PART 2

Introduction

1 We dealt in some detail with line, surface and volume integrals in an earlier programme, when we approached the subject analytically. In many practical problems, it is more convenient to express these integrals in vector form and the methods often lead to more concise working.

Line integrals

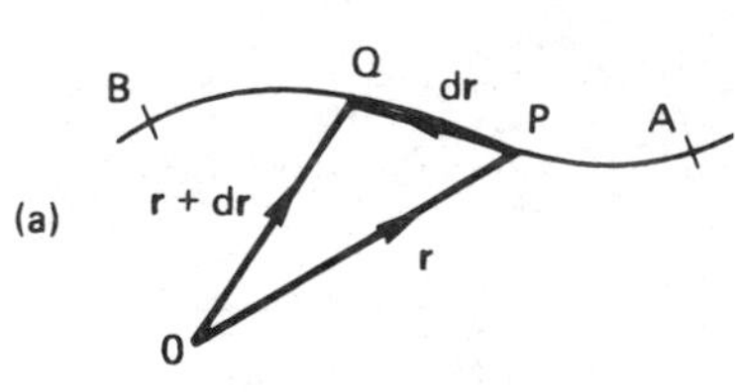

Let a point P on the curve c joining A and B be denoted by the position vector **r** with respect to a fixed origin O.

If Q is a neighbouring point on the curve with position vector $\mathbf{r} + \mathbf{dr}$, then $\overline{\mathbf{PQ}} = \mathbf{dr}$.

The curve c can be divided up into many (n) such small arcs, approximating to $\mathbf{dr}_1$, $\mathbf{dr}_2$, $\mathbf{dr}_3 \ldots \mathbf{dr}_p \ldots$ so that $\overline{\mathbf{AB}} = \sum_{p=1}^{n} \mathbf{dr}_p$ where $\mathbf{dr}_p$ is a vector representing the element of arc in both magnitude and direction.

Scalar field

If a scalar field V exists for all points on the curve, then $\sum_{p=1}^{n} V\mathbf{dr}_p$ with $\mathbf{dr} \to 0$, defines the *line integral* of V along the curve c from A to B,

$$\text{i.e. line integral} = \int_c V\mathbf{dr}.$$

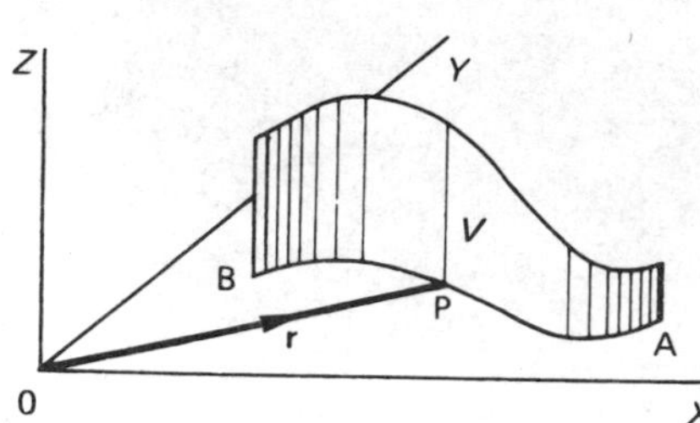

We can illustrate this integral by erecting a continuous ordinate proportional to V at each point of the curve. $\int_c V\mathbf{dr}$ is then represented by the area of the curved surface between the ends A and B of the curve c.

To evaluate a line integral, the integrand is expressed in terms of x, y, z, with $\mathbf{dr} = \ldots\ldots\ldots$

2

$$\mathbf{dr} = \mathbf{i}\,dx + \mathbf{j}\,dy + \mathbf{k}\,dz$$

In practice, x, y and z are often expressed in terms of parametric equations of a fourth variable (say u), i.e. $x = x(u)$; $y = y(u)$; $z = z(u)$. From these, dx, dy and dz can be written in terms of u and the integral evaluated in terms of this parameter u.

The following examples will show the method.

Example 1 If $V = xy^2z$, evaluate $\int_c V\,\mathbf{dr}$ along the curve c having parametric equations $x = 3u$; $y = 2u^2$; $z = u^3$ between A(0, 0, 0) and B(3, 2, 1).

$$V = xy^2z = (3u)(4u^4)(u^3) = 12u^8$$

$$\mathbf{dr} = \mathbf{i}\,dx + \mathbf{j}\,dy + \mathbf{k}\,dz = \dots\dots\dots$$

3

$$\mathbf{dr} = \mathbf{i}\,3\,du + \mathbf{j}\,4u\,du + \mathbf{k}\,3u^2\,du$$

for $x = 3u$, $\therefore dx = 3\,du$; $y = 2u^2$, $\therefore dy = 4u\,du$; $z = u^3$,

$\therefore dz = 3u^2\,du$

Limits : A(0, 0, 0) corresponds to $u = \dots\dots\dots$

B(3, 2, 1) corresponds to $u = \dots\dots\dots$

4

$$A(0,0,0) \equiv u = 0; \qquad B(3,2,1) \equiv u = 1$$

$$\therefore \int_c V\,\mathbf{dr} = \int_0^1 12u^8\,(\mathbf{i}\,3\,du + \mathbf{j}\,4u\,du + \mathbf{k}\,3u^2\,du)$$

$$= \dots\dots\dots \quad \text{Finish it off.}$$

5

$$4\mathbf{i}+\frac{24}{5}\mathbf{j}+\frac{36}{11}\mathbf{k}$$

for $\int_c V\,d\mathbf{r} = 12\int_0^1 (\mathbf{i}\,3u^8\,du + \mathbf{j}\,4u^9\,du + \mathbf{k}\,3u^{10}\,du)$ which integrates directly to give the result quoted above.

Now for another example.

6

Example 2 If $V = xy + y^2z$ evaluate $\int_c V\,d\mathbf{r}$ along the curve c defined by

$x = t^2$; $y = 2t$; $z = t + 5$ between A(0, 0, 5) and B(4, 4, 7).

As before, expressing V and $d\mathbf{r}$ in terms of the parameter t we have

$V = \dots\dots\dots$; $d\mathbf{r} = \dots\dots\dots$

7

$$V = 6t^3 + 20t^2; \quad d\mathbf{r} = \mathbf{i}\,2t\,dt + \mathbf{j}\,2\,dt + \mathbf{k}\,dt$$

since $V = xy + y^2z = (t^2)(2t) + (4t^2)(t+5) = 6t^3 + 20t^2$.

Also
$$\left.\begin{aligned} x &= t^2 \quad & dx &= 2t\,dt \\ y &= 2t & dy &= 2\,dt \\ z &= t+5 & dz &= dt \end{aligned}\right\} \quad \begin{aligned} \therefore\ d\mathbf{r} &= \mathbf{i}\,dx + \mathbf{j}\,dy + \mathbf{k}\,dz \\ &= \mathbf{i}\,2t\,dt + \mathbf{j}\,2\,dt + \mathbf{k}\,dt \end{aligned}$$

$$\therefore \int_c V\,d\mathbf{r} = \int_c (6t^3 + 20t^2)(\mathbf{i}\,2t + \mathbf{j}\,2 + \mathbf{k})\,dt$$

Limits: $A(0,0,5) \equiv t = \dots\dots\dots$

$B(4,4,7) \equiv t = \dots\dots\dots$

8

$$A(0,0,5) \equiv t=0; \quad B(4,4,7) \equiv t=2$$

$$\therefore \int_c V\,d\mathbf{r} = \int_0^2 (6t^3 + 20t^2)(\mathbf{i}\,2t + \mathbf{j}\,2 + \mathbf{k})\,dt$$

$= \ldots\ldots\ldots$ Complete the integration.

9

$$\frac{8}{15}(444\,\mathbf{i} + 290\,\mathbf{j} + 145\,\mathbf{k})$$

$$\int_c V\,d\mathbf{r} = 2\int_0^2 \{(6t^4 + 20t^3)\mathbf{i} + (6t^3 + 20t^2)\mathbf{j} + (3t^3 + 10t^2)\mathbf{k}\}\,dt.$$

The actual integration is simple enough and gives the result shown. All line integrals in scalar fields are done in the same way.

10 Vector field

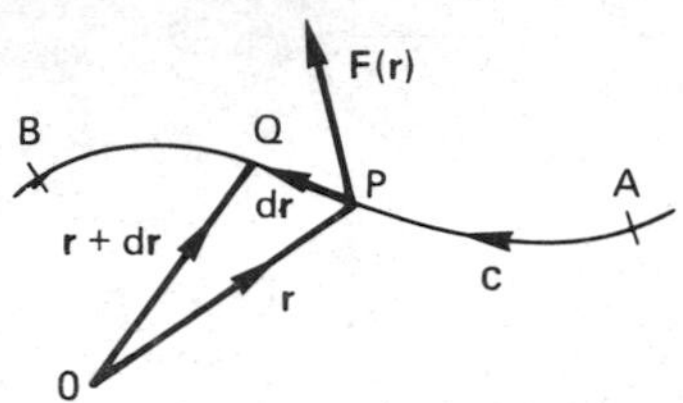

If a vector field $\mathbf{F(r)}$ exists for all points of the curve c, then for each element of arc we can form the scalar product $\mathbf{F}\cdot\mathbf{dr}$. Summing these products for all elements of arc, we have $\sum_{p=1}^{n}\mathbf{F}\cdot\mathbf{dr}_p$

Then, if $\mathbf{dr}_p \to 0$, the sum becomes the integral $\int_c \mathbf{F}\cdot\mathbf{dr}$,

i.e. the line integral of $\mathbf{F}(\mathbf{r})$ from A to B along the stated curve $= \int_c \mathbf{F}\cdot\mathbf{dr}$.

In this case, since $\mathbf{F}\cdot\mathbf{dr}$ is a scalar product, then the line integral is a scalar.

To evaluate the line integral, $\mathbf{F}$ and $\mathbf{dr}$ are expressed in terms of x, y, z and the curve in parametric form. We have

$$\mathbf{F} = F_x\mathbf{i} + F_y\mathbf{j} + F_z\mathbf{k}$$

and
$$\mathbf{dr} = \mathbf{i}\,\mathrm{d}x + \mathbf{j}\,\mathrm{d}y + \mathbf{k}\,\mathrm{d}z$$

Then
$$\mathbf{F}\cdot\mathbf{dr} = (F_x\mathbf{i} + F_y\mathbf{j} + F_z\mathbf{k})\cdot(\mathbf{i}\,\mathrm{d}x + \mathbf{j}\,\mathrm{d}y + \mathbf{k}\,\mathrm{d}z)$$
$$= F_x\,\mathrm{d}x + F_y\,\mathrm{d}y + F_z\,\mathrm{d}z$$

$$\therefore \int_c \mathbf{F}\cdot\mathbf{dr} = \int_c \mathbf{F}_x\,\mathrm{d}x + \int_c F_y\,\mathrm{d}y + \int_c F_z\,\mathrm{d}z$$

Now for an example to show it in operation.

Example 1 If $\mathbf{F}(\mathbf{r}) = x^2y\mathbf{i} + xz\mathbf{j} - 2yz\mathbf{k}$, evaluate $\int_c \mathbf{F}\cdot\mathbf{dr}$ between A(0, 0, 0) and B(4, 2, 1) along the curve having parametric equations $x = 4t$; $y = 2t^2$; $z = t^3$.

Expressing everything in terms of the parameter t, we have

$$\mathbf{F} = \ldots\ldots\ldots$$

$$\mathrm{d}x = \ldots\ldots\ldots;\quad \mathrm{d}y = \ldots\ldots\ldots;\quad \mathrm{d}z = \ldots\ldots\ldots$$

11

$$\mathbf{F} = 32t^4\,\mathbf{i} + 4t^4\,\mathbf{j} - 4t^5\,\mathbf{k}$$
$$dx = 4\,dt; \quad dy = 4t\,dt; \quad dz = 3t^2\,dt$$

$$\begin{aligned} \text{for} \quad x^2y &= (16t^2)(2t^2) = 32t^4 & x &= 4t & \therefore\ dx &= 4\,dt \\ xz &= (4t)(t^3) = 4t^4 & y &= 2t^2 & \therefore\ dy &= 4t\,dt \\ 2yz &= (4t^2)(t^3) = 4t^5 & z &= t^3 & \therefore\ dz &= 3t^2\,dt \end{aligned}$$

$$\text{Then} \int \mathbf{F}\cdot d\mathbf{r} = \int (32t^4\mathbf{i} + 4t^4\mathbf{j} - 4t^5\mathbf{k})\cdot(\mathbf{i}\,4\,dt + \mathbf{j}\,4t\,dt + \mathbf{k}\,3t^2\,dt)$$
$$= \int (128t^4 + 16t^5 - 12t^7)\,dt$$

Limits: $A(0, 0, 0) \equiv t = \ldots\ldots\ldots;\quad B(4, 2, 1) \equiv t = \ldots\ldots\ldots$

12

$$A \equiv t = 0; \quad B \equiv t = 1$$

$$\therefore \int_c \mathbf{F}\cdot d\mathbf{r} = \int_0^1 (128t^4 + 16t^5 - 12t^7)\,dt = \ldots\ldots\ldots$$

13

$$\frac{128}{5} + \frac{8}{3} - \frac{3}{2} = \frac{803}{30} = 26.77$$

If the vector field $\mathbf{F}(\mathbf{r})$ is a *force field*, then the line integral $\int_c \mathbf{F}\cdot d\mathbf{r}$ represents the work done in moving a unit particle along the prescribed curve c from A to B.

Now for another example.

Example 2 If $\mathbf{F}(\mathbf{r}) = x^2y\mathbf{i} + 2yz\mathbf{j} + 3z^2x\mathbf{k}$ evaluate $\int_c \mathbf{F}\cdot d\mathbf{r}$ between A(0, 0, 0) and B(1, 2, 3)

(a) along the straight lines c_1 from $(0, 0, 0)$ to $(1, 0, 0)$
then c_2 from $(1, 0, 0)$ to $(1, 2, 0)$
and c_3 from $(1, 2, 0)$ to $(1, 2, 3)$

(b) along the straight lines c_4 joining $(0, 0, 0)$ to $(1, 2, 3)$.

As before, we first obtain an expression for $\mathbf{F}\cdot d\mathbf{r}$ which is

. . . .

14

$$\mathbf{F}\cdot\,\mathbf{dr} = x^2y\,dx + 2yz\,dy + 3z^2x\,dz$$

for $\mathbf{F}\cdot\,\mathbf{dr} = (x^2y\,\mathbf{i} + 2yz\,\mathbf{j} + 3z^2x\,\mathbf{k})\cdot(\mathbf{i}\,dx + \mathbf{j}\,dy + \mathbf{k}\,dz)$

$$\therefore \int \mathbf{F}\cdot\,\mathbf{dr} = \int x^2y\,dx + \int 2yz\,dy + \int 3z^2x\,dz$$

(a)

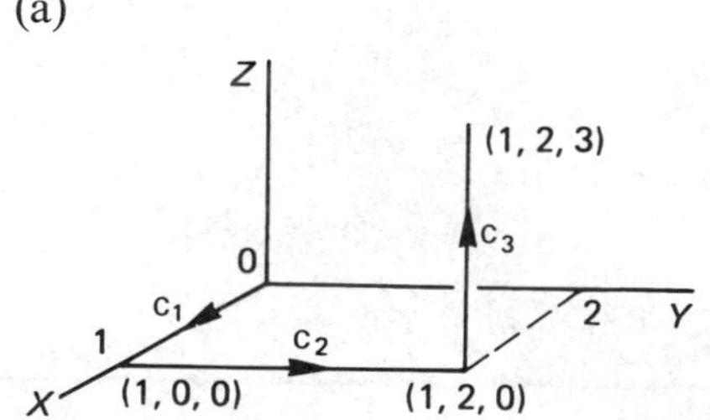

Here the integration is made in three sections, along c_1, c_2 and c_3.

(i) c_1: $y = 0$, $z = 0$, $dy = 0$, $dz = 0$

$$\therefore \int_{c_1} \mathbf{F}\cdot\,\mathbf{dr} = 0 + 0 + 0 = \underline{0}$$

(ii) c_2: The conditions along c_2 are

....

15

c_2: $x = 1$, $z = 0$, $dx = 0$, $dz = 0$

$$\therefore \int_{c_2} \mathbf{F}\cdot\,\mathbf{dr} = 0 + 0 + 0 = \underline{0}$$

(iii) c_3: $x = 1$, $y = 2$, $dx = 0$, $dy = 0$

$$\therefore \int_{c_3} \mathbf{F}\cdot\,\mathbf{dr} = \ldots\ldots\ldots$$

16

27

for $$\int_{c_3} \mathbf{F} \cdot \, d\mathbf{r} = 0 + 0 + \int_0^3 3z^2 \, dz = \underline{27}$$

Summing the three partial results

$$\int_{(0,\,0,\,0)}^{(1,\,2,\,3)} \mathbf{F} \cdot \, d\mathbf{r} = 0 + 0 + 27 = 27 \quad \therefore \underline{\int_{c_1+c_2+c_3} \mathbf{F} \cdot \, d\mathbf{r} = 27}$$

(b)

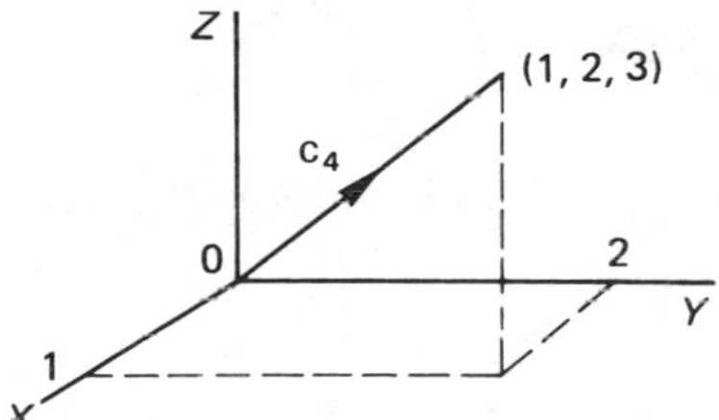

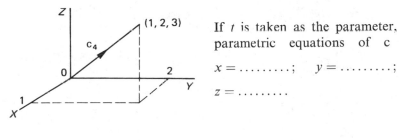

If t is taken as the parameter, the parametric equations of c are

$x = \dots\dots\dots;\quad y = \dots\dots\dots;$

$z = \dots\dots\dots$

17

$x = t;\quad y = 2t;\quad z = 3t$

and the limits of t are ....

18

$t = 0$ and $t = 1$

As in Example 1, we now express everything in terms of t and complete the integral, finally getting

$$\int_{c_4} \mathbf{F} \cdot \, d\mathbf{r} = \dots\dots\dots$$

19

$$\int_{c_4} \mathbf{F} \cdot \mathrm{d}\mathbf{r} = \frac{115}{4} = 28.75$$

for $\mathbf{F} = 2t^3\mathbf{i} + 12t^2\mathbf{j} + 27t^3\mathbf{k}$

$\mathrm{d}\mathbf{r} = \mathbf{i}\,\mathrm{d}x + \mathbf{j}\,\mathrm{d}y + \mathbf{k}\,\mathrm{d}z = \mathbf{i}\,\mathrm{d}t + \mathbf{j}\,2\,\mathrm{d}t + \mathbf{k}\,3\,\mathrm{d}t$

$$\therefore \int_{c_4} \mathbf{F} \cdot \mathrm{d}\mathbf{r} = \int_0^1 (2t^3\mathbf{i} + 12t^2\mathbf{j} + 27t^3\mathbf{k}) \cdot (\mathbf{i} + 2\mathbf{j} + 3\mathbf{k})\,\mathrm{d}t$$

$$= \int_0^1 (2t^3 + 24t^2 + 81t^3)\,\mathrm{d}t = \int_0^1 (83t^3 + 24t^2)\,\mathrm{d}t$$

$$= \left[83\frac{t^4}{4} + 8t^3\right]_0^1 = \frac{115}{4} = \underline{28.75}$$

So the value of the line integral depends on the path taken between the two end points A and B

(a) $\int \mathbf{F} \cdot \mathrm{d}\mathbf{r}$ via c_1, c_2 and $c_3 = 27$

(b) $\int \mathbf{F} \cdot \mathrm{d}\mathbf{r}$ via c_4 $= 28.75$

We shall refer to this topic later.

One further example on your own. The working is just the same as before.

Example 3 If $\mathbf{F}(\mathbf{r}) = x^2y^2\mathbf{i} + y^3z\mathbf{j} + z^2\mathbf{k}$, evaluate $\int_c \mathbf{F} \cdot \mathrm{d}\mathbf{r}$ along the curve $x = 2u^2$, $y = 3u$, $z = u^3$ between A(2, −3, −1) and B(2, 3, 1).

Proceed as before. You will have no difficulty.

$$\int_c \mathbf{F} \cdot \mathrm{d}\mathbf{r} = \dots\dots\dots$$

20

$$\int_c \mathbf{F} \cdot \, d\mathbf{r} = \frac{500}{21} = 23.8$$

Here is the working for you to check.

$$x = 2u^2; \quad y = 3u; \quad z = u^3$$

$$x^2y^2 = (4u^4)(9u^2) = 36\,u^6 \qquad dx = 4u\,du$$

$$y^3z = (27u^3)(u^3) = 27u^6 \qquad dy = 3\,du$$

$$z^2 = u^6 \qquad dz = 3u^2\,du$$

Limits : A(2, −3, −1) corresponds to $u = -1$
B(2, 3, 1) corresponds to $u = 1$

$$\therefore \int_c \mathbf{F} \cdot \, d\mathbf{r} = \int_{-1}^{1} (x^2y^2\mathbf{i} + y^3z\mathbf{j} + z^2\mathbf{k}) \cdot (\mathbf{i}\,dx + \mathbf{j}\,dy + \mathbf{k}\,dz)$$

$$= \int_{-1}^{1} (36u^6\mathbf{i} + 27u^6\mathbf{j} + u^6\mathbf{k}) \; (\mathbf{i}\,4u\,du + \mathbf{j}\,3\,du + \mathbf{k}\,3u^2\,du)$$

$$= \int_{-1}^{1} (144u^7 + 81u^6 + 3u^8)\,du$$

$$= \left[18u^8 + \frac{81u^7}{7} + \frac{u^9}{3}\right]_{-1}^{1} = \frac{500}{21} = \underline{23.8}$$

Now on to the next section.

21

Volume integrals

If V is a closed region bounded by a surface **S** and **F** is a vector field at each point of V and on its boundary surface **S**, then $\int_V \mathbf{F}\,dV$ is the *volume integral* of **F** throughout the region.

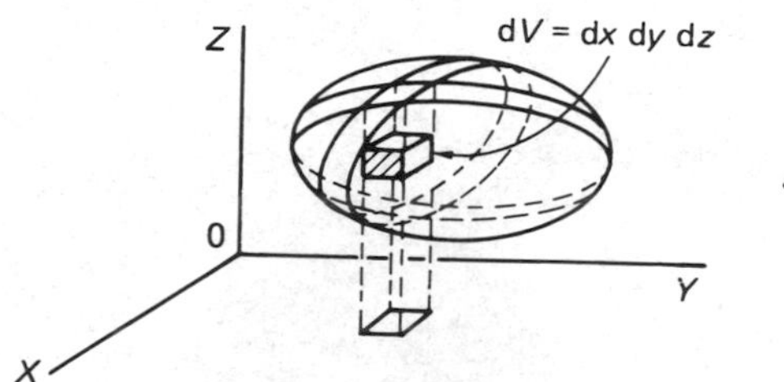

$$\int_V \mathbf{F}\,dV = \int_{x_1}^{x_2}\int_{y_1}^{y_2}\int_{z_1}^{z_2} \mathbf{F}\,dz\,dy\,dx$$

Example 1 Evaluate $\int_V \mathbf{F}\,dV$ where V is the region bounded by the planes $x=0$, $x=2$, $y=0$, $y=3$, $z=0$, $z=4$, and $\mathbf{F} = xy\mathbf{i} + z\mathbf{j} - x^2\mathbf{k}$.

We start, as in most cases, by sketching the diagram, which is

....

22

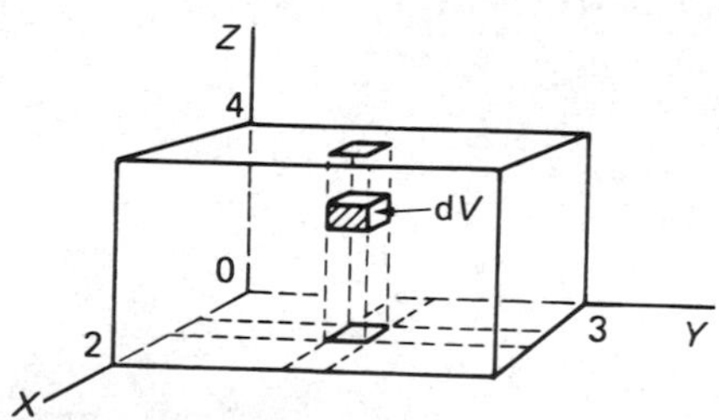

Then $\mathbf{F} = xy\,\mathbf{i} + z\,\mathbf{j} - x^2\,\mathbf{k}$ and $dV = dx\,dy\,dz$

$$\therefore \int_V \mathbf{F}\,dV = \int_0^4\int_0^3\int_0^2 (xy\mathbf{i} + z\mathbf{j} - x^2\mathbf{k})\,dx\,dy\,dz$$

$$= \int_0^4\int_0^3 \left[\frac{x^2y}{2}\mathbf{i} + xz\mathbf{j} - \frac{x^3}{3}\mathbf{k}\right]_{x=0}^{x=2} dy\,dz$$

$$= \int_0^4\int_0^3 \left(2y\mathbf{i} + 2z\mathbf{j} - \frac{8}{3}\mathbf{k}\right) dy\,dz$$

$= \ldots\ldots\ldots$ Complete the integral.

$$\int_V \mathbf{F}\,\mathrm{d}V = 4(9\mathbf{i} + 12\mathbf{j} - 8\mathbf{k})$$

since $$\int_V \mathbf{F}\,\mathrm{d}V = \int_0^4 \left[y^2\mathbf{i} + 2yz\mathbf{j} - \frac{8}{3}y\mathbf{k}\right]_{y=0}^{y=3} \mathrm{d}z$$

$$= \int_0^4 (9\mathbf{i} + 6z\mathbf{j} - 8\mathbf{k})\,\mathrm{d}z$$

$$= \left[9z\mathbf{i} + 3z^2\mathbf{j} - 8z\mathbf{k}\right]_0^4 = 36\mathbf{i} + 48\mathbf{j} - 32\mathbf{k}$$

$$= \underline{4(9\mathbf{i} + 12\mathbf{j} - 8\mathbf{k})}$$

Now another.

Example 2 Evaluate $\int_V \mathbf{F}\,\mathrm{d}V$ where V is the region bounded by the planes $x = 0$, $y = 0$, $z = 0$ and $2x + y + z = 2$, and $\mathbf{F} = 2z\mathbf{i} + y\mathbf{k}$.

To sketch the surface $2x + y + z = 2$, note that

when $z = 0$, $2x + y = 2$ i.e. $y = 2 - 2x$
when $y = 0$, $2x + z = 2$ i.e. $z = 2 - 2x$
when $x = 0$, $y + z = 2$ i.e. $z = 2 - y$

Inserting these in the planes $x = 0$, $y = 0$, $z = 0$ will help.

The diagram is therefore

....

24

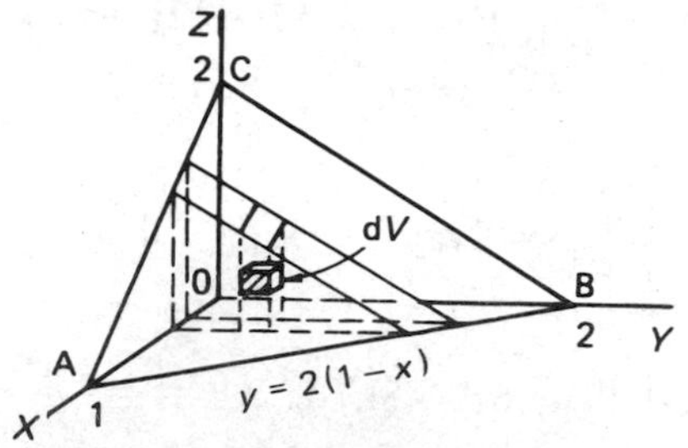

So $2x + y + z = 2$ cuts the axes at A(1, 0, 0); B(0, 2, 0); C(0, 0, 2).

Also $\mathbf{F} = 2z\mathbf{i} + y\mathbf{k}$; $\quad z = 2 - 2x - y = 2(1 - x) - y$

$$\therefore \int_V \mathbf{F}\,dV = \int_0^1 \int_0^{2(1-x)} \int_0^{2(1-x)-y} (2z\mathbf{i} + y\mathbf{k})\,dz\,dy\,dx$$

$$= \int_0^1 \int_0^{2(1-x)} \left[z^2\mathbf{i} + yz\mathbf{k} \right]_{z=0}^{z=2(1-x)-y} dy\,dx$$

$$= \int_0^1 \int_0^{2(1-x)} \{[4(1-x)^2 - 4(1-x)y + y^2]\mathbf{i}$$

$$+[2(1-x)y - y^2]\,\mathbf{k}\}\,dy\,dx$$

$$= \int_0^1 \left[\left\{ 4(1-x)^2 y - 2(1-x)y^2 + \frac{y^3}{3} \right\}\mathbf{i} \right.$$

$$\left. + \left\{ (1-x)y^2 - \frac{y^3}{3} \right\}\mathbf{k} \right]_{y=0}^{2(1-x)} dx$$

$= \ldots\ldots\ldots$ Finish the last stage.

25

$$\int_V \mathbf{F}\,\mathrm{d}V = \frac{1}{3}(2\mathbf{i} + \mathbf{k})$$

for $\displaystyle\int_V \mathbf{F}\,\mathrm{d}V = \int_0^1 \left\{\frac{8}{3}(1-x)^3\mathbf{i} + \frac{4}{3}(1-x)^3\mathbf{k}\right\}\mathrm{d}x$

$$= \left[-\frac{2}{3}(1-x)^4\mathbf{i} - \frac{1}{3}(1-x)^4\mathbf{k}\right]_0^1 = \underline{\frac{1}{3}(2\mathbf{i} + \mathbf{k})}$$

And now one more, slightly different.

Example 3 Evaluate $\displaystyle\int_V \mathbf{F}\,\mathrm{d}V$ where $\mathbf{F} = 2\mathbf{i} + 2z\mathbf{j} + y\mathbf{k}$ and V is the region bounded by the planes $z = 0$, $z = 4$ and the surface $x^2 + y^2 = 9$.

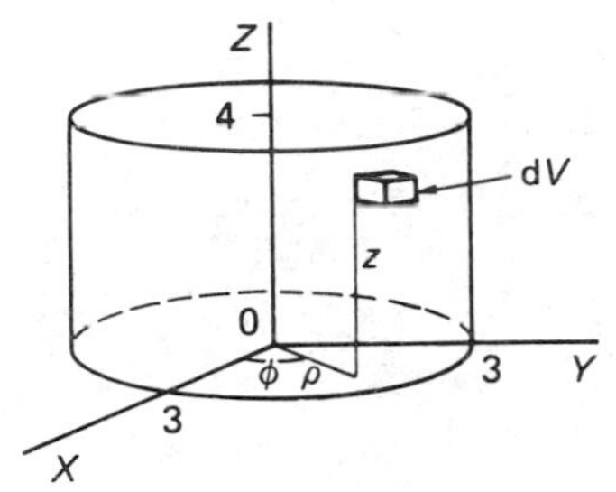

It will be convenient to use cylindrical polar coordinates (ρ, ϕ, z) so the relevant transformations are

$x = \ldots\ldots\ldots$; $\quad y = \ldots\ldots\ldots$

$z = \ldots\ldots\ldots$; $\quad \mathrm{d}V = \ldots\ldots\ldots$

26

$$x = \rho\cos\phi; \qquad y = \rho\,\sin\phi$$
$$z = z; \qquad \mathrm{d}V = \rho\,\mathrm{d}\rho\,\mathrm{d}\phi\,\mathrm{d}z$$

Then $\int_V \mathbf{F}\,\mathrm{d}V = \iiint_V (2\mathbf{i} + 2z\mathbf{j} + y\mathbf{k})\,\mathrm{d}x\,\mathrm{d}y\,\mathrm{d}z.$

Changing into cylindrical polar coordinates with appropriate change of limits this becomes

$$\int_V \mathbf{F}\,\mathrm{d}V = \int_{\phi=0}^{2\pi}\int_{\rho=0}^{3}\int_{z=0}^{4} (2\mathbf{i} + 2z\mathbf{j} + \rho\sin\phi\,\mathbf{k})\,\mathrm{d}z\,\rho\,\mathrm{d}\rho\,\mathrm{d}\phi$$

$$= \int_{\phi=0}^{2\pi}\int_{\rho=0}^{3}\Big[2z\mathbf{i} + z^2\mathbf{j} + \rho\sin\phi\,z\mathbf{k}\Big]_{z=0}^{4}\rho\,\mathrm{d}\rho\,\mathrm{d}\phi$$

$$= \int_0^{2\pi}\int_0^3 (8\mathbf{i} + 16\mathbf{j} + 4\rho\sin\phi\,\mathbf{k})\,\mathrm{d}\rho\,\mathrm{d}\phi$$

$$= 4\int_0^{2\pi}\int_0^3 (2\rho\mathbf{i} + 4\rho\mathbf{j} + \rho^2\sin\phi\,\mathbf{k})\,\mathrm{d}\rho\,\mathrm{d}\phi$$

Completing the working, we finally get

$$\int_V \mathbf{F}\,\mathrm{d}V = \dots\dots\dots$$

27

$$\boxed{72\pi(\mathbf{i} + 2\mathbf{j})}$$

for $\displaystyle\int_V \mathbf{F}\,\mathrm{d}V = 4\int_0^{2\pi}\left[\rho^2\mathbf{i} + 2\rho^2\mathbf{j} + \frac{\rho^3}{3}\sin\phi\,\mathbf{k}\right]_0^3 \mathrm{d}\phi$

$$= 4\int_0^{2\pi}(9\mathbf{i} + 18\mathbf{j} + 9\sin\phi\,\mathbf{k})\,\mathrm{d}\phi$$

$$= 36\int_0^{2\pi}(\mathbf{i} + 2\mathbf{j} + \sin\phi\,\mathbf{k})\,\mathrm{d}\phi$$

$$= 36\left[\phi\mathbf{i} + 2\phi\mathbf{j} - \cos\phi\,\mathbf{k}\right]_0^{2\pi}$$

$$= 36\{(2\pi\mathbf{i} + 4\pi\mathbf{j} - \mathbf{k}) - (-\mathbf{k})\}$$

$$= \underline{72\pi(\mathbf{i} + 2\mathbf{j})}$$

You will, of course, remember that in appropriate cases, the use of cylindrical polar coordinates or spherical polar coordinates often simplifies the subsequent calculations. So keep them in mind.

Now let us turn to surface integrals—in the next frame.

28 Surface integrals

The vector product of two vectors **A** and **B** is defined as

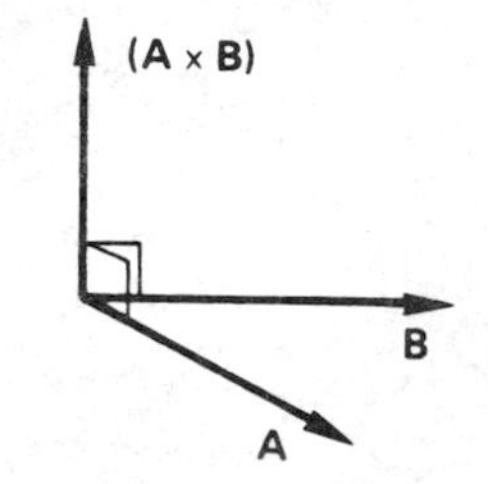

$\mathbf{A} \times \mathbf{B} = AB \sin\theta$ at right angles to the plane of **A** and **B** to form a right-handed set.

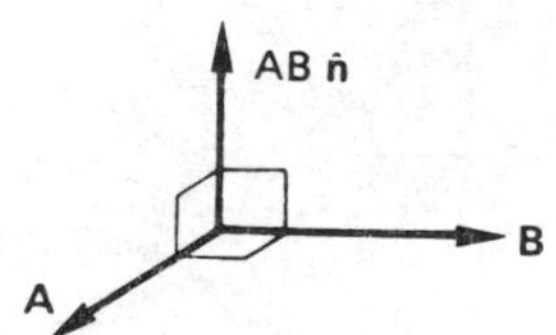

If $\theta = \frac{\pi}{2}$, then $\mathbf{A} \times \mathbf{B} = AB$ in the direction of the normal. Therefore, if $\hat{\mathbf{n}}$ is a unit normal then

$$\mathbf{A} \times \mathbf{B} = |\mathbf{A}|\,|\mathbf{B}|\hat{\mathbf{n}} = AB\,\hat{\mathbf{n}}$$

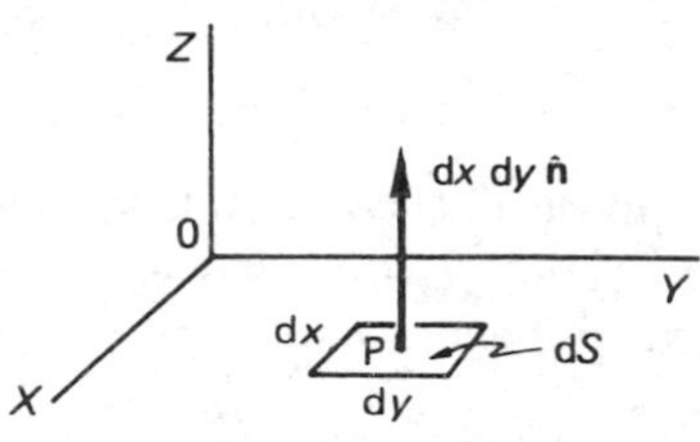

If P(x, y) is a point in the xy-plane, the element of area $\mathrm{d}x\,\mathrm{d}y$ has a vector area $\mathrm{d}\mathbf{S} = (\mathbf{i}\,\mathrm{d}x) \times (\mathbf{j}\,\mathrm{d}y)$.

i.e. $\mathrm{d}\mathbf{S} = \mathrm{d}x\,\mathrm{d}y\,\hat{\mathbf{n}}$

i.e. a vector of magnitude $\mathrm{d}x\,\mathrm{d}y$ acting in the direction of $\hat{\mathbf{n}}$ and referred to as the *vector area.*

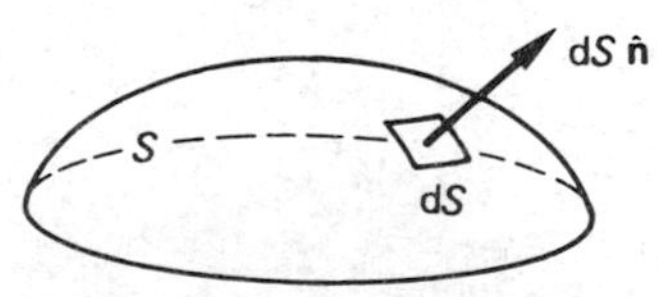

For a general surface S in space, each element of surface $\mathrm{d}S$ has a *vector area* $\mathrm{d}S$ such that $\mathrm{d}\mathbf{S} = \mathrm{d}S\,\hat{\mathbf{n}}$.

You will also remember that we established previously that the unit normal $\hat{\mathbf{n}}$ to a surface S is given by

$$\hat{\mathbf{n}} = \frac{\operatorname{grad} S}{|\operatorname{grad} S|} = \frac{\nabla S}{|\nabla S|}$$

Let us see how we can apply these results to the following examples.

29

Scalar fields

Example 1 A scalar field $V = xyz$ exists over the curved surface S defined by $x^2 + y^2 = 4$ between the planes $z = 0$ and $z = 3$ in the first octant. Evaluate $\int_S V\,\mathrm{d}\mathbf{S}$ over this surface.

We have $V = xyz \quad S: \quad x^2 + y^2 - 4 = 0, \quad z = 0$ to $z = 3$

$$\mathrm{d}\mathbf{S} = \hat{\mathbf{n}}\,\mathrm{d}S \qquad \text{where } \hat{\mathbf{n}} = \frac{\nabla S}{|\nabla S|}$$

Now $\nabla S = \dfrac{\partial S}{\partial x}\mathbf{i} + \dfrac{\partial S}{\partial y}\mathbf{j} + \dfrac{\partial S}{\partial z}\mathbf{k} = 2x\mathbf{i} + 2y\mathbf{j}$

$$|\nabla S| = \sqrt{4x^2 + 4y^2} = 2\sqrt{x^2 + y^2} = 2\sqrt{4} = 4$$

$$\therefore\ \hat{\mathbf{n}} = \frac{\nabla S}{|\nabla S|} = \frac{x\mathbf{i} + y\mathbf{j}}{2} \qquad \therefore\ \mathrm{d}\mathbf{S} = \hat{\mathbf{n}}\,\mathrm{d}S = \frac{x\mathbf{i} + y\mathbf{j}}{2}\,\mathrm{d}S$$

$$\therefore\ \int_S V\,\mathrm{d}\mathbf{S} = \int_S V\hat{\mathbf{n}}\,\mathrm{d}S = \frac{1}{2}\int_S xyz(x\mathbf{i} + y\mathbf{j})\,\mathrm{d}S$$

$$= \frac{1}{2}\int_S (x^2yz\mathbf{i} + xy^2z\mathbf{j})\,\mathrm{d}S \qquad (1)$$

We have to evaluate this integral over the prescribed surface.

Changing to cylindrical coordinates with $\rho = 2$

$x = \ldots\ldots\ldots; \qquad y = \ldots\ldots\ldots$

$z = \ldots\ldots\ldots; \qquad \mathrm{d}S = \ldots\ldots\ldots$

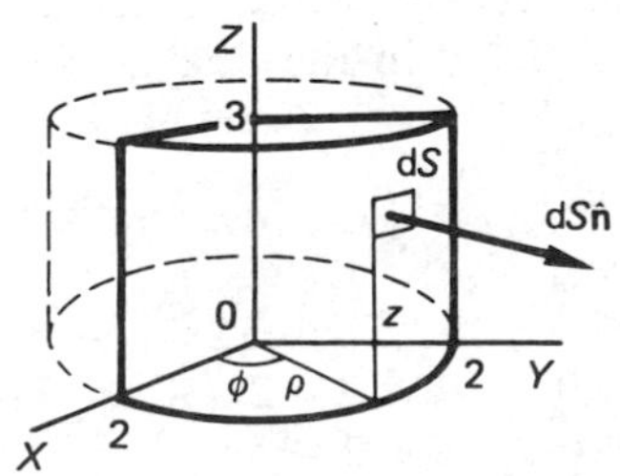

30

$$x = 2\cos\phi; \qquad y = 2\sin\phi$$
$$z = z; \qquad dS = 2\,d\phi\,dz$$

$$\therefore\ x^2yz = (4\cos^2\phi)(2\sin\phi)(z) = 8\cos^2\phi\sin\phi\ z$$
$$xy^2z = (2\cos\phi)(4\sin^2\phi)(z) = 8\cos\phi\sin^2\phi\ z$$

Then result (1) above becomes

$$\int_S V\,d\mathbf{S} = \frac{1}{2}\int_0^{\pi/2}\int_0^3 (8\cos^2\phi\sin\phi\, z\mathbf{i} + 8\cos\phi\sin^2\phi\, z\mathbf{j})\, 2\,dz\,d\phi$$

$$= 4\int_0^{\pi/2}\int_0^3 (\cos^2\phi\sin\phi\,\mathbf{i} + \cos\phi\sin^2\phi\,\mathbf{j})\, 2z\,dz\,d\phi$$

$$= 4\int_0^{\pi/2} (\cos^2\phi\sin\phi\,\mathbf{i} + \cos\phi\sin^2\phi\,\mathbf{j})\, 9\,d\phi$$

and this eventually gives

$$\int_S V\,d\mathbf{S} = \ldots\ldots\ldots$$

31

$$\int_S V\,d\mathbf{S} = 12(\mathbf{i}+\mathbf{j})$$

$$\int_S V\,d\mathbf{S} = 36\left[-\frac{\cos^3\phi}{3}\mathbf{i} + \frac{\sin^3\phi}{3}\mathbf{j}\right]_0^{\pi/2} = \underline{12(\mathbf{i}+\mathbf{j})}$$

Example 2 A scalar field $V = x + y + z$ exists over the surface S defined by $2x + 2y + z = 2$ bounded by $x = 0$, $y = 0$, $z = 0$ in the first octant. Evaluate $\int_S V\,d\mathbf{S}$ over this surface.

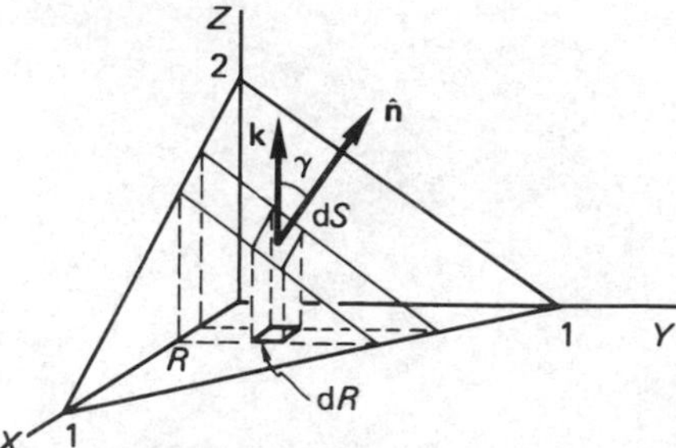

$S:\quad 2x + 2y + z = 2$

$x = 0 \qquad z = 2 - 2y$

$y = 0 \qquad z = 2 - 2x$

$z = 0 \qquad y = 1 - x$

$$\mathbf{dS} = \hat{\mathbf{n}}\,\mathrm{d}S \qquad \hat{\mathbf{n}} = \frac{\nabla S}{|\nabla S|}$$

$$\nabla S = \frac{\partial S}{\partial x}\mathbf{i} + \frac{\partial S}{\partial y}\mathbf{j} + \frac{\partial S}{\partial z}\mathbf{k} = 2\mathbf{i} + 2\mathbf{j} + \mathbf{k}$$

$$\therefore |\nabla S| = \sqrt{4+4+1} = \sqrt{9} = 3 \qquad \therefore \hat{\mathbf{n}} = \frac{\nabla S}{|\nabla S|} = \frac{1}{3}(2\mathbf{i} + 2\mathbf{j} + \mathbf{k})$$

$$\therefore \mathbf{dS} = \hat{\mathbf{n}}\,\mathrm{d}S = \frac{1}{3}(2\mathbf{i} + 2\mathbf{j} + \mathbf{k})\,\mathrm{d}S$$

If we now project $\mathrm{d}S$ on to the xy-plane, $\qquad \mathrm{d}R = \mathrm{d}S\cos\gamma$

$$\cos\gamma = \hat{\mathbf{n}}\cdot\mathbf{k} = \frac{1}{3}(2\mathbf{i} + 2\mathbf{j} + \mathbf{k})\cdot(\mathbf{k}) = \frac{1}{3}$$

$$\therefore \mathrm{d}R = \frac{1}{3}\,\mathrm{d}S \qquad \therefore \mathrm{d}S = 3\,\mathrm{d}R = 3\,\mathrm{d}x\,\mathrm{d}y$$

$$\therefore \int_S V\,\mathbf{dS} = \int_S V\hat{\mathbf{n}}\,\mathrm{d}S = \int_S\!\!\int (x+y+z)\frac{1}{3}(2\mathbf{i} + 2\mathbf{j} + \mathbf{k})3\,\mathrm{d}x\,\mathrm{d}y$$

But $z = 2 - 2x - 2y$

$$\therefore \int_S V\,\mathbf{dS} = \int_{x=0}^{1}\int_{y=0}^{1-x}(2-x-y)(2\mathbf{i} + 2\mathbf{j} + \mathbf{k})\,\mathrm{d}y\,\mathrm{d}x$$

$$= \ldots\ldots\ldots$$

32

$$\boxed{\frac{2}{3}(2\mathbf{i} + 2\mathbf{j} + \mathbf{k})}$$

since $\displaystyle\int_S V\,\mathbf{dS} = \int_0^1\left[2y - xy - \frac{y^2}{2}\right]_0^{1-x}(2\mathbf{i} + 2\mathbf{j} + \mathbf{k})\,\mathrm{d}x$

$$= \left[\frac{3}{2}x - x^2 + \frac{x^3}{6}\right]_0^1(2\mathbf{i} + 2\mathbf{j} + \mathbf{k})$$

$$= \underline{\frac{2}{3}(2\mathbf{i} + 2\mathbf{j} + \mathbf{k})}$$

33 Vector fields

Example 1 A vector field $\mathbf{F} = y\mathbf{i} + 2\mathbf{j} + \mathbf{k}$ exists over a surface S defined by $x^2 + y^2 + z^2 = 9$ bounded by $x = 0$, $y = 0$, $z = 0$ in the first octant. Evaluate $\int_S \mathbf{F} \cdot \mathrm{d}\mathbf{S}$ over the surface indicated.

$$\mathrm{d}\mathbf{S} = \hat{\mathbf{n}}\,\mathrm{d}S; \qquad \hat{\mathbf{n}} = \frac{\nabla S}{|\nabla S|} \qquad S: \quad x^2 + y^2 + z^2 - 9 = 0$$

$$\nabla S = \frac{\partial S}{\partial x}\mathbf{i} + \frac{\partial S}{\partial y}\mathbf{j} + \frac{\partial S}{\partial z}\mathbf{k} = 2x\mathbf{i} + 2y\mathbf{j} + 2z\mathbf{k}$$

$$\therefore |\nabla S| = \sqrt{4x^2 + 4y^2 + 4z^2} = 2\sqrt{x^2 + y^2 + z^2} = 2\sqrt{9} = 6$$

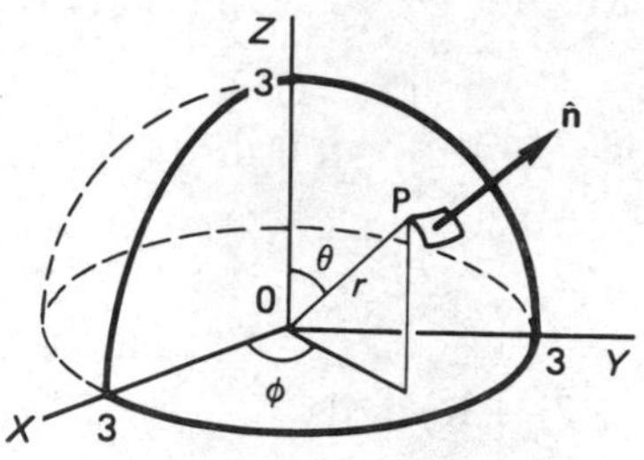

$$\therefore \hat{\mathbf{n}} = \frac{1}{6}(2x\mathbf{i} + 2y\mathbf{j} + 2z\mathbf{k})$$

$$= \frac{1}{3}(x\mathbf{i} + y\mathbf{j} + z\mathbf{k})$$

$$\int_S \mathbf{F} \cdot \mathrm{d}\mathbf{S} = \int_S \mathbf{F} \cdot \hat{\mathbf{n}}\,\mathrm{d}S = \int_S (y\mathbf{i} + 2\mathbf{j} + \mathbf{k}) \cdot \frac{1}{3}(x\mathbf{i} + y\mathbf{j} + z\mathbf{k})\,\mathrm{d}S$$

$$= \frac{1}{3}\int_S (xy + 2y + z)\,\mathrm{d}S$$

Before integrating over the surface, we convert to spherical polar coordinates.

$x = \dots\dots\dots$; $\qquad y = \dots\dots\dots$

$z = \dots\dots\dots$; $\qquad \mathrm{d}S = \dots\dots\dots$

34

$$x = 3\sin\theta\cos\phi; \qquad y = 3\sin\theta\sin\phi$$
$$z = 3\cos\theta; \qquad dS = 9\sin\theta\, d\theta\, d\phi$$

Limits of θ and ϕ are $\theta = 0$ to $\frac{\pi}{2}$; $\phi = 0$ to $\frac{\pi}{2}$.

$$\therefore \int_S \mathbf{F}\cdot d\mathbf{S} = \frac{1}{3}\int_0^{\pi/2}\int_0^{\pi/2}(9\sin^2\theta\sin\phi\cos\phi + 6\sin\theta\sin\phi + 3\cos\theta)\, 9\sin\theta\, d\theta\, d\phi$$

$$= 9\int_0^{\pi/2}\int_0^{\pi/2}(3\sin^3\theta\sin\phi\cos\phi + 2\sin^2\theta\sin\phi + \sin\theta\cos\theta)\, d\theta\, d\phi$$

$= \dots\dots\dots$ Complete the integral.

35

$$\int_S \mathbf{F}\cdot d\mathbf{S} = 9\left(1 + \frac{3\pi}{4}\right)$$

for $\int_S \mathbf{F}\cdot d\mathbf{S} = 9\int_0^{\pi/2}\left(2\sin\phi\cos\phi + \frac{\pi}{2}\sin\phi + \frac{1}{2}\right)d\phi$

$$= 9\left[\sin^2\phi - \frac{\pi}{2}\cos\phi + \frac{\phi}{2}\right]_0^{\pi/2} = 9\left(1 + \frac{3\pi}{4}\right)$$

Example 2 Evaluate $\int_S \mathbf{F}\cdot d\mathbf{S}$ where $\mathbf{F} = 2y\mathbf{j} + z\mathbf{k}$ and S is the surface $x^2 + y^2 = 4$ in the first two octants bounded by the planes $z = 0$, $z = 5$ and $y = 0$.

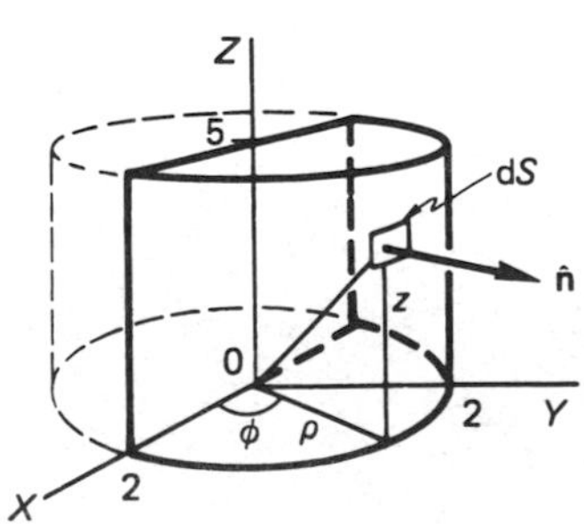

$$S:\ x^2 + y^2 - 4 = 0 \qquad \hat{\mathbf{n}} = \frac{\nabla S}{|\nabla S|}$$

$$\nabla S = \frac{\partial S}{\partial x}\mathbf{i} + \frac{\partial S}{\partial y}\mathbf{j} + \frac{\partial S}{\partial z}\mathbf{k} = 2x\mathbf{i} + 2y\mathbf{j}$$

$$\therefore |\nabla S| = \sqrt{4x^2 + 4y^2} = 2\sqrt{x^2 + y^2}$$
$$= 2\sqrt{4} = 4$$

$$\therefore \hat{\mathbf{n}} = \frac{\nabla S}{|\nabla S|} = \frac{2x\mathbf{i} + 2y\mathbf{j}}{4} = \frac{1}{2}(x\mathbf{i} + y\mathbf{j})$$

$$\therefore \int_S \mathbf{F}\cdot d\mathbf{S} = \int_S \mathbf{F}\cdot \hat{\mathbf{n}}\, dS = \dots\dots\dots$$

36

$$\boxed{\int_S y^2\,\mathrm{d}S}$$

for $\int_S \mathbf{F}\cdot\hat{\mathbf{n}}\,\mathrm{d}S = \int_S (2y\mathbf{j} + z\mathbf{k})\cdot\frac{1}{2}(x\mathbf{i} + y\mathbf{j})\,\mathrm{d}S$

$$= \frac{1}{2}\int_S (2y^2)\,\mathrm{d}S = \int_S y^2\,\mathrm{d}S$$

This is clearly a case for using cylindrical polar coordinates.

$$x = \ldots\ldots\ldots; \quad y = \ldots\ldots\ldots$$

$$z = \ldots\ldots\ldots; \quad \mathrm{d}S = \ldots\ldots\ldots$$

37

$$\boxed{\begin{array}{ll} x = 2\cos\phi; & y = 2\sin\phi \\ z = z; & \mathrm{d}S = 2\,\mathrm{d}\phi\,\mathrm{d}z \end{array}}$$

$$\therefore \int_S \mathbf{F}\cdot\mathrm{d}\mathbf{S} = \int_S y^2\,\mathrm{d}S = \int_S\int 4\sin^2\phi\, 2\,\mathrm{d}\phi\,\mathrm{d}z = 8\int_S\int \sin^2\phi\,\mathrm{d}\phi\,\mathrm{d}z$$

Limits: $\phi = 0$ to $\phi = \pi$; $z = 0$ to $z = 5$

$$\therefore \int_S \mathbf{F}\cdot\mathrm{d}\mathbf{S} = \ldots\ldots\ldots$$

$$\boxed{20\pi}$$

for $\displaystyle\int_S \mathbf{F}\cdot \mathrm{d}S = 4\int_{z=0}^{5}\int_{\phi=0}^{\pi}(1-\cos 2\phi)\,\mathrm{d}\phi\,\mathrm{d}z$

$$= 4\int_0^5\left[\phi - \frac{\sin 2\phi}{2}\right]_0^{\pi}\mathrm{d}z$$

$$= 4\int_0^5 \pi\,\mathrm{d}z = 4\pi\Big[z\Big]_0^5 = \underline{20\pi}$$

Example 3 Evaluate $\displaystyle\int_S \mathbf{F}\cdot \mathrm{d}\mathbf{S}$ where **F** is the field $x^2\mathbf{i} - y\mathbf{j} + 2z\mathbf{k}$ and S is the surface $2x + y + 2z = 2$ bounded by $x = 0,\ y = 0,\ z = 0$ in the first octant.

We can sketch the diagram by putting $x = 0,\ y = 0,\ z = 0$ in turn in the equation for S.

When	$x = 0$	$y + 2z = 2$	$z = 1 - \frac{y}{2}$
	$y = 0$	$x + z = 1$	$z = 1 - x$
	$z = 0$	$2x + y = 2$	$y = 2 - 2x$

So the diagram is

....

39

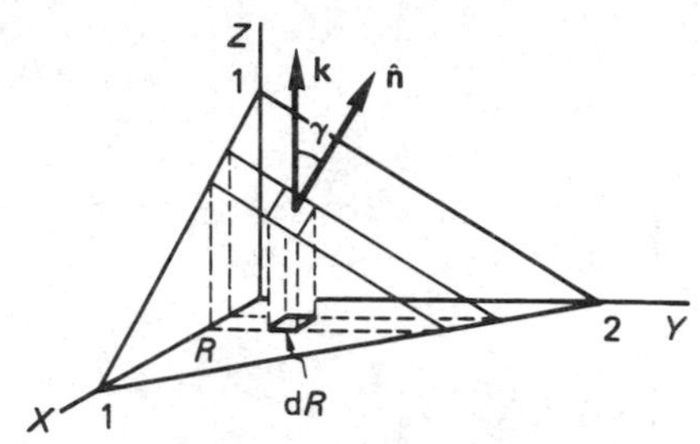

$$\mathbf{F} = x^2\mathbf{i} - y\mathbf{j} + 2z\mathbf{k}; \quad S: \quad 2x + y + 2z - 2 = 0$$

$$\nabla S = \frac{\partial S}{\partial x}\mathbf{i} + \frac{\partial S}{\partial y}\mathbf{j} + \frac{\partial S}{\partial z}\mathbf{k} = 2\mathbf{i} + \mathbf{j} + 2\mathbf{k} \qquad |\nabla S| = 3$$

$$\int_S \mathbf{F} \cdot \mathrm{d}\mathbf{S} = \int_S \mathbf{F} \cdot \hat{\mathbf{n}}\, \mathrm{d}S = \dots\dots\dots \text{ (next stage)}$$

40

$$\boxed{\frac{1}{3}\int_S (2x^2 - y + 4z)\, \mathrm{d}S}$$

since $\int_S \mathbf{F} \cdot \hat{\mathbf{n}}\, \mathrm{d}S = \int_S (x^2\mathbf{i} - y\mathbf{j} + 2z\mathbf{k}) \cdot \frac{1}{3}(2\mathbf{i} + \mathbf{j} + 2\mathbf{k})\, \mathrm{d}S$

$$= \frac{1}{3}\int_S (2x^2 - y + 4z)\, \mathrm{d}S$$

If we now project the element of surface $\mathrm{d}S$ on to the xy-plane

$$\mathrm{d}R = \mathrm{d}S \cos\gamma; \quad \cos\gamma = \hat{\mathbf{n}} \cdot \mathbf{k} \quad \therefore \mathrm{d}R = \hat{\mathbf{n}} \cdot \mathbf{k}\, \mathrm{d}S \quad \therefore \mathrm{d}S = \frac{\mathrm{d}x\, \mathrm{d}y}{\hat{\mathbf{n}} \cdot \mathbf{k}}$$

$$\therefore \hat{\mathbf{n}} \cdot \mathbf{k} = \frac{1}{3}(2\mathbf{i} + \mathbf{j} + 2\mathbf{k}) \cdot (\mathbf{k}) = \frac{2}{3} \quad \therefore \mathrm{d}S = \frac{3}{2}\, \mathrm{d}x\, \mathrm{d}y$$

Using these new relationships, $\int_S \mathbf{F} \cdot \mathrm{d}\mathbf{S} = \int_S \mathbf{F} \cdot \hat{\mathbf{n}}\, \mathrm{d}S = \dots\dots\dots$

41

$$\int_R \int \frac{1}{2}(2x^2 - y + 4z)\,dx\,dy$$

for $\int_S \mathbf{F} \cdot \hat{\mathbf{n}}\,dS = \frac{1}{3}\int_S (2x^2 - y + 4z)\,dS$

$$= \frac{1}{3}\int_R \int (2x^2 - y + 4z)\frac{3}{2}\,dx\,dy$$

$$= \frac{1}{2}\int_R \int (2x^2 - y + 4z)\,dx\,dy$$

Limits : $y = 0$ to $y = 2 - 2x$; $x = 0$ to $x = 1$

$$\therefore \int_S \mathbf{F} \cdot \hat{\mathbf{n}}\,dS = \frac{1}{2}\int_0^1 \int_0^{2-2x} (2x^2 - y + 4z)\,dy\,dx$$

But $2x + y + 2z = 2 \quad \therefore z = \frac{1}{2}(2 - 2x - y)$

$\therefore \int_S \mathbf{F} \cdot \hat{\mathbf{n}}\,dS = \ldots\ldots\ldots$ Complete the integration.

42

$$\frac{1}{2}$$

Here is the rest of the working.

$$\int_S \mathbf{F} \cdot d\mathbf{S} = \int_S \mathbf{F} \cdot \hat{\mathbf{n}}\,dS = \frac{1}{2}\int_0^1 \int_0^{2-2x} (2x^2 - y + 4 - 4x - 2y)\,dy\,dx$$

$$= \frac{1}{2}\int_0^1 \int_0^{2-2x} (2x^2 - 4x + 4 - 3y)\,dy\,dx$$

$$= \frac{1}{2}\int_0^1 \left[(2x^2 - 4x + 4)y - \frac{3y^2}{2}\right]_0^{2-2x} dx$$

$$= \frac{1}{2}\int_0^1 (4x^2 - 8x + 8 - 4x^3 + 8x^2 - 8x - 6 + 12x - 6x^2)\,dx$$

$$= \frac{1}{2}\int_0^1 (6x^2 - 4x^3 - 4x + 2)\,dx* = \int_0^1 (3x^2 - 2x^3 - 2x + 1)\,dx$$

$$= \left[x^3 - \frac{x^4}{2} - x^2 + x\right]_0^1 = \frac{1}{2}$$

While we are concerned with vector fields, let us move on to a further point of interest.

43 Conservative vector fields

In general, the value of the line integral $\int_c \mathbf{F}\cdot d\mathbf{r}$ between two stated

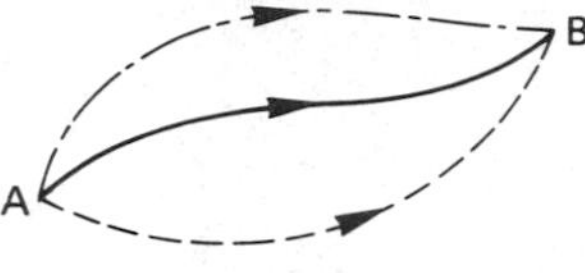

points **A** and **B** depends on the particular path of integration followed.

If, however, the line integral between A and B is independent of the path of integration between the two end points, then the vector field **F** is said to be *conservative*.

It follows that, for a closed path in a conservative field, $\oint_c \mathbf{F}\cdot d\mathbf{r} = 0$.

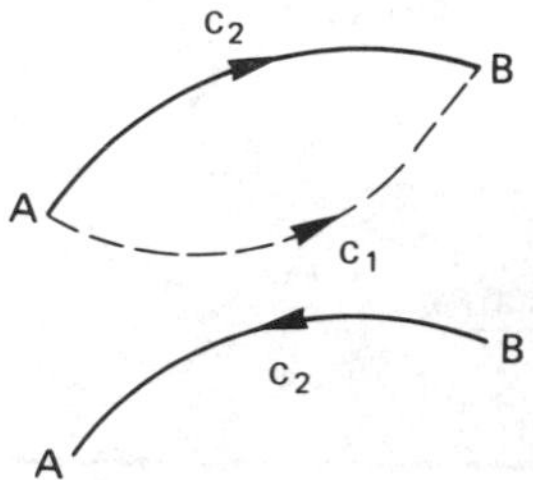

For, if the field is conservative,

$$\int_{c_1(AB)} \mathbf{F}\cdot d\mathbf{r} = \int_{c_2(AB)} \mathbf{F}\cdot d\mathbf{r}$$

$$\text{But } \int_{c_2(BA)} \mathbf{F}\cdot d\mathbf{r} = -\int_{c_2(AB)} \mathbf{F}\cdot d\mathbf{r}$$

Hence, for the closed path $\mathbf{AB}_{c_1} + \mathbf{BA}_{c_2}$, $\oint \mathbf{F}\cdot d\mathbf{r}$

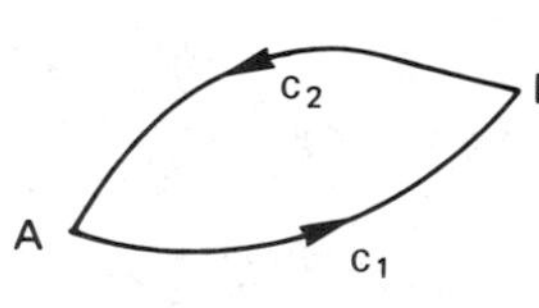

$$= \int_{c_1(AB)} \mathbf{F}\cdot d\mathbf{r} + \int_{c_2(BA)} \mathbf{F}\cdot d\mathbf{r}$$

$$= \int_{c_1(AB)} \mathbf{F}\cdot d\mathbf{r} - \int_{c_2(AB)} \mathbf{F}\cdot d\mathbf{r}$$

$$= \int_{c_1(AB)} \mathbf{F}\cdot d\mathbf{r} - \int_{c_1(AB)} \mathbf{F}\cdot d\mathbf{r} = 0$$

$$\therefore \oint \mathbf{F}\cdot d\mathbf{r} = 0$$

Note that this result holds good only for a closed curve and when the vector field is a conservative field. Now for some examples.

Example 1 If $\mathbf{F} = 2xyz\mathbf{i} + x^2z\mathbf{j} + x^2y\mathbf{k}$, evaluate the line integral $\int \mathbf{F} \cdot \mathrm{d}\mathbf{r}$ between A(0, 0, 0) and B(2, 4, 6)

(a) along the curve c whose parametric equations are $x = u$, $y = u^2$, $z = 3u$

(b) along the three straight lines c_1: (0, 0, 0) to (2, 0, 0); c_2: (2, 0, 0) to (2, 4, 0); c_3: (2, 4, 0) to (2, 4, 6).

Hence determine whether or not **F** is a conservative field.

First draw the diagram.

....

44

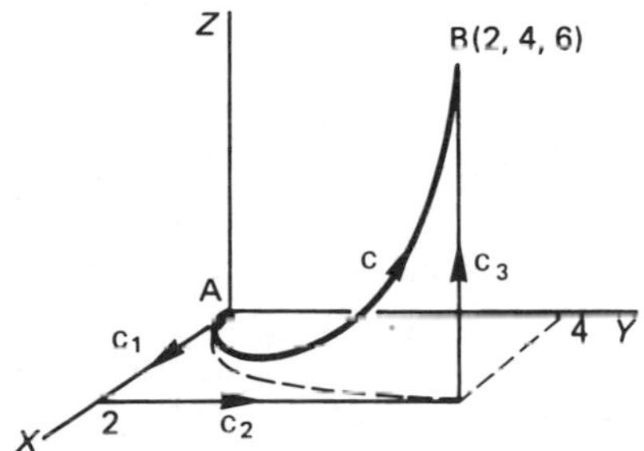

(a) $\mathbf{F} = 2xyz\mathbf{i} + x^2z\mathbf{j} + x^2y\mathbf{k}$

$$x = u; \quad y = u^2; \quad z = 3u$$
$$\therefore \mathrm{d}x = \mathrm{d}u; \quad \mathrm{d}y = 2u\,\mathrm{d}u; \quad \mathrm{d}z = 3\,\mathrm{d}u.$$

$$\mathbf{F} \cdot \mathrm{d}\mathbf{r} = (2xyz\mathbf{i} + x^2z\mathbf{j} + x^2y\mathbf{k}) \cdot (\mathbf{i}\,\mathrm{d}x + \mathbf{j}\,\mathrm{d}y + \mathbf{k}\,\mathrm{d}z)$$
$$= 2xyz\,\mathrm{d}x + x^2z\,\mathrm{d}y + x^2y\,\mathrm{d}z$$

Using the transformations shown above, we can now express $\mathbf{F} \cdot \mathrm{d}\mathbf{r}$ in terms of u.

$$\mathbf{F} \cdot \mathrm{d}\mathbf{r} = \ldots\ldots\ldots$$

45

$$\boxed{15u^4\,\mathrm{d}u}$$

for
$$2xyz\,\mathrm{d}x = (2u)(u^2)(3u)\,\mathrm{d}u = 6u^4\,\mathrm{d}u$$
$$x^2z\,\mathrm{d}y = (u^2)(3u)(2u)\,\mathrm{d}u = 6u^4\,\mathrm{d}u$$
$$x^2y\,\mathrm{d}z = (u^2)(u^2)3\,\mathrm{d}u = 3u^4\,\mathrm{d}u$$

$$\therefore\ \mathbf{F}\cdot\mathrm{d}\mathbf{r} = 6u^4\,\mathrm{d}u + 6u^4\,\mathrm{d}u + 3u^4\,\mathrm{d}u = 15u^4\,\mathrm{d}u$$

The limits of integration in u are ....

46

$$\boxed{u = 0 \quad \text{to} \quad u = 2}$$

$$\therefore \int_c \mathbf{F}\cdot\mathrm{d}\mathbf{r} = \int_0^2 15u^4\,\mathrm{d}u = \left[3u^5\right]_0^2 = 96 \qquad \underline{\int_c \mathbf{F}\cdot\mathrm{d}\mathbf{r} = 96}$$

(b) The diagram for (b) is as shown. We consider each straight line section in turn.

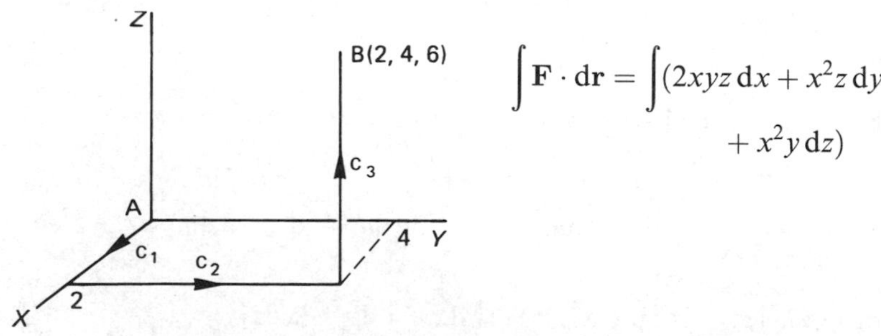

$$\int \mathbf{F}\cdot\mathrm{d}\mathbf{r} = \int (2xyz\,\mathrm{d}x + x^2z\,\mathrm{d}y + x^2y\,\mathrm{d}z)$$

c_1: (0, 0, 0) to (2, 0, 0); $y = 0$, $z = 0$, $\mathrm{d}y = 0$, $\mathrm{d}z = 0$

$$\therefore \int_{c_1} \mathbf{F}\cdot\mathrm{d}\mathbf{r} = 0 + 0 + 0 = \underline{0}$$

In the same way, we evaluate the line integral along c_2 and c_3.

$$\int_{c_2} \mathbf{F}\cdot\mathrm{d}\mathbf{r} = \text{.........};\quad \int_{c_3} \mathbf{F}\cdot\mathrm{d}\mathbf{r} = \text{.........}$$

47

$$\boxed{\int_{c_2} \mathbf{F}\cdot d\mathbf{r} = 0; \quad \int_{c_3} \mathbf{F}\cdot d\mathbf{r} = 96}$$

for we have $\int \mathbf{F}\cdot d\mathbf{r} = \int (2xyz\,dx + x^2z\,dy + x^2y\,dz)$

c_2: (2, 0, 0) to (2, 4, 0); $x = 2, \quad z = 0, \quad dx = 0, \quad dz = 0$

$$\therefore \int_{c_2} \mathbf{F}\cdot d\mathbf{r} = 0 + 0 + 0 = 0 \qquad \underline{\int_{c_2} \mathbf{F}\cdot d\mathbf{r} = 0}$$

c_3: (2, 4, 0) to (2, 4, 6); $x = 2, \quad y = 4, \quad dx = 0, \quad dy = 0$

$$\therefore \int_{c_3} \mathbf{F}\cdot d\mathbf{r} = 0 + 0 + \int_0^6 16\,dz = \Big[16z\Big]_0^6 = 96. \qquad \underline{\int_{c_3} \mathbf{F}\cdot d\mathbf{r} = 96}$$

Collecting the three results together

$$\int_{c_1+c_2+c_3} \mathbf{F}\cdot d\mathbf{r} = 0 + 0 + 96 \qquad \therefore \underline{\int_{c_1+c_2+c_3} \mathbf{F}\cdot d\mathbf{r} = 96}$$

In this particular example, the value of the line integral is independent of the two paths we have used joining the same two end points and indicates that **F** is a conservative field. It follows that

$$\int_c \mathbf{F}\cdot d\mathbf{r} - \int_{c_1+c_2+c_3} \mathbf{F}\cdot d\mathbf{r} = 0 \quad \text{i.e.} \quad \oint \mathbf{F}\cdot d\mathbf{r} = 0$$

So, <u>if **F** is a conservative field, $\oint \mathbf{F}\cdot d\mathbf{r} = 0$</u>

Make a note of this for future use.

48 Two further tests can also be applied to establish that a given vector field is conservative.

If **F** is a conservative field
(a) curl **F** $= 0$
(b) **F** can be expressed as grad V where V is a scalar field to be determined.

For example, in the work we have just completed, we showed that $\mathbf{F} = 2xyz\mathbf{i} + x^2z\mathbf{j} + x^2y\mathbf{k}$ is a conservative field.

(a) If we determine curl **F** in this case, we have

curl **F** = ...

49

curl **F** $= 0$

since $\quad \text{curl } \mathbf{F} = \begin{vmatrix} \mathbf{i} & \mathbf{j} & \mathbf{k} \\ \dfrac{\partial}{\partial x} & \dfrac{\partial}{\partial y} & \dfrac{\partial}{\partial z} \\ 2xyz & x^2z & x^2y \end{vmatrix}$

$$= (x^2 - x^2)\mathbf{i} - (2xy - 2xy)\mathbf{j} + (2xz - 2xz)\mathbf{k} = 0$$

$$\therefore \underline{\text{curl } \mathbf{F} = 0}$$

(b) We can attempt to express **F** as grad V where V is a scalar in x, y, z.

If $V = f(x, y, z)$ $\qquad$ $\text{grad } V = \dfrac{\partial V}{\partial x}\mathbf{i} + \dfrac{\partial V}{\partial y}\mathbf{j} + \dfrac{\partial V}{\partial z}\mathbf{k}$

and we have $\mathbf{F} = 2xyz\mathbf{i} + x^2z\mathbf{j} + x^2y\mathbf{k}$

$$\therefore \frac{\partial V}{\partial x} = 2xyz \qquad \therefore V = x^2yz + f(y, z)$$

$$\frac{\partial V}{\partial y} = x^2z \qquad \therefore V = \ldots\ldots\ldots$$

$$\frac{\partial V}{\partial z} = x^2y \qquad \therefore V = \ldots\ldots\ldots$$

We therefore have to find a scalar function V that satisfies the three requirements.

$V =$ ....

50

$$\boxed{V = x^2yz}$$

for $\dfrac{\partial V}{\partial x} = 2xyz \qquad \therefore\ V = x^2yz + f(y,\ z)$

$$\frac{\partial V}{\partial y} = x^2z \qquad \therefore\ V = x^2yz + g(x,\ z)$$

$$\frac{\partial V}{\partial z} = x^2y \qquad \therefore\ V = x^2yz + h(x,\ y)$$

These three are satisfied if $f(y,\ z) = g(z,\ x) = h(x,\ y) = 0$

$$\underline{\therefore\ \mathbf{F} = \text{grad}\ V \text{ where } V = x^2yz}$$

So three tests can be applied to determine whether or not a vector field is conservative. They are

(a) ....

(b) ....

(c) ....

51

$$\text{(a)} \oint \mathbf{F} \cdot d\mathbf{r} = 0$$

$$\text{(b)} \quad \text{curl } \mathbf{F} = 0$$

$$\text{(c)} \quad \mathbf{F} = \text{grad } V$$

Any one of these conditions can be applied as is convenient.
Now what about these?

Exercise Determine which of the following vector fields are conservative.

(a) $\mathbf{F} = (x + y)\mathbf{i} + (y - z)\mathbf{j} + (x + y + z)\mathbf{k}$
(b) $\mathbf{F} = (2xz + y)\mathbf{i} + (z + x)\mathbf{j} + (x^2 + y)\mathbf{k}$
(c) $\mathbf{F} = y \sin z\,\mathbf{i} + x \sin z\,\mathbf{j} + (xy \cos z + 2z)\mathbf{k}$
(d) $\mathbf{F} = 2xy\mathbf{i} + (x^2 + 4yz)\mathbf{j} + 2y^2z\mathbf{k}$
(e) $\mathbf{F} = y \cos x \cos z\,\mathbf{i} + \sin x \cos z\,\mathbf{j} - y \sin x \sin z\,\mathbf{k}$.

Complete all five and check your findings with the next frame.

52

(a) No (b) Yes (c) Yes (d) No (e) Yes

Divergence theorem (Gauss' theorem)

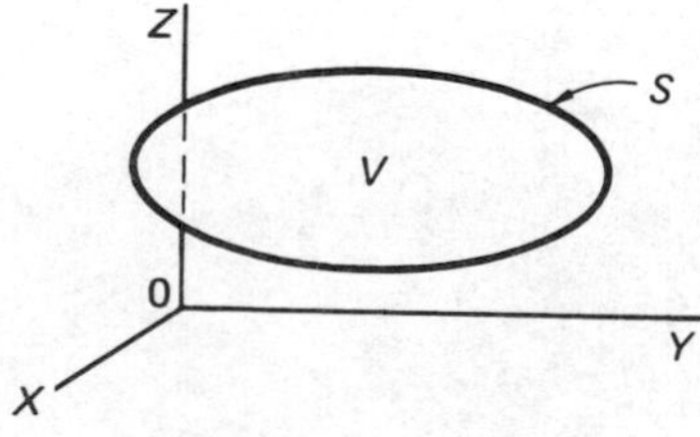

For a closed surface S, enclosing a region V in a vector field $\mathbf{F}$,

$$\int_V \text{div } \mathbf{F}\, dV = \int_S \mathbf{F} \cdot d\mathbf{S}$$

In general, this means that the volume integral (triple integral) on the left-hand side can be expressed as a surface integral (double integral) on the right-hand side. Let us work through one or two examples.

Example 1 Verify the divergence theorem for the vector field $\mathbf{F} = x^2\mathbf{i} + z\mathbf{j} + y\mathbf{k}$ taken over the region bounded by the planes $z = 0,\ z = 2,\ x = 0,\ x = 1,\ y = 0,\ y = 3$.

Start off, as always, by sketching the relevant diagram, which is

. . . .

53

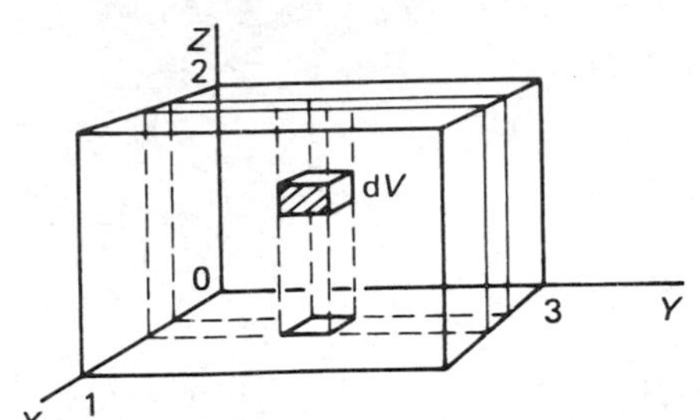

$$dV = dx\,dy\,dz$$

We have to show that

$$\int_V \text{div}\,\mathbf{F}\,dV = \int_S \mathbf{F}\cdot d\mathbf{S}$$

(a) To find $\int_V \text{div}\,\mathbf{F}\,dV$

$$\text{div}\,\mathbf{F} = \nabla\cdot\mathbf{F} = \left(\frac{\partial}{\partial x}\mathbf{i} + \frac{\partial}{\partial y}\mathbf{j}\,\frac{\partial}{\partial z}\mathbf{k}\right)\cdot(x^2\mathbf{i} + z\mathbf{j} + y\mathbf{k})$$

$$= \frac{\partial}{\partial x}(x^2) + \frac{\partial}{\partial y}(z) + \frac{\partial}{\partial z}(y) = 2x + 0 + 0 = 2x$$

$$\therefore \int_V \text{div}\,\mathbf{F}\,dV = \int_V 2x\,dV = \iiint_V 2x\,dz\,dy\,dx$$

Inserting the limits and completing the integration

$$\int_V \text{div}\,\mathbf{F}\,dV = \ldots\ldots\ldots$$

54

$$\boxed{\int_V \operatorname{div} \mathbf{F}\,dV = 6}$$

for $\displaystyle\int_V \operatorname{div} \mathbf{F}\,dV = \int_0^1\int_0^3\int_0^2 2x\,dz\,dy\,dx = \int_0^1\int_0^3 \left[2xz\right]_0^2 dy\,dx$

$$= \int_0^1 \left[4xy\right]_0^3 dx = \int_0^1 12x\,dx = \left[6x^2\right]_0^1 = \underline{6}$$

Now we have to find $\displaystyle\int_S \mathbf{F}\cdot d\mathbf{S}$

(b) To find $\displaystyle\int_S \mathbf{F}\cdot d\mathbf{S}$ i.e. $\displaystyle\int_S \mathbf{F}\cdot\hat{\mathbf{n}}\,dS$

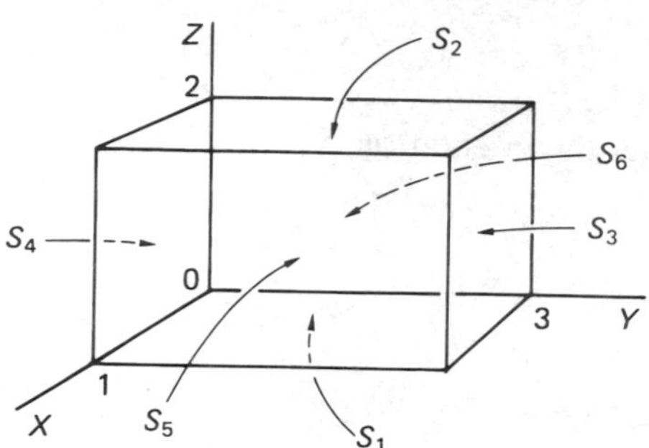

The enclosing surface S consists of six separate plane faces denoted as $S_1, S_2, \ldots S_6$ as shown. We consider each face in turn.

$$\mathbf{F} = x^2\mathbf{i} + z\mathbf{j} + y\mathbf{k}$$

(i) S_1 (base): $z = 0$; $\hat{\mathbf{n}} = -\mathbf{k}$ (outwards and downwards)

$$\therefore \mathbf{F} = x^2\mathbf{i} + y\mathbf{k} \qquad dS_1 = dx\,dy$$

$$\therefore \int_{S_1} \mathbf{F}\cdot\hat{\mathbf{n}}\,dS = \iint_{S_1} (x^2\mathbf{i} + y\mathbf{k})\cdot(-\mathbf{k})\,dy\,dx$$

$$= \int_0^1\int_0^3 (-y)\,dy\,dx = \int_0^1 \left[-\frac{y^2}{2}\right]_0^3 dx = \underline{-\frac{9}{2}}$$

(ii) S_2 (top): $z = 2$; $\hat{\mathbf{n}} = \mathbf{k}$ $dS_2 = dx\,dy$

$$\therefore \int_{S_2} \mathbf{F}\cdot\hat{\mathbf{n}}\,dS = \ldots\ldots\ldots$$

55

$$\boxed{\frac{9}{2}}$$

for $\displaystyle\int_{S_2} \mathbf{F}\cdot\hat{\mathbf{n}}\,\mathrm{d}S = \iint_{S_2} (x^2\mathbf{i} + 2\mathbf{j} + y\mathbf{k})\cdot(\mathbf{k})\,\mathrm{d}y\,\mathrm{d}x$

$$= \int_0^1\int_0^3 y\,\mathrm{d}y\,\mathrm{d}x = \underline{\frac{9}{2}}$$

So we go on.

(iii) S_3 (right-hand end): $y = 3$; $\hat{\mathbf{n}} = \mathbf{j}$ $\mathrm{d}S_3 = \mathrm{d}x\,\mathrm{d}z$

$$\mathbf{F} = x^2\mathbf{i} + z\mathbf{j} + y\mathbf{k}$$

$$\therefore \int_{S_3} \mathbf{F}\cdot\hat{\mathbf{n}}\,\mathrm{d}S = \iint_{S_3} (x^2\mathbf{i} + z\mathbf{j} + 3\mathbf{k})\cdot(\mathbf{j})\,\mathrm{d}z\,\mathrm{d}x = \int_0^1\int_0^2 z\,\mathrm{d}z\,\mathrm{d}x$$

$$= \int_0^1 \left[\frac{z^2}{2}\right]_0^2 \mathrm{d}x = \int_0^1 2\,\mathrm{d}x = \underline{2}$$

(iv) S_4 (left-hand end): $y = 0$; $\hat{\mathbf{n}} = -\mathbf{j}$ $\mathrm{d}S_4 = \mathrm{d}x\,\mathrm{d}z$

$$\therefore \int_{S_4} \mathbf{F}\cdot\hat{\mathbf{n}}\,\mathrm{d}S = \dots\dots\dots$$

56

$$\boxed{-2}$$

for $\displaystyle\int_{S_4} \mathbf{F}\cdot\hat{\mathbf{n}}\,\mathrm{d}S = \iint_{S_4} (x^2\mathbf{i} + z\mathbf{j} + y\mathbf{k})\cdot(-\mathbf{j})\,\mathrm{d}z\,\mathrm{d}x = \int_0^1\int_0^2 (-z)\,\mathrm{d}z\,\mathrm{d}x$

$$= \int_0^1 \left[-\frac{z^2}{2}\right]_0^2 \mathrm{d}x = \int_0^1 (-2)\,\mathrm{d}x = \underline{-2}$$

Now for the remaining two sides S_5 and S_6.
Evaluate these in the same manner, obtaining

$$\int_{S_5} \mathbf{F}\cdot\hat{\mathbf{n}}\,\mathrm{d}S = \dots\dots\dots; \qquad \int_{S_6} \mathbf{F}\cdot\hat{\mathbf{n}}\,\mathrm{d}S = \dots\dots\dots$$

57

$$\int_{S_5} \mathbf{F}\cdot\hat{\mathbf{n}}\,\mathrm{d}S = 6; \qquad \int_{S_6} \mathbf{F}\cdot\hat{\mathbf{n}}\,\mathrm{d}S = 0$$

Check:

(v) S_5 (front): $x = 1;\quad \hat{\mathbf{n}} = \mathbf{i}\quad \mathrm{d}S_5 = \mathrm{d}y\,\mathrm{d}z$

$$\therefore \int_{S_5} \mathbf{F}\cdot\hat{\mathbf{n}}\,\mathrm{d}S = \iint_{S_5} (\mathbf{i} + z\mathbf{j} + y\mathbf{k})\cdot(\mathbf{i})\,\mathrm{d}y\,\mathrm{d}z = \iint_{S_5} 1\,\mathrm{d}y\,\mathrm{d}z = \underline{6}$$

(vi) S_6 (back): $x = 0;\quad \hat{\mathbf{n}} = -\mathbf{i}\quad \mathrm{d}S_6 = \mathrm{d}y\,\mathrm{d}z$

$$\therefore \int_{S_6} \mathbf{F}\cdot\hat{\mathbf{n}}\,\mathrm{d}S = \iint_{S_6} (z\mathbf{j} + y\mathbf{k})\cdot(-\mathbf{i})\,\mathrm{d}y\,\mathrm{d}z = \iint_{S_6} 0\,\mathrm{d}y\,\mathrm{d}z = \underline{0}$$

Now on to the next frame where we will collect our results together.

58

For the whole surface S we therefore have

$$\int_S \mathbf{F}\cdot\mathrm{d}\mathbf{S} = -\frac{9}{2} + \frac{9}{2} + 2 - 2 + 6 + 0 = \underline{6}$$

and from our previous work in section (a) $\int_V \operatorname{div}\mathbf{F}\,\mathrm{d}V = \underline{6}$

We have therefore verified as required that, in this example

$$\underline{\int_V \operatorname{div}\mathbf{F}\,\mathrm{d}V = \int_S \mathbf{F}\cdot\mathrm{d}\mathbf{S}}$$

We have made rather a meal of this since we have set out the working in detail. In practice, the actual writing can often be considerably simplified. Let us move on to another example.

Example 2 Verify the Gauss divergence theorem for the vector field $\mathbf{F} = x\mathbf{i} + 2\mathbf{j} + z^2\mathbf{k}$ taken over the region bounded by the planes $z = 0,\ z = 4,\ x = 0,\ y = 0$ and the surface $x^2 + y^2 = 4$ in the first octant.

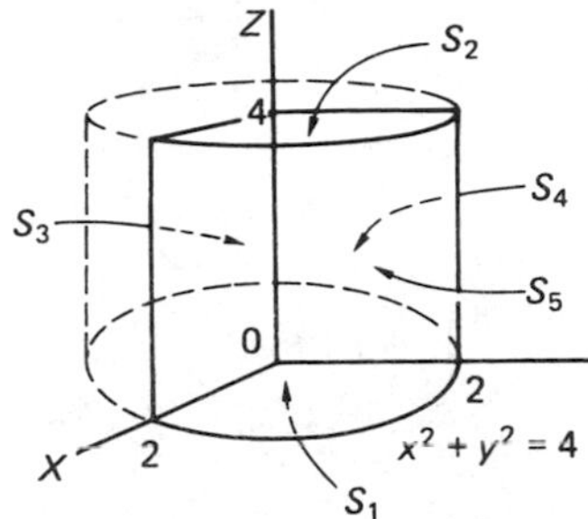

Divergence theorem

$$\int_V \operatorname{div} \mathbf{F}\, \mathrm{d}V = \int_S \mathbf{F} \cdot \mathrm{d}S$$

S consists of five surfaces $S_1,\ S_2, \dots\ S_5$ as shown.

(a) $\quad \operatorname{div} \mathbf{F} = \nabla \cdot \mathbf{F} = \left(\frac{\partial}{\partial x}\mathbf{i} + \frac{\partial}{\partial y}\mathbf{j} + \frac{\partial}{\partial z}\mathbf{k}\right) \cdot (x\mathbf{i} + 2\mathbf{j} + z^2\mathbf{k})$

$$= \dots\dots\dots$$

59

$$\boxed{1 + 2z}$$

$$\therefore \int_V \operatorname{div} \mathbf{F}\, \mathrm{d}V = \int_V \nabla \cdot \mathbf{F}\, \mathrm{d}V = \iiint_V (1 + 2z)\, \mathrm{d}x\, \mathrm{d}y\, \mathrm{d}z$$

Changing to cylindrical polar coordinates $(\rho,\ \phi,\ z)$

$$x = \rho\cos\phi; \quad y = \rho\sin\phi; \quad z = z; \quad \mathrm{d}V = \rho\, \mathrm{d}\rho\, \mathrm{d}\phi\, \mathrm{d}z$$

Transforming the variables and inserting the appropriate limits, we then have

$$\int_V \operatorname{div} \mathbf{F}\, \mathrm{d}V = \dots\dots\dots$$ Finish it.

60

$$\boxed{20\pi}$$

for $\displaystyle\int_V \text{div}\,\mathbf{F}\,\text{d}V = \int_0^{\pi/2}\int_0^2\int_0^4 (1+2z)\,\text{d}z\,\rho\,\text{d}\rho\,\text{d}\phi$

$$= \int_0^{\pi/2}\int_0^2 \left[z+z^2\right]_0^4 \rho\,\text{d}\rho\,\text{d}\phi = \int_0^{\pi/2}\int_0^2 20\rho\,\text{d}\rho\,\text{d}\phi$$

$$= \int_0^{\pi/2}\left[10\rho^2\right]_0^2 \text{d}\phi = \int_0^{\pi/2} 40\,\text{d}\phi = \underline{20\pi} \qquad (1)$$

(b) Now we evaluate $\displaystyle\int_S \mathbf{F}\cdot\text{d}\mathbf{S}$ over the closed surface.

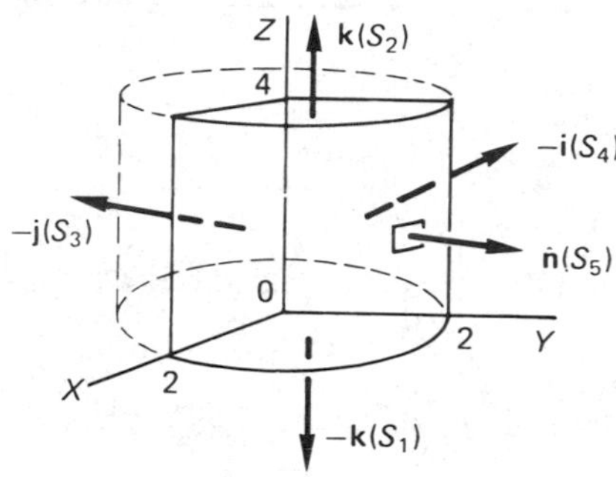

The unit normal vector for each surface is shown.

$$\mathbf{F} = x\mathbf{i} + 2\mathbf{j} + z^2\mathbf{k}$$

(i) S_1: $\quad z = 0; \quad \hat{\mathbf{n}} = -\mathbf{k} \quad \mathbf{F} = x\mathbf{i} + 2\mathbf{j}$

$$\therefore \int_{S_1} \mathbf{F}\cdot\hat{\mathbf{n}}\,\text{d}S = \int_{S_1} (x\mathbf{i} + 2\mathbf{j})\cdot(-\mathbf{k})\,\text{d}S = \underline{0}$$

(ii) S_2: $\quad z = 4; \quad \hat{\mathbf{n}} = \mathbf{k} \quad \mathbf{F} = x\mathbf{i} + 2\mathbf{j} + 16\mathbf{k}$

$$\therefore \int_{S_2} \mathbf{F}\cdot\hat{\mathbf{n}}\,\text{d}S = \int_{S_2} (x\mathbf{i} + 2\mathbf{j} + 16\mathbf{k})\cdot(\mathbf{k})\,\text{d}S = \int_{S_2} 16\,\text{d}S$$

$$= 16\left(\frac{\pi 4}{4}\right) = \underline{16\pi}$$

In the same way for S_3: $\displaystyle\int_{S_3} \mathbf{F}\cdot\hat{\mathbf{n}}\,\text{d}S = \ldots\ldots\ldots$

and for S_4: $\displaystyle\int_{S_4} \mathbf{F}\cdot\hat{\mathbf{n}}\,\text{d}S = \ldots\ldots\ldots$

61

$$\int_{S_3} \mathbf{F} \cdot \hat{\mathbf{n}} \, dS = -16; \qquad \int_{S_4} \mathbf{F} \cdot \hat{\mathbf{n}} \, dS = 0$$

for we have

(iii) S_3: $\quad y = 0; \qquad \hat{\mathbf{n}} = -\mathbf{j} \qquad \mathbf{F} = x\mathbf{i} + 2\mathbf{j} + z^2\mathbf{k}$

$$\therefore \int_{S_3} \mathbf{F} \cdot \hat{\mathbf{n}} \, dS = \int_{S_3} (x\mathbf{i} + 2\mathbf{j} + z^2\mathbf{k}) \cdot (-\mathbf{j}) \, dS$$

$$= \int_{S_3} (-2) \, dS = -2(8) = \underline{-16}$$

(iv) S_4: $\quad x = 0; \qquad \hat{\mathbf{n}} = -\mathbf{i} \qquad \mathbf{F} = 2\mathbf{j} + z^2\mathbf{k}$

$$\therefore \int_{S_4} \mathbf{F} \cdot \hat{\mathbf{n}} \, dS = \int_{S_4} (2\mathbf{j} + z^2\mathbf{k}) \cdot (-\mathbf{i}) \, dS = \underline{0}$$

Finally we have

(v) S_5: $\quad x^2 + y^2 - 4 = 0 \qquad \hat{\mathbf{n}} = \dots\dots\dots$

62

$$\hat{\mathbf{n}} = \frac{1}{2}(x\mathbf{i} + y\mathbf{j})$$

for $x^2 + y^2 - 4 = 0$ $\qquad \hat{\mathbf{n}} = \dfrac{\nabla S}{|\nabla S|} = \dfrac{2x\mathbf{i} + 2y\mathbf{j}}{\sqrt{4x^2 + 4y^2}} = \dfrac{x\mathbf{i} + y\mathbf{j}}{2}$

$$\therefore \int_{S_5} \mathbf{F} \cdot \hat{\mathbf{n}} \, dS = \int_{S_5} (x\mathbf{i} + 2\mathbf{j} + z^2\mathbf{k}) \cdot \left(\frac{x\mathbf{i} + y\mathbf{j}}{2}\right) dS = \frac{1}{2} \int_{S_5} (x^2 + 2y) \, dS$$

Converting to cylindrical polar coordinates, this gives

$$\int_{S_5} \mathbf{F} \cdot \hat{\mathbf{n}} \, dS = \dots\dots\dots$$

63

$$\boxed{4\pi + 16}$$

since we have $\displaystyle\int_{S_5} \mathbf{F} \cdot \hat{\mathbf{n}}\, \mathrm{d}S = \frac{1}{2}\int_{S_5} (x^2 + 2y)\, \mathrm{d}S$

also $x = 2\cos\phi; \quad y = 2\sin\phi$

$z = z; \quad \mathrm{d}S = 2\,\mathrm{d}\phi\,\mathrm{d}z$

$$\therefore \int_{S_5} \mathbf{F} \cdot \hat{\mathbf{n}}\, \mathrm{d}S = \frac{1}{2}\int_0^4 \int_0^{\pi/2} (4\cos^2\phi + 4\sin\phi)\, 2\,\mathrm{d}\phi\,\mathrm{d}z$$

$$= 2\int_0^4 \int_0^{\pi/2} \{(1 + \cos 2\phi) + 2\sin\phi\}\, \mathrm{d}\phi\,\mathrm{d}z$$

$$= 2\int_0^4 \left[\left(\phi = \frac{\sin 2\phi}{2}\right) - 2\cos\phi\right]_0^{\pi/2} \mathrm{d}z$$

$$= 2\int_0^4 \left(\frac{\pi}{2} + 2\right) \mathrm{d}z = \underline{4\pi + 16}$$

Therefore, for the total surface S

$$\int_S \mathbf{F} \cdot \hat{\mathbf{n}}\, \mathrm{d}S = 0 + 16\pi - 16 + 0 + 4\pi + 16 = \underline{20\pi} \qquad (2)$$

$$\underline{\therefore \int_V \operatorname{div} \mathbf{F}\, \mathrm{d}V = \int_S \mathbf{F} \cdot \mathrm{d}\mathbf{S} = 20\pi}$$

Other examples are worked in much the same way. You will remember that, for a closed surface, the normal vectors at all points are drawn in an *outward* direction.

Now we move on to a further important theorem.

64

Stokes' theorem

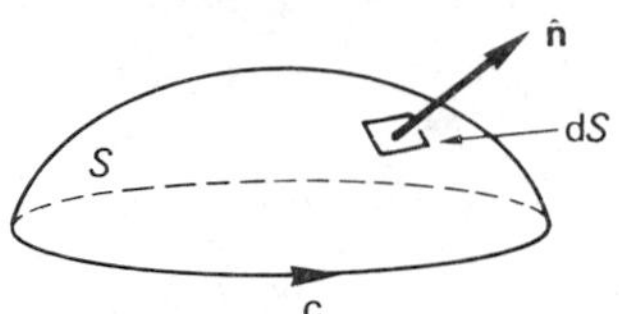

If **F** is a vector field existing over an open surface S and around its boundary, closed curve c, then

$$\int_S \operatorname{curl} \mathbf{F} \cdot d\mathbf{S} = \oint_c \mathbf{F} \cdot d\mathbf{r}$$

This means that we can express a surface integral in terms of a line integral round the boundary curve.

The proof of this theorem is rather lengthy and is to be found in the Appendix. Let us demonstrate its application in the following examples.

Example 1 A hemisphere S is defined by $x^2 + y^2 + z^2 = 4$ $(z \geq 0)$. A vector field $\mathbf{F} = 2y\mathbf{i} - x\mathbf{j} + xz\mathbf{k}$ exists over the surface and around its boundary c. Vertify Stokes' theorem that $\int_S \operatorname{curl} \mathbf{F} \cdot d\mathbf{S} = \oint_c \mathbf{F} \cdot d\mathbf{r}$.

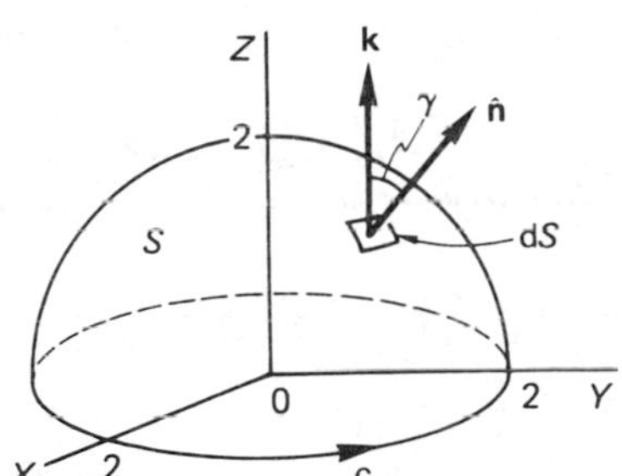

$S : x^2 + y^2 + z^2 - 4 = 0$

$\mathbf{F} = 2y\mathbf{i} - x\mathbf{j} + xz\mathbf{k}$

c is the circle $x^2 + y^2 = 4$.

(a) $$\oint_c \mathbf{F} \cdot d\mathbf{r} = \int_c (2y\mathbf{i} - x\mathbf{j} + xz\mathbf{k}) \cdot (\mathbf{i}\, dx + \mathbf{j}\, dy + \mathbf{k}\, dz)$$
$$= \int_c (2y\, dx - x\, dy + xz\, dz)$$

Converting to polar coordinates

$x = 2\cos\theta;$ $\quad y = 2\sin\theta;$ $\quad z = 0$

$dx = -2\sin\theta\, d\theta;$ $\quad dy = 2\cos\theta\, d\theta;$ $\quad$ Limits $\theta = 0$ to 2π

Making the substitutions and completing the integral

$$\oint_c \mathbf{F} \cdot d\mathbf{r} = \ldots\ldots\ldots$$

65

$$\oint_c \mathbf{F}\cdot \mathrm{d}\mathbf{r} = -12\pi$$

since $\displaystyle\oint_c \mathbf{F}\cdot \mathrm{d}\mathbf{r} = \int_0^{2\pi} (4\sin\theta[-2\sin\theta\,\mathrm{d}\theta] - 2\cos\theta\, 2\cos\theta\,\mathrm{d}\theta)$

$$= -4\int_0^{2\pi} (2\sin^2\theta + \cos^2\theta)\,\mathrm{d}\theta$$

$$= -4\int_0^{2\pi} (1+\sin^2\theta)\,\mathrm{d}\theta = -2\int_0^{2\pi} (3-\cos 2\theta)\,\mathrm{d}\theta$$

$$= -2\left[3\theta - \frac{\sin 2\theta}{2}\right]_0^{2\pi} = \underline{-12\pi} \qquad (1)$$

On to the next frame.

66 (b) Now we determine $\displaystyle\int_S \text{curl}\,\mathbf{F}\cdot \mathrm{d}\mathbf{S}$

$$\int \text{curl}\,\mathbf{F}\cdot \mathrm{d}\mathbf{S} = \int \text{curl}\,\mathbf{F}\cdot \hat{\mathbf{n}}\,\mathrm{d}S \qquad \mathbf{F} = 2y\mathbf{i} - x\mathbf{j} + xz\mathbf{k}$$

$$\therefore \text{curl}\,\mathbf{F} = \dots\dots\dots$$

67

$$\text{curl}\,\mathbf{F} = -z\mathbf{j} - 3\mathbf{k}$$

since $\text{curl}\,\mathbf{F} = \begin{vmatrix} \mathbf{i} & \mathbf{j} & \mathbf{k} \\ \dfrac{\partial}{\partial x} & \dfrac{\partial}{\partial y} & \dfrac{\partial}{\partial z} \\ 2y & -x & xz \end{vmatrix}$

$$= \mathbf{i}\,(0-0) - \mathbf{j}\,(z-0) + \mathbf{k}\,(-\mathbf{1}-\mathbf{2}) = -\mathbf{zj} - \mathbf{3k}$$

Now $\displaystyle\hat{\mathbf{n}} = \frac{\nabla S}{|\nabla S|} = \frac{2x\mathbf{i} + 2y\mathbf{j} + 2z\mathbf{k}}{\sqrt{4x^2+4y^2+4z^2}} = \frac{x\mathbf{i} + y\mathbf{j} + z\mathbf{k}}{2}$

Then $\int_S \text{curl}\,\mathbf{F}\cdot\hat{\mathbf{n}}\,dS = \int_S (-z\mathbf{j} - 3\mathbf{k})\cdot\left(\dfrac{x\mathbf{i} + y\mathbf{j} + z\mathbf{k}}{2}\right) dS$

$$= \frac{1}{2}\int_S (-yz - 3z)\,dS$$

Expressing this in spherical polar coordinates and integrating, we get

$$\int_S \text{curl}\,\mathbf{F}\cdot\hat{\mathbf{n}}\,dS = \dots\dots\dots$$

68

$$\boxed{-12\pi}$$

for $x = 2\sin\theta\,\cos\phi;\quad y = 2\sin\theta\,\sin\phi;\quad z = 2\cos\theta$

$$dS = 4\sin\theta\,d\theta\,d\phi$$

$$\therefore \int_S \text{curl}\,\mathbf{F}\cdot\hat{\mathbf{n}}\,dS = \frac{1}{2}\int_S\int(-2\sin\theta\,\sin\phi\,2\cos\theta$$

$$- 6\cos\theta)4\sin\theta\,d\theta\,d\phi$$

$$= -4\int_0^{2\pi}\int_0^{\pi/2}(2\sin^2\theta\cos\theta\sin\phi$$

$$+ 3\sin\theta\,\cos\theta)\,d\theta\,d\phi$$

$$= -4\int_0^{2\pi}\left[\frac{2\sin^3\theta\sin\phi}{3} + \frac{3\sin^2\theta}{2}\right]_0^{\pi/2} d\phi$$

$$= -4\int_0^{2\pi}\left(\frac{2}{3}\sin\phi + \frac{3}{2}\right) d\phi = \underline{-12\pi} \qquad (2)$$

So we have from our two results (1) and (2)

$$\underline{\int_S \text{curl}\,\mathbf{F}\cdot d\mathbf{S} = \oint_c \mathbf{F}\cdot d\mathbf{r}}$$

Before we proceed with another example, let us clarify a point relating to the direction of unit normal vectors now that we are dealing with surfaces.

So on to the next frame.

69

Direction of unit normal vectors to a surface S

When we were dealing with the divergence theorem, the normal vectors were drawn in a direction outward from the enclosed region.

With an open surface, as we now have, there is, in fact, no inward or outward direction. With any general surface, a normal vector can be drawn in either of two opposite directions. To avoid confusion, a convention must therefore be agreed upon and the established rule is as follows.

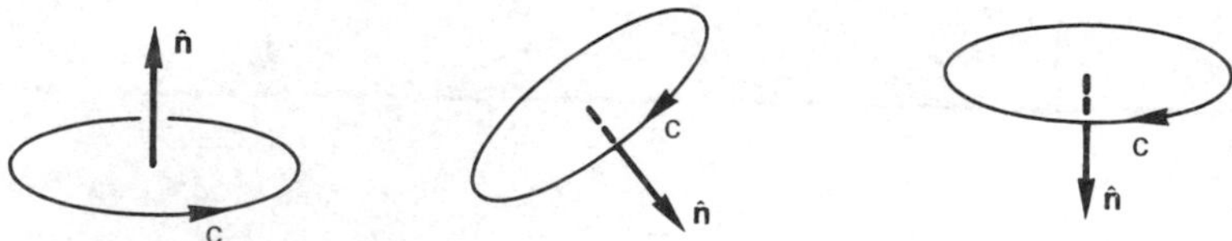

A unit normal $\hat{\mathbf{n}}$ is drawn perpendicular to the surface S at any point in the direction indicated by applying a right-handed screw sense to the direction of integration round the boundary c.

Having noted that point, we can now deal with the next example.

Example 2 A surface consists of five sections formed by the planes $x=0$, $x=1$, $y=0$, $y=3$, $z=2$ in the first octant. If the vector field $\mathbf{F}=y\mathbf{i}+z^2\mathbf{j}+xy\mathbf{k}$ exists over the surface and around its boundary, verify Stokes' theorem.

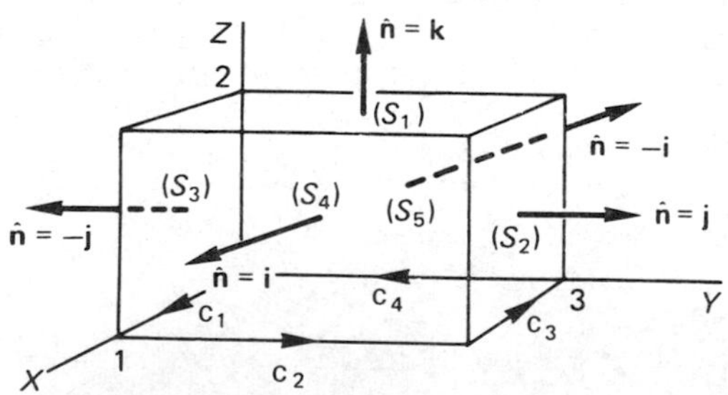

If we progress round the boundary along c_1, c_2, c_3, c_4 in an anti-clockwise manner, the normals to the surfaces will be as shown.

We have to verify that $\int_S \text{curl}\,\mathbf{F}\cdot d\mathbf{S} = \oint_c \mathbf{F}\cdot d\mathbf{r}$

(a) We will start off by finding $\oint_c \mathbf{F}\cdot d\mathbf{r}$

$$\int \mathbf{F}\cdot d\mathbf{r} = \ldots\ldots\ldots$$

70

$$\int \mathbf{F} \cdot d\mathbf{r} = \int (y\,dx + z^2\,dy + xy\,dz)$$

(i) Along c_1: $y = 0; \quad z = 0; \quad dy = 0; \quad dz = 0$

$$\therefore \int_{c_1} \mathbf{F} \cdot d\mathbf{r} = \int (0 + 0 + 0) = \underline{0}$$

(ii) Along c_2: $x = 1; \quad z = 0; \quad dx = 0; \quad dz = 0$

$$\therefore \int_{c_2} \mathbf{F} \cdot d\mathbf{r} = \int (0 + 0 + 0) = \underline{0}$$

In the same way

$\int_{c_3} \mathbf{F} \cdot d\mathbf{r} = \ldots\ldots\ldots$ and $\int_{c_4} \mathbf{F} \cdot d\mathbf{r} = \ldots\ldots\ldots$

71

$$\int_{c_3} \mathbf{F} \cdot d\mathbf{r} = -3; \qquad \int_{4} \mathbf{F} \cdot d\mathbf{r} = 0$$

for

(iii) Along c_3: $y = 3; \quad z = 0; \quad dy = 0; \quad dz = 0$

$$\therefore \int_{c_3} \mathbf{F} \cdot d\mathbf{r} = \int_1^0 (3\,dx + 0 + 0) = \left[3x\right]_1^0 = \underline{-3}$$

(iv) Along c_4: $x = 0; \quad z = 0; \quad dx = 0; \quad dz = 0$

$$\therefore \int_{c_4} \mathbf{F} \cdot d\mathbf{r} = \int (0 + 0 + 0) = \underline{0}$$

$$\therefore \oint_c \mathbf{F} \cdot d\mathbf{r} = 0 + 0 - 3 + 0 = -3 \qquad \underline{\oint_c \mathbf{F} \cdot d\mathbf{r} = -3} \quad (1)$$

(b) Now we have to find $\int_S \operatorname{curl} \mathbf{F} \cdot d\mathbf{S}$.

First we need an expression for curl $\mathbf{F}$.

$\mathbf{F} = y\mathbf{i} + z^2\mathbf{j} + xy\mathbf{k} \quad \therefore \operatorname{curl} \mathbf{F} = \ldots\ldots\ldots$

72

$$\boxed{\text{curl } \mathbf{F} = (x - 2z)\mathbf{i} - y\mathbf{j} - \mathbf{k}}$$

$$\text{for curl } \mathbf{F} = \nabla \times \mathbf{F} = \begin{vmatrix} \mathbf{i} & \mathbf{j} & \mathbf{k} \\ \dfrac{\partial}{\partial x} & \dfrac{\partial}{\partial y} & \dfrac{\partial}{\partial z} \\ y & z^2 & xy \end{vmatrix}$$

$$= \mathbf{i}(x - 2z) - \mathbf{j}(y - 0) + \mathbf{k}(0 - 1) = \underline{(x - 2z)\mathbf{i} - y\mathbf{j} - \mathbf{k}}$$

Then, for each section, we obtain $\int \text{curl } \mathbf{F} \cdot \mathrm{d}\mathbf{S} = \int \text{curl } \mathbf{F} \cdot \hat{\mathbf{n}} \, \mathrm{d}S$

(i) S_1 (top): $\quad \hat{\mathbf{n}} = \mathbf{k}$

$$\therefore \int_{S_1} \text{curl } \mathbf{F} \cdot \hat{\mathbf{n}} \, \mathrm{d}S = \dots\dots\dots$$

73

$$\boxed{-3}$$

$$\text{for } \int_{S_1} \text{curl } \mathbf{F} \cdot \hat{\mathbf{n}} \, \mathrm{d}S = \int_{S_1} \{(x - 2z)\mathbf{i} - y\mathbf{j} - \mathbf{k}\} \cdot (\mathbf{k}) \, \mathrm{d}S$$

$$= \int_{S_1} (-1) \, \mathrm{d}S = -(\text{area of } S_1) = \underline{-3}$$

Then, likewise

(ii) S_2 (right-hand end): $\quad \hat{\mathbf{n}} = \mathbf{j}$

$$\therefore \int_{S_2} \text{curl } \mathbf{F} \cdot \hat{\mathbf{n}} \, \mathrm{d}S = \int_{S_2} \{(x - 2z)\mathbf{i} - y\mathbf{j} - \mathbf{k}\} \cdot (\mathbf{j}) \, \mathrm{d}S$$

$$= \int_{S_2} (-y) \, \mathrm{d}S \qquad \text{But } y = 3 \text{ for this section}$$

$$\therefore \int_{S_2} \text{curl } \mathbf{F} \cdot \hat{\mathbf{n}} \, \mathrm{d}S = \int_{S_2} (-3) \, \mathrm{d}S = (-3)(2) = \underline{-6}$$

(iii) S_3 (left-hand end): $\quad \hat{\mathbf{n}} = -\mathbf{j}$

$$\therefore \int_{S_3} \text{curl } \mathbf{F} \cdot \hat{\mathbf{n}} \, \mathrm{d}S = \dots\dots\dots$$

74

$$\boxed{0}$$

since $\displaystyle\int_{S_3} \text{curl}\,\mathbf{F}\cdot\hat{\mathbf{n}}\,\mathrm{d}S = \int_{S_3} \{(x-2z)\mathbf{i} - y\mathbf{j} - \mathbf{k}\}\cdot(-\mathbf{j})\,\mathrm{d}S$

$$= \int_{S_3} y\,\mathrm{d}S \qquad \text{But} \quad y = 0 \quad \text{over } S_3$$

$$\therefore \int_{S_3} \text{curl}\,\mathbf{F}\cdot\hat{\mathbf{n}}\,\mathrm{d}S = \underline{0}$$

Working in the same way

$\displaystyle\int_{S_4} \text{curl}\,\mathbf{F}\cdot\hat{\mathbf{n}}\,\mathrm{d}S = \ldots\ldots\ldots;\quad \int_{S_5} \text{curl}\,\mathbf{F}\cdot\hat{\mathbf{n}}\,\mathrm{d}S = \ldots\ldots\ldots$

75

$$\boxed{\int_{S_4} \text{curl}\,\mathbf{F}\cdot\hat{\mathbf{n}}\,\mathrm{d}S = -6; \qquad \int_{S_5} \text{curl}\,\mathbf{F}\cdot\hat{\mathbf{n}}\,\mathrm{d}S = 12}$$

since

(iv) S_4 (front): $\quad \hat{\mathbf{n}} = \mathbf{i}$

$$\therefore \int_{S_4} \text{curl}\,\mathbf{F}\cdot\hat{\mathbf{n}}\,\mathrm{d}S = \int_{S_4} \{(x-2z)\mathbf{i} - y\mathbf{j} - \mathbf{k}\}\cdot(\mathbf{i})\,\mathrm{d}S$$

$$= \int_{S_4} (x-2z)\,\mathrm{d}S \qquad \text{But} \quad x = 1 \quad \text{over } S_4$$

$$\therefore \int_{S_4} \text{curl}\,\mathbf{F}\cdot\hat{\mathbf{n}}\,\mathrm{d}S = \int_0^3\int_0^2 (1-2z)\,\mathrm{d}z\,\mathrm{d}y = \int_0^3 \Big[z - z^2\Big]_0^2\,\mathrm{d}y$$

$$= \int_0^3 (-2)\,\mathrm{d}y = \Big[-2y\Big]_0^3 = \underline{-6}$$

(v) S_5 (back): $\hat{\mathbf{n}} = -\mathbf{i}$ $\quad$ with $\quad x = 0 \quad$ over S_5

Similar working to that above gives $\displaystyle\int_{S_5} \text{curl}\,\mathbf{F}\cdot\hat{\mathbf{n}}\,\mathrm{d}S = \underline{12}$

Finally, collecting the five results together gives

$$\int_S \text{curl}\,\mathbf{F}\cdot\hat{\mathbf{n}}\,\mathrm{d}S = \ldots\ldots\ldots$$

76

$$\int_S \text{curl } \mathbf{F} \cdot \hat{\mathbf{n}} \, dS = -3 - 6 + 0 - 6 + 12 = -3 \qquad (2)$$

So, referring back to our result for section (a) we see that

$$\int_S \text{curl } \mathbf{F} \cdot d\mathbf{S} = \oint_c \mathbf{F} \cdot d\mathbf{r}$$

Of course we can, on occasions, make use of Stokes' theorem to lighten the working—as in the next example.

Example 3 A surface S consists of that part of the cylinder $x^2 + y^2 = 9$ between $z = 0$ and $z = 4$ for $y \geq 0$ and the two semicircles of radius 3 in the planes $z = 0$ and $z = 4$. If $\mathbf{F} = z\mathbf{i} + xy\mathbf{j} + xz\mathbf{k}$ evaluate $\int_S \text{curl } \mathbf{F} \cdot d\mathbf{S}$ over the surface.

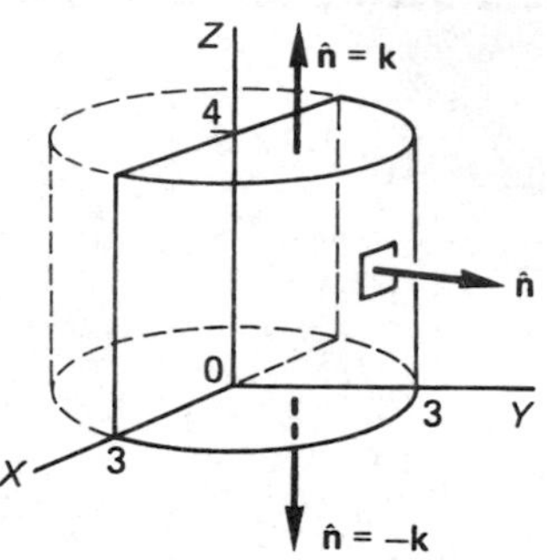

The surface S consists of three sections
(a) the curved surface of the cylinder
(b) the top and bottom semicircles.
We could therefore evaluate

$\int_S \text{curl } \mathbf{F} \cdot d\mathbf{S}$ over each of these separately.

However, we know by Stokes' theorem that $\int_S \text{curl } \mathbf{F} \cdot d\mathbf{S} = \ldots\ldots\ldots$

77

$\oint_c \mathbf{F} \cdot d\mathbf{r}$ where c is the boundary of S

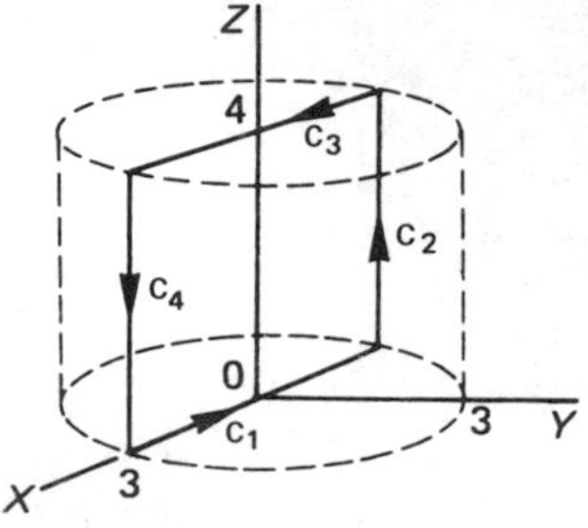

$$\mathbf{F} = z\mathbf{i} + xy\mathbf{j} + xz\mathbf{k}$$

$$\therefore \int \mathbf{F} \cdot d\mathbf{r} = \int (z\mathbf{i} + xy\mathbf{j} + xz\mathbf{k}) \cdot (\mathbf{i}\, dx + \mathbf{j}\, dy + \mathbf{k}\, dz)$$

$$= \int (z\, dx + xy\, dy + xz\, dz)$$

Now we can work through this easily enough, taking c_1, c_2, c_3, c_4 in turn, and summing the results, which gives

$$\int_S \text{curl}\,\mathbf{F}\cdot d\mathbf{S} = \oint_c \mathbf{F}\cdot d\mathbf{r} = \ldots\ldots\ldots$$

78

$$\boxed{-24}$$

Here is the working in detail. $\int \mathbf{F}\cdot d\mathbf{r} = \int (z\,dx + xy\,dy + xz\,dz)$

(i) c_1: $y = 0$; $z = 0$; $dy = 0$; $dz = 0$

$$\int_{c_1} \mathbf{F}\cdot d\mathbf{r} = \int_{c_1} (0 + 0 + 0) = \underline{0}$$

(ii) c_2: $x = -3$; $y = 0$; $dx = 0$; $dy = 0$

$$\int_{c_2} \mathbf{F}\cdot d\mathbf{r} = \int_{c_2} (0 + 0 - 3z\,dz) = \left[\frac{-3z^2}{2}\right]_0^4 = \underline{-24}$$

(iii) c_3: $y = 0$; $z = 4$; $dy = 0$; $dz = 0$

$$\int_{c_3} \mathbf{F}\cdot d\mathbf{r} = \int_{c_3} (4\,dx + 0 + 0) = \int_{-3}^{3} 4\,dx = \underline{24}$$

(iv) c_4: $x = 3$; $y = 0$; $dx = 0$; $dy = 0$

$$\int_{c_4} \mathbf{F}\cdot d\mathbf{r} \int_{c_4} (0 + 0 + 3z\,dz) = \left[\frac{3z^2}{2}\right]_4^0 = \underline{-24}$$

Totalling up these four results, we have

$$\oint_c \mathbf{F}\cdot d\mathbf{r} = 0 - 24 + 24 - 24 = -24$$

But $\int_S \text{curl}\,\mathbf{F}\cdot d\mathbf{S} = \oint_c \mathbf{F}\cdot d\mathbf{r}$ $\therefore$ $\underline{\int_S \text{curl}\,\mathbf{F}\cdot d\mathbf{S} = -24}$

This working is a good deal easier than calculating $\int_S \text{curl}\,\mathbf{F}\cdot d\mathbf{S}$ over the three separate surfaces direct.

So, if you have not already done so, make a note of Stokes' theorem:

$$\underline{\int_S \text{curl}\,\mathbf{F}\cdot d\mathbf{S} = \oint_c \mathbf{F}\cdot d\mathbf{r}}$$

Then on to the next section of the work.

79

Green's theorem

Green's theorem enables an integral over a plane area to be expressed in terms of a line integral round its boundary curve.

We showed in Programme 9 that, if P and Q are two single-valued functions of x and y, continuous over a plane surface S, and c is its boundary curve, then

$$\oint_{c}(P\,dx + Q\,dy) = \int_S\int\left(\frac{\partial Q}{\partial x} - \frac{\partial P}{\partial y}\right)dx\,dy$$

where the line integral is taken round c in an anticlockwise manner.

In vector terms, this becomes:

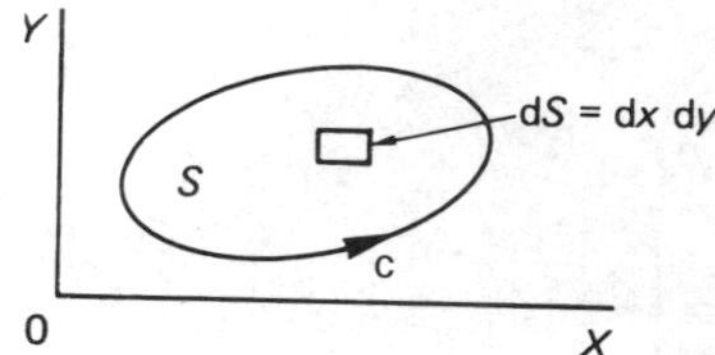

S is a two-dimensional space enclosed by a simple closed curve c.

$$dS = dx\,dy$$
$$\mathbf{dS} = \hat{\mathbf{n}}\,dS = \mathbf{k}\,dx\,dy$$

If $\mathbf{F} = P\mathbf{i} + Q\mathbf{j}$ where $P = P(x, y)$ and $Q = Q(x, y)$ then

curl $\mathbf{F}$ = ...

80

$$\boxed{\mathbf{k}\left(\frac{\partial Q}{\partial x} - \frac{\partial P}{\partial y}\right)}$$

for $\quad \text{curl}\,\mathbf{F} = \begin{vmatrix} \mathbf{i} & \mathbf{j} & \mathbf{k} \\ \dfrac{\partial}{\partial x} & \dfrac{\partial}{\partial y} & \dfrac{\partial}{\partial z} \\ P & Q & 0 \end{vmatrix}$

$$= \mathbf{i}\left(0 - \frac{\partial Q}{\partial z}\right) - \mathbf{j}\left(0 - \frac{\partial P}{\partial z}\right) + \mathbf{k}\left(\frac{\partial Q}{\partial x} - \frac{\partial P}{\partial y}\right)$$

But in the xy-plane, $\dfrac{\partial Q}{\partial z} = \dfrac{\partial P}{\partial z} = 0. \quad \therefore \text{curl}\,\mathbf{F} = \mathbf{k}\left(\dfrac{\partial Q}{\partial x} - \dfrac{\partial P}{\partial y}\right)$

So $\displaystyle\int \text{curl}\,\mathbf{F}\cdot\mathbf{dS} = \int \text{curl}\,\mathbf{F}\cdot\hat{\mathbf{n}}\,dS$ and in the xy-plane, $\hat{\mathbf{n}} = \mathbf{k}$

$$\therefore \int_S \text{curl}\,\mathbf{F}\cdot\mathbf{dS} = \int_S \mathbf{k}\left(\frac{\partial Q}{\partial x} - \frac{\partial P}{\partial y}\right)\cdot(\mathbf{k})\,dS = \int_S\int\left(\frac{\partial Q}{\partial x} - \frac{\partial P}{\partial y}\right)dx\,dy$$

$$\therefore \int_S \text{curl}\,\mathbf{F}\cdot d\mathbf{S} = \int_S\int\left(\frac{\partial Q}{\partial x} - \frac{\partial P}{\partial y}\right) dx\,dy \qquad (1)$$

Now by Stokes' theorem ....

81

$$\boxed{\int_S \text{curl}\,\mathbf{F}\cdot d\mathbf{S} = \oint_c \mathbf{F}\cdot d\mathbf{r}}$$

and, in this case, $\int \mathbf{F}\cdot d\mathbf{r} = \int (P\mathbf{i} + Q\mathbf{j})\cdot(\mathbf{i}\,dx + \mathbf{j}\,dy + \mathbf{k}\,dz)$

$$= \int (P\,dx + Q\,dy)$$

$$\therefore \oint_c \mathbf{F}\cdot d\mathbf{r} = \oint_c (P\,dx + Q\,dy) \qquad (2)$$

Therefore from (1) and (2)

Stokes' theorem $\int_S \text{curl}\,\mathbf{F}\cdot d\mathbf{S} = \oint_c \mathbf{F}\cdot d\mathbf{r}$

becomes

<u>Green's theorem $\int_S\int\left(\frac{\partial Q}{\partial x} - \frac{\partial P}{\partial y}\right) dx\,dy = \oint_c (P\,dx + Q\,dy)$</u>

Example Verify Green's theorem for the integral

$\oint_c \{(x^2 + y^2)\,dx + (x + 2y)\,dy\}$ taken round the boundary curve c defined by

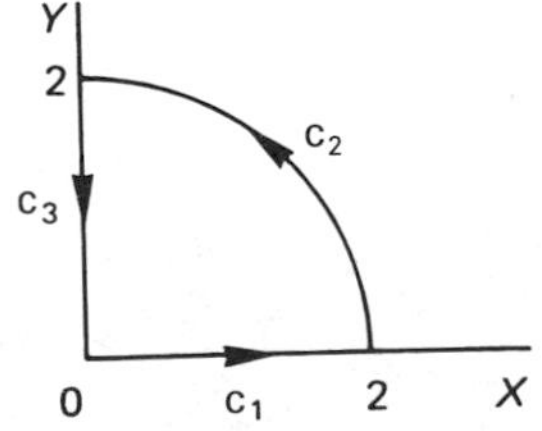

$$y = 0 \qquad 0 \le x \le 2$$
$$x^2 + y^2 = 4 \qquad 0 \le x \le 2$$
$$x = 0 \qquad 0 \le y \le 2.$$

Green's theorem: $\int_S\int\left(\frac{\partial Q}{\partial x} - \frac{\partial P}{\partial y}\right) dx\,dy = \oint_c (P\,dx + Q\,dy)$

In this case $(x^2 + y^2)\,dx + (x + 2y)\,dy = P\,dx + Q\,dy$

$$\therefore P = x^2 + y^2 \quad \text{and} \quad Q = x + 2y$$

We now take c_1, c_2, c_3 in turn.

(i) c_1: $y = 0$; $dy = 0$

$$\therefore \int_{c_1} (P\,dx + Q\,dy) = \int_0^2 x^2\,dx = \left[\frac{x^3}{3}\right]_0^2 = \underline{\frac{8}{3}}$$

(ii) c_2: $x^2 + y^2 = 4 \quad \therefore y^2 = 4 - x^2 \quad \therefore y = (4 - x^2)^{1/2}$

$$x + 2y = x + 2(4 - x^2)^{1/2}$$

$$dy = \frac{1}{2}(4 - x^2)^{-1/2}(-2x)\,dx = \frac{-x}{\sqrt{4 - x^2}}\,dx$$

$$\therefore \int_{c_2} (P\,dx + Q\,dy) = \ldots\ldots\ldots$$

Make any necessary substitutions and evaluate the line integral for c_2.

82

$$\boxed{\pi - 4}$$

for we have

$$\int_{c_2} (P\,dx + Q\,dy) = \int_{c_2} \left\{4 + (x + 2\sqrt{4 - x^2})\left(\frac{-x}{\sqrt{4 - x^2}}\right)\right\} dx$$

$$= \int_{c_2} \left\{4 - 2x - \frac{x^2}{\sqrt{4 - x^2}}\right\} dx$$

Putting $x = 2\sin\theta$, $\quad \sqrt{4 - x^2} = 2\cos\theta \quad dx = 2\cos\theta\,d\theta$

$$\text{Limits}: x = 2,\ \theta = \frac{\pi}{2}; \quad x = 0,\ \theta = 0.$$

$$\therefore \int_{c_2} (P\,dx + Q\,dy) = \int_{\pi/2}^{0} \left\{4 - 4\sin\theta - \frac{4\sin^2\theta}{2\cos\theta}\right\} 2\cos\theta\,d\theta$$

$$= 4\left[2\sin\theta - \sin^2\theta - \frac{1}{2}\left(\theta - \frac{\sin 2\theta}{2}\right)\right]_{\pi/2}^{0}$$

$$= 4\left[-\left(2 - 1 - \frac{\pi}{4}\right)\right] = \underline{\pi - 4}$$

Finally

(iii) c_3 : $x = 0$; $\quad dx = 0$

$$\therefore \int_{c_3} (P\,dx + Q\,dy) = \int_2^0 2y\,dy = \left[y^2\right]_2^0 = \underline{-4}$$

∴ Collecting our three partial results

$$\oint_c (P\,dx + Q\,dy) = \frac{8}{3} + \pi - 4 - 4 = \underline{\pi - \frac{16}{3}} \qquad (1)$$

That is one part done. Now we have to evaluate $\int_S\int \left(\frac{\partial Q}{\partial x} - \frac{\partial P}{\partial y}\right) dx\,dy$

$$P = x^2 + y^2 \qquad \therefore \frac{\partial P}{\partial y} = 2y$$

$$Q = x + 2y \qquad \therefore \frac{\partial Q}{\partial x} = 1$$

$$\therefore \int_S\int \left(\frac{\partial Q}{\partial x} - \frac{\partial P}{\partial y}\right) dx\,dy = \int_S\int (1 - 2y)\,dy\,dx$$

It will be more convenient to work in polar coordinates, so we make the substitutions

$$x = r\cos\theta; \quad y = r\sin\theta; \quad dS = dx\,dy = r\,dr\,d\theta$$

$$\therefore \int_S\int \left(\frac{\partial Q}{\partial x} - \frac{\partial P}{\partial y}\right) dx\,dy = \int_0^{\pi/2}\int_0^2 (1 - 2r\sin\theta)r\,dr\,d\theta$$

$$= \ldots\ldots\ldots \quad \text{Complete it.}$$

83

$$\boxed{\pi - \frac{16}{3}}$$

Here it is:

$$\int_S\int\left(\frac{\partial Q}{\partial x} - \frac{\partial P}{\partial y}\right) \mathrm{d}x\,\mathrm{d}y = \int_0^{\pi/2}\int_0^2 (r - 2r^2 \sin\theta)\, \mathrm{d}r\,\mathrm{d}\theta$$

$$= \int_0^{\pi/2}\left[\frac{r^2}{2} - \frac{2r^3}{3}\sin\theta\right]_0^2 \mathrm{d}\theta$$

$$= \int_0^{\pi/2}\left\{2 - \frac{16}{3}\sin\theta\right\} \mathrm{d}\theta$$

$$= \left[2\theta + \frac{16}{3}\cos\theta\right]_0^{\pi/2} = \underline{\pi - \frac{16}{3}} \qquad (2)$$

So we have established once again that

$$\underline{\oint_c (P\,\mathrm{d}x + Q\,\mathrm{d}y) = \int_S\int\left(\frac{\partial Q}{\partial x} - \frac{\partial P}{\partial y}\right) \mathrm{d}x\,\mathrm{d}y}$$

And that brings us to the end of this particular programme. We have covered a number of important sections, so check carefully down the Revision Summary and then work through the Test Exercise that follows.

84

REVISION SUMMARY

1. *Line integrals*

   (a) Scalar field V: $\int_c V\,\mathrm{d}\mathbf{r}$

   The curve c is expressed in parametric form.

   $\mathrm{d}\mathbf{r} = \mathbf{i}\,\mathrm{d}x + \mathbf{j}\,\mathrm{d}y + \mathbf{k}\,\mathrm{d}z$

   (b) Vector field $\mathbf{F}(\mathbf{r})$: $\int_c \mathbf{F}(\mathbf{r}) \cdot \mathrm{d}\mathbf{r}$

$$\mathbf{F}(\mathbf{r}) = F_x\mathbf{i} + F_y\mathbf{j} + F_z\mathbf{k}$$

$$\mathrm{d}\mathbf{r} = \mathbf{i}\,\mathrm{d}x + \mathbf{j}\,\mathrm{d}y + \mathbf{k}\,\mathrm{d}z$$

$$\mathbf{F} \cdot \mathrm{d}\mathbf{r} = F_x\,\mathrm{d}x + F_y\,\mathrm{d}y + F_z\,\mathrm{d}z$$

2. *Volume integrals*

   $\mathbf{F}$ is a vector field; V a closed region with boundary surface S.

$$\int_V \mathbf{F}\,\mathrm{d}V = \int_{x_1}^{x_2}\int_{y_1}^{y_2}\int_{z_1}^{z_2} \mathbf{F}\,\mathrm{d}z\,\mathrm{d}y\,\mathrm{d}x$$

3. *Surface integrals*

   (a) Scalar field $V(x, y, z)$:

$$\int_S V\,\mathrm{d}\mathbf{S} = \int_S V\hat{\mathbf{n}}\,\mathrm{d}S; \qquad \hat{\mathbf{n}} = \frac{\nabla S}{|\nabla S|} = \frac{\text{grad } S}{|\text{grad } S|}$$

   (b) Vector field $\mathbf{F} = F_x\mathbf{i} + F_y\mathbf{j} + F_z\mathbf{k}$:

$$\int_S \mathbf{F} \cdot \mathrm{d}\mathbf{S} = \int_S \mathbf{F} \cdot \hat{\mathbf{n}}\,\mathrm{d}S; \qquad \hat{\mathbf{n}} = \frac{\nabla S}{|\nabla S|}$$

4. *Polar coordinates*

(a) Plane polar coordinates (r, θ)

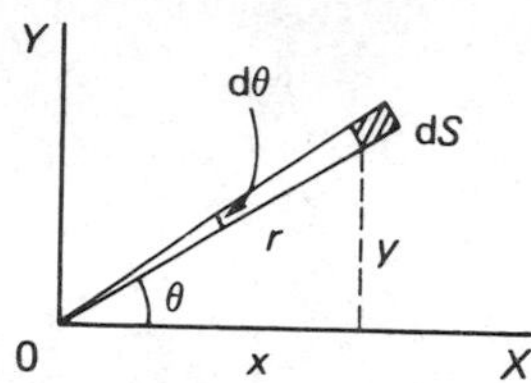

$$x = r\cos\theta; \qquad y = r\sin\theta$$
$$dS = r\,dr\,d\theta$$

(b) Cylindrical polar coordinates (ρ, ϕ, z)

$$x = \rho\cos\phi; \qquad y = \rho\sin\phi; \qquad z = z$$

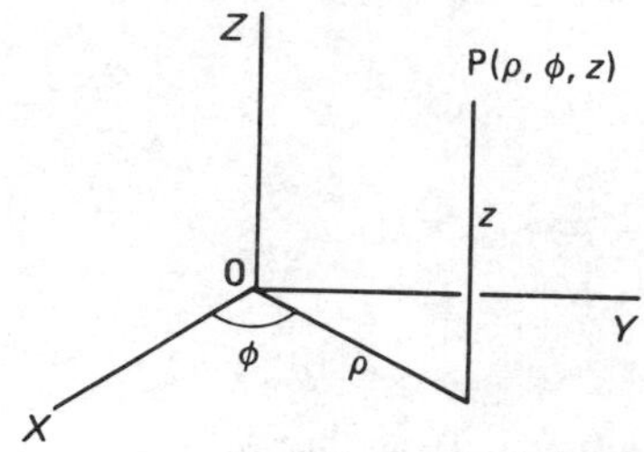

$$dS = \rho\,d\phi\,dz$$
$$dV = \rho\,d\rho\,d\phi\,dz$$

(c) Spherical polar coordinates (r, θ, ϕ)

$$x = r\sin\theta\cos\phi; \qquad y = r\sin\theta\sin\phi$$
$$z = r\cos\theta$$

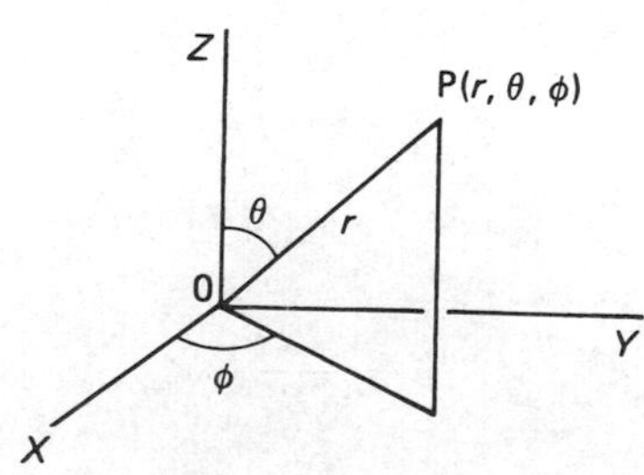

$$dS = r^2\sin\theta\,d\theta\,d\phi$$
$$dV = r^2\sin\theta\,dr\,d\theta\,d\phi$$

5. *Conservative vector fields*

A vector field **F** is conservative if

(a) $\oint_c \mathbf{F}\cdot d\mathbf{r} = 0$

(b) curl $\mathbf{F} = 0$

(c) $\mathbf{F} = \text{grad}\,V$ where V is a scalar.

6. *Divergence theorem* (Gauss' theorem)

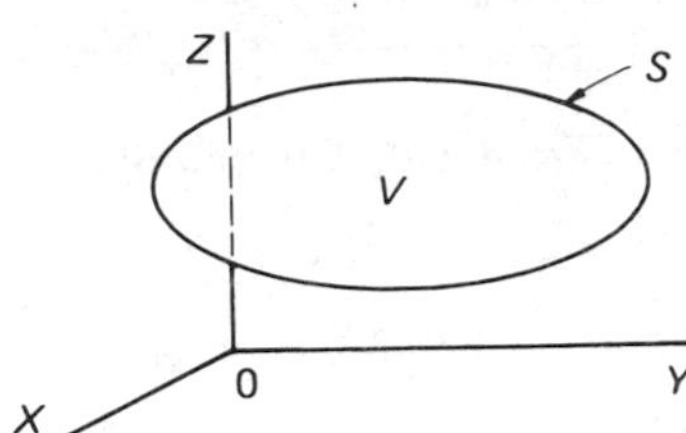

Closed surface S enclosing a region V in a vector field $\mathbf{F}$.

$$\int_V \operatorname{div} \mathbf{F} \, \mathrm{d}V = \int_S \mathbf{F} \cdot \mathrm{d}\mathbf{S}$$

7. *Stokes' theorem*

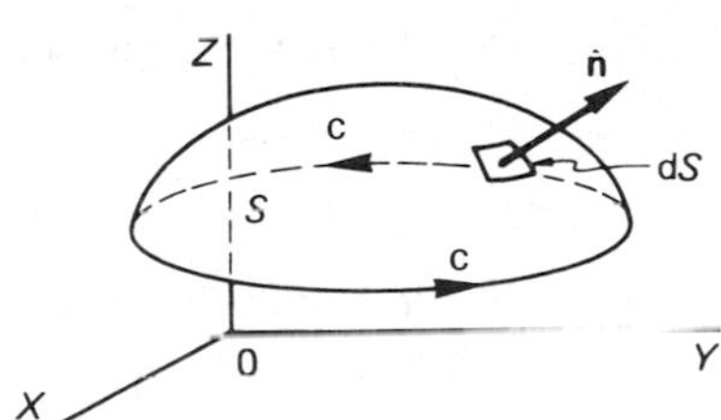

An open surface S bounded by a simple closed curve c, then

$$\int_S \operatorname{curl} \mathbf{F} \cdot \mathrm{d}\mathbf{S} = \oint_c \mathbf{F} \cdot \mathrm{d}\mathbf{r}$$

8. *Green's theorem*

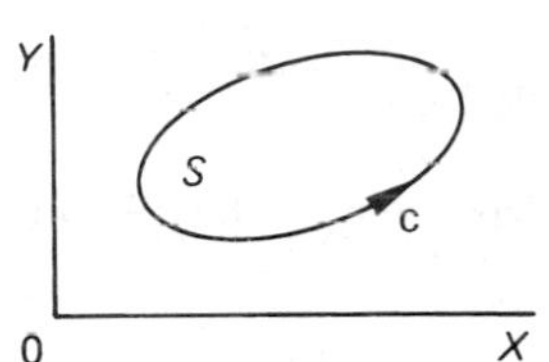

The curve c is a simple closed curve enclosing a plane space S in the xy-plane. P and Q are functions of both x and y.

$$\text{Then } \int_S \int \left(\frac{\partial Q}{\partial x} - \frac{\partial P}{\partial y} \right) \mathrm{d}x \, \mathrm{d}y = \oint_c (P \, \mathrm{d}x + Q \, \mathrm{d}y).$$

85 TEST EXERCISE XIII

1. If $V = x^3y + 2xy^2 + yz$, evaluate $\int_c V\,d\mathbf{r}$ between A(0, 0, 0) and B(2, 1, –3) along the curve with parametric equations $x = 2t$, $y = t^2$, $z = -3t^3$.
2. If $\mathbf{F}(\mathbf{r}) = x^2y^3\,\mathbf{i} + yz^2\,\mathbf{j} + zx^2\,\mathbf{k}$, evaluate $\int_c \mathbf{F}\cdot d\mathbf{r}$ along the curve $x = 3u^2$, $y = u$, $z = 2u^3$ between A(3, –1, –2) and B(3, 1, 2).
3. Evaluate $\int_V \mathbf{F}\,dV$ where $\mathbf{F} = 3\mathbf{i} + z\mathbf{j} + 2y\mathbf{k}$ and V is the region bounded by the planes $z = 0$, $z = 3$ and the surface $x^2 + y^2 = 4$.
4. If V is the scalar field $V = xyz^2$, evaluate $\int_S V\,d\mathbf{S}$ over the surface S defined by $x^2 + y^2 = 9$ between $z = 0$ and $z = 2$ in the first octant.
5. Evaluate $\int_S \mathbf{F}\cdot d\mathbf{S}$ over the surface S defined by $x^2 + y^2 + z^2 = 4$ for $z \geq 0$ and bounded by $x = 0$, $y = 0$, $z = 0$ in the first octant. $\mathbf{F} = x\mathbf{i} + 2z\mathbf{j} + y\mathbf{k}$.
6. Determine which of the following vector fields are conservative.
   (a) $\mathbf{F} = (2xy + z)\mathbf{i} + (x^2 + 2yz)\mathbf{j} + (x + y^2)\mathbf{k}$
   (b) $\mathbf{F} = (yz + 2y)\mathbf{i} + (xz + 2x)\mathbf{j} + (xy + 3)\mathbf{k}$
   (c) $\mathbf{F} = (yz^2 + 3)\mathbf{i} + (xz^2 + 2)\mathbf{j} + (2xyz + 4)\mathbf{k}$.
7. By the use of the divergence theorem, determine $\int_S \mathbf{F}\cdot d\mathbf{S}$ where $\mathbf{F} = x\mathbf{i} + xy\mathbf{j} + 2\mathbf{k}$, taken over the region bounded by the planes $z = 0$, $z = 4$, $x = 0$, $y = 0$ and the surface $x^2 + y^2 = 9$ in the first octant.
8. A surface consists of the planes $x = 0$, $x = 2$, $y = 0$, $y = 2$ and $z = 3 - y$. Apply Stokes' theorem to evaluate $\int_S \text{curl}\,\mathbf{F}\cdot d\mathbf{S}$ over the surface where $\mathbf{F} = 2x\mathbf{i} + xz\mathbf{j} + yz\mathbf{k}$.

FURTHER PROBLEMS XIII

1. If $V = x^2yz$, evaluate $\int_c V\,\mathbf{dr}$ between A(0, 0, 0) and B(6, 2, 4)

   (a) along the straight lines c_1: (0, 0, 0) to (6, 0, 0)

   c_2: (6, 0, 0) to (6, 2, 0)

   c_3: (6, 2, 0) to (6, 2, 4)

   (b) along the path c_4 having parametric equations

   $$x = 3t,\; y = t,\; z = 2t.$$

2. If $V = xy^2 + yz$, evaluate $\int_c V\,\mathbf{dr}$ along the curve c having parametric equations $x = 2t^2$, $y = 4t$, $z = 3t + 5$ between A(0, 0, 5) and B(8, 8, 11).

3. Evaluate the integral $\int_c (xyz + 4x^2y)\,\mathbf{dr}$ along the curve c with parametric equations $x = 2u$, $y = u^2$, $z = 3u^3$ between A(2, 1, 3) and B(4, 4, 24).

4. If $\mathbf{F}(\mathbf{r}) = xy\mathbf{i} + yz\mathbf{j} + 3xyz\mathbf{k}$, evaluate $\int_c \mathbf{F}\cdot\mathbf{dr}$ between A(0, 2, 0) and B(3, 6, 1) where c has the parametric equations $x = 3u$, $y = 4u + 2$, $z = u^2$.

5. $\mathbf{F}(\mathbf{r}) = x^2\mathbf{i} - 2xy\mathbf{j} + yz\mathbf{k}$. Evaluate $\int_c \mathbf{F}\cdot\mathbf{dr}$ between A(2, 1, 2) and B(4, 4, 5) where c is the path with parametric equations $x = 2u$, $y = u^2$, $z = 3u - 1$.

6. A unit particle is moved in an anticlockwise manner round a circle with centre (0, 0, 4) and radius 2 in the plane $z = 4$ in a force field defined as $\mathbf{F} = (xy + z)\mathbf{i} + (2x + y)\mathbf{j} + (x + y + z)\mathbf{k}$. Find the work done.

7. Evaluate $\int_V \mathbf{F}\,dV$ where $\mathbf{F} = \mathbf{i} - y\mathbf{j} + \mathbf{k}$ and V is the region bounded by the plane $z = 0$ and the hemisphere $x^2 + y^2 + z^2 = 4$, for $z \geq 0$.

8. V is the region bounded by the planes $x = 0$, $y = 0$, $z = 0$ and the surfaces $y = 4 - x^2$ $(z \geq 0)$ and $y = 4 - z^2$ $(y \geq 0)$.
   If $\mathbf{F} = 2\mathbf{i} + y^2\mathbf{j} - \mathbf{k}$, evaluate $\int_V \mathbf{F}\,dV$ throughout the region.

9. If $\mathbf{F} = 3\mathbf{i} + 2\mathbf{j} - 2x\mathbf{k}$, evaluate $\int_V \mathbf{F}\,dV$ where V is the region bounded by the planes $y = 0$, $z = 0$, $z = 4 - y$ $(z \geq 0)$ and the surface $x^2 + y^2 = 16$.

10. A scalar field $V = x + y$ exists over a surface S defined by $x^2 + y^2 + z^2 = 9$, bounded by the planes $x = 0$, $y = 0$, $z = 0$ in the first octant. Evaluate $\int_S V\,\mathbf{dS}$ over the curved surface.

11. A surface S is defined by $y^2 + z = 4$ and is bounded by the planes $x = 0,\ x = 3,\ y = 0,\ z = 0$ in the first octant. Evaluate $\int_S V\,\mathbf{dS}$ over this curved surface where V denotes the scalar field $V = x^2yz$.

12. Evaluate $\int_S \text{curl}\,\mathbf{F}\cdot\mathbf{dS}$ over the surface S defined by $2x + 2y + z = 2$ and bounded by $x = 0,\ y = 0,\ z = 0$ in the first octant and where $\mathbf{F} = y^2\mathbf{i} + 2yz\mathbf{j} + xy\mathbf{k}$.

13. Evaluate $\int_S \mathbf{F}\cdot\mathbf{dS}$ over the hemisphere defined by $x^2 + y^2 + z^2 = 25$ with $z \geq 0$, where $\mathbf{F} = (x + y)\mathbf{i} - 2z\mathbf{j} + y\mathbf{k}$.

14. A vector field $\mathbf{F} = 2x\mathbf{i} + z\mathbf{j} + y\mathbf{k}$ exists over a surface S defined by $x^2 + y^2 + z^2 = 16$, bounded by the planes $z = 0,\ z = 3,\ x = 0,\ y = 0$. Evaluate $\int_S \mathbf{F}\cdot\mathbf{dS}$ over the stated curved surface.

15. Evaluate $\int_S \mathbf{F}\cdot\mathbf{dS}$, where $\mathbf{F}$ is the vector field $x^2\mathbf{i} + 2z\mathbf{j} - y\mathbf{k}$, over the curved surface S defined by $x^2 + y^2 = 25$ and bounded by $z = 0,\ z = 6,\ y \geq 3$.

16. A region V is defined by the hemisphere $x^2 + y^2 + z^2 = 16,\ z \geq 0,\ y \geq 0$ and the planes $z = 0,\ y = 0$. A vector field $\mathbf{F} = xy\mathbf{i} + y^2\mathbf{j} + \mathbf{k}$ exists throughout and on the boundary of the region. Verify the Gauss divergence theorem for the region stated.

17. A surface consists of parts of the planes $x = 0,\ x = 1,\ y = 0,\ y = 2,\ z = 1$ in the first octant. If $\mathbf{F} = y\mathbf{i} + x^2z\mathbf{j} + xy\mathbf{k}$, verify Stokes' theorem.

18. S is the surface $z = x^2 + y^2$ bounded by the planes $z = 0$ and $z = 4$. Verify Stokes' theorem for a vector field $\mathbf{F} = xy\mathbf{i} + x^3\mathbf{j} + xz\mathbf{k}$.

19. A vector field $\mathbf{F} = xy\mathbf{i} + z^2\mathbf{j} + xyz\mathbf{k}$ exists over the surfaces $x^2 + y^2 + z^2 = a^2,\ x = 0$ and $y = 0$ in the first octant. Verify Stokes' theorem that $\int_S \text{curl}\,\mathbf{F}\cdot\mathbf{dS} = \oint_c \mathbf{F}\cdot\mathbf{dr}$.

20. A surface is defined by $z^2 = 4(x^2 + y^2)$ where $0 \leq z \leq 6$. If a vector field $\mathbf{F} = z\mathbf{i} + xy^2\mathbf{j} + x^2z\mathbf{k}$ exists over the surface and on the boundary circle c, show that $\oint_c \mathbf{F}\cdot\mathbf{dr} = \int_S \text{curl}\,\mathbf{F}\cdot\mathbf{dS}$. (Evaluate the line integral in a clockwise manner, in this instance.)

Programme 14

Vector Analysis

PART 3

Orthogonal Curvilinear Coordinates

1 **Introduction**

This short programme is an extension of the two previous ones and may not be required for all courses. It can well be bypassed without adversely affecting the rest of the work.

Curvilinear coordinates

Let us consider two variables u and v, each of which is a function of x and y

i.e $$u = f(x, y); \qquad v = g(x, y)$$

If u and v are each assigned a constant value a and b, the equations will, in general, define two intersecting curves.

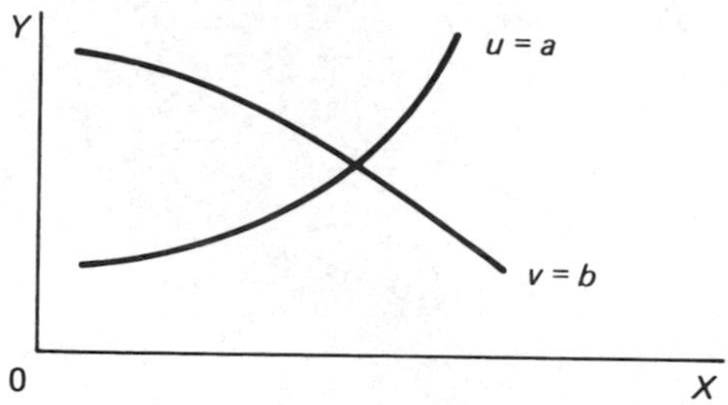

If u and v are each given several such values, the equations define a network of curves covering the xy-plane

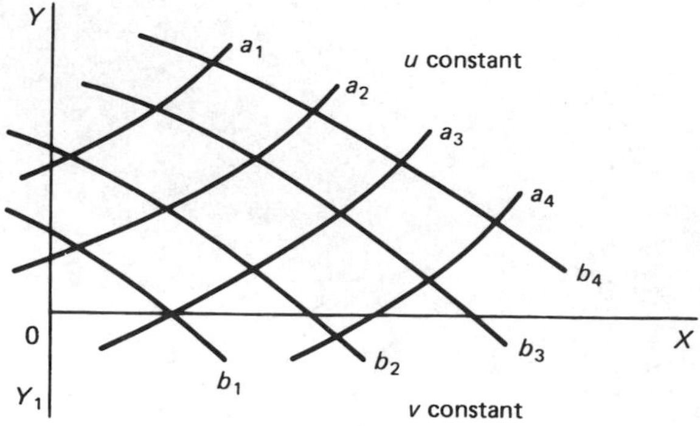

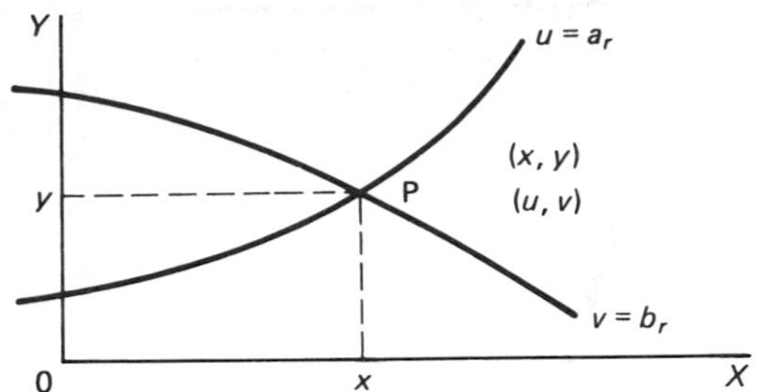

A pair of curves $u = a_r$ and $v = b_r$ pass through each point in the plane. Hence, any point in the plane can be expressed in *rectangular coordinates* (x, y) or in *curvilinear coordinates* (u, v).

Let us see how this works out in an example, so move on.

2

Example 1 Let us consider the case where $u = xy$ and $v = x^2 - y$.

(a) *With* $u = xy$, if we put $u = 4$, then $y = \dfrac{4}{x}$ and we can plot y against x to obtain the relevant curve.

Similarly, putting $u = 8,\ 16,\ 32,\ \ldots$ we can build up a family of curves, all of the pattern $u = xy$.

x		0.5	1.0	2.0	3.0	4.0
y	$u = 4$	8	4	2	1.33	1.0
	$u = 8$	16	8	4	2.67	2
	$u = 16$	32	16	8	5.33	4
	$u = 32$	64	32	16	10.67	8

If we plot these on graph paper between $x = 0$ and $x = 4$ with a range of y from $y = 0$ to $y = 20$, we obtain

. . . .

3

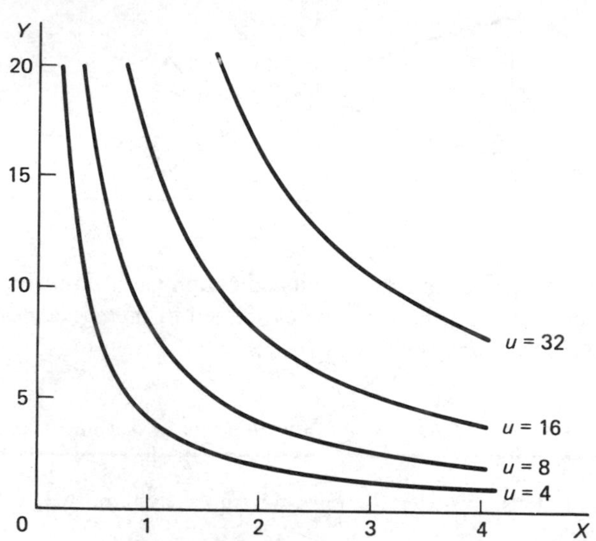

Note that each graph is labelled with its individual u-value.

(b) *With* $v = x^2 - y$, we proceed in just the same way. We rewrite the equation as $y = x^2 - v$; assign values such as 8, 4, 0, –4, –8, –12, –16 ... to v; and draw the relevant curve in each case. If we do that for $x = 0$ to $x = 4$ and limit the y-values to the range $y = 0$ to $y = 20$, we obtain the family of curves

....

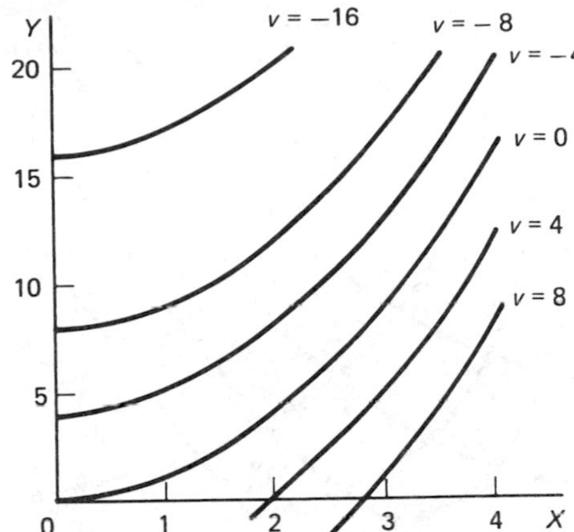

The table of function values is as follows.

x		0	1	2	3	4
y	$v = 8$	−8	−7	−4	1	8
	$v = 4$	−4	−3	0	5	12
	$v = 0$	0	1	4	9	16
	$v = -4$	4	5	8	13	20
	$v = -8$	8	9	12	17	24
	$v = -12$	12	13	16	21	28
	$v = -16$	16	17	20	25	32

Note again that we label each graph with its own v-value.

This again is a family of curves with the common pattern $v = x^2 - y$, the members being distinguished from each other by the value assigned to v in each case.

Now we draw both sets of curves on a common set of xy-axes, taking

the range of x from $x = 0$ to $x = 4$

and the range of y from $y = 0$ to $y = 20$.

It is worthwhile taking a little time over it—and good practice!

When you have the complete picture, move on to the next frame.

5

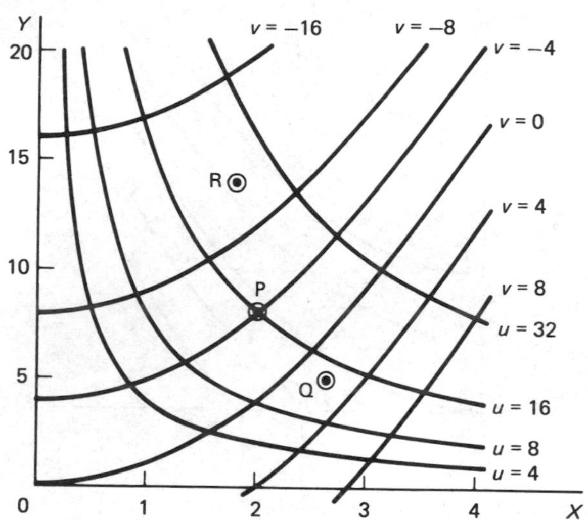

The position of any point in the plane can now be stated in two ways. For example, the point P has cartesian rectangular coordinates $x = 2,\ y = 8$. It can also be stated in curvilinear coordinates $u = 16,\ v = -4$, for it is at the point of intersection of the two curves corresponding to $u = 16$ and $v = -4$.

Likewise, for the point Q, the position in rectangular coordinates is $x = 2.65,\ y = 5.0$ and for its position in curvilinear coordinates we must estimate it within the network. Approximate values are $u = 13,\ v = 2$.

Similarly, the curvilinear coordinates of $R(x = 1.8,\ y = 14)$ are approximately $u = \dots\dots\dots$; $v = \dots\dots\dots$

6

$$u = 26; \quad v = -11$$

Their actual values are in fact $u = 25.2$ and $v = -10.76$.

Now let us deal with another example.

7

Example 2 If $u = x^2 + 2y$ and $v = y - (x+1)^2$, these can be re-written as $y = \frac{1}{2}(u - x^2)$ and $y = v + (x+1)^2$. We can now plot the family of curves, say between $x = 0$ and $x = 4$, with $u = 5(5)30$ and $v = -20(5)5$, i.e. values of u from 5 to 30 at intervals of 5 units and values of v from −20 to 5 at intervals of 5 units.

The resulting network is easily obtained and appears as

....

8

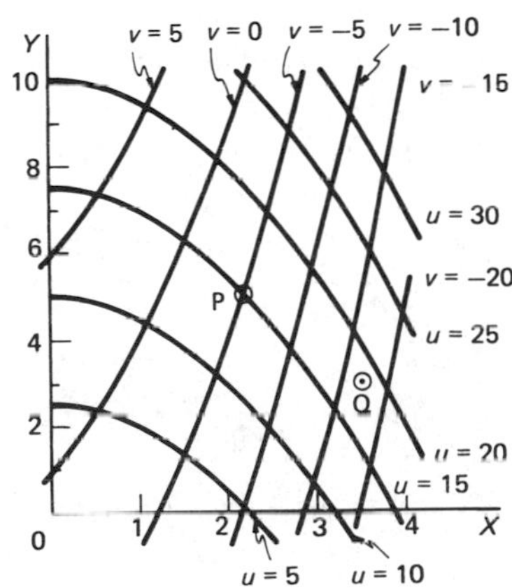

For P, the rectangular coordinates are $(x = 2.18,\ y = 5.1)$

and the curvilinear coordinates are $(u = 15,\ v = -5)$.

For Q, the rectangular coordinates are ....

and the curvilinear coordinates are ....

9

Q: $(x = 3.5,\ y = 3.0)$; $(u = 18.5;\ v = -17)$

Orthogonal curvilinear coordinates

If the coordinate curves for u and v forming the network cross at right angles, the system of curvilinear coordinates is said to be *orthogonal*.

Let us extend these ideas to three dimensions. Move on.

10 Orthogonal coordinate systems in space

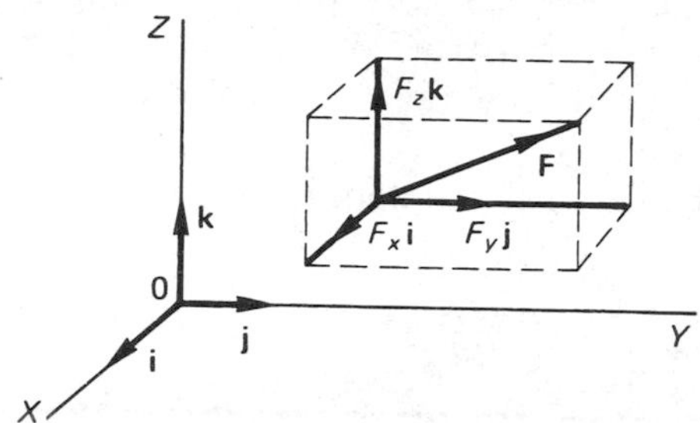

Any vector **F** can be expressed in terms of its components in three mutually perpendicular directions, which have normally been the directions of the coordinate axes, i.e.

$$\mathbf{F} = F_x\mathbf{i} + F_y\mathbf{j} + F_z\mathbf{k}$$

where **i, j, k** are the unit vectors parallel to the x, y, z axes respectively.

Situations can arise, however, where the directions of the unit vectors do not remain fixed, but vary from point to point in space according to prescribed conditions. Examples of this occur in cylindrical and spherical polar coordinates, with which we are already familiar.

1. *Cylindrical polar coordinates* (ρ, ϕ, z)

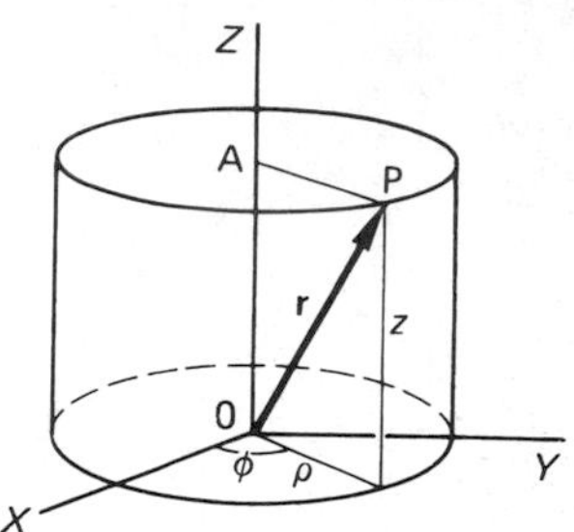

Let P be a point with cylindrical coordinates (ρ, ϕ, z) as shown. The position of P is a function of the three variables ρ, ϕ, z.

(a) If ϕ and z remain constant and ρ varies, then P will move out along AP by an amount $\dfrac{\partial \mathbf{r}}{\partial \rho}$ and the unit vector **I** in this direction will be given by

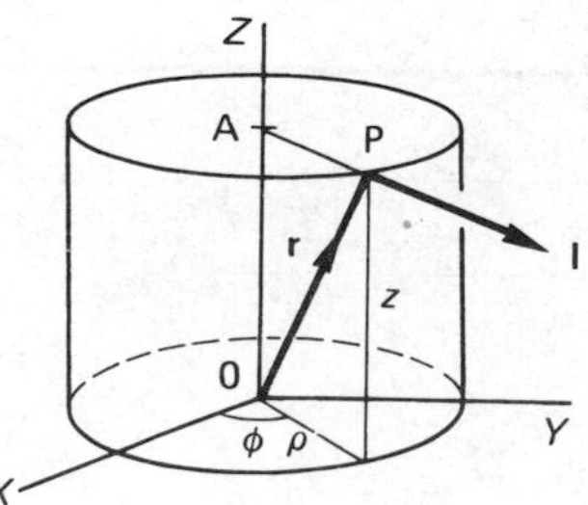

$$\mathbf{I} = \frac{\partial \mathbf{r}}{\partial \rho} \Big/ \left|\frac{\partial \mathbf{r}}{\partial \rho}\right|$$

(b) If, instead, ρ and z remain constant and ϕ varies, P will move

. . . .

round the circle with AP as radius

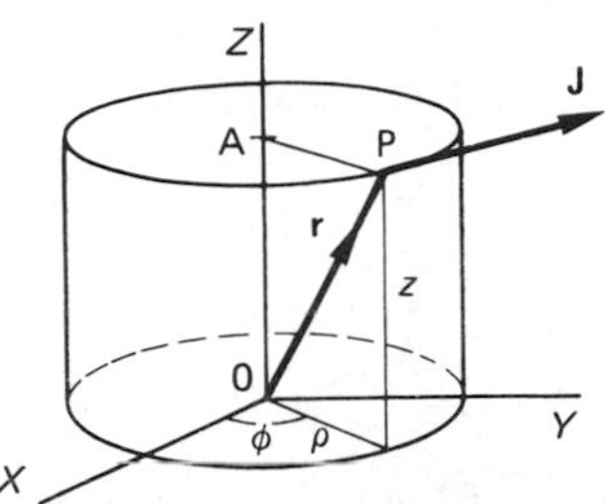

$\dfrac{\partial \mathbf{r}}{\partial \phi}$ is therefore a vector along the tangent to the circle at P and the unit vector **J** at P will be given by

$$\mathbf{J} = \frac{\partial \mathbf{r}}{\partial \phi} \Big/ \left| \frac{\partial \mathbf{r}}{\partial \phi} \right|$$

(c) Finally, if ρ and ϕ remain constant and z increases, the vector $\dfrac{\partial \mathbf{r}}{\partial z}$ will be parallel to the z-axis and the unit vector **K** in this direction will be given by

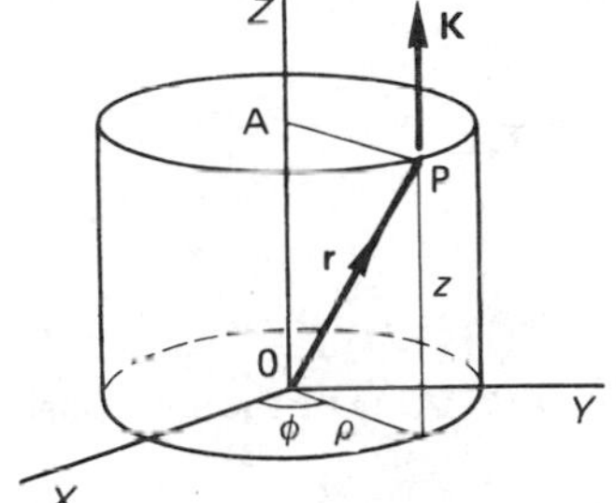

$$\mathbf{K} = \frac{\partial \mathbf{r}}{\partial z} \Big/ \left| \frac{\partial \mathbf{r}}{\partial z} \right|$$

Putting our three unit vectors on to one diagram, we have

....

12

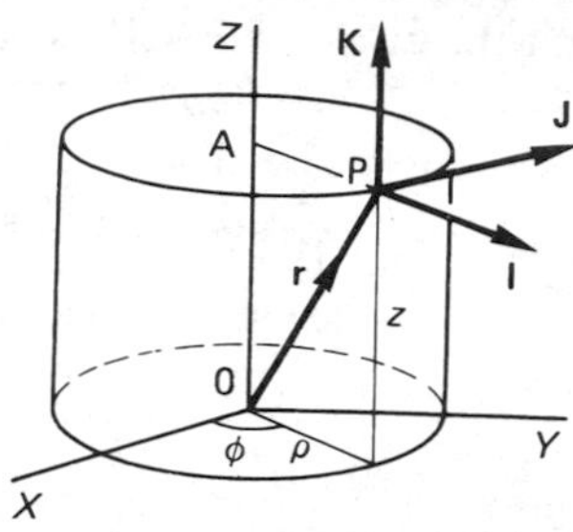

Note that **I, J, K** are mutually perpendicular and form a right-handed set. But note also that, unlike the unit vectors **i, j, k** in the cartesian system, the unit vectors **I, J, K**, or *base vectors* as they are called, are not fixed in directions, but change as the position of P changes.

So we have, for cylindrical polar coordinates

$$\mathbf{I}=\frac{\partial \mathbf{r}}{\partial \rho}\Bigg/\left|\frac{\partial \mathbf{r}}{\partial \rho}\right|;\quad \mathbf{J}=\frac{\partial \mathbf{r}}{\partial \phi}\Bigg/\left|\frac{\partial \mathbf{r}}{\partial \phi}\right|;\quad \mathbf{K}=\frac{\partial \mathbf{r}}{\partial z}\Bigg/\left|\frac{\partial \mathbf{r}}{\partial z}\right|$$

If **F(r)** is a vector associated with P, then $\mathbf{F(r)} = F_\rho\mathbf{I}+F_\phi\mathbf{J}+F_z\mathbf{K}$ where F_ρ, F_ϕ, F_z are the components of **F** in the directions of the unit base vectors $\mathbf{I}, \mathbf{J}, \mathbf{K}$.

Now let us attend to spherical coordinates in the same way.

2. *Spherical polar coordinates* (r, θ, ϕ)

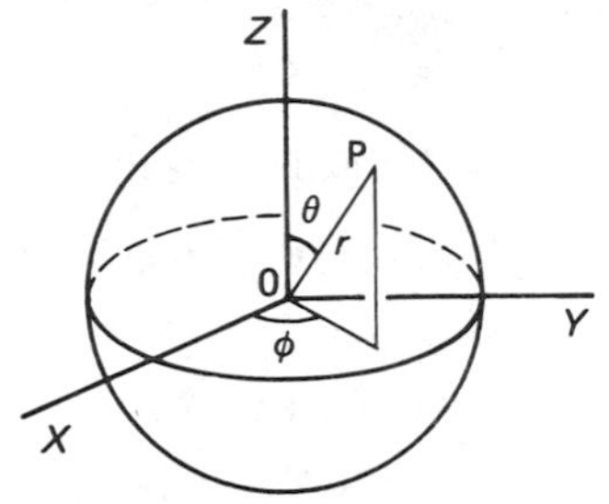

P is a function of the three variables r, θ, ϕ.

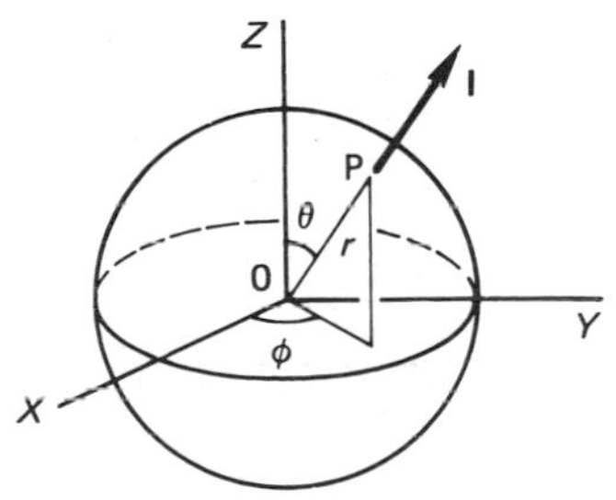

(a) If θ and ϕ remain constant and r increases, P moves outwards in the direction OP. $\dfrac{\partial \mathbf{r}}{\partial r}$ is thus a vector normal to the surface of the sphere at P and the unit vector **I** in that direction is therefore

$$\mathbf{I} = \frac{\partial \mathbf{r}}{\partial r} \bigg/ \left|\frac{\partial \mathbf{r}}{\partial r}\right|$$

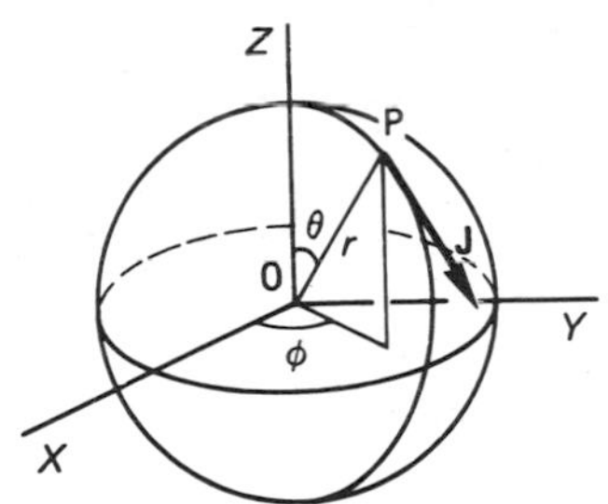

(b) If r and ϕ remain constant and θ increases, P will move along the 'meridian' through P, i.e. $\dfrac{\partial \mathbf{r}}{\partial \theta}$ is a tangent vector to this circle at P and the unit vector **J** is given by

$$\mathbf{J} = \frac{\partial \mathbf{r}}{\partial \theta} \bigg/ \left|\frac{\partial \mathbf{r}}{\partial \theta}\right|$$

(c) If r and θ remain constant and ϕ increases, P will move

. . . .

14

along the circle through P perpendicular to the z-axis

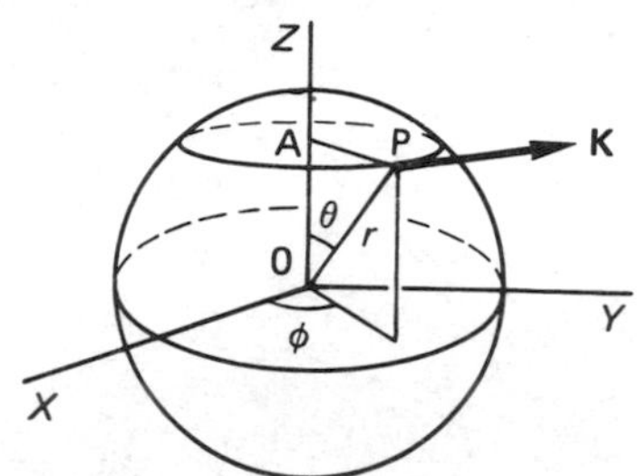

$\dfrac{\partial \mathbf{r}}{\partial \phi}$ is therefore a tangent vector at P and the unit vector **K** in this direction is given by

$$\mathbf{K} = \frac{\partial \mathbf{r}}{\partial \phi} \bigg/ \left| \frac{\partial \mathbf{r}}{\partial \phi} \right|$$

So, putting the three results on one diagram, we have

....

15

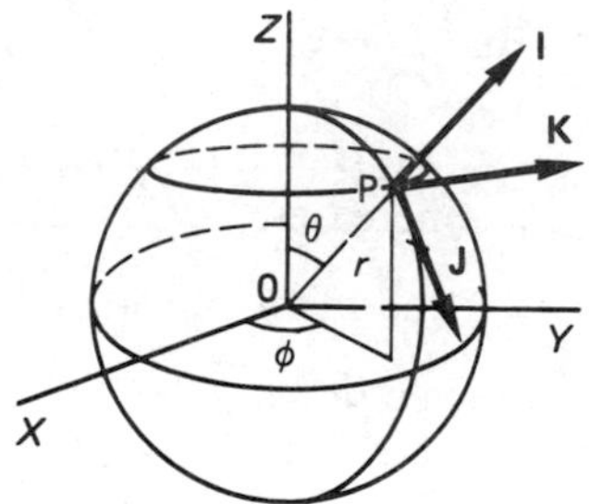

Once again, the three unit vectors at **P** (base vectors) are mutually perpendicular and form a right-handed set. Their directions in space, however, change as the position of **P** changes.

A vector $\mathbf{F}(\mathbf{r})$ associated with P can therefore be expressed as $\mathbf{F}(\mathbf{r}) = F_r\mathbf{I} + F_\theta\mathbf{J} + F_\phi\mathbf{K}$ where F_r, F_θ, F_ϕ are the components of **F** in the directions of the base vectors **I, J, K**.

Both cylindrical and spherical polar coordinate systems are

....

16

orthogonal

Scale factors

Collecting the recent results together, we have:

1. For cylindrical polar coordinates, the unit base vectors are

$$\mathbf{I} = \frac{\partial \mathbf{r}}{\partial \rho} \bigg/ \left|\frac{\partial \mathbf{r}}{\partial \rho}\right| = \frac{1}{h_\rho}\frac{\partial \mathbf{r}}{\partial \rho} \qquad \text{where } h_\rho = \left|\frac{\partial \mathbf{r}}{\partial \rho}\right|$$

$$\mathbf{J} = \frac{\partial \mathbf{r}}{\partial \phi} \bigg/ \left|\frac{\partial \mathbf{r}}{\partial \phi}\right| = \frac{1}{h_\phi}\frac{\partial \mathbf{r}}{\partial \phi} \qquad \text{where } h_\phi = \left|\frac{\partial \mathbf{r}}{\partial \phi}\right|$$

$$\mathbf{K} = \frac{\partial \mathbf{r}}{\partial z} \bigg/ \left|\frac{\partial \mathbf{r}}{\partial z}\right| = \frac{1}{h_z}\frac{\partial \mathbf{r}}{\partial z} \qquad \text{where } h_z = \left|\frac{\partial \mathbf{r}}{\partial z}\right|$$

2. For spherical polar coordinates, the unit base vectors are

$$\mathbf{I} = \frac{\partial \mathbf{r}}{\partial r} \bigg/ \left|\frac{\partial \mathbf{r}}{\partial r}\right| = \frac{1}{h_r}\frac{\partial \mathbf{r}}{\partial r} \qquad \text{where } h_r = \left|\frac{\partial \mathbf{r}}{\partial r}\right|$$

$$\mathbf{J} = \frac{\partial \mathbf{r}}{\partial \theta} \bigg/ \left|\frac{\partial \mathbf{r}}{\partial \theta}\right| = \frac{1}{h_\theta}\frac{\partial \mathbf{r}}{\partial \theta} \qquad \text{where } h_\theta = \left|\frac{\partial \mathbf{r}}{\partial \theta}\right|$$

$$\mathbf{K} = \frac{\partial \mathbf{r}}{\partial \phi} \bigg/ \left|\frac{\partial \mathbf{r}}{\partial \phi}\right| = \frac{1}{h_\phi}\frac{\partial \mathbf{r}}{\partial \phi} \qquad \text{where } h_\phi = \left|\frac{\partial \mathbf{r}}{\partial \phi}\right|$$

In each case, h is called the *scale factor*.

Move on.

17

Scale factors for coordinate systems

1. *Rectangular coordinates* (x, y, z)

With rectangular coordinates, $h_x = h_y = h_z = 1$.

2. *Cylindrical coordinates* (ρ, ϕ, z)

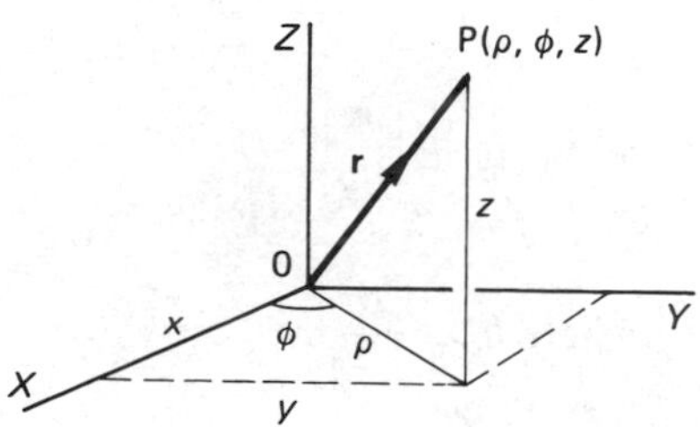

$$\mathbf{r} = x\mathbf{i} + y\mathbf{j} + z\mathbf{k}$$
$$x = \rho \cos \phi$$
$$y = \rho \sin \phi$$
$$z = z$$

$$\therefore \mathbf{r} = \rho \cos \phi \, \mathbf{i} + \rho \sin \phi \, \mathbf{j} + z\mathbf{k}$$

$$\mathbf{I} = \frac{\partial \mathbf{r}}{\partial \rho} \bigg/ \left|\frac{\partial \mathbf{r}}{\partial \rho}\right| = \frac{1}{h_\rho}\frac{\partial \mathbf{r}}{\partial \rho} \qquad h_\rho = \left|\frac{\partial \mathbf{r}}{\partial \rho}\right| = |\cos \phi \, \mathbf{i} + \sin \phi \, \mathbf{j}|$$
$$= (\cos^2\phi + \sin^2\phi)^{1/2} = 1 \quad \therefore \underline{h_\rho = 1}$$

$$\mathbf{J} = \frac{\partial \mathbf{r}}{\partial \phi} \bigg/ \left|\frac{\partial \mathbf{r}}{\partial \phi}\right| = \frac{1}{h_\phi}\frac{\partial \mathbf{r}}{\partial \phi} \qquad h_\phi = \left|\frac{\partial \mathbf{r}}{\partial \phi}\right| = |-\rho \sin \phi \, \mathbf{i} + \rho \cos \phi \, \mathbf{j}|$$
$$= (\rho^2 \sin^2 \phi + \rho^2 \cos^2 \phi)^{1/2} = \rho \quad \therefore \underline{h_\phi =}$$

$$\mathbf{K} = \frac{\partial \mathbf{r}}{\partial z} \bigg/ \left|\frac{\partial \mathbf{r}}{\partial z}\right| = \frac{1}{h_z}\frac{\partial \mathbf{r}}{\partial z} \qquad h_z = \left|\frac{\partial \mathbf{r}}{\partial z}\right| = |\mathbf{k}| = 1 \qquad \therefore \underline{h_z = 1}$$

$$\therefore \underline{h_\rho = 1; \;\; h_\phi = \rho; \;\; h_z = 1}$$

3. *Spherical coordinates* (r, θ, ϕ)

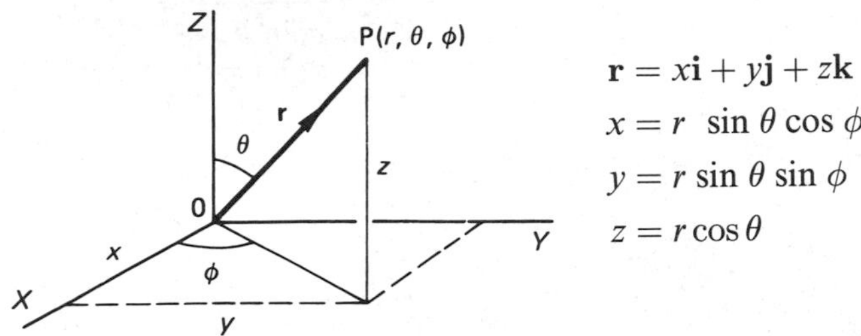

$$\mathbf{r} = x\mathbf{i} + y\mathbf{j} + z\mathbf{k}$$
$$x = r\sin\theta\cos\phi$$
$$y = r\sin\theta\sin\phi$$
$$z = r\cos\theta$$

$$\therefore\ \mathbf{r} = r\sin\theta\cos\phi\,\mathbf{i} + r\sin\theta\sin\phi\,\mathbf{j} + r\cos\theta\,\mathbf{k}$$

Then working as before

$$h_r = \ldots\ldots\ldots;\quad h_\theta = \ldots\ldots\ldots;\quad h_\phi = \ldots\ldots\ldots$$

18

$$\boxed{h_r = 1; \quad h_\theta = r; \quad h_\phi = r\sin\theta}$$

for $\quad \mathbf{r} = r\sin\theta\cos\phi\,\mathbf{i} + r\sin\theta\sin\phi\,\mathbf{j} + r\cos\theta\,\mathbf{k}$

$$\mathbf{I} = \frac{\partial \mathbf{r}}{\partial r} \Big/ \left|\frac{\partial \mathbf{r}}{\partial r}\right| = \frac{1}{h_r}\frac{\partial \mathbf{r}}{\partial r}$$

$$\begin{aligned} h_r = \left|\frac{\partial \mathbf{r}}{\partial r}\right| &= |\sin\theta\cos\phi\,\mathbf{i} + \sin\theta\sin\phi\,\mathbf{j} + \cos\theta\,\mathbf{k}| \\ &= (\sin^2\theta\cos^2\phi + \sin^2\theta\sin^2\phi + \cos^2\theta)^{1/2} \\ &= (\sin^2\theta + \cos^2\theta)^{1/2} = 1 \qquad \therefore \underline{h_r = 1} \end{aligned}$$

$$\mathbf{J} = \frac{\partial \mathbf{r}}{\partial \theta} \Big/ \left|\frac{\partial \mathbf{r}}{\partial \theta}\right| = \frac{1}{h_\theta}\frac{\partial \mathbf{r}}{\partial \theta}$$

$$\begin{aligned} h_\theta = \left|\frac{\partial \mathbf{r}}{\partial \theta}\right| &= |r\cos\theta\cos\phi\,\mathbf{i} + r\cos\theta\sin\phi\,\mathbf{j} - r\sin\theta\,\mathbf{k}| \\ &= (r^2\cos^2\theta\cos^2\phi + r^2\cos^2\theta\sin^2\phi + r^2\sin^2\theta)^{1/2} \\ &= (r^2\cos^2\theta + r^2\sin^2\theta)^{1/2} = r \qquad \therefore \underline{h_\theta = r} \end{aligned}$$

$$\mathbf{K} = \frac{\partial \mathbf{r}}{\partial \phi} \Big/ \left|\frac{\partial \mathbf{r}}{\partial \phi}\right| = \frac{1}{h_\phi}\frac{\partial \mathbf{r}}{\partial \phi}$$

$$\begin{aligned} h_\phi = \left|\frac{\partial \mathbf{r}}{\partial \phi}\right| &= |-r\sin\theta\sin\phi\,\mathbf{i} + r\sin\theta\cos\phi\,\mathbf{j}| \\ &= (r^2\sin^2\theta\sin^2\phi + r^2\sin^2\theta\cos^2\phi)^{1/2} \\ &= (r^2\sin^2\theta)^{1/2} = r\sin\theta \qquad \therefore \underline{h_\phi = r\sin\theta} \\ &\therefore \underline{h_r = 1; \quad h_\theta = r; \quad h_\phi = r\sin\theta} \end{aligned}$$

So: (a) for cylindrical coordinates

$$\mathbf{I} = \frac{\partial \mathbf{r}}{\partial \rho}; \quad \mathbf{J} = \frac{1}{\rho}\frac{\partial \mathbf{r}}{\partial \phi}; \quad \mathbf{K} = \frac{\partial \mathbf{r}}{\partial z}$$

(b) for spherical coordinates

$$\mathbf{I} = \frac{\partial \mathbf{r}}{\partial r}; \quad \mathbf{J} = \frac{1}{r}\frac{\partial \mathbf{r}}{\partial \theta}; \quad \mathbf{K} = \frac{1}{r\sin\theta}\frac{\partial \mathbf{r}}{\partial \phi}$$

19

General curvilinear coordinate system (u, v, w)

Any system of coordinates can be treated in like manner to obtain expressions for the appropriate unit vectors **I, J, K**.

$$\mathbf{I} = \frac{\partial \mathbf{r}}{\partial u} \Big/ \left|\frac{\partial \mathbf{r}}{\partial u}\right|; \quad \mathbf{J} = \frac{\partial \mathbf{r}}{\partial v} \Big/ \left|\frac{\partial \mathbf{r}}{\partial v}\right|; \quad \mathbf{K} = \frac{\partial \mathbf{r}}{\partial w} \Big/ \left|\frac{\partial \mathbf{r}}{\partial w}\right|$$

These unit vectors are not always at right angles to each other. If they are mutually perpendicular, the coordinate system is

....

20

orthogonal

Unit vectors **I, J, K** are orthogonal if

$$\mathbf{I} \cdot \mathbf{J} = \mathbf{J} \cdot \mathbf{K} = \mathbf{K} \cdot \mathbf{I} = 0$$

Exercise

Determine the unit base vectors in the directions of the following vectors and determine whether the vectors are orthogonal.

1. $\mathbf{i} - 2\mathbf{j} + 4\mathbf{k}$
   $2\mathbf{i} + 3\mathbf{j} + \mathbf{k}$
   $-2\mathbf{i} + \mathbf{j} + \mathbf{k}$

2. $2\mathbf{i} - 3\mathbf{j} + 2\mathbf{k}$
   $\mathbf{i} + 2\mathbf{j} + 2\mathbf{k}$
   $-10\mathbf{i} - 2\mathbf{j} + 7\mathbf{k}$

3. $4\mathbf{i} + 2\mathbf{j} - \mathbf{k}$
   $3\mathbf{i} - 5\mathbf{j} + 2\mathbf{k}$
   $\mathbf{i} + 2\mathbf{j} + 6\mathbf{k}$

4. $3\mathbf{i} + 2\mathbf{j} + \mathbf{k}$
   $\mathbf{i} - 3\mathbf{j} + 3\mathbf{k}$
   $6\mathbf{i} + \mathbf{j} - \mathbf{k}$

21 The results are as follows:

1. $\mathbf{I} = \dfrac{1}{\sqrt{21}}(\mathbf{i} - 2\mathbf{j} + 4\mathbf{k})$; $\mathbf{J} = \dfrac{1}{\sqrt{14}}(2\mathbf{i} + 3\mathbf{j} + \mathbf{k})$; $\mathbf{K} = \dfrac{1}{\sqrt{6}}(-2\mathbf{i} + \mathbf{j} + \mathbf{k})$

$$\mathbf{I} \cdot \mathbf{J} = 0; \quad \mathbf{J} \cdot \mathbf{K} = 0; \quad \mathbf{K} \cdot \mathbf{I} = 0 \quad \therefore \ \underline{\text{orthogonal}}$$

2. $\mathbf{I} = \dfrac{1}{\sqrt{17}}(2\mathbf{i} - 3\mathbf{j} + 2\mathbf{k})$; $\mathbf{J} = \dfrac{1}{3}(\mathbf{i} + 2\mathbf{j} + 2\mathbf{k})$; $\mathbf{K} = \dfrac{1}{\sqrt{153}}(-10\mathbf{i} + 2\mathbf{j} + 7\mathbf{k})$

$$\mathbf{I} \cdot \mathbf{J} = 0; \quad \mathbf{J} \cdot \mathbf{K} = 0; \quad \mathbf{K} \cdot \mathbf{I} = 0 \quad \therefore \ \underline{\text{orthogonal}}$$

3. $\mathbf{I} = \dfrac{1}{\sqrt{21}}(4\mathbf{i} + 2\mathbf{j} - \mathbf{k})$; $\mathbf{J} = \dfrac{1}{\sqrt{38}}(3\mathbf{i} - 5\mathbf{j} + 2\mathbf{k})$; $\mathbf{K} = \dfrac{1}{\sqrt{41}}(\mathbf{i} + 2\mathbf{j} + 6\mathbf{k})$

$$\mathbf{I} \cdot \mathbf{J} = 0; \quad \mathbf{J} \cdot \mathbf{K} \neq 0 \qquad \therefore \ \underline{\text{not orthogonal}}$$

4. $\mathbf{I} = \dfrac{1}{\sqrt{14}}(3\mathbf{i} + 2\mathbf{j} + \mathbf{k})$; $\mathbf{J} = \dfrac{1}{\sqrt{19}}(\mathbf{i} - 3\mathbf{j} + 3\mathbf{k})$; $\mathbf{K} = \dfrac{1}{\sqrt{38}}(6\mathbf{i} + \mathbf{j} - \mathbf{k})$

$$\mathbf{I} \cdot \mathbf{J} = 0; \quad \mathbf{J} \cdot \mathbf{K} = 0; \quad \mathbf{K} \cdot \mathbf{I} \neq 0 \quad \therefore \ \underline{\text{not orthogonal}}$$

Transformation equations

In general coordinates, the transformation equations are of the form

$$x = f(u, v, w); \quad y = g(u, v, w); \quad z = h(u, v, w)$$

where the functions f, g, h are continuous and single-valued, and whose partial derivatives are continuous.

Then $\mathbf{r} = x\mathbf{i} + y\mathbf{j} + z\mathbf{k} = f(u, v, w)\mathbf{i} + g(u, v, w)\mathbf{j} + h(u, v, w)\mathbf{k}$ and co-ordinate curves can be formed by keeping two of the three variables constant.

$$\text{Now } \mathbf{r} = x\mathbf{i} + y\mathbf{j} + z\mathbf{k} \qquad \therefore\ d\mathbf{r} = \frac{\partial \mathbf{r}}{\partial u} du + \frac{\partial \mathbf{r}}{\partial v} dv + \frac{\partial \mathbf{r}}{\partial w} dw \qquad (1)$$

$\dfrac{\partial \mathbf{r}}{\partial u}$ is a tangent vector to the u-coordinate curve at P

$\dfrac{\partial \mathbf{r}}{\partial v}$ is a tangent vector to the v-coordinate curve at P

$\dfrac{\partial \mathbf{r}}{\partial w}$ is a tangent vector to the w-coordinate curve at P

$$\mathbf{I} = \frac{\partial \mathbf{r}}{\partial u} \Big/ \left|\frac{\partial \mathbf{r}}{\partial u}\right| \quad \therefore\ \frac{\partial \mathbf{r}}{\partial u} = h_u \mathbf{I} \text{ where } h_u = \left|\frac{\partial \mathbf{r}}{\partial u}\right|$$

$$\mathbf{J} = \frac{\partial \mathbf{r}}{\partial v} \Big/ \left|\frac{\partial \mathbf{r}}{\partial v}\right| \quad \therefore\ \frac{\partial \mathbf{r}}{\partial v} = h_v \mathbf{J} \text{ where } h_v = \left|\frac{\partial \mathbf{r}}{\partial v}\right|$$

$$\mathbf{K} = \frac{\partial \mathbf{r}}{\partial w} \Big/ \left|\frac{\partial \mathbf{r}}{\partial w}\right| \quad \therefore\ \frac{\partial \mathbf{r}}{\partial w} = h_w \mathbf{K} \text{ where } h_w = \left|\frac{\partial \mathbf{r}}{\partial w}\right|$$

Then (1) above becomes

$$\underline{d\mathbf{r} = h_u\, du\, \mathbf{I} + h_v\, dv\, \mathbf{J} + h_w\, dw\, \mathbf{K}}$$

where, as before, h_u, h_v, h_w are the scale factors.

23

Element of arc ds and element of volume dV in orthogonal curvilinear coordinates

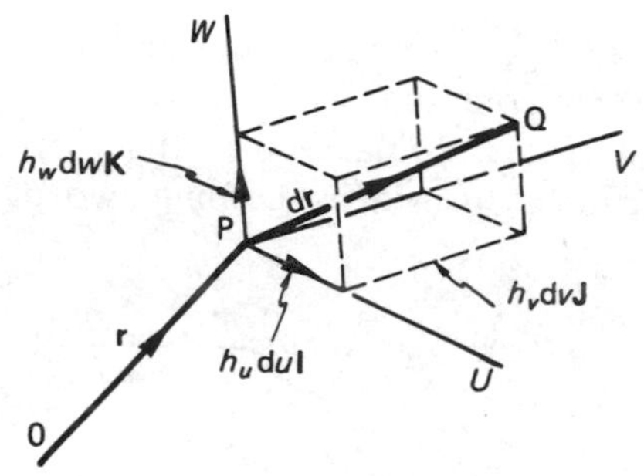

(a) *Element of arc* ds

Element of arc ds from P to Q is given by

$$\mathrm{d}\mathbf{r} = h_u\,\mathrm{d}u\,\mathbf{I} + h_v\,\mathrm{d}v\,\mathbf{J} + h_w\,\mathrm{d}w\,\mathbf{K}$$

$$\therefore\ \mathrm{d}\mathbf{r}\cdot\mathrm{d}\mathbf{r} = (h_u\,\mathrm{d}u\,\mathbf{I} + h_v\,\mathrm{d}v\,\mathbf{J} + h_w\,\mathrm{d}w\,\mathrm{K})\cdot(h_u\,\mathrm{d}u\,\mathbf{I} + h_v\,\mathrm{d}v\,\mathbf{J} + h_w\,\mathrm{d}w\,\mathbf{K})$$

$$\therefore\ \mathrm{d}s^2 = h_u^2\,\mathrm{d}u^2 + h_v^2\,\mathrm{d}v^2 + h_w^2\,\mathrm{d}w^2$$

$$\therefore\ \underline{\mathrm{d}s = (h_u^2\,\mathrm{d}u^2 + h_v^2\,\mathrm{d}v^2 + h_w^2\,\mathrm{d}w^2)^{1/2}}$$

(b) *Element of volume* dV

$$\mathrm{d}V = (h_u\,\mathrm{d}u\,\mathbf{I}).(h_v\,\mathrm{d}v\,\mathbf{J} \times h_w\,\mathrm{d}w\,\mathbf{K})$$

$$= (h_u\,\mathrm{d}u\,\mathbf{I}).(h_v\,\mathrm{d}v\,h_w\,\mathrm{d}w\,\mathbf{I}) = h_u\,\mathrm{d}u\,h_v\,\mathrm{d}v\,h_w\,\mathrm{d}w$$

$$\therefore\ \underline{\mathrm{d}V = h_u\,h_v\,h_w\,\mathrm{d}u\,\mathrm{d}v\,\mathrm{d}w}$$

Note also that

$$\mathrm{d}V = \left|\frac{\partial\mathbf{r}}{\partial u}\cdot\left(\frac{\partial\mathbf{r}}{\partial v}\times\frac{\partial\mathbf{r}}{\partial w}\right)\right|\mathrm{d}u\,\mathrm{d}v\,\mathrm{d}w$$

$$= \frac{\partial(x,\ y,\ z)}{\partial(u,\ v,\ w)}\,\mathrm{d}u\,\mathrm{d}v\,\mathrm{d}w$$

where $\dfrac{\partial(x,\ y,\ z)}{\partial(u,\ v,\ w)}$ is the Jocabian of the transformation.

Grad, div and curl in orthogonal curvilinear coordinates 24

(a) *Grad V* (∇V)

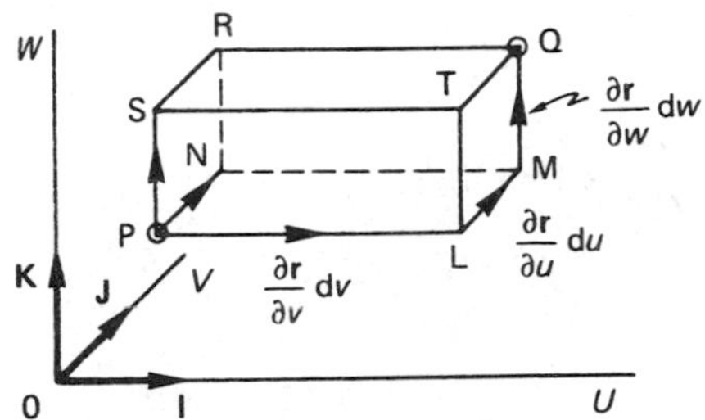

Let a scalar field V exist in space and let $\mathrm{d}V$ be the change in V from P to Q. If the position vector of P is $\mathbf{r}$ then that of Q is $\mathbf{r} + \mathrm{d}\mathbf{r}$.

$$\text{Then } \mathrm{d}V = \frac{\partial V}{\partial u}\,\mathrm{d}u + \frac{\partial V}{\partial v}\,\mathrm{d}v + \frac{\partial V}{\partial w}\,\mathrm{d}w$$

$$\text{Let grad } V = \nabla V = (\nabla V)_u \mathbf{I} + (\nabla V)_v \mathbf{J} + (\nabla V)_w \mathbf{K}$$

where $(\nabla V)_{u,v,w}$ are the components of grad V in the u, v, w directions.

$$\text{Also } \mathrm{d}\mathbf{r} = \frac{\partial \mathbf{r}}{\partial u}\,\mathrm{d}u + \frac{\partial \mathbf{r}}{\partial v}\,\mathrm{d}v + \frac{\partial \mathbf{r}}{\partial w}\,\mathrm{d}w$$

$$\text{But } \frac{\partial \mathbf{r}}{\partial u} = \left|\frac{\partial \mathbf{r}}{\partial u}\right| \mathbf{I} = h_u\,\mathbf{I}; \qquad \frac{\partial \mathbf{r}}{\partial v} = \left|\frac{\partial \mathbf{r}}{\partial v}\right| \mathbf{J} = h_v\,\mathbf{J};$$

$$\text{and } \frac{\partial \mathbf{r}}{\partial w} = \left|\frac{\partial \mathbf{r}}{\partial w}\right| \mathbf{K} = h_w\,\mathbf{K}.$$

$$\therefore\ \mathrm{d}\mathbf{r} = h_u\,\mathrm{d}u\,\mathbf{I} + h_v\,\mathrm{d}v\,\mathbf{J} + h_w\,\mathrm{d}w\,\mathrm{K}$$

We have previously established that $\mathrm{d}V = \text{grad } V \cdot \mathrm{d}\mathbf{r}$

$$\therefore\ \mathrm{d}V = \{(\nabla V)_u\mathbf{I} + (\nabla V)_v\mathbf{J} + (\nabla V)_w\mathbf{K}\}.\{h_u\mathrm{d}u\mathbf{I} + h_v\mathrm{d}v\mathbf{J} + h_w\mathrm{d}w\mathbf{K}\}$$
$$= (\nabla V)_u h_u\,\mathrm{d}u + (\nabla V)_v h_v\,\mathrm{d}v + (\nabla V)_w h_w\,\mathrm{d}w$$

$$\text{But } \mathrm{d}V = \frac{\partial V}{\partial u}\mathrm{d}u + \frac{\partial V}{\partial v}\mathrm{d}v + \frac{\partial V}{\partial w}\mathrm{d}w$$

$\therefore$ Equating coefficients, we then have

$$\frac{\partial V}{\partial u} = (\nabla V)_u h_u \qquad \therefore (\nabla V)_u = \frac{1}{h_u}\frac{\partial V}{\partial u}$$

$$\frac{\partial V}{\partial v} = (\nabla V)_v h_v \qquad \therefore (\nabla V)_v = \frac{1}{h_v}\frac{\partial V}{\partial v}$$

$$\frac{\partial V}{\partial w} = (\nabla V)_w h_w \qquad \therefore (\nabla V)_w = \frac{1}{h_w}\frac{\partial V}{\partial w}$$

$$\therefore \text{grad } V = \nabla V = \frac{1}{h_u}\frac{\partial V}{\partial u}\mathbf{I} + \frac{1}{h_v}\frac{\partial V}{\partial v}\mathbf{J} + \frac{1}{h_w}\frac{\partial V}{\partial w}\mathbf{K}$$

$$\text{i.e. grad operator } \nabla = \frac{\mathbf{I}}{h_u}\frac{\partial}{\partial u} + \frac{\mathbf{J}}{h_v}\frac{\partial}{\partial v} + \frac{\mathbf{K}}{h_w}\frac{\partial}{\partial w}$$

---

25 (b) *Div* **F** $(\nabla \cdot \mathbf{F})$

$$\text{div } \mathbf{F} = \nabla \cdot \mathbf{F} = \left\{\frac{\mathbf{I}}{h_u}\frac{\partial}{\partial u} + \frac{\mathbf{J}}{h_v}\frac{\partial}{\partial v} + \frac{\mathbf{K}}{h_w}\frac{\partial}{\partial w}\right\} \cdot \{F_u\mathbf{I} + F_v\mathbf{J} + F_w\mathbf{K}\}$$

$$= \frac{1}{h_u}\frac{\partial F_u}{\partial u} + \frac{1}{h_v}\frac{\partial F_v}{\partial v} + \frac{1}{h_w}\frac{\partial F_w}{\partial w}$$

$$\therefore \text{div } \mathbf{F} = \frac{1}{h_u h_v h_w}\left\{\frac{\partial}{\partial u}(h_v h_w F_u) + \frac{\partial}{\partial v}(h_w h_u F_v) + \frac{\partial}{\partial w}(h_u h_v F_w)\right\}$$

(c) *Curl* **F** $(\nabla \times \mathbf{F})$

Applying the usual definition for curl of a vector, we have

$$\text{curl } \mathbf{F} = \nabla \times \mathbf{F} = \ldots\ldots\ldots$$

26

$$\text{curl } \mathbf{F} = \begin{vmatrix} \mathbf{I} & \mathbf{J} & \mathbf{K} \\ \frac{1}{h_u}\frac{\partial}{\partial u} & \frac{1}{h_v}\frac{\partial}{\partial v} & \frac{1}{h_w}\frac{\partial}{\partial w} \\ F_u & F_v & F_w \end{vmatrix}$$

If we multiply the first column by h_u, the second by h_v and the third by h_w, this can be written

$$\text{curl } \mathbf{F} = \frac{1}{h_u\, h_v\, h_w} = \begin{vmatrix} h_u\mathbf{I} & h_v\mathbf{J} & h_w\mathbf{K} \\ \frac{\partial}{\partial u} & \frac{\partial}{\partial v} & \frac{\partial}{\partial w} \\ h_u F_u & h_v F_v & h_w F_w \end{vmatrix}$$

We could, of course, expand the determinant in the usual way, but the expression becomes somewhat tedious in symbolic form, so the result is best remembered in the determinant form as stated.

27

(d) *Div grad V* ($\nabla \cdot \nabla V$ i.e. $\nabla^2 V$)

In orthogonal curvilinear coordinates, we can combine operations as we did with rectangular coordinates. As an example, let us consider div grad V where V is a scalar field.

We have already established that

$$\text{grad } V = \nabla V = \frac{1}{h_u}\frac{\partial V}{\partial u}\mathbf{I} + \frac{1}{h_v}\frac{\partial V}{\partial v}\mathbf{J} + \frac{1}{h_w}\frac{\partial V}{\partial w}\mathbf{K}$$

and that

$$\text{div } \mathbf{F} = \nabla \cdot \mathbf{F} = \frac{1}{h_u}\frac{\partial F_u}{\partial u} + \frac{1}{h_v}\frac{\partial F_v}{\partial v} + \frac{1}{h_w}\frac{\partial F_w}{\partial w}$$

$\therefore$ div grad V = ...

28

$$\text{div grad } V = \frac{1}{h_u}\frac{\partial}{\partial u}\left(\frac{1}{h_u}\frac{\partial V}{\partial u}\right) + \frac{1}{h_v}\frac{\partial}{\partial v}\left(\frac{1}{h_v}\frac{\partial V}{\partial v}\right) + \frac{1}{h_w}\frac{\partial}{\partial w}\left(\frac{1}{h_w}\frac{\partial V}{\partial w}\right)$$

div grad $V = \nabla \cdot (\text{grad } V) = \nabla \cdot \nabla V$, often written as $\nabla^2 V$.

The result we obtained is usually expressed in the form

$$\nabla^2 V = \frac{1}{h_u h_v h_w}\left\{\frac{\partial}{\partial u}\left(\frac{h_v h_w}{h_u}\cdot\frac{\partial V}{\partial u}\right) + \frac{\partial}{\partial v}\left(\frac{h_u h_w}{h_v}\cdot\frac{\partial V}{\partial v}\right) + \frac{\partial}{\partial w}\left(\frac{h_u h_v}{h_w}\cdot\frac{\partial V}{\partial w}\right)\right\}$$

Example 1 If $V(u, v, w) = u + v^2 + w^3$ with scale factors $h_u = 2$, $h_v = 1$, $h_w = 1$, find $\nabla^2 V$ at the point $(5, 3, 4)$.

There is very little to it. All we have to do is to determine the various partial derivatives and substitute in the expression above with relevant values.

div grad $V = \ldots\ldots\ldots$

29

$$\boxed{26}$$

for

$$\nabla^2 V = \frac{1}{h_u\, h_v\, h_w}\left\{\frac{\partial}{\partial u}\left(\frac{h_v\, h_w}{h_u}\cdot\frac{\partial V}{\partial u}\right) + \frac{\partial}{\partial v}\left(\frac{h_u\, h_w}{h_v}\cdot\frac{\partial V}{\partial v}\right) + \frac{\partial}{\partial w}\left(\frac{h_u\, h_v}{h_w}\cdot\frac{\partial V}{\partial w}\right)\right\}$$

In this case, $V = u + v^2 + w^3 \quad \therefore \quad \frac{\partial V}{\partial u} = 1; \quad \frac{\partial V}{\partial v} = 2v; \quad \frac{\partial V}{\partial w} = 3w^2$

Also $\qquad h_u = 2, \; h_v = 1, \; h_w = 1$

$$\therefore \nabla^2 V = \frac{1}{2}\left\{\frac{\partial}{\partial u}\left(\frac{1}{2}\right) + \frac{\partial}{\partial v}(4v) + \frac{\partial}{\partial w}(6w^2)\right\}$$

$$= \tfrac{1}{2}\{0 + 4 + 12w\} \quad \therefore \text{ At } w = 4, \quad \underline{\nabla^2 V = 26}$$

That is all there is to it. Here is another.

Example 2 If $V = (u^2 + v^2)w^3$ with $h_u = 3$, $h_v = 1$, $h_w = 2$, find div grad V at the point $(2, -2, 1)$.

$\nabla^2 V = \ldots\ldots\ldots$

30

$$\boxed{14\tfrac{2}{9}}$$

since $V = (u^2 + v^2)w^3 \quad \therefore \dfrac{\partial V}{\partial u} = 2uw^3; \quad \dfrac{\partial V}{\partial v} = 2vw^3; \quad \dfrac{\partial V}{\partial w} = 3(u^2 + v^2)w^2$

also $h_u = 3, \quad h_v = 1, \quad h_w = 2$

$$\therefore \nabla^2 V = \frac{1}{6}\left\{\frac{\partial}{\partial u}\left(\frac{2}{3}\frac{\partial V}{\partial u}\right) + \frac{\partial}{\partial v}\left(6\frac{\partial V}{\partial v}\right) + \frac{\partial}{\partial w}\left(\frac{3}{2}\frac{\partial V}{\partial w}\right)\right\}$$

$$= \frac{1}{6}\left\{\frac{\partial}{\partial u}\left(\frac{4}{3}uw^3\right) + \frac{\partial}{\partial v}(12vw^3) + \frac{\partial}{\partial w}\left(\frac{9}{2}(u^2 + v^2)w^2\right)\right\}$$

$\therefore$ at (2, –2, 1)

$$\nabla^2 V = \tfrac{1}{6}\{(\tfrac{4}{3}w^3) + (12w^3) + 9(u^2 + v^2)w\}$$

$$= \tfrac{1}{6}\{\tfrac{4}{3} + 12 + 72\} = \frac{256}{18} = \underline{14\tfrac{2}{9}}$$

Particular orthogonal systems

We can apply the general results for div, grad and curl to special coordinate systems by inserting the appropriate scale factors—as we shall now see.

31

(a) *Cartesian rectangular coordinate system*

If we replace u, v, w by x, y, z and insert values of $h_x = h_y = h_z = 1$, we obtain expressions for grad, div and curl in rectangular coordinates, so that

grad $V = \ldots\ldots\ldots$; div $\mathbf{F} = \ldots\ldots\ldots$; curl $\mathbf{F} = \ldots\ldots\ldots$

32

$$\text{grad } V = \frac{\partial V}{\partial x}\mathbf{i} + \frac{\partial V}{\partial y}\mathbf{j} + \frac{\partial V}{\partial z}\mathbf{k}$$

$$\text{div } \mathbf{F} = \frac{\partial F_x}{\partial x} + \frac{\partial F_y}{\partial y} + \frac{\partial F_z}{\partial z}$$

$$\text{curl } \mathbf{F} = \begin{vmatrix} \mathbf{i} & \mathbf{j} & \mathbf{k} \\ \dfrac{\partial}{\partial x} & \dfrac{\partial}{\partial y} & \dfrac{\partial}{\partial z} \\ F_x & F_y & F_z \end{vmatrix}$$

all of which you will surely recognise.

(b) *Cylindrical polar coordinate system*

Here we simply replace u, v, w with ρ, ϕ, z and insert $h_u = h_\rho = 1$, $h_v = h_\phi = \rho$, $h_w = h_z = 1$ giving

grad V = ... ; div $\mathbf{F}$ = ... ; curl $\mathbf{F}$ = ...

33

$$\text{grad } V = \frac{\partial V}{\partial \rho}\mathbf{I} + \frac{1}{\rho}\frac{\partial V}{\partial \phi}\mathbf{J} + \frac{\partial V}{\partial z}\mathbf{K}$$

$$\text{div } \mathbf{F} = \frac{1}{\rho}\left\{\frac{\partial}{\partial \rho}(\rho F_\rho) + \frac{\partial}{\partial \phi}(F_\phi) + \frac{\partial}{\partial z}(\rho F_z)\right\}$$

$$\text{curl } \mathbf{F} = \frac{1}{\rho}\begin{vmatrix} \mathbf{I} & \rho\mathbf{J} & \mathbf{K} \\ \dfrac{\partial}{\partial \rho} & \dfrac{\partial}{\partial \phi} & \dfrac{\partial}{\partial z} \\ F_\rho & \rho F_\phi & F_z \end{vmatrix}$$

Spherical polar coordinate system

Replacing u, v, w with r, θ, ϕ with $h_r = 1$, $h_\theta = r$, $h_\phi = r\sin\theta$,

grad V = ... ; div $\mathbf{F}$ = ... ; curl $\mathbf{F}$ = ...

34

$$\text{grad } V = \frac{\partial V}{\partial r}\mathbf{I} + \frac{1}{r}\frac{\partial V}{\partial \theta}\mathbf{J} + \frac{1}{r\sin\theta}\frac{\partial V}{\partial \phi}\mathbf{K}$$

$$\text{div } \mathbf{F} = \frac{1}{r^2\sin\theta}\frac{\partial}{\partial r}(r^2\sin\theta\, F_r) + \frac{\partial}{\partial \theta}(r^2\sin\theta\, F_\theta) + \frac{\partial}{\partial \phi}(rF_\phi)$$

$$\text{curl } \mathbf{F} = \frac{1}{r^2\sin\theta}\begin{vmatrix} \mathbf{I} & r\mathbf{J} & r\sin\theta\,\mathbf{K} \\ \dfrac{\partial}{\partial r} & \dfrac{\partial}{\partial \theta} & \dfrac{\partial}{\partial \phi} \\ F_r & rF_\theta & r\sin\theta\, F_\phi \end{vmatrix}$$

The results we have compiled are sometimes written in slightly different forms, but they are, of course, equivalent.

That brings us to the end of this programme which is designed as an introduction to the topic of curvilinear coordinates. It has considerable applications, but these are beyond the scope of this present course of study.

The Revision Summary follows as usual. Make any further notes as necessary: then you can work through the Test Exercise without difficulty.

35 REVISION SUMMARY

1. *Curvilinear coordinates in two dimensions*

$$u = f(x, y); \qquad v = g(x, y)$$

2. *Orthogonal coordinate system in space*

   (a) *Cartesian rectangular coordinates* (x, y, z)

   $\mathbf{F} = F_x\mathbf{i} + F_y\,\mathbf{j} + F_z\,\mathbf{k}$ Scale factors $h_x = h_y = h_z = 1$

   (b) *Cylindrical polar coordinates* (ρ, ϕ, z)

   $\mathbf{r} = \rho\cos\phi\,\mathbf{i} + \rho\,\sin\phi\,\mathbf{j} + z\,\mathbf{k}$

   Base unit vectors: Scale factors:

$$\mathbf{I} = \frac{\partial\mathbf{r}}{\partial\rho}\Bigg/\left|\frac{\partial\mathbf{r}}{\partial\rho}\right| \qquad h_\rho = \left|\frac{\partial\mathbf{r}}{\partial\rho}\right| = 1$$

$$\mathbf{J} = \frac{\partial\mathbf{r}}{\partial\phi}\Bigg/\left|\frac{\partial\mathbf{r}}{\partial\phi}\right| \qquad h_\phi = \left|\frac{\partial\mathbf{r}}{\partial\phi}\right| = \rho$$

$$\mathbf{K} = \frac{\partial\mathbf{r}}{\partial z}\Bigg/\left|\frac{\partial\mathbf{r}}{\partial z}\right| \qquad h_z = \left|\frac{\partial\mathbf{r}}{\partial z}\right| = 1$$

   $\mathbf{F} = F_\rho\,\mathbf{I} + F_\phi\,\mathbf{J} + F_z\,\mathbf{K}$

   (c) *Spherical polar coordinates* (r, θ, ϕ)

   $\mathbf{r} = r\sin\theta\,\cos\phi\,\mathbf{i} + r\sin\theta\,\sin\phi\,\mathbf{j} + r\cos\theta\,\mathbf{k}$

   Base unit vectors: Scale factors:

$$\mathbf{I} = \frac{\partial\mathbf{r}}{\partial r}\Bigg/\left|\frac{\partial\mathbf{r}}{\partial r}\right| \qquad h_r = \left|\frac{\partial\mathbf{r}}{\partial r}\right| = 1$$

$$\mathbf{J} = \frac{\partial\mathbf{r}}{\partial\theta}\Bigg/\left|\frac{\partial\mathbf{r}}{\partial\theta}\right| \qquad h_\theta = \left|\frac{\partial\mathbf{r}}{\partial\theta}\right| = r$$

$$\mathbf{K} = \frac{\partial\mathbf{r}}{\partial\phi}\Bigg/\left|\frac{\partial\mathbf{r}}{\partial\phi}\right| \qquad h_\phi = \left|\frac{\partial\mathbf{r}}{\partial\phi}\right| = r\sin\theta$$

   $\mathbf{F} = F_r\,\mathbf{I} + F_\theta\,\mathbf{J} + F_\phi\,\mathbf{K}$

3. *General orthogonal curvilinear coordinates* (u, v, w)

$$x = f(u, v, w); \quad y = g(u, v, w); \quad w = h(u, v, w)$$

$$\mathbf{r} = x\mathbf{i} + y\mathbf{j} + z\mathbf{k}$$

$$\frac{\partial \mathbf{r}}{\partial u} = h_u\mathbf{I} \quad \text{where} \quad h_u = \left|\frac{\partial \mathbf{r}}{\partial u}\right|$$

$$\frac{\partial \mathbf{r}}{\partial v} = h_v\mathbf{J} \quad \text{where} \quad h_v = \left|\frac{\partial \mathbf{r}}{\partial v}\right|$$

$$\frac{\partial \mathbf{r}}{\partial w} = h_w\mathbf{K} \quad \text{where} \quad h_w = \left|\frac{\partial \mathbf{r}}{\partial w}\right|$$

Element of arc: $\mathrm{d}s = (h_u^2\,\mathrm{d}u^2 + h_v^2\,\mathrm{d}v^2 + h_w^2\,\mathrm{d}w^2)^{1/2}$

Element of volume: $\mathrm{d}V = h_u h_v h_w\,\mathrm{d}u\,\mathrm{d}v\,\mathrm{d}w$

$$= \frac{\partial(x, y, z)}{\partial(u, v, w)}\,\mathrm{d}u\,\mathrm{d}v\,\mathrm{d}w$$

4. *Grad, div and curl in orthogonal curvilinear coordinates*

(a) Grad $V = \nabla V = \dfrac{1}{h_u}\dfrac{\partial V}{\partial u}\mathbf{I} + \dfrac{1}{h_v}\dfrac{\partial V}{\partial v}\mathbf{J} + \dfrac{1}{h_w}\dfrac{\partial V}{\partial w}\mathbf{K}$

grad operator $= \nabla = \dfrac{\mathbf{I}}{h_u}\dfrac{\partial}{\partial u} + \dfrac{\mathbf{J}}{h_v}\dfrac{\partial}{\partial v} + \dfrac{\mathbf{K}}{h_w}\dfrac{\partial}{\partial w}$

(b) Div $\mathbf{F} = \dfrac{1}{h_u h_v h_w}\left\{\dfrac{\partial}{\partial u}(h_v h_w F_u) + \dfrac{\partial}{\partial v}(h_w h_u F_v) + \dfrac{\partial}{\partial w}(h_u h_v F_w)\right\}$

(c) Curl $\mathbf{F} = \dfrac{1}{h_u h_v h_w}\begin{vmatrix} h_u\mathbf{I} & h_v\mathbf{J} & h_w\mathbf{K} \\ \dfrac{\partial}{\partial u} & \dfrac{\partial}{\partial v} & \dfrac{\partial}{\partial w} \\ h_u F_u & h_v F_v & h_w F_w \end{vmatrix}$

(d) Div grad $V = \nabla \cdot \nabla V = \nabla^2 V$

$$= \frac{1}{h_u h_v h_w}\left\{\frac{\partial}{\partial u}\left(\frac{h_v h_w}{h_u}\cdot\frac{\partial V}{\partial u}\right) + \frac{\partial}{\partial v}\left(\frac{h_u h_w}{h_v}\cdot\frac{\partial V}{\partial v}\right) + \frac{\partial}{\partial w}\left(\frac{h_u h_v}{h_w}\cdot\frac{\partial V}{\partial w}\right)\right\}$$

5. *Grad, div and curl in cylindrical and spherical coordinates*

(a) *Cylindrical coordinates* (ρ, ϕ, z)

grad $V = \dfrac{\partial V}{\partial \rho}\mathbf{I} + \dfrac{1}{\rho}\dfrac{\partial V}{\partial \phi}\mathbf{J} + \dfrac{\partial V}{\partial z}\mathbf{K}$

$$\text{div } \mathbf{F} = \frac{1}{\rho}\left\{\frac{\partial(\rho F_\rho)}{\partial \rho}\right\} + \frac{1}{\rho}\left\{\frac{\partial F_\phi}{\partial \phi}\right\} + \frac{\partial F_z}{\partial z}$$

$$\text{curl } \mathbf{F} = \frac{1}{\rho}\begin{vmatrix} \mathbf{I} & \rho\mathbf{J} & \mathbf{K} \\ \dfrac{\partial}{\partial \rho} & \dfrac{\partial}{\partial \phi} & \dfrac{\partial}{\partial z} \\ F_\rho & \rho F_\phi & F_z \end{vmatrix}$$

(b) *Spherical coordinates* (r, θ, ϕ)

$$\text{grad } V = \frac{\partial V}{\partial r}\mathbf{I} + \frac{1}{r}\frac{\partial V}{\partial \theta}\mathbf{J} + \frac{1}{r\sin\theta}\frac{\partial V}{\partial \phi}\mathbf{K}$$

$$\text{div } \mathbf{F} = \frac{1}{r^2}\frac{\partial}{\partial r}(r^2 F_r) + \frac{1}{r\sin\theta}\frac{\partial}{\partial \theta}(\sin\theta\, F_\theta) + \frac{1}{\sin\theta}\frac{\partial}{\partial \phi}(F_\phi)$$

$$\text{curl } \mathbf{F} = \frac{1}{r^2\sin\theta}\begin{vmatrix} \mathbf{I} & r\mathbf{J} & r\sin\theta\,\mathbf{K} \\ \dfrac{\partial}{\partial \mathbf{r}} & \dfrac{\partial}{\partial \theta} & \dfrac{\partial}{\partial \phi} \\ F_r & rF_\theta & r\sin\theta\, F_\phi \end{vmatrix}$$

36 TEST EXERCISE XIV

1. Determine the unit vectors in the directions of the following three vectors and test whether they form an orthogonal set.

$$\begin{aligned} 3\mathbf{i} - 2\mathbf{j} + \mathbf{k} \\ \mathbf{i} + 2\mathbf{j} + \mathbf{k} \\ -2\mathbf{i} - \mathbf{j} + 4\mathbf{k}. \end{aligned}$$

2. If $\mathbf{r} = u\sin 2\theta\,\mathbf{I} + u\cos 2\theta\,\mathbf{J} + v^2\,\mathbf{K}$, determine the scale factors h_u, h_v, h_θ.
3. If $P(\mathbf{r})$ is a point $\mathbf{r} = \rho\cos\phi\,\mathbf{i} + \rho\sin\phi\,\mathbf{j} + z\,\mathbf{k}$ and a scalar field $V = \rho^2 z\sin 2\phi$ exists in space, using cylindrical polar coordinates (ρ, ϕ, z) determine grad V at the point at which $\rho = 1$, $\phi = \pi/4$, $z = 2$.
4. A vector field $\mathbf{F}$ is given in cylindrical coordinates by

$$\mathbf{F} = \rho\cos\phi\,\mathbf{I} + \rho\sin 2\phi\,\mathbf{J} + z\mathbf{K}$$

Determine (a) div $\mathbf{F}$; (b) curl $\mathbf{F}$.

5. Using spherical coordinates (r, θ, ϕ) determine expressions for (a) an element of arc $\mathrm{d}s$; (b) an element of volume $\mathrm{d}V$.
6. If V is a scalar field such that $V = u^2vw^3$ and scale factors are $h_u = 1$, $h_v = 2$, $h_w = 4$, determine $\nabla^2 V$ at the point (2, 3, –1).

FURTHER PROBLEMS XIV

1. Determine whether the following sets of three vectors are orthogonal.

(a) $4\mathbf{i} - 2\mathbf{j} - \mathbf{k}$
$3\mathbf{i} + 5\mathbf{j} + 2\mathbf{k}$
$\mathbf{i} - 11\mathbf{j} + 26\mathbf{k}$

(b) $2\mathbf{i} + 3\mathbf{j} - \mathbf{k}$
$4\mathbf{i} - 2\mathbf{j} + 2\mathbf{k}$
$\mathbf{i} + 4\mathbf{j} + 2\mathbf{k}$

2. If $V(u, v, w) = v^3 w^2 \sin 2u$ with scale factors $h_u = 3,\ h_v = 1,\ h_w = 2$, determine div grad V at the point $(\pi/4,\ -1,\ 3)$.

3. A scalar field $V = \dfrac{u^2 e^{2w}}{v}$ exists in space. If the relevant scale factors are $h_u = 2,\ h_v = 3,\ h_w = 1$, determine the value of $\nabla^2 V$ at the point (1, 2, 0).

4. If $\mathbf{r} = x\mathbf{i} + y\mathbf{j} + z\mathbf{k}$ and $x = r\sin\theta\cos\phi,\ y = r\sin\theta\sin\phi,\ z = r\cos\theta$ in spherical polar coordinates $(r,\ \theta,\ \phi)$, prove that, for any vector field $\mathbf{F}$ where

$$\mathbf{F} = F_x\,\mathbf{i} + F_y\,\mathbf{j} + F_z\,\mathbf{k} = F_r\,\mathbf{I} + F_\theta\,\mathbf{J} + F_\phi\,\mathbf{K}$$

then
$$F_x = F_r\sin\theta\cos\phi + F_\theta\cos\theta\cos\phi - F_\phi\sin\phi$$
$$F_y = F_r\sin\theta\sin\phi + F_\theta\cos\theta\sin\phi + F_\phi\cos\phi$$
$$F_z = F_r\cos\theta - F_\theta\sin\theta\ .$$

5. If V is a scalar field, determine an expression for $\nabla^2 V$

(a) in cylindrical polar coordinates,

(b) in spherical polar coordinates.

6. Transformation equations from rectangular coordinates (x, y, z) to parabolic cylindrical coordinates (u, v, w) are

$$x = \frac{u^2 - v^2}{2};\quad y = uv;\quad z = w$$

V is a scalar field and $\mathbf{F}$ a vector field.

(a) Prove that the (u, v, w) system is orthogonal

(b) Determine the scale factors

(c) Find div $\mathbf{F}$

(d) Obtain an expression for $\nabla^2 V$.

Programme 15

Complex Variable
PART 1

Prerequisites: *Engineering Mathematics* (fourth edition)
Programmes 1, 2, 3

Introduction

1 The foundations of complex numbers and their application to hyperbolic functions were treated fully in Programmes 1, 2 and 3 of *Engineering Mathematics* and these provide valuable revision should you feel it to be necessary before embarking on the new work.

It will be assumed that you are already familiar with the material covered in those previous programmes and it would be a wise move to work through the relevant Test Exercises to refresh your memory on this all-important part of the course.

Functions of a complex variable

We are familiar enough with the method of graphical representation of a function $f(x)$ of a single real variable x, where we normally denote the function by y, i.e. $y = f(x)$. For various real values assigned to x, we can therefore plot specific points in the xy-plane and so obtain the graph of $y = f(x)$.

When the independent variable is complex, i.e. $x + jy$, there are problems. We normally denote the complex variable by the symbol z, i.e. $z = x + jy$ and a function of z, $f(z)$, will in general also be complex. Now z already consists of two real variables x and y, so that $f(z)$ will also be a function of x and y. For example, if $z = x + jy$ and $f(z) = z^2$ then $f(z) = (x + jy)^2 = x^2 + j2xy - y^2 = (x^2 - y^2) + j2xy$.

We can plot values of x, y and z on one set of axes, as in an Argand diagram, but we are unable to plot values of x, y and $f(z)$ on one set of axes. We, therefore, represent values of $f(z)$ on a separate plane. We let $w = f(z)$ where $w = u + jv$. So in our example above

$$w = u + jv = f(z) = z^2 = (x + jy)^2 = (x^2 - y^2) + j2xy$$

Equating real and imaginary parts

$$u = x^2 - y^2 \quad \text{and} \quad v = 2xy$$

So, for a particular value of z, e.g. $z = 4 + j3$

$$u = \ldots\ldots\ldots\ldots\ldots; \quad v = \ldots\ldots\ldots\ldots\ldots$$

2

$$u = 7; \quad v = 24$$

For with $z = 4 + j3$, $x = 4$ and $y = 3$. Then $u = 16 - 9 = 7$ and $v = 24$.

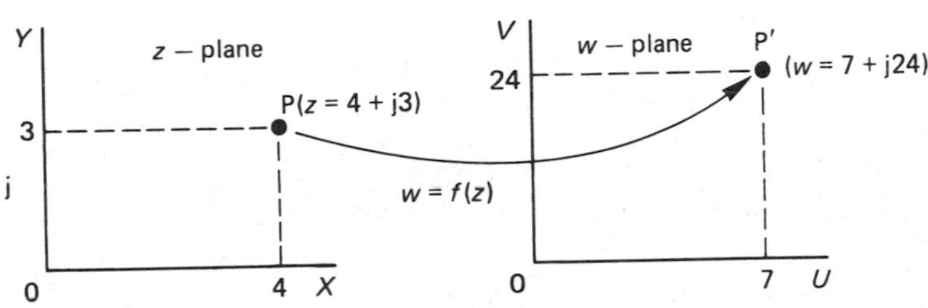

Therefore, z (where $z = x + jy$) and w (where $w = u + jv$) are two complex variables related by the equation $w = f(z)$.

Any other point in the z-plane will similarly be transformed into a corresponding point in the w-plane, the resulting position P′ depending on

(a) the initial position of P
(b) the relationship $w = f(z)$, called the *transformation equation* or *transformation function.*

Complex mapping

The transformation of P in the z-plane on to P′ in the w-plane is said to be a *mapping* of P on to P′ under the transformation $w = f(z)$ and P′ is sometimes referred to as the *image* of P.

Example 1 Determine the image of the point P, $z = 3 + j2$, on the w-plane under the transformation $w = 3z + 2 - j$.

$$\begin{aligned} w = u + jv = f(z) &= 3z + 2 - j \\ &= 3(x + jy) + 2 - j \end{aligned}$$

so that, for this example, $u = \ldots\ldots\ldots;\quad v = \ldots\ldots\ldots$

3

$$u = 3x + 2; \quad v = 3y - 1$$

Then the point P $(z = 3 + j2)$ transforms on to ....

4

$$w = 11 + j5$$

for $z = 3 + j2 \quad \therefore x = 3, \ y = 2$

$u = 3x + 2 = 11; \quad v = 3y - 1 = 5; \quad \therefore w = 11 + j5$

We can illustrate the transformation thus:

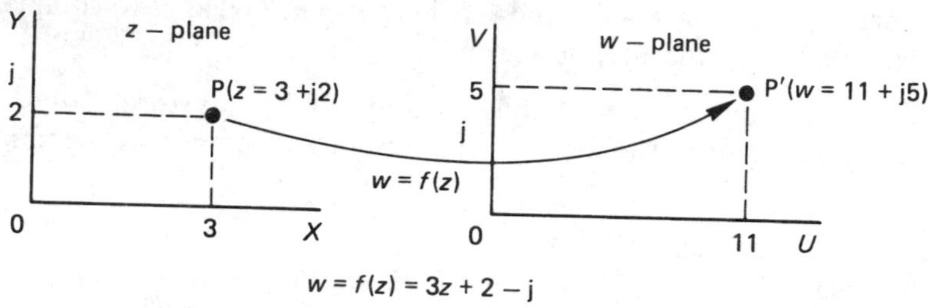

Here is another.

Example 2 Map the points A $(z = -2 + j)$ and B $(z = 3 + j4)$ on to the w-plane under the transformation $w = j2z + 3$ and illustrate the transformation on a diagram.

This is no different from the previous example. Complete the job and check with the next frame.

5

$$\text{A}'\ (w = 1 - \text{j}4);\quad \text{B}'\ (w = -5 + \text{j}6)$$

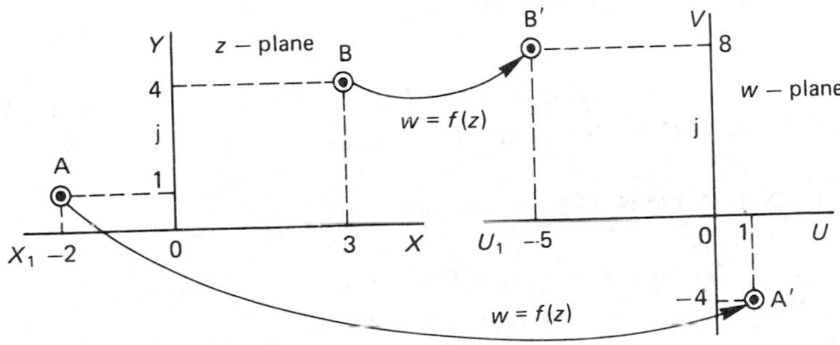

since $w = f(z) = \text{j}2z + 3 = \text{j}2(x + \text{j}y) + 3 = (3 - 2y) + \text{j}2x$

$w = u + \text{j}v \quad \therefore u = 3 - 2y; \quad v = 2x$

A: $x = -2,\ y = 1 \quad \therefore \text{A}'\!: u = 3 - 2 = 1;\ v = -4 \quad \therefore \text{A}'\!: w = 1 - \text{j}4$

B: $x = 3,\ y = 4 \quad \therefore \text{B}'\!: u = 3 - 8 = -5;\ v = 6 \quad \therefore \text{B}'\!: w = -5 + \text{j}6$

There now follows a short practice exercise. Work all four of the items before you check the results. There is no need to illustrate the transformation in each case.

So move on.

6

Exercise Map the following points in the z-plane on to the w-plane under the transformation $w = f(z)$ stated in each case.

1. $z = 4 - \text{j}2$ under $w = \text{j}3z + \text{j}2$
2. $z = -2 - \text{j}$ under $w = \text{j}z + 3$
3. $z = 3 + \text{j}2$ under $w = (1 + \text{j})z - 2$
4. $z = 2 + \text{j}$ under $w = z^2$.

7

1. $w = 6 + j14$	2. $w = 4 - j2$
3. $w = -1 + j5$	4. $w = 3 + j4$

That was easy enough. Now let us extend the ideas.

Mapping of a straight line in the z-plane on to the w-plane under the transformation $w = f(z)$

A typical example will show the method.

Example 1 To map the straight line joining A $(-2 + j)$ and B $(3 + j6)$ in the z-plane on to the w-plane when $w = 3 + j2z$.

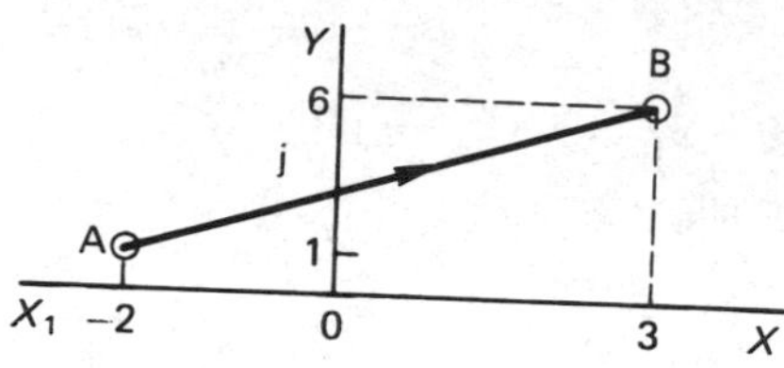

We first of all map the end points A and B on to the w-plane to obtain A′ and B′ as in the previous cases.

A′: $w =$ ....; B′: $w =$ ....

8

A′: $w = 1 - j4$; B′: $w = -9 + j6$

for (i) A: $z = -2 + j$ $w = 3 + j2z$

$$\therefore \text{A}': \; w = 3 + j2(-2 + j) = 3 - j4 - 2 = \underline{1 - j4}$$

(ii) B: $z = 3 + j6$

$$\therefore \text{B}': \; w = 3 + j2(3 + j6) = 3 + j6 - 12 = \underline{-9 + j6}$$

Then, if we illustrate the transformations on a diagram, as before, we get

....

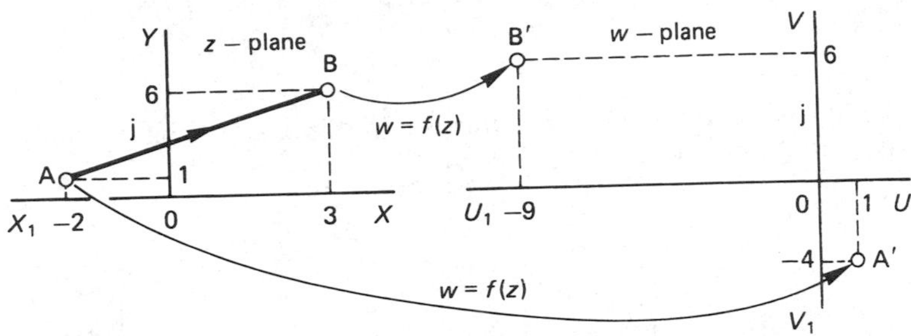

As z moves along the line A to B in the z-plane, we cannot assume that its images in the w-plane travels along a straight line from A′ to B′. As yet, we have no evidence of what the path is. We therefore have to find a general point $w = u + \text{j}v$ in the w-plane corresponding to a general point $z = x + \text{j}y$ in thc z-plane.

$$w = u + \text{j}v = f(z) = 3 + \text{j}2z$$

$$= \ldots\ldots\ldots\ldots\ldots$$

10

$$\boxed{w = u + \text{j}v = (3 - 2y) + \text{j}2x}$$

for $w = 3 + \text{j}2(x + \text{j}y) = 3 + \text{j}2x - 2y = (3 - 2y) + \text{j}2x$

$$\therefore\ u = 3 - 2y \quad \text{and} \quad v = 2x$$

Rearranging these results, we also have $y = \dfrac{3 - u}{2}$; $\quad x = \dfrac{v}{2}$.

Now the cartesian equation of AB is $y = x + 3$ and substituting from the previous line, we have $\dfrac{3 - u}{2} = \dfrac{v}{2} + 3$

which simplifies to . . . .

11

$$v = -u - 3$$

which is the equation of a straight line, so, in this case, the path joining A′ and B′ *is* in fact a straight line.

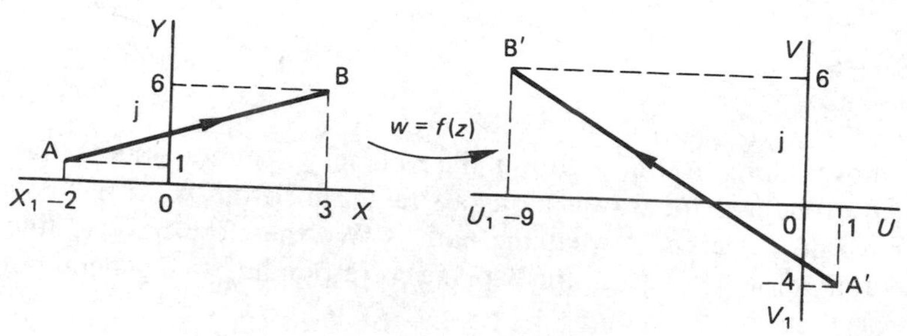

Note that it is useful to attach arrow heads to show the corresponding direction of progression in the transformation.

On to the next.

12

Example 2 If $w = z^2$, find the path traced out by w as z moves along the straight line joining A $(2 + j0)$ and B $(0 + j2)$.

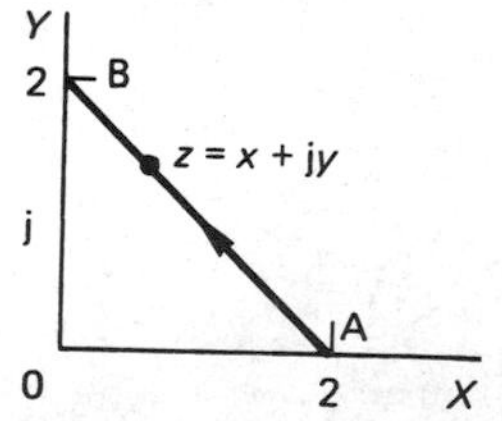

Cartesian equation of AB is

$$y = 2 - x$$

First we transform the two end points A and B on to A′ and B′ in the w-plane.

A′: ....; B′: ....

13

$$A': \ w = 4 + j0; \quad B': \ w = -4 + j0$$

for $w = z^2$ A: $z = 2$ $\therefore$ A′: $w = 2^2 = 4$

B: $z = j2$ $\therefore$ B′: $w = (j2)^2 = -4$

So we have

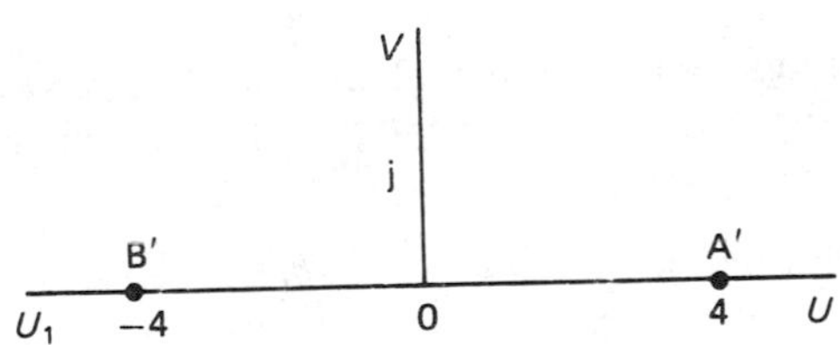

Now we have to find the path from A′ to B′.
The cartesian equation of AB in the z-plane is $y = 2 - x$.

Also $w = z^2 = (x + jy)^2 = (x^2 - y^2) + j2xy$

$$\therefore u = x^2 - y^2 \quad \text{and} \quad v = 2xy$$

Substituting $y = 2 - x$ in these results we can express u and v in terms of x.

$$u = \ldots\ldots\ldots; \quad v = \ldots\ldots\ldots$$

14

$$u = 4x - 4; \quad v = 4x - 2x^2$$

So, from the first of these $x = \dfrac{u + 4}{4}$

Substituting in the second $v = 4\left(\dfrac{u+4}{4}\right) - 2\left(\dfrac{u+4}{4}\right)^2$

$$= u + 4 - \frac{1}{8}(u^2 + 8u + 16)$$

$$= -\frac{1}{8}(u^2 - 16)$$

Therefore the path is $v = -\frac{1}{8}(u^2 - 16)$ which is a parabola for which at $u = 0,\ v = 2$.

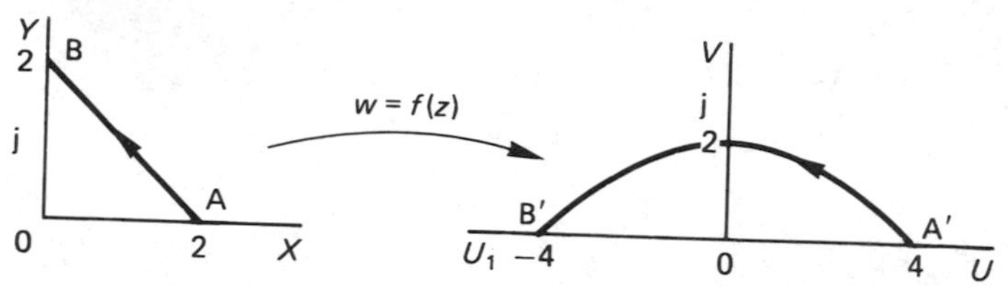

Note that a straight line in the z-plane does not always map on to a straight line in the w-plane. It depends on the particular transformation equation $w = f(z)$.

If the transformation is a *linear equation*, $w = f(z) = az + b$, where a and b may themselves be real or complex, then a straight line in the z-plane maps on to a corresponding straight line in the w-plane.

Example 3 A triangle in the z-plane has vertices at A $(z = 0)$; B $(z = 3)$ and C $(z = 3 + j2)$. Determine the image of this triangle in the w-plane under the transformation equation $w = (2 + j)z$.

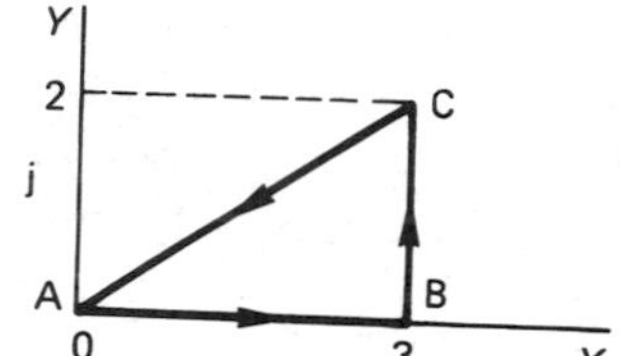

$$w = u + jv = f(z) = (2 + j)z = (2 + j)(x + jy) = (2x - y) + j(2y + x)$$

$$\therefore\ u = 2x - y; \qquad v = 2y + x$$

We now transform each vertex in turn on to the w-plane to determine A′, B′, C′.

These are A′ : . . . . ; B′ : . . . . ; C′ : . . . .

15

$$\text{A}': w = 0; \quad \text{B}': w = 6 + \text{j}3; \quad \text{C}': w = 4 + \text{j}7$$

The transformation is linear (of the form $w = az$) so A′B′, B′C′ and C′A′ are straight lines and the transformation can be illustrated in the diagram

. . . .

16

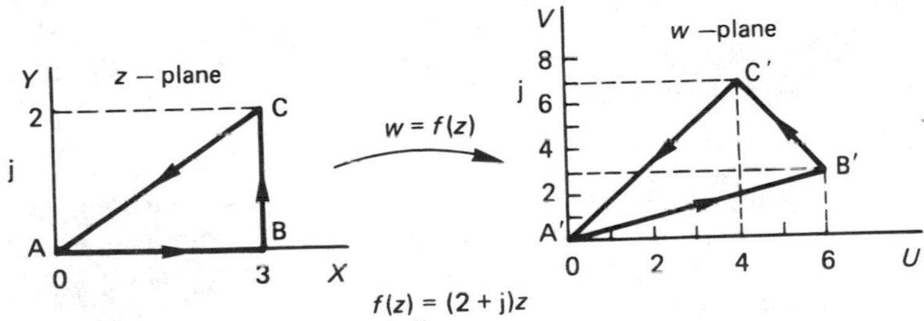

All very straightforward. Let us now take a more detailed look at linear transformations.

Types of transformation of the form $w = az + b$ where the constants a and b may be real or complex.

1. *Translation*

   Let $a = 1$ and $b = 2 - \text{j}$ i.e. $w = z + (2 - \text{j})$.

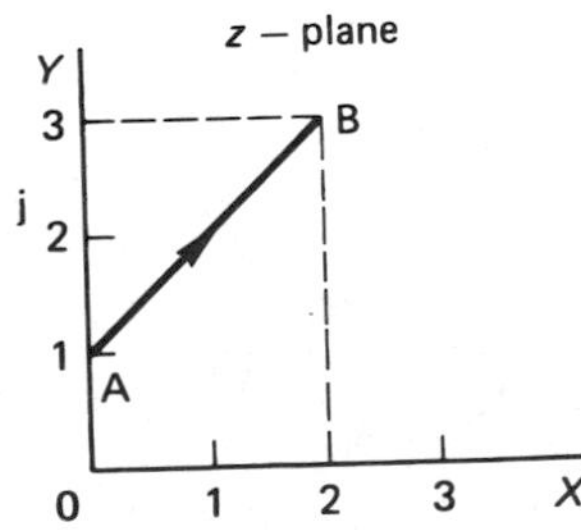

If we apply this to the straight line joining A $(0 + \text{j})$ and B $(2 + \text{j}3)$ in the z-plane, then

$$w = x + \text{j}y + 2 - \text{j} = (x + 2) + \text{j}(y - 1).$$

so the corresponding end points A′ and B′ in the w-plane are

A′ : . . . .; B′ : . . . .

17

$$A': w = 2; \quad B': w = 4 + j2$$

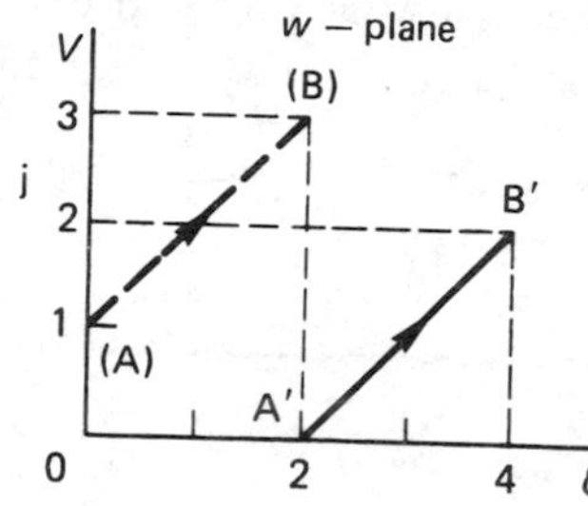

The transformed line A′B′ is then as shown. The broken line (A)(B) indicates the position of the original line AB in the z-plane.

Note that the whole line AB has moved two units to the right and one unit downwards, while retaining its original magnitude (length) and direction.

Such a transformation is called a *translation* and occurs whenever the transformation equation is of the form $w = z + b$. The degree of translation is given by the value of b—in this case $(2 - j)$, i.e. 2 units along the positive real axis and 1 unit in the direction of the negative imaginary axis.

On to the next frame.

2. *Magnification* 18

Consider now $w = az + b$ where $b = 0$ and a is real, e.g. $w = 2z$.

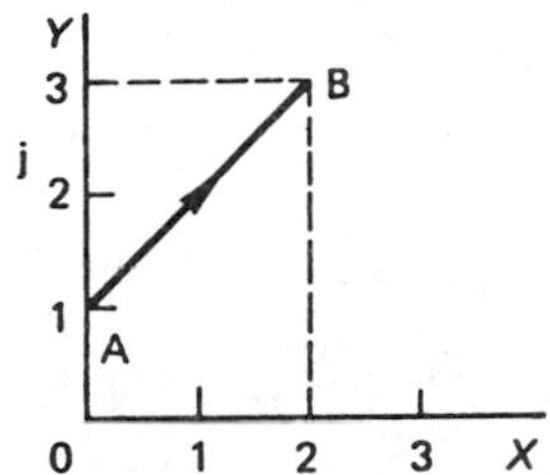

Applying the transformation to the same line AB as before, we have

$$w = u + \text{j}v = 2z = 2(x + \text{j}y)$$

$$\therefore u = 2x \quad \text{and} \quad v = 2y$$

Transforming the end points A $(0 + \text{j})$ and B $(2 + \text{j}3)$ on to A′ and B′ in the w-plane, we have A′ : ....; B′ : .... and the w-plane diagram becomes

....

19

A′: $w = \text{j}2$; B′: $w = 4 + \text{j}6$

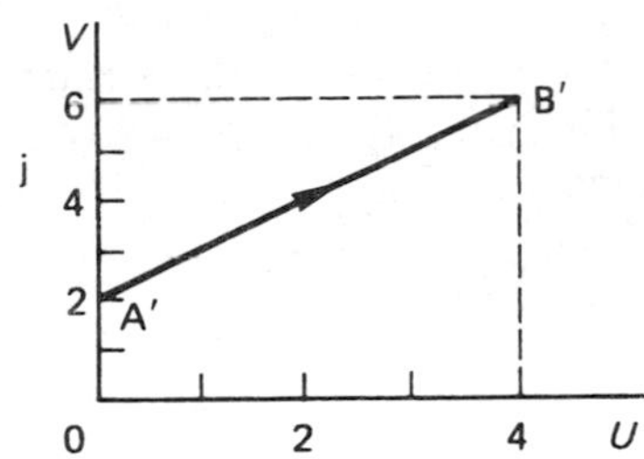

Note that (a) all distances in the z-plane are magnified by a factor 2, and (b) the direction of A′B′ is that of AB unchanged. Any such transformation $w = az$ where a is real, is said to be a *magnification* by the factor a.

So, if we apply the transformation $w = z/2$ to AB shown here, we can map AB on to A′B′ in the w-plane and obtain

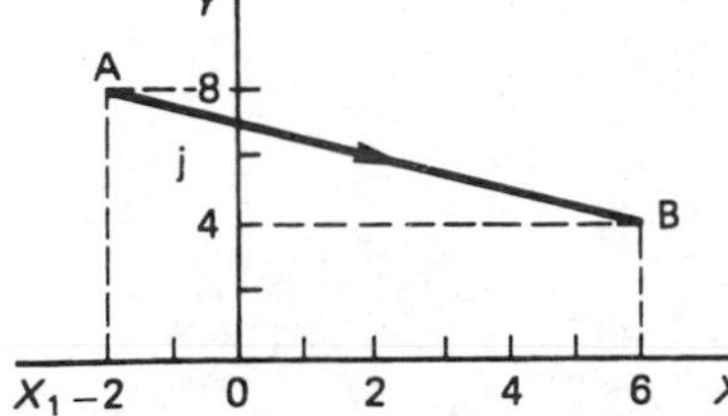

....

(Sketch the result)

20

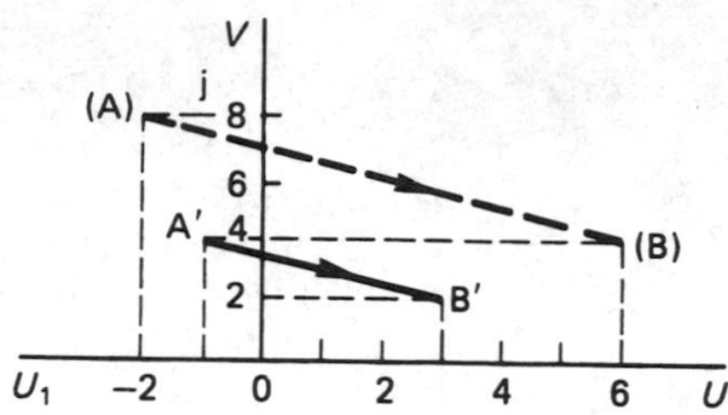

3. *Rotation*

Finally let us consider $w = az + b$ with $b = 0$ and a complex, e.g.

$$w = jz.$$

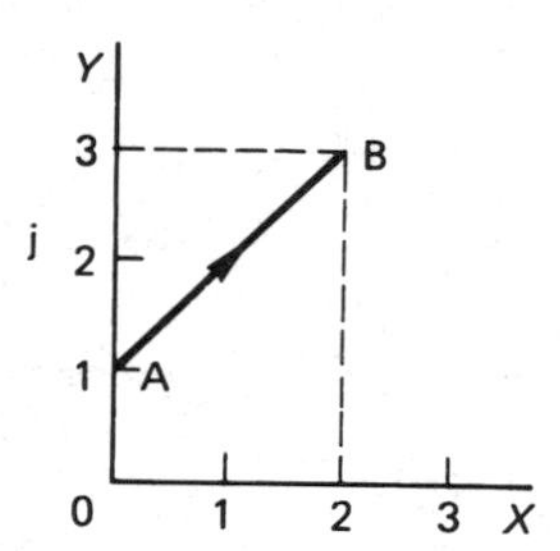

$$w = u + jv = jz = j(x + jy)$$
$$= -y + jx$$

$$\therefore\ u = -y \quad \text{and} \quad v = x$$

Transforming the end points as usual, we can sketch the original line AB and the mapping A′B′, which gives

. . . .

21

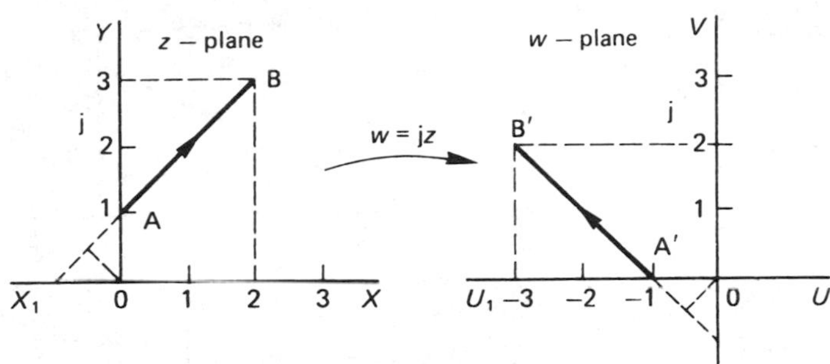

A′ is the point $w = -1 + \mathrm{j}0$; B′ is the point $w = -3 + \mathrm{j}2$

Note $\quad \mathrm{AB} = 2\sqrt{2}$ $\qquad \mathrm{A'B'} = 2\sqrt{2}$

Slope of AB $= m = 1$ $\qquad$ Slope of A′B′ $= m_1 = -1$

$$mm_1 = 1(-1) = -1$$

Therefore in transformation by $w = \mathrm{j}z$, AB retains its original length but is rotated about the origin, in this case, through 90° in a positive (anticlockwise) direction.

Some degree of rotation always occurs when the transformation equation is of the form $w = az + b$ with a complex.

Move on to the next frame.

22

4. *Combined magnification and rotation*

If $w = (a + \mathrm{j}b)z$, the effect of transformation is

(a) magnification $|\, a + \mathrm{j}b \,| = \sqrt{a^2 + b^2}$

(b) rotation anticlockwise through $\arg(a + \mathrm{j}b)$, i.e. $\arctan \dfrac{b}{a}$.

Let us see this with an example.

Example Map the straight line joining A $(0 + \mathrm{j}2)$ and B $(4 + \mathrm{j}6)$ in the z-plane on to the w-plane under the transformation $w = (3 + \mathrm{j}2)z$.

The working is just as before. Draw the z-plane and w-plane diagrams, which give

. . . .

23

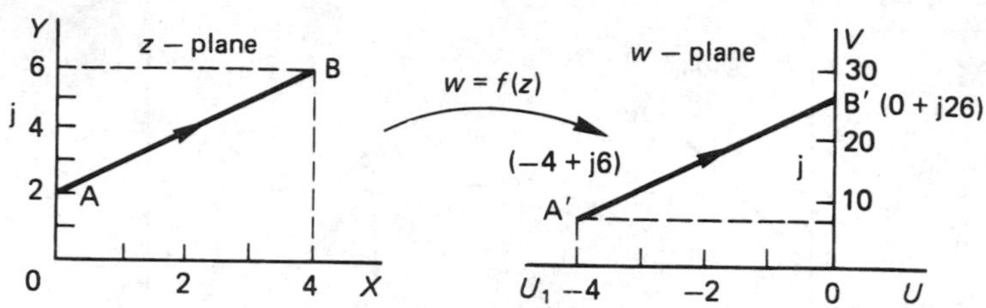

$$w = (3 + j2)z$$

$$\therefore\ u + jv = (3 + j2)(x + jy) = (3x - 2y) + j(2x + 3y)$$

$$\therefore\ u = 3x - 2y \quad \text{and} \quad v = 2x + 3y$$

A: $z = 0 + j2$, i.e. $x = 0$, $y = 2$ $\therefore$ A′: $u = -4$, $v = 6$ $\therefore$ A′: $(-4 + j6)$

B: $z = 4 + j6$, i.e. $x = 4$, $y = 6$ $\therefore$ B′: $u = 0$, $v = 26$ $\therefore$ B′: $(0 + j26)$

By a simple application of Pythagoras, we can now calculate the lengths of AB and A′B′, and then determine the magnification factor (A′B′)/(AB).

AB = ...; A′B′ = ...; magnification = ...

24

$$\boxed{AB = 4\sqrt{2};\quad A'B' = 4\sqrt{26};\quad \text{mag} = \sqrt{13}}$$

for $AB = \sqrt{16+16} = \sqrt{32} = 4\sqrt{2}$

$A'B' = \sqrt{16+400} = \sqrt{416} = 4\sqrt{26}$

$$\therefore \text{ magnification} = \frac{4\sqrt{26}}{4\sqrt{2}} = \sqrt{13}$$

Also $|a+jb| = |3+j2| = \sqrt{9+4} = \sqrt{13}$ $\quad\therefore$ <u>mag $= |a+jb|$</u>

Now let us check the rotation.

For AB $\quad \tan\theta_1 = 1 \quad \therefore\ \theta_1 = 45° = 0.7854$

For A′B′ $\quad \tan\theta_2 = 5 \quad \therefore\ \theta_2 = 78°\ 41' = 1.3733$

$$\therefore \text{ rotation} = \theta_2 - \theta_1 = 1.3733 - 0.7854 = 0.5879$$

i.e. <u>rotation = 0.5879 radians</u>

Also $\arg(a+jb) = \arg(3+j2) = \dots\dots\dots$

25

$$\boxed{0.5879}$$

for $\arg(3+j2) = \arctan\frac{2}{3} = 33°\ 41' = 0.5879$.

So, in transformation $w = (a+jb)z = (3+j2)z$

(a) AB is magnified by $|a+jb|$ i.e. $\sqrt{13}$

(b) AB is rotated anticlockwise through $\arg(a+jb)$, i.e. $\arg(3+j2)$ i.e. 0.5879 radians.

5. *Combined magnification, rotation and translation*

The work we have just done can be extended to include all three effects of transformation.

In general, a transformation equation $w = az + b$, where a and b are each real or complex, results in

magnification $|a|$; rotation $\arg a$; translation b.

Therefore, if $w = (3+j)z + 2 - j$

magnification = ...; rotation = ...; translation = ...

26

> mag $= \sqrt{10} = 3.162$; rotation $= 18^\circ\ 26' = 0.3218$;
> translation $=$ 2 units to right, 1 unit downwards

for (a) magnification $= |3 + \mathrm{j}| = \sqrt{9+1} = \sqrt{10} = 3.162$

(b) rotation $= \arg(3 + \mathrm{j}) = \arctan\frac{1}{3} = 18^\circ\ 26' = 0.3218$ radians

(c) translation $= 2 - \mathrm{j}$, i.e. 2 to the right, 1 downwards.

Let us work through an example in detail.

Example 1 The straight line joining A $(-2 - \mathrm{j}3)$ and B $(3 + \mathrm{j})$ in the z-plane is subjected to the linear transformation equation

$$w = (1 + \mathrm{j}2)z + 3 - \mathrm{j}4$$

Illustrate the mapping on to the w-plane and state the resulting magnification, rotation and translation involved.

The first part is just like examples we have already done. So,

(a) transform the end points A and B on to A′ and B′ in the w-plane

(b) join A′ and B′ with a straight line, since AB is a straight line and the transformation equation is linear.

That can be done without trouble, the final diagram being

....

27

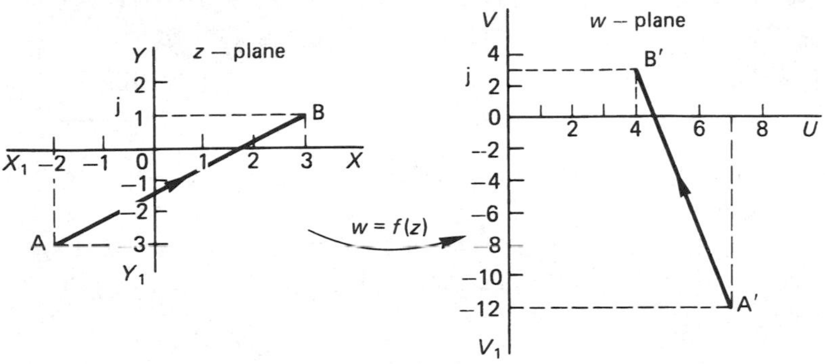

Check the working. $w = (1 + j2)z + 3 - j4$

A: $z = x + jy = -2 - j3$

A′: $w = u + jv = (1 + j2)(-2 - j3) + 3 - j4 = -2 - j7 + 6 + 3 - j4$

$$= \underline{7 - j11}$$

B: $z = x + jy = 3 + j$

B′: $w = u + jv = (1 + j2)(3 + j) + 3 - j4 = 3 + j7 - 2 + 3 - j4 = \underline{4 + j3}$

Now for the second part of the problem, we have to state the magnification, rotation and translation when $w = (1 + j2)z + 3 - j4$. We remember that the 'tailpiece' i.e. $3 - j4$, independent of z, represents the ....

28

translation

So, for the moment, we concentrate on $w = (1 + j2)z$, which determines the magnification and rotation. This tells us that

magnification = ... ; rotation = ...

29

$$\text{mag} = |a| = |1 + j2| = \sqrt{1 + 4} = \sqrt{5} = 2.236$$
$$\text{rotation} = \arg a = \arctan \tfrac{2}{1} = 63^\circ\ 26' = 1.107 \text{ radians}$$

The translation is given by $(3 - j4)$, i.e. 3 units to the right, 4 units downwards.

We can in fact see the intermediate steps if we deal first with the transformation $w = (1 + j2)z$ and subsequently with the translation $w = 3 - j4$.

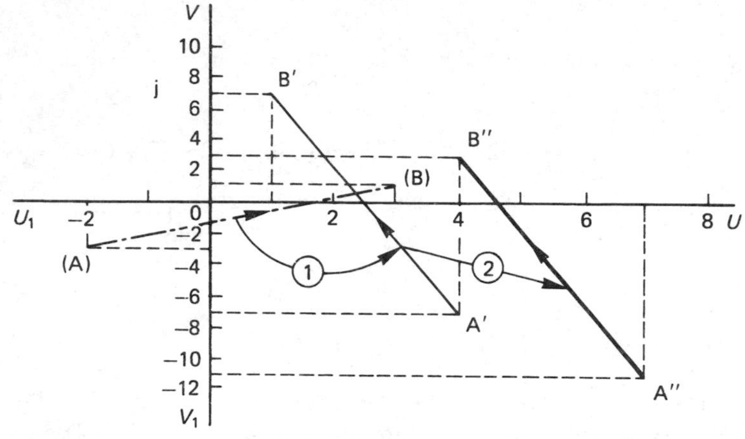

Under $w = (1 + j2)z$, A and B map on to A′ and B′ where A′ is $w = 4 - j7$ and B′ is $w = 1 + j7$.

Then the translation $w = 3 - j4$ moves all points 3 units to the right and 4 units downwards, so that A′ and B′ now map on to A″ and B″ where A″ is $w = 7 - j11$ and B″ is $w = 4 + j3$.

Normally, there is no need to analyse the transformation into intermediate steps.

Now for—

Example 2 Map the straight line joining A $(1 + j2)$ and B $(4 + j)$ in the z-plane on to the w-plane using the transformation equation

$$w = (2 - j3)z - 4 + j5$$

and state the magnification, rotation and translation involved.

There are no snags. Complete the working and check with the next frame.

30

Here is the complete working.

$$w = (2 - \mathrm{j}3)z - 4 + \mathrm{j}5$$

A: $z = 1 + \mathrm{j}2$

B: $z = 4 + \mathrm{j}$

A: $\quad z = 1 + \mathrm{j}2$

A′: $\quad w = (2 - \mathrm{j}3)(1 + \mathrm{j}2) - 4 + \mathrm{j}5 = 2 + \mathrm{j} + 6 - 4 + \mathrm{j}5 = \underline{4 + \mathrm{j}6}$

B: $\quad z = 4 + \mathrm{j}$

B′: $\quad w = (2 - \mathrm{j}3)(4 + \mathrm{j}) - 4 + \mathrm{j}5 = 8 - \mathrm{j}10 + 3 - 4 + \mathrm{j}5 = \underline{7 - \mathrm{j}5}$

So we have

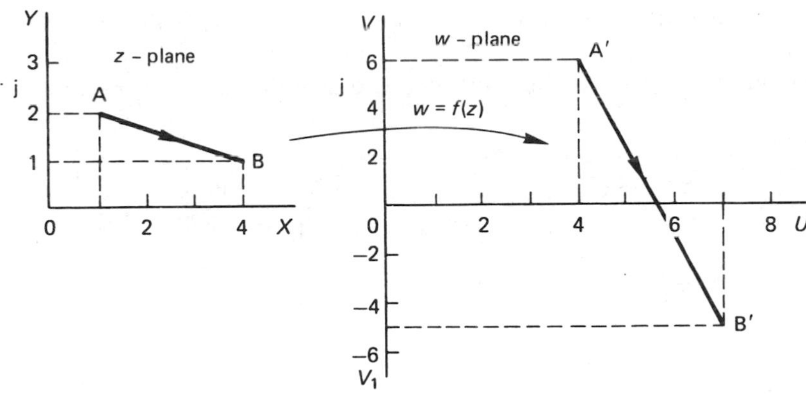

Also we have

(a) magnification $= |2 - \mathrm{j}3| = \sqrt{4 + 9} = \sqrt{13} = \underline{3.606}$

(b) rotation $= \arg(2 - \mathrm{j}3) = \arctan\left(\frac{-3}{2}\right) = -56° \; 19'$

$\qquad = \underline{0.9828 \text{ clockwise}}$

(c) translation $= -4 + \mathrm{j}5$ i.e. $\underline{\text{4 units to left, 5 units upwards}}$

All very straightforward. Before we move on, here is a short revision exercise.

Exercise

Calculate (a) the magnification, (b) the rotation, (c) the translation involved in each of the following transformations.

1. $w = (1 - j2)z + 2 - j3$
2. $w = (4 + j3)z - 2 + j5$
3. $w = (2 - j3)z - 1 - j$
4. $w = (j - 4)z + j2 - 3$
5. $w = j2z + 4 - j$
6. $w = (5 + j2)z + j(j3 - 4)$.

Complete all six and then check the results with the next frame.

31

Results:

1. $$w = (1 - j2)z + 2 - j3$$

(a) magnitude $= |\,1 - j2\,| = \sqrt{1+4} = \sqrt{5} = \underline{2.236}$

(b) rotation $= \arg(1 - j2) = \arctan(-2) = -63°\ 26'$
$= \underline{1.107 \text{ clockwise}}$

(c) translation $= 2 - j3$, i.e. $\underline{\text{2 units to right, 3 units downwards.}}$

The others are done in the same way and give the following results.

No.	Magnitude	Rotation	Translation
2.	5	0.6435 ac	2L, 5U
3.	3.606	0.9828 c	1L, 1D
4.	4.123	0.2450 c	3L, 2U
5.	2	1.5708 ac	4R, 1D
6.	5.385	0.3805 ac	3L, 4D

Now let us start a new section, so on to the next frame.

32

Non-linear transformations

So far, we have concentrated on linear transformations of the form $w = az + b$. We can now proceed to something rather more interesting.

1. *Transformation* $w = z^2$

   The general principles are those we have used before. An example will show the development.

Example 1 The straight line AB in the z-plane as shown is mapped on to the w-plane by $w = z^2$.

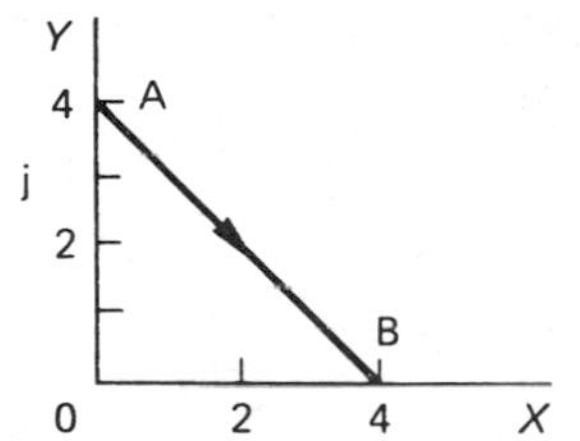

As before, we start by transforming the end points on to A′ and B′ in the w plane.

A′: $w = \ldots\ldots\ldots$; B′: $w = \ldots\ldots\ldots$

33

A′: $w = -16$; B′: $w = 16$

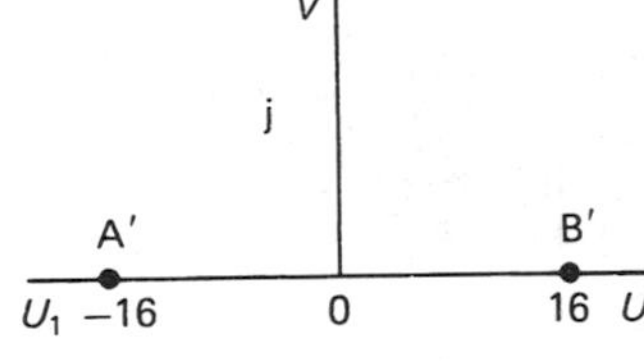

We cannot however assume that AB maps on to the straight line A′B′, since the transformation is not linear. We therefore have to deal with a general point.

$$w = u + \mathrm{j}v = z^2 = (x + \mathrm{j}y)^2 = x^2 + \mathrm{j}2xy - y^2 = (x^2 - y^2) + \mathrm{j}2xy$$

$$\therefore\ u = x^2 - y^2 \quad \text{and} \quad v = 2xy$$

The cartesian equation of AB in the z-plane is $y = 4 - x$. So, substituting in the results of the previous line, we can express u and v in terms of x.

$u = \ldots\ldots\ldots$; $v = \ldots\ldots\ldots$

34

$$u = 8x - 16; \qquad v = 8x - 2x^2$$

The first gives $x = \dfrac{u+16}{8}$ and substituting this in the expression for v gives

....

35

$$v = -\tfrac{1}{32}u^2 + 8$$

for $v = 8\left(\dfrac{u+16}{8}\right) - 2\left(\dfrac{u+16}{8}\right)^2 = u + 16 - \dfrac{u^2}{32} - u - 8$

$$\therefore\ v = -\frac{u^2}{32} + 8$$

which is an 'inverted' parabola, symmetrical about the v-axis, with $v = 8$ at $u = 0$. The mapping is therefore

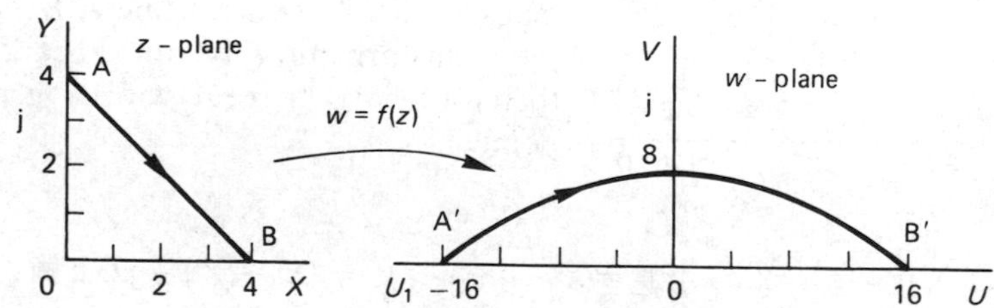

36

Example 2 AB is a straight line in the z-plane joining the origin A to the point B$(a + \text{j}b)$. Obtain the mapping of AB on to the w-plane under the transformation $w = z^2$.

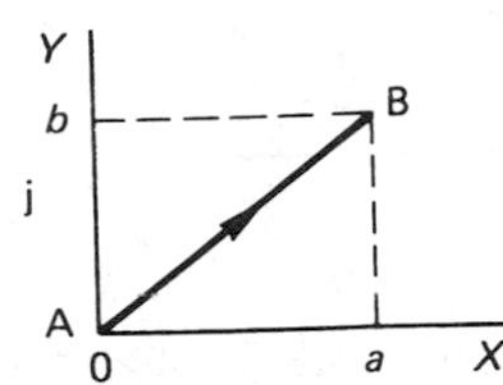

As always, first map the end points.

$$\text{A}': \; w = 0$$

$$\text{B}': \; w = (a + \text{j}b)^2 = (a^2 - b^2) + \text{j}2ab$$

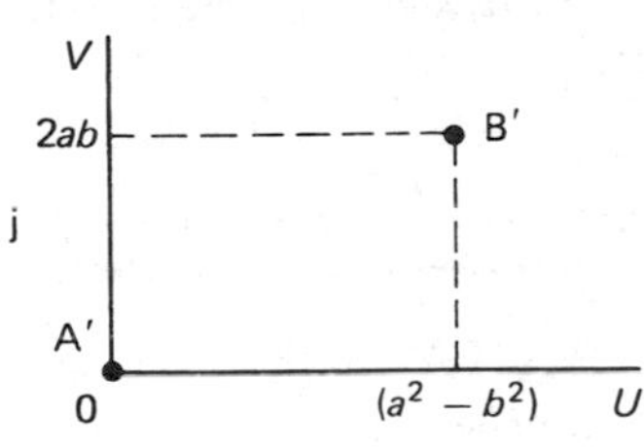

Now to find the path joining A′ and B′, we consider a general point $z = x + \text{j}y$.

$$w = u + \text{j}v = z^2 = (x + \text{j}y)^2 - (x^2 - y^2) + \text{j}2xy$$

$$\therefore \; u = x^2 - y^2 \quad \text{and} \quad v = 2xy$$

The equation of AB is $y = \dfrac{b}{a}x$. We can therefore find u and v in terms of x and hence v in terms of u

$$u = \ldots\ldots\ldots; \quad v = \ldots\ldots\ldots; \quad v = f(u) = \ldots\ldots\ldots$$

37

$$u = \left(\frac{a^2 - b^2}{a^2}\right)x^2; \; v = \left(\frac{2b}{a}\right)x^2; \; v = \left(\frac{2ab}{a^2 - b^2}\right)u$$

for $u = x^2 - y^2 = x^2 - \left(\dfrac{b^2}{a^2}\right)x^2 = \left(\dfrac{a^2 - b^2}{a^2}\right)x^2$

$$v = 2xy = 2x\left(\frac{b}{a}\right)x = \left(\frac{2b}{a}\right)x^2$$

From the expression for u, $x^2 = \left(\dfrac{a^2}{a^2 - b^2}\right)u \quad \therefore \; v = \dfrac{2b}{a}\left(\dfrac{a^2}{a^2 - b^2}\right)u$

$\therefore \; v = \left(\dfrac{2ab}{a^2 - b^2}\right)u$ which is of the form $v = ku$.

A′B′ is therefore a straight line through the origin.

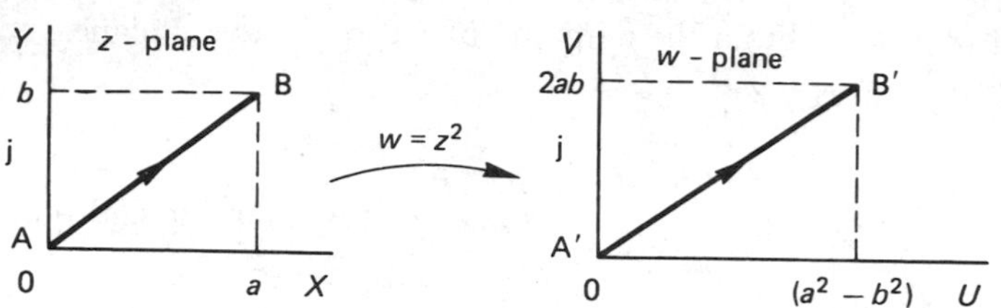

Therefore, under the transformation $w = z^2$, a straight line through the origin in the z-plane maps on to a straight line through the origin in the w-plane.

This is worth remembering, so make a note of it.

---

38 *Example 3* A triangle consisting of AB, BC, CA in the z-plane is mapped on to the w-plane by the transformation $w = z^2$.

The transformation is $w = z^2$.

$$\therefore w = (x + \text{j}y)^2 = (x^2 - y^2) + \text{j}2xy$$
$$= u + \text{j}v$$

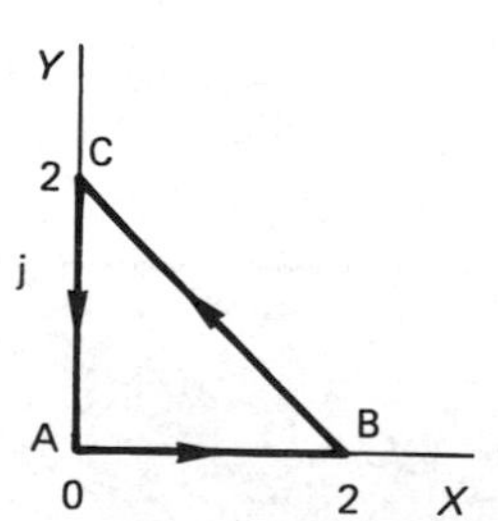

$$\therefore u = x^2 - y^2 \quad \text{and} \quad v = 2xy$$

First we can map the end points A, B, C on to A′, B′, C′ in the w-plane

A′: ....; B′: ....; C′: ....

39

$$\boxed{A': w = 0; \quad B': w = 4; \quad C': w = -4}$$

So we establish

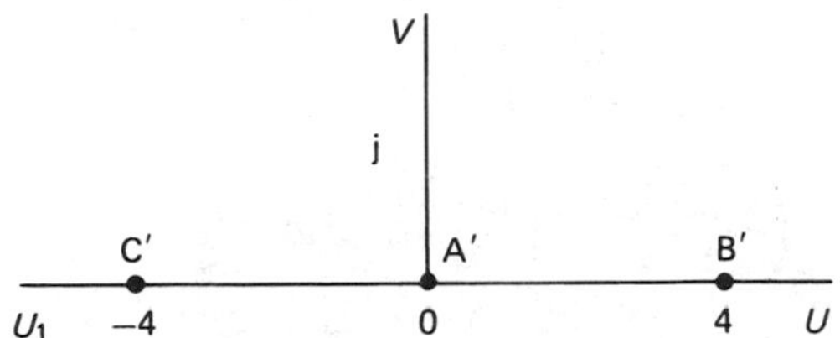

To find the paths joining these three transformed end points, we consider each of the sides of the triangle in turn.

(a) AB: Equation of AB is $y = 0 \quad \therefore u = x^2; \quad v = 0$
$\therefore$ Each point in AB maps on to a point between A′ and B′ for which $v - 0$, i.c. part of thc u-axis.

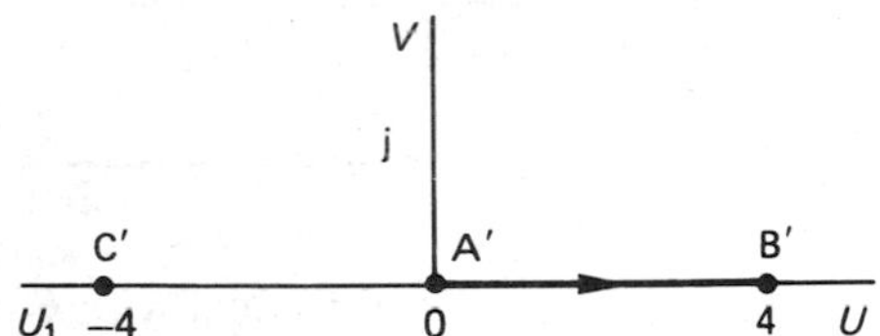

(b) BC: Equation of BC is $y = 2 - x$
Substitute in $u = x^2 - y^2$ and $v = 2xy$ and determine v as a function of u.

$u = \ldots\ldots\ldots; \quad v = \ldots\ldots\ldots; \quad v = f(u) = \ldots\ldots\ldots$

40

$$u = 4x - 4; \quad v = 4x - 2x^2; \quad v = 2 - \frac{u^2}{8}$$

for $\quad u = x^2 - y^2 = x^2 - (2 - x)^2 = 4x - 4 \quad \therefore x = \dfrac{u+4}{4}$

$$v = 2xy = 2x(2 - x) = 4x - 2x^2$$

$$\therefore v = 4\left(\frac{u+4}{4}\right) - 2\left(\frac{u+4}{4}\right)^2 = 2 - \frac{u^2}{8}$$

Therefore, the path joining B′ to C′ is an ....

41

inverted parabola

$v = 2 - \dfrac{u^2}{8} \quad \therefore$ at $u = 0$, $v = 2$ and the w-plane diagram now becomes

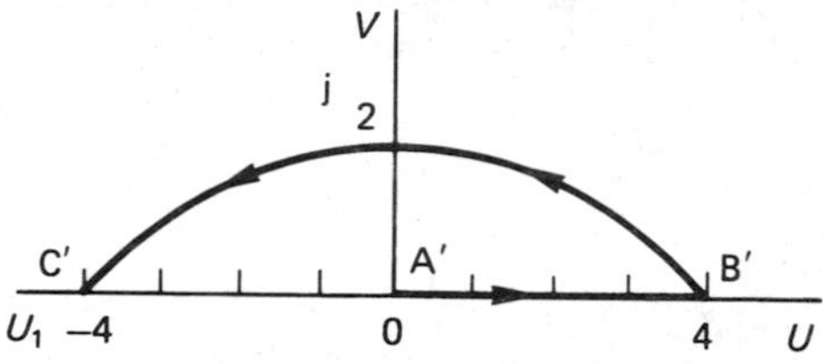

To complete the mapping, we have still to deal with CA. This transforms on to

the u-axis between C′ and A′

(c) CA: Equation of CA is $x = 0$ $\therefore$ $u = -y^2$, $v = 0$

$\therefore$ Each point between C and A maps on to the negative part of the u-axis between C′ and A′.

So finally we have

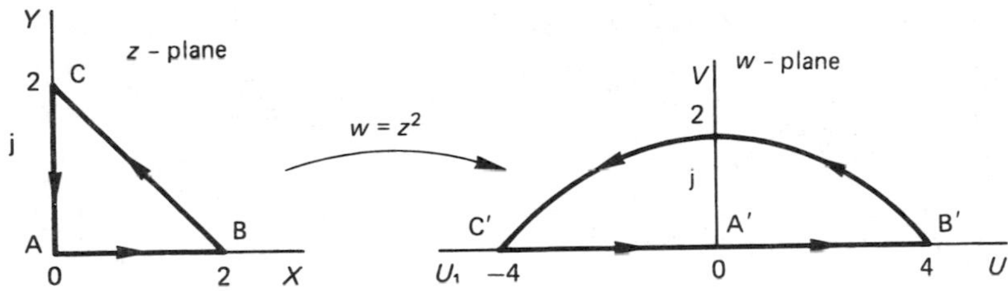

Note: Mapping of regions

In this last example, the three lines AB, BC, CA form the boundary of the triangular region. In the transformed figure in the w-plane, the boundary A′B′C′A′ could be regarded as the boundary of the internal region or of the external region.

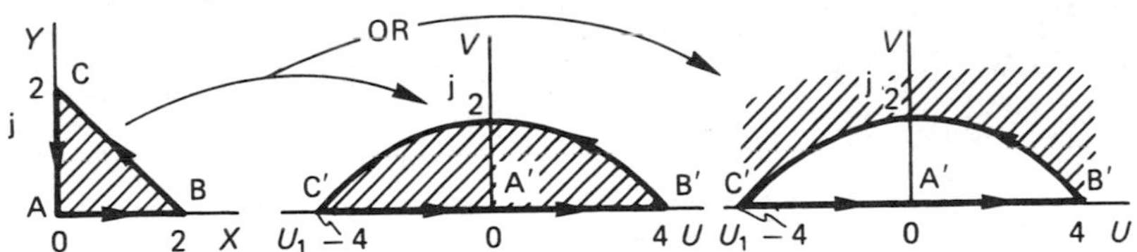

In the z-plane, the region is on the left-hand side as we proceed round the figure in the direction of the arrows ABCA. The region on the left-hand side as we proceed round the figure A′B′C′A′ in the w-plane determines that the transformed region in this case is, in fact, the internal region.

So

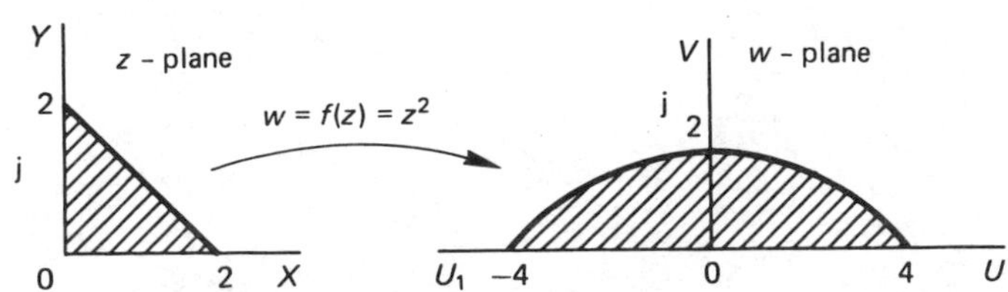

Therefore, every point in the region shaded in the z-plane maps on to a corresponding point in the region shaded in the w-plane.

43

2. *Transformation* $w = \frac{1}{z}$ (inversion)

Example 1 A straight line joining A $(-j)$ and B $(2 + j)$ in the z-plane is mapped on to the w-plane by the transformation equation $w = \frac{1}{z}$.

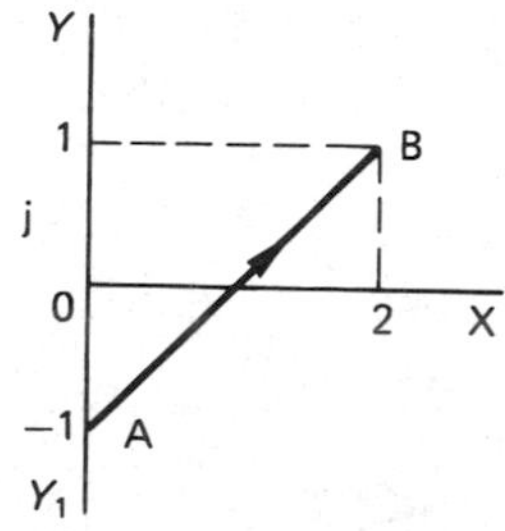

Proceeding as before

$$w = \frac{1}{z} \quad \therefore u + jv = \frac{1}{x + jy} = \frac{x - jy}{x^2 + y^2}$$

$$\therefore u = \frac{x}{x^2 + y^2}; \quad v = \frac{-y}{x^2 + y^2}$$

First we map the end points A and B on to the w-plane.

A′: $w =$ ....; B′: $w =$ ....

44

$$\boxed{A': \ w = j; \quad B': \ w = \tfrac{2}{5} - j\tfrac{1}{5}}$$

since A: $x = 0, \ y = -1 \quad \therefore A': \ u = 0, \ v = 1 \qquad \therefore A'$ is $\underline{w = j}$

B: $x = 2, \ y = 1 \quad \therefore B': \ u = \frac{2}{5}, \ v = -\frac{1}{5} \quad \therefore B'$ is $\underline{w = \frac{2}{5} - j\frac{1}{5}}$

So far then we have

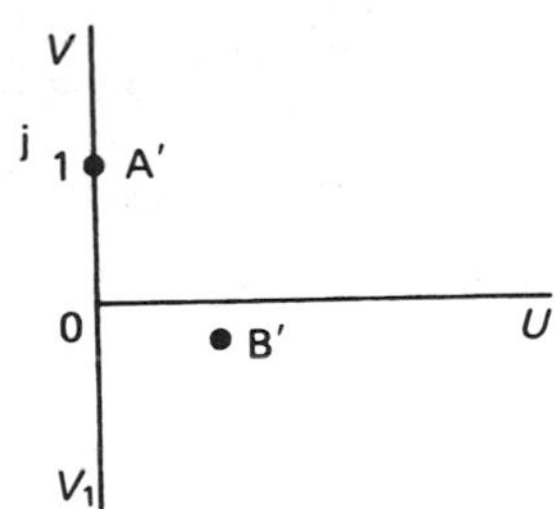

To determine the path A′B′, we can proceed as follows

$$w = \frac{1}{z} \qquad \therefore z = \frac{1}{w} \quad \text{i.e.} \quad x + jy = \frac{1}{u + jv} = \frac{u - jv}{u^2 + v^2}$$

$$\therefore x = \frac{u}{u^2 + v^2} \quad \text{and} \quad y = \frac{-v}{u^2 + v^2}$$

The equation of AB is $\qquad y = x - 1$

$$\therefore \frac{-v}{u^2 + v^2} = \frac{u}{u^2 + v^2} - 1$$

which simplifies into ....

45

$$\boxed{u^2 + v^2 - u - v = 0}$$

for $\dfrac{-v}{u^2 + v^2} = \dfrac{u}{u^2 + v^2} - 1 \quad \therefore \; -v = u - u^2 - v^2$

$$\therefore \; \underline{u^2 + v^2 - u - v = 0}$$

We can write this as $(u^2 - u) + (v^2 - v) = 0$

and completing the square in each bracket this becomes

$$\left(u - \frac{1}{2}\right)^2 + \left(v - \frac{1}{2}\right)^2 = \frac{1}{2}$$

which we recognise as the equation of a ....

46

$$\boxed{\text{circle with centre } \left(\frac{1}{2}, \frac{1}{2}\right) \text{ and radius } \frac{1}{\sqrt{2}}}$$

The path joining A′ and B′ is therefore an arc of this circle.

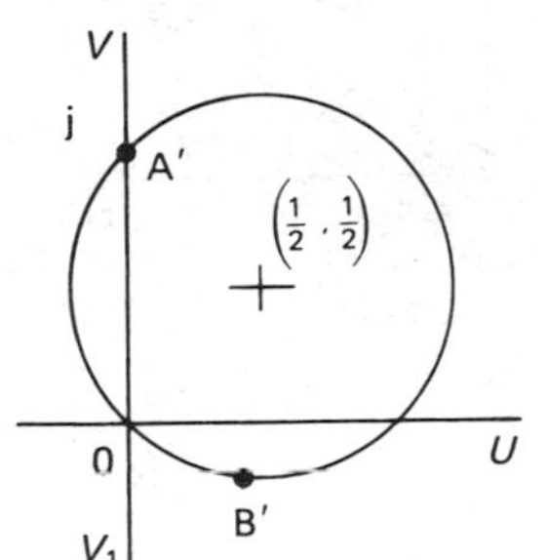

But we still have problems, for it could be the minor arc or the major arc.

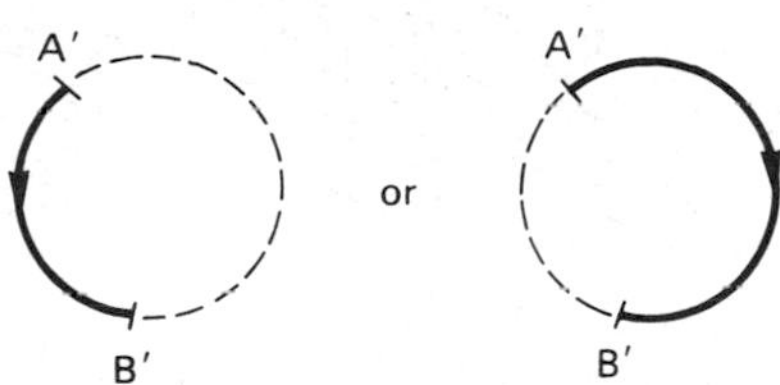

To decide which is correct, we take a further convenient point on the original line AB and determine its image on the w-plane.

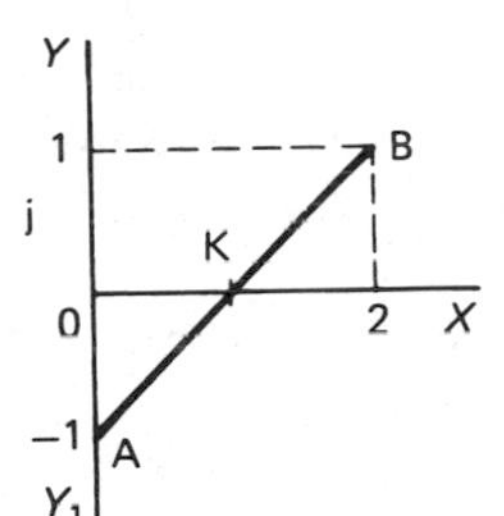

For instance, for K, $x = 1,\ y = 0$

$$\therefore \text{ For K}',\quad u = \frac{x}{x^2 + y^2} = 1$$

$$v = \frac{-y}{x^2 + y^2} = 0$$

$\therefore$ K′ is the point $w = 1$

The path is, therefore, the major arc A′K′B′ developed in the direction indicated.

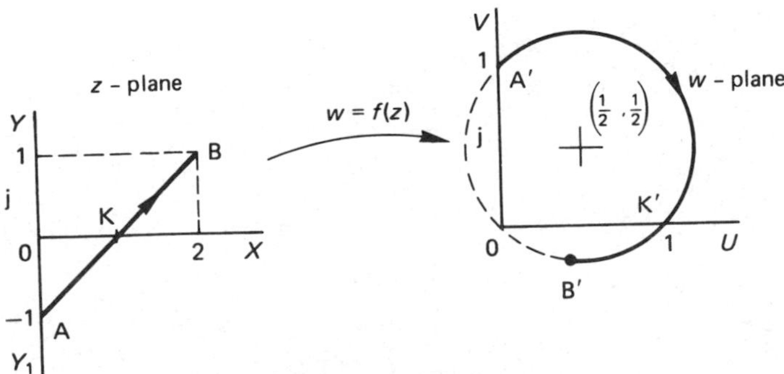

47 If we consider the line AB of our last example extended to infinity in each direction, its image in the w-plane would then be the complete circle.

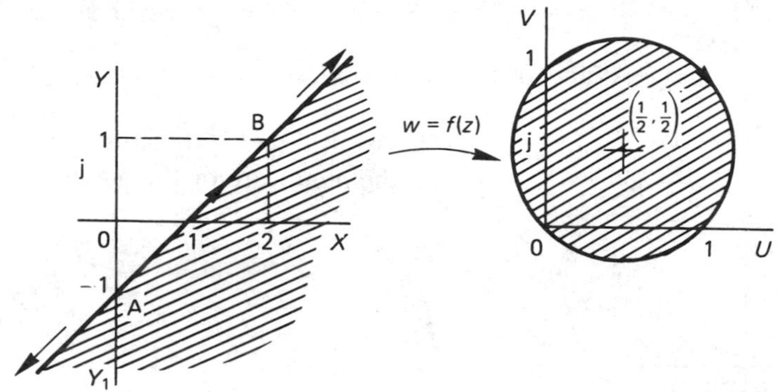

Furthermore, the line AB cuts the entire z-plane into two regions and

(a) the region on the right-hand side of the line relative to the arrowed direction maps on to the region inside the circle in the w-plane

(b) the region on the left-hand side of the line maps on to

. . . .

48

> the region outside the circle in the w-plane

Let us now consider a general case.

Example 2 Determine the image in the w-plane of a circle in the z-plane under the inversion transformation $w = \frac{1}{z}$.

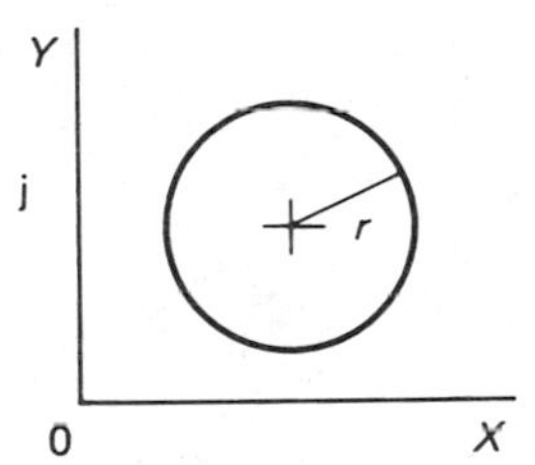

The general equation of a circle is

$$x^2 + y^2 + 2gx + 2fy + c = 0$$

with centre $(-g, -f)$

and radius $\sqrt{g^2 + f^2 - c}$.

It is convenient at times to write this as

$$A(x^2 + y^2) + Dx + Ey + F = 0$$

in which case

centre is .... and radius is ....

49

> centre $\left(-\frac{D}{2A}, -\frac{E}{2A}\right)$; radius $= \frac{1}{2A}\sqrt{D^2 + E^2 - 4AF}$

since $g = \frac{D}{2A}$, $f = \frac{E}{2A}$, $c = \frac{F}{A}$.

As before we have $w = \frac{1}{z}$ $\therefore z = \frac{1}{w}$

$$\therefore x + jy = \frac{1}{u + jv} = \frac{u - jv}{u^2 + v^2} \quad \therefore x = \frac{u}{u^2 + v^2}; \quad y = \frac{-v}{u^2 + v^2}$$

Then $A(x^2 + y^2) + Dx + Ey + F = 0$

becomes ....

(Simplify it as far as possible.)

50

$$\boxed{A + Du - Ev + F(u^2 + v^2) = 0}$$

for we have $\dfrac{A(u^2+v^2)}{(u^2+v^2)^2} + \dfrac{Du}{u^2+v^2} - \dfrac{Ev}{u^2+v^2} + F = 0$

$$\therefore\ A + Du - Ev + F(u^2 + v^2) = 0$$

and changing the order of terms, this can be written

$$F(u^2 + v^2) + Du - Ev + A = 0$$

which is the equation of a circle with

centre ....; radius ....

51

$$\boxed{\text{centre } \left(-\frac{D}{2F}, \frac{E}{2F}\right);\ \text{radius } \frac{1}{2F}\sqrt{D^2 + E^2 - 4FA}}$$

Thus any circle in the z-plane transforms, with $w = \dfrac{1}{z}$, on to another circle in the w-plane.

We have already seen previously that, under inversion, a straight line also maps on to a circle. This may be regarded as a special case of the general result, if we accept a straight line as the circumference of a circle of .... radius.

52

infinite

for $$A(x^2 + y^2) + Dx + Ey + F = 0$$

If $A = 0$, this becomes $Dx + Ey + F = 0$ i.e. a straight line and also the

centre $\left(-\frac{D}{2A}, -\frac{E}{2A}\right)$ becomes infinite

and the radius $\frac{1}{2A}\sqrt{D^2 + E^2 - 4AF}$ becomes infinite.

Therefore, combining the results we have obtained, we have this conclusion:

Under inversion, $w = \frac{1}{z}$, a circle or a straight line in the z-plane transforms on to a circle in the w-plane.

Now for one more example.

Example 3 A circle in the z-plane has its centre at $z = 3$ and a radius of 2 units. Determine its image in the w-plane when transformed by $w = \frac{1}{z}$.

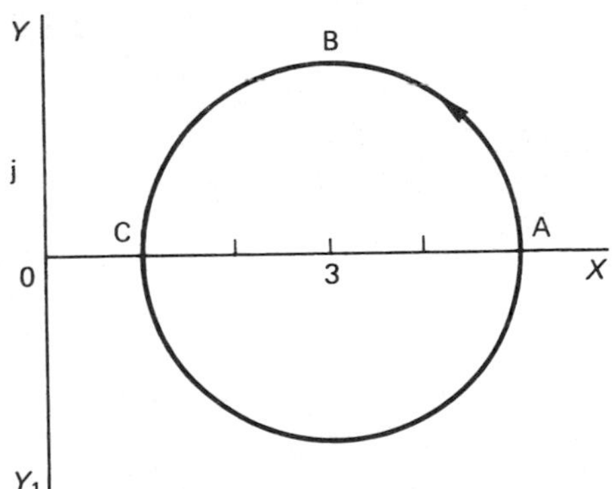

Equation of the circle is

$$(x - 3)^2 + y^2 = 4$$
$$x^2 - 6x + 9 + y^2 = 4$$
$$x^2 + y^2 - 6x + 5 = 0.$$

Using $w = \frac{1}{z}$, we can obtain x and y in terms of u and v.

$x = \ldots\ldots\ldots$; $y = \ldots\ldots\ldots$

53

$$x = \frac{u}{u^2 + v^2}; \quad y = \frac{-v}{u^2 + v^2}$$

since $w = \dfrac{1}{z}$, $\quad \therefore z = \dfrac{1}{w} \quad \therefore x + jy = \dfrac{1}{u + jv} = \dfrac{u - jv}{u^2 + v^2}$

$$\therefore x = \frac{u}{u^2 + v^2}; \quad y = \frac{-v}{u^2 + v^2}$$

Substituting these in the equation of the circle, we get a relationship between u and v, which is ....

54

$$5(u^2 + v^2) - 6u + 1 = 0$$

for the circle is $\quad x^2 + y^2 - 6x + 5 = 0$

$$\therefore \frac{u^2}{(u^2 + v^2)^2} + \frac{v^2}{(u^2 + v^2)^2} - \frac{6u}{u^2 + v^2} + 5 = 0$$

$$\frac{1}{u^2 + v^2} - \frac{6u}{u^2 + v^2} + 5 = 0$$

$$\underline{5(u^2 + v^2) - 6u + 1 = 0}$$

This is of the form $A(u^2 + v^2) + Du + Ev + F = 0$

where $A = 5$, $D = -6$, $E = 0$, $F = 1$.

Therefore, the centre is .... and the radius is ....

55

$$\text{centre} = \left(\frac{3}{5}, 0\right); \quad \text{radius} = \frac{2}{5}$$

since the centre is $\left(-\dfrac{D}{2A}, -\dfrac{E}{2A}\right) = \left(\dfrac{6}{10}, 0\right)$ i.e. $\left(\dfrac{3}{5}, 0\right)$

and the radius $= \dfrac{1}{2A}\sqrt{D^2 + E^2 - 4AF} = \dfrac{1}{10}\sqrt{36 + 0 - 20} = \dfrac{2}{5}$.

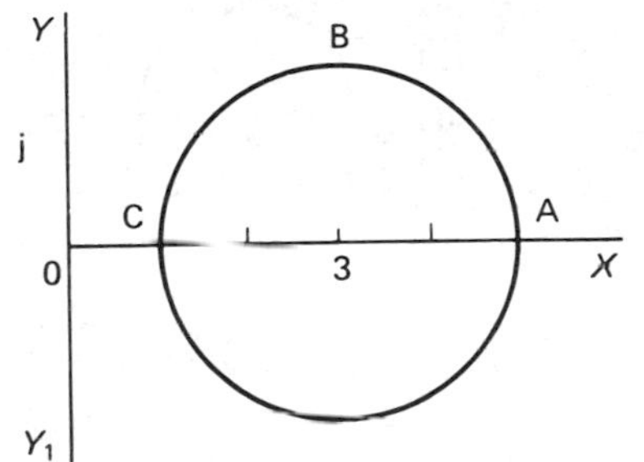

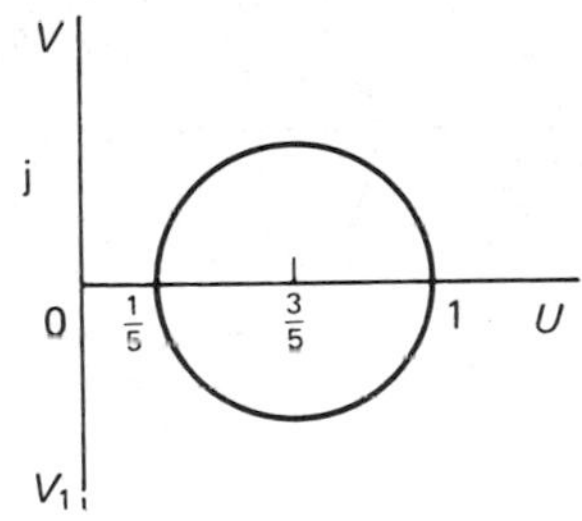

Taking three sample points A, B, C as shown, we can map these on to the w-plane using $u = \dfrac{x}{x^2 + y^2}$ and $v = \dfrac{-y}{x^2 + y^2}$

A′ : ....; B′ : ....; C′ : ....

56

$$\text{A}': \left(\frac{1}{5}, 0\right); \quad \text{B}': \left(\frac{3}{13}, -\frac{2}{13}\right); \quad \text{C}': (1, 0)$$

So we finally have

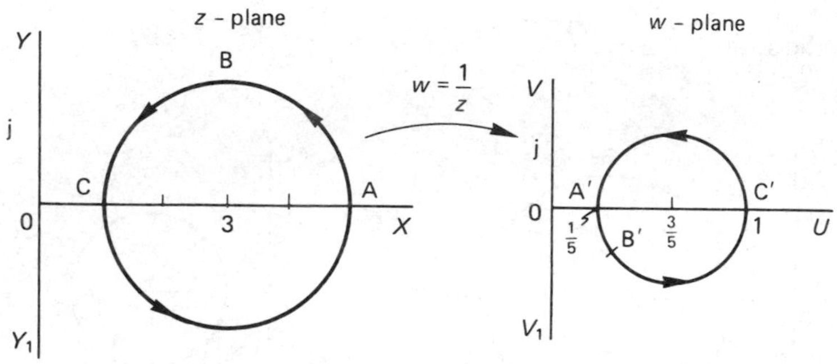

3. *Transformation* $w = \dfrac{1}{z-a}$

An extension of the method we have just applied occurs with transformations of the form $w = \dfrac{1}{z-a}$ where a is real or complex.

Example 4 A circle $|z| = 1$ in the z-plane is mapped on to the w-plane by $w = \dfrac{1}{z-2}$.

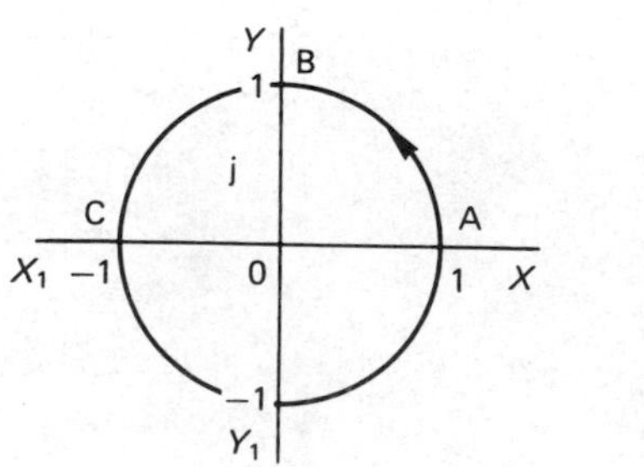

$$w = \frac{1}{z-2} \quad \therefore z - 2 = \frac{1}{w}$$

$$x + \text{j}y - 2 = \frac{1}{u + \text{j}v}$$

$$(x-2) + \text{j}y = \frac{u - \text{j}v}{u^2 + v^2}$$

$$\therefore x = \frac{u}{u^2+v^2} + 2; \quad y = \frac{-v}{u^2+v^2}$$

Cartesian equation of the circle is $x^2 + y^2 = 1$.
We then substitute the expressions for x and y in terms of u and v and obtain the relationship between u and v, which is

....

57

$$\boxed{3(u^2 + v^2) + 4u + 1 = 0}$$

We have $$\left\{\frac{u + 2(u^2 + v^2)}{u^2 + v^2}\right\}^2 + \left\{\frac{-v}{u^2 + v^2}\right\}^2 = 1$$

$$\{u + 2(u^2 + v^2)\}^2 + v^2 = (u^2 + v^2)^2$$

$$1 + 4u + 4(u^2 + v^2) = u^2 + v^2$$

$$3(u^2 + v^2) + 4u + 1 = 0$$

This can be expressed as

$$u^2 + \frac{4}{3}u + v^2 + \frac{1}{3} = 0$$

$$\left(u + \frac{2}{3}\right)^2 + v^2 = \left(\frac{1}{3}\right)^2$$

which is a circle with centre $\left(-\frac{2}{3}, 0\right)$ and radius $\frac{1}{3}$.

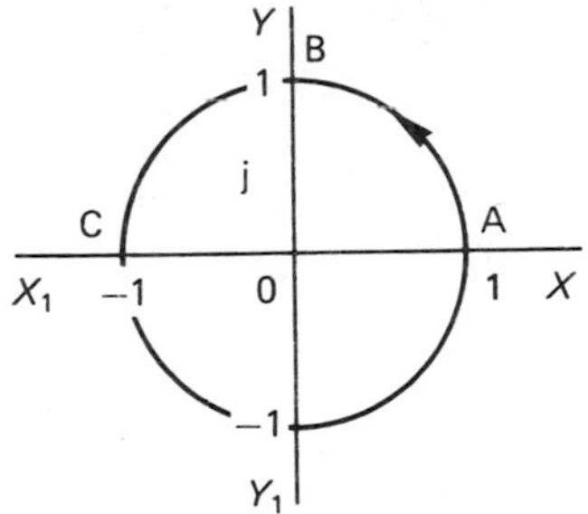

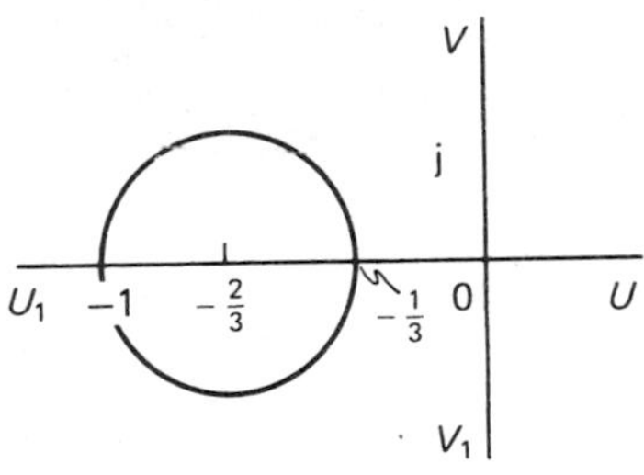

To determine the direction of development relative to the arrowed direction in the z-plane, we consider the mapping of three sample points A, B, C as shown on to the w-plane, giving A′, B′, C′.

A′ : ....; B′ : ....; C′ : ....

58

$$\text{A}': \quad w = (-1,\ 0); \quad \text{B}': \quad w = \left(-\frac{2}{5},\ -\frac{1}{5}\right); \quad \text{C}': \quad w = \left(\frac{1}{3},\ 0\right)$$

for $\text{A}: z = 1 \quad \therefore w = \dfrac{1}{z-2} = -1 \qquad \therefore \text{A}' = (-1,\ 0)$

$\text{B}: z = \text{j} \quad \therefore w = \dfrac{1}{\text{j}-2} = \dfrac{\text{j}+2}{-5} \qquad \therefore \text{B}' = \left(-\dfrac{2}{5},\ -\dfrac{1}{5}\right)$

$\text{C}: z = -1 \quad \therefore w = -\dfrac{1}{3} \qquad \text{C}' = \left(-\dfrac{1}{3},\ 0\right)$

Where upon we have

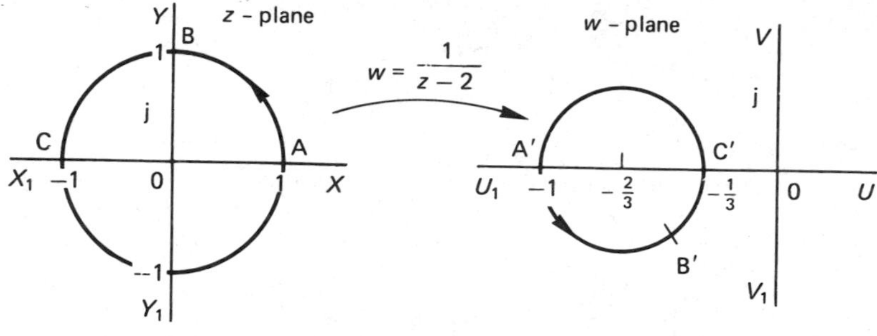

We now have one further transformation which is important, so move on to the next frame for a fresh start.

59

4. *Bilinear transformation* $w = \dfrac{az + b}{cz + d}$

Transformation of the form $w = \dfrac{az + b}{cz + d}$ where a, b, c, d are, in general, complex.

Note that (a) if $cz + d = 1$, $w = az + b$, i.e. the general linear transformation

(b) if $az + b = 1$, $w = \dfrac{1}{cz + d}$, i.e. the form of inversion just considered.

Example Determine the image in the w-plane of the circle $|z| = 2$ in the z-plane under the transformation $w = \dfrac{z + j}{z - j}$ and show the region in the w-plane on to which the region within the circle is mapped.

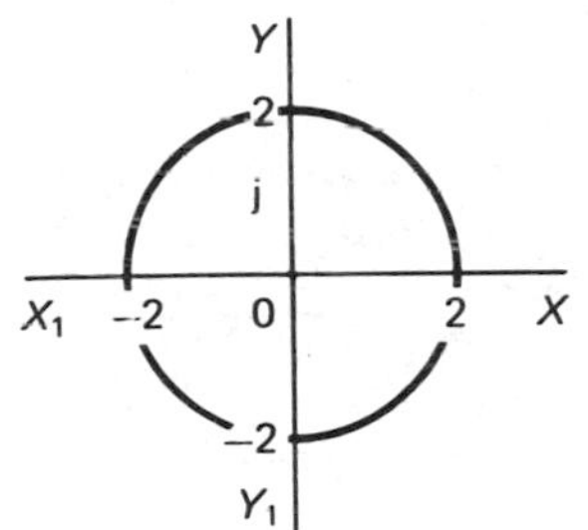

We begin in very much the same way as before by expressing u and v in terms of x and y.

u . . . . ; $v =$ . . . .

60

$$u = \frac{x^2 + y^2 - 1}{x^2 + y^2 - 2y + 1}; \quad v = \frac{2x}{x^2 + y^2 - 2y + 1}$$

for $w = u + jv = \dfrac{z + j}{z - j} = \dfrac{x + j(y+1)}{x + j(y-1)}$

$$= \frac{\{x + j(y+1)\}\{x - j(y-1)\}}{\{x + j(y-1)\}\{x - j(y-1)\}}$$

$$= \frac{x^2 + jx(y + 1 - y + 1) + y^2 - 1}{x^2 + (y-1)^2}$$

$$= \frac{x^2 + y^2 - 1 + j2x}{x^2 + y^2 - 2y + 1}$$

$$\therefore u = \frac{x^2 + y^2 - 1}{x^2 + y^2 - 2y + 1} \quad \text{and} \quad v = \frac{2x}{x^2 + y^2 - 2y + 1}$$

But the equation of the circle is $x^2 + y^2 = 4$, so these expressions simplify to $u = \ldots\ldots\ldots$ and $v = \ldots\ldots\ldots$

61

$$u = \frac{3}{5 - 2y}; \quad v = \frac{2x}{5 - 2y}$$

From these, we can form expressions for x and y in terms of u and v.

$$x = \ldots\ldots\ldots; \quad y = \ldots\ldots\ldots$$

62

$$x = \frac{3v}{2u}; \quad y = \frac{5u - 3}{2u}$$

for, from the first, $y = \dfrac{5u - 3}{2u}$ and substituting in the second gives $x = \dfrac{3v}{2u}$.

But $x^2 + y^2 = 4 \quad \therefore \dfrac{9v^2}{4u^2} + \dfrac{(5u-3)^2}{4u^2} = 4$

which can be simplified to $\ldots\ldots\ldots$

63

$$\boxed{9(u^2 + v^2) - 30u + 9 = 0}$$

for $9v^2 + 25u^2 - 30u + 9 = 16u^2 \quad \therefore 9(u^2 + v^2) - 30u + 9 = 0.$

Dividing through by 9, we can now rearrange this to

$$\left(u^2 - \frac{30}{9}u\right) + v^2 + 1 = 0$$

i.e.

$$\left(u - \frac{5}{3}\right)^2 + v^2 + 1 - \frac{25}{9} = 0$$

$$\left(u - \frac{5}{3}\right)^2 + v^2 = \left(\frac{4}{3}\right)^2$$

which, you will recognise, is a circle in the w-plane with

centre .... and radius ....

64

$$\boxed{\text{centre} = \left(\frac{5}{3}, 0\right); \quad \text{radius} = \frac{4}{3}}$$

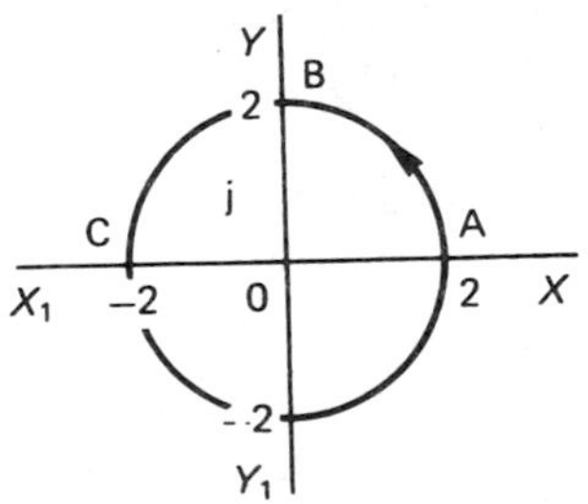

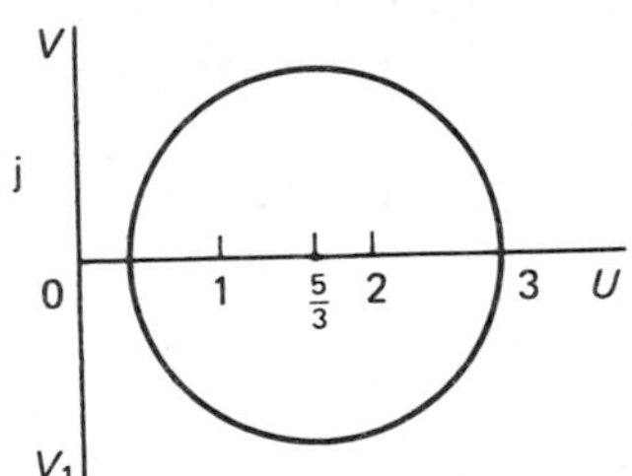

To find the direction of development, we map three sample points A, B, C on to A′, B′, C′ as usual.

A′ : ....; B′ : ....; C′ : ....

65

$$A': \; w = \tfrac{3}{5} + j\tfrac{4}{5}; \quad B': \; w = 3; \quad C': \; w = \tfrac{3}{5} - j\tfrac{4}{5}$$

for A: $z = 2 \qquad \therefore w = \dfrac{2+j}{2-j} = \dfrac{(2+j)^2}{5} = \dfrac{4+j4-1}{5} = \underline{\tfrac{3}{5} + j\tfrac{4}{5}}$ i.e. A′

B: $z = j2 \qquad \therefore w = \dfrac{j2+j}{j2-j} = \dfrac{j3}{j} = 3 \qquad \therefore \underline{w = 3}$ i.e. B′

C: $z = -2 \qquad \therefore w = \dfrac{-2+j}{-2-j} = \dfrac{2-j}{2+j} = \dfrac{(2-j)^2}{5} = \underline{\tfrac{3}{5} - j\tfrac{4}{5}}$ i.e. C′

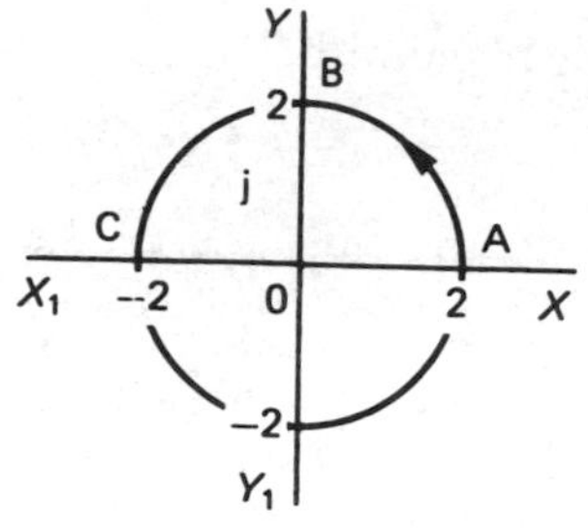

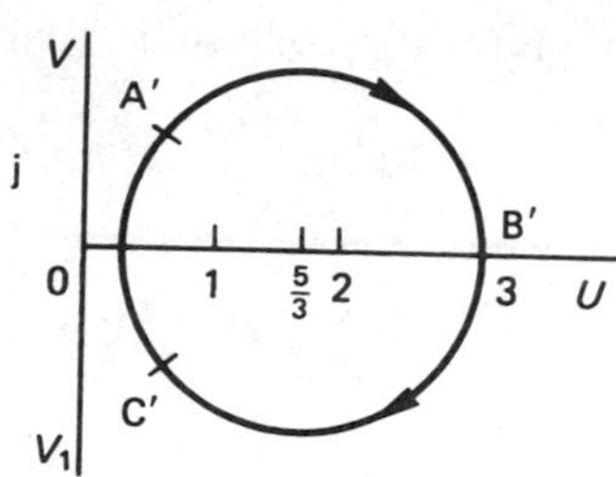

So an anticlockwise progression in the z-plane becomes a clockwise progression in the w-plane with this particular example.

Now we can complete the problem, for the region inside the circle in the z-plane maps on to .... in the w-plane.

66

the region outside the circle

for the enclosed region in the z-plane is on the left-hand side of the direction of progression. The region on the left-hand side of the direction of progression in the w-plane is thus the region outside the transformed circle.

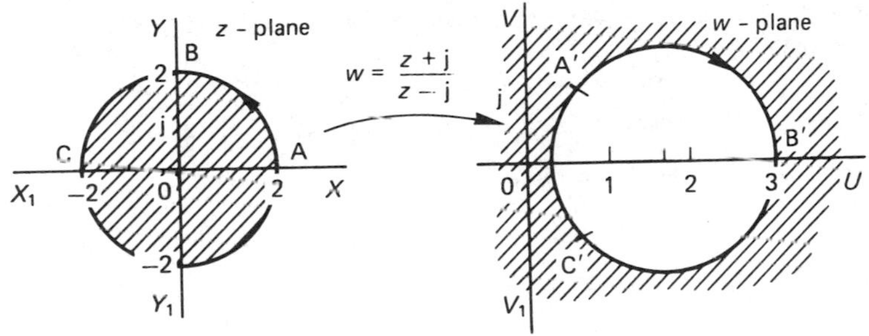

And that brings us successfully to the end of this programme. We shall pursue the topic further in the succeeding programme. Meanwhile, all that remains is to check down the Revision Summary before working through the Test Exercise. All very straightforward.

67 REVISION SUMMARY

1. *Transformation equation*

$$z = x + jy \qquad w = u + jv$$

The transformation equation is the relationship between z and w, i.e. $w = f(z)$.

2. *Linear transformation* $w = az + b$ where a and b are real or complex. A straight line in the z-plane maps on to a corresponding straight line in the w-plane.

3. *Types of transformation* $w = az + b$

   (a) *magnification*—given by $|a|$

   (b) *rotation*—given by arg a

   (c) *translation*—given by b.

4. *Non-linear transformation*

   (a) $w = z^2$

   A straight line through the origin maps on to a corresponding straight line through the origin in the w-plane.

   (b) $w = \dfrac{1}{z}$ (inversion)

   A straight line or a circle maps on to a circle in the w-plane.

   A straight line may be regarded as a circle of infinite radius.

   (c) $w = \dfrac{az + b}{cz + d}$ (bilinear transformation)—with a, b, c, d real or complex.

5. *Mapping of a region* depends on the direction of development. Right-hand regions map on to right-hand regions: left-hand regions on to left-hand regions.

68

TEST EXERCISE XV

1. Map the following points in the z-plane on to the w-plane under the transformation $w = f(z)$

(a) $z = 3 + j2$; $w = 2z - j6$
(b) $z = -2 + j$; $w = 4 + jz$
(c) $z = j(1 - j)$; $w = (2 + j)z - 1$
(d) $z = j - 2$; $w = (1 - j)(z + 3)$.

2. Map the straight line joining A $(2 - j)$ and B $(4 - j3)$ in the z-plane on to the w-plane using the transformation $w = (1 + j2)z + 1 - j3$. State the magnification, rotation and translation involved.

3. A triangle ABC in the z-plane as shown is mapped on to the w-plane under the transformation $w = z^2$.

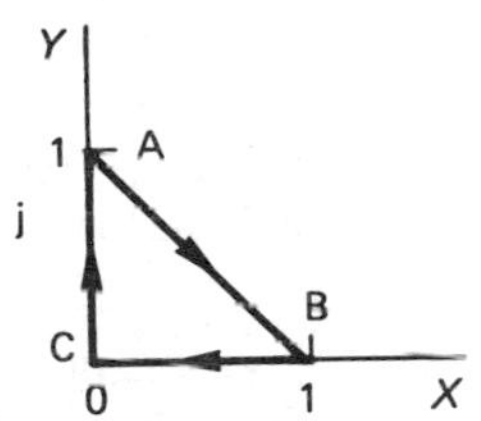

Determine the image in the w-plane and indicate the mapping of the interior triangular region ABC.

4. Map the straight line joining A $(z = j)$ and B $(z = 3 + j4)$ in the z-plane on to the w-plane under the inversion transformation $w = \dfrac{1}{z}$.
Sketch the image of AB in the w-plane.

5. The unit circle $|z| = 1$ in the z-plane is mapped on to the w-plane by $w = \dfrac{1}{z - j2}$. Determine (a) the position of the centre and (b) the radius of the circle obtained.

6. The circle $|z| = 2$ is mapped on to the w-plane by the transformation $w = \dfrac{z + j2}{z + j}$. Determine the centre and radius of the resulting circle in the w-plane.

FURTHER PROBLEMS XV

1. A triangle ABC in the z-plane with vertices A $(-1-j)$, B $(2+j2)$, C $(-1+j2)$ is mapped on to the w-plane under the transformation $w=(1-j)z+(1+j2)$. Determine the image A′B′C′ of ABC in the w-plane.

2. The straight line joining A $(1+j2)$ and B $(4-j3)$ in the z-plane is mapped on to the w-plane by the transformation equation $w=(2+j5)z$. Determine (a) the images of A and B, (b) the magnification, rotation and translation involved.

3. Map the straight line joining A $(-2+j3)$ and B $(1+j2)$ in the z-plane on to the w-plane using the transformation equation $w=(-3+j)z+2+j4$. State the magnification, rotation and translation occurring in the process.

4.

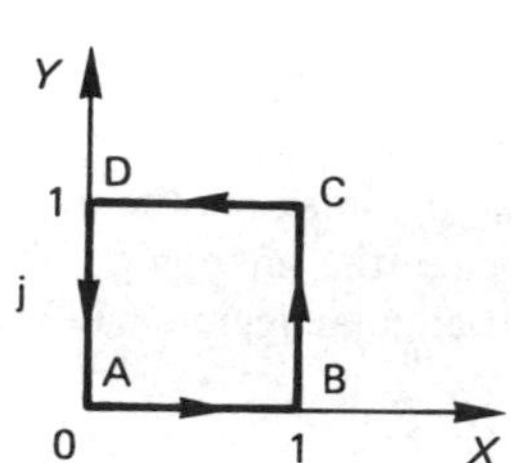

Transform the square ABCD in the z-plane on to the w-plane under the transformation $w=z^2$.

5.

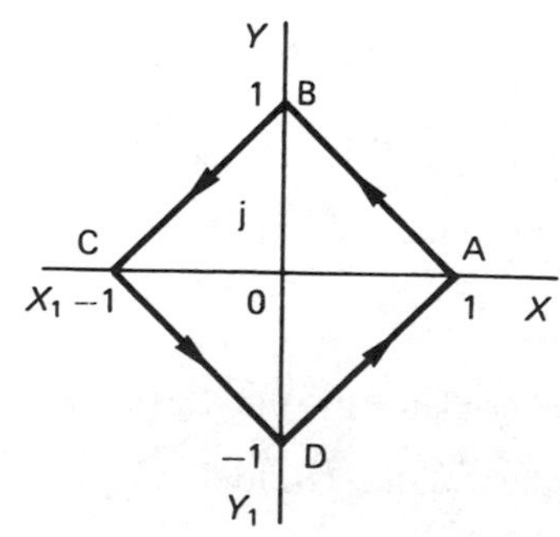

Map the square ABCD in the z-plane on to the w-plane using the transformation $w=2z^2+2$.

6.

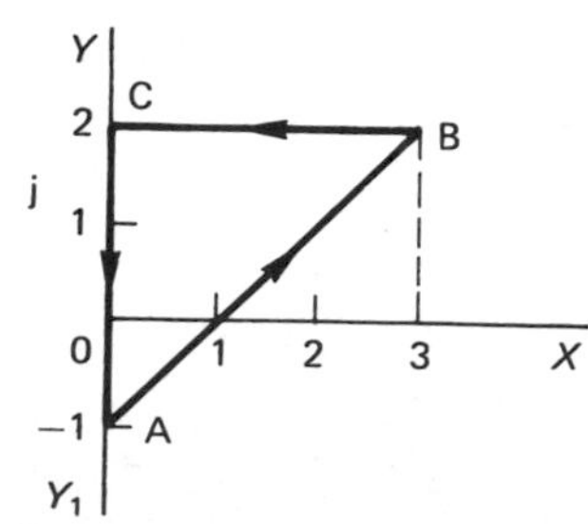

The triangle ABC in the z-plane is mapped on to the w-plane by the transformation $w=j2z^2+1$. Determine the image of ABC in the w-plane.

7. A circle in the z-plane has its centre at the point $(-\frac{3}{4} - j)$ and radius $\frac{7}{4}$. Show that its cartesian equation can be expressed as

$$2(x^2 + y^2) + 3x + 4y - 3 = 0$$

Determine the image of the circle in the w-plane under the inversion transformation $w = \dfrac{1}{z}$.

8. The transformation $w = \dfrac{1}{z - 1}$ is applied to the circle $|z| = 2$ in the z-plane. Determine

   (a) the image of the circle in the w-plane

   (b) the region in the w-plane on to which the region enclosed within the circle in the z-plane is mapped.

9. The circle $|z| = 4$ is described in the z-plane in an anticlockwise manner. Obtain its image in the w-plane under the transformation $w = \dfrac{z + 1}{z - 2}$ and state the direction of development.

10. The bilinear transformation $w = \dfrac{z - j}{z + j2}$ is applied to the circle $|z| = 3$ in the z-plane. Determine the equation of the image in the w-plane and state its centre and radius.

11. The unit circle $|z| = 1$ in the z-plane is mapped on to the w-plane under the transformation $w = \dfrac{z - 1}{z - 3}$. Determine the equation of its image and the region on to which the region within the circle is mapped.

12. Obtain the image of the unit circle $|z| = 1$ in the z-plane under the transformation $w = \dfrac{z + j3}{z - j2}$.

13. The circle $|z| = 2$ is mapped on to the w-plane by the transformation $w = \dfrac{z + j}{2z - j}$. Determine

    (a) the image of the circle in the w-plane

    (b) the mapping of the region enclosed by $|z| = 2$.

14. Show that the transformation equation $w = \dfrac{z - a}{z - b}$ where $z = x + jy$, $a = 1 + j4$ and $b = 2 + j3$, transforms the circle $(x - 3)^2 + (y - 5)^2 = 5$ into a straight line through the origin in the w-plane.

Programme 16

Complex Variable

PART 2

Introduction

1 In the previous programme we introduced the ideas of mapping from one complex plane to another and considered some of the more common transformation functions. Now we pursue our consideration of the complex variable a little further.

Differentiation of a complex function

In differentiation of a function of a single real variable, $y = f(x)$, the differential coefficient of y with respect to x can be defined as the limiting value of $\dfrac{(y + \delta y) - y}{\delta x}$ as δx tends to zero.

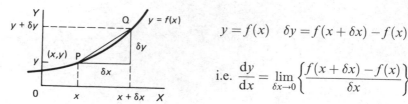

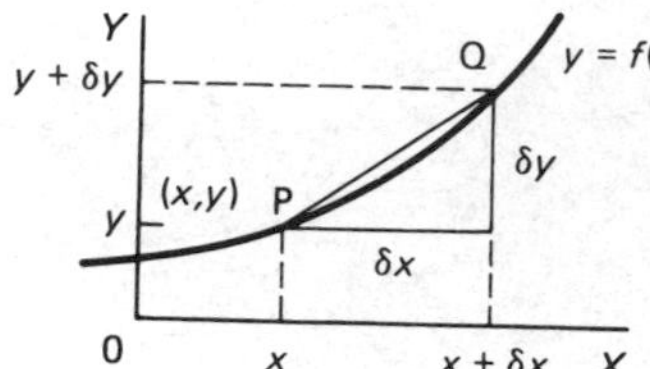

$$y = f(x) \quad \delta y = f(x + \delta x) - f(x)$$

$$\text{i.e. } \frac{dy}{dx} = \lim_{\delta x \to 0}\left\{\frac{f(x + \delta x) - f(x)}{\delta x}\right\}$$

In considering the differentiation of a function of a complex variable, $w = f(z)$, the differential coefficient of w with respect to z can similarly be defined as the limiting value of .... as z tends to zero.

2

$$\frac{(w+\delta w)-w}{\delta z} \quad \text{i.e.} \quad \frac{f(z+\delta z)-f(z)}{\delta z}$$

Now, of course, we are dealing in vectors.

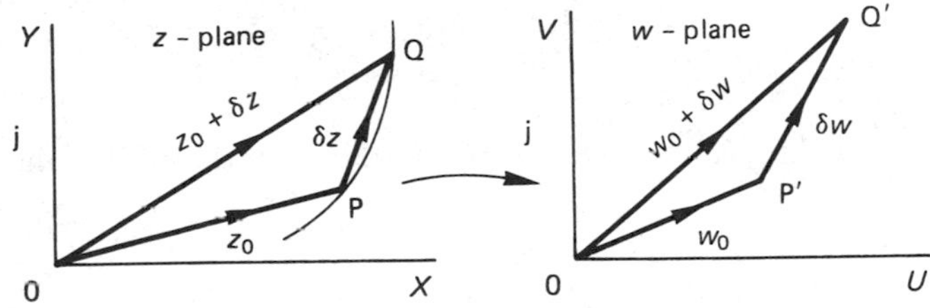

If P and Q in the z-plane map on to P′ and Q′ in the w-plane, then

$$P'Q' = \delta w = (w_0 + \delta w) - w_0 = f(z_0 + \delta z) - f(z_0)$$

Therefore, the derivative of w at P′ $(z = z_0)$ is the limiting value of $\frac{\delta w}{\delta z}$ as $\delta z \to 0$, i.e. $\left[\frac{dw}{dz}\right]_{z_0} = \lim_{\delta z \to 0}\left\{\frac{f(z_0 + \delta z) - f(z_0)}{\delta z}\right\} = \lim_{Q \to P}\left(\frac{P'Q'}{PQ}\right)$

If this limiting value exists—which is not always the case as we shall see—the function $f(z)$ is said to be *differentiable at* P.

Also, if $w = f(z)$ and $f'(z)$ has a limit for all points z_0 within a given region for which $w = f(z)$ is defined, then $f(z)$ is said to be differentiable in that region. From this, it follows that the limit exists whatever the path of approach from Q $(z = z_0 + \delta z)$ to P $(z = z_0)$.

Regular function

A function $w = f(z)$ is said to be *regular* (or analytic) at a point $z = z_0$, if it is defined and single-valued, and has a derivative at every point at and around z_0. Points in a region where $f(z)$ ceases to be regular are called *singular points*, or *singularities*.

We have introduced quite a few new definitions, so let us pause here while you make a note of them. We shall be meeting the various terms quite often.

3

In those cases where a derivative exists, the usual rules of differentiation apply. For example, the derivative of $w = z^2$ can be found from first principles in the normal way.

$$w = z^2 \quad \therefore\ w + \delta w = (z + \delta z)^2 = z^2 + 2z\delta z + \delta z^2$$

$$\therefore\ \delta w = 2z\delta z + \delta z^2 \quad \therefore\ \frac{\delta w}{\delta z} = 2z + \delta z$$

$\therefore\ \dfrac{dw}{dz} = \lim\limits_{\delta z \to 0}(2z + \delta z) = 2z$ and this does not depend on the path along which δz tends to zero.

That was elementary. Here is a rather different one.

Example To find the derivative of $w = z\bar{z}$ where $z = x + jy$ and $\bar{z} = x - jy$.

We have $w = z\bar{z} \quad \therefore\ w + \delta w = (z + \delta z)(\bar{z} + \delta\bar{z})$

from which $\dfrac{\delta w}{\delta z} = \ldots\ldots\ldots$

4

$$\boxed{\frac{\delta w}{\delta z} = \bar{z} + z\frac{\delta\bar{z}}{\delta z} + \delta\bar{z}}$$

for $w + \delta w = (z + \delta z)(\bar{z} + \delta\bar{z}) = z\bar{z} + \bar{z}\delta z + z\delta\bar{z} + \delta z\delta\bar{z}$

$$\therefore\ \delta w = \bar{z}\delta z + z\delta\bar{z} + \delta z\delta\bar{z} \quad \therefore\ \frac{\delta w}{\delta z} = \bar{z} + z\frac{\delta\bar{z}}{\delta z} + \delta\bar{z}$$

Now since $z = x + jy$ and $\bar{z} = x - jy$, we can express $\dfrac{\delta w}{\delta z}$ in terms of x and y. $\dfrac{\delta w}{\delta z} = \ldots\ldots\ldots$

5

$$\boxed{\frac{\delta w}{\delta z} = (x - jy) + (x + jy)\left\{\frac{\delta x - j\delta y}{\delta x + j\delta y}\right\} + \delta x - j\delta y}$$

$$\left.\begin{aligned} \text{for}\quad z &= x + jy \quad \therefore \delta z = \delta x + j\delta y \\ \bar{z} &= x - jy \quad \therefore \delta \bar{z} = \delta x - j\delta y \end{aligned}\right\} \quad \therefore \frac{\delta \bar{z}}{\delta z} = \frac{\delta x - j\delta y}{\delta x + j\delta y}$$

Then $\dfrac{\delta w}{\delta z} = \bar{z} + z\dfrac{\delta \bar{z}}{\delta z} + \delta \bar{z}$ gives the expression quoted above.

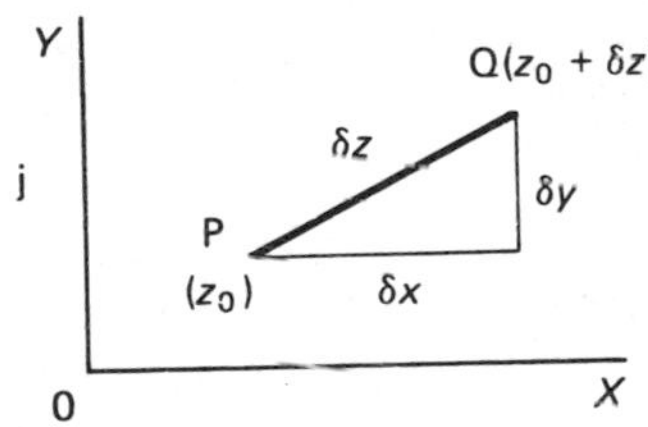

The next step is to reduce δz to zero. But δz consists of $\delta x + j\delta y$ and so reducing δz to zero can be done in one of two ways.

(i) First let $\delta y \to 0$ and afterwards let $\delta x \to 0$.

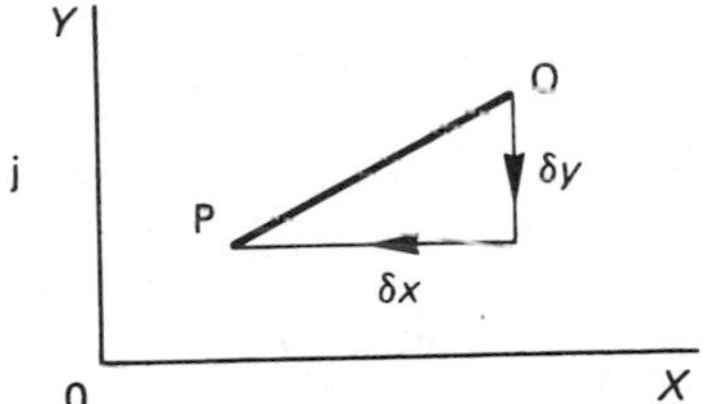

If $\delta y \to 0$, $\quad \dfrac{\delta w}{\delta z} = x - jy + (x + jy)\dfrac{\delta x}{\delta x} + \delta x$

Then $\dfrac{dw}{dz} = \lim\limits_{\delta x \to 0} \{x - jy + x + jy + \delta x\} = \ldots\ldots\ldots$

6

$$\boxed{\frac{dw}{dz} = 2x}$$

On the other hand, we could have reduced δz to zero in the second way.

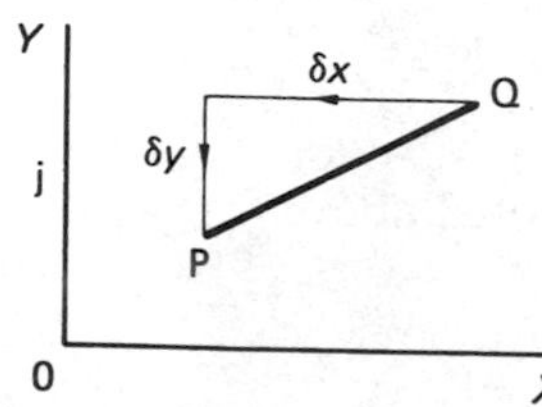

(ii) First let $\delta x \to 0$ and afterwards let $\delta y \to 0$.

We have $$\frac{\delta w}{\delta z} = x - jy + (x + jy)\left\{\frac{\delta x - j\delta y}{\delta x + j\delta y}\right\} + \delta x - j\delta y$$

If $\delta x \to 0$ $$\frac{\delta w}{\delta z} = x - jy + (x + jy)(-1) - j\delta y = -j2y - j\delta y$$

Then $$\frac{dw}{dz} = \lim_{\delta y \to 0}\{-j2j - j\delta y\} = -j2y$$

So, in the first case, $\underline{\dfrac{dw}{dz} = 2x}$ and in the second case $\underline{\dfrac{dw}{dz} = -j2y}$.

These two results are clearly not the same for all values of x and y—with one exception, i.e.

....

7

$$\boxed{\text{when } x = y = 0}$$

Therefore $w = z\bar{z}$ is a function that has no specific derivative, except at $z = 0$—and there are others. It would be convenient, therefore, to have some form of test to see whether a particular function $w = f(z)$ has a derivative $f'(z)$ at $z = z_0$. This useful tool is provided by the Cauchy–Riemann equations.

Cauchy–Riemann equations

The development is very much along the same lines as in the last example. If $w = f(z) = u + \mathrm{j}v$, we have to establish conditions for $w = f(z)$ to have a derivative at a given point $z = z_0$.

$$w = u + \mathrm{j}v \;\therefore\; \delta w = \delta u + \mathrm{j}\delta v; \qquad z = x + \mathrm{j}y \;\therefore\; \delta z = \delta x + \mathrm{j}\delta y$$

Then $$f'(z) = \frac{\mathrm{d}w}{\mathrm{d}z} = \lim_{\delta z \to 0}\left\{\frac{\delta u + \mathrm{j}\delta v}{\delta z}\right\} = \lim_{\substack{\delta x \to 0 \\ \delta y \to 0}}\left\{\frac{\delta u + \mathrm{j}\delta v}{\delta x + \mathrm{j}\delta y}\right\} \qquad (1)$$

(a) Let $\delta x \to 0$, followed by $\delta y \to 0$

Then from (1) above, $f'(z) = \dfrac{\mathrm{d}w}{\mathrm{d}z} = \ldots\ldots\ldots$

8

$$\boxed{\frac{\mathrm{d}w}{\mathrm{d}z} = \frac{\partial v}{\partial y} - \mathrm{j}\frac{\partial u}{\partial y}}$$

for $$f'(z) = \lim_{\delta y \to 0}\left\{\frac{\delta u + \mathrm{j}\delta v}{\mathrm{j}\delta y}\right\} = \lim_{\delta y \to 0}\left\{\frac{\delta v}{\delta y} - \mathrm{j}\frac{\delta u}{\delta y}\right\} = \underline{\frac{\partial v}{\partial y} - \mathrm{j}\frac{\partial u}{\partial y}} \qquad (2)$$

We use the 'partial' notation since u and v are functions of both x and y.

Or (b) Let $\delta y \to 0$, followed by $\delta x \to 0$.

This gives $\ldots\ldots\ldots$

9

$$\frac{dw}{dz} = \frac{\partial u}{\partial x} + j\frac{\partial v}{\partial x}$$

for $f'(z) = \lim_{\delta x \to 0}\left\{\frac{\delta u + j\delta v}{\delta x}\right\} = \lim_{\delta x \to 0}\left\{\frac{\delta u}{\delta x} + j\frac{\delta v}{\delta x}\right\} = \frac{\partial u}{\partial x} + j\frac{\partial v}{\partial x}$ (3)

If the results (2) and (3) are to have the same value for $f'(z)$ irrespective of the path chosen for δz to tend to zero, then

....

10

$$\frac{\partial u}{\partial x} + j\frac{\partial v}{\partial x} = \frac{\partial v}{\partial y} - j\frac{\partial u}{\partial y}$$

Equating real and imaginary parts, this gives

$$\underline{\frac{\partial u}{\partial x} = \frac{\partial v}{\partial y}} \quad \text{and} \quad \underline{\frac{\partial v}{\partial x} = -\frac{\partial u}{\partial y}}$$

These are the *Cauchy–Riemann equations.*

So, to sum up:

A necessary condition for $w = f(z) = u + jv$ to be regular at $z = z_0$ is that u, v and their partial derivatives are continuous and that in the neighbourhood of $z = z_0$

$$\underline{\frac{\partial u}{\partial x} = \frac{\partial v}{\partial y}} \quad \text{and} \quad \underline{\frac{\partial v}{\partial x} = -\frac{\partial u}{\partial y}}$$

Make a note of this important result—then move on to the next frame.

11

We said earlier that where a function fails to be regular, a *singular point*, or *singularity* occurs, i.e. where $w = f(z)$ is not continuous or where the Cauchy–Riemann test fails.

Exercise Determine where each of the following functions fails to be regular, i.e. where singularities occur.

1. $w = z^2 - 4$
2. $w = \dfrac{z}{z-2}$
3. $w = \dfrac{z+5}{z+1}$
4. $w = \dfrac{1}{(z-2)(z-3)}$
5. $w = z\bar{z}$
6. $w = \dfrac{x + jy}{x^2 + y^2}$

Finish all six: then check with the next frame.

12

Conclusions:

1. Putting $z = x + jy$, the Cauchy–Riemann conditions are satisfied. Therefore, no singularity in $w = z^2 - 4$.
2. The function becomes discontinuous at $z = 2$. Singularity at $z = 2$.
3. The function is discontinuous at $z = -1$. Singularity at $z = -1$.
4. Sigularities at $z = 2$ and $z = 3$.
5. We have already seen that $w = z\bar{z}$ has no derivative for all values of z apart from $z = 0$. All points on $w = z\bar{z}$ are singularities.
6. Singularity occurs where $x^2 + y^2 = 0$, i.e. $x = 0$ and $y = 0 \;\therefore\; z = 0$.

13 Complex integration

At the beginning of this programme, we defined differentiation with respect to z in the case of a complex function, since z is a function of two independent variables x and y, i.e. $z = x + jy$. Complex integration is approached in the same way.

$z = x + jy$ and $w = f(z) = u + jv$ where u and v are also functions of x and y.

Also $\mathrm{d}z = \mathrm{d}x + j\,\mathrm{d}y$ and $\mathrm{d}w = \mathrm{d}u + j\,\mathrm{d}v$

$$\therefore \int w\,\mathrm{d}z = \int f(z)\,\mathrm{d}z = \int (u + jv)(\mathrm{d}x + j\,\mathrm{d}y)$$

$$= \int \{(u\,\mathrm{d}x - v\,\mathrm{d}y) + j(v\,\mathrm{d}x + u\,\mathrm{d}y)\}$$

$$\therefore \int f(z)\,\mathrm{d}z = \int (u\,\mathrm{d}x - v\,\mathrm{d}y) + j\int (v\,\mathrm{d}x + u\,\mathrm{d}y)$$

That is, the integral reduces to two real-variable integrals

$$\int (u\,\mathrm{d}x - v\,\mathrm{d}y) \quad \text{and} \quad \int (v\,\mathrm{d}x + u\,\mathrm{d}y)$$

Note that each of these two integrals is of the general form $\int (P\,\mathrm{d}x + Q\,\mathrm{d}y)$ which we met before during our work on *line integrals* and, in the complex plane, this rather neatly leads us *into contour integration*.

Let us make a fresh start.

14 Contour integration—line integrals in the z-plane

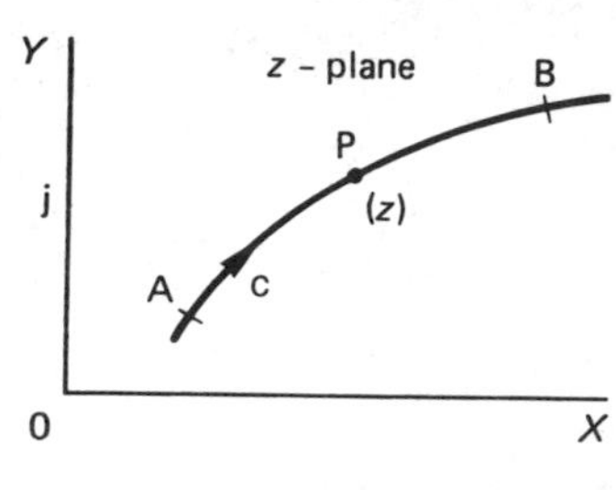

If z moves along the curve c in the z-plane and at each position z has associated with it a function of z, i.e. $f(z)$, then summing up $f(z)$ for all such points between A and B means that we are evaluating a line integral in the z-plane between A $(z = z_1)$ and B $(z = z_2)$ along the curve c, i.e. we are evaluating $\int_c f(z)\,\mathrm{d}z$ where c is the particular path joining A to B.

The evaluation of line integrals in the complex plane is known as *contour integration*. Let us see how it works in practice.

15

Example Evaluate the integral $\int_c f(z)\,dz$ where $f(z) = (z - j)^2$ and c is the straight line joining A $(z = 0)$ to B $(z = 1 + j2)$.

$$z = x + jy;\ dz = dx + j\,dy$$

$$f(z) = (z - j)^2 = \{x + j(y - 1)\}^2 = x^2 - (y - 1)^2 + j\,2x(y - 1)$$

$$\therefore\ I = \int\{(x^2 - y^2 + 2y - 1) + j(2xy - 2x)\}\{dx + j\,dy\}$$

$$= \int\{(x^2 - y^2 + 2y - 1)\,dx - (2xy - 2x)\,dy\}$$

$$+ j\int\{(2xy - 2x)\,dx + (x^2 - y^2 + 2y - 1)\,dy\}$$

Now the equation of AB is $y = 2x$. $\therefore$ $dy = 2\,dx$ and substituting these in the expression for I, between the limits $x = 0$ and $x = 1$ gives

$I = \ldots\ldots\ldots$ Finish it.

16

$$\boxed{I = \tfrac{1}{3}(-2 + j)}$$

$$\text{for } I = \int_0^1\{(x^2 - 4x^2 + 4x - 1)\,dx - (4x^2 - 2x)2\,dx\}$$

$$+ j\int_0^1\{(4x^2 - 2x)\,dx + (2x^2 - 8x^2 + 8x - 2)\,dx\}$$

$$= \int_0^1(-11x^2 + 8x - 1)\,dx + j\int_0^1(-2x^2 + 6x - 2)\,dx$$

and this, by elementary integration, gives $I = \frac{1}{3}(-2 + j)$

Now you will remember that, in general, the value of a line integral depends on the path of integration between the end points, but that the line integral $\int(P\,dx + Q\,dy)$ is independent of the path of integration in a simply connected region if $\dfrac{\partial P}{\partial y} = \dfrac{\partial Q}{\partial x}$ throughout the region.

In our example

$$I = \int \{(x^2 - y^2 + 2y - 1)\,dx - (2xy - 2x)\,dy\}$$

$$+ j\int \{(2xy - 2x)\,dx + (x^2 - y^2 + 2y - 1)\,dy\} \equiv I_1 + j\,I_2$$

If we apply the test to I_1, we get ....

17

$$\boxed{\frac{\partial P}{\partial y} = \frac{\partial Q}{\partial x}}$$

for $I_1 = \int \{(x^2 - y^2 + 2y - 1)\,dx - (2xy - 2x)\,dx\} \equiv \int (P\,dx + Q\,dy)$

$$\left.\begin{aligned} \therefore P &= x^2 - y^2 + 2y - 1 \quad \therefore \frac{\partial P}{\partial y} = -2y + 2 \\ Q &= -2xy + 2x \quad\quad \therefore \frac{\partial Q}{\partial x} = -2y + 2 \end{aligned}\right\} \therefore \underline{\frac{\partial P}{\partial y} = \frac{\partial Q}{\partial x}}$$

Similarly

for $I_2 = \int \{(2xy - 2x)\,dx + (x^2 - y^2 + 2y - 1)dy\} \equiv \int (Pdx + Qdy)$

$$\left.\begin{aligned} \therefore P &= 2xy - 2x \quad\quad \therefore \frac{\partial P}{\partial y} = 2x \\ Q &= x^2 - y^2 + 2y - 1 \;\therefore \frac{\partial Q}{\partial x} = 2x \end{aligned}\right\} \quad \therefore \underline{\frac{\partial P}{\partial y} = \frac{\partial Q}{\partial x}}$$

Therefore, in this example, the value of the line integral is independent of the path of integration.

Just to satisfy our conscience, determine the value of the line integral between the same two end points, but along the parabola $y = 2x^2$.

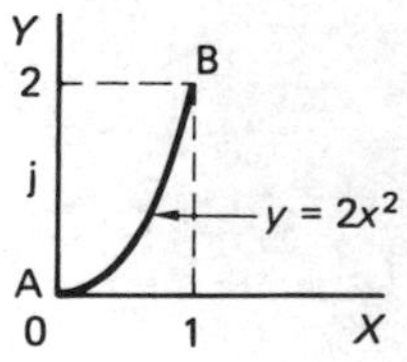

$$f(z) = (z - j)^2$$
$$y = 2x^2 \quad \therefore dy = 4x\,dx$$

As before we have

$$I = \int \{(x^2 - y^2 + 2y - 1)\,dx - (2xy - 2x)\,dy\}$$

$$+ j\int (2xy - 2x)\,dx + (x^2 - y^2 + 2y - 1)\,dy\}$$

Substituting $y = 2x^2$ and $dy = 4x\,dx$, the evaluation gives

$$I = \ldots\ldots\ldots$$

18

$$I = \tfrac{1}{3}(-2 + \mathrm{j})$$

We have

$$I = \int_0^1 \{(x^2 - 4x^4 + 4x^2 - 1)\,\mathrm{d}x - (4x^3 - 2x)4x\,\mathrm{d}x\}$$
$$+\mathrm{j}\int_0^1 \{(4x^3 - 2x)\,\mathrm{d}x + (x^2 - 4x^4 + 4x^2 - 1)4x\,\mathrm{d}x\}$$
$$= \int_0^1 (-20x^4 + 13x^2 - 1)\,\mathrm{d}x + \mathrm{j}\int_0^1 (-16x^5 + 24x^3 - 6x)\,\mathrm{d}x$$

The rest is easy enough, giving $\underline{I = \tfrac{1}{3}(-2 + \mathrm{j})}$ which is, of course, the same result as before.

Now on to the next frame.

19

Cauchy's theorem

We have already seen that if $w = f(z)$ where, as usual, $w = u + \mathrm{j}v$ and $z = x + \mathrm{j}y$, then $\mathrm{d}z = \mathrm{d}x + \mathrm{j}\mathrm{d}y$ and

$$\int f(z)\,\mathrm{d}z = \int (u + \mathrm{j}v)(\mathrm{d}x + \mathrm{j}\,\mathrm{d}y)$$
$$= \int (u\,\mathrm{d}x - v\,\mathrm{d}y) + \mathrm{j}\int (v\,\mathrm{d}x + u\,\mathrm{d}y)$$

If c is a simply connected closed curve as the path of integration,

then $\displaystyle\oint_c f(z)\,\mathrm{d}z = \oint_c (u\,\mathrm{d}x - u\,\mathrm{d}y) + \mathrm{j}\oint_c (v\,\mathrm{d}x + u\,\mathrm{d}y)$

Applying Green's theorem to each of the two integrals on the right-hand side in turn, we have

(a) $\displaystyle\oint_c (u\,\mathrm{d}x - v\,\mathrm{d}y) = \int_S\int \left(-\frac{\partial v}{\partial x} - \frac{\partial u}{\partial y}\right)\mathrm{d}x\,\mathrm{d}y$

where S is the region enclosed by the curve c.
Also, if $f(z)$ is regular at every point within and on c, then

$$\frac{\partial u}{\partial y} = -\frac{\partial v}{\partial x} \text{ and therefore } -\frac{\partial v}{\partial x} - \frac{\partial u}{\partial y} = 0$$

$$\therefore \oint_c (u\,\mathrm{d}x - v\,\mathrm{d}y) = 0 \qquad (1)$$

(b) Similarly, with the second integral, we have

. . . .

20

$$\oint_c (v\,dx + u\,dy) = 0$$

since $\oint_c (v\,dx + u\,dy) = \int_s\int \left(\frac{\partial u}{\partial x} - \frac{\partial v}{\partial y}\right) dx\,dy$

Again, if $f(z)$ is regular at every point within and on c, then

$$\frac{\partial u}{\partial x} = \frac{\partial v}{\partial y} \text{ and therefore } \frac{\partial u}{\partial x} - \frac{\partial v}{\partial y} = 0$$

$$\therefore \oint_c (v\,dx + u\,dy) = 0 \qquad (2)$$

Combining the two results (1) and (2) we have the following result.

If $f(z)$ is regular at every point within and on a simply connected closed curve c, then $\oint_c f(z)\,dz = 0$

This is Cauchy's theorem. Make a note of the result; then we can see an example.

21

Example 1 Verify Cauchy's theorem by evaluating the integral $\oint_c f(z)\,\mathrm{d}z$ where $f(z) = z^2$ around the square formed by joining the points $z = 1, \quad z = 2, \quad z = 2 + \mathrm{j}, \quad z = 1 + \mathrm{j}$.

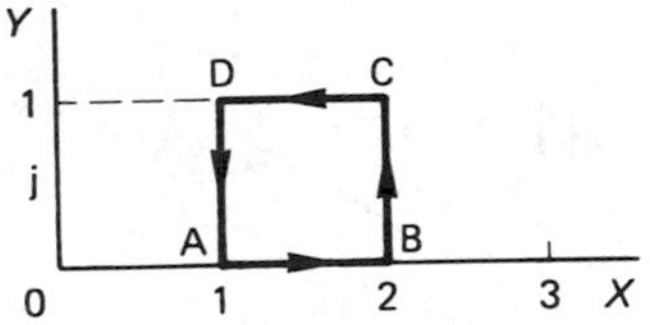

Y, 1, j, D, C, A, B, 0, 1, 2, 3 X

$$z = x + \mathrm{j}y$$
$$z^2 = x^2 - y^2 + \mathrm{j}\,2xy$$
$$\mathrm{d}z = \mathrm{d}x + \mathrm{j}\,\mathrm{d}y$$

$$\oint_c f(z)\,\mathrm{d}z = \oint_c z^2\,\mathrm{d}z = \oint_c \{x^2 - y^2 + \mathrm{j}\,2xy\}\{\mathrm{d}x + \mathrm{j}\,\mathrm{d}y\}$$
$$= \oint_c \{(x^2 - y^2)\,\mathrm{d}x - 2xy\,\mathrm{d}y\} + \mathrm{j}\oint_c \{2xy\,\mathrm{d}x + (x^2 - y^2)\,\mathrm{d}y\}$$

We now take each of the sides in turn.

(a) AB: $y = 0 \quad \therefore \mathrm{d}y = 0$

$$\therefore \int_{AB} f(z)\,\mathrm{d}z = \int_1^2 x^2\mathrm{d}x = \left[\frac{x^3}{3}\right]_1^2 = \frac{8}{3} - \frac{1}{3} = \underline{\frac{7}{3}}$$

(b) BC: $x = 2 \quad \therefore \mathrm{d}x = 0$

$$\therefore \int_{BC} f(z)\,\mathrm{d}z = \int_0^1 (-4y\mathrm{d}y) + \mathrm{j}\int_0^1 (4 - y^2)\,\mathrm{d}y$$
$$= \left[-2y^2\right]_0^1 + \mathrm{j}\left[4y - \frac{y^3}{3}\right]_0^1$$
$$= -2 + \mathrm{j}\left(4 - \frac{1}{3}\right) = \underline{-2 + \mathrm{j}\frac{11}{3}}$$

Continuing in the same way, the results for the remaining two sides are

.... and ....

22

$$\text{CD: } -\tfrac{4}{3} - \text{j}3; \quad \text{DA: } 1 - \text{j}\tfrac{2}{3}$$

since

(c) CD: $y = 1 \quad \therefore \mathrm{d}y = 0$

$$\therefore \int_{CD} f(z)\,\mathrm{d}z = \int_2^1 (x^2 - 1)\,\mathrm{d}x + \text{j}\int_2^1 2x\,\mathrm{d}x$$

$$= \left[\frac{x^3}{3} - x\right]_2^1 + \text{j}\left[x^2\right]_2^1 = -\tfrac{4}{3} - \text{j}\,3$$

(d) DA: $x = 1 \quad \therefore \mathrm{d}x = 0$

$$\therefore \int_{DA} f(z)\,\mathrm{d}z = \int_1^0 (-2y\mathrm{d}y) + \text{j}\int_1^0 (1 - y^2)\,\mathrm{d}y$$

$$= \left[-y^2\right]_1^0 + \text{j}\left[y - \frac{y^3}{3}\right]_1^0 = 1 - \text{j}\tfrac{2}{3}$$

So, collecting the four results, $\oint_c f(z)\,\mathrm{d}z = \ldots\ldots\ldots$

23

$$\oint_c f(z)\,\mathrm{d}z = 0$$

for $\oint_c f(z)\,\mathrm{d}z = \frac{7}{3} + \left(-2 + \text{j}\frac{11}{3}\right) + \left(-\frac{4}{3} - \text{j}3\right) + \left(1 - \text{j}\frac{2}{3}\right) = 0.$

Example 2 A region in the z-plane has a boundary c consisting of

(a) OA joining $z = 0$ to $z = 2$
(b) AB a quadrant of the circle $|z| = 2$ from $z = 2$ to $z = \text{j}2$
(c) BO joining $z = \text{j}2$ to $z = 0$.

Verify Cauchys theorem by evaluating the integral $\int_c (z^2 + 1)\,\mathrm{d}z$

(i) along the arc from A to B,
(ii) along BO and OA.

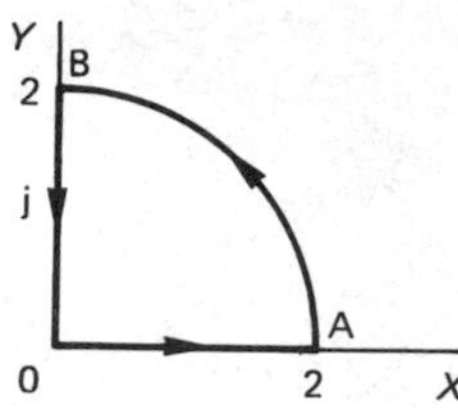

$$f(z) = z^2 + 1 = (x + \text{j}y)^2 + 1$$
$$= (x^2 - y^2 + 1) + \text{j}\,2xy$$
$$z = x + \text{j}y \quad \therefore \mathrm{d}z = \mathrm{d}x + \text{j}\,\mathrm{d}y$$

So the general expression for $\int f(z)\,\mathrm{d}z = \ldots\ldots\ldots$

24

$$\int \{(x^2 - y^2 + 1) + \mathrm{j}\,2xy\}\{\mathrm{d}x + \mathrm{j}\,\mathrm{d}y\}$$
$$= \int \{(x^2 - y^2 + 1)\,\mathrm{d}x - 2xy\,\mathrm{d}y\} + \mathrm{j}\int \{2xy\,\mathrm{d}x + (x^2 - y^2 + 1)\,\mathrm{d}y\}$$

(i) Arc AB: $x^2 + y^2 = 4 \qquad \therefore\ y^2 = 4 - x^2 \qquad \therefore\ y = \sqrt{4 - x^2}$

$$\mathrm{d}y = \frac{1}{2}(4 - x^2)^{-1/2}(-2x)\,\mathrm{d}x \qquad \therefore\ \mathrm{d}y = \frac{-x}{\sqrt{4 - x^2}}\,\mathrm{d}x$$

$$\therefore \int_{AB} f(z)\,\mathrm{d}z = \int_2^0 \left\{(x^2 - 4 + x^2 + 1)\,\mathrm{d}x - 2x\sqrt{4 - x^2}\left(\frac{(-x)}{\sqrt{4 - x^2}}\right)\mathrm{d}x\right\}$$
$$+\mathrm{j}\int_2^0 \left\{2x\sqrt{4 - x^2}\,\mathrm{d}x + (x^2 - 4 + x^2 + 1)\left(\frac{(1 - x)}{\sqrt{4 - x^2}}\right)\mathrm{d}x\right\}$$
$$= \int_2^0 (4x^2 - 3)\,\mathrm{d}x + \mathrm{j}\int_2^0 \frac{11x - 4x^3}{\sqrt{4 - x^2}}\,\mathrm{d}x = \underline{-\frac{14}{3} + \mathrm{j}\,I_1}$$

Now we must attend to $I_1 = \displaystyle\int_2^0 \frac{11x - 4x^3}{\sqrt{4 - x^2}}\,\mathrm{d}x.$

Substituting $x = 2\sin\theta$ and $\mathrm{d}x = 2\cos\theta\,\mathrm{d}\theta$ with appropriate limits we have

25

$$I_1 = -\tfrac{2}{3}$$

$$\text{for } I_1 = \int_{\pi/2}^0 \left(\frac{22\sin\theta - 32\sin^3\theta}{2\cos\theta}\right)2\cos\theta\,\mathrm{d}\theta$$
$$= \int_0^{\pi/2} (32\sin^3\theta - 22\sin\theta)\,\mathrm{d}\theta$$
$$= 32\frac{2}{(3)(1)} + \Big[22\cos\theta\Big]_0^{\pi/2} = \frac{64}{3} - 22 = -\frac{2}{3}$$

$$\therefore \int_{AB} f(z)\,\mathrm{d}z = -4\frac{2}{3} - \mathrm{j}\frac{2}{3} = \underline{-\frac{2}{3}(7 + \mathrm{j})}$$

(ii) Along BO and OA. Complete this section on your own in the same way.

$$\int_{BO} f(z)\,\mathrm{d}z = \ldots\ldots\ldots; \qquad \int_{OA} f(z)\,\mathrm{d}z = \ldots\ldots\ldots$$

26

$$\boxed{\int_{BO} f(z)\,dz = j\frac{2}{3};\quad \int_{OA} f(z)\,dz = 4\frac{2}{3}}$$

for we have

BO: $x = 0 \quad \therefore dx = 0$

$$\therefore \int_{BO} f(z)\,dz = j\int_{2}^{1}(1-y^2)\,dy = j\left[y - \frac{y^3}{3}\right]_2^0 = \underline{j\frac{2}{3}}$$

OA: $y = 0 \quad \therefore dy = 0$

$$\therefore \int_{OA} f(z)\,dz = \int_{0}^{2}(x^2+1)\,dx = \left[\frac{x^3}{3} + x\right]_0^2 = \underline{4\frac{2}{3}}$$

Collecting the results together, therefore

$$\int_{AB} f(z)\,dz = -\frac{14}{3} - j\frac{2}{3}$$

$$\int_{BO+OA} f(z)\,dz = j\frac{2}{3} + 4\frac{2}{3} = \frac{14}{3} + j\frac{2}{3}$$

$$\therefore \oint_c f(z)\,dz = \int_{AB} f(z)\,dz + \int_{BO+OA} f(z)\,dz = \underline{0}$$

which, once again, verifies Cauchy's theorem.

Just by way of revision, Cauchy's theorem actually states that

. . . .

If $f(z)$ is *regular* at every point within and on a simply connected closed curve c, then $\oint_c f(z)\,dz = 0$

In our examples so far, $f(z)$ has been regular and no problems have arisen. Let us now consider a case where one or more singularities occur within the region enclosed by the curve c.

Deformation of contours at singularities

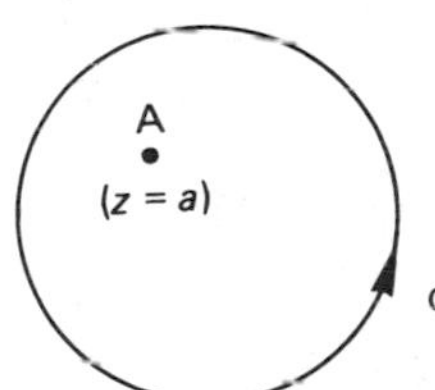

If c is the boundary curve (or *contour*) of a region and $f(z)$ is regular for all points within and on the contour, then the evaluation of $\oint_c f(z)\,dz$ around the contour is straightforward.

However, if $f(z) = \dfrac{1}{z-a}$, where a is a complex constant, and point A corresponds to $z = a$, then at A, $f(z)$ ceases to be regular and a singularity occurs at that point.

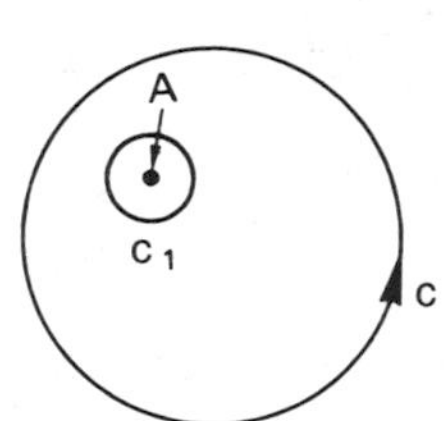

We can isolate A in a very small region within a contour c_1 and then $f(z)$ will be regular at all points within the region c and outside c_1. But the original region is now no longer simply connected (it now has a 'hole' in it) and this was one of our initial conditions.

However, all is not lost! We select a suitable point B on the contour c and join it to the inner contour c_1. If we now consider the integration $\int f(z)\,dz$ starting from a point K and proceeding anticlockwise, the path of integration can be taken as K B L M N B D E K which is now a simply connected closed curve.

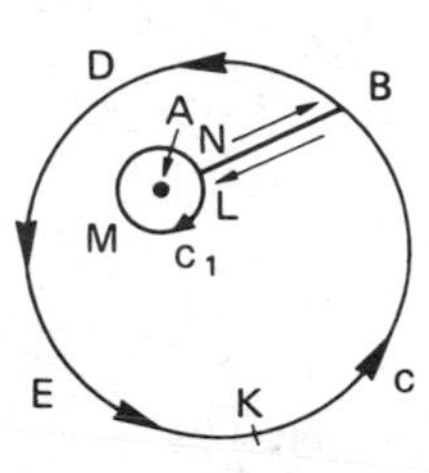

Therefore

$$\int f(z)\,dz = I = I_{KB} + I_{BL} + I_{LMN} + I_{NB} + I_{BDEK} = \dots\dots\dots$$

28

$$\boxed{0}$$

The function $f(z)$ is now regular at all points within and on the deformed contour. Remember that the inner contour c_1 can be made as small as we wish.

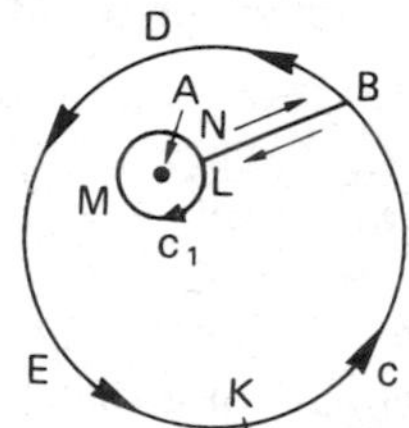

Note that $I_{NB} = -I_{BL}$ being in opposite directions and these therefore cancel out.
The previous result then becomes

$$I_{KB} + I_{LMN} + I_{BDEK} = 0 \qquad \text{i.e.} \qquad I_{KB} + I_{BDEK} + I_{LMN} = 0$$

But $I_{KB} + I_{BDEK} = \oint_c f(z)\,dz$ and $I_{LMN} = \ointclockwise_{c_1} f(z)\,dz$

$$\therefore \oint_c f(z)\,dz + \ointclockwise_{c_1} f(z)\,dz = 0$$

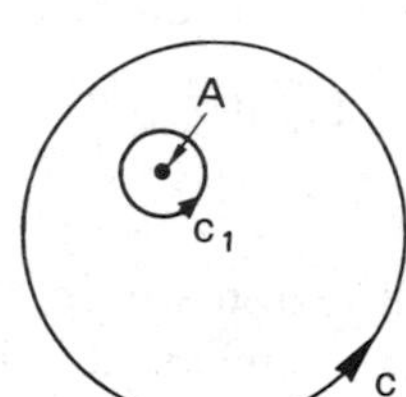

$$\therefore \oint_c f(z)\,dz - \oint_{c_1} f(z)\,dz = 0$$

$$\therefore \underline{\oint_c f(z)\,dz = \oint_{c_1} f(z)\,dz}$$

The process can, of course, be extended to cases with more than one such singularity.

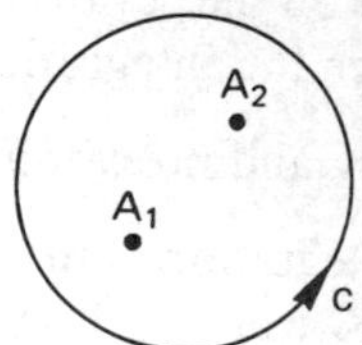

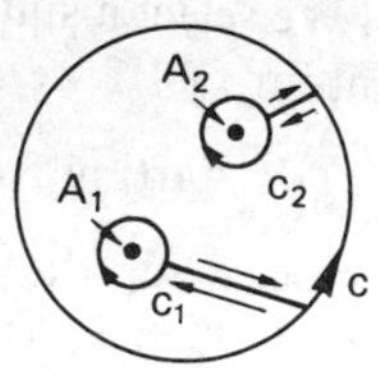

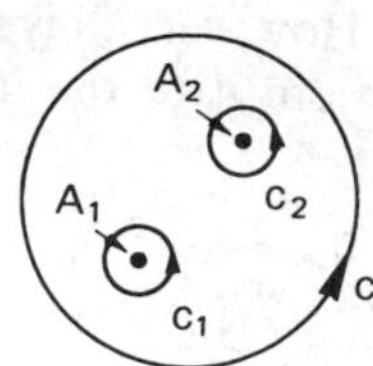

The corresponding result then becomes

$$\underline{\oint_c f(z)\,dz = \oint_{c_1} f(z)\,dz + \oint_{c_2} f(z)\,dz \ldots \text{etc.}}$$

Now let us apply these ideas to an example.

Example 1 Consider the integral $\oint_c f(z)\,dz$ where $f(z) = \frac{1}{z}$, evaluated round a closed contour in the z-plane.

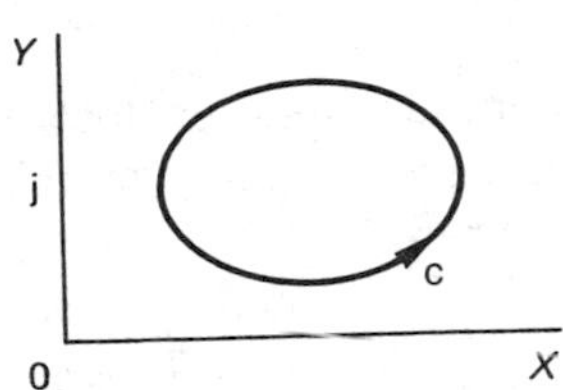

We first check the function $f(z) = \frac{1}{z}$ for singularities and find at once that

....

29

At $z = 0$, $f(z) = \frac{1}{z}$ ceases to be regular and a singularity occurs at that point

The actual position of the closed contour is not specified in the problem, so there are two possibilities: either the contour does enclose the origin, or it does not. Let us consider them in turn.

(a) The contour does not enclose the origin.

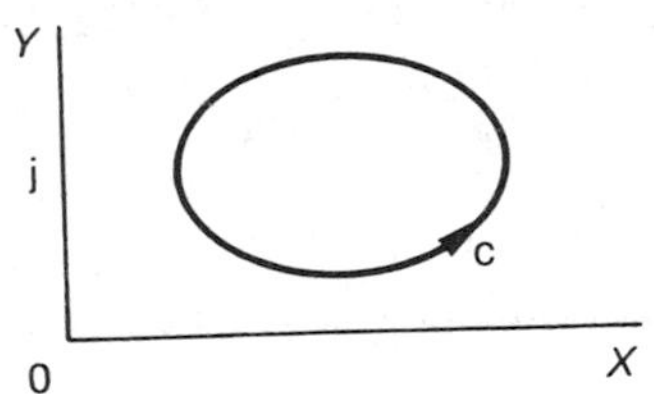

No difficulty arises here and by Cauchy's theorem

....

30

$$\oint_c f(z)\,dz = 0$$

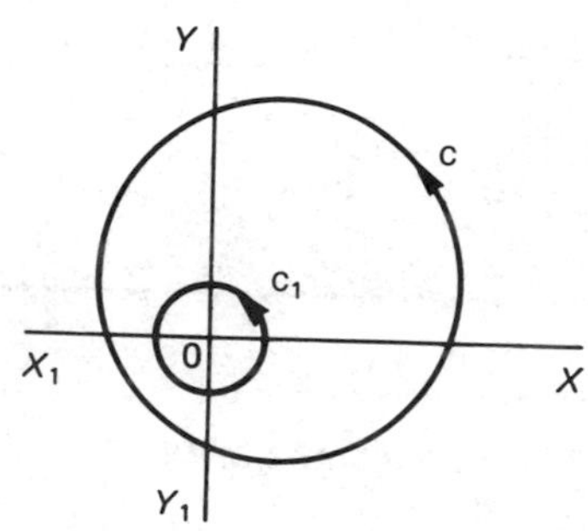

(b) If the contour *does* enclose the origin, the singularity must be taken into account. Then

$$\oint_c f(z)\,dz = \oint_{c_1} f(z)\,dz = \oint_{c_1} \frac{1}{z}\,dz$$

and we attend to evaluating $\oint_{c_1} \frac{1}{z}\,dz$

where c_1 is a small circle of radius r entirely within the region bounded by c.

If we take an enlarged view of the small circle c_1, we have

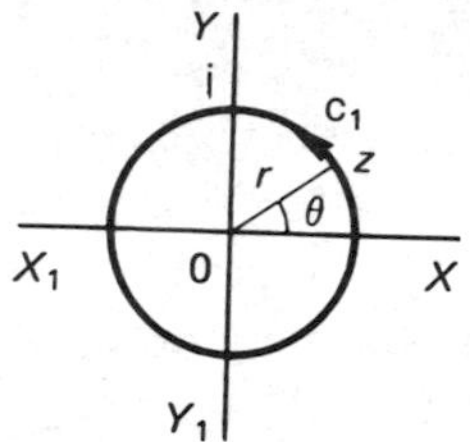

$z = x + jy$ which can be expressed in polar form .... and in exponential form ....

31

$$z = r\,(\cos\theta + j\sin\theta)$$
$$z = re^{j\theta}$$

Using $z = re^{j\theta}$ then $dz = jre^{j\theta}d\theta$

and $\oint_{c_1} \frac{1}{z}\,dz = \ldots\ldots\ldots$ Complete it.

32

$$\boxed{j\,2\pi}$$

for $\quad \oint_{c_1} \frac{1}{z}\,dz = \int_0^{2\pi} \frac{1}{re^{j\theta}}\{jre^{j\theta}\}d\theta = \int_0^{2\pi} j\,d\theta = j2\pi$

$$\therefore \oint_c \frac{1}{z}\,dz = \oint_{c_1} \frac{1}{z}\,dz = j2\pi$$

So we have:

(a) $\oint_c \frac{1}{z}\,dz = 0$ if the contour c does not enclose the origin

(b) $\oint_c \frac{1}{z}\,dz = j2\pi$ if the contour c does enclose the origin.

These two constitute an important result, so note them well.

33

Example 2 Consider the integral $\oint_c f(z)\,dz$ where $f(z) = \frac{1}{z^n}$ $(n = 2, 3, 4 \ldots)$.

Again, a singularity clearly occurs at $z = 0$ and again also we have two possible cases.

(a) If the contour c does not enclose the origin, then by Cauchy's theorem $\oint_c f(z)\,dz = 0$.

(b) If the contour c does enclose the origin, then we proceed very much as before.

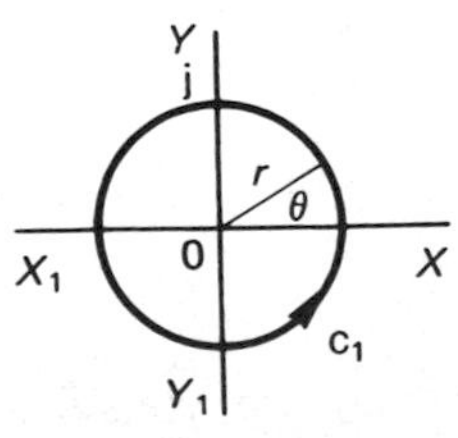

Using $z = re^{j\theta}$, $\quad dz = jre^{j\theta}d\theta$

and $z^n = r^n e^{jn\theta}$

Then $\oint_c f(z)\,dz = \oint_{c_1} f(z)\,dz = \int_0^{2\pi} \frac{1}{r^n e^{jn\theta}}\{jre^{j\theta}\}\,d\theta$

$$= \frac{j}{r^{n-1}} \int_0^{2\pi} e^{-j(n-1)\theta} d\theta$$

$$= \frac{-1}{(n-1)r^{n-1}} \left[e^{-j(n-1)\theta}\right]_0^{2\pi}$$

$= \ldots\ldots\ldots$ Finish it off.

34

$$\boxed{0}$$

since $\displaystyle\oint_c \frac{1}{z^n}\,dz = \frac{-1}{(n-1)r^{n-1}}\{e^{-j(n-1)2\pi} - 1\}$

$$= \frac{-1}{(n-1)r^{n-1}}\{\cos(n-1)2\pi - j\sin(n-1)2\pi - 1\}$$

$$= 0 \quad \text{since } \left.\begin{array}{l}\cos(n-1)2\pi = 1\\ \sin(n-1)2\pi = 0\end{array}\right\} \quad n = 2, 3, 4, \ldots$$

So $\displaystyle\oint_c \frac{1}{z^n}\,dz = 0$ for all positive integer values of n other than $n = 1$, where c is any simply connected closed contour.

The particular case when $n = 1$ we have seen in Example 1.

Now we can easily cope with this next example.

Example 3 Consider $\displaystyle\oint_c f(z)\,dz$ where $f(z) = \dfrac{1}{(z-a)^n}$ for $n = 1, 2, 3, \ldots$

This is a simple extension of the last piece of work. Here we see that a singularity occurs at $z = a$ and yet again we have two cases to consider.

(a)

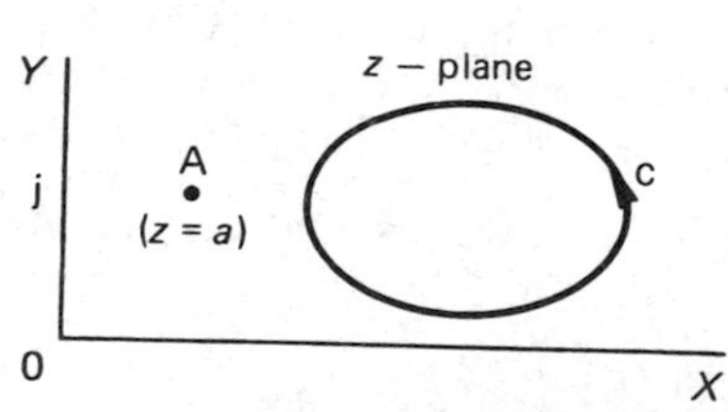

If the contour c does not enclose $z = a$, then by Cauchy's theorem

$$\oint_c f(z)\,dz = 0$$

(b)

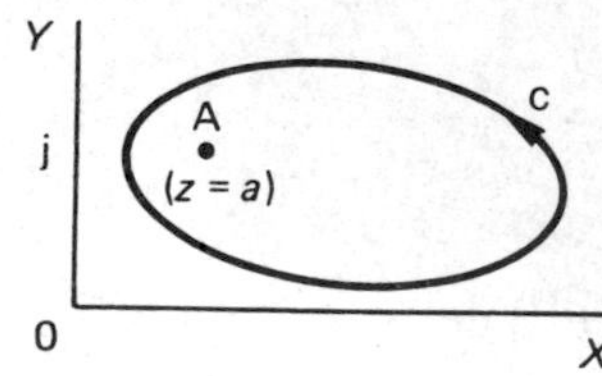

If c encloses A $(z = a)$ we consider separately the cases when

(i) $n = 1$ and (ii) $n > 1$.

(i) If $n = 1$
$$\oint_c f(z)\,dz = \oint_c \frac{1}{z-a}\,dz$$

Putting $z - a = w$ $\therefore dz = dw$ $\therefore$ $\displaystyle\oint_c \frac{1}{z-a}\,dz = \oint_c \frac{1}{w}\,dw$

and this we have already established has a value ....

35

$$\boxed{j\,2\pi}$$

(ii) If $n > 1$, $\displaystyle\oint_c f(z)\,dz = \oint_c \frac{1}{(z-a)^n}\,dz = \oint_c \frac{1}{w^n}\,dw = 0$ for $n \neq 1$.

So collecting our results together, we have the following

For $\displaystyle\oint_c f(z)\,dz$, where $f(z) = \dfrac{1}{(z-a)^n}$, $n = 1, 2, 3, \ldots$ and c is a simply connected closed contour

$$\begin{aligned}\oint_c \frac{1}{(z-a)^n}\,dz &= 0 && n \neq 1\\ &= 0 && n = 1 \text{ and c does not enclose } z = a\\ &= j2\pi && n = 1 \text{ and c does enclose } z = a.\end{aligned}$$

You will notice that this is a more general result and includes the results obtained from Examples 1 and 2. *Make a note of it, therefore: it is quite important.*

Then on to Example 4.

36

Example 4 Finally, we can go one stage further and consider the contour integral of functions such as $f(z) = \dfrac{z-j-4}{(z+j)(z-2)}$.

First we express $f(z)$ in partial fractions

$$\frac{z-j-4}{(z+j)(z-2)} = \frac{A}{z+j} + \frac{B}{z-2}$$

One quick way of finding A and B is by the 'cover up' method.

(a) *To find A*, temporarily cover up the denominator $(z+j)$ in the partial fraction $\dfrac{A}{[z+j]}$ and in the function $\dfrac{z-j-4}{[z+j](z-2)}$ and substitute $z+j=0$, i.e. $z=-j$ in the remainder of the function.

$$A = \frac{-j-j-4}{-j-2} = \frac{4+j2}{2+j} = 2 \quad \therefore \underline{A=2}$$

(b) *To find B*, cover up the denominator $(z-2)$ in the partial fraction $\dfrac{B}{[z-2]}$ and in the function $\dfrac{z-j-4}{(z+j)[z-2]}$ and substitute $z-2=0$, i.e. $z=2$ in the remainder of the function.

$$B = \dots\dots\dots$$

37

$$\boxed{B=-1}$$

for $$B = \frac{2-j-4}{2+j} = \frac{-2-j}{2+j} = -1.$$

Therefore the function $f(z)$ becomes

$$f(z) = \frac{z-j-4}{(z+j)(z-2)} \equiv \frac{2}{z+j} - \frac{1}{z-2}$$

Now we can see that there are singularities at . . . .

38

$$\boxed{z = -\text{j} \quad \text{and} \quad z = 2}$$

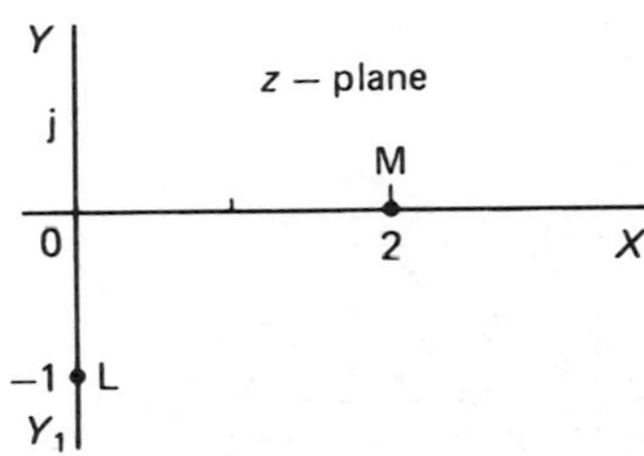

Denote the singularities by L and M

$$\therefore \oint_c \frac{z - \text{j} - 4}{(z + \text{j})(z - 2)}\,\text{d}z = \oint_c \left\{ \frac{2}{z + \text{j}} - \frac{1}{z - 2} \right\} \text{d}z$$

$$= \oint_c \left\{ 2\left(\frac{1}{z + \text{j}}\right) - \frac{1}{z - 2} \right\} \text{d}z$$

So we now have *four* cases to consider, depending on whether L, M, neither, or both, are enclosed within the contour c.

(a) *Neither L nor M enclosed*

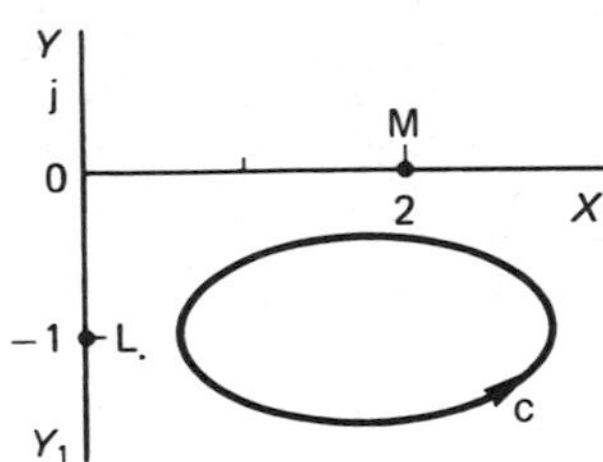

Then, once again, by Cauchy's theorem

$$\oint_c f(z)\,\text{d}z = \underline{0}$$

(b) *L enclosed but not M*

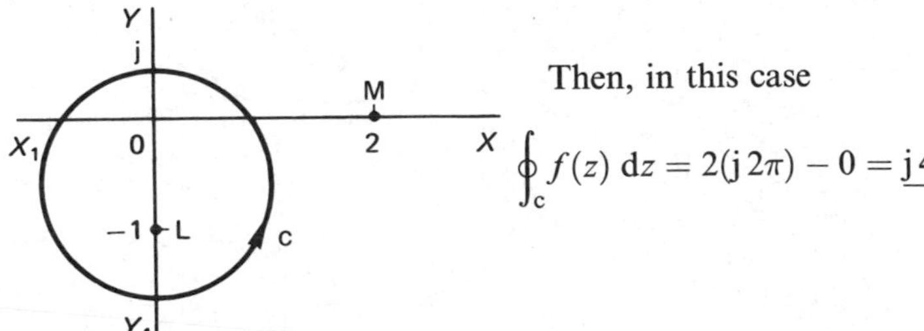

Then, in this case

$$\oint_c f(z)\,\text{d}z = 2(\text{j}\,2\pi) - 0 = \underline{\text{j}\,4\pi}$$

(c) *M enclosed but not L*

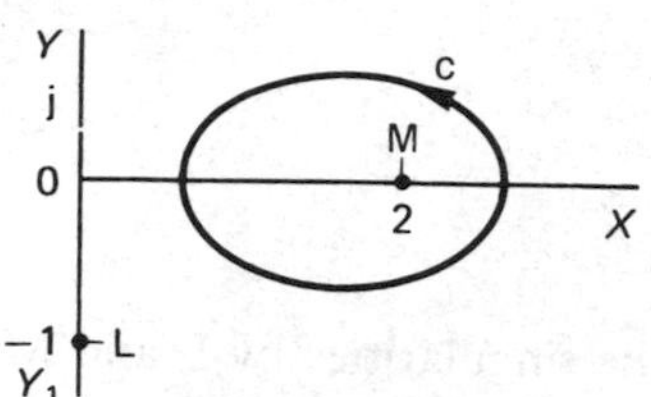

Here

$$\oint_c f(z)\,dz = 0 - (j\,2\pi) = \underline{-j\,2\pi}$$

(d) *Both L and M enclosed*

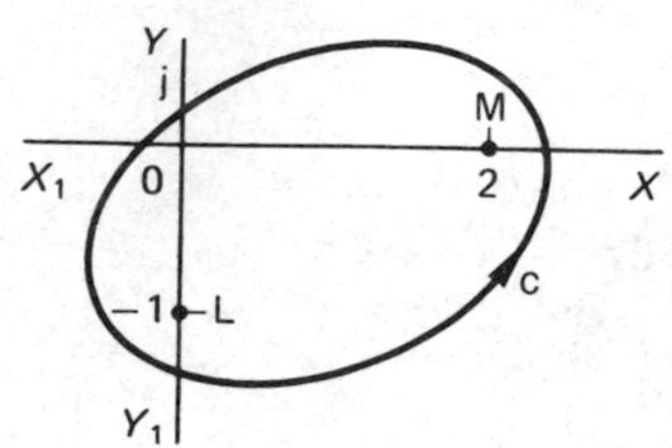

In this case

$$\oint_c f(z)\,dz = \ldots\ldots\ldots$$

39

$$\boxed{j\,2\pi}$$

for, when both L and M are enclosed

$$\oint_c f(z)\,dz = \oint_c \left\{2\left(\frac{1}{z+j}\right) - \frac{1}{z-2}\right\}dz$$

$$= 2(j\,2\pi) - j\,2\pi = \underline{j\,2\pi}$$

The key is provided by the results we established earlier.

$$\oint_c \frac{1}{(z-a)^n}\,dz = \ldots\ldots\ldots \text{ if } \ldots\ldots\ldots$$

$$= \ldots\ldots\ldots \text{ if } \ldots\ldots\ldots$$

$$= \ldots\ldots\ldots \text{ if } \ldots\ldots\ldots$$

40

$$\oint_c \frac{1}{(z-a)^n}\,dz = 0 \qquad \text{if } n \neq 1$$
$$= 0 \qquad \text{if } n = 1 \text{ and c does not enclose } z = a$$
$$= j2\pi \qquad \text{if } n = 1 \text{ and c does enclose } z = a.$$

Now for something somewhat different.

Conformal Transformation (conformal mapping)

A mapping from the z-plane on to the w-plane is said to be *conformal* if the angles between lines in the z-plane are preserved both in magnitude and in sense of rotation when transformed on to the corresponding lines

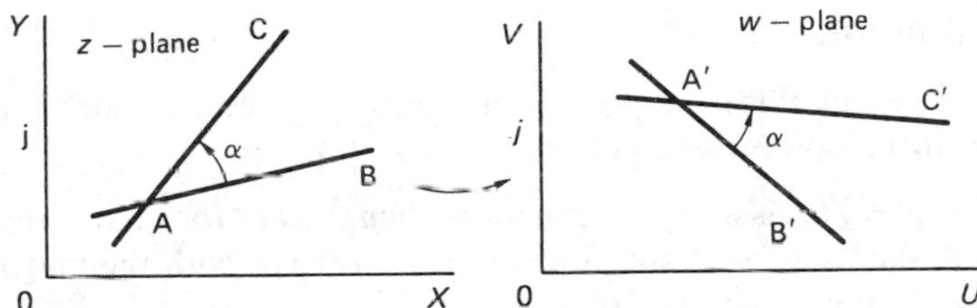

in the w-plane.

The angle between two intersecting curves in the z-plane is defined by the

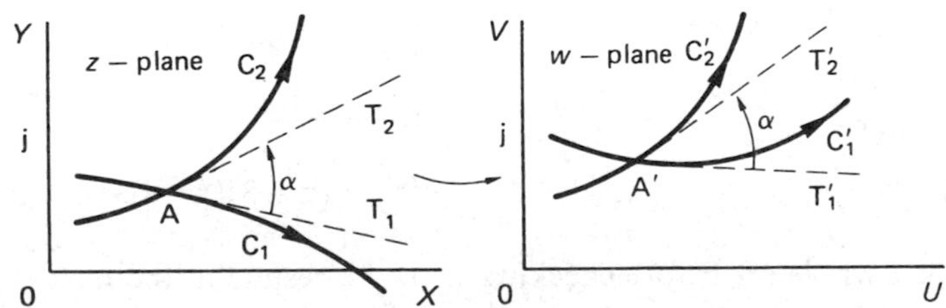

angle α $(0 \leq \alpha \leq \pi)$ between their two tangents at the point of intersection, and this is preserved.

The essential characteristic of a conformal mapping is that

....

41

angles are preserved both in magnitude and in sense of rotation

Conditions for conformal transformation

The conditions necessary in order that a transformation shall be conformal are as follows

1. The transformation function $w = f(z)$ must be a regular function of z. That is, it must be defined and single-valued, have a continuous derivative at every point in the region and satisfy the Cauchy–Riemann equations.

2. The derivative $\frac{dw}{dz}$ must not be zero, i.e. $f'(z) \neq 0$ at a point of intersection.

Critical points

A point at which $f'(z) = 0$ is called a *critical point* and, at such a point, the transformation is not conformal.

So, if $w = f(z)$ is a regular function, then, except for points at which $f'(z) = 0$, the transformation function will preserve both the magnitude of the angle and its sense of rotation.

Now for a short exercise by way of practice.

Exercise Determine critical points (if any) which occur in the following transformations $w = f(z)$.

1. $f(z) = (z-1)^2$
2. $f(z) = e^z$
3. $f(z) = \frac{1}{z^2}$
4. $f(z) = z + \frac{1}{z}$
5. $f(z) = (2z+3)^3$
6. $f(z) = z^3 + 6z + 9$
7. $f(z) = \frac{z-j}{z+j}$
8. $f(z) = (z+3)(z-j)$.

Finish the whole set before checking with the results in the next frame.

42

1. $z = 1$	5. $z = -\frac{3}{2}$
2. none	6. $z = \pm j\sqrt{2}$
3. none	7. none
4. $z = \pm 1$	8. $z = \frac{1}{2}(j - 3)$

All that is required is to differentiate each function and to find for which values of z, $f'(z) = 0$.

Now one or two simple examples on conformal mapping.

Example 1 Linear transformation $w = az + b$, $a \neq 0$, a and b complex.

(i) Cauchy–Riemann conditions satisfied.

(ii) $f'(z) = a$ i.e. not zero $\therefore$ no critical points.

Therefore, the transformation $w = az + b$ provides conformal mapping throughout the entire z-plane.

Example 2 Non-linear transformation $w = z^2$.

First check for singularities and critical points. These, if any, occur at

....

43

no singularities; critical point at $z = 0$

for $$f'(z) = 2z \quad \therefore f'(z) = 0 \text{ at } z = 0.$$

Therefore, the transformation is not conformal at the origin.

If we choose to express z in exponential form $z = x + \mathrm{j}y = r\mathrm{e}^{\mathrm{j}\theta}$, then $w = z^2 = r^2\mathrm{e}^{\mathrm{j}2\theta}$, i.e. r is squared and the angle doubled.

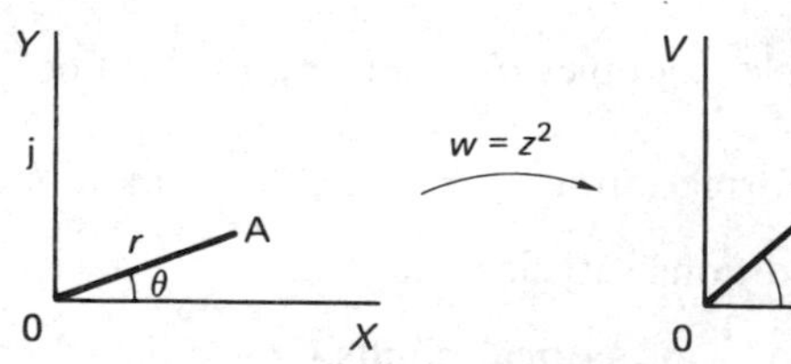

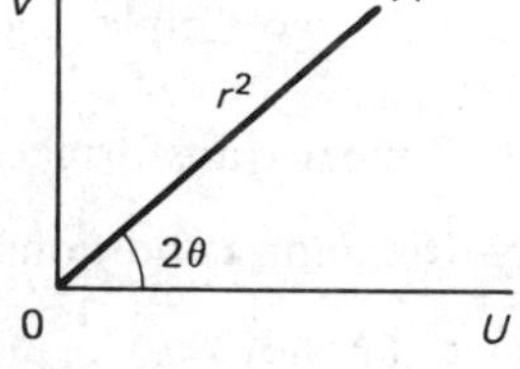

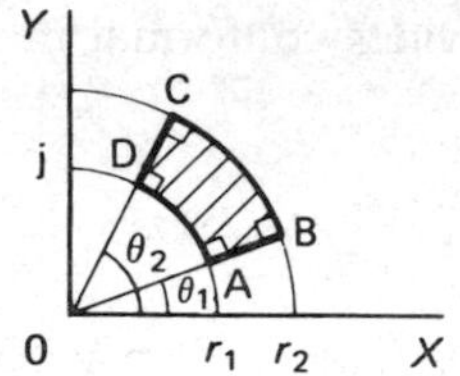

So ABCD, a section of an annulus of inner and outer radii r_1 and r_2 respectively, will be mapped on to

. . . .

44

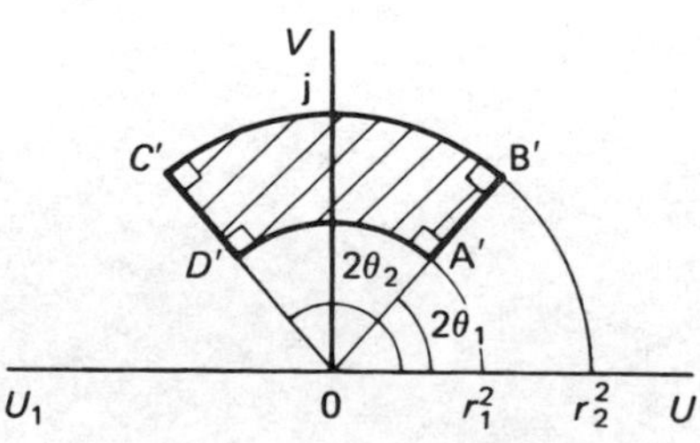

The angles at the origin are doubled, but notice that the right angles at A, B, C, D are preserved at A′, B′, C′, D′, i.e. the transformation there is conformal.

Example 3 Consider the mapping of the circle $|z| = 1$ under the transformation $w = z + \dfrac{4}{z}$ on to the w-plane.

First, as always, check for singularities and critical points. We find

. . . .

45

> singularity at $z = 0$; critical points at $z = \pm 2$

A singularity occurs at $z = 0$, i.e. $f'(z)$ does not exist at $z = 0$. Also

$f(z) = z + \frac{4}{z}$ $\therefore f'(z) = 1 - \frac{4}{z^2}$ $\therefore f'(z) = 0$ at $z = \pm 2$.

Therefore the transformation is not conformal at $z = 0$ and at $z = \pm 2$.

In fact, if we carry out the transformation $w = z + \frac{4}{z}$ on the unit circle $|z| = 1$, we get .... (Complete it: it is good revision)

46

> the ellipse $\frac{u^2}{5^2} + \frac{v^2}{3^2} = 1$

$$\text{for } w = u + \mathrm{j}v = z + \frac{4}{z} = x + \mathrm{j}y + \frac{4}{x + \mathrm{j}y} = x + \mathrm{j}y + \frac{4(x - \mathrm{j}y)}{x^2 + y^2}$$

$$\therefore u = x + \frac{4x}{x^2 + y^2}; \qquad v = y - \frac{4y}{x^2 + y^2}$$

$$|z| = 1 \quad \therefore x^2 + y^2 = 1 \quad \therefore u = x(1 + 4) = 5x; \quad v = y(1 - 4) = -3y$$

$$\therefore x = \frac{u}{5} \quad \text{and} \quad y = -\frac{v}{3}$$

Then $x^2 + y^2 = 1$ gives $\underline{\frac{u^2}{5^2} + \frac{v^2}{3^2} = 1}$

The image of the unit circle is therefore an ellipse with centre at the origin; semi major axis 5; semi minor axis 3.

Now let us move on to a new section.

47

Schwarz–Christoffel transformation

Example 1 Consider a semi-infinite strip on BC as base, the arrows at A and D indicating that the ordinate boundaries extend to infinity in the positive y direction and that progression round the boundary is to be taken in the direction indicated.

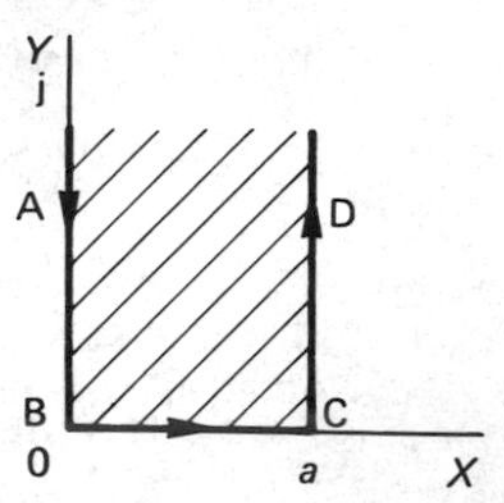

Let us apply the transformation $w = -\cos\dfrac{\pi z}{a}$ to the shaded region.

$$\text{Then } w = u + \mathrm{j}v = -\cos\frac{\pi z}{a} = -\cos\frac{\pi(x+\mathrm{j}y)}{a}$$

$$= -\left\{\cos\frac{\pi x}{a}\cos\frac{\mathrm{j}\pi y}{a} - \sin\frac{\pi x}{a}\sin\frac{\mathrm{j}\pi y}{a}\right\}$$

Now $\cos \mathrm{j}\theta = \cosh\theta$ and $\sin \mathrm{j}\theta = \mathrm{j}\sinh\theta$.

$$\therefore\ w = u + \mathrm{j}v = -\cos\frac{\pi x}{a}\cosh\frac{\pi y}{a} + \mathrm{j}\sin\frac{\pi x}{a}\sinh\frac{\pi y}{a}$$

$$\therefore\ u = -\cos\frac{\pi x}{a}\cosh\frac{\pi y}{a}; \qquad v = \sin\frac{\pi x}{a}\sinh\frac{\pi y}{a}$$

So B and C map on to B′ and C′ where B′ = ...; C′ = ...

48

$$\text{B}': u = -1,\ v = 0; \quad \text{C}': u = 1,\ v = 0$$

for (i) at B, $x = 0,\ y = 0\ \therefore u = -(1)(1) = -1;\ v = (0)(0) = 0$
and (ii) at C, $x = a,\ y = 0\ \therefore u = -(-1)(1) = 1;\ v = (0)(0) = 0$

So we have

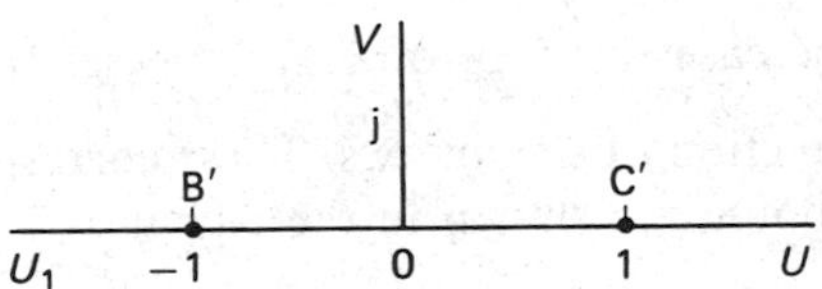

Now we map AB, BC, CD on to the w-plane giving A′B′, B′C′, C′D′.

(a) AB: $x = 0\ \therefore$ A′B′: $u = -\cosh\dfrac{\pi y}{a}$; $v = 0$

$\therefore$ As y decreases from ∞ to 0, u increases from $-\infty$ to -1.

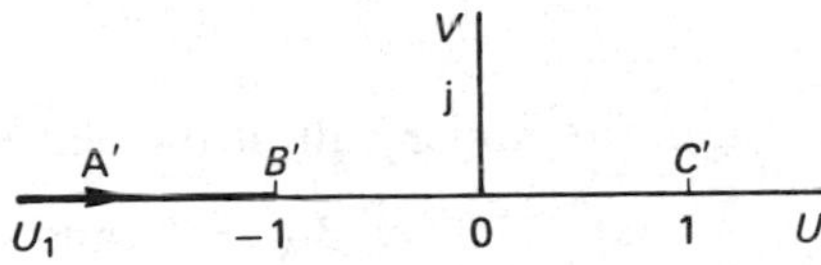

(b) BC: $y = 0\ \therefore$ B′C′: $u = -\cos\dfrac{\pi x}{a}$; $v = 0$

$\therefore$ As x increases from 0 to a, u increases from -1 to 1.

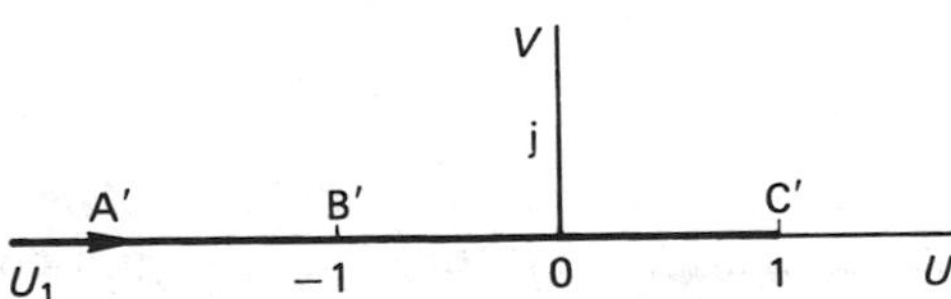

(c) CD: In the same way you can map CD and C′D′ in the w-plane and the mapping then becomes ....

49

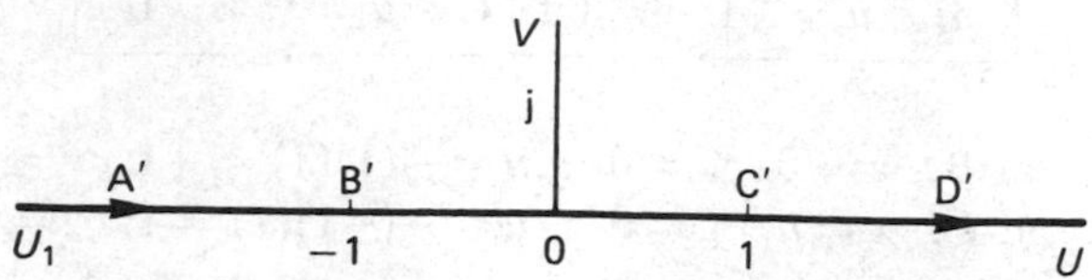

since CD: $\quad x = a \;\therefore\;$ C′D′: $\quad u = \cosh \dfrac{\pi y}{a}; \quad v = 0.$

Therefore, as y increases from 0 to ∞, u increases from 1 to ∞.

Notice the direction of the arrows. These correspond to the directed travel round the boundary shown in the z-plane.

The shaded region in the z-plane is on the left-hand side of the boundary as traversed. This maps on to the left-hand side of the image

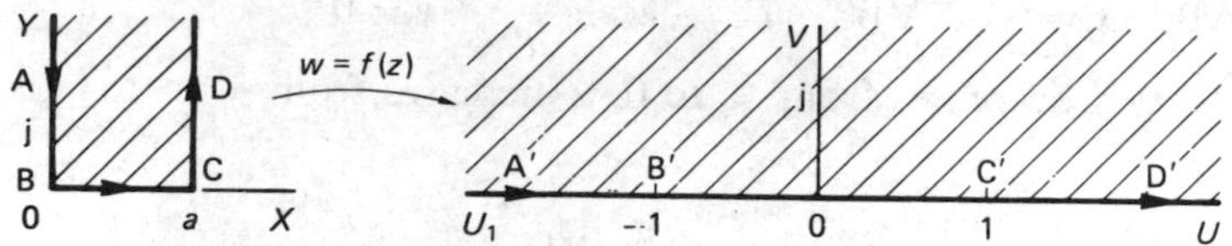

on the w-plane, i.e. the entire upper half of the plane.

Note that $\dfrac{dw}{dz} = \dfrac{\pi}{a} \sin \dfrac{\pi z}{a} \quad \therefore$ at B ($z = 0$) and C ($z = a$), $\dfrac{dw}{dz} = 0.$

Therefore, the conformal property does not hold at these points. The internal angle at B and at C is $\dfrac{\pi}{2}$, while at B′ and C′ it is π .

Example 2 Consider an infinite strip in the z-plane bounded by the real axis and $z = \text{j}a$

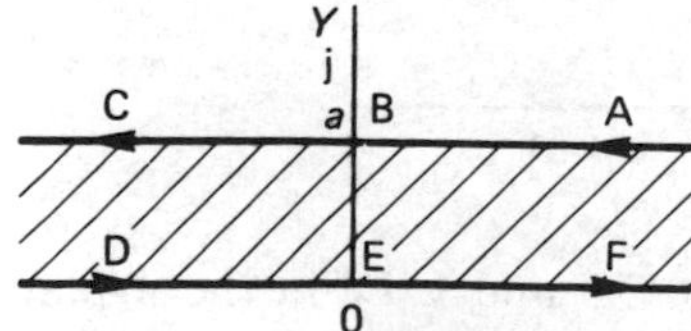

Note the arrows. The boundary comes from $+\infty$ (A) and continues to $-\infty$ (C); then returns from $-\infty$ (D) to $+\infty$ (F).

The strip can be considered as a closed figure with the left- and right-hand vertices at infinity.

We now map the infinite strip on to the w-plane by the transformation $w = e^{\pi z/a}$.

$$\therefore\; w = u + \text{j}v = e^{\pi z/a}, \text{ from which}$$

$$u = \ldots\ldots\ldots; \quad v = \ldots\ldots\ldots$$

50

$$u = e^{\pi x/a} \cos \frac{\pi y}{a}; \quad v = e^{\pi x/a} \sin \frac{\pi y}{a}$$

for $u + jv = e^{\pi z/a} = e^{\pi(x+jy)/a} = e^{\pi x/a} e^{j\pi y/a}$

$$= e^{\pi x/a}\left(\cos \frac{\pi y}{a} + j \sin \frac{\pi y}{a}\right)$$

$$\therefore u = e^{\pi x/a} \cos \frac{\pi y}{a}; \qquad v = e^{\pi x/a} \sin \frac{\pi y}{a}$$

Now we map points B and E on to B′ and E′.

(i) B: $x = 0, \ y = a$ $\qquad \therefore$ B′: $u = -1, \ v = 0$
(ii) E: $x = 0, \ y = 0$ $\qquad \therefore$ E′: $u = 1, \ v = 0$

i.e.

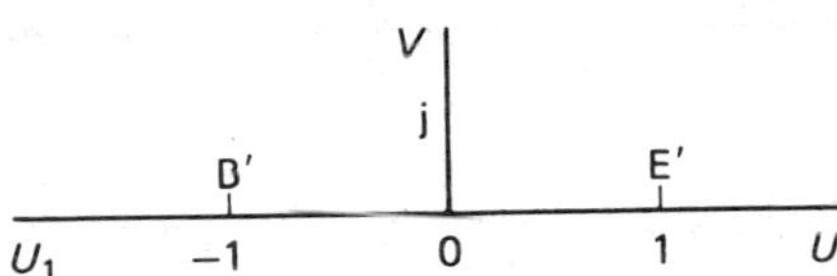

Now we map the lines AB, BC, DE, EF on to the w-plane.

(a) AB: $y = a \quad \therefore u = -e^{\pi x/a}, \quad v = 0$

$\therefore$ As x decreases from $+\infty$ to 0, u increases from $-\infty$ to -1.

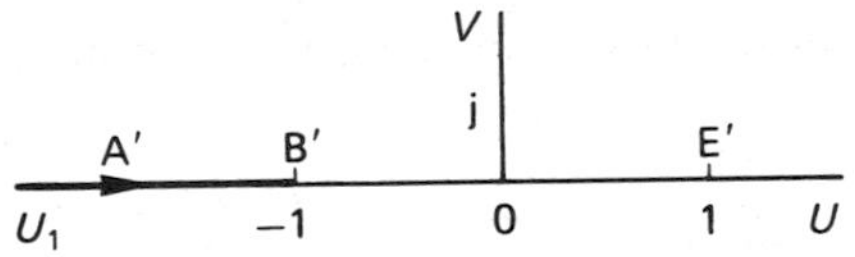

(b) BC: $y = a \quad \therefore u = -e^{\pi x/a}, \quad v = 0$ (as for AB)

$\therefore$ As x decreases from 0 to $-\infty$, u increases from -1 to 0.

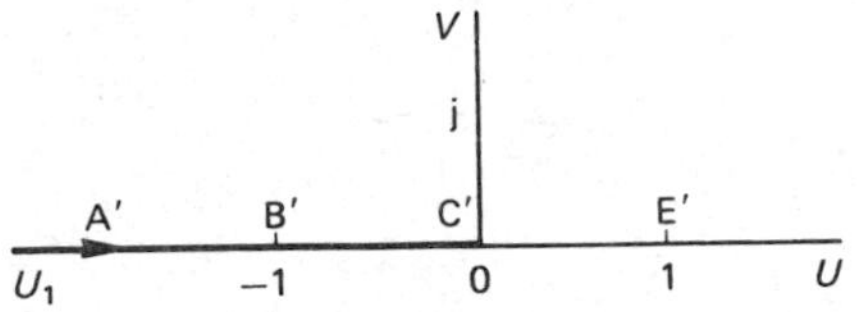

(c) Now there is DE which maps on to

....

51

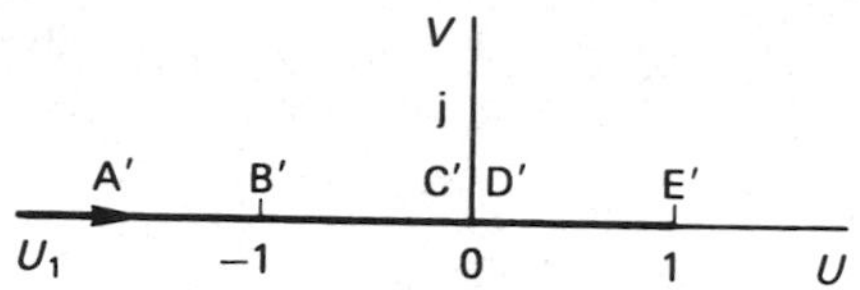

Since

(c) DE: $y = 0 \quad \therefore u = e^{\pi x/a}, \quad v = 0$

$\therefore$ As x increases from $-\infty$ to 0, u increases from 0 to 1.

(d) EF: $y = 0 \quad \therefore u = e^{\pi x/a}, \quad v = 0$ (as for DE)

$\therefore$ As x increases from 0 to $+\infty$, u increases from 1 to $+\infty$.

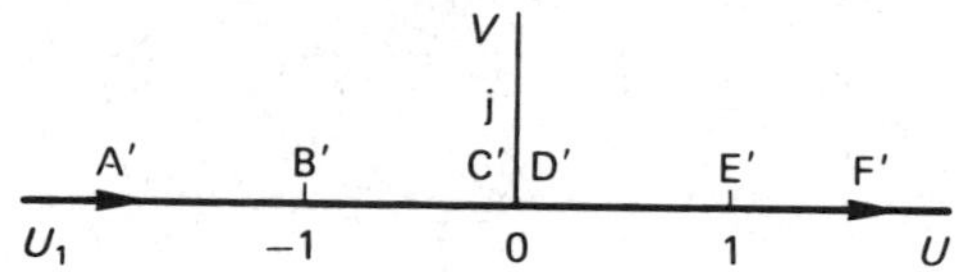

Finally, what about the shaded region in the z-plane? This maps on to

....

52

the upper half of the w-plane

since it is on the left-hand side of the directed boundary in the z-plane.

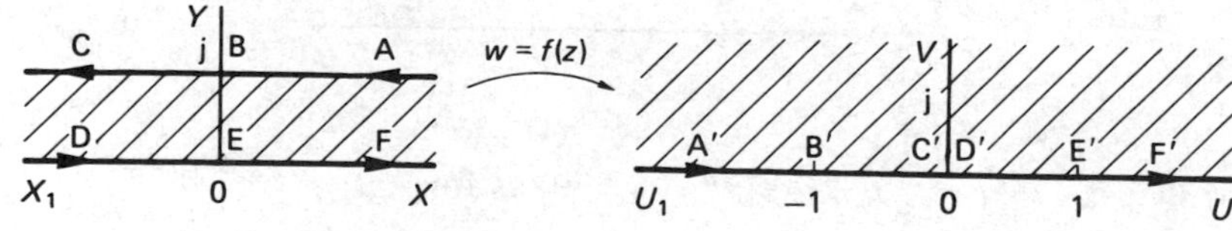

Schwarz–Christoffel transformation

The last two examples have been simple cases of the application of the Schwarz–Christoffel transformation under which any polygon in the z-plane can be made to map on to the entire *upper half* of the w-plane and the boundary of the polygon on to the *real axis* of the w-plane.

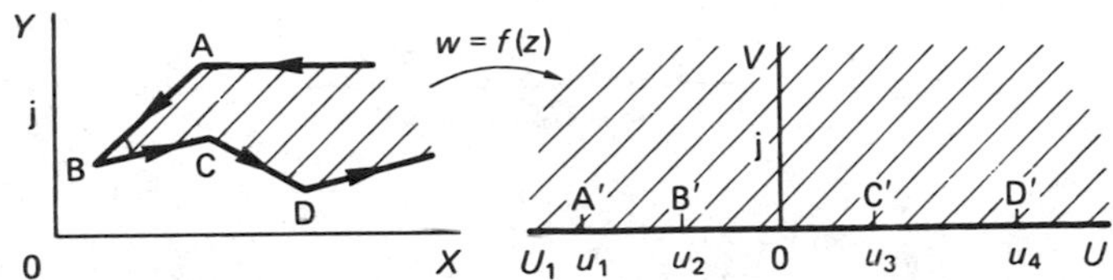

The process depends, of course, on the right choice of transformation function for any particular polygon, which can be defined by its vertices and the internal angle at each vertex.

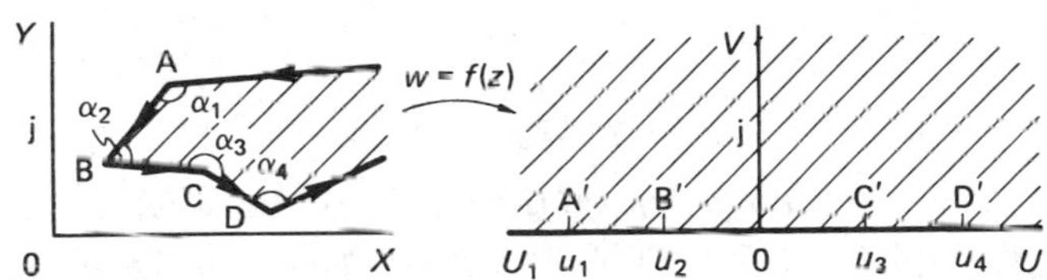

The Schwarz–Christoffel transformation function is given by

$$\frac{dz}{dw} = A(w-u_1)^{\alpha_1/\pi-1}(w-u_2)^{\alpha_2/\pi-1}(w-u_3)^{\alpha_3/\pi-1}\dots$$

$$\therefore\ z = A\int (w-u_1)^{\alpha_1/\pi-1}(w-u_2)^{\alpha_2/\pi-1}\dots(w-u_n)^{\alpha_n/\pi-1}\mathrm{d}w + B$$

where A and B are complex constants, determined by the physical properties of the polygon.

This is not as bad as it looks!

Make a careful note of it: then we will apply it.

53 Here it is again.

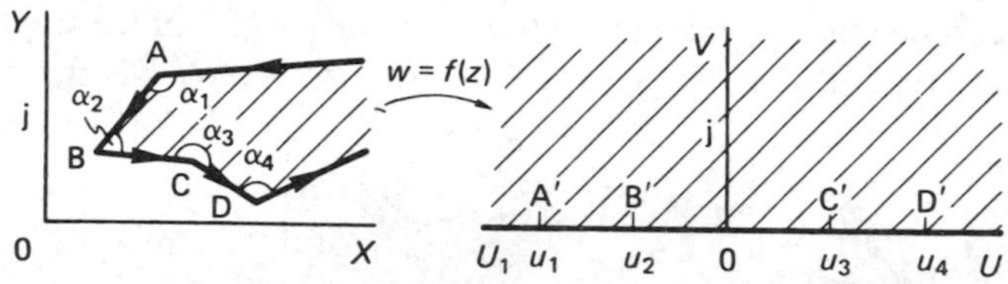

$$\frac{dz}{dw} = A(w - u_1)^{\alpha_1/\pi - 1}(w - u_2)^{\alpha_2/\pi - 1}(w - u_3)^{\alpha_3/\pi - 1}\dots$$

$$\therefore z = A\int(w - u_1)^{\alpha_1/\pi - 1}(w - u_2)^{\alpha_2/\pi - 1}\dots(w - u_n)^{\alpha_n/\pi - 1}dw + B$$

where A and B are complex constants.

Three other points also have to be noted.

1. Any three points u_1, u_2, u_3 on the u-axis can be selected as required.
2. It is convenient to choose one such point, u_n, at infinity, in which case the relevant factor in the integral above does not occur.
3. Infinite open polygons are regarded as limiting cases of closed polygons where one (or more) vertex is taken to infinity.

Open polygons

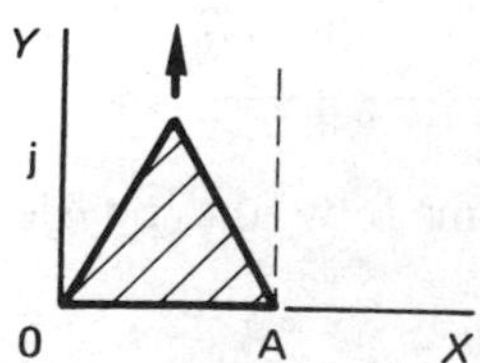

We have already introduced these in Examples 1 and 2 of this section.

In Example 1, the semi-infinite strip is a case of a triangle with one vertex

....

54

taken to infinity in the y-direction

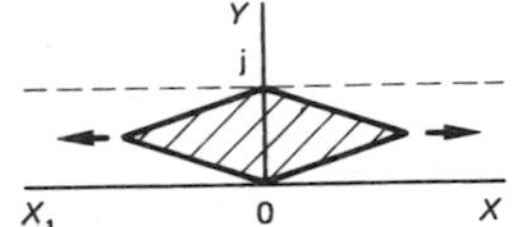

In Example 2, the infinite strip is a case of a double triangle, or quadrilateral, with two vertices taken to infinity.

An open polygon with n sides with one vertex at infinity will have $(n-1)$ internal angles.
An open polygon with n sides with two vertices at infinity will have $(n-2)$ internal angles.

Now for an example to see how all this works.

55

Example 3 To determine the transformation that will map the semi-infinite strip ABCD on to the w-plane so that the images of B and C occur at $u=-1$ and $u=1$, respectively, and the shaded region maps on to the upper half of the w-plane.

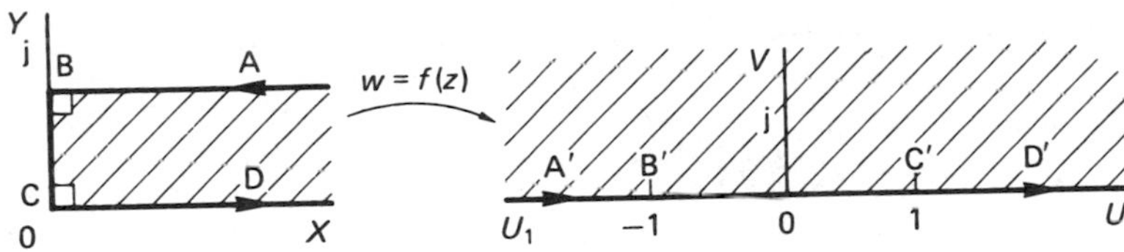

In this case, B′ is $u_1=-1$ and C′ is $u_2=1$.

The corresponding internal angles are: at B, $\alpha_1=\dfrac{\pi}{2}$ and at C, $\alpha_2=\dfrac{\pi}{2}$.

So we have

$$\frac{dz}{dw}=A(w+1)^{(\pi/2)/\pi-1}(w-1)^{(\pi/2)/\pi-1} \quad A \text{ a complex constant}$$

$$=A(w+1)^{-1/2}(w-1)^{-1/2}=A(w^2-1)^{-1/2}$$

$$=K(1-w^2)^{-1/2}=\frac{K}{\sqrt{1-w^2}}$$

$$\therefore z=\int\frac{K}{\sqrt{1-w^2}}dw=\ldots\ldots\ldots$$

56

$$\boxed{z = K \arcsin w + B}$$

$$\therefore \arcsin w = \frac{z - B}{K} \qquad \therefore \underline{w = \sin\frac{z - B}{K}}$$

Now we have to find B and K.

(a) We require B $(z = \mathrm{j}a)$ to map on to B′ $(w = -1)$

$$\therefore -1 = \sin\frac{\mathrm{j}a - B}{K}$$

$$\therefore \frac{\mathrm{j}a - B}{K} = -\frac{\pi}{2} \qquad \therefore 2\mathrm{j}a - 2B = -K\pi \qquad (1)$$

(b) We also require C $(z = 0)$ to map on to C′ $(w = 1)$ $\therefore 1 = \sin\dfrac{0 - B}{K}$

$$\therefore -\frac{B}{K} = \frac{\pi}{2} \qquad \therefore -2B = K\pi \qquad (2)$$

Then, from (1) and (2), $B = \dots\dots\dots$; $K = \dots\dots\dots$;

57

$$\boxed{B = \frac{\mathrm{j}a}{2}; \quad K = -\frac{\mathrm{j}a}{\pi}}$$

$$\therefore w = \sin\left\{\frac{z - (\mathrm{j}a)/2}{-\mathrm{j}a/\pi}\right\} = \sin\left\{\mathrm{j}z\frac{\pi}{a} + \frac{\pi}{2}\right\} = \cos\frac{\mathrm{j}z\pi}{a}$$

But $\cos \mathrm{j}\theta = \cosh\theta$ $\qquad \therefore \underline{w = \cosh\dfrac{\pi z}{a}}$

To verify that this is the required transformation, let us apply it to the figure given in the z-plane. We will do that in the next frame.

58

We have:

$$w = u + jv = \cosh\frac{\pi z}{a} = \cosh\frac{(x + jy)\pi}{a}$$

$$\therefore\ u + jv = \cosh\frac{x\pi}{a}\cosh\frac{jy\pi}{a} + \sinh\frac{x\pi}{a}\sinh\frac{jy\pi}{a}$$

But $\cosh j\theta = \cosh\theta$ and $\sinh j\theta = j\sin\theta$

$$\therefore\ u + jv = \cosh\frac{x\pi}{a}\cos\frac{y\pi}{a} + j\sinh\frac{x\pi}{a}\sin\frac{y\pi}{a}$$

$$\therefore\ u = \cosh\frac{x\pi}{a}\cos\frac{y\pi}{a}; \quad v = \sinh\frac{x\pi}{a}\sin\frac{y\pi}{a}$$

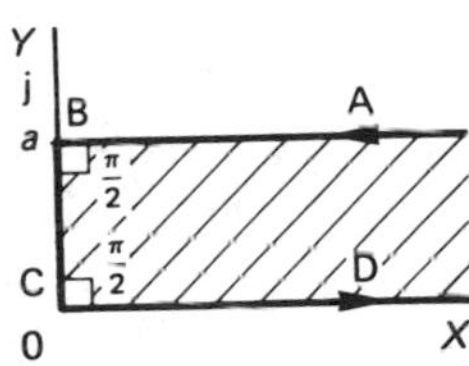

First map the points B and C on to B′ and C′ in the w-plane.

B′: ....; C′: ....

59

$$\text{B}': u = -1, v = 0; \quad \text{C}': u = 1, v = 0$$

for B: $x = 0,\ y = a\ \therefore\ \text{B}':\ u = \cos\pi = -1,\ v = 0\ \therefore\ \text{B}':\ u = -1,\ v = 0$
C: $x = 0,\ y = 0\ \therefore\ \text{C}':\ u = 1,\ v = 0 \qquad \therefore\ \text{C}':\ u = 1,\ v = 0.$

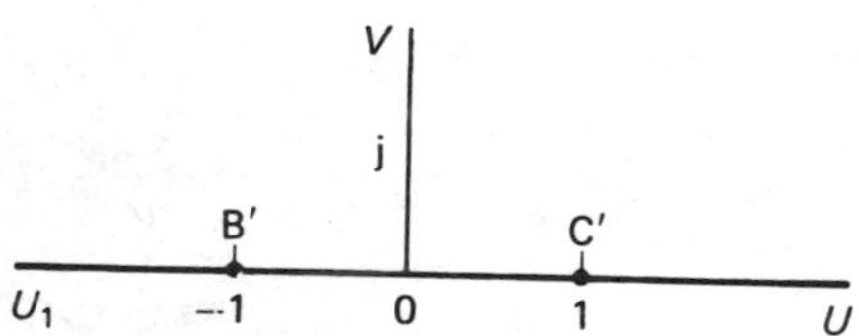

Now we map AB, BC, CD in turn.

(a) AB: $y = a \quad \therefore\ u = -\cosh\dfrac{x\pi}{a}, \quad v = 0$

$\therefore$ As x decreases from ∞ to 0, u increases from $-\infty$ to -1.

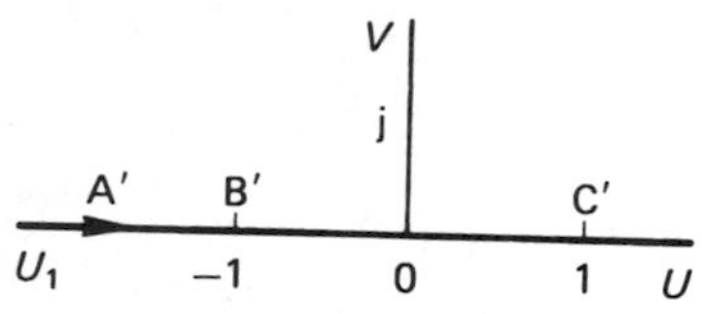

(b) BC:
(c) CD: } Complete the working and show the mapped region which is

. . . .

60

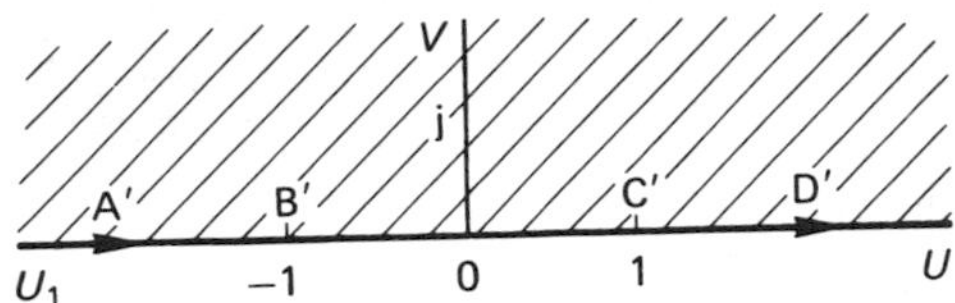

for we have

(b) BC: $x = 0 \quad \therefore\ u = \cos\dfrac{y\pi}{a}, \quad v = 0$

$\therefore$ As y decreases from a to 0, u increases from –1 to 1.

CD: $y = 0 \quad \therefore\ u = \cosh\dfrac{x\pi}{a}, \quad v = 0$

$\therefore$ As x increases from 0 to ∞, u increases from 1 to ∞.

In each plane, the shaded region is on the left-hand side of the boundary.

We will now finish up with one further example.

So move on.

61

Example 4

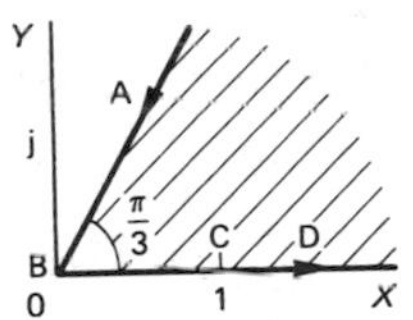

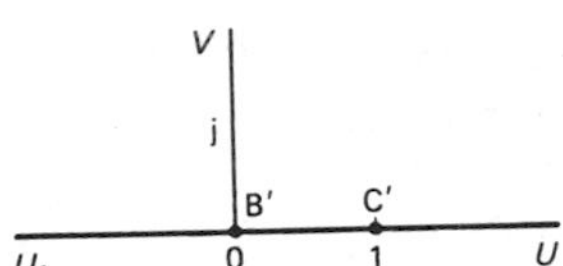

Determine the transformation function $w = f(z)$ that maps the infinite sector in the z-plane on to the upper half of the w-plane with points B and C mapping on to B′ and C′ as shown.

The transformation function $w = f(z)$ is given by

. . . .

62

$$\frac{dz}{dw} = A(w - u_1)^{\alpha_1/\pi - 1}(w - u_2)^{\alpha_2/\pi - 1} \ldots (w - u_n)^{\alpha_n/\pi - 1}$$

At B, $\alpha_1 = \dfrac{\pi}{3}$. At C, $\alpha_2 = \pi$.

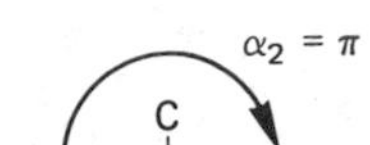

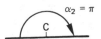

With that reminder, you can now work through on your own, just as we did before, finally obtaining $w = \ldots\ldots\ldots$

63

$$\boxed{w = z^3}$$

Check with the working.

$$\frac{dz}{dw} = A(w-0)^{(\pi/3)/\pi - 1}(w-1)^{\pi/\pi - 1}$$

$$= Aw^{-2/3}(w-1)^0 = Aw^{-2/3} \quad \therefore z = 3Aw^{1/3} + B$$

$$\therefore z = Kw^{1/3} + B \quad \therefore w = \left(\frac{z-B}{K}\right)^3$$

To find B and K

(a) At B: $z = 0$ At B′: $w = 0 \therefore 0 = \left(\frac{-B}{K}\right)^3 \therefore B = 0 \therefore w = \left(\frac{z}{K}\right)^3$

(b) At C: $z = 1$ At C′: $w = 1 \therefore 1 = \left(\frac{1}{K}\right)^3 \therefore K = 1 \therefore w = z^3$

$\therefore$ the transformation function is $w = z^3$

Finally, as a check—and a little more valuable practice—apply the function $w = z^3$ to the region shaded in the z-plane.

$$w = u + jv = (x + jy)^3 = x^3 + 3x^2(jy) + 3x(jy)^2 + (jy)^3$$

$$\therefore u = \ldots\ldots\ldots; \quad v = \ldots\ldots\ldots$$

64

$$u = x^3 - 3xy^2; \quad v = 3x^2y - y^3$$

At B: $x = 0, \ y = 0 \quad \therefore u = 0, \ v = 0 \qquad \therefore B': u = 0, \ v = 0$

At C: $x = 1, \ y = 0 \quad \therefore u = 1, \ v = 0 \qquad \therefore C': u = 1, \ v = 0$

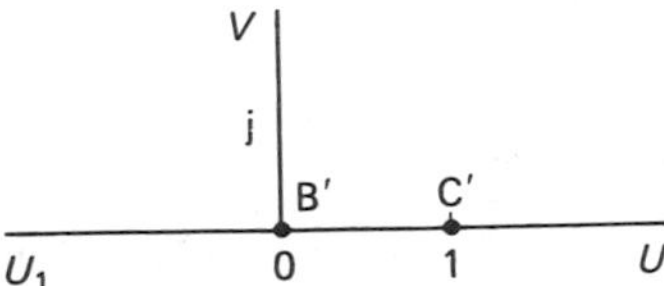

Now we map AB, BC, CD on to A′B′, B′C′, C′D′

AB: $y = \sqrt{3}x \quad \therefore u = x^3 - 9x^3 = -8x^3, \quad v = 0$

$\therefore$ As x decreases from ∞ to 0, u increases from $-\infty$ to 0.

You can now deal with BC and CD in the same way and finally show the transformed region. So we get . . . .

65

Here is the remaining working.

BC: $y = 0 \ \therefore u = x^3, \ v = 0$

$\therefore$ As x increases from 0 to 1, u increases from 0 to 1.

CD: $y = 0 \ \therefore u = x^3, \ v = 0$

$\therefore$ As x increases from 1 to ∞, u increases from 1 to ∞.

So we have

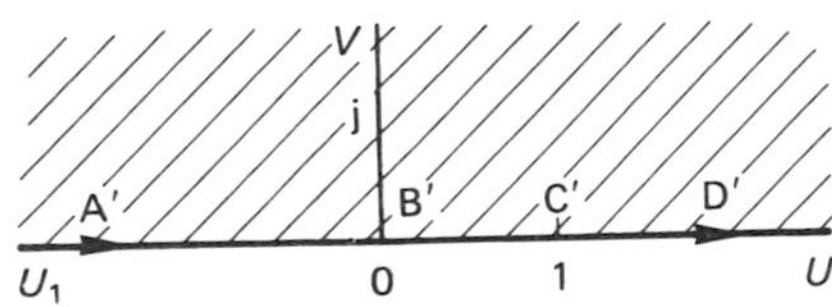

The shaded region is to the left of the directed boundary in the z-plane. This therefore maps on to the region to the left of the directed real axis in the w-plane, i.e. the upper half of the plane.

We have just touched on the fringe of the work on Schwarz–Christoffel transforamtion. The whole topic of mapping between planes has applications in fluid mechanics, heat conduction, electromagnetic theory, etc. and it is at times convenient to solve a problem relating to the z-plane by transforming to the upper half of the w-plane and later to transform back to the z-plane. The transformation function can be operated in either direction.

And that is it. The Revision Summary follows: then on to the Test Exercise.

66 REVISION SUMMARY

1. *Differentiation of a complex function*

$$w = f(z) \qquad \frac{dw}{dz} = f'(z) = \lim_{\delta z \to 0}\left\{\frac{f(z_0 + \delta z) - f(z_0)}{\delta z}\right\}$$

2. *Regular (or analytic) function*

   $w = f(z)$ is *regular* at z_0 if it is defined, single-valued and has a derivative at every point at and around $z = z_0$.

3. *Singularities* or singular points—points at which $f(z)$ ceasese to be regular.

4. *Cauchy–Riemann equations* test whether $w = f(z)$ has a derivative $f'(z)$ at $z = z_0$. $\quad w = u + jv = f(z)$ where $z = x + jy$.

   Then $\quad \dfrac{\partial u}{\partial x} = \dfrac{\partial v}{\partial y}$ and $\dfrac{\partial v}{\partial x} = -\dfrac{\partial u}{\partial y}$

5. *Complex integration*

$$\int w\mathrm{d}z = \int f(z)\mathrm{d}z = \int (u\mathrm{d}x - v\mathrm{d}y) + \mathrm{j}\int (v\mathrm{d}x + u\mathrm{d}y)$$

6. *Contour integration*—evaluation of line integrals in the z-plane.

7. *Cauchy's theorem* If $f(z)$ is regular at every point within and on a simply connected closed curve c, then $\oint_c f(z)\,\mathrm{d}z = 0$.

8. *Deformation of contours*

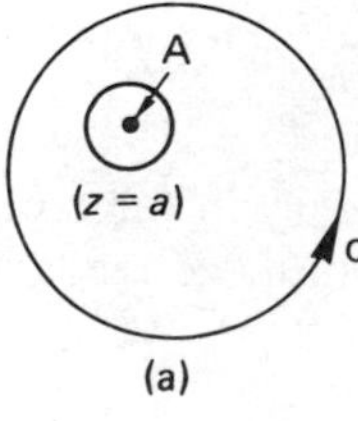

(a)

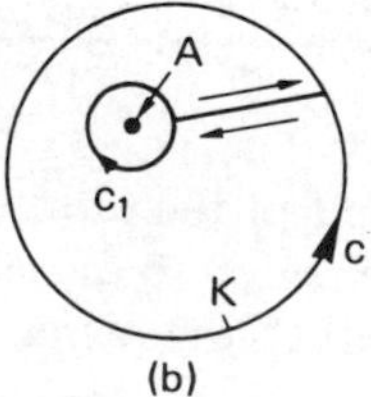

(b)

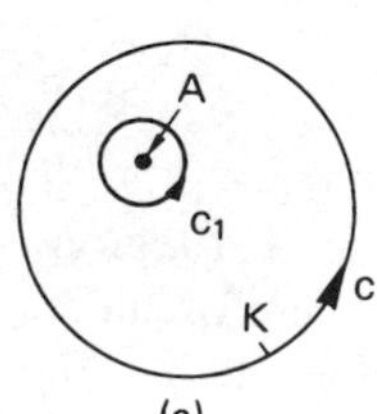

(c)

(a) Singularity at A

(b) Restored to simply connected closed curve

(c) $\oint_c f(z)\,\mathrm{d}z = \oint_{c_1} f(z)\,\mathrm{d}z.$

For $\oint_c f(z)\, dz$ where $f(z) = \dfrac{1}{(z-a)^n}$ $n = 1, 2, 3 \ldots$

$$\oint_c \frac{1}{(z-a)^n} dz = 0 \text{ if } n \neq 1$$

$$= 0 \text{ if } n = 1 \text{ and c does not enclose } z = a$$

$$= j2\pi \text{ if } n = 1 \text{ and c does enclose } z = a$$

.

9. *Conformal transformation*—mapping in which angles are preserved in size and sense of rotation.

Conditions

1. $w = f(z)$ must be a regular function of z.
2. $f'(z)$, i.e. $\dfrac{dw}{dz}$, $\neq 0$ at the point of intersection.

   If $f'(z) = 0$ at $z = z_0$, then z_0 is a *critical point.*

10. *Schwarz–Christoffel transformation* maps any polygon in the z-plane on to the entire *upper half* of the w-plane and the boundary of the polygon on to the *real axis* of the w-plane.

$$\frac{dz}{dw} = A(w - u_1)^{\alpha_1/\pi - 1}(w - u_2)^{\alpha_2/\pi - 1} \ldots (w - u_n)^{\alpha_n/\pi - 1}$$

1. Any three points u_1, u_2, u_3 can be selected on the u-axis.
2. One such point can be chosen at infinity.
3. Infinite open polygons are regarded as limiting cases of closed polygons.

67 TEXT EXERCISE XVI

1. Determine where each of the following functions fails to be regular

(a) $w = z^3 + 4$

(b) $w = \dfrac{z}{z+5}$

(c) $w = e^{2z+4}$

(d) $w = \dfrac{z-2}{(z-4)(z+1)}$

(e) $w = \dfrac{x - jy}{x^2 + y^2}$.

2. Verify Cauchy's theorem by evaluating $\oint_c f(z)\,dz$ where $f(z) = z^2$ round the rectangle formed by joining the points $z = 2 + j$, $z = 2 + j4$, $z = j4$, $z = j$.

3. Evaluate the integral $\oint_c f(z)dz$ where $f(z) = \dfrac{3z - 6 - j}{(z - j)(z - 3)}$ round the contour $|z| = 2$.

4. Determine critical points, if any, at which the following transformation functions $w = f(z)$ fail to be conformal.

(a) $w = z^4$

(b) $w = z^3 - 3z$

(c) $w = e^{1-z}$

(d) $w = z + \dfrac{2}{z}$

(e) $w = e^{(z^2)}$

(f) $w = \dfrac{z + j}{z - j}$.

5. Determine the Schwarz–Christoffel transformation function $w = f(z)$ that will map the semi-infinite strip shaded in the z-plane on to the upper half of the w-plane, so that the image of B is B′ ($w = -1$) and that of C is C′ ($w = 0$). Obtain the image of the point D.

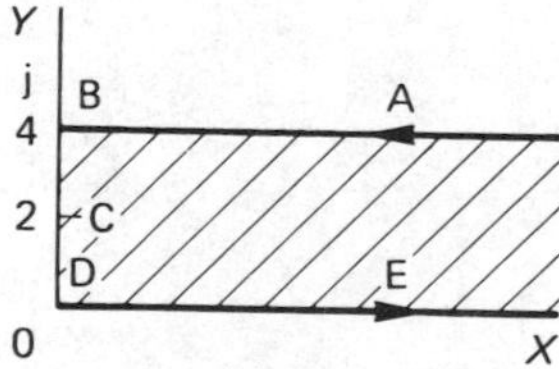

FURTHER PROBLEMS XVI

1. Verify Cauchy's theorem for the closed path c consisting of three straight lines joining A $(1+\mathrm{j})$, B $(3+\mathrm{j}3)$, C $(-1+\mathrm{j}3)$ where $f(z)=z-1+\mathrm{j}$.

2. If $z=2+\mathrm{j}y$ is mapped on to the w-plane under the transformation $w=f(z)=\dfrac{1}{z}$, show that the locus of w is a circle with centre $w=0.25$ and radius 0.25.

3. Determine the image in the w-plane of the circle $|z-2|=1$ in the z-plane under the transformation $w=(1-\mathrm{j})z+3$.

4. The unit circle $|z|=1$ in the z-plane is generated in an anticlockwise manner from the point A $(z=1)$ and is transformed on to the w-plane by $w=\dfrac{z}{z-2}$. Determine the locus of w and the direction in which it is generated.

5. Evaluate $\oint_c f(z)\,\mathrm{d}z$ where $f(z)=\dfrac{5z-2-\mathrm{j}3}{(z-\mathrm{j})(z-1)}$ around the closed contour c for the two cases when

   (a) c is the path $|z|=2$

   (b) c is the path $|z-1|=1$.

6. If $f(z)=\dfrac{5z+\mathrm{j}}{(z-\mathrm{j})(z+\mathrm{j}2)}$, evaluate $\oint_c f(z)\mathrm{d}z$ along the contours

   (a) $|z-1|=1$; (b) $|z|=\dfrac{3}{2}$; (c) $|z|=3$.

7. If $z=x+\mathrm{j}y$ and $w=f(z)$, show that, if $\dfrac{\mathrm{j}(w+z)}{w-z}$ is entirely real, then $|w|=|z|$.

8. Evaluate $\oint_c f(z)\,\mathrm{d}z$, where $f(z)=\dfrac{3z-\mathrm{j}5}{(z+1-\mathrm{j}2)(z-2-\mathrm{j})}$, around the perimeter of the rectangle formed by the lines $z=1$, $z=\mathrm{j}3$, $z=-2$, $z=-\mathrm{j}$.

9. If $f(z)=\dfrac{8z^2-2}{z(z-1)(z+1)}$, evaluate $\oint_c f(z)\,\mathrm{d}z$ along the contour c where c is the triangle joining the points $z=2$, $z=\mathrm{j}$, $z=-1-\mathrm{j}$.

10. (a) For the transformation $w=z+\dfrac{1}{z}$, state (i) singularities, (ii) critical points.

    (b) Apply $w=z+\dfrac{1}{z}$ to map the circle $|z|=2$ on to the w-plane.

11. Find the images in the w-plane of (a) the line $y = 0$ and (b) the line $y = x$ that result from the mapping $w = \dfrac{z - \text{j}}{z + \text{j}}$. Show that the curves intersect at the points $(\pm 1, 0)$ in the w-plane and determine the angle at which they intersect.

12. Use the transformation $w = \dfrac{\text{j}(1 + z)}{1 - z}$ to map the unit circle $|z| = 1$ in the z-plane on to the w-plane. Determine also the image in the w-plane of the region bounded by $|z| = 1$ and inside the circle.

13. Determine the transformation that will map the semi-infinite strip shown, on to the upper half of the w-plane, where the image of B is B′ $(w = -1)$ and that of C is C′ $(w = 1)$.

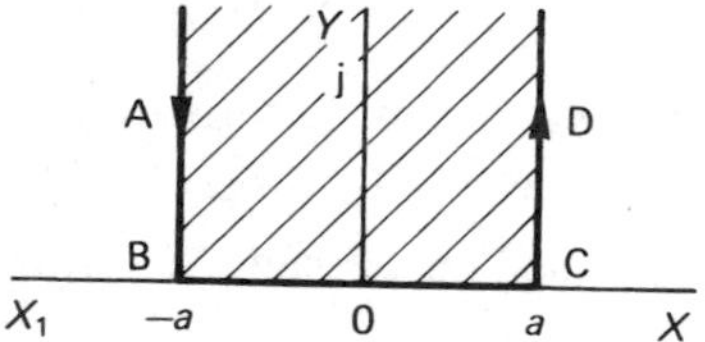

Programme 17

Fourier Series

Prerequisites: Engineering Mathematics (fourth edition)
Programmes 15, 17

Introduction

1

We have seen earlier that many functions can be expressed in the form of infinite series. Problems involving various forms of oscillations are common in fields of modern technology and *Fourier series*, with which we shall now be concerned, enable us to represent a periodic function as an infinite trigonometrical series in sine and cosine terms. One important advantage of Fourier series is that it can represent a function containing discontinuities, whereas Maclaurin's and Taylor's series require the function to be continuous throughout.

Periodic functions

A function $f(x)$ is said to be *periodic* if its function values repeat at regular intervals of the independent variable. The regular interval between repetitions is the *period* of the oscillations.

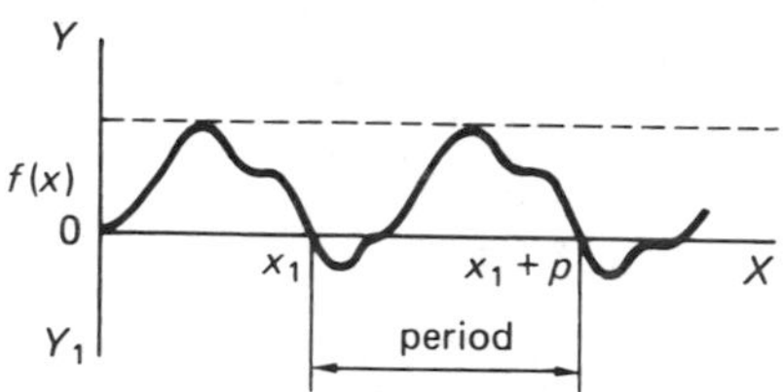

Graphs of $y = A \sin nx$

(a) $y = \sin x$ The obvious example of a periodic function is $y = \sin x$, which goes through its complete range of values while x increases from 0° to 360°. The period is therefore 360° or 2π radians and the amplitude, the maximum displacement from the position of rest, is 1.

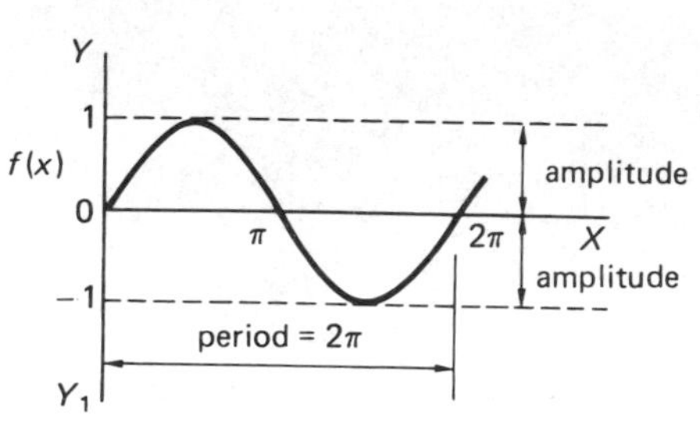

(b) $y = 5\sin 2x$

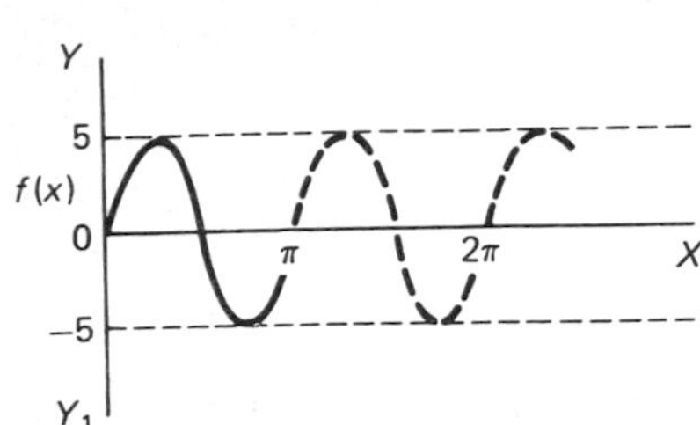

The amplitude is 5.
The period is 180° and there are thus 2 complete cycles in 360°.

(c) $y = A\sin nx$
Thinking along the same lines, the function $y = A\sin nx$ has amplitude ....; period ....; and will have .... complete cycles in 360°.

2

$$\text{amplitude} = A; \quad \text{period} = \frac{360°}{n} = \frac{2\pi}{n}; \quad n \text{ cycles in } 360°$$

Graphs of $y = A\cos nx$ have the same characteristics.

By way of revising earlier work, then, complete the following short exercise.

Exercise In each of the following state (a) the amplitude and (b) the period.

1. $y = 3\sin 5x$
2. $y = 2\cos 3x$
3. $y = \sin\dfrac{x}{2}$
4. $y = 4\sin 2x$
5. $y = 5\cos 4x$
6. $y = 2\sin x$
7. $y = 3\cos 6x$
8. $y = 6\sin\dfrac{2x}{3}$

Deal with all eight. They will not take much time.

3

No.	Amplitude	Period	No.	Amplitude	Period
1	3	72°	5	5	90°
2	2	120°	6	2	360°
3	1	720°	7	3	60°
4	4	180°	8	6	540°

Harmonics

A function $f(x)$ is sometimes expressed as a series of a number of different sine components. The component with the largest period is the *first harmonic*, or *fundamental* of $f(x)$.

$y = A_1 \sin x$ is the first harmonic or fundamental
$y = A_2 \sin 2x$ is the second harmonic
$y = A_3 \sin 3x$ is the third harmonic, etc.

and in general

$y = A_n \sin nx$ is the .... harmonic, with amplitude .... and period ....

4

nth harmonic; amplitude A_n; period $= \dfrac{360°}{n} = \dfrac{2\pi}{n}$

Non-sinusoidal periodic functions

Although we introduced the concept of a periodic function via a sine curve, a function can be periodic without being obviously sinusoidal in appearance.

Example In the following cases, the x-axis carries a scale of t in milliseconds.

(a) period $= 8$ ms

(b)

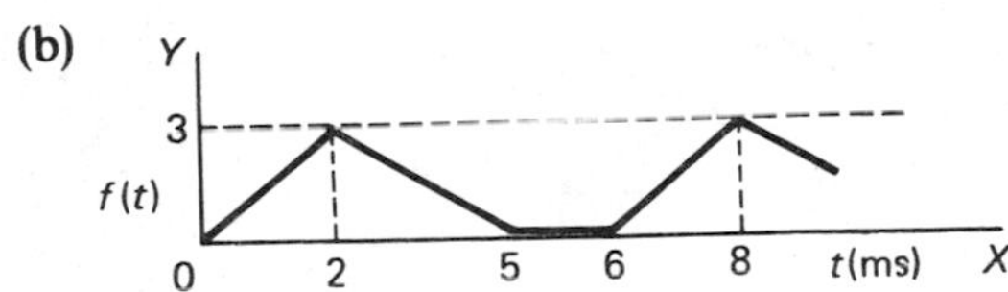

period $= \dots$

(c)

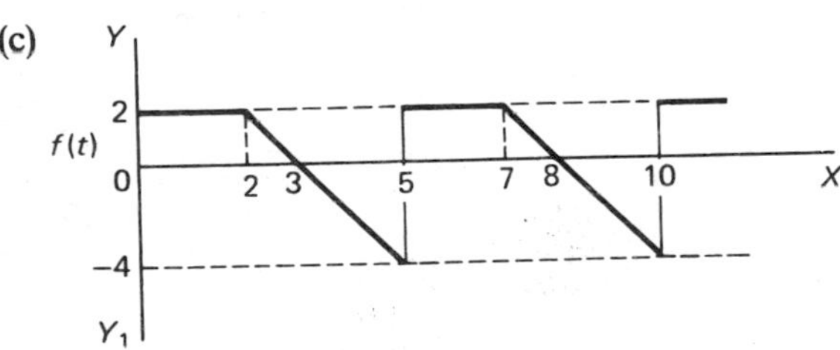

period $= \dots\dots$

5

(b) period $= 6$ ms; (c) period $= 5$ ms

Analytical description of a periodic function

A periodic function can be defined analytically in many cases.

Example 1

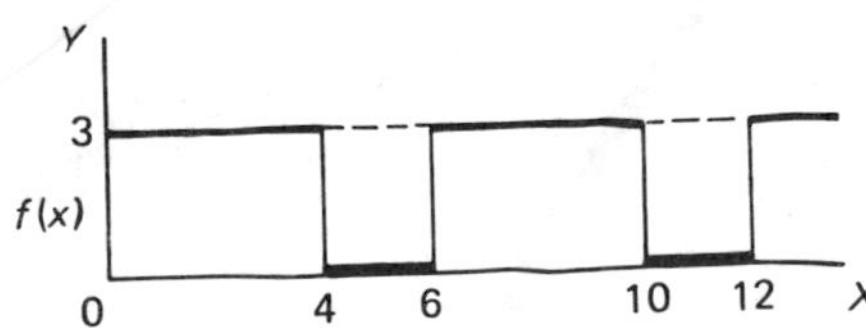

(a) Between $x = 0$ and $x = 4$, $y = 3$, i.e. $f(x) = 3$ $\quad 0 < x < 4$
(b) Between $x = 4$ and $x = 6$, $y = 0$, i.e. $f(x) = 0$ $\quad 4 < x < 6$
So we could define the function by

$$f(x) = 3 \qquad 0 < x < 4$$
$$f(x) = 0 \qquad 4 < x < 6$$
$$f(x) = f(x + 6)$$

the last line indicating that $\dots\dots\dots$

6

the function is periodic with period 6 units

Example 2

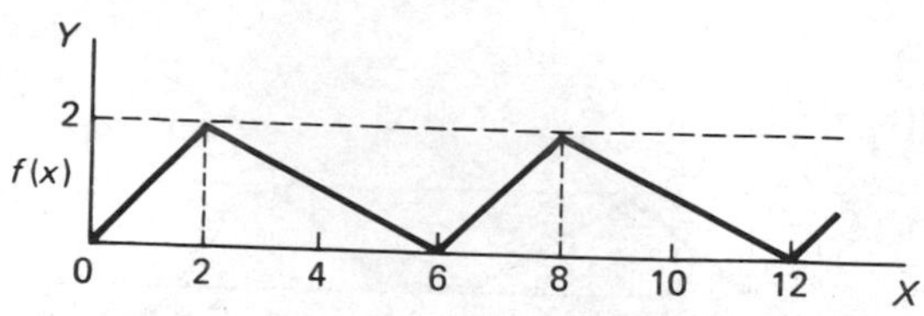

In this case

(a) between $x = 0$ and $x = 2$, $y = x$ i.e. $f(x) = x \quad 0 < x < 2$

(b) between $x = 2$ and $x = 6$, $y = -\dfrac{x}{2} + 3$, i.e. $f(x) = 3 - \dfrac{x}{2} \quad 2 < x < 6$

(c) the period is 6 units i.e. $f(x) = f(x + 6)$.

So we have

$$f(x) = x \qquad 0 < x < 2$$
$$f(x) = 3 - \frac{x}{2} \qquad 2 < x < 6$$
$$f(x) = f(x + 6).$$

Example 3

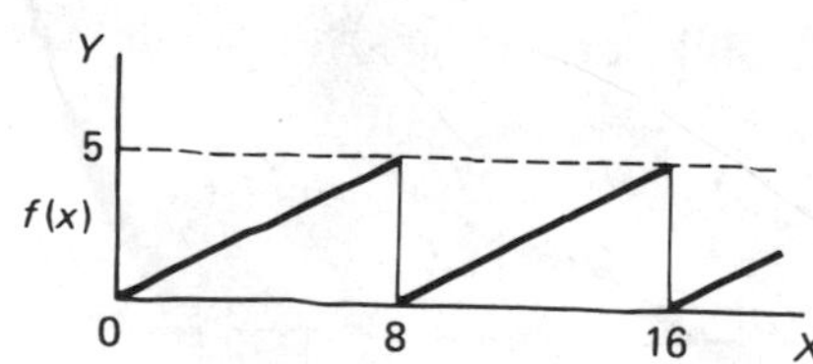

In this case ....

....

7

$$f(x) = \frac{5x}{8} \qquad 0 < x < 8$$
$$f(x) = f(x+8)$$

Here is a short exercise.

Exercise Define analytically the periodic functions shown.

1.

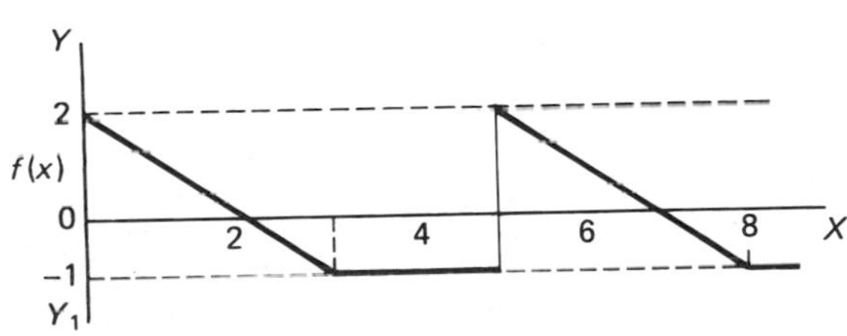

2.

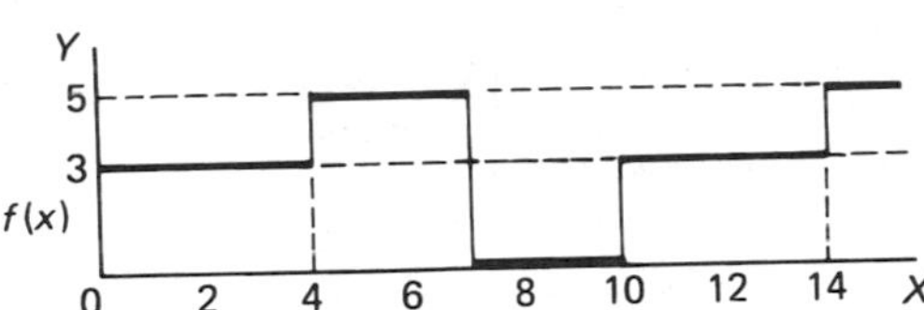

3.

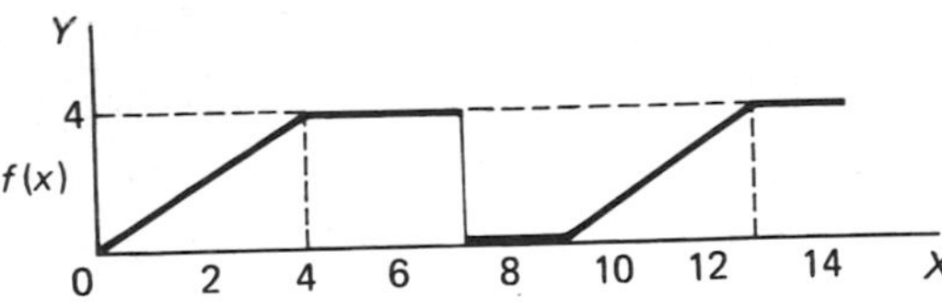

4.

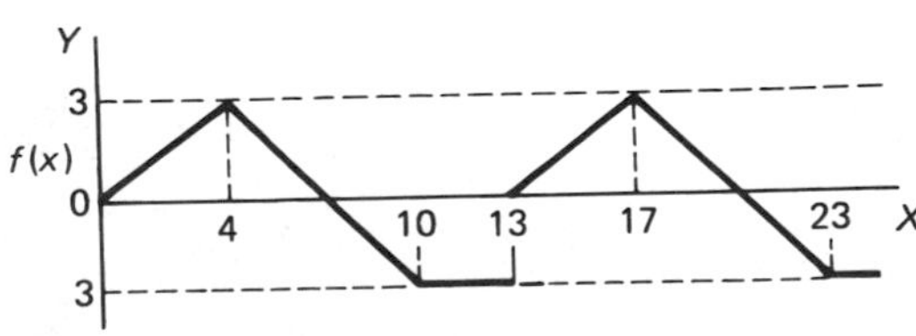

5.

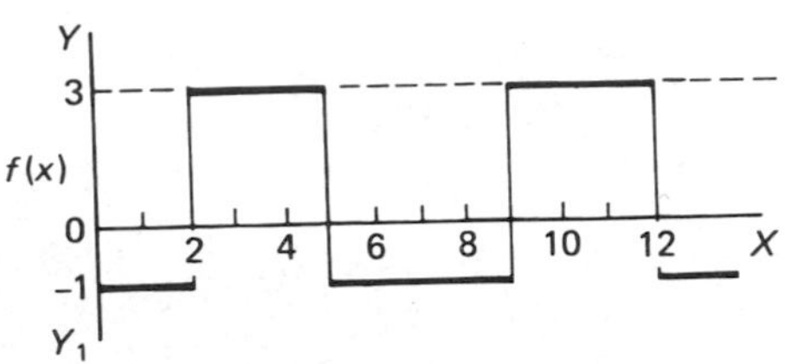

Finish all five and then check the results.

8 Here are the details.

1. $f(x) = 2 - x \qquad 0 < x < 3$
   $f(x) = -1 \qquad 3 < x < 5$
   $f(x) = f(x + 5).$

2. $f(x) = 3 \qquad 0 < x < 4$
   $f(x) = 5 \qquad 4 < x < 7$
   $f(x) = f(x + 10).$

3. $f(x) = x \qquad 0 < x < 4$
   $f(x) = 4 \qquad 4 < x < 7$
   $f(x) = 0 \qquad 7 < x < 9$
   $f(x) = f(x + 9).$

4. $f(x) = \dfrac{3x}{4} \qquad 0 < x < 4$
   $f(x) = 7 - x \qquad 4 < x < 10$
   $f(x) = -3 \qquad 10 < x < 13$
   $f(x) = f(x + 13).$

5. $f(x) = -1 \qquad 0 < x < 2$
   $f(x) = 3 \qquad 2 < x < 5$
   $f(x) = -1 \qquad 5 < x < 7$
   $f(x) = f(x + 7).$

Now we have the same thing in reverse.

Exercise Sketch the graphs of the following, inserting relevant values.

1. $f(x) = 4 \qquad 0 < x < 5$
   $f(x) = 0 \qquad 5 < x < 8$
   $f(x) = f(x + 8).$

2. $f(x) = 3x - x^2 \qquad 0 < x < 3$
   $f(x) = f(x + 3).$

3. $f(x) = 2 \sin x \qquad 0 < x < \pi$
   $f(x) = 0 \qquad \pi < x < 2\pi$
   $f(x) = f(x + 2\pi).$

4. $f(x) = \dfrac{x}{2} \qquad 0 < x < \pi$

$f(x) = \pi - \dfrac{x}{2} \qquad \pi < x < 2\pi$

$f(x) = f(x + 2\pi).$

5. $f(x) = \dfrac{x^2}{4} \qquad 0 < x < 4$

$f(x) = 4 \qquad 4 < x < 6$

$f(x) = 0 \qquad 6 < x < 8$

$f(x) = f(x + 8).$

9 Here they are: check carefully.

1.

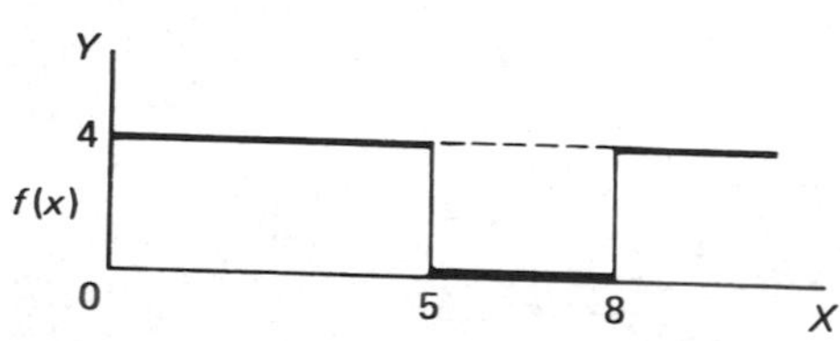

2.

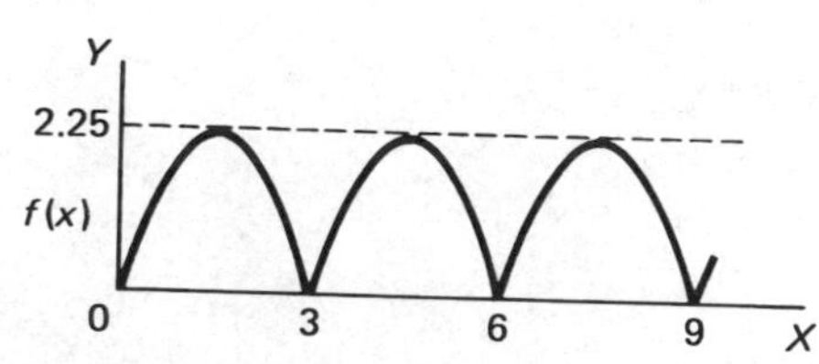

3.

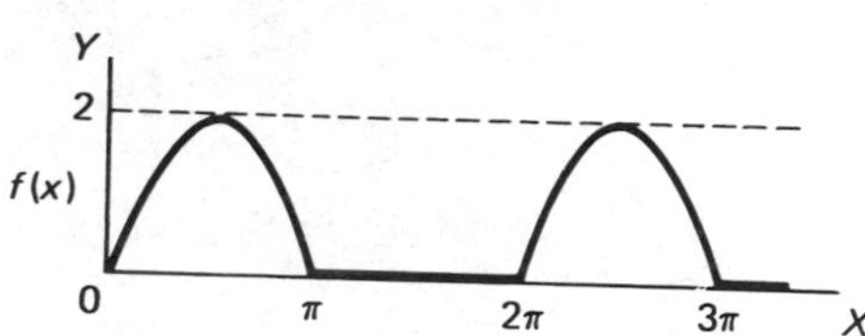

4.

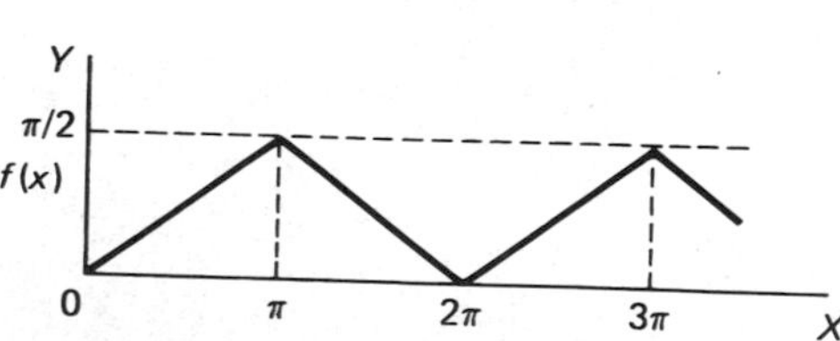

5.

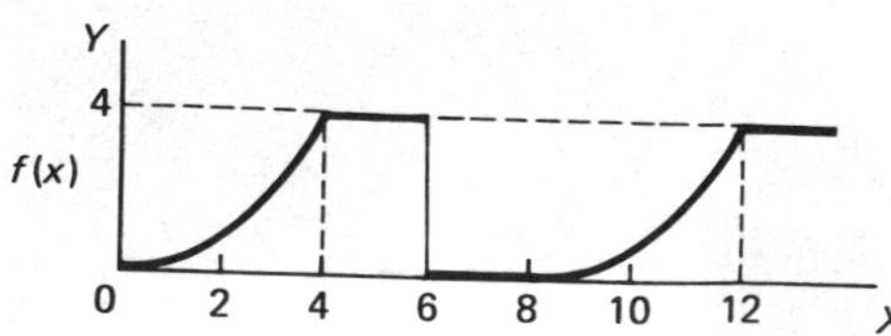

All this is in preparation for what is to come, so let us now consider Fourier series.

Move on then to the next frame.

10

Fourier series Periodic functions of period 2π

The basis of a Fourier series is to represent a periodic function by a trigonometrical series of the form

$$f(x) = A_0 + c_1 \sin(x + \alpha_1) + c_2 \sin(2x + \alpha_2) + c_3 \sin(3x + \alpha_3) + \dots + c_n \sin(nx + \alpha_n) + \dots$$

where A_0 is a constant term
$c_1, c_2, c_3 \dots c_n$ denote the amplitudes of the compound sine terms
$\alpha_1, \alpha_2, \alpha_3 \dots$ are constant auxiliary angles.

The *percentage nth harmonic* is the ratio of its amplitude to that of the fundamental, the result being expressed as a percentage.

$$\therefore \text{ percentage } n\text{th harmonic} = \frac{c_n}{c_1} \times 100 \text{ per cent}$$

Note that, in the definition of the series given above

$c_1 \sin(x + \alpha_1)$ is the fundamental or first harmonic of $f(x)$
$c_n \sin(nx + \alpha_n)$ is the nth harmonic of $f(x)$.

Each sine term, $c_n \sin(nx + \alpha_n)$ can be expanded thus:

$$\begin{aligned} c_n \sin(nx + \alpha_n) &= c_n\{\sin nx \cos \alpha_n + \cos nx \sin \alpha_n\} \\ &= (c_n \sin \alpha_n) \cos nx + (c_n \cos \alpha_n) \sin nx \\ &= a_n \cos nx + b_n \sin nx \end{aligned}$$

Also, for convenience in calculation, we write A_0 as $\frac{1}{2}a_0$ and then, putting $n = 1, 2, 3, \dots$, the whole series becomes

$$f(x) = \tfrac{1}{2}a_0 + a_1 \cos x + a_2 \cos 2x + a_3 \cos 3x + \dots + a_n \cos nx + \dots + b_1 \sin x + b_2 \sin 2x + b_3 \sin 3x + \dots + b_n \sin nx + \dots$$

It is in this form that we use the series and our main concern is to find the values of the coefficients $a_0, a_1, a_2, a_3, \dots$ and $b_1, b_2, b_3, \dots$ so that the series shall accurately represent the given periodic function. We can, of course, always combine sine and cosine terms into compound sine terms if necessary.

e.g. $$a_1 \cos x + b_1 \sin x = c_1 \sin(x + \alpha_1)$$

where $c_1 = \dots\dots\dots$ and $\alpha_1 = \dots\dots\dots$

11

$$c_1 = \sqrt{a_1^2 + b_1^2}; \quad \alpha_1 = \arctan\left(\frac{a_1}{b_1}\right)$$

Dirichlet conditions

If the Fourier series is to represent a function $f(x)$, then putting $x = x_1$ will give an infinite series in x_1 and the value of this should converge to the value of $f(x_1)$ as more and more terms of the series are evaluated. For this to happen, the following conditions must be fulfilled.

(a) The function $f(x)$ must be defined and single-valued.
(b) $f(x)$ must be continuous or have a finite number of finite discontinuities within a periodic interval.
(c) $f(x)$ and $f'(x)$ must be piecewise continuous in the periodic interval.

If these *Dirichlet conditions* are satisfied, the Fourier series converges to $f(x_1)$, if $x = x_1$ is a point of continuity.

In most practical cases, these conditions are met and the series can be taken as representing the particular function. Normally also the series converges fairly quickly and only the first few terms are required to give a result of adequate accuracy.

Exercise If the following functions are defined over the interval $-\pi < x < \pi$ and $f(x) = f(x + 2\pi)$, state whether or not each function can be represented by a Fourier series.

1. $f(x) = x^3$
2. $f(x) = 4x - 5$
3. $f(x) = \dfrac{2}{x}$
4. $f(x) = \dfrac{1}{x - 5}$
5. $f(x) = \tan x$
6. $f(x) = y$ where $x^2 + y^2 = 9$.

12

1. Yes	4. Yes
2. Yes	5. No: infinite discontinuity at $x = \pi/2$
3. No: infinite discontinuity at $x = 0$	6. No: two valued

We shall not get very far in this work without meeting a number of specific integrals, so let us establish a few results once and for all.

On then.

13

Useful integrals

The following integrals appear frequently in our work on Fourier series, so it will help if we obtain the results in readiness. In each case, m and n are integers other than zero.

(a) $\displaystyle\int_{-\pi}^{\pi} \sin nx \, dx = \left[\frac{-\cos nx}{n}\right]_{-\pi}^{\pi} = \frac{1}{n}\{-\cos n\pi + \cos n\pi\} = \underline{0}$

(b) $\displaystyle\int_{-\pi}^{\pi} \cos nx \, dx = \left[\frac{\sin nx}{n}\right]_{-\pi}^{\pi} = \frac{1}{n}\{\sin n\pi + \sin n\pi\} = \underline{0}$

(c) $\displaystyle\int_{-\pi}^{\pi} \sin^2 nx \, dx = \frac{1}{2}\int_{-\pi}^{\pi} (1 - \cos 2nx) dx = \frac{1}{2}\left[x - \frac{\sin 2nx}{2n}\right]_{-\pi}^{\pi}$

$$= \underline{\pi} \quad (n \neq 0)$$

(d) $\displaystyle\int_{-\pi}^{\pi} \cos^2 nx \, dx = \ldots\ldots\ldots$

14

$$\int_{-\pi}^{\pi} \cos^2 nx \, dx = \pi \quad (n \neq 0)$$

for $\displaystyle\int_{-\pi}^{\pi} \cos^2 nx \, dx = \frac{1}{2}\int_{-\pi}^{\pi} (1 + \cos 2nx) \, dx = \frac{1}{2}\left[x + \frac{\sin 2nx}{2n}\right]_{-\pi}^{\pi}$

$$= \underline{\pi} \quad (n \neq 0)$$

(e) $\displaystyle\int_{-\pi}^{\pi} \sin nx \cos mx \, dx = \frac{1}{2}\int_{-\pi}^{\pi} \{\sin(n+m)x + \sin(n-m)x\} \, dx$

$$= \frac{1}{2}\{0 + 0\} = \underline{0} \quad \text{from result (a) with } n \neq m$$

(f) $\displaystyle\int_{-\pi}^{\pi} \cos nx \cos mx \, dx = \frac{1}{2}\int_{-\pi}^{\pi} \{\cos(n+m)x + \cos(n-m)x\} \, dx$

$$= \frac{1}{2}\{0 + 0\} = \underline{0} \quad \text{from result (b) with } n \neq m$$

(g) $\displaystyle\int_{-\pi}^{\pi} \sin nx \sin mx \, dx = \dots\dots\dots$

(This is good revision of trig. identities!)

15

$$\int_{-\pi}^{\pi} \sin nx \sin mx \, dx = 0 \quad n \neq m$$

for $\displaystyle\int_{-\pi}^{\pi} \sin nx \sin mx \, dx = \frac{1}{2}\int_{-\pi}^{\pi} \{\cos(n-m)x - \cos(n+m)x\} \, dx$

$$= \frac{1}{2}\{0 - 0\} \underline{= 0} \quad \text{from result (b) with } n \neq m.$$

Note: If $n = m$, then (g) becomes $\displaystyle\int_{-\pi}^{\pi} \sin^2 nx \, dx = \pi$ from (c) above

and (f) becomes $\displaystyle\int_{-\pi}^{\pi} \cos^2 nx \, dx = \pi$ from (d) above.

We have evaluated the integrals between $-\pi$ and π, but, provided integration is carried out over a complete periodic interval of 2π, the results are the same. Thus, the limits could just as well be $-\pi$ to π, 0 to 2π, $-\pi/2$ to $3\pi/2$, etc. We can therefore choose the limits to suit the particular problem.

Let us therefore summarise these important results.

16

Summary of integrals

Collecting the results together, we have

(a) $\int_{-\pi}^{\pi} \sin nx \, dx = 0$

(b) $\int_{-\pi}^{\pi} \cos nx \, dx = 0$

(c) $\int_{-\pi}^{\pi} \sin^2 nx \, dx = \pi \quad (n \neq 0)$

(d) $\int_{-\pi}^{\pi} \cos^2 nx \, dx = \pi \quad (n \neq 0)$

(e) $\int_{-\pi}^{\pi} \sin nx \cos mx \, dx = 0$

(f) $\int_{-\pi}^{\pi} \cos nx \cos mx \, dx = 0 \quad (n \neq m)$

$= \pi \quad (n = m)$

(g) $\int_{-\pi}^{\pi} \sin nx \sin mx \, dx = 0 \quad (n \neq m)$

$= \pi \quad (n = m).$

If you have not already done so, make a note of this list of integrals, for we shall certainly have to use the results from time to time.

17

Fourier coefficients

We have defined Fourier series in the form

$$\begin{aligned} f(x) = \tfrac{1}{2}a_0 &+ a_1 \cos x + a_2 \cos 2x + a_3 \cos 3x + \ldots \\ &+ b_1 \sin x + b_2 \sin 2x + b_3 \sin 3x + \ldots \end{aligned}$$

This is sometimes written in the form

$$f(x) = \tfrac{1}{2}a_0 + \sum_{n=1}^{\infty} \{a_n \cos nx + b_n \sin nx\}; \quad n \text{ a positive integer}$$

Now, at last, we are in a position to obtain expressions for the various coefficients.

(a) *To find* a_0, we integrate $f(x)$ with respect to x from $-\pi$ to π.

$$\int_{-\pi}^{\pi} f(x)\mathrm{d}x = \frac{1}{2}\int_{-\pi}^{\pi} a_0\,\mathrm{d}x + \sum_{n=1}^{\infty}\left\{\int_{-\pi}^{\pi} a_n \cos nx\,\mathrm{d}x + \int_{-\pi}^{\pi} b_n \sin nx\,\mathrm{d}x\right\}$$

$$= \frac{1}{2}\Big[a_0 x\Big]_{-\pi}^{\pi} + \sum_{n=1}^{\infty}\{0+0\} \text{ from (a) and (b) in the list above.}$$

$$= a_0\pi \qquad \therefore\ a_0 = \frac{1}{\pi}\int_{-\pi}^{\pi} f(x)\,\mathrm{d}x$$

We see the convenience of having evaluated the first two listed integrals. The others will also appear in due course.

Now we find an expression for a_n, *so move on to the next frame.*

18

(b) *To find* a_n we multiply $f(x)$ by $\cos mx$ and integrate from $-\pi$ to π.

$$\int_{-\pi}^{\pi} f(x)\cos mx\,\mathrm{d}x = \frac{1}{2}\int_{-\pi}^{\pi} a_0 \cos mx\,\mathrm{d}x + \sum_{n=1}^{\infty}\left\{\int_{-\pi}^{\pi} a_n \cos nx\ \cos mx\,\mathrm{d}x\right.$$

$$\left. + \int_{-\pi}^{\pi} b_n \sin nx\ \cos mx\,\mathrm{d}x\right\}$$

Making use of our listed integrals, we can therefore say

$$\int_{-\pi}^{\pi} f(x)\cos\ mx\,\mathrm{d}x = \dots\dots\dots$$

19

$$\int_{-\pi}^{\pi} f(x)\cos mx\,\mathrm{d}x = 0 \quad \text{for } n \neq m$$
$$= a_n\pi \quad \text{for } n = m$$

for $$\int_{-\pi}^{\pi} f(x)\cos mx\,\mathrm{d}x = \frac{1}{2}\int_{-\pi}^{\pi} a_0 \cos mx\,\mathrm{d}x$$

$$+\sum_{n=1}^{\infty}\left\{\int_{-\pi}^{\pi} a_n \cos nx\ \cos mx\,\mathrm{d}x\right.$$

$$\left. + \int_{-\pi}^{\pi} b_n \sin nx\ \cos mx\,\mathrm{d}x\right\}$$

$$= \frac{1}{2}a_0\{0\} + \sum_{n=1}^{\infty}\{a_n(0) + b_n(0)\} = 0 \quad \text{for } n \neq m$$

or
$$= 0 + a_n\pi + 0 = a_n\pi \quad \text{for } n = m$$

$$\therefore\ a_n = \frac{1}{\pi}\int_{-\pi}^{\pi} f(x)\cos nx\,dx$$

(c) *To find* b_n we multiply $f(x)$ by $\sin mx$ and integrate from $-\pi$ to π. Work through it, as before, so that

$$b_n = \ldots\ldots\ldots$$

20

$$\boxed{b_n = \frac{1}{\pi}\int_{-\pi}^{\pi} f(x)\sin nx\,dx}$$

for
$$\int_{-\pi}^{\pi} f(x)\sin mx\,dx = \frac{1}{2}\int_{-\pi}^{\pi} a_0 \sin mx\,dx$$
$$+ \sum_{n=1}^{\infty}\left\{\int_{-\pi}^{\pi} a_n\cos nx\ \sin mx\,dx\right.$$
$$\left.+ \int_{-\pi}^{\pi} b_n\sin nx\ \sin mx\,dx\right\}$$

$$= \frac{1}{2}a_0\{0\} + \sum_{n=1}^{\infty}\{a_n(0) + b_n(0)\} = 0 \quad \text{for } n \neq m$$

or
$$= 0 + 0 + b_n\pi = b_n\pi \qquad \text{for } n = m$$

$$\therefore\ b_n = \frac{1}{\pi}\int_{-\pi}^{\pi} f(x)\sin nx\,dx$$

Let us now record these three results together. Then we can use them.

---

21 Here they are:

(a) $a_0 = \dfrac{1}{\pi}\displaystyle\int_{-\pi}^{\pi} f(x)\,dx$ $= 2 \times$ mean value of $f(x)$ over a period

(b) $a_n = \dfrac{1}{\pi}\displaystyle\int_{-\pi}^{\pi} f(x)\cos nx\,dx$ $= 2 \times$ mean value of $f(x)\cos nx$ over a period

(c) $b_n = \dfrac{1}{\pi}\displaystyle\int_{-\pi}^{\pi} f(x)\sin nx\,dx$ $= 2 \times$ mean value of $f(x)\sin nx$ over a period.

In each case, $n = 1, 2, 3, \ldots$

Make a careful note of these three results. They are the key to all that follows. Then we will see some examples.

22 *Example 1*

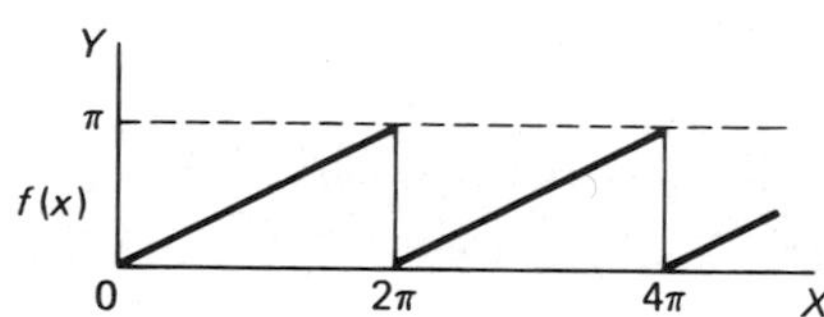

Determine the Fourier series to represent the periodic function shown.

It is more convenient here to take the limits as 0 to 2π.

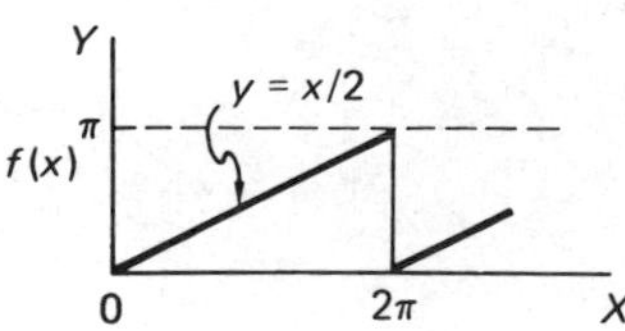

The function can be defined as

$$f(x) = \frac{x}{2} \qquad 0 < x < 2\pi$$

$$f(x) = f(x + 2\pi) \quad \text{i.e. period} = 2\pi.$$

Now to find the coefficients.

(a) $a_0 = \dfrac{1}{\pi}\displaystyle\int_0^{2\pi} f(x)\,dx = \dfrac{1}{\pi}\int_0^{2\pi}\left(\dfrac{x}{2}\right)dx = \dfrac{1}{4\pi}\left[x^2\right]_0^{2\pi}$

$= \pi$ $\qquad\qquad \therefore\ \underline{a_0 = \pi}$

(b) $a_n = \dfrac{1}{\pi}\displaystyle\int_0^{2\pi} f(x)\cos nx\,\mathrm{d}x = \dfrac{1}{\pi}\int_0^{2\pi}\left(\dfrac{x}{2}\right)\cos nx\,\mathrm{d}x$

$$= \frac{1}{2\pi}\int_0^{2\pi} x\cos nx\,\mathrm{d}x$$

$= \ldots\ldots\ldots$ (integrating by parts)

23

$$\boxed{a_n = 0}$$

since $a_n = \dfrac{1}{2\pi}\displaystyle\int_0^{2\pi} x\cos nx\,\mathrm{d}x = \dfrac{1}{2\pi}\left\{\left[\dfrac{x\sin nx}{n}\right]_0^{2\pi} - \dfrac{1}{n}\int_0^{2\pi}\sin nx\,\mathrm{d}x\right\}$

$$= \frac{1}{2\pi}\left\{(0-0) - \frac{1}{n}(0)\right\} = 0 \qquad \therefore\ \underline{a_n = 0}$$

(c) $b_n = \dfrac{1}{\pi}\displaystyle\int_0^{2\pi} f(x)\sin nx\,\mathrm{d}x = \ldots\ldots\ldots$

24

$$\boxed{b_n = -\frac{1}{n}}$$

Straightforward integration by parts, as for a_n, gives the result stated.

So we now have $a_0 = \ldots\ldots\ldots$; $a_n = \ldots\ldots\ldots$; $b_n = \ldots\ldots\ldots$

25

$$\boxed{a_0 = \pi; \quad a_n = 0; \quad b_n = -\frac{1}{n}}$$

Now the general expression for a Fourier series is

$$\ldots\ldots\ldots$$

26

$$f(x) = \frac{1}{2}a_0 + \sum_{n=1}^{\infty}\{a_n \cos nx + b_n \sin nx\}$$

Therefore in this case

$$f(x) = \frac{\pi}{2} + \sum_{n=1}^{\infty}\{b_n \sin nx\} \qquad \text{since } a_n = 0$$

$$= \frac{\pi}{2} + \left\{-\frac{1}{1}\sin x - \frac{1}{2}\sin 2x - \frac{1}{3}\sin 3x - \ldots\right\}$$

$$\therefore f(x) = \frac{\pi}{2} - \left\{\sin x + \frac{1}{2}\sin 2x + \frac{1}{3}\sin 3x + \ldots\right\}$$

Note that in this example, the series contains a constant term and sine terms only.

Example 2

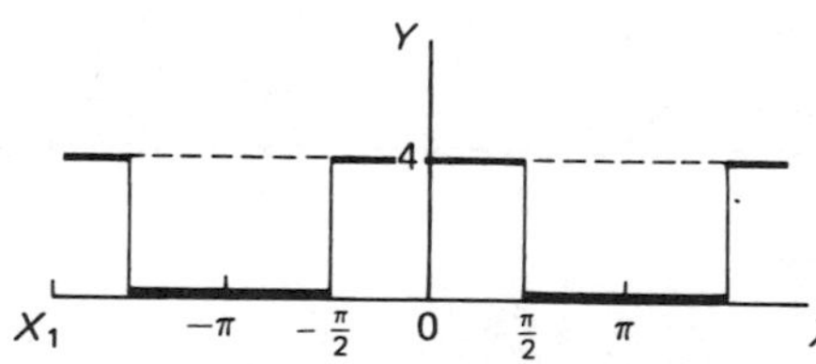

Find the Fourier series for the function shown.

Consider one cycle between $x = -\pi$ and $x = \pi$.

The function can be defined by
$$\begin{aligned} f(x) &= 0 && -\pi < x < -\frac{\pi}{2} \\ f(x) &= 4 && -\frac{\pi}{2} < x < \frac{\pi}{2} \\ f(x) &= 0 && \frac{\pi}{2} < x < \pi \\ f(x) &= f(x + 2\pi). \end{aligned}$$

As before $f(x) = \frac{1}{2}a_0 + \sum_{n=1}^{\infty}\{a_n \cos nx + b_n \sin nx\}$

The expression for a_0 is ....

27

$$\boxed{a_0 = \frac{1}{\pi}\int_{-\pi}^{\pi} f(x)\,\mathrm{d}x}$$

This gives

$$a_0 = \frac{1}{\pi}\left\{\int_{-\pi}^{-\pi/2} 0\,\mathrm{d}x + \int_{-\pi/2}^{\pi/2} 4\,\mathrm{d}x + \int_{\pi/2}^{\pi} 0\,\mathrm{d}x\right\}$$

$$= \frac{1}{\pi}\Big[4x\Big]_{\pi/2}^{\pi/2} \qquad \therefore\ \underline{a_0 = 4}$$

(b) *To find* a_n $\qquad a_n = \dfrac{1}{\pi}\displaystyle\int_{-\pi}^{\pi} f(x)\cos nx\,\mathrm{d}x$

$$\therefore\ a_n = \frac{1}{\pi}\left\{\int_{-\pi}^{-\pi/2} (0)\cos nx\,\mathrm{d}x + \int_{-\pi/2}^{\pi/2} 4\cos nx\,\mathrm{d}x + \int_{\pi/2}^{\pi} (0)\cos nx\,\mathrm{d}x\right\}$$

$$\therefore\ a_n = \ldots\ldots\ldots$$

28

$$\boxed{a_n = \frac{8}{\pi n}\sin\frac{n\pi}{2}}$$

Then considering different integer values of n, we have

If n is even	$a_n = 0$
If $n = 1,\ 5,\ 9,\ \ldots$	$a_n = \dfrac{8}{n\pi}$
If $n = 3,\ 7,\ 11,\ \ldots$	$a_n = -\dfrac{8}{n\pi}$

We keep these in mind while we find b_n.

(c) *To find* b_n

$$b_n = \frac{1}{\pi}\int_{-\pi}^{\pi} f(x)\ \sin\ nx\,\mathrm{d}x = \ldots\ldots\ldots$$

29

$$\boxed{b_n = 0}$$

for we have

$$b_n = \frac{1}{\pi}\left\{\int_{-\pi}^{-\pi/2}(0)\sin nx\,dx + \int_{-\pi/2}^{\pi/2}4\sin nx\,dx + \int_{\pi/2}^{\pi}(0)\sin nx\,dx\right\}$$

$$= \frac{4}{\pi}\int_{-\pi/2}^{\pi/2}\sin nx\,dx = \frac{4}{\pi}\left[\frac{-\cos nx}{n}\right]_{-\pi/2}^{\pi/2}$$

$$= -\frac{4}{n\pi}\left\{\cos\frac{n\pi}{2} - \cos\left(\frac{-n\pi}{2}\right)\right\} = 0 \qquad \therefore\ \underline{b_n = 0}$$

So with $a_0 = 4$; a_n as stated above; $b_n = 0$; the Fourier series is

$$\underline{f(x) = 2 + \frac{8}{\pi}\left\{\cos x - \frac{1}{3}\cos 3x + \frac{1}{5}\cos 5x - \frac{1}{7}\cos 7x + \ldots\right\}}$$

In this particular example, there are, in fact, no sine terms.

Example 3 Find the Fourier series for the function defined by

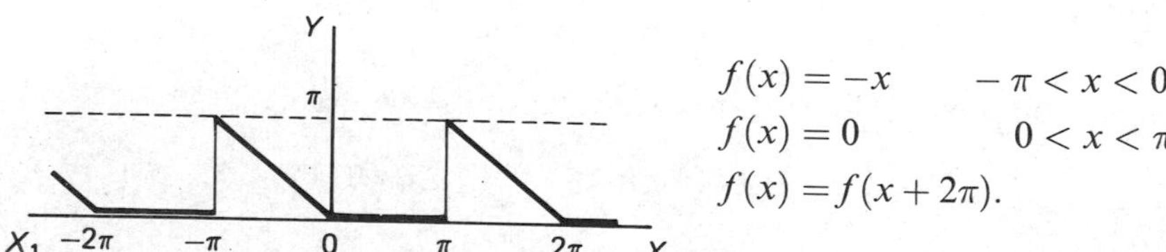

$$f(x) = -x \qquad -\pi < x < 0$$
$$f(x) = 0 \qquad 0 < x < \pi$$
$$f(x) = f(x + 2\pi).$$

The general expressions for a_0, a_n, b_n are

$a_0 = \ldots\ldots\ldots$

$a_n = \ldots\ldots\ldots$

$b_n = \ldots\ldots\ldots$

30

$$a_0 = \frac{1}{\pi}\int_{-\pi}^{\pi} f(x)\,dx$$
$$a_n = \frac{1}{\pi}\int_{-\pi}^{\pi} f(x)\cos nx\,dx$$
$$b_n = \frac{1}{\pi}\int_{-\pi}^{\pi} f(x)\sin nx\,dx$$

With that reminder, in this example $a_0 = \dots\dots\dots$

31

$$a_0 = \frac{\pi}{2}$$

for $a_0 = \frac{1}{\pi}\int_{-\pi}^{\pi} f(x)\,dx = \frac{1}{\pi}\int_{-\pi}^{0}(-x)\,dx = \frac{1}{\pi}\left[-\frac{x^2}{2}\right]_{-\pi}^{0} = \underline{\frac{\pi}{2}}$

(b) *To find* a_n

$$a_n = \frac{1}{\pi}\int_{-\pi}^{\pi} f(x)\cos nx\,dx = \dots\dots\dots$$

32

$$a_n = -\frac{2}{\pi n^2}\ (n \text{ odd});\qquad 0\ (n \text{ even})$$

since $a_n = \frac{1}{\pi}\int_{-\pi}^{\pi} f(x)\cos nx\,dx = \frac{1}{\pi}\int_{-\pi}^{0}(-x)\cos nx\,dx$

$$= -\frac{1}{\pi}\int_{-\pi}^{0} x\cos nx\,dx$$

$$= -\frac{1}{\pi}\left\{\left[x\frac{\sin nx}{n}\right]_{-\pi}^{0} - \frac{1}{n}\int_{-\pi}^{0}\sin nx\,dx\right\}$$

$$= -\frac{1}{\pi}\left\{(0-0) - \frac{1}{n}\left[\frac{-\cos nx}{n}\right]_{-\pi}^{0}\right\} = -\frac{1}{\pi n^2}\{1-\cos n\pi\}$$

But $\cos n\pi = 1$ (n even) or -1 (n odd)

$$\therefore\ \underline{a_n = -\frac{2}{\pi n^2}\ (n \text{ odd}) \text{ or } 0\ (n \text{ even})}$$

(c) *Now to find* b_n. Working as for a_n, we obtain

$$b_n = \dots\dots\dots$$

33

$$b_n = -\frac{1}{n}\ (n\text{ even})\ \text{or}\ \frac{1}{n}\ (n\text{ odd})$$

for $b_n = \frac{1}{\pi}\int_{-\pi}^{\pi} f(x)\sin nx\,dx = \frac{1}{\pi}\int_{-\pi}^{0}(-x)\sin nx\,dx$

$$= -\frac{1}{\pi}\int_{-\pi}^{0} x\sin nx\,dx$$

$$= -\frac{1}{\pi}\left\{\left[x\left(\frac{-\cos nx}{n}\right)\right]_{-\pi}^{0} + \frac{1}{n}\int_{-\pi}^{0}\cos nx\,dx\right\}$$

$$= -\frac{1}{\pi}\left\{\frac{\pi\cos n\pi}{n} + \frac{1}{n}\left[\frac{\sin nx}{n}\right]_{-\pi}^{0}\right\} = -\frac{\cos n\pi}{n}$$

$$\therefore\ b_n = -\frac{1}{n}\ (n\text{ even});\ \frac{1}{n}\ (n\text{ odd})$$

So we have $a_0 = \frac{\pi}{2}$; $a_n = 0\ (n\text{ even})$ or $-\frac{2}{\pi n^2}\ (n\text{ odd})$

$$b_n = -\frac{1}{n}\ (n\text{ even})\ \text{or}\ \frac{1}{n}\ (n\text{ odd})$$

$\therefore f(x) = \ldots\ldots\ldots$ Complete the series.

$$f(x) = \frac{\pi}{4} - \frac{2}{\pi}\left(\cos x + \frac{1}{9}\cos 3x + \frac{1}{25}\cos 5x + \ldots\right) + \left(\sin x - \frac{1}{2}\sin 2x + \frac{1}{3}\sin 3x - \frac{1}{4}\sin 4x + \ldots\right)$$

It is just a case of substituting $n = 1, 2, 3$, etc.
In this particular example, we have a constant term and both sine and cosine terms.

Effect of harmonics

It is interesting to see just how accurately the Fourier series represents the function with which it is associated. The complete representation requires an infinite number of terms, but we can, at least, see the effect of including the first few terms of the series.

Let us consider the waveform shown. We established earlier that the function

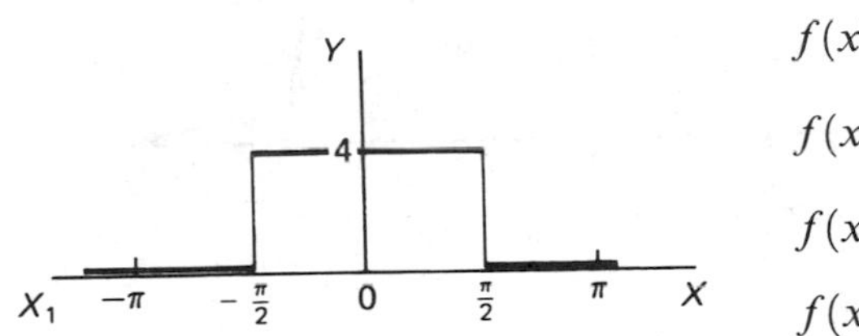

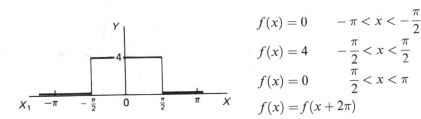

$$f(x) = 0 \qquad -\pi < x < -\frac{\pi}{2}$$
$$f(x) = 4 \qquad -\frac{\pi}{2} < x < \frac{\pi}{2}$$
$$f(x) = 0 \qquad \frac{\pi}{2} < x < \pi$$
$$f(x) = f(x + 2\pi)$$

gives the Fourier series

$$f(x) = 2 + \frac{8}{\pi}\left\{\cos x - \frac{1}{3}\cos 3x + \frac{1}{5}\cos 5x - \frac{1}{7}\cos 7x + \ldots\right\}$$

If we start with just one cosine term, we can then see the effect of including subsequent harmonics. Let us restrict our attention to just the right-hand half of the symmetrical waveform. Detailed plotting of points gives the following development.

(1) $f(x) = 2 + \frac{8}{\pi}\cos x$

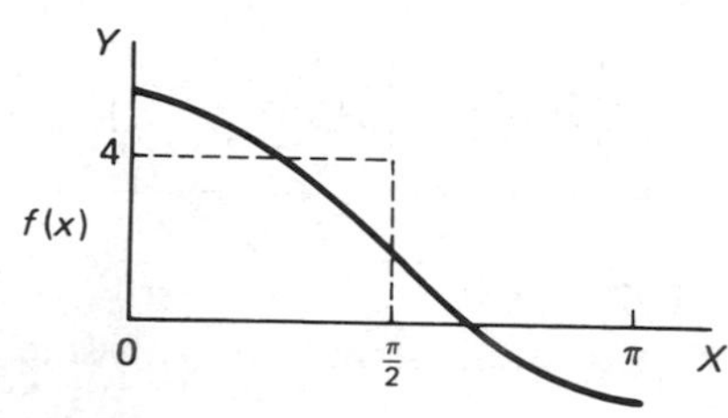

(2) $f(x) = 2 + \frac{8}{\pi}\left\{\cos x - \frac{1}{3}\cos 3x\right\}$

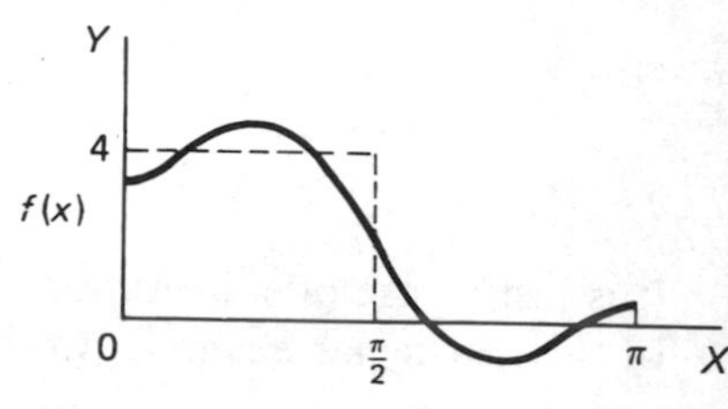

(3) $f(x) = 2 + \frac{8}{\pi}\left\{\cos x - \frac{1}{3}\cos 3x + \frac{1}{5}\cos 5x\right\}$

(4) $f(x) = 2 + \frac{8}{\pi}\left\{\cos x - \frac{1}{3}\cos 3x + \frac{1}{5}\cos 5x - \frac{1}{7}\cos 7x\right\}$

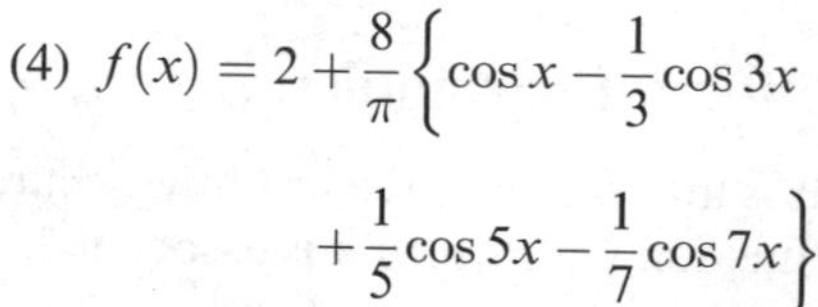

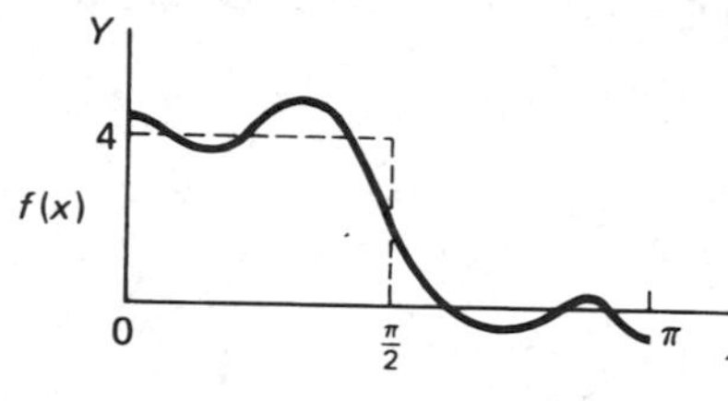

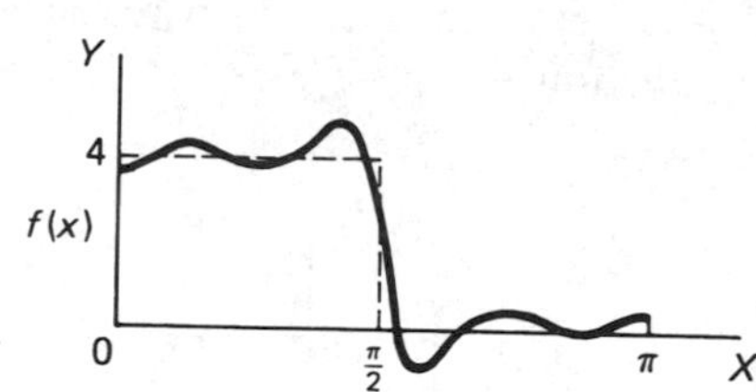

As the number of terms is increased, the graph gradually approaches the shape of the original square waveform. The ripples increase in number and decrease in amplitude, but a perfectly square waveform is unattainable in practice. For practical purposes, the first few terms normally suffice to give an accuracy of acceptable level.

Sum of a Fourier series at a point of discontinuity

35

$$f(x) = \tfrac{1}{2}a_0 + \sum_{n=1}^{\infty}\{a_n \cos nx + b_n \sin nx\}$$

At $x = x_1$, the series converges to the value $f(x_1)$ as the number of terms included increases to infinity. A particular point of interest occurs at a point of finite discontinuity or 'jump' of the function $y = f(x)$.

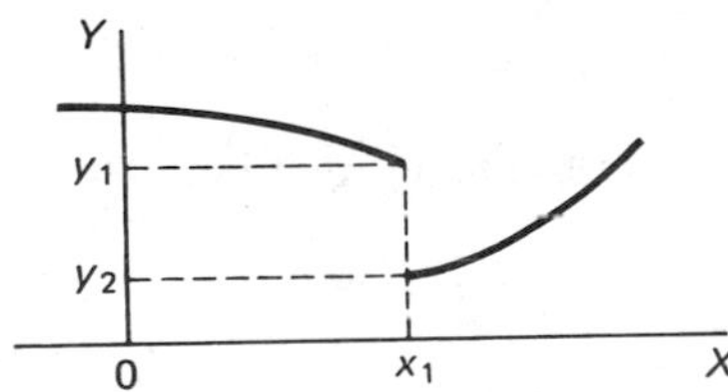

At $x = x_1$, the function appears to have two distinct values y_1 and y_2.

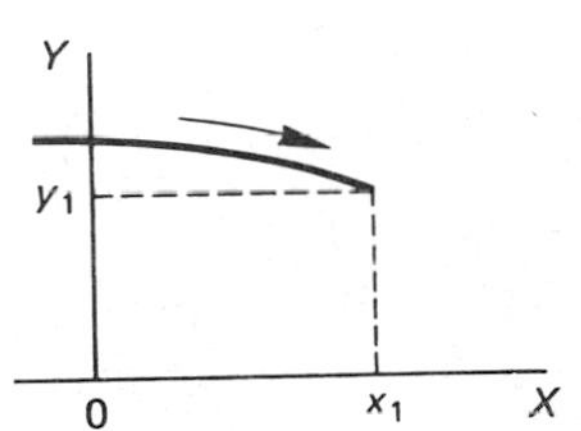

If we approach $x = x_1$ from below that value, the limiting value of $f(x)$ is y_1.

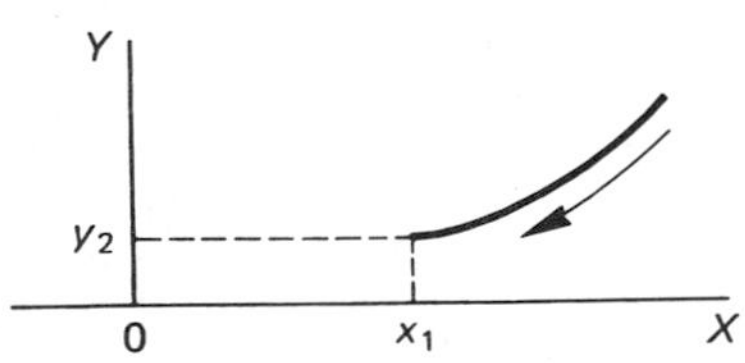

If we approach $x = x_1$ from above that value, the limiting value of $f(x)$ is y_2.

To distinguish between these two values we write

$y_1 = f(x_1 - 0)$ denoting immediately before $x = x_1$

$y_2 = f(x_1 + 0)$ denoting immediately after $x = x_1$.

In fact, if we substitute $x = x_1$ in the Fourier series for $f(x)$, it can be shown that the series converges to the value $\frac{1}{2}\{f(x_1 - 0) + f(x_1 + 0)\}$ i.e. $\frac{1}{2}(y_1 + y_2)$, the average of y_1 and y_2.

Example Consider the function

$$f(x) = 0 \qquad -\pi < x < \pi$$
$$f(x) = a \qquad \pi < x < 2\pi$$
$$f(x) = f(x + 2\pi).$$

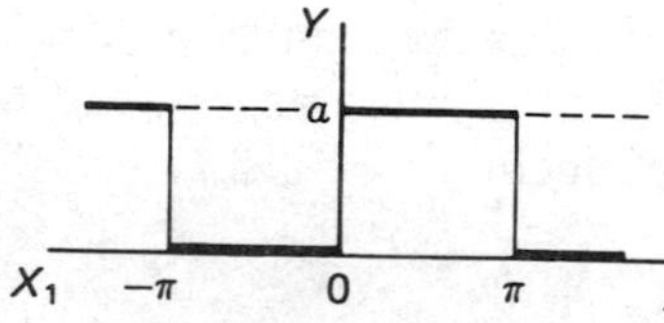

First of all, determine the Fourier series to represent the function. There are no snags. $f(x) = \dots\dots\dots$

36

$$f(x) = \frac{a}{2} + \frac{2a}{\pi}\{\sin x + \tfrac{1}{3}\sin 3x + \tfrac{1}{5}\sin 5x + \ldots\}$$

Check the working

(a) $a_0 = \frac{1}{\pi}\int_{-\pi}^{\pi} f(x)\,dx = \frac{1}{\pi}\int_0^{\pi} a\,dx = \frac{1}{\pi}\Big[ax\Big]_0^{\pi} = a \qquad \therefore \underline{a_0 = a}$

(b) $a_n = \frac{1}{\pi}\int_{-\pi}^{\pi} f(x)\cos nx\,dx = \frac{1}{\pi}\int_0^{\pi} a\cos nx\,dx$

$= \frac{a}{\pi}\left[\frac{\sin nx}{n}\right]_0^{\pi} = 0 \qquad \therefore \underline{a_n = 0}$

(c) $b_n = \frac{1}{\pi}\int_{-\pi}^{\pi} f(x)\sin nx\,dx = \frac{1}{\pi}\int_0^{\pi} a\sin nx\,dx$

$= \frac{a}{\pi}\left[\frac{-\cos nx}{n}\right]_0^{\pi} = \frac{a}{n\pi}(1 - \cos n\pi)$

But
$\cos n\pi = 1$ (n even) and -1 (n odd) $\qquad \therefore \underline{b_n = 0 \ (n \text{ even});\ \frac{2a}{n\pi}\ (n \text{ odd})}$

$$\therefore f(x) = \frac{1}{2}a_0 + \sum_{n=1}^{\infty} b_n \sin nx$$

$$\therefore f(x) = \frac{a}{2} + \frac{2a}{\pi}\left\{\sin x + \frac{1}{3}\sin 3x + \frac{1}{5}\sin 5x + \ldots\right\}$$

A finite discontinuity, or 'jump', occurs at $x = 0$. If we substitute $x = 0$ in the series obtained, all the sine terms vanish and we get $f(x) = a/2$, which is, in fact, the average of the two function values at $x = 0$.

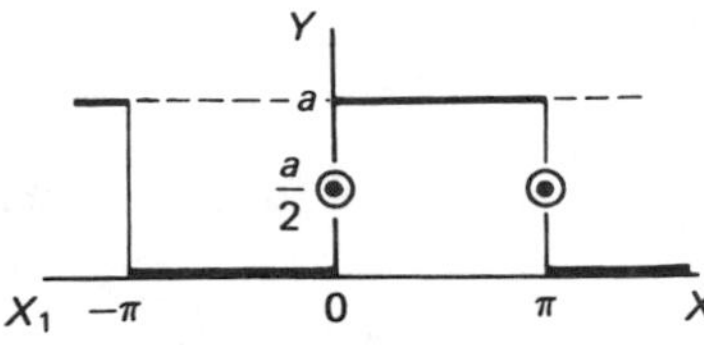

Note also that at $x = \pi$, another finite discontinuity occurs and substituting $x = \pi$ in the series gives the same result.

Because of this ambiguity, the function is said to be 'undefined' at $x = 0$, $x = \pi$, etc.

Now on to something new.

37

Odd and even functions

(a) *Even functions* A function $f(x)$ is said to be *even* if

$$f(-x) = f(x)$$

i.e. the function value for a particular negative value of x is the same as that for the corresponding positive value of x. The graph of an even function is therefore *symmetrical about the y-axis.*

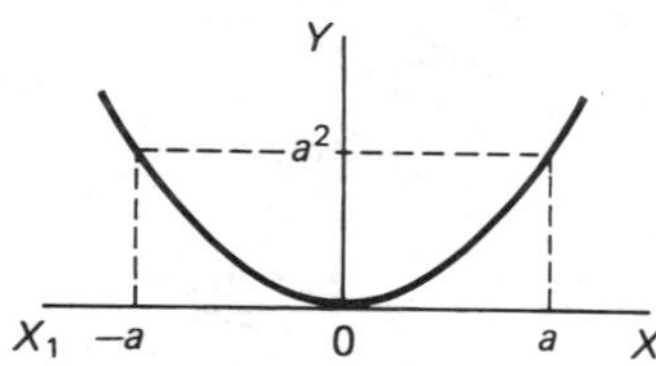

$y = f(x) = x^2$ is an even function since

$$f(-2) = 4 = f(2)$$
$$f(-3) = 9 = f(3) \quad \text{etc.}$$

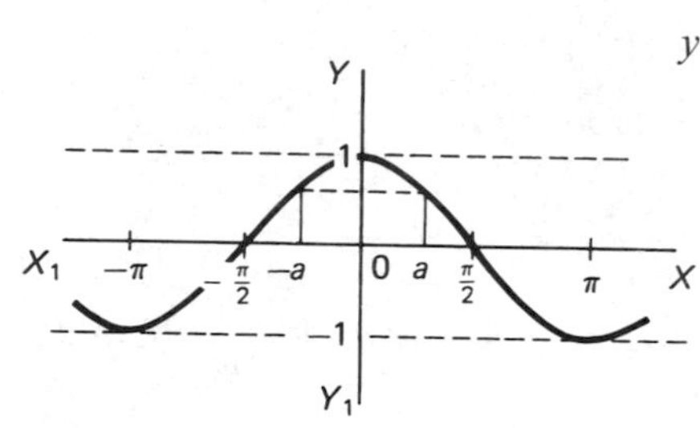

$y = f(x) = \cos x$ is an even function

since $\cos(-x) = \cos x$

$$f(-a) = \cos a = f(a).$$

(b) *Odd functions* A function $f(x)$ is said to be *odd* if

$$f(-x) = -f(x)$$

i.e. the function value for a particular negative value of x is numerically equal to that for the corresponding positive value of x but opposite in sign. The graph of an odd function is thus *symmetrical about the origin.*

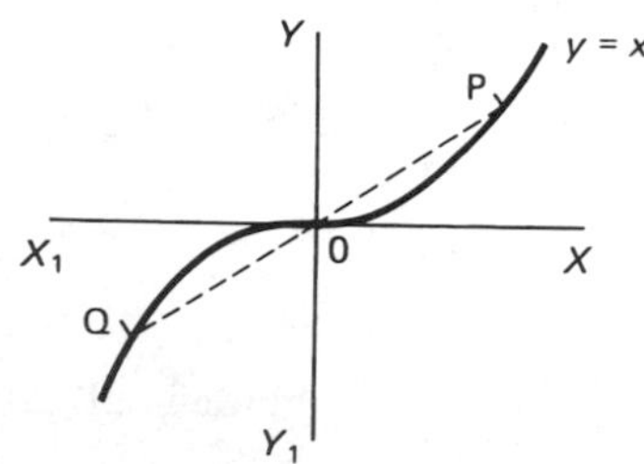

$y = f(x) = x^3$ is an odd function since

$$f(-2) = -8 = -f(2)$$
$$f(-5) = -125 = -f(5) \quad \text{etc.}$$

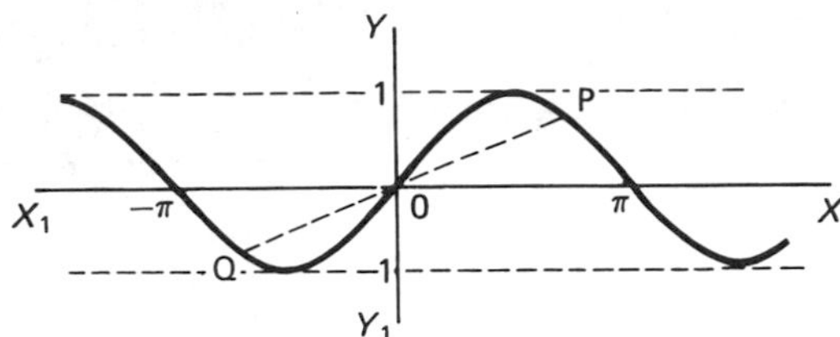

$y = f(x) = \sin x$ is an odd function since $\sin(-x) = -\sin x$

$$f(-a) = -f(a).$$

So, for an even function $f(-x) = f(x)$ symmetrical about the y-axis
for an odd function $f(-x) = -f(x)$ symmetrical about the origin.

Example 1

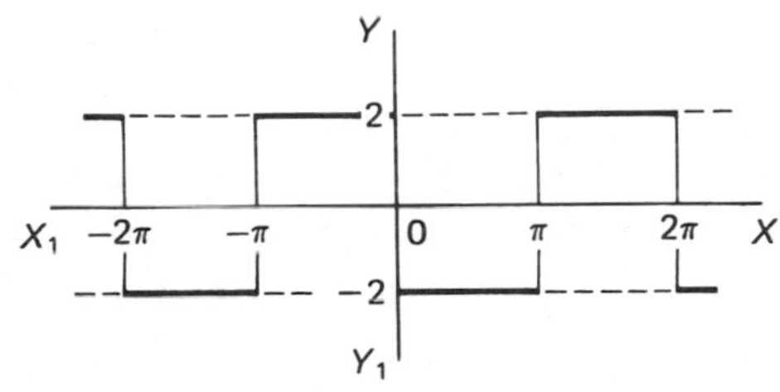

$f(x)$ shown by the waveform is therefore an .... function

since it is ....

38

odd; symmetrical about the origin, i.e. $f(-x) = -f(x)$

Example 2

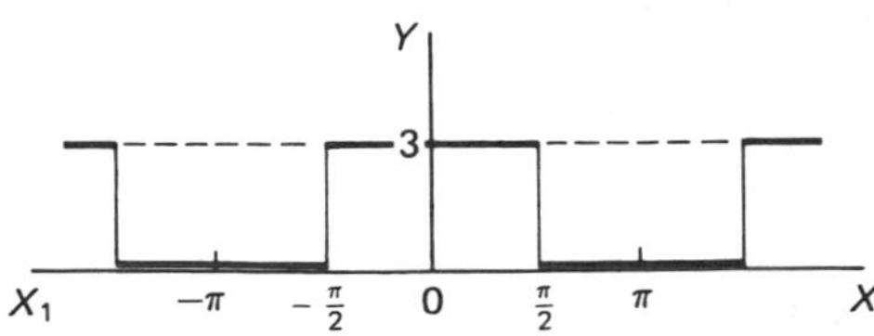

Hence the waveform of $y = f(x)$ depicts an .... function,

since it is ....

39

even; symmetrical about the y-axis, i.e. $f(-x) = f(x)$

Example 3

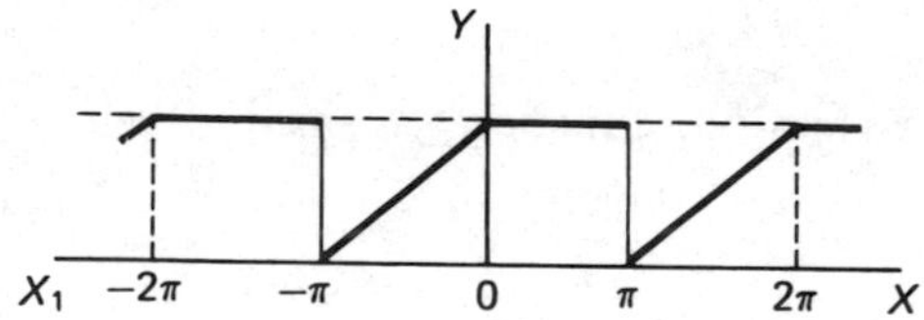

In this case, the waveform shows a function that is ....

since ....

40

neither even nor odd; not symmetrical about either the y-axis or the origin

Exercise State whether each of the following functions is odd, even, or neither.

1.

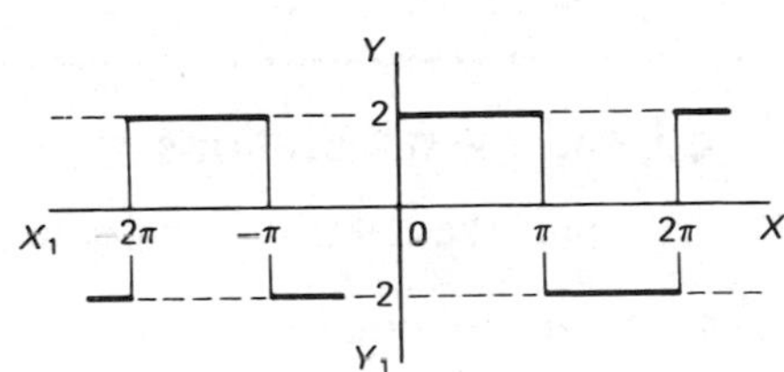

2.

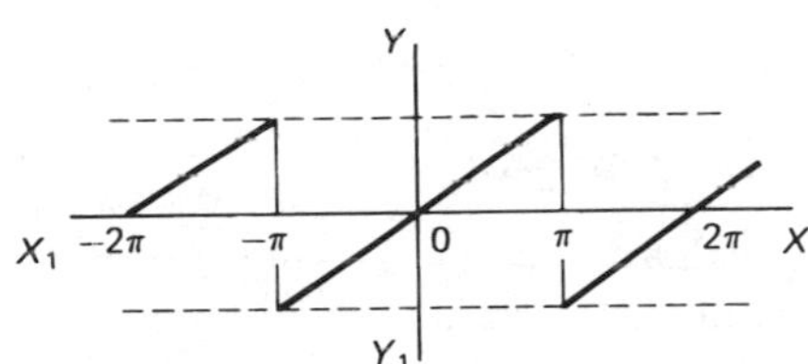

3.

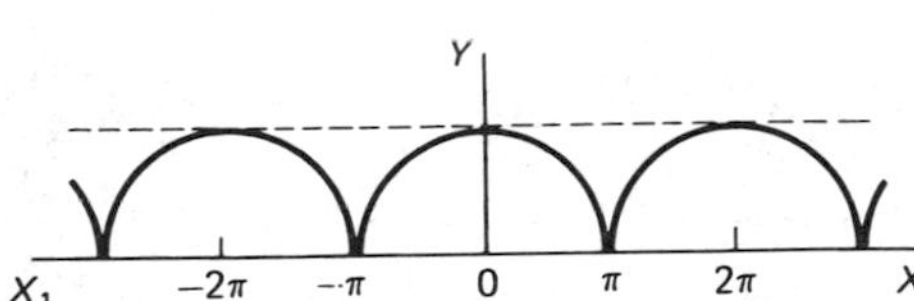

4.

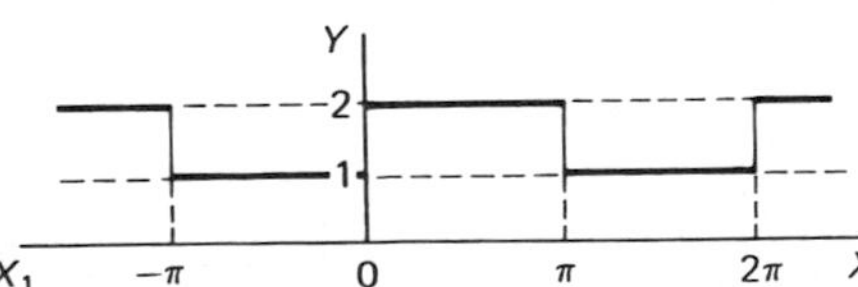

5.

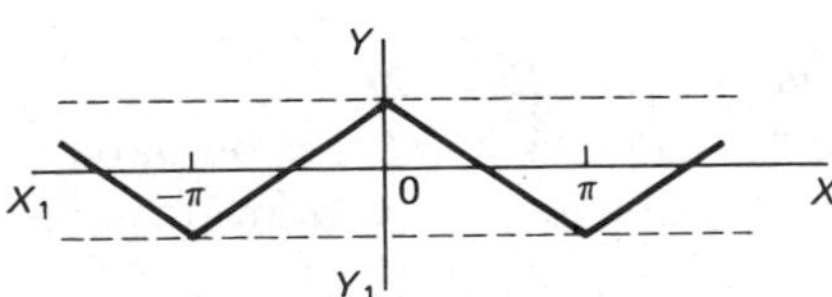

6.

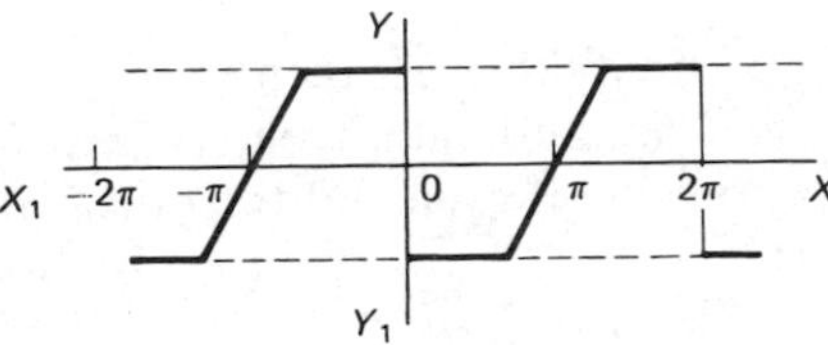

41

1. odd	2. odd	3. even
4. neither	5. even	6. odd

We shall shortly see that a knowledge of odd and even functions can save us a lot of unnecessary calculation. *First, however, let us consider products of odd and even functions in the next frame.*

42

Products of odd and even functions

The rules closely resemble the elementary rules of signs.

$$\begin{aligned} (\text{even}) \times (\text{even}) &= (\text{even}) \quad \text{like} \quad (+) \times (+) = (+) \\ (\text{odd}) \times (\text{odd}) &= (\text{even}) \qquad\qquad (-) \times (-) = (+) \\ (\text{odd}) \times (\text{even}) &= (\text{odd}) \qquad\qquad\; (-) \times (+) = (-). \end{aligned}$$

The results can easily be proved.

(a) *Two even functions*
Let $F(x) = f(x)g(x)$ where $f(x)$ and $g(x)$ are even functions.
Then $F(-x) = f(-x)g(-x) = f(x)g(x)$ since $f(x)$ and $g(x)$ are even.

$$\therefore F(-x) = F(x) \qquad \text{i.e. } \underline{F(x) \text{ is even}}$$

(b) *Two odd functions*
Let $F(x) = f(x)g(x)$ where $f(x)$ and $g(x)$ are odd functions.
Then $F(-x) = f(-x)g(-x) = \{-f(x)\}\{-g(x)\}$ since $f(x)$ and $g(x)$ are odd.

$$= f(x)g(x) = F(x)$$

$$\therefore F(-x) = F(x) \qquad \text{i.e. } \underline{F(x) \text{ is even}}$$

Finally
(c) *One odd and one even function*
Let $F(x) = f(x)g(x)$ where $f(x)$ is odd and $g(x)$ even.
Then $F(-x) = f(-x)g(-x) = -f(x)g(x) = -F(x)$

$$\therefore F(-x) = -F(x) \quad \text{i.e. } \underline{F(x) \text{ is odd}}$$

So if $f(x)$ and $g(x)$ are both even, then $f(x)g(x)$ is even
and if $f(x)$ and $g(x)$ are both odd, then $f(x)g(x)$ is even
but if either $f(x)$ or $g(x)$ is even and the other odd,
then $f(x)g(x)$ is odd.

Now for a short exercise, so move on.

43

Exercise State whether each of the following products is odd, even, or neither.

1. $x^2 \sin 2x$
2. $x^3 \cos x$
3. $\cos 2x \cos 3x$
4. $x \sin nx$
5. $3 \sin x \cos 4x$
6. $(2x+3) \sin 4x$
7. $\sin^2 x \cos 3x$
8. $x^3 e^x$
9. $(x^4+4) \sin 2x$
10. $\dfrac{1}{x+2} \cosh x$

Finish all ten and then check with the next frame.

44

1. odd (E)(O) = (O)	6. neither (N)(O) = (N)
2. odd (O)(E) = (O)	7. even (E)(E) = (E)
3. even (E)(E) = (E)	8. neither (O)(N) = (N)
4. even (O)(O) = (E)	9. odd (E)(O) = (O)
5. odd (O)(E) = (O)	10. neither (N)(E) = (N)

Two useful facts emerge from odd and even functions. Thinking in terms of areas under the graphs,

(a)

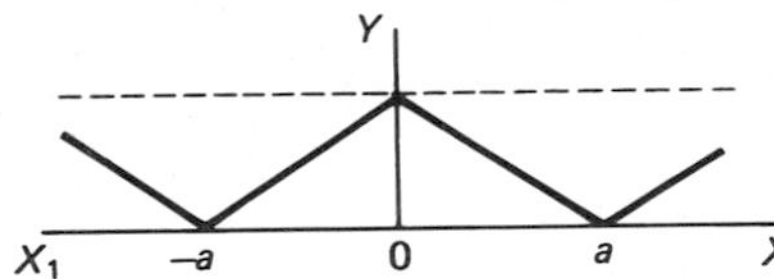

For an *even* function

$$\int_{-a}^{a} f(x)\mathrm{d}x = 2\int_{0}^{a} f(x)\mathrm{d}x$$

(b)

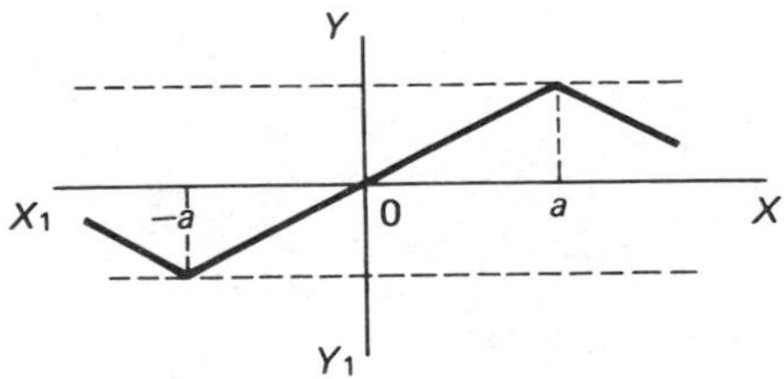

For an *odd* function

$$\int_{-a}^{a} f(x)\mathrm{d}x = 0$$

We can now look at two important theorems concerning odd and even functions.

Theorem 1 If $f(x)$ is defined over the interval $-\pi < x < \pi$ and $f(x)$ is *even*, then the Fourier series for $f(x)$ *contains cosine terms* only. Included in this is a_0 which may be regarded as $a_n \cos nx$ with $n = 0$.

Proof: Since $f(x)$ is even, $\int_{-\pi}^{0} f(x)\,dx = \int_{0}^{\pi} f(x)\,dx$

(a) $a_0 = \frac{1}{\pi}\int_{-\pi}^{\pi} f(x)\,dx = \frac{2}{\pi}\int_{0}^{\pi} f(x)\,dx$ $\qquad \therefore\ a_0 = \frac{2}{\pi}\int_{0}^{a} f(x)\,dx$

(b) $a_n = \frac{1}{\pi}\int_{-\pi}^{\pi} f(x)\cos nx\,dx$. But $f(x)\cos nx$ is the product of two even functions and therefore itself even. $\therefore\ a_n = \dots\dots\dots$

45

$$a_n = \frac{2}{\pi}\int_{0}^{\pi} f(x)\cos nx\,dx$$

for, because the integrand is even,

$$a_n = \frac{1}{\pi}\int_{-\pi}^{\pi} f(x)\cos nx\,dx = \frac{2}{\pi}\int_{0}^{\pi} f(x)\cos nx\,dx.$$

(c) $b_n = \frac{1}{\pi}\int_{-\pi}^{\pi} f(x)\sin nx\,dx$

Arguing along similar lines, this gives $b_n = \dots\dots\dots$

46

$$\boxed{b_n = 0}$$

Since $f(x)$ sin nx is the product of an even function and an odd function, it is itself odd. $\therefore\ b_n = \dfrac{1}{\pi}\displaystyle\int_{-\pi}^{\pi} f(x)\sin nx\,\mathrm{d}x = 0.$ $\therefore\ \underline{b_n = 0}$

Therefore, there are no sine terms in the Fourier series for $f(x)$.

Now for an example.

Example The waveform shown is symmetrical about the y-axis. The function is therefore even and there will be no sine terms in the series.

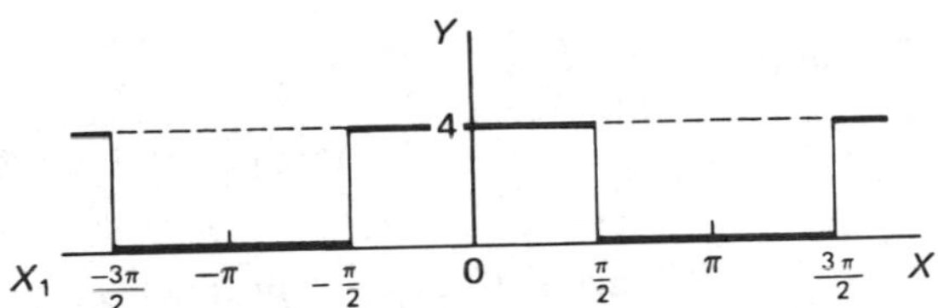

$$\therefore f(x) = \frac{1}{2}a_0 + \sum_{n=1}^{\infty} a_n \cos nx$$

(a) $a_0 = \dfrac{1}{\pi}\displaystyle\int_{-\pi}^{\pi} f(x)\,\mathrm{d}x = \dfrac{2}{\pi}\int_0^{\pi} f(x)\,\mathrm{d}x = \dfrac{2}{\pi}\int_0^{\pi/2} 4\,\mathrm{d}x = \dfrac{2}{\pi}\Big[4x\Big]_0^{\pi/2} = \underline{4}$

(b) $a_n = \dfrac{1}{\pi}\displaystyle\int_{-\pi}^{\pi} f(x)\cos nx\,\mathrm{d}x = \dfrac{2}{\pi}\int_0^{\pi} f(x)\cos nx\,\mathrm{d}x$

$= \ldots\ldots\ldots$ Finish the integration.

47

$$a_n = 0 \ (n \text{ even}); \quad a_n = \frac{8}{\pi n} \ (n = 1, 5, 9, \ldots);$$
$$a_n = -\frac{8}{\pi n} (n = 3, 7, 11, \ldots)$$

for $a_n = \dfrac{2}{\pi}\displaystyle\int_0^{\pi} f(x) \cos nx \, dx = \dfrac{2}{\pi}\displaystyle\int_0^{\pi/2} 4 \cos nx \, dx$

$$= \frac{8}{\pi}\left[\frac{\sin nx}{n}\right]_0^{\pi/2} = \frac{8}{\pi n} \sin\frac{n\pi}{2}$$

But $\sin\dfrac{n\pi}{2} = 0$ for n even

$= 1$ for $n = 1, 5, 9, \ldots$

$= -1$ for $n = 3, 7, 11, \ldots$ Hence the result stated.

(c) We know that $b_n = 0$, since $f(x)$ is an even function.

Therefore, the required series is

$f(x) = \ldots\ldots\ldots\ldots\ldots\ldots\ldots\ldots$

48

$$f(x) = 2 + \frac{8}{\pi}\left\{\cos x - \frac{1}{3}\cos 3x + \frac{1}{5}\cos 5x - \frac{1}{7}\cos 7x + \ldots\right\}$$

If you care to look back to Example 2 on page 844, you will see how much time and effort we have saved by not having to evaluate b_n.

A similar theorem applies to odd functions.

Theorem 2 If $f(x)$ is an *odd* function defined over the interval $-\pi < x < \pi$, then the Fourier series for $f(x)$ contains *sine terms* only.

Proof: Since $f(x)$ is an odd function, $\int_{-\pi}^{0} f(x)\,dx = -\int_{0}^{\pi} f(x)\,dx$.

(a) $a_0 = \frac{1}{\pi}\int_{-\pi}^{\pi} f(x)\,dx$. But $f(x)$ is odd $\quad \therefore \underline{a_0 = 0}$

(b) $a_n = \frac{1}{\pi}\int_{-\pi}^{\pi} f(x)\cos nx\,dx$

Remembering that $f(x)$ is odd and $\cos nx$ is even, the product $f(x)\cos nx$ is ....

49

odd

$$\therefore a_n = \frac{1}{\pi}\int_{-\pi}^{\pi} f(x)\cos nx\,dx = \frac{1}{\pi}\int_{-\pi}^{\pi} (\text{odd function})\,dx = 0 \quad \therefore \underline{a_n = 0}$$

Now for b_n we have

(c) $b_n = \frac{1}{\pi}\int_{-\pi}^{\pi} f(x)\sin nx\,dx$ and since $f(x)$ and $\sin nx$ are each odd, the product $f(x)\sin nx$ is ....

50

even

Then $b_n = \frac{1}{\pi}\int_{-\pi}^{\pi} f(x)\sin nx\,dx = \frac{1}{\pi}\int_{-\pi}^{\pi}(\text{even function})\,dx$

$= \frac{2}{\pi}\int_0^{\pi} f(x)\sin nx\,dx$

$\therefore\ b_n = \frac{2}{\pi}\int_0^{\pi} f(x)\sin nx\,dx$

So, if $f(x)$ is odd, $a_0 = 0$; $a_n = 0$; $b_n = \frac{2}{\pi}\int_0^{\pi} f(x)\sin nx\,dx$ i.e. the Fourier series contains sine terms only.

Example Consider the function shown.

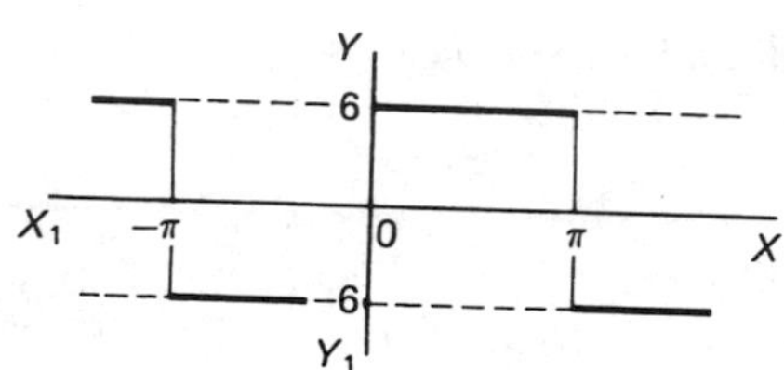

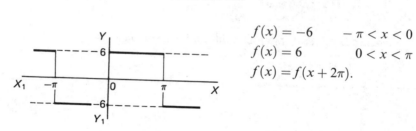

$f(x) = -6 \qquad -\pi < x < 0$
$f(x) = 6 \qquad 0 < x < \pi$
$f(x) = f(x + 2\pi).$

Before we do any evaluation, we can see that this is .... and therefore ....

51

an odd function; sine terms only, i.e. $a_0 = 0$ and $a_n = 0$

$b_n = \frac{1}{\pi}\int_{-\pi}^{\pi} f(x)\sin nx\,dx.$

$f(x)\sin nx$ is a product of two odd functions and is therefore even.

$\therefore\ b_n = \frac{2}{\pi}\int_0^{\pi} f(x)\sin nx\,dx = \ldots\ldots\ldots$

52

$$b_n = 0 \ (n \text{ even}) \text{ or } \frac{24}{\pi n} \ (n \text{ odd})$$

for $b_n = \frac{2}{\pi}\int_0^{\pi} 6 \sin nx \, dx = \frac{12}{\pi}\left[\frac{-\cos nx}{n}\right]_0^{\pi} = \frac{12}{\pi n}(1 - \cos n\pi)$.

Hence the result stated above.

So the series is $f(x) = \ldots\ldots\ldots$

53

$$f(x) = \frac{24}{\pi}\left\{\sin x + \frac{1}{3}\sin 3x + \frac{1}{5}\sin 5x + \ldots\right\}$$

Of course, if $f(x)$ is neither an odd nor an even function, then we must obtain expressions for a_0, a_n and b_n in full.
One more example

Example 3 Determine the Fourier series for the function shown.

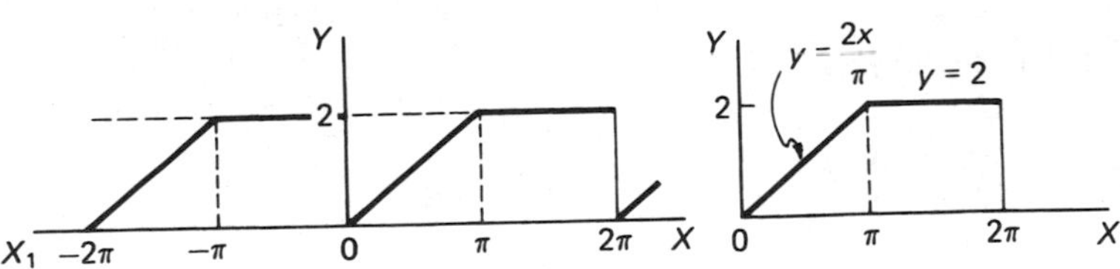

This is neither odd nor even. Therefore we must find a_0, a_n and b_n.

$$f(x) = \frac{1}{2}a_0 + \sum_{n=1}^{\infty}\{a_n \cos nx + b_n \sin nx\}$$

(a) $a_0 = \frac{1}{\pi}\int_0^{2\pi} f(x) \, dx = \frac{1}{\pi}\left\{\int_0^{\pi} \frac{2}{\pi} x dx + \int_{\pi}^{2\pi} 2 dx\right\}$

$= \frac{1}{\pi}\left\{\left[\frac{x^2}{\pi}\right]_0^{\pi} + \left[2x\right]_{\pi}^{2\pi}\right\} = \frac{1}{\pi}\{\pi + 4\pi - 2\pi\} = 3 \qquad \therefore \underline{a_0 = 3}$

(b) $a_n = \dfrac{1}{\pi}\displaystyle\int_0^{2\pi} f(x)\cos nx\,\mathrm{d}x = \dfrac{1}{\pi}\left\{\int_0^{\pi}\left(\dfrac{2}{\pi}x\right)\cos nx\,\mathrm{d}x\right.$

$$\left. + \int_0^{2\pi} 2\cos nx\,\mathrm{d}x\right\}$$

$$= \frac{2}{\pi}\left\{\frac{1}{\pi}\left[\frac{x\sin nx}{n}\right]_0^{\pi} - \frac{1}{\pi n}\int_0^{\pi}\sin nx\,\mathrm{d}x + \int_{\pi}^{2\pi}\cos nx\,\mathrm{d}x\right\}$$

$= \ldots\ldots\ldots$ Finish it off.

54

$$a_n = 0\ (n\text{ even});\quad a_n = \frac{-4}{\pi^2 n^2}\ (n\text{ odd})$$

(c) To find b_n, we proceed in the same general manner

$b_n = \ldots\ldots\ldots$ Complete it on your own.

55

$$b_n = -\frac{2}{\pi n}$$

Here is the working

$$b_n = \frac{1}{\pi}\int_0^{2\pi} f(x)\sin nx\,\mathrm{d}x = \frac{1}{\pi}\left\{\int_0^{\pi}\left(\frac{2}{\pi}x\right)\sin nx\,\mathrm{d}x + \int_{\pi}^{2\pi} 2\sin nx\,\mathrm{d}x\right\}$$

$$= \frac{2}{\pi}\left\{\frac{1}{\pi}\left[\frac{-x\cos nx}{n}\right]_0^{\pi} + \frac{1}{\pi n}\int_0^{\pi}\cos nx\,\mathrm{d}x + \int_{\pi}^{2\pi}\sin nx\,\mathrm{d}x\right\}$$

$$= \frac{2}{\pi}\left\{\frac{1}{\pi n}(-\pi\cos n\pi) + \frac{1}{\pi n}\left[\frac{\sin nx}{n}\right]_0^{\pi} + \left[\frac{-\cos nx}{n}\right]_{\pi}^{2\pi}\right\}$$

$$= \frac{2}{\pi}\left\{-\frac{1}{n}\cos n\pi + (0-0) - \frac{1}{n}(\cos 2\pi n - \cos n\pi)\right\}$$

$$= \frac{2}{\pi}\left\{-\frac{1}{n}\cos 2n\pi\right\} = -\frac{2}{\pi n}\cos 2n\pi$$

But $\cos 2n\pi = 1$. $\quad\therefore\ \underline{b_n = -\dfrac{2}{\pi n}}$

So the required series is $f(x) = \ldots\ldots\ldots$

56

$$f(x) = \frac{3}{2} - \frac{4}{\pi^2}\left\{\cos x + \frac{1}{9}\cos 3x + \frac{1}{25}\cos 5x + \ldots\right\}$$
$$- \frac{2}{\pi}\left\{\sin x + \frac{1}{2}\sin 2x + \frac{1}{3}\sin 3x + \frac{1}{4}\sin 4x \ldots\right\}$$

At this stage, let us take stock of our findings so far.

If a function $f(x)$ is defined over the range $-\pi$ to π, or any other periodic interval of 2π, then the Fourier series for $f(x)$ is of the form

$$f(x) = \frac{1}{2}a_0 + \sum_{n=1}^{\infty}\{a_n \cos nx + b_n \sin nx\}$$

where $a_0 = \frac{1}{\pi}\int_{-\pi}^{\pi} f(x)\,dx \qquad = 2 \times$ mean value of $f(x)$ over a period

$a_n = \frac{1}{\pi}\int_{-\pi}^{\pi} f(x)\cos nx\,dx = 2 \times$ mean value of $f(x)\cos nx$ over a period

$b_n = \frac{1}{\pi}\int_{-\pi}^{\pi} f(x)\sin nx\,dx = 2 \times$ mean value of $f(x)\sin nx$ over a period.

We also know that

(a) if $f(x)$ is an *even* function, the series will contain *no sine terms*

(b) if $f(x)$ is an *odd* function, the series will contain *only sine terms*

(c) if $f(x)$ is *neither odd nor even*, the series will, in general, contain a constant term, cosine terms and sine terms.

57 Half-range series

Sometimes a function of period 2π is defined over the range 0 to π, instead of the normal $-\pi$ to π, or 0 to 2π. We then have a choice of how to proceed.

For example, if we are told that between $x = 0$ and $x = \pi$, $f(x) = 2x$, then, since the period is 2π, we have no evidence of how the function behaves between $x = -\pi$ and $x = 0$.

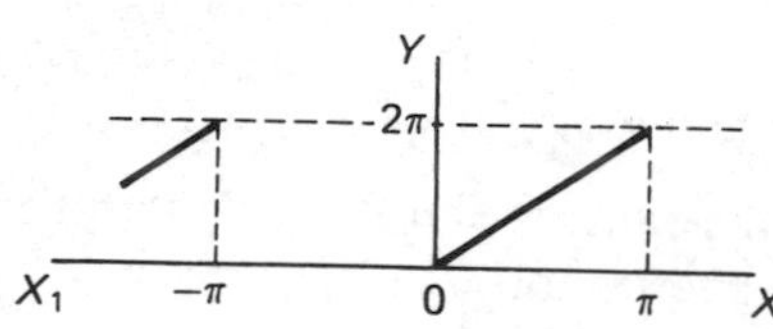

(a)

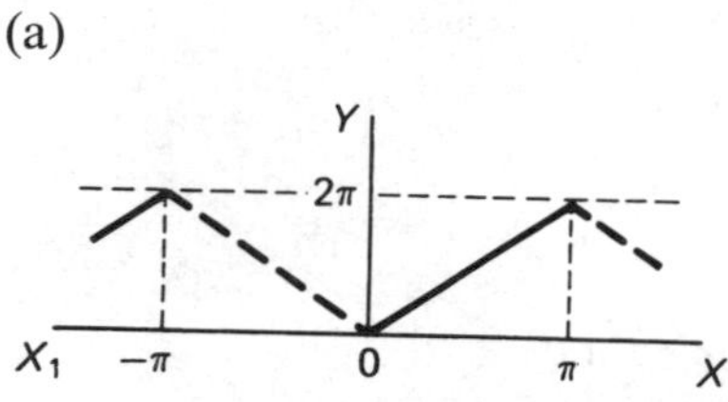

If the waveform were as shown in (a), the function would be an even function, symmetrical about the y-axis and the series would have *only cosine terms* (including possibly a_0).

(b)

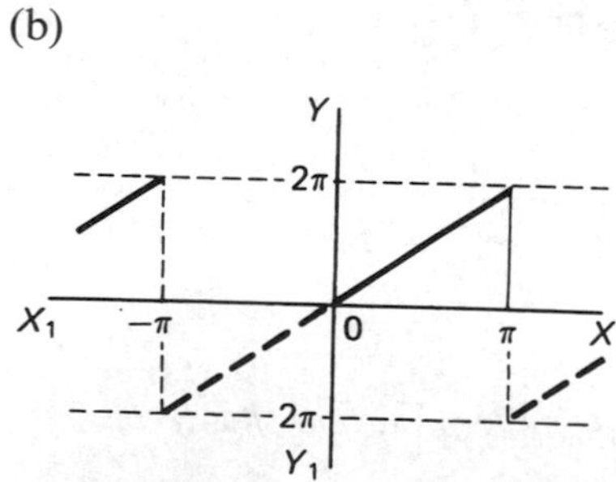

On the other hand, if the waveform were as shown in (b), the function would be odd, being symmetrical about the origin and the series would have *only sine terms.*

(c)

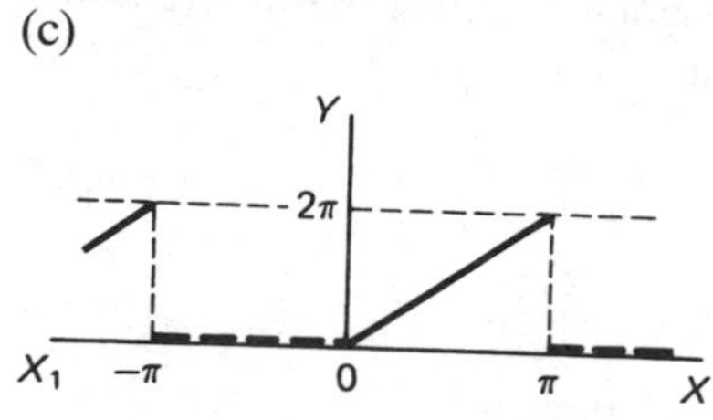

Of course, if we choose something quite different for the waveform between $x = -\pi$ and $x = 0$, then $f(x)$ will be neither odd nor even and the series will then contain ....

58

both sine and cosine terms (including a_0)

In each case, we are making an assumption on how the function behaves between $x = -\pi$ and $x = 0$, and the resulting Fourier series will therefore apply only to $f(x)$ between $x = 0$ and $x = \pi$ for which it is defined. For this reason, such series are called *half-range series.*

Example 1 A function $f(x)$ is defined by $f(x) = 2x \quad 0 < x < \pi$

$$f(x) = f(x + 2\pi).$$

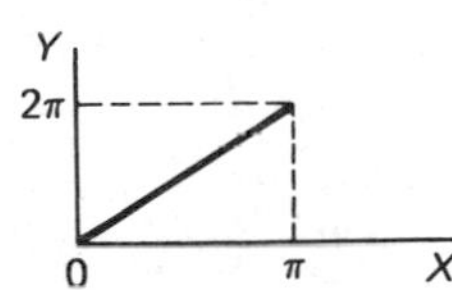

Obtain a half-range cosine series to represent the function.

To obtain a cosine series, i.e. a series with no sine terms, we need an

.... function.

59

even

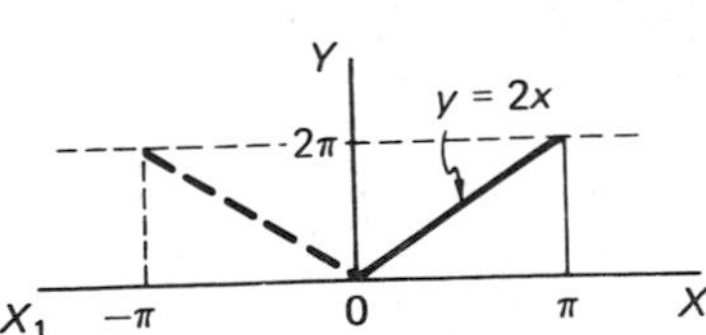

Therefore, we assume the waveform between $x = -\pi$ and $x = 0$ to be as shown, making the total graph symmetrical about the y-axis.

Now we can find expressions for the Fourier coefficients as usual.

$$a_0 = \ldots\ldots\ldots$$

60

$a_0 = 2\pi$

for $a_0 = \dfrac{2}{\pi}\displaystyle\int_0^{\pi} f(x)\,dx = \dfrac{2}{\pi}\int_0^{\pi} 2x\,dx = \dfrac{2}{\pi}\Big[x^2\Big]_0^{\pi} = 2\pi \qquad \therefore\ \underline{a_0 = 2\pi}$

Then we need a_n which is ....

61

$$a_n = 0\ (n \text{ even}) = -\frac{8}{\pi n^2}\ (n \text{ odd})$$

since $a_n = \frac{2}{\pi}\int_0^{\pi} 2x\cos nx\,dx = \frac{4}{\pi}\int_0^{\pi} x\cos nx\,dx$

$$= \frac{4}{\pi}\left\{\left[\frac{x\sin nx}{n}\right]_0^{\pi} - \frac{1}{n}\int_0^{\pi}\sin nx\,dx\right\}$$

$$= \frac{4}{\pi}\left\{(0-0) - \frac{1}{n}\left[\frac{-\cos nx}{n}\right]_0^{\pi}\right\} = \frac{4}{\pi n^2}(\cos n\pi - 1)$$

$$\cos n\pi = 1\ (n \text{ even}) = -1\ (n \text{ odd})$$

$$\therefore\ a_n = 0\ (n \text{ even}) \quad \text{and} \quad a_n = -\frac{8}{\pi n^2}\ (n \text{ odd})$$

All that now remains is b_n which is ....

62

zero, since $f(x)$ is an even function, i.e. $b_n = 0$

So $a_0 = 2\pi$, $a_n = 0$ (n even) or $-\frac{8}{\pi n^2}$ (n odd), $b_n = 0$.
Therefore $f(x) =$ ....

63

$$f(x) = \pi - \frac{8}{\pi}\left\{\cos x + \frac{1}{9}\cos 3x + \frac{1}{25}\cos 5x + \dots\right\}$$

Let us look at a further example, so move on to the next frame.

64

Example 2 Determine a half-range sine series to represent the function $f(x)$ defined by

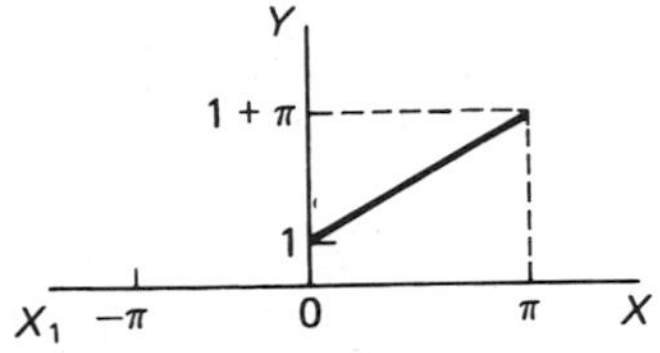

$$f(x) = 1 + x \qquad 0 < x < \pi$$
$$f(x) = f(x + 2\pi).$$

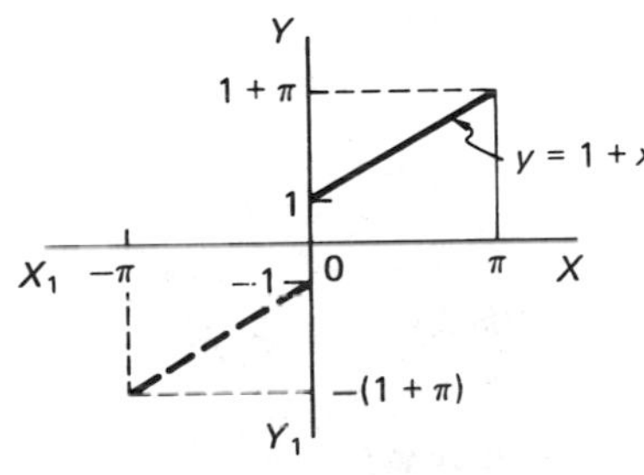

We choose the waveform between $x = -\pi$ and $x = 0$ so that the graph is symmetrical about the origin. The function is then an odd function and the series will contain only sine terms.

$$\therefore\ a_0 = 0 \quad \text{and} \quad a_n = 0$$

b_n can now easily be determined and the required series obtained

$$f(x) = \ldots\ldots\ldots$$

65

$$f(x) = \left(\frac{4}{\pi} + 2\right)\left\{\sin x + \frac{1}{3}\sin 3x + \frac{1}{5}\sin 5x + \dots\right\}$$
$$-2\left\{\frac{1}{2}\sin 2x + \frac{1}{4}\sin 4x + \frac{1}{6}\sin 6x + \dots\right\}$$

Check the working.

$$b_n = \frac{2}{\pi}\int_0^{\pi}(1+x)\sin nx\,\mathrm{d}x = \frac{2}{\pi}\left\{\left[(1+x)\frac{-\cos nx}{n}\right]_0^{\pi} + \frac{1}{n}\int_0^{\pi}\cos nx\,\mathrm{d}x\right\}$$

$$= \frac{2}{\pi}\left\{-\frac{1+\pi}{n}\cos n\pi + \frac{1}{n} + \frac{1}{n}\left[\frac{\sin nx}{n}\right]_0^{\pi}\right\}$$

$$= \frac{2}{\pi}\left\{\frac{1}{n} - \frac{1+\pi}{n}\cos n\pi\right\} = \frac{2}{\pi n}\{1 - (1+\pi)\cos n\pi\}$$

$$\cos n\pi = 1 \ (n \text{ even}) \quad = -1 \ (n \text{ odd})$$

$$\therefore\ b_n = -\frac{2}{n}\ (n \text{ even}) = \frac{4+2\pi}{\pi n}\ (n \text{ odd})$$

Substituting in the general expression $f(x) = \sum_{n=1}^{\infty} b_n \sin nx$ we have

$$f(x) = \frac{4+2\pi}{\pi}\left\{\sin x + \frac{1}{3}\sin 3x + \frac{1}{5}\sin 5x + \dots\right\}$$
$$-2\left\{\frac{1}{2}\sin 2x + \frac{1}{4}\sin 4x + \frac{1}{6}\sin 6x + \dots\right\}$$

So a knowledge of odd and even functions and of half-range series saves a deal of unnecessary work on occasions.

Now let us consider the presence of odd or even harmonics, so move on.

66

Series containing only odd harmonics or only even harmonics

$$f(x) = \tfrac{1}{2}a_0 + a_1 \cos x + a_2 \cos 2x + a_3 \cos 3x + \dots$$
$$+ b_1 \sin x + b_2 \sin 2x + b_3 \sin 3x + \dots$$

If we replace x by $(x + \pi)$, this becomes

$$f(x + \pi) = \tfrac{1}{2}a_0 + \sum_{n=1}^{\infty}\{a_n \cos n(x + \pi) + b_n \sin n(x + \pi)\}$$

Now $\cos(nx + n\pi) = \cos nx \cos n\pi - \sin nx \sin n\pi$.
But for $n = 1, 2, 3, \dots \quad \sin n\pi = 0$

$$\therefore \cos n(x + \pi) = \cos nx \cos n\pi$$

Also for $n = 1, 2, 3, \dots \quad \cos n\pi = 1 \ (n \text{ even}) = -1 \ (n \text{ odd})$.

$$\therefore \cos n(x + \pi) = \cos nx \ (n \text{ even}) = -\cos nx \ (n \text{ odd}) \qquad (1)$$

Similarly, $\sin(nx + n\pi) = \sin nx \cos n\pi + \cos nx \sin n\pi$.
Therefore, as before

$$\sin n(x + \pi) = \sin nx \ (n \text{ even}) = -\sin nx \ (n \text{ odd}) \qquad (2)$$

$$\therefore f(x + \pi) = \tfrac{1}{2}a_0 - a_1 \cos x + a_2 \cos 2x - a_3 \cos 3x + \dots$$
$$- b_1 \sin x + b_2 \sin 2x - b_3 \sin 3x + \dots$$

But $$f(x) = \tfrac{1}{2}a_0 + a_1 \cos x + a_2 \cos 2x + a_3 \cos 3x + \dots$$
$$+ b_1 \sin x + b_2 \sin 2x + b_3 \sin 3x + \dots$$

If $f(x) = f(x + \pi)$, these two series are equal and the odd harmonics that you see differ in sign must be zero.

$$\therefore f(x) = f(x + \pi) = \tfrac{1}{2}a_0 + a_2 \cos 2x + a_4 \cos 4x + \dots$$
$$+ b_2 \sin 2x + b_4 \sin 4x + \dots$$

$\therefore$ If $f(x) = f(x + \pi)$, the Fourier series for $f(x)$ contains even harmonics only.

Similarly, from the same two series above

if $f(x) = -f(x + \pi)$, the Fourier series for $f(x)$ contains odd harmonics only.

Make a note of these two results: you will find them useful.

67 *Example 1*

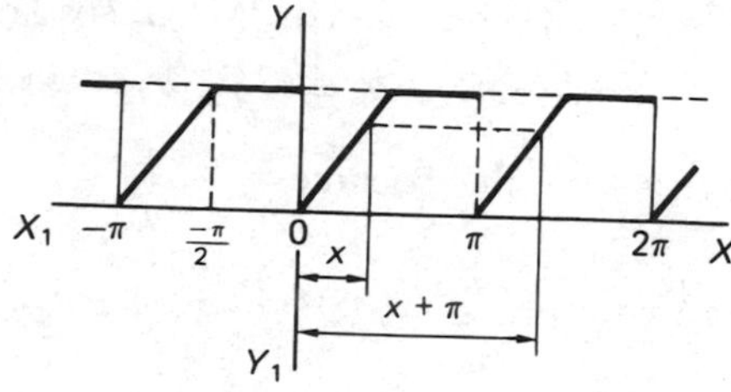

Here $f(x) = f(x + \pi)$

Therefore, the series contains ....

68

even harmonics only

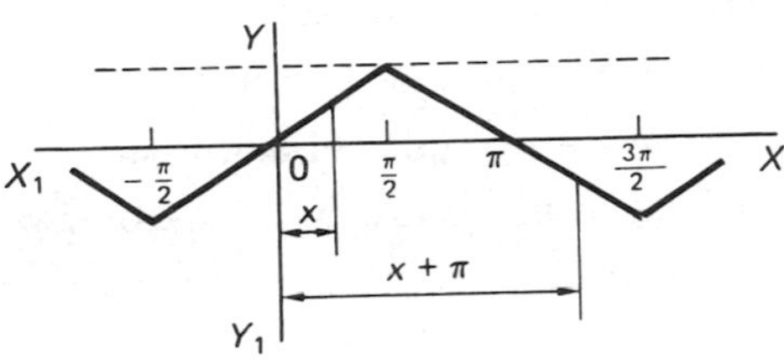

Example 2

Here we see that $f(x) = -f(x + \pi)$
Therefore, the series contains odd harmonics only.

Now we can apply our knowledge to date to the following exercise.

Exercise From each of the following waveforms, we can describe the nature of the terms in the relevant Fourier series.

1.

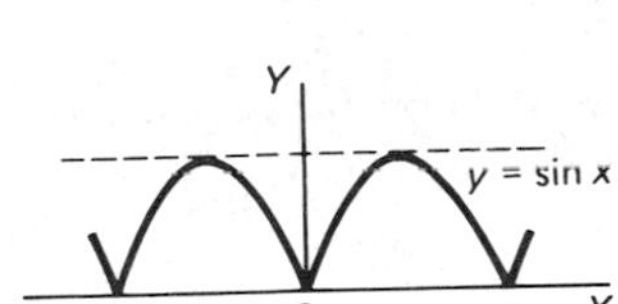

2.

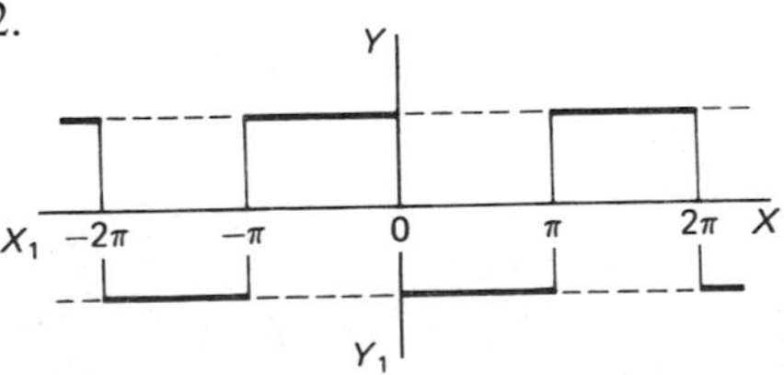

3.

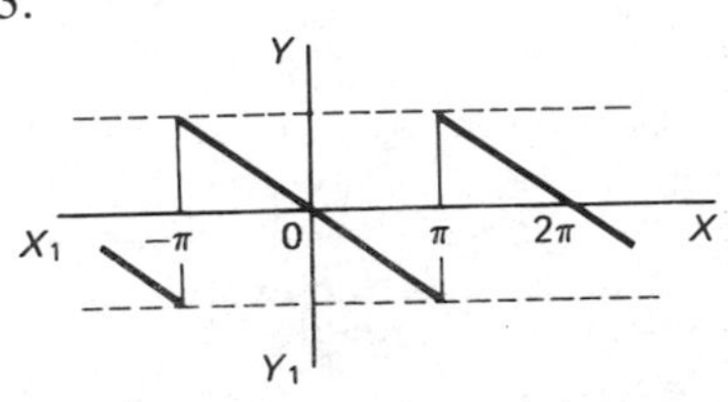

4.

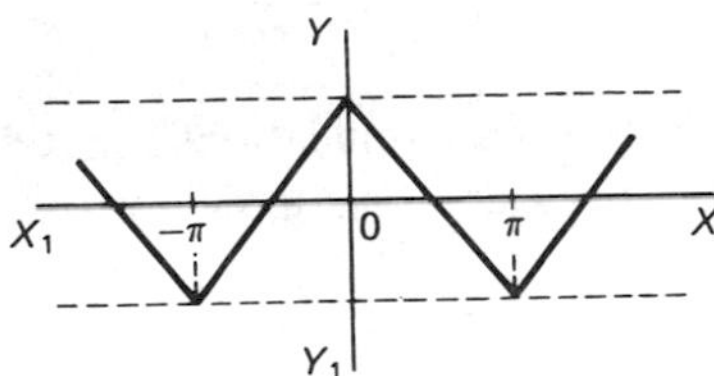

5.

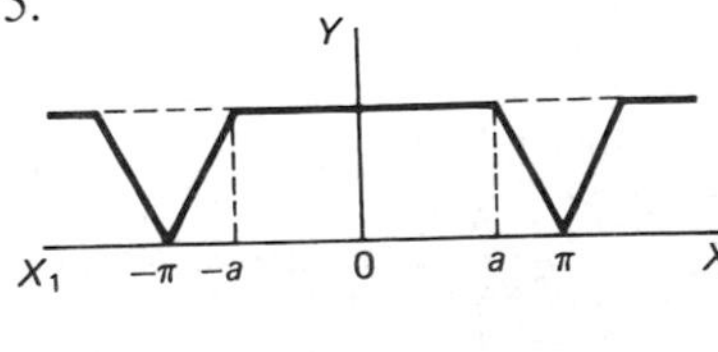

6.

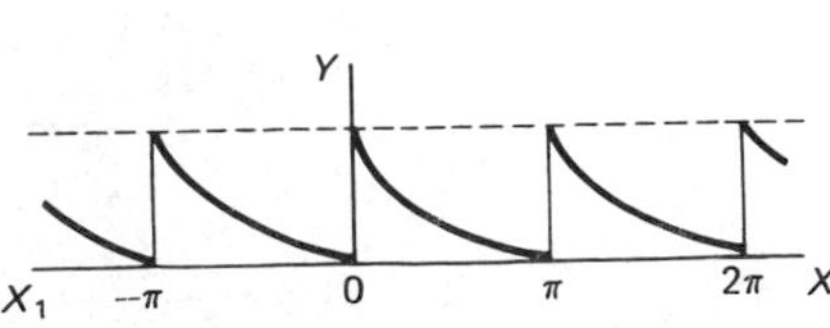

69

1. cosine terms $(+a_0)$ only; even harmonics only
2. sine terms only; odd harmonics only
3. sine terms only; all harmonics
4. cosine terms $(+a_0)$ only; odd harmonics only
5. cosine terms $(+a_0)$ only; all harmonics
6. a_0, sine and cosine terms; even harmonics only.

On we go.

70

Significance of the constant term $\frac{1}{2}a_0$

We might, at this point, note that the effect of the constant term $\frac{1}{2}a_0$ is to raise, or lower, the whole waveform on the y-axis.

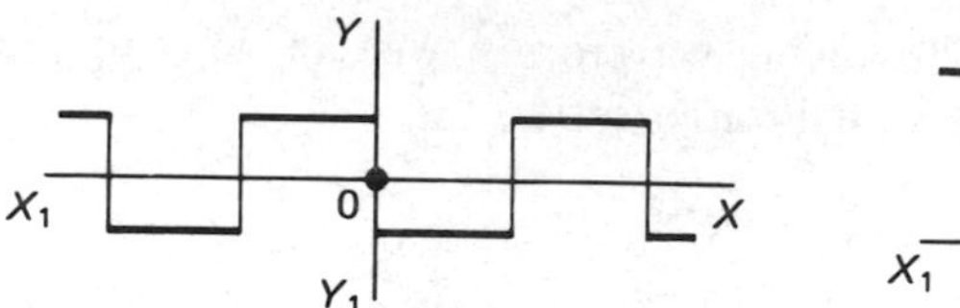

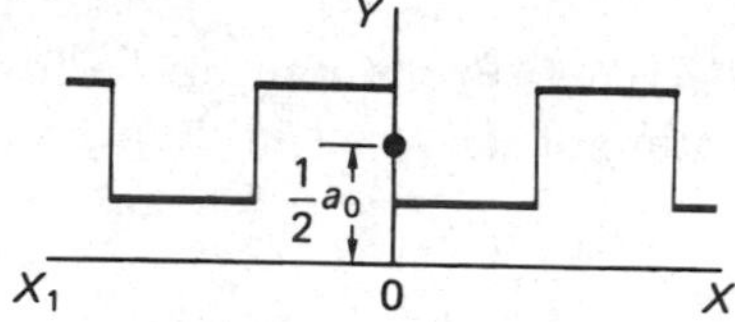

In electrical applications to alternating currents, the constant term $\frac{1}{2}a_0$ of the Fourier series indicates the d.c. component.

71

Functions with periods other than 2π

So far, we have considered functions $f(x)$ with period 2π. In practice, we often encounter functions defined over periodic intervals other than 2π, e.g. from 0 to T, $-\frac{T}{2}$ to $\frac{T}{2}$, etc.

Functions with period T

If $y = f(x)$ is defined in the range $-\frac{T}{2}$ to $\frac{T}{2}$, i.e. has a period T, we can convert this to an interval of 2π by changing the units of the independent variable.

In many practical cases involving physical oscillations, the independent variable is time (t) and the periodic interval is normally denoted by T, i.e.

$$f(t) = f(t + T)$$

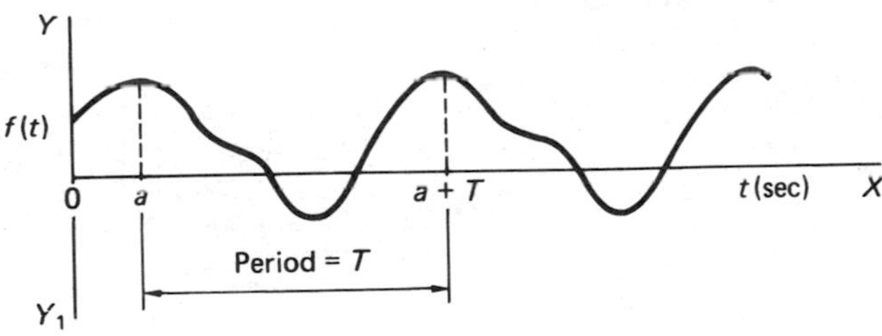

Each cycle is therefore completed in T seconds and the frequency f hertz (oscillations per second) of the periodic function is therefore given by $f = \frac{1}{T}$. If the angular velocity, ω radians per second, is defined by $\omega = 2\pi f$, then

$$\omega = \frac{2\pi}{T} \quad \text{and} \quad T = \frac{2\pi}{\omega}.$$

The angle, x radians, at any time t is therefore $x = \omega t$ and the Fourier series to represent the function can be expressed as

$$f(t) = \tfrac{1}{2}a_0 + \sum_{n=1}^{\infty}\{a_n \cos n\omega t + b_n \sin n\omega t\}$$

which can also be written in the form

$$f(t) = \frac{1}{2}A_0 + \sum_{n=1}^{\infty} B_n \sin(n\omega t + \phi_n) \qquad n = 1, 2, 3 \ldots$$

Comparing these two expressions, we see that

$A_0 = \ldots\ldots\ldots;\quad B_n \sin\phi_n = \ldots\ldots\ldots;\quad B_n \cos\phi_n = \ldots\ldots\ldots$

72

$$\boxed{A_0 = a_0;\ B_n \sin\phi_n = a_n;\ B_n \cos\phi_n = b_n}$$

and from this it follows that

$$B_n = \ldots\ldots\ldots;\quad \phi_n = \ldots\ldots\ldots$$

73

$$B_n = \sqrt{a_n^2 + b_n^2}; \quad \phi_n = \arctan\left(\frac{a_n}{b_n}\right)$$

So:

$B_1 \sin(\omega t + \phi_1)$ is the first harmonic or fundamental (lowest frequency)
$B_2 \sin(2\omega t + \phi_2)$ is the second harmonic (frequency twice that of the fundamental)

$B_n \sin(n\omega t + \phi_n)$ is the nth harmonic (frequency n times that of the fundamental).

And for the series to converge, the values of B_n must eventually decrease with higher-order harmonics, i.e. $B_n \to 0$ as $n \to \infty$.

Fourier coefficients

With the new variable, the Fourier coefficients become:

$$f(t) = \tfrac{1}{2}a_0 + \sum_{n=1}^{\infty}\{a_n \cos n\omega t + b_n \sin n\omega t\}$$

$$a_0 = \frac{2}{T}\int_0^T f(t)\,dt \qquad = \frac{\omega}{\pi}\int_0^{2\pi/\omega} f(t)\,dt$$

$$a_n = \frac{2}{T}\int_0^T f(t)\cos n\omega t\,dt = \frac{\omega}{\pi}\int_0^{2\pi/\omega} f(t)\cos n\omega t\,dt$$

$$b_n = \frac{2}{T}\int_0^T f(t)\sin n\omega t\,dt = \frac{\omega}{\pi}\int_0^{2\pi/\omega} f(t)\sin n\omega t\,dt.$$

We can see that there is very little difference between these expressions and those that have gone before. The limits can, of course, be 0 to T, $-\dfrac{T}{2}$ to $\dfrac{T}{2}$, $-\dfrac{\pi}{\omega}$ to $\dfrac{\pi}{\omega}$, 0 to $\dfrac{2\pi}{\omega}$ etc. as is convenient, so long as they cover a complete period.

Example Determine the Fourier series for a periodic function defined by

$$f(t) = 2(1 + t) \qquad -1 < t < 0$$
$$f(t) = 0 \qquad 0 < t < 1$$
$$f(t) = f(t + 2)$$

The first step is to sketch the waveform which is ....

74

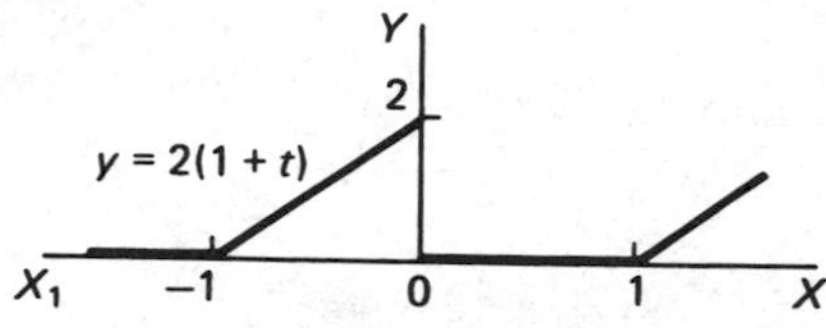

We have $$f(t) = \tfrac{1}{2}a_0 + \sum_{n=1}^{\infty}\{a_n \cos n\omega t + b_n \sin n\omega t\} \qquad T = 2$$

$$a_0 = \frac{2}{T}\int_{-T/2}^{T/2} f(t)\,dt = \frac{2}{2}\int_{-1}^{1} f(t)\,dt = \int_{-1}^{0} 2(1+t)\,dt + \int_{0}^{1}(0)\,dt$$

$$= \left[2t + t^2\right]_{-1}^{0} = 1 \qquad \therefore \underline{a_0 = 1}$$

$$a_n = \frac{2}{T}\int_{-T/2}^{T/2} f(t)\cos n\omega t\,dt = \frac{2}{2}\int_{-1}^{1} f(t)\cos n\omega t\,dt$$

$$= \int_{-1}^{0} 2(1+t)\cos n\omega t\,dt = \dots\dots\dots$$

Finish this part.

75

$$\boxed{a_n = 0\ (n \text{ even}) = \frac{4}{n^2\omega^2}\ (n \text{ odd})}$$

for $$a_n = 2\left\{\left[(1+t)\frac{\sin n\omega t}{n\omega}\right]_{-1}^{0} - \frac{1}{n\omega}\int_{-1}^{0}\sin n\omega t\,dt\right\}$$

$$= 2\left\{(0-0) - \frac{1}{n\omega}\left[-\frac{\cos n\omega t}{n\omega}\right]_{-1}^{0}\right\} = \frac{2}{n^2\omega^2}(1 - \cos n\omega)$$

Now $T = \dfrac{2\pi}{\omega}$ $\therefore\ \omega = \dfrac{2\pi}{T} = \dfrac{2\pi}{2} = \pi$ $\quad\therefore\ a_n = \dfrac{2}{n^2\omega^2}(1 - \cos n\pi)$

$$\therefore\ \underline{a_n = 0\ (n \text{ even}) = \frac{4}{n^2\omega^2}\ (n \text{ odd})}$$

Now for b_n $$b_n = \frac{2}{T}\int_{-T/2}^{T/2} f(t)\sin n\omega t\,dt = \dots\dots\dots$$

76

$$\boxed{b_n = -\frac{2}{n\omega}}$$

for $b_n = \frac{2}{2}\int_{-1}^{0} 2(1+t)\sin n\omega t\, dt$

$$= 2\left\{\left[(1+t)\frac{-\cos n\omega t}{n\omega}\right]_{-1}^{0} + \frac{1}{n\omega}\int_{-1}^{0}\cos n\omega t\, dt\right\}$$

$$= 2\left\{-\frac{1}{n\omega} + \frac{1}{n\omega}\left[\frac{\sin n\omega t}{n\omega}\right]_{-1}^{0}\right\}$$

$$= 2\left\{-\frac{1}{n\omega} + \frac{1}{n^2\omega^2}(\sin n\omega)\right\} \qquad \text{As before } \omega = \pi$$

$$\therefore b_n = -\frac{2}{n\omega}$$

So the first few terms of the series give

$f(t) = \dots\dots\dots$

77

$$\boxed{\begin{aligned} f(t) = \frac{1}{2} + \frac{4}{\omega^2}&\left\{\cos\omega t + \frac{1}{9}\cos 3\omega t + \frac{1}{25}\cos 5\omega t + \dots\right\} \\ -\frac{2}{\omega}&\left\{\sin\omega t + \frac{1}{2}\sin 2\omega t + \frac{1}{3}\sin 3\omega t + \frac{1}{4}\sin 4\omega t \dots\right\}\end{aligned}}$$

Half-range series

The theory behind the half-range sine and cosine series still applies with the new variable.

(a) *Even function* Half-range cosine series

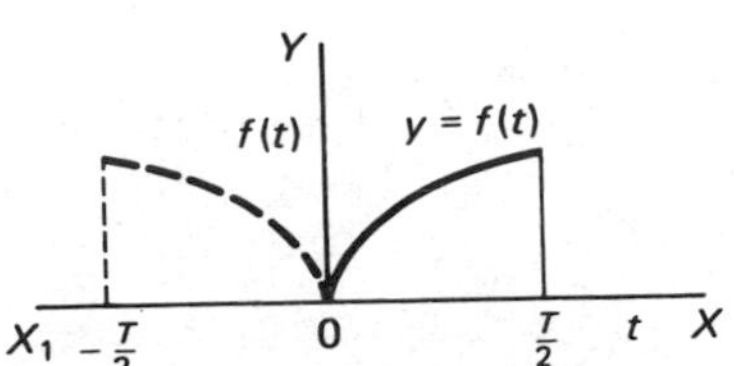

$y = f(t) \qquad 0 < t < \frac{T}{2}$

$f(t) = f(t+T)$

symmetrical about the y-axis.

With an even function, we know that $b_n = 0$

$$\therefore f(t) = \frac{1}{2}a_0 + \sum_{n=1}^{\infty} a_n \cos n\omega t$$

where $$a_0 = \frac{4}{T}\int_0^{T/2} f(t)\,\mathrm{d}t$$

and $$a_n = \frac{4}{T}\int_0^{T/2} f(t)\cos n\omega t\,\mathrm{d}t$$

(b) *Odd function* Half-range sine series

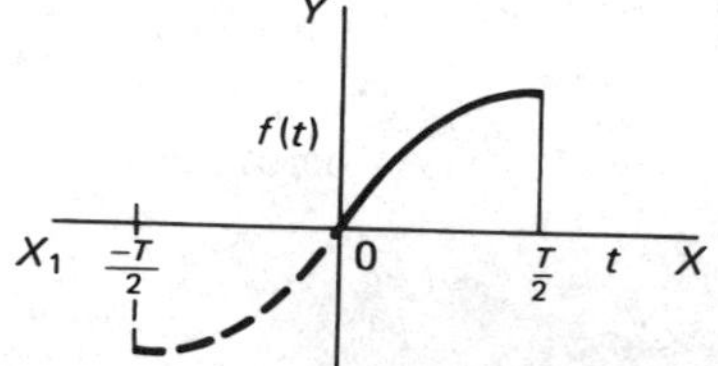

$y = f(t) \qquad 0 < t < \dfrac{T}{2}$

$f(t) = f(t + T)$

symmetrical about the origin.

$\therefore\ a_0 = 0$ and $a_n = 0$

Then $\quad f(t) = \ldots\ldots\ldots\ldots\ldots\ldots\ldots$

and $\quad b_n = \ldots\ldots\ldots\ldots\ldots\ldots\ldots$

78

$$f(t) = \sum_{n=1}^{\infty} b_n \sin n\omega t; \quad b_n = \frac{4}{T}\int_0^{T/2} f(t) \sin n\omega t \, dt$$

Now for an example or two with the new variable.

So turn on.

79

Example 1 A function $f(t)$ is defined by $f(t) = 4 - t, \quad 0 < t < 4.$

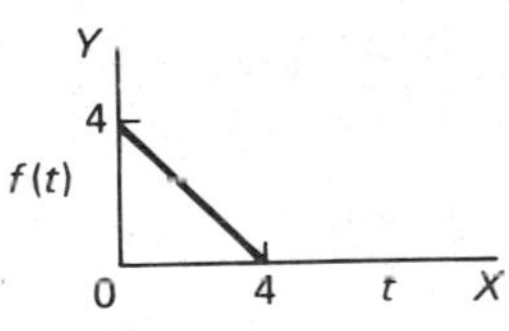

We have to form a half-range cosine series to represent the function in this interval.

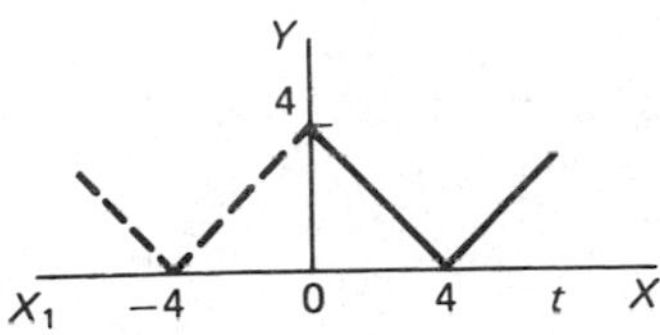

First we form an even function, i.e. symmetrical about the y-axis.

Now for a useful little trick. If we lower the waveform 2 units, i.e. to its 'average' position, balanced above and below the x-axis, then in this new position $\frac{1}{2}a_0 = 0$ and we have been saved one set of calculations.

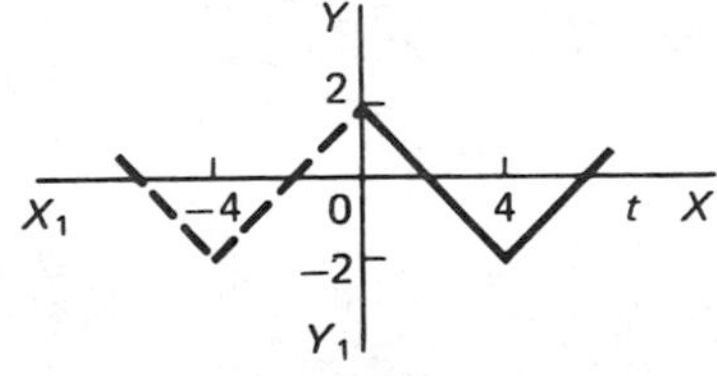

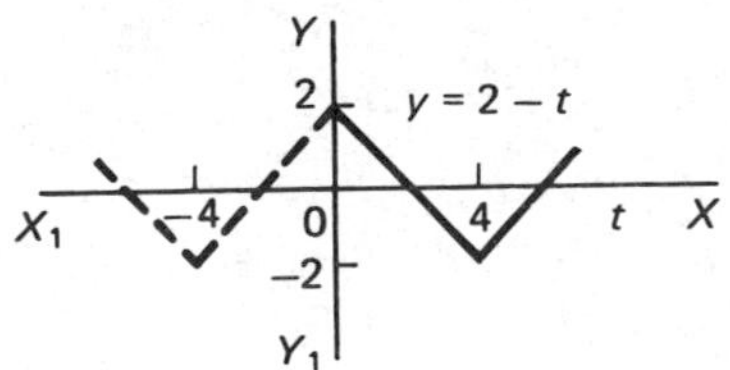

The function is now $y = f_1(t) = 2 - t$ and, for the moment $\frac{1}{2}a_0 = 0$. Also, being an even function $b_n = 0$. All we need to do is to evaluate a_n.

$$\text{So } a_n = \frac{4}{T}\int_0^{T/2} f_1(t) \cos n\omega t \, dt = \frac{4}{8}\int_0^4 (2 - t) \cos n\omega t \, dt$$

$$= \dots\dots\dots$$

80

$$a_n = 0 \text{ } (n \text{ even}) = \frac{1}{n^2\omega^2} \text{ } (n \text{ odd})$$

Simple integration by parts gives

$$a_n = \frac{1}{2}\left\{-\frac{2\sin\ 4n\omega}{n\omega} - \frac{1}{n^2\omega^2}(\cos\ 4n\omega - 1)\right\}$$

But $\omega = \frac{2\pi}{T} = \frac{2\pi}{8} = \frac{\pi}{4}$

$$\therefore\ a_n = \frac{1}{2}\left\{-\frac{2\sin n\pi}{n\omega} - \frac{1}{n^2\omega^2}(\cos n\pi - 1)\right\} \qquad n = 1,\ 2,\ 3,\ \ldots$$

$\sin n\pi = 0$; $\quad \cos n\pi = 1$ (n even); $\quad \cos n\pi = -1$ (n odd)

$$\therefore\ a_n = 0 \text{ } (n \text{ even}) \text{ and } a_n = \frac{1}{n^2\omega^2} \text{ } (n \text{ odd})$$

$$\therefore\ f_1(t) = \ldots\ldots\ldots$$

81

$$f_1(t) = \frac{1}{\omega^2}\left\{\cos \omega t + \frac{1}{9}\cos 3\omega t + \frac{1}{25}\cos 5\omega t + \dots\right\}$$

Now if we finally lift the waveform back to its original position by restoring the 2 units (i.e. $\frac{1}{2}a_0 = 2$), the original function is now regained with $f(t) = f_1(t) + 2$.

$$\therefore f(t) = 2 + \frac{1}{\omega^2}\left\{\cos \omega t + \frac{1}{9}\cos 3\omega t + \frac{1}{25}\cos 5\omega t + \dots\right\}$$

where $\omega = \frac{\pi}{4}$.

Example 2 A function $f(t)$ is defined by $f(t) = 3 + t \quad 0 < t < 2$

$f(t) = f(t = 4)$.

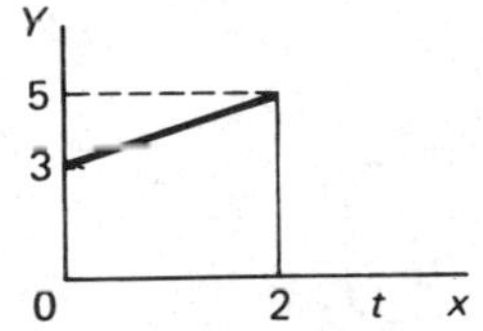

Obtain the half-range sine series for the function in this range.

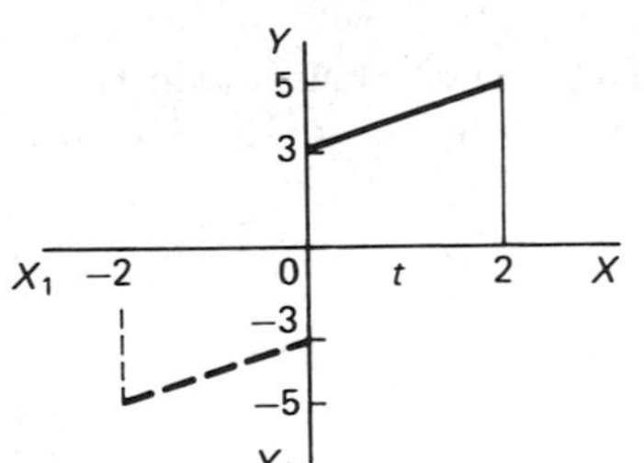

Sine series required. Therefore, we form an odd function, symmetrical about the origin

$$a_0 = 0;\ a_n = 0;\ T = 4$$

$$f(t) = \sum_{n=1}^{\infty} b_n \sin n\omega t$$

$$\therefore b_n = \frac{4}{T}\int_0^2 f(t)\ \sin n\omega t\,dt = \int_0^2 (3 + t)\sin n\omega t\ dt$$

This you can easily evaluate and then, putting $n = 1, 2, 3, \dots$ obtain the series $f(t) = \dots\dots\dots$

82

$$f(t) = \frac{2}{\omega}\left\{4\sin\omega t - \frac{1}{2}\sin 2\omega t + \frac{4}{3}\sin 3\omega t - \frac{1}{4}\sin 4\omega t \dots\right\}$$

Straightforward integration by parts gives

$$b_n = \frac{1}{n\omega}(3 - 5\cos 2n\omega) + \frac{1}{n^2\omega^2}(\sin 2n\omega)$$

But $T = \dfrac{2\pi}{\omega}$ $\quad\therefore\ \omega = \dfrac{2\pi}{T} = \dfrac{\pi}{2}$

$$\therefore\ b_n = \frac{1}{n\omega}(3 - 5\cos n\pi) + \frac{1}{n^2\omega^2}\sin n\pi$$

$$= -\frac{2}{n\omega}\ (n \text{ even}) = \frac{8}{n\omega}\ (n \text{ odd})$$

$$\therefore\ f(t) = \frac{2}{\omega}\left\{4\sin\omega t - \frac{1}{2}\sin 2\omega t + \frac{4}{3}\sin 3\omega t - \frac{1}{4}\sin 4\omega t \dots\right\}$$

And that just about brings this particular programme to an end. Fourier series have wide applications so it is very worthwhile paying considerable attention to them. We shall approach them again from a different viewpoint in the next programme when we deal with numerical harmonic analysis.

Meanwhile, the Revision Summary now follows, after which you will have no trouble with the Test Exercise.

REVISION SUMMARY

83

1. *Graphs of $y = A \sin nx$ and $A \cos nx$*

$$\text{Amplitude} = A;\ \text{period} = \frac{360^\circ}{n} = \frac{2\pi}{n} \text{ radians}$$

2. *Harmonics*

$y = A_1 \sin x$ is the first harmonic or fundamental

$y = A_n \sin nx$ is the nth harmonic.

3. *Periodic function*

$$f(x) = f(x + P) \qquad P = \text{period}$$

4. *Fourier series—functions of period 2π*

$$f(x) = \tfrac{1}{2}a_0 + a_1 \cos x + a_2 \cos 2x + a_3 \cos 3x + \ldots + a_n \cos nx \ldots$$
$$+ b_1 \sin x + b_2 \sin 2x + b_3 \sin 3x + \ldots + b_n \sin nx \ldots$$
$$= \tfrac{1}{2}a_0 + \sum_{n=1}^{\infty} \{a_n \cos nx + b_n \sin nx\}$$

Can also be expressed in compound sine terms

$$f(x) = \tfrac{1}{2}a_0 + \sum_{n=1}^{\infty} \{c_n \sin(nx + \alpha_n)\}$$

where $c_n = \sqrt{a_n^2 + b_n^2}$ and $\alpha_n = \arctan\left(\dfrac{a_n}{b_n}\right)$.

5. *Dirichlet conditions*

(a) The function $f(x)$ must be defined and single-valued.

(b) $f(x)$ must be continuous or have a finite number of finite discontinuities within a periodic interval.

(c) $f(x)$ and $f'(x)$ must be piecewise continuous in the periodic interval.

6. *Fourier coefficients*

$$a_0 = \frac{1}{\pi}\int_{-\pi}^{\pi} f(x)\,dx \qquad = 2 \times \text{mean value of } f(x) \text{ over a period}$$

$$a_n = \frac{1}{\pi}\int_{-\pi}^{\pi} f(x)\cos nx\,dx = 2 \times \text{mean value of } f(x)\cos nx$$

$$b_n = \frac{1}{\pi}\int_{-\pi}^{\pi} f(x)\sin nx\,dx = 2 \times \text{ mean value of } f(x)\sin\, nx$$

where, in each case, $n = 1, 2, 3, \ldots$

7. *Sum of Fourier series at a finite discontinuity*

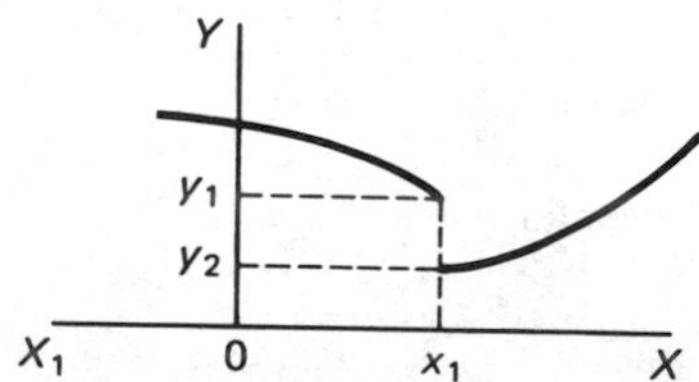

At $x = x_1$, series for $f(x)$ converges to the value

$\frac{1}{2}\{f(x_1 - 0) + f(x_1 + 0)\} = \frac{1}{2}(y_1 + y_2)$

8. *Odd and even functions*

(a) Even function: $f(-x) = f(x)$: symmetrical about y-axis.

(b) Odd function: $f(-x) = -f(x)$: symmetrical about the origin.

Product of odd and even functions

$$\begin{aligned}(\text{even}) \times (\text{even}) &= (\text{even})\\ (\text{odd}) \times (\text{odd}) &= (\text{even})\\ (\text{odd}) \times (\text{even}) &= (\text{odd}).\end{aligned}$$

9. *Sine series and cosine series*

If $f(x)$ is *even*, the series contains *cosine terms only* (including a_0).

If $f(x)$ is *odd*, the series contains *sine terms only*.

10. *Half-range series*

Function of period 2π, defined over the range 0 to π. Can be considered as half of an even function, or half of an odd function.

11. *Series containing only odd harmonics or only even harmonics*

If $f(x) = f(x + \pi)$ the Fourier series contains *even harmonics* only. If $f(x) = -f(x + \pi)$ the Fourier series contains *odd harmonics* only.

12. *Functions with period T*

$$f(t) = \frac{1}{2}a_0 + \sum_{n=1}^{\infty}\{a_n \cos\, n\omega t + b_n \sin\, n\omega t\}$$

$$a_0 = \frac{2}{T}\int_0^T f(t)\,dt \qquad = \frac{\omega}{\pi}\int_0^{2\pi/\omega} f(t)\,dt$$

$$a_n = \frac{2}{T}\int_0^T f(t)\cos n\omega t\, dt \quad = \frac{\omega}{\pi}\int_0^{2\pi/\omega} f(t)\cos n\omega t\, dt$$

$$b_n = \frac{2}{T}\int_0^T f(t)\sin n\omega t\, dt \quad = \frac{\omega}{\pi}\int_0^{2\pi/\omega} f(t)\sin n\omega t\, dt$$

$$\text{where } \omega = \frac{2\pi}{T} \quad \text{i.e. } T = \frac{2\pi}{\omega}.$$

13. *Half-range series—period T*

(a) Even function: half-range cosine series

$$a_0 = \frac{4}{T}\int_0^{T/2} f(t)\, dt$$

$$a_n = \frac{4}{T}\int_0^{T/2} f(t)\cos n\omega t\, dt$$

$$b_n = 0.$$

(b) Odd function: half-range sine series

$$a_0 = 0$$

$$a_n = 0$$

$$b_n = \frac{4}{T}\int_0^{T/2} f(t)\sin n\omega t\, dt.$$

TEST EXERCISE XVII

84

1. If $f(x)$ is defined in the interval $-\pi < x < \pi$ and $f(x) = f(x + 2\pi)$, state whether or not each of the following functions can be represented by a Fourier series.

(a) $f(x) = x^4$

(b) $f(x) = 3 - 2x$

(c) $f(x) = \dfrac{1}{x}$

(d) $f(x) = e^{2x}$

(e) $f(x) = \operatorname{cosec} x$

(f) $f(x) = \pm\sqrt{4x}$.

2. Determine the Fourier series for the function defined by

$$f(x) = 2x \qquad 0 < x < 2\pi$$
$$f(x) = f(x + 2\pi).$$

3. State whether each of the following products is odd, even, or neither.

(a) $x^3 \cos 2x$ (d) $x^2 e^{2x}$

(b) $x^2 \sin 3x$ (e) $(x+5)\cos 2x$

(c) $\sin 2x \sin 3x$ (f) $\sin^2 x \cos x$.

4. A function $f(x)$ is defined by $f(x) = \pi - x \quad 0 < x < \pi$

$$f(x) = f(x + 2\pi).$$

Express the function (a) as a half-range cosine series

(b) as a half-range sine series.

5. Comment on the nature of the terms in the Fourier series for the following functions.

(a)

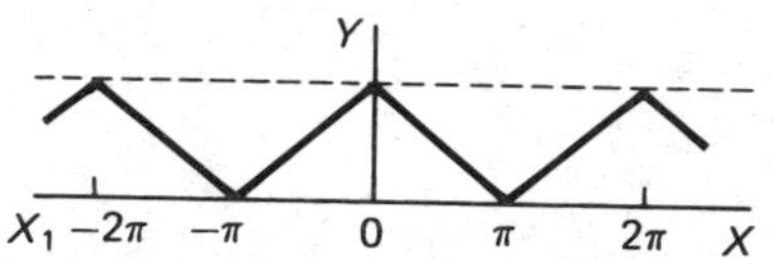

(c)

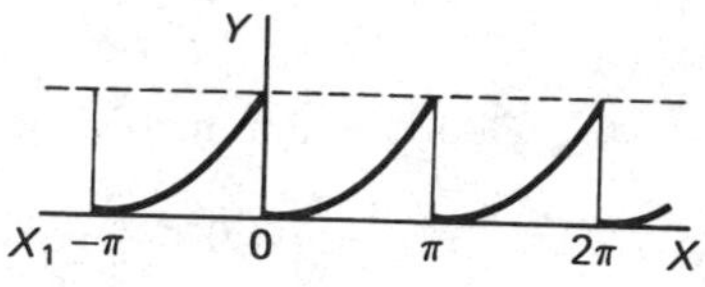

(b)

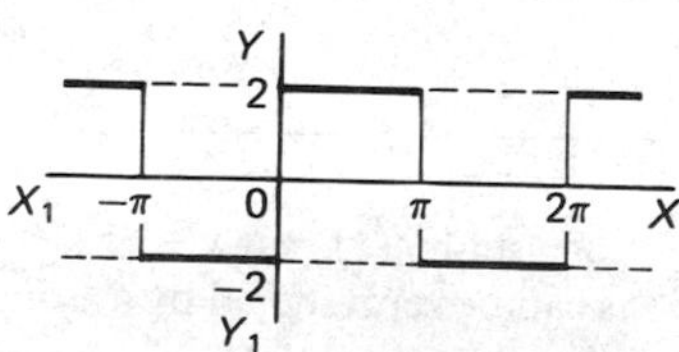

(d)

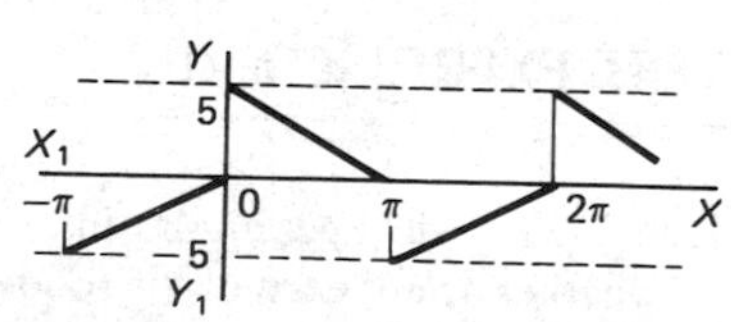

6. A function $f(t)$ is defined by

$$f(t) = 0 \quad -2 < t < 0$$
$$f(t) = t \quad 0 < t < 2$$
$$f(t) = f(t + 4).$$

Determine its Fourier series.

FURTHER PROBLEMS XVII

1. A periodic function $f(x)$ is defined by

$$f(x) = 1 - \frac{x}{\pi} \qquad 0 < x < 2\pi$$
$$f(x) = f(x + 2\pi).$$

Determine the Fourier series up to and including the third harmonic.

2. Determine the Fourier series representation of the function $f(t)$ defined by

$$f(t) = 3 \qquad -2 < t < 0$$
$$f(t) = -5 \qquad 0 < t < 2$$
$$f(t) - f(t + 4).$$

3. Determine the half-range cosine series for the function $f(x) = \sin x$ defined in the range $0 < x < \pi$.

4. A function is defined by $f(x) = \pi + x \qquad -\pi < x < 0$
$$f(x) = \pi - x \qquad 0 < x < \pi$$
$$f(x) = f(x + 2\pi).$$

Obtain the Fourier series.

5. A periodic function is defined by

$$f(x) = A \sin x \qquad 0 < x < \pi$$
$$f(x) = -A \sin x \qquad \pi < x < 2\pi$$
$$f(x) = f(x + 2\pi).$$

Determine its Fourier series up to and including the fourth harmonic.

6. If $\left.\begin{array}{ll} f(x) = 0 & -\pi < x < 0 \\ f(x) = x & 0 < x < \pi \end{array}\right\}$ and $f(x) = f(x + 2\pi)$

obtain the Fourier series and determine the percentage second harmonic.

7. Determine the Fourier series to represent a half-wave rectifier output current, i amperes, defined by

$$i = f(t) = A \sin \omega t \qquad 0 < t < \frac{T}{2}$$
$$f(t) = 0 \qquad \frac{T}{2} < t < T$$
$$f(t) = f(t + T).$$

8. A function $f(x)$ is defined by

$$\begin{aligned} f(x) &= a & 0 < x < \frac{\pi}{3} \\ f(x) &= 0 & \frac{\pi}{3} < x < \frac{2\pi}{3} \\ f(x) &= -a & \frac{2\pi}{3} < x < \pi \\ f(x) &= f(x+\pi). \end{aligned}$$

Obtain the Fourier series to represent the function.

9. If $f(x)$ is defined by $f(x) = x(\pi - x) \quad 0 < x < \pi$, express the function as

(a) a half-range cosine series

(b) a half-range sine series.

10. Determine the Fourier cosine series to represent the function $f(x)$ where

$$\begin{aligned} f(x) &= \cos x & 0 < x < \frac{\pi}{2} \\ f(x) &= 0 & \frac{\pi}{2} < x < \pi \\ f(x) &= f(x+2\pi). \end{aligned}$$

11. If $\left.\begin{aligned} f(x) &= 0 & 0 < x < \frac{\pi}{2} \\ f(x) &= \cos x & \frac{\pi}{2} < x < \pi \end{aligned}\right\} f(x) = f(x+2\pi),$ obtain

the Fourier cosine series for $f(x)$ in the range $x = 0$ to $x = \pi$.

12. A function $f(x)$ is defined over the interval $0 < x < \pi$ by

$$\begin{aligned} f(x) &= x & 0 < x < \frac{\pi}{2} \\ f(x) &= \pi - x & \frac{\pi}{2} < x < \pi \end{aligned}$$

For the range $x = 0$ to $x = \pi$, determine the Fourier series.

13. A function $f(t)$ is defined by

$$\begin{aligned} f(t) &= -1 & -1 < t < 0 \\ f(t) &= 2t & 0 < t < 1 \\ f(t) &= f(t+2). \end{aligned}$$

Obtain the Fourier series up to and including the third harmonic.

14. If $f(x) = x^2 \quad -\pi < x < \pi$ and $f(x) = f(x+2\pi)$, determine the Fourier series for $f(x)$.

15. A function $f(x)$ is given by $f(x) = 7 - \dfrac{3x}{\pi}$ for $-\pi < x < \pi$ with $f(x) = f(x+2\pi)$. Obtain the Fourier series up to and including the fourth harmonic.

16. A function $f(t)$ is defined by $f(t) = 1 - t^2 \qquad -1 < t < 1$

$$f(t) = f(t + 2).$$

Determine its Fourier series.

17. A function $f(x)$ is such that

$$f(x) = \frac{x + \pi}{2} \qquad -\pi < x < 0$$

$$f(x) = \frac{x - \pi}{2} \qquad 0 < x < \pi$$

$$f(x) = f(x + 2\pi).$$

Obtain the Fourier series.

18. Determine the Fourier series for a periodic function such that

$$f(t) = 1 \qquad -2 < t < -1$$

$$f(t) = 0 \qquad -1 < t < 1$$

$$f(t) = -1 \qquad 1 < t < 2$$

$$f(t) = f(t + 4).$$

19. A function is defined by $f(x) = x^2 \qquad 0 < x < 2\pi$

$$f(x) = f(x + 2\pi).$$

Determine (a) its Fourier series

(b) its percentage second harmonic.

20. Determine the Fourier series for the function $f(t)$ defined by

$$f(t) = 0 \qquad -2 < t < 0$$

$$f(t) = \frac{3t}{4} \qquad 0 < t < 4$$

$$f(t) = f(t + 6).$$

Calculate also its percentage third harmonic.

Programme 18

Numerical Harmonic Analysis

Introduction

1 As we have seen in the previous programme, the determination of the coefficients in the Fourier series representation of a periodic function entails the evaluation of various integrals involving the function of the waveform.

In experimental work, an output function is often expressed in the form of a set of readings or as a graph, and direct integration to obtain the coefficients of the relevant Fourier series is then not possible. Some form of approximate method must then be used—and that is what this programme is all about.

Approximate integration

You will, of course, recall that there are several methods of approximate integration, each of which is equivalent to finding the area under a curve. The most common are by the application of the mid-ordinate rule, the trapezoidal rule or Simpson's rule. Let us refresh our ideas about one of these.

Trapezoidal rule

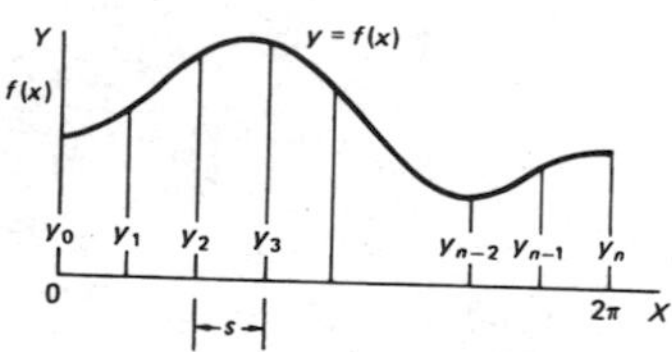

Consider one cycle of a periodic function of period 2π.
If we divide the figure into n equal-width strips, the width of each strip is $\dfrac{2\pi}{n}$.

We denote the ordinates as $y_0,\ y_1,\ y_2,\ \dots y_{n-1},\ y_n$, as shown.
Treating each strip as a trapezium, the area of the whole figure is given by

$$\begin{aligned}A &\approx \tfrac{1}{2}(y_0+y_1)s+\tfrac{1}{2}(y_1+y_2)s+\dots+\tfrac{1}{2}(y_{n-2}+y_{n-1})s\\ &\qquad\qquad\qquad\qquad\qquad\qquad +\tfrac{1}{2}(y_{n-1}+y_n)s\\ &\approx \tfrac{1}{2}s\{(y_0+y_1)+(y_1+y_2)+\dots+(y_{n-2}+y_{n-1})+(y_{n-1}+y_n)\}\\ &\approx s\{\tfrac{1}{2}(y_0+y_n)+y_1+y_2+\dots+y_{n-1}\}\end{aligned}$$

Now, since $f(x)=f(x+2\pi)$, then $y_n=y_0$.

$$\therefore\ A=\int_0^{2\pi} f(x)\,\mathrm{d}x\ \dots\dots\dots$$

2

$$\int_0^{2\pi} f(x)\,dx \approx s\{y_0 + y_1 + y_2 + \dots\dots\dots + y_{n-1}\}$$

where $n =$ number of equal-width strips

$s =$ width of each strip

Note that the series within the brackets stops at y_{n-1}. We regard y_n as the first ordinate in the next cycle.

The values of y_0, y_1, y_2, ... are often available as a given table of values at regular intervals. If the function values are not given at regular intervals, we simply draw a graph of y against x and read off a fresh set of values of y at regular intervals of x and carry on from there.

Example Evaluate $\int_0^{2\pi} f(x)\,dx$ from the following set of function values

$x°$	0	30	60	90	120	150	180	210	240	270	300	330	360
$f(x)$	1.4	1.6	2.0	2.1	1.9	1.1	0.4	0.4	0.7	0.6	0.5	1.0	1.4

Width of strip $= s = \dfrac{\pi}{6}$ radians.

So in this case $\int_0^{2\pi} f(x)\,dx = \dots\dots\dots$

3

7.173

for $\int_0^{2\pi} f(x)\,dx = s(y_0 + y_1 + y_2 + \dots + y_{n-1})$

$$= \frac{\pi}{6}\sum_{r=0}^{n-1} y_r = \frac{\pi}{6}(13.7) = \underline{7.173}$$

The accuracy of the result depends on the number of strips into which the figure is divided. As usual, greater accuracy entails a larger number of thinner strips and that means more work!

Fourier series

By now we are certainly familiar with the following expressions

$$f(x) = \tfrac{1}{2}a_0 + \sum_{n=1}^{\infty}\{a_n \cos nx + b_n \sin nx\}$$

where $a_0 = \dfrac{1}{\pi}\displaystyle\int_0^{2\pi} f(x)\,\mathrm{d}x = 2 \times$ mean value of $f(x)$ over a period

$a_n = \dfrac{1}{\pi}\displaystyle\int_0^{2\pi} f(x)\cos nx\,\mathrm{d}x = 2 \times$ mean value of $f(x)\cos nx$ over a period

$b_n = \dfrac{1}{\pi}\displaystyle\int_0^{2\pi} f(x)\sin nx\,\mathrm{d}x = 2 \times$ mean value of $f(x)\sin nx$ over a period.

If $f(x)$ is given as a set of evenly spaced function values, then we can evaluate each of these integrals by finding twice times the mean value of $f(x)$, $f(x)\cos nx$, $f(x)\sin nx$, for successive values of n, over a complete cycle—and that is precisely what we have just done with the trapezoidal rule.

Let us make a new start.

4

Twelve point analysis

In practice, one complete cycle of the function is divided up into twelve equal-width strips, i.e. at intervals of $\dfrac{2\pi}{12} = \dfrac{\pi}{6} = 30°$ and the ordinates (function values) $y_0, y_1, y_2, \ldots y_{n-1}$ tabulated. Remember that the final boundary ordinate y_n is omitted since this is regarded as ....

5

the first ordinate of the next cycle

Example 1 Values of $f(x)$, a periodic function of period 2π, at intervals of 30° from $x = 0°$ to $x = 360°$ are given.

$x°$	0	30	60	90	120	150	180	210	240	270	300	330	360
y_r	y_0	y_1	y_2	y_3	y_4	y_5	y_6	y_7	y_8	y_9	y_{10}	y_{11}	y_{12}
$f(x)$	1.4	1.6	2.0	2.1	1.9	1.1	0.4	0.4	0.7	0.6	0.5	1.0	1.4

Now $a_0 = 2 \times \text{mean value of } f(x) \text{ over a cycle}$

$$= 2 \times \frac{\text{sum of the twelve ordinates}}{12}$$

$$= \frac{1}{6}(y_0 + y_1 + y_2 + \ldots + y_{11}) = \ldots\ldots\ldots$$

6

$a_0 = 2.283$

Now $a_n = 2 \times$ mean value of $f(x) \cos nx$ over a period

$\therefore\ a_1 = 2 \times$ mean value of $f(x) \cos x$ over a period

$a_2 = 2 \times$ mean value of $f(x) \cos 2x$ over a period.

First let us concentrate on finding a_1.

New frame.

7

To find a_1

$a_1 = 2 \times$ mean value of $f(x)\cos x$ over a period.

We therefore multiply each ordinate by the cosine of its corresponding angle (x) and calculate twice times the mean value of the products.

$x°$	0	30	60	90	120	150	180	210	240	270	300	330	360
y_r	y_0	y_1	y_2	y_3	y_4	y_5	y_6	y_7	y_8	y_9	y_{10}	y_{11}	–
$f(x)$	1.4	1.6	2.0	2.1	1.9	1.1	0.4	0.4	0.7	0.6	0.5	1.0	–
$\cos x$	1.0	0.866	0.5	0	−0.5	−0.866	−1.0	−0.866	−0.5	0	0.5	0.866	–
$f(x)\cos x$	1.4	1.386	1.0	0	−0.95	−0.953	−0.4	−0.346	−0.35	0	0.25	0.866	–

Then $a_1 = 2 \times \dfrac{\sum f(x)\cos x}{12} = \dfrac{1}{6}(1.903) = 0.3172 \qquad \therefore \underline{a_1 = 0.3172}$

To find a_2

$a_2 = 2 \times$ mean value of $f(x)\cos 2x$ over a period.

This means that we compile a similar table, but this time we multiply the values of $f(x)$ by the cosine of twice the relevant angle.

$x°$	0	30	60	90	120	150	180	210	240	270	300	330	360
y_r	y_0	y_1	y_2	y_3	y_4	y_5	y_6	y_7	y_8	y_9	y_{10}	y_{11}	–
$f(x)$	1.4	1.6	2.0	2.1	1.9	1.1	0.4	0.4	0.7	0.6	0.5	1.0	–
$\cos 2x$													
$f(x)\cos 2x$													

Complete the table and finish the calculation, then $a_2 = \dots\dots\dots$

8

$$a_2 = -0.2333$$

for the last line of the table reads

x	0	30	60	90	120	150	180	210	240	270	300	330
$f(x)\cos 2x$	1.4	0.8	−1.0	−2.1	−0.95	0.55	0.4	0.2	−0.35	−0.6	−0.25	0.5

Then $a_2 = 2 \times \dfrac{\sum f(x)\cos 2x}{12} = \dfrac{1}{6}(-1.40) = -0.2333 \qquad \therefore \underline{a_2 = -0.2333}$

On to the next frame.

9

To find b_1

$$b_1 = 2 \times \text{ mean value of } f(x)\sin x \text{ over a period.}$$

Here we compile a similar table, but multiply each ordinate by the sine of the corresponding angle. Then we total up the products and finish off the calculation as before, getting $b_1 = \ldots\ldots\ldots$

10

$$\boxed{b_1 = 0.748}$$

$x°$	0	30	60	90	120	150	180	210	240	270	300	330	360
$f(x)$	1.4	1.6	2.0	2.1	1.9	1.1	0.4	0.4	0.7	0.6	0.5	1.0	–
$\sin x$	0	0.5	0.866	1.0	0.866	0.5	0	−0.5	−0.866	−1.0	−0.866	−0.5	–
$f(x)\sin x$	0	0.8	1.732	2.1	1.645	0.55	0	−0.2	−0.606	−0.6	−0.433	−0.5	–

Then $b_1 = 2 \times \dfrac{\sum f(x)\sin x}{12} = \dfrac{1}{6}(4.488) = 0.748$ $\qquad \therefore \underline{b_1 = 0.748}$

To find b_2 $\qquad b_2 = 2 \times$ mean value of $f(x)\sin 2x$ over a period.

So, forming a table as before and using the sine of the double angles in the products, we finally obtain $b_2 = \dots\dots\dots$

11

$$\boxed{b_2 = 0.0288}$$

If we now collect our results together, we have

$$a_0 = 2.283; \qquad a_1 = 0.3172; \qquad b_1 = 0.748$$
$$a_2 = -0.2333; \qquad b_2 = 0.0288$$

Substituting these in the general expression for Fourier series

$$f(x) = \tfrac{1}{2}a_0 + a_1\cos x + a_2\cos 2x + \dots$$
$$+ b_1\sin x + b_2\sin 2x + \dots$$

we have $\underline{f(x) = 1.142 + 0.317\cos x - 0.233\cos 2x + \dots}$

$$\underline{+ 0.748\sin x + 0.029\sin 2x + \dots}$$

Using $a\sin\theta + b\cos\theta = c\sin(\theta + \phi)$ we can write the result in terms of compound sines, i.e. $f(x) = \dots\dots\dots$

$$\boxed{f(x) = 1.142 + 0.812\sin(x + 0.401) + 0.235\sin(2x - 1.447)}$$

since $0.748\sin x + 0.317\cos x = c_1\sin(x + \phi_1)$

where $c_1 = \sqrt{0.748^2 + 0.317^2} = \sqrt{0.66} = 0.8124$

$$\phi_1 = \arctan\frac{0.317}{0.748} = \arctan 0.4238 = 22^\circ\ 58' = 0.4008$$

$$\therefore\ 0.748\sin x + 0.317\cos x = \underline{0.812\sin(x + 0.401)}$$

Also $\quad 0.029\sin 2x - 0.233\cos 2x = c_2\sin(2x - \phi_2)$

where $c_2 = \sqrt{0.029^2 + 0.233^2} = \sqrt{0.05513} = 0.2348.$

$$\phi_2 = \arctan\frac{0.233}{0.029} = \arctan 8.0345 = 82^\circ\ 54' = 1.4469$$

$$\therefore\ 0.029\sin 2x - 0.233\cos 2x = \underline{0.235\sin(2x - 1.447).}$$

$$\underline{\therefore f(x) = 1.142 + 0.812\sin(x + 0.401) + 0.235\sin(2x - 1.447)}$$

We could, of course, continue in the same manner and obtain higher harmonics. Setting each table out in detail while explaining the method has nevertheless meant a lot of repetition and in practice we can incorporate much of the work in one re-designed table which is far more convenient.

Let us look at the same problem again.

x°	0	30	60	90	120	150	180	210	240	270	300	330	360
$f(x)$	1.4	1.6	2.0	2.1	1.9	1.1	0.4	0.4	0.7	0.6	0.5	1.0	–

$x°$	$f(x)$	$\cos x$	$y \cos x$	$\cos 2x$	$y \cos 2x$	$\sin x$	$y \sin x$	$\sin 2x$	$y \sin 2x$
0	1.4	1.0	1.4	1.0	1.4	0	0	0	0
30	1.6	0.866	1.386	0.5	0.8	0.5	0.8	0.866	1.386
60	2.0	0.5	1.0	−0.5	−1.0	0.866	1.732	0.866	1.732
90	2.1	0	0	−1.0	−2.1	1.0	2.1	0	0
120	1.9	−0.5	−0.95	−0.5	−0.95	0.866	1.645	−0.866	−1.645
150	1.1	−0.866	−0.953	0.5	0.55	0.5	0.55	−0.866	−0.953
180	0.4	−1.0	−0.4	1.0	0.4	0	0	0	0
210	0.4	−0.866	−0.346	0.5	0.2	−0.5	−0.2	0.866	0.346
240	0.7	−0.5	−0.35	−0.5	−0.35	−0.866	−0.606	0.866	0.606
270	0.6	0	0	−1.0	−0.6	−1.0	−0.6	0	0
300	0.5	0.5	0.25	−0.5	−0.25	−0.866	−0.433	−0.866	−0.433
330	1.0	0.866	0.866	0.5	0.5	−0.5	−0.5	−0.866	−0.866
$\sum y = 13.7$		$\sum y \cos x = 1.903$		$\sum y \cos 2x = -1.40$		$\sum y \sin x = 4.488$		$\sum y \sin 2x = 0.173$	
$\therefore a_0 = 2.283$		$\therefore a_1 = 0.3172$		$\therefore a_2 = -0.2333$		$\therefore b_1 = 0.748$		$\therefore b_2 = 0.0288$	

$$f(x) = 1.142 + 0.317 \cos x - 0.233 \cos 2x + \dots$$
$$+ 0.748 \sin x + 0.029 \sin 2x + \dots$$

The composite table has real advantages

(a) The whole development can be seen at a glance.
(b) It takes up less space.
(c) Continual repetition is avoided: 'copying' errors are minimised.
(d) It means less work!

Example 2 Determine the Fourier series up to and including the third harmonic to represent the periodic function $y = f(x)$ defined by the table of values given below. $y = f(x) = f(x + 2\pi)$.

$x°$	0	30	60	90	120	150	180	210	240	270	300	330	360
$f(x)$	1.8	1.7	1.5	1.0	0.6	0.4	0.5	1.0	1.6	2.0	2.1	1.9	–

The basis of the calculation is, as always,

$$f(x) = \frac{1}{2}a_0 + \sum_{t=1}^{\infty}\{a_n \cos nx + b_n \sin nx\}$$

where $a_0 = 2 \times$ mean value of $f(x)$ over a period

$a_n = 2 \times$ mean value of $f(x) \cos nx$ over a period

$b_n = 2 \times$ mean value of $f(x) \sin nx$ over a period.

First we set up a table to determine the coefficients of the cosine terms and insert the given data.

$x°$	$f(x)$	$\cos x$	$y \cos x$	$\cos 2x$	$y \cos 2x$	$\cos 3x$	$y \cos 3x$
0	1.8						
30	1.7						
60	1.5						
90	1.0						
120	0.6						
150	0.4						
180	0.5						
210	1.0						
240	1.6						
270	2.0						
300	2.1						
330	1.9						
$\sum y =$		$\sum y \cos x =$		$\sum y \cos 2x =$		$\sum y \cos 3x =$	
$a_0 =$		$a_1 =$		$a_2 =$		$a_3 =$	

Now we insert the values of $\cos x$ and multiply each by the relevant value of y, i.e. $f(x)$. Do just that and then move on to the next step.

14 So we have

$x°$	$f(x)$	$\cos x$	$y\cos x$	$\cos 2x$	$y\cos 2x$
0	1.8	1.0	1.8		
30	1.7	0.866	1.472		
60	1.5	0.5	0.75		
90	1.0	0	0		
120	0.6	−0.5	−0.3		
150	0.4	−0.866	−0.346		
180	0.5	−1.0	−0.5		
210	1.0	−0.866	−0.866		
240	1.6	−0.5	−0.8		
270	2.0	0	0		
300	2.1	0.5	1.05		
330	1.9	0.866	1.645		
$\sum y =$		$\sum y\cos x =$			
$a_0 =$		$a_1 =$			

Now we can insert the values of $\cos 2x$ and complete the sixth column with values of $y\cos 2x$.

The table then becomes ....

15

Now we have reached this stage

$x°$	$f(x)$	$\cos x$	$y\cos x$	$\cos 2x$	$y\cos 2x$	$\cos 3x$	$y\cos 3x$
0	1.8	1.0	1.8	1.0	1.8		
30	1.7	0.866	1.472	0.5	0.85		
60	1.5	0.5	0.75	−0.5	−0.75		
90	1.0	0	0	−1.0	−1.0		
120	0.6	−0.5	−0.3	−0.5	−0.3		
150	0.4	−0.866	−0.346	0.5	0.2		
180	0.5	−1.0	−0.5	1.0	0.5		
210	1.0	−0.866	−0.866	0.5	0.5		
240	1.6	−0.5	−0.8	−0.5	−0.8		
270	2.0	0	0	−1.0	−1.0		
300	2.1	0.5	1.05	−0.5	−1.05		
330	1.9	0.866	1.645	0.5	0.95		
$\sum y =$		$\sum y\cos x =$		$\sum y\cos 2x =$			
$a_0 =$		$a_1 =$		$a_2 =$			

In the same way, we now insert values of $\cos 3x$ and then values of $y\cos 3x$ in the eighth column.

So then we have ....

16 We finally arrive at

$x°$	$f(x)$	$\cos x$	$y \cos x$	$\cos 2x$	$y \cos 2x$	$\cos 3x$	$y \cos 3x$
0	1.8	1.0	1.8	1.0	1.8	1.0	1.8
30	1.7	0.866	1.472	0.5	0.85	0	0
60	1.5	0.5	0.75	−0.5	−0.75	−1.0	−1.5
90	1.0	0	0	−1.0	−1.0	0	0
120	0.6	−0.5	−0.3	−0.5	−0.3	1.0	0.6
150	0.4	−0.866	−0.346	0.5	0.2	0	0
180	0.5	−1.0	−0.5	1.0	0.5	−1.0	−0.5
210	1.0	−0.866	−0.866	0.5	0.5	0	0
240	1.6	−0.5	−0.8	−0.5	−0.8	1.0	1.6
270	2.0	0	0	−1.0	−1.0	0	0
300	2.1	0.5	1.05	−0.5	−1.05	−1.0	−2.1
330	1.9	0.866	1.645	0.5	0.95	0	0
$\sum y =$		$\sum y \cos x =$		$\sum y \cos 2x =$		$\sum y \cos 3x =$	
$a_0 =$		$a_1 =$		$a_2 =$		$a_3 =$	

All that now remains is to total up the appropriate columns and, in each case, to divide the sum by 6 (i.e. $2 \times \frac{1}{12}$) which gives the values of a_0, a_1, a_2.

The bottom of the table then becomes ....

17

$\sum y = 16.1$	$\sum y \cos x = 3.905$	$\sum y \cos 2x = -0.10$	$\sum y \cos 3x = -0.10$
$a_0 = 2.683$	$a_1 = 0.651$	$a_2 = -0.0167$	$a_3 = -0.0167$

Now we have to find the coefficients of the sine terms, i.e. b_1, b_2, b_3. So we have now to construct a similar table to the last, using $\sin x$, $\sin 2x$, $\sin 3x$, instead of the cosines.

The lay-out and method are exactly the same as for the cosine coefficients, so it is very straightforward.

Complete the table then for b_1, b_2, b_3 and then check with the results in the next frame.

$x°$	$f(x)$	$\sin x$	$y \sin x$	$\sin 2x$	$y \sin 2x$	$\sin 3x$	$y \sin 3x$
0	1.8	0	0	0	0	0	0
30	1.7	0.5	0.85	0.866	1.472	1.0	1.7
60	1.5	0.866	1.299	0.866	1.299	0	0
90	1.0	1.0	1.0	0	0	−1.0	−1.0
120	0.6	0.866	0.520	−0.866	−0.520	0	0
150	0.4	0.5	0.2	−0.866	−0.346	1.0	0.4
180	0.5	0	0	0	0	0	0
210	1.0	−0.5	−0.5	0.866	0.866	−1.0	−1.0
240	1.6	−0.866	−1.386	0.866	1.386	0	0
270	2.0	−1.0	−2.0	0	0	1.0	2.0
300	2.1	−0.866	−1.819	−0.866	−1.819	0	0
330	1.9	−0.5	−0.95	−0.866	−1.645	−1.0	−1.9
		$\sum y \sin x = -2.786$		$\sum y \sin 2x = 0.693$		$\sum y \sin 3x = 0.20$	
		$b_1 = -0.464$		$b_2 = 0.116$		$b_3 = 0.033$	

From our two tables of results then we have

$$a_0 = 2.683; \quad a_1 = 0.651; \quad a_2 = -0.0167; \quad a_3 = -0.0167$$
$$b_1 = -0.464; \quad b_2 = 0.116; \quad b_3 = 0.0333$$

$$\therefore f(x) = 1.342 + 0.651 \cos x - 0.017 \cos 2x - 0.017 \cos 3x + \ldots$$

$$- 0.464 \sin x + 0.116 \sin 2x + 0.033 \sin 3x + \ldots$$

In practice, we go one step further and compile both sets of calculations in one ‘broadsheet’ with the cosine and sine tables side by side. It is then necessary for the values of x and $f(x)$ to appear only once on the left-hand side.

Example 3 A function $f(x)$ has a period 2π and function values over one cycle as shown in the table.

$x°$	0	30	60	90	120	150	180	210	240	270	300	330	360
$f(x)$	0.5	0.8	1.4	2.0	1.9	1.4	1.2	1.4	1.1	0.5	0.3	0.4	–

Determine the Fourier series up to and including the third harmonic.

So build up the relevant double table and obtain the series. *Finish it completely before checking the results with the next frame.*

$x°$	$f(x)$	$\cos x$	$y\cos x$	$\cos 2x$	$y\cos 2x$	$\cos 3x$	$y\cos 3x$	$\sin x$	$y\sin x$	$\sin 2x$	$y\sin 2x$	$\sin 3x$	$y\sin 3x$
0	0.5	1.0	0.5	1.0	0.5	1.0	0.5	0	0	0	0	0	0
30	0.8	0.866	0.693	0.5	0.4	0	0	0.5	0.4	0.866	0.693	1.0	0.8
60	1.4	0.5	0.7	−0.5	−0.7	−1.0	−1.4	0.866	1.212	0.866	1.212	0	0
90	2.0	0	0	−1.0	−2.0	0	0	1.0	2.0	0	0	−1.0	−2.0
120	1.9	−0.5	−0.95	−0.5	−0.95	1.0	1.9	0.866	1.645	−0.866	−1.645	0	0
150	1.4	−0.866	−1.212	0.5	0.7	0	0	0.5	0.7	−0.866	−1.212	1.0	1.4
180	1.2	−1.0	−1.2	1.0	1.2	−1.0	−1.2	0	0	0	0	0	0
210	1.4	−0.866	−1.212	0.5	0.7	0	0	−0.5	−0.7	0.866	1.212	−1.0	−1.4
240	1.1	−0.5	−0.55	−0.5	−0.55	1.0	1.1	−0.866	−0.953	0.866	0.953	0	0
270	0.5	0	0	−1.0	−0.5	0	0	−1.0	−0.5	0	0	1.0	0.5
300	0.3	0.5	0.15	−0.5	−0.15	−1.0	−0.3	−0.866	−0.260	−0.866	−0.260	0	0
330	0.4	0.866	0.346	0.5	0.2	0	0	−0.5	−0.2	−0.866	−0.346	−1.0	−0.4
$\sum y = 12.9$		$\sum y\cos x = -2.735$		$\sum y\cos 2x = -1.15$		$\sum y\cos 3x = 0.60$		$\sum y\sin x = 3.344$		$\sum y\sin 2x = 0.607$		$\sum y\sin 3x = -1.10$	
$a_0 = 2.15$		$a_1 = -0.456$		$a_2 = -0.192$		$a_3 = 0.10$		$b_1 = 0.557$		$b_2 = 0.101$		$b_3 = -0.183$	

$$\therefore \underline{f(x) = 1.08 - 0.456\cos x - 0.192\cos 2x + 0.100\cos 3x + \ldots}$$

$$\underline{+\, 0.557\sin x + 0.101\sin 2x - 0.183\sin 3x + \ldots}$$

20

If we combine corresponding sine and cosine terms, we can express the series as compound sine terms, the series then being written in the form

$$f(x) = \dots\dots\dots$$

21

$$f(x) = 1.08 + 0.720\sin(x - 0.686) + 0.217\sin(2x - 1.086)$$
$$-0.209\sin(3x - 0.500) + \dots$$

From this result, we can determine that the percentage second harmonic is ………

22

30.1%

$$\text{for second harmonic} = \frac{0.217}{0.720} \times 100\% = \underline{30.1\%}$$

If the period of the given function is other than 2π, we convert the period into a value of 2π by changing the units of the independent variable.

Example 4 A current waveform $i = f(t)$ amperes is given by the following set of values over a period of 6 ms.

t (ms)	0	0.5	1.0	1.5	2.0	2.5	3.0	3.5	4.0	4.5	5.0	5.5	6.0
i (A)	7	13	17	13	5	6	11	17	21	15	5	3	–

We are required to find (a) the Fourier series for $f(t)$
(b) the percentage third harmonic.

For the current waveform $i = f(t) = f(t + 6)$, i.e. period $= T = 6$ ms.

If we put $x = \dfrac{2\pi t}{T} = \dfrac{2\pi t}{6} = \dfrac{\pi t}{3}$ i.e. $x = \dfrac{\pi t}{3}$

then when $t = 0$, $x = 0$ and when $t = 6$, $x = 2\pi$.

Now we can re-write the table, changing the values of t into their corresponding values of x. The values of i, of course, remain unaltered.

The table then appears as ....

23

t (ms)	0	0.5	1.0	1.5	2.0	2.5	3.0	3.5	4.0	4.5	5.0	5.5	6.0
x	0	$\frac{\pi}{6}$	$\frac{\pi}{3}$	$\frac{\pi}{2}$	$\frac{2\pi}{3}$	$\frac{5\pi}{6}$	π	$\frac{7\pi}{6}$	$\frac{4\pi}{3}$	$\frac{3\pi}{2}$	$\frac{5\pi}{3}$	$\frac{11\pi}{6}$	2π
	0	30	60	90	120	150	180	210	240	270	300	330	360°
i (A)	7	13	17	13	5	6	11	17	21	15	5	3	–

Before we set forth on the calculations, it might be of interest to draw the waveform with scales of both t and x on the x-axis.

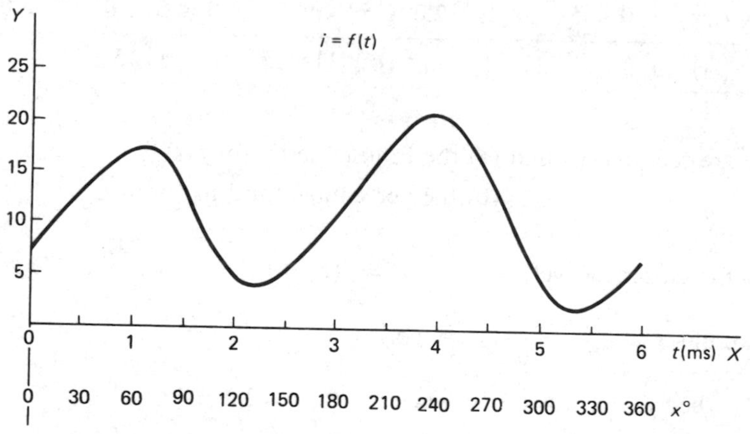

Now we proceed as in the previous example, this time using values of i as y and so obtain the series for $y = F(x)$. Substitution of the original symbols and units then gives the required series for $i = f(t)$.

There is nothing special to say about the waveform, so the series will contain both sine and cosine terms. This will require a double table as we had in the last example, so allow plenty of room. When you have obtained the series, check the results and the working with the table given in full in the next frame.

x°	y	$\cos x$	$y\cos x$	$\cos 2x$	$y\cos 2x$	$\cos 3x$	$y\cos 3x$	$\sin x$	$y\sin x$	$\sin 2x$	$y\sin 2x$	$\sin 3x$	$y\sin 3x$
0	7	1.0	7.0	1.0	7.0	1.0	7.0	0	0	0	0	0	0
30	13	0.866	11.258	0.5	6.5	0	0	0.5	6.5	0.866	11.258	1.0	13
60	17	0.5	8.5	−0.5	−8.5	−1.0	−17	0.866	14.722	0.866	14.722	0	0
90	13	0	0	−1.0	−13	0	0	1.0	13	0	0	−1.0	−13
120	5	−0.5	−2.5	−0.5	−2.5	1.0	5.0	0.866	4.33	−0.866	−4.33	0	0
150	6	−0.866	−5.196	0.5	3.0	0	0	0.5	3.0	−0.866	−5.196	1.0	6
180	11	−1.0	−11.0	1.0	11	−1.0	−11.0	0	0	0	0	0	0
210	17	−0.866	−14.722	0.5	8.5	0	0	−0.5	−8.5	0.866	14.722	−1.0	−17
240	21	−0.5	−10.5	−0.5	−10.5	1.0	21.0	−0.866	−18.186	0.866	18.186	0	0
270	15	0	0	−1.0	−15.0	0	0	−1.0	−15.0	0	0	1.0	15
300	5	0.5	2.5	−0.5	−2.5	−1.0	−5.0	−0.866	−4.33	−0.866	−4.33	0	0
330	3	0.866	2.598	0.5	1.5	0	0	−0.5	−1.5	−0.866	−2.598	−1.0	−3
$\sum y = 133$		$\sum y\cos x = -12.062$		$\sum y\cos 2x = -14.5$		$\sum y\cos 3x = 0$		$\sum y\sin x = -5.964$		$\sum y\sin 2x = 42.434$		$\sum y\sin 3x = 1.0$	
$a_0 = 22.167$		$a_1 = -2.010$		$a_2 = -2.417$		$a_3 = 0$		$b_1 = -0.994$		$b_2 = 7.072$		$b_3 = 0.167$	

$$\therefore\ F(x) = 11.08 - 2.010\cos x - 2.417\cos 2x + 0\cos 3x + \dots$$

$$-0.994\sin x + 7.072\sin 2x + 0.167\sin 3x + \dots$$

25

From the table, we have

$$F(x) = 11.08 - 2.010\cos x - 2.417\cos 2x + 0\cos 3x + \dots$$
$$- 0.994\sin x + 7.072\sin 2x + 0.167\sin 3x + \dots$$

But $x = \dfrac{\pi t}{3} = \omega t$ where $\omega = \dfrac{\pi}{3}$ $\quad \therefore x = \omega t.$

$$\therefore \underline{i = f(t) = 11.08 - 2.010\cos\omega t - 2.417\cos 2\omega t + 0\cos 3\omega t + \dots}$$
$$\underline{- 0.994\sin\omega t + 7.072\sin 2\omega t + 0.167\sin 3\omega t + \dots}$$
$$\text{where } \omega = \frac{\pi}{3}.$$

We were also required in the question to determine the percentage third harmonic. This is ....

26

7.45%

for the amplitude of the fundamental $= \sqrt{2.010^2 + 0.994^2} = 2.242$
and the amplitude of the third harmonic $= 0.167$.

$$\therefore \text{ percentage third harmonic} = \frac{0.167}{2.242} \times 100\% = \underline{7.45\%}$$

Of course, we can still draw on our previous knowledge of Fourier series as the occasion demands. For instance, we remember that

(a) if $f(x) = f(x + \pi)$, the series contains only .... harmonics

(b) if $f(x) = -f(x + \pi)$, the series contains only .... harmonics.

27

$f(x) = f(x + \pi)$, even harmonics; $f(x) = -f(x + \pi)$, odd harmonics

Example 5 A periodic function $y = f(x)$, of period 2π, is defined between $x = 0$ and $x = \pi$ by the function values given in the table. If the function is known to contain odd harmonics only, determine the Fourier series up to and including the third harmonic and obtain the percentage third harmonic.

$x°$	0	30	60	90	120	150	180
y	0	8.0	11.5	6.0	4.0	5.4	0

The complete table from $x = 0$ to $x = 2\pi$ can be written as

....

28

$x°$	0	30	60	90	120	150	180	210	240	270	300	330	360
y	0	8.0	11.5	6.0	4.0	5.4	0	−8.0	−11.5	−6.0	−4.0	−5.4	–

We are told there are only odd harmonics. $\therefore f(x) = -f(x + \pi)$. So from $x = 180°$ onwards the values of y repeat with the signs changed.

Now we can proceed as on previous occasions. Total up the usual products and determine the series.

$f(x) = \ldots\ldots\ldots$

29

$$y = f(x) = 2.0\cos x - 2.5\cos 3x + \ldots$$
$$+ 8.71\sin x + 2.47\sin 3x + \ldots$$

To determine the percentage third harmonic, we need the amplitudes of the fundamental and the third harmonic.

$$A_1 = \ldots\ldots\ldots; \qquad A_3 = \ldots\ldots\ldots$$

30

$$A_1 = 8.937; \quad A_3 = 3.514$$

$\therefore$ percentage third harmonic $= \ldots\ldots\ldots$

31

39.3%

i.e. Percentage harmonic $= \dfrac{3.514}{8.937} \times 100\% = \underline{39.3\%}$

That brings this programme to an end. The Revision Summary now follows; then the Test Exercise as usual. All the problems are done in much the same way: careful lay-out of the tables helps to avoid numerical slips.

REVISION SUMMARY

1. *Trapezoidal rule for approximate integration*

$$\int_0^{2\pi} f(x)\,dx = s\{y_0 + y_1 + y_2 + \ldots + y_{n-1}\}$$

where $s =$ width of each strip

$y_0, y_1, y_2 \ldots$ equally spaced ordinates.

2. *Fourier series*

$$f(x) = \frac{1}{2}a_0 + \sum_{n=1}^{\infty}\{a_n \cos nx + b_n \sin nx\}$$

$a_0 = 2 \times$ mean value of $f(x)$ over a period

$a_n = 2 \times$ mean value of $f(x) \cos nx$ over a period

$b_n = 2 \times$ mean value of $f(x) \sin nx$ over a period.

3. *Twelve point analysis*

(a) *Functions of period* 2π

Divide periodic interval into 12 equal parts.

Ordinates $y_0, y_1, y_2, \ldots y_{11}$ at intervals of $\dfrac{2\pi}{12} = \dfrac{\pi}{6} = 30°$.

$$a_0 = 2 \times \frac{1}{12}\sum_{r=0}^{11} y_r \qquad = \frac{1}{6}\sum_{r=0}^{11} y_r$$

$$a_n = 2 \times \frac{1}{12}\sum_{r=0}^{11} y_r \cos nx = \frac{1}{6}\sum_{r=0}^{11} y_r \cos nx$$

$$b_n = 2 \times \frac{1}{12}\sum_{r=0}^{11} y_r \sin nx = \frac{1}{6}\sum_{r=0}^{11} y_r \sin nx.$$

Expressed as compound sine terms

$$b_n \sin nx + a_n \cos nx = A_n \sin(nx + \phi_n)$$

where $A_n = \sqrt{a_n^2 + b_n^2}$; $\quad \phi_n = \arctan\left(\dfrac{a_n}{b_n}\right)$.

(b) *Functions of period other than* 2π e.g. $y = f(t)$ with period T.

Convert to period of 2π, by change of units.

Put $x = \dfrac{2\pi t}{T}$. $\therefore$ when $t = 0$, $x = 0$; when $t = T$, $x = 2\pi$.

4. *Percentage nth harmonic* $= \dfrac{A_n}{A_1} \times 100\%$

where A_n = amplitude of nth harmonic.

A_1 = amplitude of the fundamental.

TEST EXERCISE XVIII

1. A periodic function $y = f(x)$, of period 2π, is defined between $x = 0°$ and $x = 360°$ by the following table of values.

$x°$	0	30	60	90	120	150	180	210	240	270	300	330	360
y	3.0	4.0	4.6	4.8	3.6	2.8	2.2	1.0	0.6	1.0	1.6	2.0	–

Determine the Fourier series up to and including the third harmonic.

2. An alternating current, $i = f(t)$ amperes, has a period of 24 ms and a waveform defined by the following table of values.

t (ms)	0	2	4	6	8	10	12	14	16	18	20	22	24
i (A)	6.5	9.0	9.0	5.5	7.5	4.3	1.5	3.2	6.2	5.8	3.5	4.2	–

Determine the Fourier series up to and including the third harmonic.

3. A periodic function $y = f(x)$ of period 2π is known to have odd harmonics only. Values of y from $x = 0$ to $x = \pi$ at intervals of $\frac{\pi}{6}$ are given.

$x°$	0	30	60	90	120	150	180
y	0	9	8	10	17	14	0

Determine (a) the Fourier series up to and including the third harmonic

(b) the percentage third harmonic.

FURTHER PROBLEMS XVIII

Note: In problems 1 to 6 inclusive, the values tabled relate to periodic functions of period 2π. In each case, determine the Fourier series up to and including the third harmonic.

1.

$x°$	0	30	60	90	120	150	180	210	240	270	300	330	360
y	3.2	4.3	4.9	4.7	3.6	2.5	2.7	3.8	3.9	2.3	1.7	2.1	–

2.

$x°$	0	30	60	90	120	150	180	210	240	270	300	330	360
y	5.6	3.2	3.2	4.6	6.6	7.6	7.6	8.0	9.0	9.6	9.4	8.2	–

3.

$x°$	0	30	60	90	120	150	180	210	240	270	300	330	360
y	6.8	7.8	7.8	6.4	6.8	6.8	4.0	2.2	2.8	4.2	4.4	5.2	–

4.

$x°$	0	30	60	90	120	150	180	210	240	270	300	330	360
y	4.0	1.0	−1.0	−1.7	−1.5	0	1.7	3.2	4.0	4.2	5.3	5.5	–

5.

$x°$	0	30	60	90	120	150	180	210	240	270	300	330	360
y	6.0	8.0	8.3	6.5	6.0	7.5	8.3	7.5	5.0	2.3	2.0	3.3	–

6.

$x°$	0	30	60	90	120	150	180	210	240	270	300	330	360
y	1.5	2.2	4.3	7.0	9.0	9.0	5.2	4.5	5.7	6.0	4.5	2.2	–

7. The vertical movement y of a cam follower in one revolution of the shaft is recorded at angular intervals of $\frac{\pi}{6}$.

$x°$	0	30	60	90	120	150	180	210	240	270	300	330	360
y	1.4	1.3	0.9	0.4	0.5	0.9	1.5	1.9	1.9	1.6	1.3	1.3	–

Determine the Fourier series to represent the movement up to and including the third harmonic.

8. The waveform of a periodic function $y = f(x)$, from $x = 0$ to $x = 2\pi$, is defined as follows.

$x°$	0	30	60	90	120	150	180	210	240	270	300	330	360
y	3.0	4.0	7.0	11.0	14.0	13.0	10.0	13.0	15.0	9.0	5.0	3.5	–

If $y = f(x) = f(x + 2\pi)$, determine

(a) the Fourier series to represent $f(x)$ up to and including the third harmonic

(b) the percentage second harmonic.

9. A function $y = f(u)$ where $f(u) = f(u + 60)$ is defined over the range $u = 0$ to $u = 60$ by the following table of function values.

u	0	5	10	15	20	25	30	35	40	45	50	55	60
y	14.0	6.5	4.0	7.6	13.0	14.4	10.5	9.0	10.6	16.5	19.6	19.6	–

Determine (a) the Fourier series up to and including the third harmonic

(b) the percentage third harmonic.

10. A function $y = f(u)$ where $f(u) = f(u + 12)$ has the following function values from $u = 0$ to $u = 12$.

u	0	1	2	3	4	5	6	7	8	9	10	11	12
y	6.0	5.5	3.6	1.7	2.6	5.5	8.5	9.2	6.5	4.0	4.0	5.5	–

Determine the Fourier series to represent $f(u)$ involving terms up to and including the third harmonic.

11. A current waveform $i = f(t)$ is defined by the following set of values, where i is in milliamperes and t is in milliseconds. $f(t) = f(t + 24)$.

t (ms)	0	2	4	6	8	10	12	14	16	18	20	22	24
i (mA)	6.0	7.1	9.0	8.7	7.5	5.8	5.0	4.3	2.5	1.5	1.7	3.6	–

Determine the Fourier series representing i up to and including the third harmonic.

12. A periodic function $y = f(x)$ of period 2π is known to have even harmonics only. Values of y for values of x at intervals of $\frac{\pi}{6}$ between $x = 0$ and $x = \pi$ are given.

$x°$	0	30	60	90	120	150	180
y	4.0	5.4	9.5	10.0	5.0	3.0	4.0

Determine the amplitude A_2 and the phase angle ϕ_2 of the second harmonic $A_2 \sin(2x + \phi_2)$.

13. A periodic function $y = f(x)$, of period 2π, is known to have odd harmonics only. Values of $f(x)$ from $x = 0$ to $x = \pi$ are given.

$x°$	0	30	60	90	120	150	180
y	0	7.0	11.0	10.0	6.0	5.0	0

Determine the Fourier series for the function and express the result in compound sine terms.

Programme 19

Partial Differential Equations

Prerequisites: Engineering Mathematics (fourth edition)
Programmes 24, 25

Introduction

1 The formation of ordinary linear differential equations and their solution by various methods were covered in some detail in Programme 24, 25, 26 of the previous year's work as presented in *Engineering Mathematics* (fourth edition) and reference to these sections before undertaking the new work of this programme could be beneficial—especially Programme 25 which dealt with second-order equations. Working through the Test Exercise of that programme would provide worthwhile revision.

The main results obtained are listed here for convenience and easy reference.

1. Equations of the form $a\dfrac{d^2y}{dx^2}+b\dfrac{dy}{dx}+cy=0$

Auxiliary equation $am^2+bm+c=0$. Solutions depend on the roots of this equation.

(a) Real and different roots: $m=m_1$ and $m=m_2$

$$\text{Solution } y=Ae^{m_1x}+Be^{m_2x} \qquad (1)$$

(b) Real and equal roots: $m=m_1$ (twice)

$$\text{Solution } y=e^{m_1x}(A+Bx) \qquad (2)$$

(c) Complex roots: $m=\alpha\pm j\beta$

$$\text{Solution } y=e^{\alpha x}(A\cos\beta x+B\sin\beta x) \qquad (3)$$

2. Equations of the form $\dfrac{d^2y}{dx^2}\pm n^2y=0$

If we take the general equation $a\dfrac{d^2y}{dx^2}+b\dfrac{dy}{dx}+cy=0$ and consider the case when $b=0$, then dividing through by a, we have $\dfrac{d^2y}{dx^2}+\dfrac{c}{a}y=0$ which we write as $\dfrac{d^2y}{dx^2}\pm n^2y=0$ to cover separately the two cases when $\dfrac{c}{a}$ is positive or $\dfrac{c}{a}$ is negative.

(a) $\dfrac{d^2y}{dx^2}+n^2y=0 \quad \therefore\ m^2+n^2=0 \quad \therefore\ m^2=-n^2 \quad \therefore\ m=\pm jn$

$$\text{Solution } y = A\cos nx + B\sin nx \tag{4}$$

(b) $\dfrac{d^2y}{dx^2}-n^2y=0 \quad \therefore\ m^2-n^2=0 \quad \therefore\ m^2=n^2 \quad \therefore\ m=\pm n$

$$\left.\begin{aligned}\text{Solution } y &= A\cosh nx + B\sinh nx\\ \text{or } y &= Ae^{nx}+Be^{-nx}\\ \text{or } y &= A\sinh n(x+\phi)\end{aligned}\right\} \tag{5}$$

In each case, A and B are arbitrary constants depending on the initial conditions.

2

Partial differential equations

A partial differential equation is a relationship between a dependent variable u and two or more independent variables $(x, y, t, \ldots)$ and partial differential coefficients of u with respect to these independent variables. The solution is therefore of the form $u = f(x, y, t, \ldots)$.

Solution by direct integration

The simplest form of partial differential equation is such that a solution can be determined by direct partial integration.

Example 1 Solve the equation $\dfrac{\partial^2 u}{\partial x^2} = 12x^2(t+1)$ given that at $x = 0$, $u = \cos 2t$ and $\dfrac{\partial u}{\partial x} = \sin t$.

$\dfrac{\partial^2 u}{\partial x^2} = 12x^2(t+1)$. Integrating partially with respect to x, we have

$\dfrac{\partial u}{\partial x} = 4x^3(t+1)+\phi(t)$ where the arbitrary function $\phi(t)$ takes the place of the normal arbitrary constant in ordinary integration. Integrating partially again with respect to x gives

$$u = \ldots\ldots\ldots$$

3

$$u = x^4(t+1) + x\phi(t) + \theta(t)$$

where $\theta(t)$ is a second arbitrary function.

To find the two arbitrary functions $\phi(t)$ and $\theta(t)$, we apply the given initial conditions that at $x = 0$, $\dfrac{\partial u}{\partial x} = \sin t$ and $u = \cos\, 2t$.
Substituting these in the relevant equations gives

$$\phi(t) = \dots\dots\dots; \quad \theta(t) = \dots\dots\dots$$

4

$$\phi(t) = \sin t; \quad \theta(t) = \cos 2t$$

$$\therefore\ u = x^4(t+1) + x\sin t + \cos 2t$$

Example 2 Solve the equation $\dfrac{\partial^2 u}{\partial x \partial y} = \sin(x+y)$, given that at $y = 0$, $\dfrac{\partial u}{\partial x} = 1$ and at $x = 0$, $u = (y-1)^2$.

In just the same way as before $u = \dots\dots\dots$

5

$$u = -\sin(x+y) + x + \sin x + (y-1)^2$$

for $\dfrac{\partial^2 u}{\partial x \partial y} = \sin(x+y) \quad \therefore\ \dfrac{\partial u}{\partial x} = -\cos(x+y) + \phi(x)$.

At $y = 0$, $\dfrac{\partial u}{\partial x} = 1 \quad \therefore\ 1 = -\cos x + \phi(x) \quad \therefore\ \phi(x) = 1 + \cos x$

$$\therefore\ \frac{\partial u}{\partial x} = -\cos(x+y) + 1 + \cos x$$

Integrating again partially, this time with respect to x, we have

$$u = -\sin(x+y) + x + \sin x + \theta(y)$$

But at $x = 0$, $u = (y-1)^2$. $\therefore\ (y-1)^2 = -\sin y + \theta(y)$

$$\therefore\ \theta(y) = (y-1)^2 + \sin y$$

$$\therefore\ u = -\sin(x+y) + x + \sin x + \sin y + (y-1)^2$$

Initial conditions and boundary conditions

As with any differential equation, the arbitrary constants or arbitrary functions in any particular case are determined from the additional information given concerning the variables of the equation. These extra facts are called the *initial conditions* or, more generally, the *boundary conditions* since they do not always refer to zero values of the independent variables.

Example 3 Solve the equation $\dfrac{\partial^2 u}{\partial x \partial y} = \sin x \cos y$, subject to the boundary conditions that at $y = \dfrac{\pi}{2}$, $\dfrac{\partial u}{\partial x} = 2x$ and at $x = \pi$, $u = 2\sin y$.

Work through it: it is easy enough. $u = \dots\dots\dots$

6

$$u = x^2 + \cos x(1 - \sin y) + \sin y + 1 - \pi^2$$

for $\dfrac{\partial^2 u}{\partial x \partial y} = \sin x \cos y \qquad \therefore \dfrac{\partial u}{\partial x} = \sin x \sin y + \phi(x)$

But $\dfrac{\partial u}{\partial x} = 2x$ at $y = \dfrac{\pi}{2} \qquad \therefore \phi(x) = 2x - \sin x$

$\therefore \dfrac{\partial u}{\partial x} = 2x - \sin x(1 - \sin y) \quad \therefore u = x^2 + \cos x(1 - \sin y) + \theta(y)$

But $u = 2\sin y$ at $x = \pi$. $\therefore \theta(y) = 1 - \pi^2 + \sin y$

$$\underline{u = x^2 + \cos x(1 - \sin y) + \sin y + 1 - \pi^2}$$

On to the next frame.

7 Before we take a closer look at some of the more important partial differential equations occurring in branches of technology, let us recall the fact that if $u = u_1$, $u = u_2$, $u = u_3$, ... are different solutions of a linear partial differential equation, so also is

$$u = c_1 u_1 + c_2 u_2 + c_3 u_3 + \ldots$$

where c_1, c_2, c_3, ... are arbitrary constants.

There are many types of partial differential equations, some requiring special treatment in their solution. In this programme we are concerned with a restricted number of such equations that occur in branches of science and technology, which can be solved by the method of separating the variables, and which also link up with the work we have done on Fourier series techniques.

Let us make a new start.

8

The wave equation

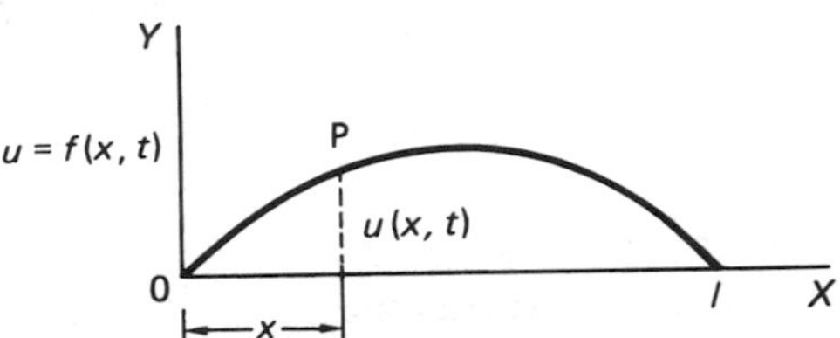

If we consider a perfectly flexible elastic string stretched between two points at $x = 0$ and $x = l$ with uniform tension T and the string is displaced slightly from its initial position of rest and released, with the end points remaining fixed, then the string will vibrate. The position of any point P in the string will then depend on its distance from one end and on the instant in time. Its displacement u at any time t can thus be expressed as $u = f(x, t)$ where x is its distance from the left-hand end.

The equation of motion is given by $\dfrac{\partial^2 u}{\partial x^2} = \dfrac{1}{c^2} \cdot \dfrac{\partial^2 u}{\partial t^2}$, where $c^2 = \dfrac{T}{\rho}$ in which T is the tension in the string and ρ the mass per unit length of the string. The displacement of the string is regarded as small so that T and ρ remain constant.

Now let us deal with the solution of this equation.

On to the next frame.

9

Solution of the wave equation

The new equation $\dfrac{\partial^2 u}{\partial x^2} = \dfrac{1}{c^2} \cdot \dfrac{\partial^2 u}{\partial t^2}$ has a solution $u = f(x, t)$ written $u(x, t)$.

Boundary conditions:

(a) The string is fixed at both ends, i.e. at $x = 0$ and at $x = l$ for all values of time t. Therefore $u\ (x, t)$ becomes

$$\left.\begin{aligned} u(0, t) &= 0 \\ u(l, t) &= 0 \end{aligned}\right\} \quad \text{for all values of } t \geq 0$$

Initial conditions:

(b) If the initial deflection of P at $t = 0$ is denoted by $f(x)$, then

$$u(x, 0) = f(x)$$

(c) Let the initial velocity of P be $g(x)$, then

$$\left[\frac{\partial u}{\partial t}\right]_{t=0} = g(x)$$

So now we have listed all the information available from the question. Next we turn to solving the equation.

Solution by separating the variables

We assume a trial solution of the form $u(x, t) = X(x)T(t)$ where

$X(x)$ is a function of x only

$T(t)$ is a function of t only.

If we simplify the symbols to $u = XT$ and denote differential coefficients with respect to their own independent variables by primes, we have

$$u = XT \qquad \therefore \frac{\partial u}{\partial x} = X'T \quad \text{and} \quad \frac{\partial^2 u}{\partial x^2} = X''T$$

$$\frac{\partial u}{\partial t} = XT' \quad \text{and} \quad \frac{\partial^2 u}{\partial t^2} = XT''$$

The wave equation $\dfrac{\partial^2 u}{\partial x^2} = \dfrac{1}{c^2} \cdot \dfrac{\partial^2 u}{\partial t^2}$ can then be written as

....

$$X''T = \frac{1}{c^2}XT''$$

and this can be transposed into $\dfrac{X''}{X} = \dfrac{1}{c^2} \cdot \dfrac{T''}{T}$

Notice that the left-hand side expression involves functions of x only and that the right-hand side expression involves functions of t only. Therefore, if these two expressions are to be equal for all values of the separate variables, then both expressions must be equal to

. . . .

11

a constant

Denote this arbitrary constant by k. Then we have

$$\frac{X''}{X} = k \quad \text{and} \quad \frac{1}{c^2} \cdot \frac{T''}{T} = k$$

$$\therefore X'' - kX = 0 \quad \text{and} \quad T'' - c^2 kT = 0$$

Let us consider the first of these two equations for different values of k.

(i) *If* $k = 0$, $\quad X'' = 0 \quad \therefore X' = a \quad \therefore X = ax + b.$

$$\left.\begin{array}{lll} \text{But } X = 0 \text{ at } x = 0 & \therefore b = 0 & \therefore X = ax \\ \text{and } X = 0 \text{ at } x = 1 & \therefore a = 0 & \end{array}\right\} \therefore a = b = 0$$

$\therefore X = 0$ which is not oscillatory as the problem requires.

(ii) *If k is positive*, let $k = p^2 \quad \therefore X'' - p^2 X = 0.$

The auxiliary equation is therefore $m^2 - p^2 = 0 \quad \therefore m^2 = p^2$

$$m = \pm p$$

$$\therefore X = Ae^{px} + Be^{-px}$$

But $X = 0$ at $x = 0 \quad \therefore 0 = A + B \quad \therefore B = -A$

and $X = 0$ at $x = l \quad \therefore 0 = Ae^{pl} - Ae^{-pl} \quad \therefore 0 = A(e^{pl} - e^{-pl})$

$\therefore A = 0 \qquad \therefore A = B = 0$

Here again $X = 0$ which is not oscillatory.

(iii) *If k is negative*, let $k = -p^2 \quad \therefore X'' + p^2 X = 0.$

This is one of the standard equations listed at the beginning of the programme and gives a solution

$$\underline{X = A\cos px + B\sin px} \qquad (1)$$

which fits the requirements.

The second equation $T'' - c^2 kT = 0$ therefore now becomes

....

$$\boxed{T'' + c^2p^2T = 0}$$

since the same value for k must apply. This equation is of the same form as before and gives the solution

$$\underline{T = C\cos cpt + D\sin cpt} \tag{2}$$

So our suggested solution $u = XT$ now becomes

$$u(x, t) = (A\cos px + B\sin px)\,(C\cos cpt + D\sin cpt)$$

and, if we put $cp = \lambda \qquad \therefore p = \dfrac{\lambda}{c}$, this becomes

$$\underline{u(x, t) = \left(A\cos\frac{\lambda}{c}x + B\sin\frac{\lambda}{c}x\right)(C\cos\lambda t + D\sin\lambda t)} \tag{3}$$

where A, B, C, D are arbitrary constants.

The result, of course, must satisfy the set of boundary conditions which we now turn to.

(a) $u = 0$ when $x = 0$ for all values of t. From this, we get

. . . .

13

$$\boxed{A = 0}$$

for substituting $u = 0$ and $x = 0$ in result (3) above

$0 = A(C\cos\lambda t + D\sin\lambda t)$for all t $\quad\therefore\ \underline{A = 0}$

$$\therefore\ u(x, t) = B\sin\frac{\lambda}{c}x(C\cos\lambda t + D\sin\lambda t)$$

(b) $u = 0$ when $x = l$ for all t $\quad\therefore\ 0 = B\ \sin\dfrac{\lambda l}{c}(C\cos\lambda t + D\sin\lambda t)$

Now $B \neq 0$ or $u(x, t)$ would be identically zero. $\therefore\ \sin\dfrac{\lambda l}{c} = 0.$

$$\therefore\ \frac{\lambda l}{c} = n\pi \text{ where } n = 1,\ 2,\ 3,\ \ldots\quad \therefore\ \lambda = \frac{nc\pi}{l} \text{ for } n = 1,\ 2,\ 3,\ \ldots$$

Note that we exclude $n = 0$ since this would also make $u(x,\ t)$ identically zero.

As we can see, there is an infinite set of values of λ and each separate value gives a particular solution for $u(x, t)$. The values of λ are called the *eigenvalues* and each corresponding solution the *eigenfunction*.

Putting $n = 1,\ 2,\ 3,\ \ldots$ we therefore have

	Eigenvalues	Eigenfunctions
n	$\lambda = \dfrac{nc\pi}{l}$	$u(x, t) = B\ \sin\dfrac{\lambda x}{c}\ \{C\cos\lambda t + D\sin\lambda t\}$
1	$\lambda_1 = \dfrac{c\pi}{l}$	$u_1 = \sin\dfrac{\pi x}{l}\left\{C_1\cos\dfrac{c\pi t}{l} + D_1\sin\dfrac{c\pi t}{l}\right\}$
2	$\lambda_2 = \dfrac{2c\pi}{l}$	$u_2 = \sin\dfrac{2\pi x}{l}\left\{C_2\cos\dfrac{2c\pi t}{l} + D_2\sin\dfrac{2c\pi t}{l}\right\}$
3	$\lambda_3 = \dfrac{3c\pi}{l}$	$u_3 = \sin\dfrac{3\pi x}{l}\left\{C_3\cos\dfrac{3c\pi t}{l} + D_3\sin\dfrac{3c\pi t}{l}\right\}$
$\vdots$	$\vdots$	$\vdots$
r	$\lambda_r = \dfrac{rc\pi}{l}$	$u_r = \sin\dfrac{r\pi x}{l}\left\{C_r\cos\dfrac{rc\pi t}{l} + D_r\sin\dfrac{rc\pi t}{l}\right\}$

where $C_1,\ C_2,\ C_3,\ \ldots$ and $D_1,\ D_2,\ D_3,\ \ldots$ are arbitrary constants.

Since the original wave equation is linear in form, we have already noted that if $u = u_1,\ u = u_2,\ u = u_3 \ldots$ are particular solutions, a more general solution is ....

$$u = u_1 + u_2 + u_3 + \ldots$$

The more general solution is therefore

$$u(x,t) = \sum_{r=1}^{\infty} u_r = \sum_{r=1}^{\infty}\left\{\sin\frac{r\pi x}{l}\left(C_r\cos\frac{rc\pi t}{l} + D_r\sin\frac{rc\pi t}{l}\right)\right\} \tag{4}$$

We still have to find C_r and D_r and for this we use the initial conditions which we have not yet taken into account.

(c) At $t = 0$, $u(x,\ 0) = f(x)$ for $0 \leq x \leq l$

Therefore from (4), $u(x,\ 0) = f(x) = \sum\limits_{r=1}^{\infty} C_r\ \sin\dfrac{r\pi x}{l}.$

(d) Also at $t = 0$, $\left[\dfrac{\partial u}{\partial t}\right]_{t=0} = g(x)$ for $0 \leq x \leq l$

We therefore differentiate (4) with respect to t and put $t = 0$, which gives

....

15

$$g(x) = \frac{c\pi}{l}\sum_{r=1}^{\infty} D_r\, r \sin\frac{r\pi x}{l}$$

for $\dfrac{\partial u}{\partial t} = \displaystyle\sum_{r=1}^{\infty} \sin\frac{r\pi x}{l}\left\{-C_r\frac{rc\pi}{l}\sin\frac{rc\pi t}{l} + D_r\frac{rc\pi}{l}\cos\frac{rc\pi t}{l}\right\}$

$\therefore$ With $t = 0$, $\quad \dfrac{\partial u}{\partial t} = g(x) = \displaystyle\sum_{r=1}^{\infty} D_r\frac{rc\pi}{l}\sin\frac{r\pi x}{l}$

$$\therefore\ g(x) = \frac{c\pi}{l}\sum_{r=1}^{\infty} D_r\, r \sin\frac{r\pi x}{l}$$

Finally we can draw on our knowledge of Fourier series techniques to determine the coefficients C_r and D_r.

$C_r = 2 \times$ mean value of $f(x)\sin\dfrac{r\pi x}{l}$ between $x = 0$ and $x = l$

$$\therefore\ C_r = \frac{2}{l}\int_0^l f(x)\sin\frac{r\pi x}{l}\,dx \qquad r = 1,\ 2,\ 3,\ \ldots$$

and $D_r\dfrac{rc\pi}{l} = 2 \times$ mean value of $g(x)\sin\dfrac{r\pi x}{l}$ between $x = 0$ and $x = l$

$$\therefore\ D_r = \frac{2}{rc\pi}\int_0^l g(x)\sin\frac{r\pi x}{l}\,dx \qquad r = 1,\ 2,\ 3,\ \ldots$$

The general solution (4) then becomes

$$u(x,\ t) = \sum_{r=1}^{\infty}\left\{\left[\frac{2}{l}\int_0^l f(w)\sin\frac{r\pi w}{l}\,dw\right]\cos\frac{rc\pi t}{l}\sin\frac{r\pi x}{l}\right.$$

$$\left. + \left[\frac{2}{rc\pi}\int_0^l g(w)\sin\frac{r\pi w}{l}\,dw\right]\sin\frac{rc\pi t}{l}\sin\frac{r\pi x}{l}\right\} \qquad (5)$$

The variable w has been introduced in the evaluation of the definite integrals to distinguish it from the independent variable x of the function $u(x, t)$.

At first sight, the solution seems very involved, but it can be analysed into a definite sequence of logical steps. Given the equation and relevant initial and boundary conditions, we go through the following stages.

(a) Assume a solution of the form $u = XT$ and express the equation in terms of X and T and their derivatives.
(b) Transpose the equation by separation of the variables and equate each side to a constant, so obtaining two separate equations, one in x and the other in t.
(c) Choose $k = -p^2$ to give an oscillatory solution.
(d) The two solution are of the form

$$X = A\cos px + B\sin px$$
$$T = C\cos cpt + D\sin cpt$$

Then $u(x, t) = \{A\cos px + B\sin px\}\{C\cos cpt + D\sin cpt\}$.

(e) Putting $cp = \lambda$, i.e. $p = \dfrac{\lambda}{c}$, this becomes

$$u(x, t) = \left\{A\cos\frac{\lambda}{c}x + B\sin\frac{\lambda}{c}x\right\}\{C\cos\lambda t + D\sin\lambda t\}.$$

(f) Apply boundary conditions to determine A and B.
(g) List the eigenvalues and eigenfunctions for $n = 1, 2, 3, \ldots$ and determine the general solution in t.
(h) Apply the remaining initial or boundary conditions.
(i) Determine the coefficients C_r and D_r by Fourier series techniques.

Make a list of these steps: then we can follow them with an example.

16 *Example* A stretched string of length 20 cm is set oscillating by displacing its mid-point a distance 1 cm from its rest position and releasing it with zero initial velocity. Solve the wave equation $\dfrac{\partial^2 u}{\partial x^2} = \dfrac{1}{c^2} \cdot \dfrac{\partial^2 u}{\partial t^2}$ where $c^2 = 1$ to determine the resulting motion, $u(x, t)$.

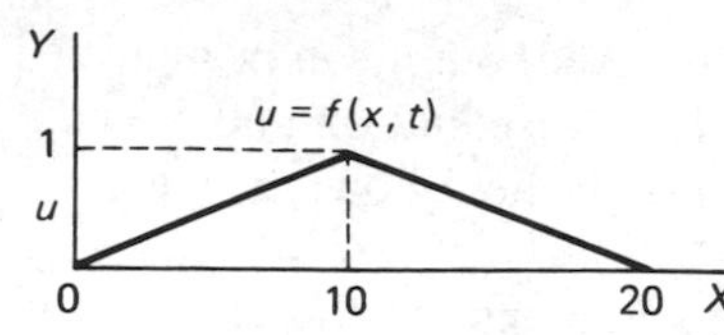

First we make a list of the boundary conditions from the data given in the question.

$$u(0, t) = \dots\dots\dots; \quad u(20, t) = \dots\dots\dots$$

$$u(x, 0) = \dots\dots\dots\dots\dots$$

$$\dots\dots\dots\dots\dots$$

$$\left[\frac{\partial u}{\partial t}\right]_{t=0} = \dots\dots\dots$$

17

$$u(0, t) = 0; \quad u(20, t) = 0 \quad \text{[fixed end points]}$$

$$u(x, 0) = f(x) = \frac{x}{10} \qquad 0 \le x \le 10$$

$$= \frac{20 - x}{10} \qquad 10 \le x \le 20$$

$$\left[\frac{\partial u}{\partial t}\right]_{t=0} = 0 \qquad \text{[zero initially velocity]}$$

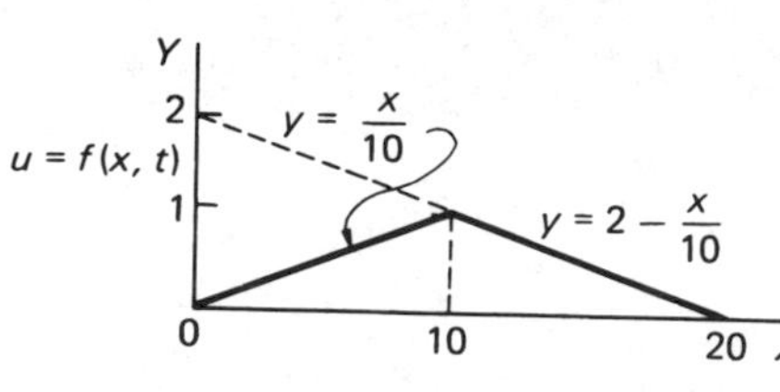

Now we can apply our sequence of operations which we listed.

So move on.

18 (a) Assume a solution $u = XT$ where X is a function of x only and T is a function of t only. Then the equation $\dfrac{\partial^2 u}{\partial x^2} = \dfrac{\partial^2 u}{\partial t^2}$ (since $c = 1$) becomes ....

19

$$X''T = XT''$$

for $u = XT$ $\quad\therefore\ \dfrac{\partial u}{\partial x} = X'T \qquad \dfrac{\partial^2 u}{\partial x^2} = X''T$

and $\quad \dfrac{\partial u}{\partial t} = XT' \qquad \dfrac{\partial^2 u}{\partial t^2} = XT''$

$$\frac{\partial^2 u}{\partial x^2} = \frac{\partial^2 u}{\partial t^2} \qquad \therefore\ X''T = XT''$$

(b) Next we rearrange the equation to separate the variables, giving

....

20

$$\frac{X''}{X} = \frac{T''}{T}$$

(c) Since the two sides are equal for all values of the variables, each must be equal to a constant k and to give an oscillatory solution we put $k = -p^2$. The two separate equations then are written

.... and ....

21

$$X'' + p^2X = 0 \text{ and } T'' + p^2T = 0$$

(d) These have solution $X =$ ....

$T =$ ....

so that $u(x, t) =$ ....

22

$$X = A\cos px + B\sin px; \quad T = C\cos pt + D\sin pt$$
$$\therefore\ u(x, t) = \{A\cos px + B\sin px\}\{C\cos pt + D\sin pt\}$$

(e) We normally now put $cp = \lambda$, but in this case $c = 1$ $\quad\therefore\ p = \lambda$ and

$u(x, t) =$ ....

23

$$u(x,t) = \{A\cos\lambda x + B\sin\lambda x\}\{C\cos\lambda t + D\sin\lambda t\}$$

(f) Now we determine A and B from the boundary conditions.

(i) $u(0,t) = 0 \quad \therefore 0 = A(C\cos\lambda t + D\sin\lambda t) \quad \therefore \underline{A = 0}$

$$\therefore u(x,t) = B\sin\lambda x(C\cos\lambda t + D\sin\lambda t)$$

(ii) $u(20,t) = 0 \quad \therefore 0 + B\sin 20\lambda(C\cos\lambda t + D\sin\lambda t)$

$B \neq 0$ or u would be identically zero. $\therefore \sin 20\lambda = 0$.

$$\therefore 20\lambda = n\pi \quad \therefore \lambda = \frac{n\pi}{20}$$

$$\therefore u(x,t) = \sin\frac{n\pi}{20}x\left\{P\cos\frac{n\pi}{20}t + Q\sin\frac{n\pi}{20}t\right\}$$

where $P = B \times C$ and $Q = B \times D$.

(g) The next step is to list the eigenvalues and eigenfunctions.

	Eigenvalues	Eigenfunctions
n	$\lambda = \frac{n\pi}{20}$	$u(x,t) = \sin\lambda x\{P\cos\lambda t + Q\sin\lambda t\}$
1	$\lambda_1 = \frac{\pi}{20}$	$u_1 = \sin\frac{\pi x}{20}\left\{P_1\cos\frac{\pi t}{20} + Q_1\sin\frac{\pi t}{20}\right\}$
2	$\lambda_2 = \frac{2\pi}{20}$	$u_2 = \sin\frac{2\pi x}{20}\left\{P_2\cos\frac{2\pi t}{20} + Q_2\sin\frac{2\pi t}{20}\right\}$
3	$\lambda_3 = \frac{3\pi}{20}$	$u_3 = \sin\frac{3\pi x}{20}\left\{P_3\cos\frac{3\pi t}{20} + Q_3\sin\frac{3\pi t}{20}\right\}$
$\vdots$	$\vdots$	$\vdots$
r	$\lambda_r = \frac{r\pi}{20}$	$u_r = \sin\frac{r\pi x}{20}\left\{P_r\cos\frac{r\pi t}{20} + Q_r\sin\frac{r\pi t}{20}\right\}$

$$u = u_1 + u_2 + u_3 + \dots$$

$$\therefore u(x,t) = \sum_{r=1}^{\infty}\sin\frac{r\pi x}{20}\left\{P_r\cos\frac{r\pi t}{20} + Q_r\sin\frac{r\pi t}{20}\right\}$$

(h) Now we apply the remaining initial conditions

(i) $u(x,0) = f(x) = \dfrac{x}{10} \qquad 0 \leq x \leq 10$

$= \dfrac{20-x}{10} \qquad 10 \leq x \leq 20$

Also $u(x,0) = \ldots\ldots\ldots$

24

$$u(x,0) = \sum_{r=1}^{\infty} P_r \sin\frac{r\pi x}{20}$$

Then $P_r = 2 \times$ mean value of $f(x)\,\sin\dfrac{r\pi x}{20}$ between $x = 0$ and $x = 20$

$$= \frac{2}{20}\int_0^{20} f(x)\,\sin\frac{r\pi x}{20}\,dx$$

$$\therefore\ 10P_r = \int_0^{10} \frac{x}{10}\,\sin\frac{r\pi x}{20}\,dx + \int_{10}^{20} \frac{20-x}{10}\sin\frac{r\pi x}{20}\,dx$$

$$= \qquad I_1 \qquad + \qquad I_2$$

$$I_1 = \int_0^{10} \frac{x}{10}\sin\frac{r\pi x}{20}\,dx = \ldots\ldots\ldots$$

25

$$I_1 = -\frac{20}{r\pi}\cos\frac{r\pi}{2} + \frac{40}{r^2\pi^2}\sin\frac{r\pi}{2}$$

Similarly, integrating by parts

$$I_2 = \int_{10}^{20} \frac{20-x}{10}\,\sin\frac{r\pi x}{20}\,dx = \ldots\ldots\ldots$$

26

$$I_2 = \frac{20}{r\pi}\cos\frac{r\pi}{2} - \frac{40}{r^2\pi^2}\sin r\pi$$

Then $10\,P_r = -\frac{20}{r\pi}\cos\frac{r\pi}{2} + \frac{40}{r^2\pi^2}\sin\frac{r\pi}{2} + \frac{20}{r\pi}\cos\frac{r\pi}{2} - \frac{40}{r^2\pi^2}\sin r\pi.$

$$\therefore \text{ For } r = 1,\ 2,\ 3,\ \dots\ P_r = \frac{4}{r^2\pi^2}\sin\frac{r\pi}{2}$$

$$\therefore\ u(x,t) = \sum_{r=1}^{\infty}\sin\frac{r\pi x}{20}\left\{\frac{4}{r^2\pi^2}\sin\frac{r\pi}{2}\cos\frac{r\pi t}{20} + Q_r\sin\frac{r\pi t}{20}\right\}$$

(ii) Also at $t = 0$, $\dfrac{\partial u}{\partial t} = 0$.

$$\frac{\partial u}{\partial t} = \dots\dots\dots$$

27

$$\frac{\partial u}{\partial t} = \sum_{r=1}^{\infty}\sin\frac{r\pi x}{20}\left\{\left(\frac{4}{r^2\pi^2}\sin\frac{r\pi}{2}\right)\left(-\frac{r\pi}{20}\sin\frac{r\pi t}{20}\right) + Q_r\frac{r\pi}{20}\cos\frac{r\pi t}{20}\right\}$$

$\therefore$ At $t = 0$, $\quad 0 = \sum_{r=1}^{\infty}\sin\frac{r\pi x}{20}Q_r\frac{r\pi}{20} \qquad \therefore\ \underline{Q_r = 0}$

So finally we have $\quad u\,(x,t) = \dots\dots\dots$

$$u(x, t) = \frac{4}{\pi^2} \sum_{r=1}^{\infty} \frac{1}{r^2} \sin \frac{r\pi x}{20} \sin \frac{r\pi}{2} \cos \frac{r\pi t}{20}$$

And that is it.

Now let us turn to a slightly different equation, but one for which the solution is very much along the same lines.

The heat conduction equation for a uniform finite bar

The conduction of heat in a uniform bar depends on the initial distribution of temperature and on the physical properties of the bar, i.e. the thermal conductivity and specific heat of the material, and the mass per unit length of the bar.

With a uniform bar insulated except at its ends, any heat flow is along the bar and, at any instant, the temperature u at a point P is a function of its distance x from one end and of the time t.

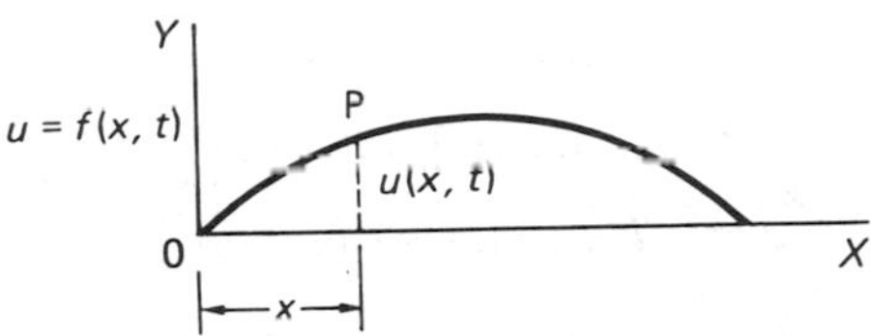

The one-dimensional heat equation is then of the form

$$\frac{\partial^2 u}{\partial x^2} = \frac{1}{c^2} \cdot \frac{\partial u}{\partial t} \qquad (1)$$

where $c^2 = \dfrac{k}{\sigma\rho}$ in which $k =$ thermal conductivity of the material; $\sigma =$ specific heat of the material; $\rho =$ mass per unit length of the bar.

You will already have noticed that the heat equation differs from the wave equation only in the fact that the right-hand side contains the first partial derivative instead of the second. It is not surprising therefore that the method of solution is very much like that of our previous examples.

Solutions of the heat conduction equation

Consider the case where

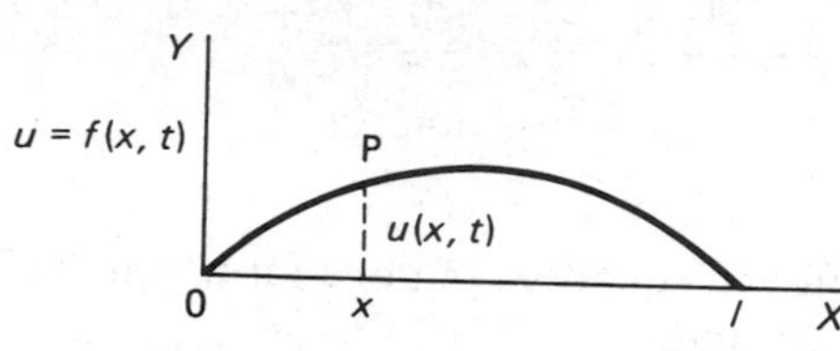

(a) the bar extends from $x = 0$ to $x = l$
(b) the temperature of the ends of the bar is maintained at zero
(c) the initial temperature distribution along the bar is defined by $f(x)$.

The boundary conditions can be expressed as ....

29

$$u(0,\ t) = 0 \quad \text{and} \quad u(l,\ t) = 0 \quad \text{for all } t \geq 0$$
$$u(x,\ 0) = f(x) \quad \text{for } 0 \leq x \leq 1$$

As before, we assume a solution of the form $u(x,\ t) = X(x)T(t)$ where

X is a function of x only

T is a function of t only.

Then, starting with $u = XT$ we can write the equation $\dfrac{\partial^2 u}{\partial x^2} = \dfrac{1}{c^2} \cdot \dfrac{\partial u}{\partial t}$ in terms of X and T, and separating the variables, we obtain

....

30

$$\frac{X''}{X} = \frac{1}{c^2} \cdot \frac{T'}{T}$$

Arguing as before, since the left-hand side is a function of x only and the right-hand side a function of t only, for these to be equal each side must equal the same constant. Let this be $(-p^2)$ as before.

$$\therefore \frac{X''}{X} = -p^2 \quad \therefore X'' + p^2 X = 0 \text{ giving } X = A\cos px + B\sin px$$

and $\dfrac{1}{c^2} \cdot \dfrac{T'}{T} = -p^2 \quad \therefore T' + p^2c^2T = 0$ giving $T =$ ....

31

$$\boxed{T = Ce^{-p^2c^2t}}$$

for $\dfrac{T'}{T} = -p^2c^2 \quad \therefore \; \ln T = -p^2c^2t + c_1 \quad \therefore \; T = C\,e^{-p^2c^2t}$

$$u(x,t) = XT = \{A\cos px + B\sin px\}\,Ce^{-p^2c^2t}$$

$$\therefore \; u(x,t) = \{P\cos px + Q\sin px\}e^{-p^2c^2t}$$

Now put $pc = \lambda \quad \therefore \; p = \dfrac{\lambda}{c}$

$$\therefore \; u(x,t) = \left\{P\cos\frac{\lambda}{c}x + Q\sin\frac{\lambda}{c}x\right\}e^{-\lambda^2t}$$

Applying the boundary condition $u(0,t) = 0$ gives

.... and ....

32

$$\boxed{P = 0 \text{ and } u(x,t) = Qe^{-\lambda^2t}\sin\frac{\lambda}{c}x}$$

Also $u(l,t) = 0$ and from this we get

....

33

$$\boxed{\lambda = \frac{nc\pi}{l} \text{ for } n = 1,\ 2,\ 3,\ \ldots}$$

for, if $u = 0$ when $x = l$, $\quad 0 = Q\,\mathrm{e}^{-\lambda^2 t} \sin \dfrac{\lambda l}{c}$

$Q \neq 0$ or $u(x, t)$ would be identically zero $\therefore\ \sin\dfrac{\lambda l}{c} = 0$

$$\therefore\ \frac{\lambda l}{c} = n\pi \qquad \therefore\ \lambda = \frac{nc\pi}{l} \qquad n = 1,\ 2,\ 3,\ \ldots$$

Now we can compile the table of eigenfunctions.

n	$\lambda = \dfrac{nc\pi}{l}$	$u(x,t) = Q\,\mathrm{e}^{-\lambda^2 t}\sin\dfrac{n\pi x}{l}$
1	$\lambda_1 = \dfrac{c\pi}{l}$	$u_1 = Q_1\,\mathrm{e}^{-\lambda_1^2 t}\sin\dfrac{\pi x}{l}$
2	$\lambda_2 = \dfrac{2c\pi}{l}$	$u_2 = Q_2\,\mathrm{e}^{-\lambda_2^2 t}\sin\dfrac{2\pi x}{l}$
3	$\lambda_3 = \dfrac{3c\pi}{l}$	$u_3 = Q_3\,\mathrm{e}^{-\lambda_3^2 t}\sin\dfrac{3\pi x}{l}$
$\vdots$	$\vdots$	$\vdots$
r	$\lambda_r = \dfrac{rc\pi}{l}$	$u_r = Q_r\,\mathrm{e}^{-\lambda_r^2 t}\sin\dfrac{r\pi x}{l}$

$$u = u_1 + u_2 + u_3 + \ldots$$

$$\therefore\ u(x,t) = \sum_{r=1}^{\infty}\left\{Q_r\,\mathrm{e}^{-\lambda_r^2 t}\sin\frac{r\pi x}{l}\right\}$$

The remaining boundary condition still to be applied is that

when $t = 0$, $\quad u(x, 0) = f(x) \qquad 0 \leq x \leq l$

This gives $\quad f(x) = \ldots\ldots\ldots$

34

$$f(x) = \sum_{r=1}^{\infty} \left\{ Q_r \sin \frac{r\pi x}{l} \right\}$$

and from our knowledge of Fourier series techniques,

$Q_r = \ldots\ldots\ldots$

35

$$Q_r = 2 \times \text{ mean value of } f(x) \sin \frac{r\pi x}{l} \text{ from } x = 0 \text{ to } x = l$$

$\therefore\ Q_r = \dfrac{2}{l} \displaystyle\int_0^l f(x) \sin \frac{r\pi x}{l} \mathrm{d}x$ and the final solution becomes

$$u(x, t) = \frac{2}{l} \sum_{r=1}^{\infty} \left\{ \left[\int_0^l f(w) \sin \frac{r\pi w}{l} \mathrm{d}w \right] \mathrm{e}^{-\lambda_r^2 t} \sin \frac{r\pi x}{l} \right\}$$

where $\lambda = \dfrac{rc\pi}{l}$ $\quad r = 1, 2, 3, \ldots$

Now on to the next frame for an example.

36

Example A bar length 2 m is fully insulated along its sides. It is initially at a uniform temperature of 10°C and at $t - 0$ the ends are plunged into ice and maintained at a temperature of 0°C. Determine an expression for the temperature at a point P at a distance x from one end at any subsequent time t seconds after $t = 0$.

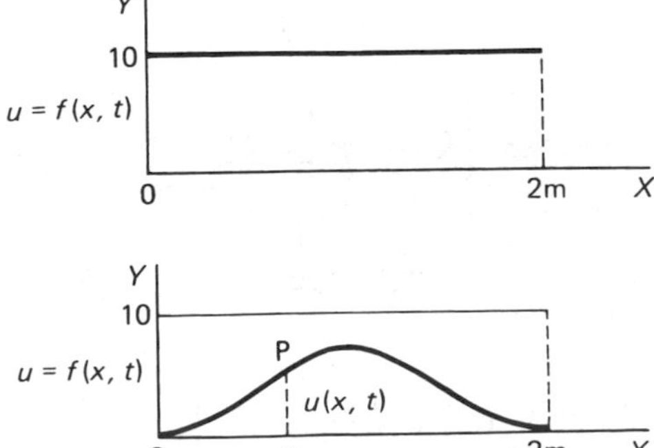

We have the heat equation $\dfrac{\partial^2 u}{\partial x^2} = \dfrac{1}{c^2} \cdot \dfrac{\partial u}{\partial t}$ with the boundary conditions

$\ldots\ldots\ldots;$ $\ldots\ldots\ldots;$ and $\ldots\ldots\ldots$

37

$$u(0, t) = 0; \quad u(2, t) = 0; \quad u(x, 0) = 10$$

Assuming a solution of the form $u = XT$, we know that this gives for this equation $\quad X = A\cos px + B\sin px$

and $\quad T = Ce^{-p^2c^2t}$

so that the general solution is

$$u(x, t) = \{P\cos px + Q\sin px\}\,e^{-p^2c^2t}$$

If we now write $pc = \lambda$, $\quad p = \dfrac{\lambda}{c}$ and the solution becomes

$$u(x, t) = \left\{P\cos\frac{\lambda}{c}x + Q\sin\frac{\lambda}{c}x\right\}e^{-\lambda^2 t}$$

Applying the first two of the boundary conditions gives us

....

38

$$P = 0 \text{ and } u(x, t) = \left\{Q\sin\frac{n\pi x}{2}\right\}e^{-\lambda^2 t}$$

for $\quad u(0, t) = 0 \quad \therefore 0 = P\,e^{-\lambda^2 t} \quad \therefore \underline{P = 0}$

$$\therefore u(x, t) = \left\{Q\sin\frac{\lambda}{c}x\right\}e^{-\lambda^2 t}$$

Also $u(2, t) = 0 \quad \therefore 0 = \left\{Q\sin\dfrac{2\lambda}{c}\right\}e^{-\lambda^2 t}$

$$Q \neq 0 \quad \therefore \sin\frac{2\lambda}{c} = 0 \quad \therefore \frac{2\lambda}{c} = n\pi \quad \therefore \lambda = \frac{nc\pi}{2} \quad n = 1, 2, 3\ldots$$

$$\therefore u(x, t) = \left\{Q\sin\frac{n\pi x}{2}\right\}e^{-\lambda^2 t}$$

There is, of course, an infinite number of such solutions with different values of n. We can write the solution so far therefore as

$$u(x, t) = \ldots\ldots\ldots$$

39

$$u(x,t)=\sum_{r=1}^{\infty}Q_r\sin\frac{r\pi x}{2}e^{-\lambda^2 t}$$

Finally, there is the remaining initial condition that at $t=0,\ u=10$.

$$\therefore u(x,0)=f(x)=10 \quad \therefore 10=\sum_{r=1}^{\infty}Q_r\sin\frac{r\pi x}{2}$$

where $Q_r=2\times$ mean value of $10\ \sin\dfrac{r\pi x}{2}$ from $x=0$ to $x=2$.

$$\therefore Q_r=\dots\dots\dots$$

40

$$0\ (r\text{ even});\quad \frac{40}{\pi r}\ (r\text{ odd})$$

$$\text{for } Q_r=\frac{2}{2}\int_0^2 10\sin\frac{r\pi x}{20}\,dx=10\int_0^2\sin\frac{r\pi x}{2}\,dx$$

$$=-\frac{20}{\pi r}\left[\cos\frac{r\pi x}{2}\right]_0^2=\frac{20}{\pi r}\{1-\cos r\pi\}$$

$$=0\ (r\text{ even})\text{ and }\frac{40}{r\pi}\ (r\text{ odd})$$

Therefore the required solution is

$$u(x,\ t)=\dots\dots\dots$$

41

$$u(x,t) = \frac{40}{\pi} \sum_{r\,(\text{odd})=1}^{\infty} \frac{1}{r} \sin \frac{r\pi x}{2} e^{-\lambda^2 t} \quad r = 1, 3, 5, \ldots$$
$$\text{where } \lambda = \frac{rc\pi}{2}$$

By now you will appreciate that the approach to all these problems is very much the same, as indeed it still is with the next important equation.

Laplace's equation

The Laplace equation concerns the distribution of a field, e.g. temperature, potential, etc., over a plane area subject to certain boundary conditions.

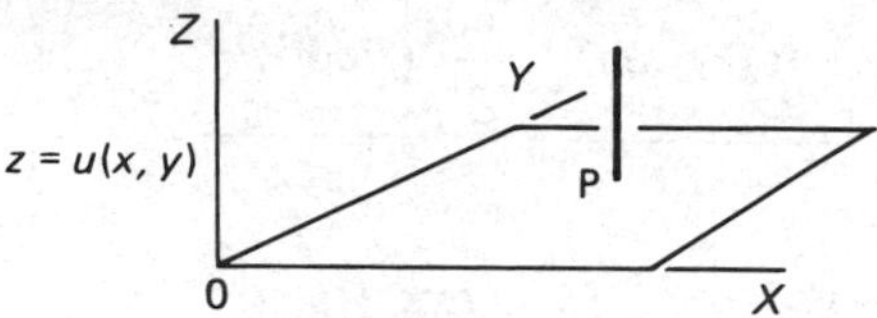

The potential at a point P in a plane can be indicated by an ordinate axis and is a function of its position, i.e. $z = u(x, y)$ where $u(x, y)$ is the solution of the Laplace two-dimensional equation $\dfrac{\partial^2 u}{\partial x^2} + \dfrac{\partial^2 u}{\partial y^2} = 0$.

Let us consider the situation in the next frame.

42

Solution of the Laplace equation

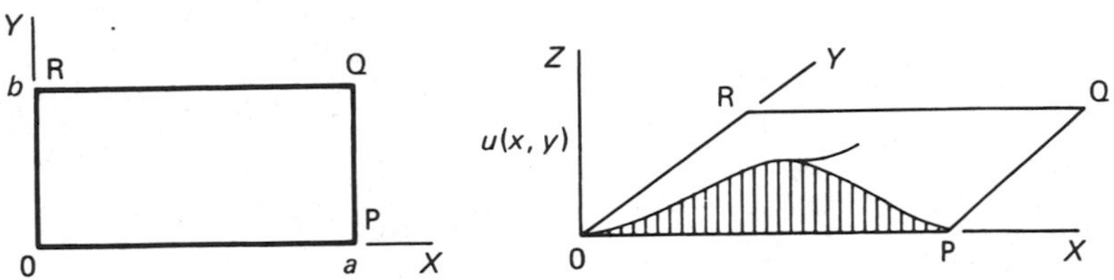

We are required to determine a solution of the equation $\dfrac{\partial^2 u}{\partial x^2} + \dfrac{\partial^2 u}{\partial y^2} = 0$

for the rectangle bounded by the lines $x = 0$, $y = 0$, $x = a$, $y = b$, subject to the following boundary conditions.

$$\begin{array}{llll} u = 0 & \text{when} & x = 0 & 0 \le y \le b \\ u = 0 & \text{when} & x = a & 0 \le y \le b \\ u = 0 & \text{when} & y = b & 0 \le x \le a \\ u - f(x) & \text{when} & y = 0 & 0 \le x \le a \end{array}$$

i.e. $u(0, y) = 0$ and $u(a, y) = 0$ for $0 \le y \le b$

and $u(x, b) = 0$ and $u(x, 0) = f(x)$ for $0 \le x \le a$.

The solution $z = u(x, y)$ will give the potential at any point within the rectangle OPQR.

We start off, as usual, by assuming a solution of the form $u(x, y) = X(x)Y(y)$ where X is a function of x only and Y is a function of y only. We now express the equation in terms of X and Y and separate the variables to give

....

43

$$\frac{X''}{X} = -\frac{Y''}{Y}$$

for $u = XY$ $\quad \therefore \dfrac{\partial u}{\partial x} = X'Y \quad$ and $\quad \dfrac{\partial^2 u}{\partial x^2} = X''Y$

$\dfrac{\partial u}{\partial y} = XY' \quad$ and $\quad \dfrac{\partial^2 u}{\partial y^2} = XY''$

The equation is then $X''Y = -XY'' \quad \therefore \dfrac{X''}{X} = -\dfrac{Y''}{Y}$

Putting each side equal to a constant $(-p^2)$ gives two equations

$$X'' + p^2 X = 0 \quad \text{and} \quad Y'' - p^2 Y = 0$$

$X'' + p^2 X = 0$ has a solution $X = \dots\dots\dots$

44

$$X = A\cos px + B\sin px$$

In the introduction to this programme we said that the equation $Y'' - p^2 Y = 0$ has a solution of the form $Y = C\cosh py + D\sinh py$ which can also be expressed as $Y = E\sinh p(y + \phi)$

$$\therefore u(x, y) = \{A\cos px + B\sin px\}E\sinh p(y + \phi)$$
$$\therefore u(x, y) = \{P\cos px + Q\sin px\}\sinh p(y + \phi)$$

Now we apply the first of the boundary conditions.

$u(0, y) = 0 \quad \therefore 0 = P\sinh p(y + \phi) \quad \therefore P = 0$

$$\therefore u(x, y) = Q\sin px \sinh p(y + \phi)$$

From the second boundary condition, we have

$$u(a, y) = 0 \quad \therefore 0 = Q\sin pa \sinh p(y + \phi) \quad \therefore \sin pa = 0$$

$$\therefore pa = n\pi \qquad \text{for } n = 1, 2, 3, \dots$$

If we write $\lambda = p$ then $\lambda = \dfrac{n\pi}{a}$ and $u(x, y) = Q\sin \lambda x \sinh \lambda(y + \phi)$

Now from the third condition

$u(x, b) = 0 \quad$ from which we have $\dots\dots\dots$

45

$$u(x,\ y) = Q \sin \lambda x \sinh \lambda(b - y)$$

for $\quad 0 = Q \sin \lambda x \sinh \lambda(b + \phi) \quad \therefore\ \sinh \lambda(b + \phi) = 0 \quad \therefore\ \phi = -b.$

$$\therefore\ u(x,\ y) = Q \sin \lambda x \sinh \lambda(y - b)$$

$\sinh \lambda(y - b) = -\sinh \lambda(b - y) \quad \therefore\ u(x,\ y) = Q \sin \lambda x \sinh \lambda(b - y),$

the minus sign being absorbed in the symbol Q whose value has yet to be found. Now $\lambda = \dfrac{n\pi}{a}$ with $n = 1, 2, 3, \ldots$ and there is therefore an infinite number of values for λ and hence an infinite number of solutions for $u(x,\ y)$. Therefore, again using $u = u_1 + u_2 + u_3 + \ldots$ we have $u(x,\ y) = \ldots\ldots\ldots$

46

$$u(x,\ y) = \sum_{r=1}^{\infty} Q_r \sin \lambda x \sinh \lambda(b - y)$$

Now there remains the fourth boundary condition to be applied.

$u(x,\ 0) = f(x) \quad \therefore\ f(x) = \sum_{r=1}^{\infty} Q_r \sin \lambda x\ \sinh\ \lambda b$

$$\begin{aligned}
\therefore\ Q_r \sinh \lambda b &= 2 \times \text{ mean value of } f(x) \sin \lambda x \text{ from } x = 0 \text{ to } x = a \\
&= \frac{2}{a} \int_0^a f(x) \sin \lambda x \,\mathrm{d}x \\
&= \frac{2}{a} \int_0^a f(x) \sin \frac{r\pi x}{a} \,\mathrm{d}x
\end{aligned}$$

from which the coefficients Q_r can be found.

Let us work through an example with numerical values.

Example Determine a solution $u(x,\ y)$ of the Laplace equation $\dfrac{\partial^2 u}{\partial x^2} + \dfrac{\partial^2 u}{\partial y^2} = 0$ subject to the following boundary conditions.

$$\begin{aligned}
&u = 0 \text{ when } x = 0; \qquad &&u = 0 \text{ when } x = \pi \\
&u \to 0 \text{ when } y \to \infty; &&u = 3 \text{ when } y = 0
\end{aligned}$$

As always, we begin with $u(x,\ y) = X(x)Y(y)$, re-write the equation in terms of X and Y and separate the variables. The equation then becomes

$\ldots\ldots\ldots$

47

$$\frac{X''}{X} = -\frac{Y''}{Y}$$

Equating each side to $-p^2$, we have $X'' + p^2X = 0$ and $Y'' - p^2Y = 0$.

$X'' + p^2X = 0$ has a solution ....

48

$$X = A\cos px + B\sin px$$

The solution of $Y'' - p^2Y = 0$ can be stated in three different forms

$Y = C\cosh py + D\sinh py$; $\quad Y = C\,e^{py} + D\,e^{-py}$; $\quad Y = C\sinh p(y + \phi)$

On this occasion, we will use the second one

$$Y = C\,e^{py} + D\,e^{-py}$$

Then $\qquad u(x, y) = \{A\cos px + B\sin px\}\{C\,e^{py} + D\,e^{-py}\}$

Application of the first boundary condition $u(0, y) = 0$ gives

.... and ....

49

$$A = 0 \text{ and } u(x, y) = \sin px\{P\,e^{py} + Q\,e^{-py}\}$$

for $\quad 0 = A\{C\,e^{py} + D\,e^{-py}\} \qquad \therefore A = 0$

and $u(x, y) = B\sin px\{C\,e^{py} + D\,e^{-py}\} = \sin px\{P\,e^{py} + Q\,e^{-py}\}$.

The second boundary condition $u(\pi, y) = 0$ then gives

....

50

$$u(x,\ y) = \sin nx\ \{P\ \mathrm{e}^{ny} + Q\ \mathrm{e}^{-ny}\} \qquad n = 1,\ 2,\ 3,\ \ldots$$

for $u = 0$ when $x = \pi \ \therefore\ 0 = \sin p\pi\{P\ \mathrm{e}^{py} + Q\ \mathrm{e}^{-py}\}$

$$\therefore\ \sin p\pi = 0\ \ \therefore\ p\pi = n\pi\ \ \therefore\ p = n \qquad n = 1,\ 2,\ 3,\ \ldots$$

$$\therefore\ u(x,\ y) = \sin\ nx\ \{P\ \mathrm{e}^{ny} + Q\ \mathrm{e}^{-ny}\}$$

The third condition is that $u \to 0$ as $y \to \infty$.

$$\therefore\ 0 = \sin nx\ \{P\ \mathrm{e}^{ny}\}\ \text{since } \mathrm{e}^{-ny} \to 0 \qquad \therefore\ P = 0$$

$$\therefore\ u(x,\ y) = Q\ \mathrm{e}^{-ny} \sin nx$$

But n can have an infinite number of values giving an infinite number of solutions

$$u_1 = Q_1\ \mathrm{e}^{-y} \sin x$$

$$u_2 = Q_2\ \mathrm{e}^{-2y} \sin 2x$$

$$u_3 = Q_3\ \mathrm{e}^{-3y} \sin 3x$$

$$\vdots \qquad \vdots$$

$$u_r = Q_r\ \mathrm{e}^{-ry} \sin rx$$

So the solution at this stage can be written as

$$u(x,\ y) = \ldots\ldots\ldots$$

51

$$u(x, y) = \sum_{r=1}^{\infty} Q_r \, \mathrm{e}^{-ry} \sin rx$$

Now we turn to the final boundary condition that $u = 3$ when $y = 0$.

$\therefore 3 = \sum_{r=1}^{\infty} Q_r \sin rx$ from which we obtain

$$Q_r = \dots\dots\dots$$

52

$$Q_r = 0 \ (r \text{ even}); \quad Q_r = \frac{12}{r\pi} (r \text{ odd})$$

for $Q_r = 2 \times$ mean value of $3 \sin rx$ between $x = 0$ and $x = \pi$

$$= \frac{2}{\pi} \int_0^{\pi} 3 \sin rx \, \mathrm{d}x = \frac{6}{\pi} \left[-\frac{\cos rx}{r} \right]_0^{\pi} = \frac{6}{r\pi} (1 - \cos r\pi)$$

$$\therefore Q_r = 0 \ \ (r \text{ even}) \text{ and } \frac{12}{r\pi} \ \ (r \text{ odd})$$

$$\therefore u(x, y) = \sum_{r\,(\text{odd})=1}^{\infty} \frac{12}{r\pi} \mathrm{e}^{-ry} \sin rx \qquad r = 1, 3, 5, \dots$$

$$\therefore u(x, y) = \frac{12}{\pi} \left\{ \mathrm{e}^{-y} \sin x + \frac{1}{3} \mathrm{e}^{-3y} \sin 3x + \frac{1}{5} \mathrm{e}^{-5y} \sin 5x + \dots \right\}$$

That covers the main steps in the method of solving second-order partial differential equations by separation of the variables applied specifically to the wave equation, the heat conduction equation and the Laplace equation. The same approach can be made with other similar equations.

The Revision Summary now follows: then the Test Exercise with problems just like those we have considered. Although the solutions take rather more steps than with other forms of equations, the method is straightforward and follows a clear pattern.

REVISION SUMMARY

1. *Ordinary second-order linear differential equations*

(a) Equation of the form $a\dfrac{d^2y}{dx^2} + b\dfrac{dy}{dx} + cy = 0$

Auxiliary equation $am^2 + bm + c = 0$

(i) Real and different roots: $m = m_1$ and $m = m_2$

$$y = A\,e^{m_1 x} + B\,e^{m_2 x}$$

(ii) Real and equal roots: $m = m_1$ (twice)

$$y = e^{m_1 x}(A + Bx)$$

(iii) Complex roots: $m = \alpha \pm j\beta$

$$y = e^{\alpha x}\{A\cos\beta x + B\sin\beta x\}.$$

(b) Equations of the form $\dfrac{d^2y}{dx^2} \pm n^2 y = 0$

(i) $\dfrac{d^2y}{dx^2} + n^2 y = 0$; $\quad y = A\cos nx + B\sin nx$

(ii) $\dfrac{d^2y}{dx^2} - n^2 y = 0$; $\quad y = A\cosh nx + B\sinh nx$

or $\quad y = A\,e^{nx} + B\,e^{-nx}$

or $\quad y = A\sinh n(x + \phi)$.

2. *Partial differential equations* Solution $u = f(x, y, t, \ldots)$

Linear equations: If $u = u_1,\ u = u_2,\ u = u_3,\ \ldots$ are solutions, so also is

$$u = u_1 + u_2 + u_3 + \ldots + u_r + \ldots = \sum_{r=1}^{\infty} u_r.$$

(a) *Wave equation*—transverse vibrations of an elastic string

$$\frac{\partial^2 u}{\partial x^2} = \frac{1}{c^2}\cdot\frac{\partial^2 u}{\partial t^2} \quad \text{where } c^2 = \frac{T}{\rho},$$

$T =$ tension of string

$\rho =$ mass per unit length.

(b) *Heat conduction equation*—heat flow in uniform finite bar

$$\frac{\partial^2 u}{\partial x^2} = \frac{1}{c^2} \cdot \frac{\partial u}{\partial t} \qquad \text{where } c^2 = \frac{k}{\sigma \rho}$$

$k =$ thermal conductivity of material

$\sigma =$ specific heat of the material

$\rho =$ mass per unit length of bar.

(c) *Laplace equation* — distribution of a field over a plane area

$$\frac{\partial^2 u}{\partial x^2} + \frac{\partial^2 u}{\partial y^2} = 0.$$

3. *Separating the variables*

Let $u(x, y) = X(x)Y(y)$ where $X(x)$ is a function of x only
and $Y(y)$ is a function of y only.

Then $\dfrac{\partial u}{\partial x} = X'Y$; $\quad \dfrac{\partial^2 u}{\partial x^2} = X''Y$

$\dfrac{\partial u}{\partial y} = XY'$; $\quad \dfrac{\partial^2 u}{\partial y^2} = XY''$

Substitute in the given partial differential equation and form separate differential equations to give $X(x)$ and $Y(y)$ by introducing a common constant $(-p^2)$. Determine arbitrary functions by use of the initial and boundary conditions.

54

TEST EXERCISE XIX

1. Solve the following equations

(a) $\dfrac{\partial^2 u}{\partial x^2} = 24x^2(t-2)$, given that at $x = 0$, $u = e^{2t}$ and $\dfrac{\partial u}{\partial x} = 4t$.

(b) $\dfrac{\partial^2 u}{\partial x \partial y} = 4\,e^y \cos 2x$, given that at $y = 0$, $\dfrac{\partial u}{\partial x} = \cos x$

and at $x = \pi$, $u = y^2$.

2. A perfectly elastic string is stretched between two points 10 cm apart. Its centre point is displaced 2 cm from its position of rest at right angles to the original direction of the string and then released with zero velocity. Applying the equation $\dfrac{\partial^2 u}{\partial x^2} = \dfrac{1}{c^2} \cdot \dfrac{\partial^2 u}{\partial t^2}$ with $c^2 = 1$, determine the subsequent motion $u(x, t)$.

3. One end A of an insulated metal bar AB of length 2 m is kept at 0°C while the other end B is maintained at 50°C until a steady state of temperature along the bar is achieved. At $t = 0$, the end B is suddenly reduced to 0°C and kept at that temperature. Using the heat conduction equation $\dfrac{\partial^2 u}{\partial x^2} = \dfrac{1}{c^2} \cdot \dfrac{\partial u}{\partial t}$, determine an expression for the temperature at any point in the bar distance x from A at any time t.

4. A square plate is bounded by the lines $x = 0$, $y = 0$, $x = 2$, $y = 2$. Apply the Laplace equation $\dfrac{\partial^2 u}{\partial x^2} + \dfrac{\partial^2 u}{\partial y^2} = 0$ to determine the potential distribution $u(x, y)$ over the plate, subject to the following boundary conditions.

$$\begin{array}{lll} u = 0 & \text{when } x = 0 & 0 \le y \le 2 \\ u = 0 & \text{when } x = 2 & 0 \le y \le 2 \\ u = 0 & \text{when } y = 0 & 0 \le x \le 2 \\ u = 5 & \text{when } y = 2 & 0 \le x \le 2. \end{array}$$

FURTHER PROBLEMS XIX

1. Show that the equation $\frac{\partial^2 u}{\partial x^2} - \frac{1}{c^2} \cdot \frac{\partial^2 u}{\partial t^2} = 0$ is satisfied by $u = f(x + ct) + F(x - ct)$ where f and F are arbitrary functions.

2. If $\frac{\partial^2 u}{\partial x^2} = \frac{1}{c^2} \cdot \frac{\partial^2 u}{\partial t^2}$ and $c = 3$, determine the solution $u = f(x, t)$ subject to the boundary conditions

$$u(0, t) = 0 \text{ and } u(2, t) = 0 \text{ for } t \geq 0$$
$$u(x, 0) = x(2 - x) \text{ and } \left[\frac{\partial u}{\partial t}\right]_{t=0} = 0 \quad 0 \leq x \leq 2.$$

3. The centre point of a perfectly elastic string stretched between two points A and B, 4 m apart, is deflected a distance 0.01 m from its position of rest perpendicular to AB and released initially with zero velocity. Apply the wave equation $\frac{\partial^2 u}{\partial x^2} = \frac{1}{c^2} \cdot \frac{\partial^2 u}{\partial t^2}$ where $c = 10$ to determine the subsequent motion of a point P distant x from A at time t.

4. An elastic string is stretched between two points 10 cm apart. A point P on the string 2 cm from the left-hand end, i.e. the origin, is drawn aside 1 cm from its position of rest and released with zero velocity. Solve the one-dimensional wave equation to determine the displacement of any point at any instant.

5. An insulated uniform metal bar, 10 units long, has the temperature of its ends maintained at 0°C and at $t = 0$ the temperature distribution $f(x)$ along the bar is defined by $f(x) = x(10 - x)$. Solve the heat conduction equation $\frac{\partial^2 u}{\partial x^2} = \frac{1}{c^2} \cdot \frac{\partial u}{\partial t}$ with $c^2 = 4$ to determine the temperature u of any point in the bar at time t.

6. The ends of an insulated rod AB, 10 units long, are maintained at 0°C. At $t = 0$, the temperature within the rod rises uniformly from each end reaching 2°C at the mid-point of AB. Determine an expression for the temperature $u(x, t)$ at any point in the rod, distant x from the left-hand end at any subsequent time t.

7. A rectangular plate OPQR is bounded by the lines $x = 0, y = 0, x = 4, y = 2$. Determine the potential distribution $u(x, y)$ over the rectangle using the Laplace equation $\frac{\partial^2 u}{\partial x^2} + \frac{\partial^2 u}{\partial y^2} = 0$, subject to the following boundary conditions.

$$\begin{array}{ll} u(0, y) = 0 & 0 \leq y \leq 2 \\ u(4, y) = 0 & 0 \leq y \leq 2 \\ u(x, 2) = 0 & 0 \leq x \leq 4 \\ u(x, 0) = x(4 - x) & 0 \leq x \leq 4. \end{array}$$

8. Two sides AB and AD of a rectangular plate ABCD lie along the x and y axes respectively. The remaining two sides are the lines $x = 5$ and $y = 2$. The sides BC, CD and DA are maintained at zero temperature. The temperature distribution along AB is defined by $f(x) = x(x - 5)$. Determine an expression for the steady-state temperature at any point in the plate.

Programme 20

Linear Optimisation (Linear Programming)

Optimisation

1

An *optimisation problem* is one requiring the determination of the *optimal (maximum or minimum) value* of a given function, called the *objective function*, subject to a set of stated restrictions, or *constraints*, placed on the variables concerned.

In practice, for example, we may need to maximise an objective function representing units of output in a manufacturing situation, subject to constraints reflecting the availability of labour, machine time, stocks of raw materials, transport conditions, etc.

Linear programming (or linear optimisation)

Linear programming is a method of solving an optimisation problem when the objective function is a *linear function* and the constraints are *linear equations* or *linear inequalities*.

In this programme, we shall restrict our considerations to problems of this type which form an important introduction to the much wider study of operational research.

Let us consider a simple example, so move on to the next frame.

2

A simple linear programming problem may look like this:

$$\begin{aligned} &\text{Maximise } P = x + 2y \quad &&(\textit{objective function})\\ &\text{subject to} \quad \left.\begin{aligned} y &\leq 3\\ x+y &\leq 5\\ x-2y &\leq 2\\ x \geq 0;\ y &\geq 0 \end{aligned}\right\} &&(\textit{constraints}) \end{aligned}$$

The last two constraints, i.e. $x \geq 0$ and $y \geq 0$, apply to all linear programming problems and indicate that the problem variables, x and y, are restricted to non-negative values: they may have zero or positive values, but NOT negative values. These two constraints are often combined and written $x, y \geq 0$ — or omitted altogether since they are taken for granted in all LP problems.

Before we proceed, we will take a brief look at linear inequalities in general.

On, then, to Frame 3.

3

Linear inequalities

In most respects, *linear inequalities* can be manipulated in the same manner as can equations.

(a) Both sides may be increased or decreased by a common term, e.g.

$$2x \leq y + 4 \quad \therefore\ 2x - y \leq 4$$

(b) Both sides may be multiplied or divided by a positive factor, e.g.

$$4x + 6y \geq 12 \quad \therefore\ 2x + 3y \geq 6$$

But NOTE this:

(c) If both sides are multiplied or divided by a negative factor, e.g. (–1), then the inequality sign must be reversed, i.e. $\geq$ becomes $\leq$ and vice versa.

Here, then, is a short exercise.

Exercise Simplify the following inequalities so that each right-hand side consists of a positive constant term only.

(a) $3x - 5 \leq 4y$ (b) $2(x + 2y) \leq -8$
(c) $4x - 6y \leq -10$ (d) $2x + 3 \geq -(y + 4)$
(e) $-(x - 3y + 5) \geq 2x + 4y - 6$

Check the results in the next frame.

4

(a) $3x - 4y \leq 5$	(b) $-x - 2y \geq 4$
(c) $-2x + 3y \geq 5$	(d) $-2x - y \leq 7$
(e) $3x + y \leq 1$	

Graphical representation of linear inequalities

Consider the inequality $y - 2x \leq 3$. We can add $2x$ to each side, so that $y \leq 2x + 3$.

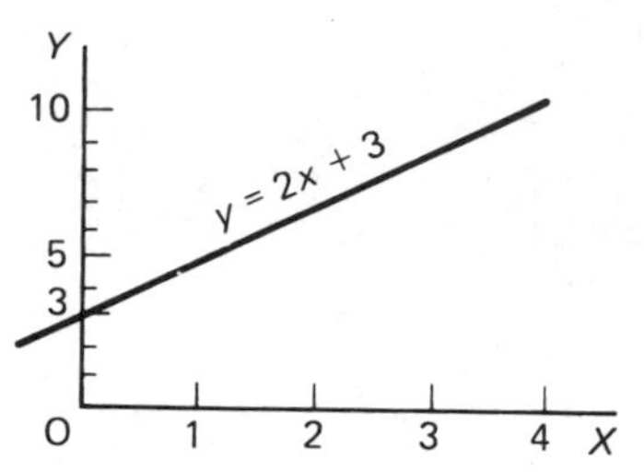

The equation $y = 2x + 3$ can be represented by a straight line, dividing the xy-plane into two parts.

For all points on the line, $y = 2x + 3$.

For all points below the line,

....

5

$y < 2x + 3$

$\therefore$ $y \leq 2x + 3$ indicated all points on or below the straight line, but excludes all points above it. We can indicate this exclusion zone by

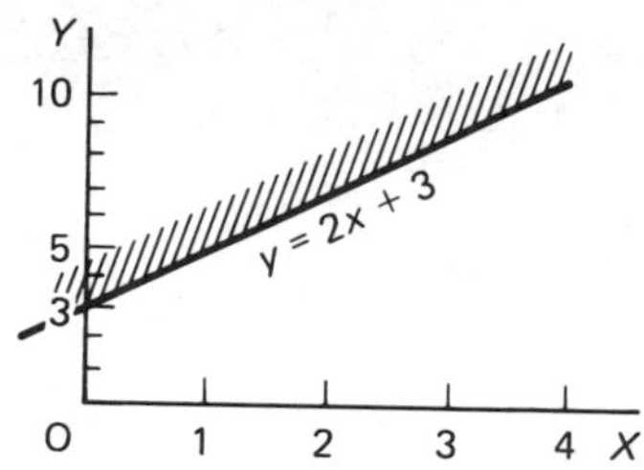

shading the upper side of the line.

Arguing in much the same way, $x - 2y \leq 2$ can be re-written as $y \geq \dfrac{x}{2} - 1$ and we can draw the line $y = \dfrac{x}{2} - 1$ and shade in the exclusion zone

....

6

below the line

i.e.

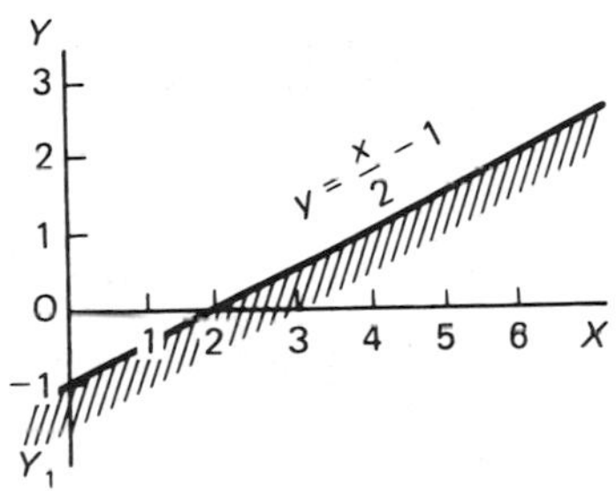

Example 1

The problem we quoted earlier in frame 2 was

$$\begin{aligned} \text{Maximise } P = \; & x + 2y && \textit{(objective function)} \\ \text{subject to } \; & y \le 3 \\ & x + y \le 5 \\ & x - 2y \le 2 && \textit{(constraints)} \\ & x \ge 0; \; y \ge 0 \end{aligned}$$

Now, on a common pair of x and y axes, we can represent the five constraints and shade in the exclusion zone for each. We then have the composite diagram

. . . .

7

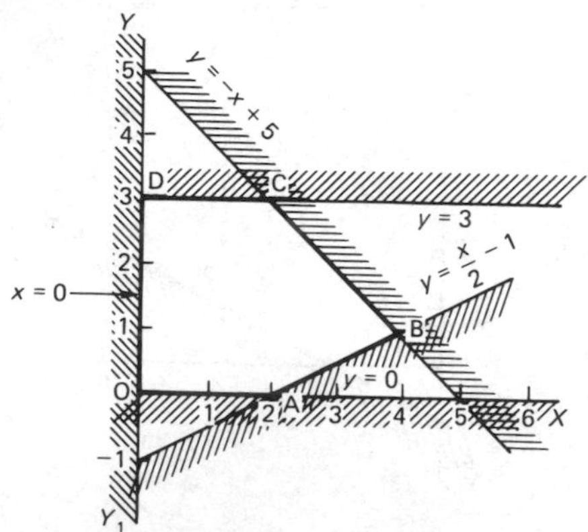

The coordinates of all points on the boundary of the polygon OABCD, or within the figure so formed, satisfy the system of constraints. The set of variables for each such point is called a *feasible point* or *feasible solution* and the figure OABCD is the *feasible domain* or *feasible polygon.*

Note these definitions.

8

Our problem now is to find the particular point within this domain that makes the objective function $P = x + 2y$ a maximum. The equation can be re-written as $y = -\dfrac{x}{2} + \dfrac{P}{2}$ and this represents a set of parallel lines with different values of the intercept $\dfrac{P}{2}$.

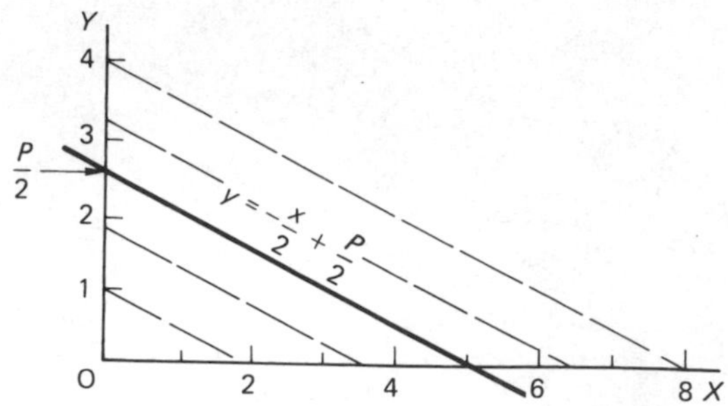

If we draw one sample line of this set to cross the feasible polygon we have just obtained, we get

. . . .

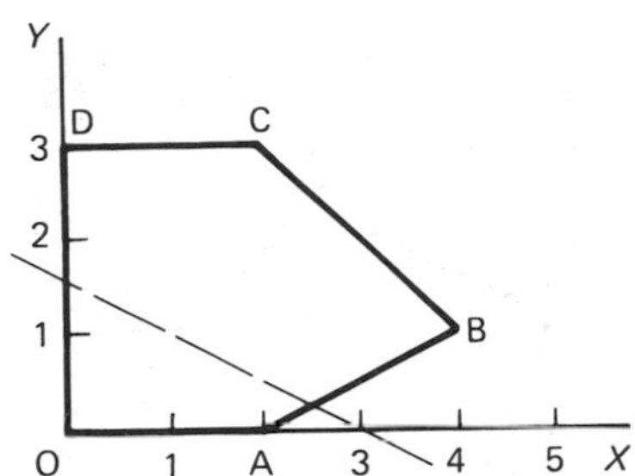

We can increase the value of P and hence raise the objective line up the page until it passes through the extreme point C. Any further increase in the value of P would take the line outside the feasible polygon and hence fail to conform to the given set of constraints.

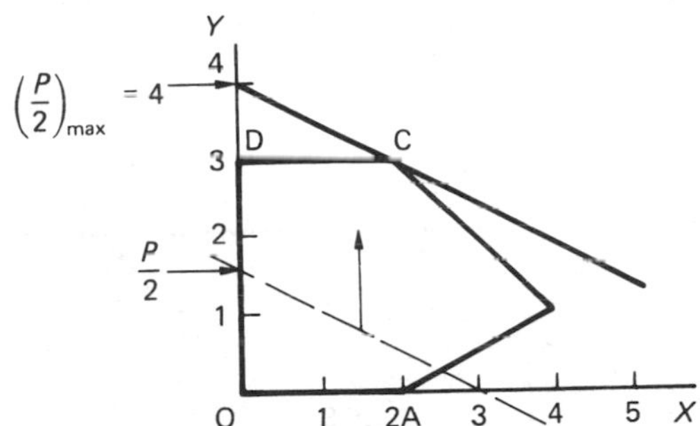

In this example, then, point C gives the optimal solution.

$$\left(\frac{P}{2}\right)_{\text{max}} = 4 \qquad \therefore\ P_{\text{max}} = 8$$

and this occurs when $x = 2,\ y = 3$ (the coordinates of C).

A graphical method of solution is clearly limited to linear programming problems involving two variables only. However, it is a useful introduction to other techniques, so let us deal with another example.

Example 2

$$\begin{aligned} \text{Maximise} \quad & P = x + 4y \\ \text{subject to} \quad & -x + 2y \leq 6 \\ & 5x + 4y \leq 40 \\ & x, y \geq 0 \end{aligned}$$

First of all, plot the appropriate straight line graphs to obtain the feasible polygon. This gives

. . . .

10

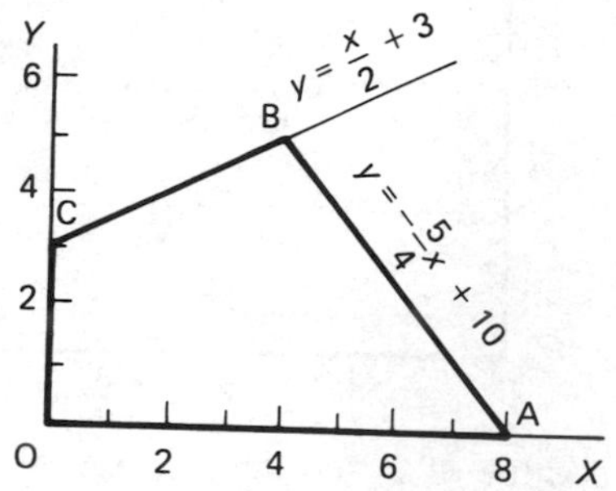

The objective function $P = x + 4y$ can be expressed in the form $y = -\frac{x}{4} + \frac{P}{4}$ and its graph added to the feasible polygon, as before. We then obtain

....

11

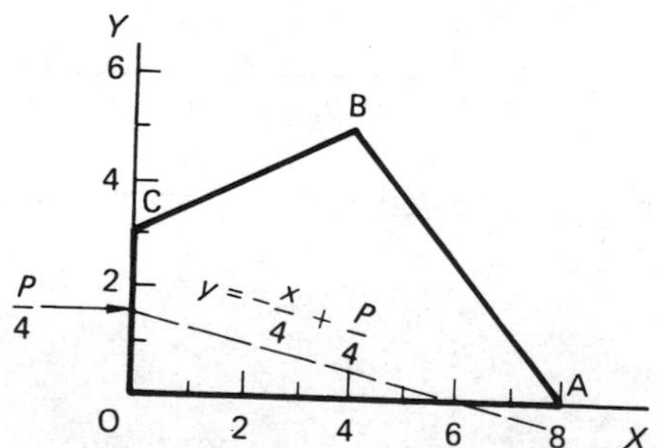

The line $y = -\frac{x}{4} + \frac{P}{4}$ is then raised to give the optimal solution, which is

....

$$\boxed{P_{max} = 24 \quad \text{with} \quad x = 4,\ y = 5}$$

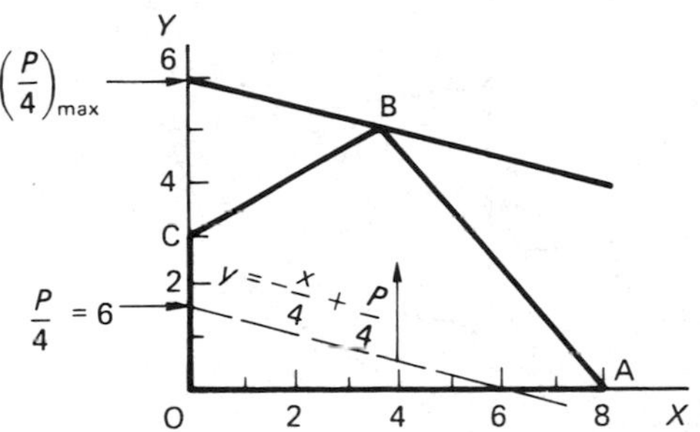

$\left(\frac{P}{4}\right)_{max} = 6 \quad \therefore \underline{P_{max} = 24 \quad \text{with} \quad x = 4,\ y = 5.}$

As easy as that.

Now this one.

Example 3

$$\begin{aligned} &\text{Minimise} && P = -4x + 6y \\ &\text{subject to} && -x + 6y \geq 24 \\ &&& 2x - y \leq 7 \\ &&& x + 8y \leq 80 \\ &&& x, y \geq 0 \end{aligned}$$

It is very much as before. Complete it on your own.

13

$$P_{\min} = 6 \quad \text{with} \quad x = 6, \; y = 5$$

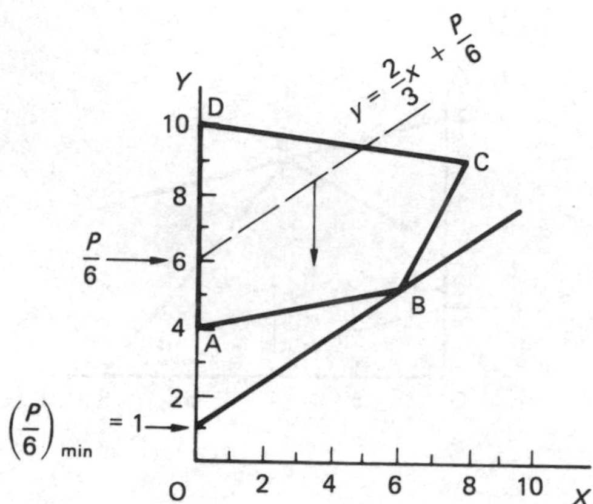

To obtain the minimum optimal value of P, the graph of the objective function is, of course, lowered to the appropriate extreme point.

In practice, linear programming problems usually contain many more variables than the two we have so far considered and a computational method is then required. One such technique is the *simplex method* and the remainder of this programme will be devoted to the steps necessary to put it into practice.

So move on to frame 14.

14

The simplex method

The first step in the *simplex method* is to ensure that each constraint is written with a *positive* right-hand side constant term. Then we express all inequalities as equations by the introduction of *slack variables*.

For example, $-x + 2y \leq 6$ can be written $-x + 2y + w_1 \qquad = 6$

and $\qquad 5x + 4y \leq 40$ can be written $5x + 4y \qquad + w_2 = 40$

where w_1 and w_2 are positive (or zero) variables with unit coefficients, required to make up the left-hand side to the value of the right hand side constant term. The new variables, w_1 and w_2, are called *slack variables*.

Let us look again at the problem we solved earlier.

Example 1

Maximise $P = x + 4y$

subject to $-x + 2y \leq 6$

$5x + 4y \leq 40 \qquad$ (As always, $x, y \geq 0$.)

The constraints now become

$$-x + 2y + w_1 \qquad = 6$$
$$5x + 4y \qquad + w_2 = 40$$

and the objective function

$$P - x - 4y \qquad = 0$$

From this, we can now begin to form the simplex tableau (or table).

So make a note of the above information — and then move on.

15

Setting up the simplex tableau

(a) *Framework* First construct a framework with the headings shown.

x	y	w_1	w_2	b	*check*

Next, we enter in the framework, the coefficients of the problem variables and of the slack variables in the constraints, together with the right-hand side constants in the column headed b. (Ignore the *check* column for the time being.)

So we have

. . . .

16

Problem variables		Slack variables		Const.	
x	y	w_1	w_2	b	*check*
-1	2	1	0	6	
5	4	0	1	40	
body		unity matrix			

(b) *Check column* The right-hand side column is included to provide a check on the numerical calculations as we develop the simplex, so, for each row, total up the entries in that row, including the constant column, and enter the sum in the check column.

Do that

17

Basis	x	y	w_1	w_2	b	check
w_1	−1	2	1	0	6	8
w_2	5	4	0	1	40	50

(c) *Starting basic solution* The two constraints now contain four variables, but if we start by letting x and y each be zero, then we have the temporary solutions, $w_1 = 6$ and $w_2 = 40$, and we indicate there variables in the extra left-hand side column, as shown.

Note (i) The columns with the slack variables form a unity matrix.
(ii) There are now four variables, x, y, w_1, w_2 $(n = 4)$.
(iii) There are two constraints $(m = 2)$.
(iv) We put $(n - m)$ variables, i.e. two variables (x and y), equal to zero as a start.

Finally, we have to deal with the objective function, so move on to the next frame.

18

(d) *The objective function* The objective function, $P = x + 4y$, is written $P - x - 4y = 0$ and this forms the bottom row, or *index row*, of the tableau, thus:

Basis	x	y	w_1	w_2	b	check
w_1	−1	2	1	0	6	8
w_2	5	4	0	1	40	50
P	−1	−4	0	0	0	−5

Complete your tableau, if you have not already done so, and then we will see how the computation is carried out.

Computation of the simplex

(i) First we select the column containing the most negative entry in the index row: in this case –4. This is called the *key column* and we enclose it as shown.

Basis	x	y	w_1	w_2	b	*check*
w_1	– 1	2	1	0	6	8
w_2	5	4	0	1	40	50
P	– 1	–4	0	0	0	– 5

key column

(ii) In each row, we now divide the value in the b column by the positive entry in the key column: the smaller ratio determines the *key row*.

For row 1 (w_1), $r = \frac{6}{2} = 3$

row 2 (w_2), $r = \frac{40}{4} = 10$

$\therefore$ row 1 is the key row. Enclose it as shown

Basis	x	y (pivot)	w_1	w_2	b	*check*
w_1	–1	2	1	0	6	8 ← key row
w_2	5	4	0	1	40	50
P	–1	–4	0	0	0	– 5

The number at the intersection of the key column and key row is the *key number* or *pivot*: in this case the number 2.

(iii) We now divide all entries in the key row by the pivot to reduce the pivot to *unity* – which we then circle. The new version of the key row is sometimes called the *main row*. The rest of the tableau remains unchanged, so we then get

....

20

unit pivot (y column, w_1 row)

Basis	x	y	w_1	w_2	b	check	
w_1	$-\frac{1}{2}$	①	$\frac{1}{2}$	0	3	4	← main row
w_2	5	4	0	1	40	50	
P	-1	-4	0	0	0	-5	← index row

↑ key column (y)

So far, so good. Now we deal with the actual calculations.

Next frame.

21

(iv) Using the main row, we now operate on the remaining rows of the tableau, including the index row, to reduce the other entries in the key column to zero. Note that the main row remains unaltered. The new value in any position in the other rows, including the b column and the check column, can be calculated as follows:

New number = old number − the product of the corresponding entries in the main row and key column

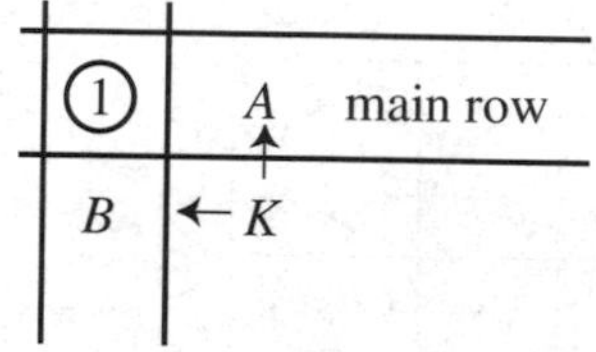

K is replaced by $K-(A\times B)$

For example, in the second row (w_2):

$$5 \text{ is replaced by } 5-(-\tfrac{1}{2})(4)=5+2=7$$
$$4 \text{ is replaced by } 4-(1)(4)=4-4=0$$
$$0 \text{ is replaced by } 0-(\tfrac{1}{2})(4)=0-2=-2$$
$$50 \text{ is replaced by } 50-(4)(4)=50-16=34, \text{ etc.}$$

and, in the third row (P):

-1 is replaced by $-1-(-\frac{1}{2})(-4)=-1-2=-3$.

Completing the operations for rows (w_2) and (P), we have

. . . .

22

Basis	x	y	w_1	w_2	b	*check*
w_1	$-\frac{1}{2}$	1	$\frac{1}{2}$	0	3	4
w_2	7	0	-2	1	28	34
P	-3	0	2	0	12	11

Now confirm that the new values in the check column are, indeed, the sums of the entries in the corresponding rows. If not, there is a mistake somewhere in the working to be corrected before we proceed.

If all is well, move on to the next frame.

23

(v) *Change of basic variables* In its final form, the key column consists of a single 1 and the remaining entries zero. This is in the column headed y which indicates that the basic variable w_1 in the main row can be replaced by y.

Basis	x	y	w_1	w_2	b	*check*
y ~~w_1~~	$-\frac{1}{2}$	1	$\frac{1}{2}$	0	3	4
w_2	7	0	-2	1	28	34
P	-3	0	2	0	12	11

Note that there are two columns containing a single 1 and the rest 0. These are headed y and w_2 which are now also the basic variables in the left-hand side column. Reading the values in the b column therefore gives a basic solution $y = 3$ and $w_2 = 28$. Any variable not listed in the basis column is zero. One basic solution at this stage is, therefore $x = 0$, $y = 3$, $w_2 = 28$. However, we are not finished.

The index row (P) still contains another negative entry, so we have to repeat the simplex process using the same steps as before.

Basis	x	y	w_1	w_2	b	*check*	
y	$-\frac{1}{2}$	1	$\frac{1}{2}$	0	3	4	
w_2	7	0	-2	1	28	34	← key row
P	-3	0	2	0	12	11	
	↑ key column						

Now divide the key row by the key number (7) to reduce the pivot to a unit pivot. This gives

....

24

Basis	x	y	w_1	w_2	b	*check*	
y	$-\frac{1}{2}$	1	$\frac{1}{2}$	0	3	4	
w_2	(1)	0	$-\frac{2}{7}$	$\frac{1}{7}$	4	$\frac{34}{7}$	← main row
P	-3	0	2	0	12	11	

Using the main row, operate on the remaining rows (including the index row) to reduce the other entries in the key column to zero. Complete that stage and we have

....

25

Basis	x	y	w_1	w_2	b	*check*
y	0	1	$\frac{5}{14}$	$\frac{1}{14}$	5	$\frac{45}{7}$
w_2	1	0	$-\frac{2}{7}$	$\frac{1}{7}$	4	$\frac{34}{7}$
P	0	0	$\frac{8}{7}$	$\frac{3}{7}$	24	$\frac{179}{7}$

Again, at this stage, check your working by totalling up the entries in each row and satisfy yourself that the sum agrees with the value in the check column.

Note that w_2 in the basis column can now be replaced by x which was the heading of the column containing the last unit pivot.

So finally, we have:

Basis	x	y	w_1	w_2	b	*check*
y	0	1	$\frac{5}{14}$	$\frac{1}{14}$	5	$\frac{45}{7}$
x	1	0	$-\frac{2}{7}$	$\frac{1}{7}$	4	$\frac{34}{7}$
P	0	0	$\frac{8}{7}$	$\frac{3}{7}$	24	$\frac{179}{7}$

A new basic solution now emerges as $x = 4$, $y = 5$.

Furthermore, since there is no further negative entry in the index row, this is also the optimal solution and the optimal value of P is given in the b column, i.e $\underline{P_{max} = 24}$.

For interest, you may wish to compare this result with that obtained in frame 12.

We have been through the simplex operation in some detail by way of explanation. Many problems involve more than just two variables, but the method of computation is basically the same, being an iterative process which is repeated until the index row contains no negative entry, at which point the optimal value of the objective function has been attained. **26**

The problem we have just solved would normally look like this:

Maximise $P = x + 4y$
subject to $-x + 2y \leq 6$
$5x + 4y \leq 40 \quad (x,\ y \geq 0)$

Entering slack variables, etc., this is written

$$-x + 2y + w_1 \qquad = 6$$
$$5x + 4y \qquad + w_2 = 40$$
$$P - x - 4y \qquad = 0$$

The complete tableau is given in the next frame.

27

Basis	x	y	w_1	w_2	b	*check*
w_1	-1	2	1	0	6	8
w_2	5	4	0	1	40	50
P	-1	-4	0	0	0	-5
y ~~w_1~~	$-\frac{1}{2}$	(1)	$\frac{1}{2}$	0	3	4
w_2	5	4	0	1	40	50
P	-1	-4	0	0	0	-5
y ~~w_1~~	$-\frac{1}{2}$	1	$\frac{1}{2}$	0	3	4
w_2	7	0	-2	1	28	34
P	-3	0	2	0	12	11
y	$-\frac{1}{2}$	1	$\frac{1}{2}$	0	3	4
w_2	(1)	0	$-\frac{2}{7}$	$\frac{1}{7}$	4	$\frac{34}{7}$
P	-3	0	2	0	12	11
y	0	1	$\frac{5}{14}$	$\frac{1}{14}$	5	$\frac{45}{7}$
x ~~w_2~~	1	0	$-\frac{2}{7}$	$\frac{1}{7}$	4	$\frac{34}{7}$
P	0	0	$\frac{8}{7}$	$\frac{3}{7}$	24	$\frac{179}{7}$

$$\underline{P_{max} = 24 \text{ with } x = 4,\ y = 5}$$

Now for another example — so move on to frame 28.

28

Here is one for you to do on your own. The method is just the same as before so you will have no difficulty.

Example 2

$$\begin{aligned} &\text{Maximise} \quad P = 4x + 3y \\ &\text{subject to} \quad -x + y \le 4 \\ &\qquad\qquad x + 2y \le 14 \\ &\qquad\qquad 2x + y \le 16 \quad (x,\ y \ge 0) \end{aligned}$$

We have three inequalities this time, so we shall need to introduce three slack variables. Converting the inequalities into equations, we obtain

. . . .

29

$$\begin{aligned} -x + y + w_1 \qquad\qquad &= 4 \\ x + 2y \quad + w_2 \qquad &= 14 \\ 2x + y \qquad\quad + w_3 &= 16 \end{aligned}$$

Then we set out the simplex framework with appropriate headings, i.e.

. . . .

30

Basis	x	y	w_1	w_2	w_3	b	*check*

Remembering that the index row uses $P - 4x - 3y = 0$, we can set out the first tableau. Choosing x and y, as usual, to be zero for a start, we have

. . . .

31

Basis	x	y	w_1	w_2	w_3	b	*check*
w_1	−1	1	1	0	0	4	5
w_2	1	2	0	1	0	14	18
w_3	2	1	0	0	1	16	20
P	−4	−3	0	0	0	0	−7

Carry on now and complete the working on this first tableau.

Check with the next frame.

32

Here is the working so far:

Basis	x	y	w_1	w_2	w_3	b	*check*
w_1	-1	1	1	0	0	4	5
w_2	1	2	0	1	0	14	18
w_3	2	1	0	0	1	16	20
P	-4	-3	0	0	0	0	-7
w_1	-1	1	1	0	0	4	5
w_2	1	2	0	1	0	14	18
w_3	(1)	$\frac{1}{2}$	0	0	$\frac{1}{2}$	8	10
P	-4	-3	0	0	0	0	-7
w_1	0	$\frac{3}{2}$	1	0	$\frac{1}{2}$	12	15
w_2	0	$\frac{3}{2}$	0	1	$-\frac{1}{2}$	6	8
x ~~w_3~~	1	$\frac{1}{2}$	0	0	$\frac{1}{2}$	8	10
P	0	-1	0	0	2	32	33

(a) The basic variable (w_3) of the unit pivot can now be replaced by the variable at the heading of the unit pivot (x).

(b) We see there is still a negative value in the index row, so we repeat the process for this second tableau.

Now you can finish it off.

33 Check to see if you agree.

Basis	x	y	w_1	w_2	w_3	b	*check*
w_1	0	$\frac{3}{2}$	1	0	$\frac{1}{2}$	12	15
w_2	0	$\frac{3}{2}$	0	1	$-\frac{1}{2}$	6	8
x	1	$\frac{1}{2}$	0	0	$\frac{1}{2}$	8	10
P	0	-1	0	0	2	32	33
w_1	0	$\frac{3}{2}$	1	0	$\frac{1}{2}$	12	15
w_2	0	(1)	0	$\frac{2}{3}$	$-\frac{1}{3}$	4	$\frac{16}{3}$
x	1	$\frac{1}{2}$	0	0	$\frac{1}{2}$	8	10
P	0	-1	0	0	2	32	33
w_1	0	0	1	-1	1	6	7
y ~~w_2~~	0	1	0	$\frac{2}{3}$	$-\frac{1}{3}$	4	$\frac{16}{3}$
x	1	0	0	$-\frac{1}{3}$	$\frac{2}{3}$	6	$\frac{22}{3}$
P	0	0	0	$\frac{2}{3}$	$\frac{5}{3}$	36	$\frac{115}{3}$

The basic variable (w_2) can now be replaced by y, being the heading of the unit pivot column.

There is no further negative entry in the index row: therefore, the optimal value of P has been attained.

$$\therefore\; P_{\max} = 36 \quad \text{with} \quad x = 6,\; y = 4.$$

Note: We also see that $w_1 = 6$, since the unity matrix has headings x, y, w_1. The full result, therefore, is

$$P_{\max} = 36 \quad \text{with} \quad x = 6,\; y = 4,\; w_1 = 6,\; w_2 = 0,\; w_3 = 0$$

though we do not normally require this extra information.

Now we will extend the simplex method to an example involving three problem variables.

Next frame.

34

Simplex with three problem variables

$$\begin{aligned} \text{Maximise} \quad & P = p_1x + p_2y + p_3z \\ \text{subject to} \quad & a_{11}x + a_{12}y + a_{13}z \leq b_1 \\ & a_{21}x + a_{22}y + a_{23}z \leq b_2 \\ & a_{31}x + a_{32}y + a_{33}z \leq b_3 \\ & x, y, z \geq 0 \end{aligned}$$

Introducing slack variables we have

$$\begin{aligned} a_{11}x + a_{12}y + a_{13}z + w_1 \qquad\qquad &= b_1 \\ a_{21}x + a_{22}y + a_{23}z \qquad + w_2 \qquad &= b_2 \\ a_{31}x + a_{32}y + a_{33}z \qquad\qquad + w_3 &= b_3 \\ P - p_1x - p_2y - p_3z \qquad\qquad &= 0 \end{aligned}$$

If there is now a total of n variables and m constraints, then at least $(n - m)$ variables are equated to zero. The remainder form the basic variable column entries. Equating x, y, z to zero, then, the basic variables are w_1, w_2, w_3.

Basis	x	y	z	w_1	w_2	w_3	b	*check*
w_1	a_{11}	a_{12}	a_{13}	1	0	0	b_1	
w_2	a_{21}	a_{22}	a_{23}	0	1	0	b_2	
w_3	a_{31}	a_{32}	a_{33}	0	0	1	b_3	
P	$-p_1$	$-p_2$	$-p_3$	0	0	0	0	

The variables in the basis column are the variables heading the unity matrix. The method is exactly as before.

(a) Select the most negative entry in the index row to determine the *key column*.
(b) Divide the entries in the constant column (b) by the corresponding positive entries in the key column. The smallest positive ratio determines the *key row*.
(c) The entry at the intersection of the key column and the key row is the *key number* or *pivot*.
(d) Divide each entry in the key row by the pivot to reduce the key number to a *unit pivot*. The revised key row is now called the *main row*.
(e) Use the main row to operate on the remaining rows to reduce all other entries in the key column to zero.
(f) Repeat steps (a) to (e) until no negative entry remains in the index row.

Now for an example.

35 *Example 1*

$$\begin{aligned} \text{Maximise} \quad & P = 2x + 6y + 4z \\ \text{subject to} \quad & 2x + 5y + 2z \le 38 \\ & 4x + 2y + 3z \le 57 \\ & x + 3y + 5z \le 57 \\ & x, y, z \ge 0 \end{aligned}$$

Re-writing the inequalities as equations, gives

....

36

$$\begin{aligned} 2x + 5y + 2z + w_1 \quad \quad \quad \quad &= 38 \\ 4x + 2y + 3z \quad + w_2 \quad \quad &= 57 \\ x + 3y + 5z \quad \quad \quad + w_3 &= 57 \end{aligned}$$

We also have $P - 2x - 6y - 4z = 0$, so we can now set up the simplex tableau ready for solution. That is ....

37

Basis	x	y	z	w_1	w_2	w_3	b	*check*
w_1	2	5	2	1	0	0	38	48
w_2	4	2	3	0	1	0	57	67
w_3	1	3	5	0	0	1	57	67
P	−2	−6	−4	0	0	0	0	−12

Now we just apply the normal simplex routine until there is no negative entry in the index row.

Remember:

(i) to replace the basic variables as the problem variables become available at each stage, and
(ii) any variable not appearing in the basis column has zero value.

Now you can work the solution right through and then check the result with the next frame. Take your time: there are no snags.

38

Basis	x	y	z	w_1	w_2	w_3	b	*check*
w_1	2	5	2	1	0	0	38	48
w_2	4	2	3	0	1	0	57	67
w_3	1	3	5	0	0	1	57	67
P	-2	-6	-4	0	0	0	0	-12
w_1	$\frac{2}{5}$	(1)	$\frac{2}{5}$	$\frac{1}{5}$	0	0	$\frac{38}{5}$	$\frac{48}{5}$
w_2	4	2	3	0	1	0	57	67
w_3	1	3	5	0	0	1	57	67
P	-2	-6	-4	0	0	0	0	-12
y ~~w_1~~	$\frac{2}{5}$	1	$\frac{2}{5}$	$\frac{1}{5}$	0	0	$\frac{38}{5}$	$\frac{48}{5}$
w_2	$\frac{16}{5}$	0	$\frac{11}{5}$	$-\frac{2}{5}$	1	0	$\frac{209}{5}$	$\frac{239}{5}$
w_3	$-\frac{1}{5}$	0	$\frac{19}{5}$	$-\frac{3}{5}$	0	1	$\frac{171}{5}$	$\frac{191}{5}$
P	$\frac{2}{5}$	0	$-\frac{8}{5}$	$\frac{6}{5}$	0	0	$\frac{228}{5}$	$\frac{228}{5}$
y	$\frac{2}{5}$	1	$\frac{2}{5}$	$\frac{1}{5}$	0	0	$\frac{38}{5}$	$\frac{48}{5}$
w_2	$\frac{16}{5}$	0	$\frac{11}{5}$	$-\frac{2}{5}$	1	0	$\frac{209}{5}$	$\frac{239}{5}$
w_3	$-\frac{1}{19}$	0	(1)	$-\frac{3}{19}$	0	$\frac{3}{19}$	9	$\frac{191}{19}$
P	$\frac{2}{5}$	0	$-\frac{8}{5}$	$\frac{6}{5}$	0	0	$\frac{228}{5}$	$\frac{228}{5}$
y	$\frac{8}{19}$	1	0	$\frac{5}{19}$	0	$-\frac{2}{19}$	4	$\frac{106}{19}$
w_2	$\frac{63}{19}$	0	0	$-\frac{1}{19}$	1	$-\frac{11}{19}$	22	$\frac{448}{19}$
z ~~w_3~~	$-\frac{1}{19}$	0	1	$-\frac{3}{19}$	0	$\frac{5}{19}$	9	$\frac{191}{19}$
P	$\frac{16}{19}$	0	0	$\frac{18}{19}$	0	$\frac{8}{19}$	60	$\frac{1172}{19}$

$$\therefore\ P_{max} = 60 \quad \text{with} \quad x = 0,\ y = 4,\ z = 9.$$

Do you agree?

If so, on to the next frame.

39 *Example 2*

$$\begin{aligned} \text{Maximise} \quad & P = 3x + 4y + 5z \\ \text{subject to} \quad & 2x + 4y + 3z \le 80 \\ & 4x + 2y + z \le 48 \\ & x + y + 2z \le 40 \\ & x, y, z \ge 0 \end{aligned}$$

It is much the same as before. Work through it carefully and then check with the next frame.

40

$$\begin{aligned} 2x + 4y + 3z + w_1 \qquad\qquad & = 80 \\ 4x + 2y + z \qquad + w_2 \qquad & = 48 \\ x + y + 2z \qquad\qquad + w_3 & = 40 \\ P - 3x - 4y - 5z \qquad\qquad\qquad & = 0 \end{aligned}$$

Basis	x	y	z	w_1	w_2	w_3	b	*check*
w_1	2	4	3	1	0	0	80	90
w_2	4	2	1	0	1	0	48	56
w_3	1	1	2	0	0	1	40	45
P	-3	-4	-5	0	0	0	0	-12
w_1	2	4	3	1	0	0	80	90
w_2	4	2	1	0	1	0	48	56
w_3	$\frac{1}{2}$	$\frac{1}{2}$	(1)	0	0	$\frac{1}{2}$	20	$\frac{45}{2}$
P	-3	-4	-5	0	0	0	0	-12
w_1	$\frac{1}{2}$	$\frac{5}{2}$	0	1	0	$-\frac{3}{2}$	20	$\frac{45}{2}$
w_2	$\frac{7}{2}$	$\frac{3}{2}$	0	0	1	$-\frac{1}{2}$	28	$\frac{67}{2}$
z ~~w_3~~	$\frac{1}{2}$	$\frac{1}{2}$	1	0	0	$\frac{1}{2}$	20	$\frac{45}{2}$
P	$-\frac{1}{2}$	$\frac{3}{2}$	0	0	0	$\frac{5}{2}$	100	$\frac{201}{2}$
w_1	$\frac{1}{5}$	(1)	0	$\frac{2}{5}$	0	$-\frac{3}{5}$	8	9
w_2	$\frac{7}{2}$	$\frac{3}{2}$	0	0	1	$-\frac{1}{2}$	28	$\frac{67}{2}$
z	$\frac{1}{2}$	$\frac{1}{2}$	1	0	0	$\frac{1}{2}$	20	$\frac{45}{2}$
P	$-\frac{1}{2}$	$-\frac{3}{2}$	0	0	0	$\frac{5}{2}$	100	$\frac{201}{2}$
y ~~w_1~~	$\frac{1}{5}$	1	0	$\frac{2}{5}$	0	$-\frac{3}{5}$	8	9
w_2	$\frac{16}{5}$	0	0	$-\frac{3}{5}$	1	$\frac{2}{5}$	16	20
z	$\frac{2}{5}$	0	1	$-\frac{1}{5}$	0	$\frac{4}{5}$	16	18
P	$-\frac{1}{5}$	0	0	$\frac{3}{5}$	0	$\frac{8}{5}$	112	114
y	$\frac{1}{5}$	1	0	$\frac{2}{5}$	0	$-\frac{3}{5}$	8	9
w_2	(1)	0	0	$-\frac{3}{16}$	$\frac{5}{16}$	$\frac{1}{8}$	5	$\frac{25}{4}$
z	$\frac{2}{5}$	0	1	$-\frac{1}{5}$	0	$\frac{4}{5}$	16	18
P	$-\frac{1}{5}$	0	0	$\frac{3}{5}$	0	$\frac{8}{5}$	112	114
y	0	1	0	$\frac{7}{16}$	$-\frac{1}{16}$	$-\frac{5}{8}$	7	$\frac{31}{4}$
x ~~w_2~~	1	0	0	$-\frac{3}{16}$	$\frac{5}{16}$	$\frac{1}{8}$	5	$\frac{25}{4}$
z	0	0	1	$-\frac{1}{8}$	$-\frac{1}{8}$	$\frac{3}{4}$	14	$\frac{31}{2}$
P	0	0	0	$\frac{9}{16}$	$\frac{1}{16}$	$\frac{13}{8}$	113	$\frac{461}{4}$

$\therefore\ \underline{P_{max} = 113 \quad \text{with } x = 5,\ y = 7,\ z = 14.}$

Now let us meet a further complication. *Next frame.*

41 Artificial variables

So far, our approach to each problem has been the same.

(a) We first of all convert the 'less than' inequalities into equations by the inclusion of slack variables.
(b) If there are now n variables and m constraints, then at least $(n - m)$ variables are equated to zero – usually x, y, z, etc. – so that the initial basic solution is given by the slack variables, the coefficients of which form the unity matrix in the tableau.
(c) We then proceed by the simplex method to convert the basic solution to one containing the problem variables, the tableau entries for which now form a new unity matrix.
(d) The method is repeated as necessary. When no negative entry remains in the index row, the value of P denoted in the constant column is the optimal value of the objective function.

Now let us look at this example.

Example 1

$$\begin{aligned} \text{Maximise} \quad & P = 7x + 4y \\ \text{subject to} \quad & 2x + \ y \leq 150 \\ & 4x + 3y \leq 350 \\ & x + \ y \geq \ 80 \quad (x, y \geq 0) \end{aligned}$$

Converting the inequalities to equations, we have

....

42

$$\begin{aligned} 2x + y + w_1 \qquad\qquad &= 150 \\ 4x + 3y \qquad + w_2 \qquad &= 350 \\ x + y \qquad\qquad - w_3 &= \ 80 \end{aligned}$$

Also, of course, $P - 7x - 4y = 0$.

NOTE that since the third constraint is a 'greater than' statement, we must subtract the slack positive variable (w_3) to form the equation.

Forming the first tableau, in the usual manner, we obtain

....

43

Basis	x	y	w_1	w_2	w_3	b	*check*
w_1	2	1	1	0	0	150	154
w_2	4	3	0	1	0	350	358
w_3	1	1	0	0	−1	80	81
P	−7	−4	0	0	0	0	−11

Now we are stuck, for we do not have a unity matrix to start off with. The entry in the w_3 column is −1 and not the necessary +1, and no amount of manipulation will help since the entries in the constant column (b) are, by definition, positive.

So how can we find a starting technique?

Let us restate the problem.

44

Maximise $P - 7x + 4y$

subject to

$$\begin{aligned} 2x + y + w_1 \qquad\qquad &= 150 \\ 4x + 3y \qquad + w_2 \qquad &= 350 \\ x + y \qquad\qquad - w_3 &= 80 \end{aligned}$$

The trouble comes in the third constraint by virtue of the negative sign of the slack variable. To save the situation, we introduce a new small positive variable (w_4) so that w_1, w_2 and w_4 will give rise to a unity matrix and the simplex computation can then proceed. Of course, w_4 is ficticious, is extremely small and cannot appear in the final basic solution. To establish this, we include in the objective function a new term $-Mw_4$, where M is an extremely large positive value which will ensure that w_4 will ultimately vanish. So we now write

$$P = 7x + 4y - Mw_4$$

The new variable, w_4, is called an *artificial variable*: it is introduced solely so that the simplex procedure can be carried out; and it must not appear in the final basic solution listed in the basis column.

The third constraint above now becomes

$$x + y \qquad - w_3 + w_4 = 80$$

Next frame.

45 We now have:

$$\begin{aligned} 2x + y + w_1 &= 150 \\ 4x + 3y + w_2 &= 350 \\ x + y - w_3 + w_4 &= 80 \\ P - 7x - 4y + Mw_4 &= 0 \end{aligned}$$

Forming the tableau in the usual way:

Basis	x	y	w_1	w_2	w_3	w_4	b	*check*
w_1	2	1	1	0	0	0	150	154
w_2	4	3	0	1	0	0	350	358
w_3	1	1	0	0	−1	1	80	82
P	−7	−4	0	0	0	M	0	$M - 11$

Note that:

(i) The columns headed w_1, w_2, w_4 now form the unity matrix.
(ii) There are now 6 variables and 3 constraints, i.e. $n = 6$ and $m = 3$. At least $(n - m)$, i.e. $6 - 3 = 3$, variables are put equal to zero. We start off with x, y, w_3 as zero and w_1, w_2, w_4 form the first basic solution with the values given in the b column.

We now proceed in the normal way. Solve the first tableau and check the results so far in the following frame.

46

Basis	x	y	w_1	w_2	w_3	w_4	b	*check*
w_1	2	1	1	0	0	0	150	154
w_2	4	3	0	1	0	0	350	358
w_4	1	1	0	0	−1	1	80	82
P	−7	−4	0	0	0	M	0	$M-11$
w_1	(1)	$\frac{1}{2}$	$\frac{1}{2}$	0	0	0	75	77
w_2	4	3	0	1	0	0	350	358
w_4	1	1	0	0	−1	1	80	82
P	−7	−4	0	0	0	M	0	$M-11$
x ~~w_1~~	1	$\frac{1}{2}$	$\frac{1}{2}$	0	0	0	75	77
w_2	0	1	−2	1	0	0	50	50
w_4	0	$\frac{1}{2}$	$-\frac{1}{2}$	0	−1	1	5	5
P	0	$-\frac{1}{2}$	$\frac{7}{2}$	0	0	M	525	$M+528$

The basic variable w_1 can be replaced by x and we continue as before to remove the further negative entry in the index row.

Do that.

47

Basis	x	y	w_1	w_2	w_3	w_4	b	check
x	1	$\frac{1}{2}$	$\frac{1}{2}$	0	0	0	75	77
w_2	0	1	-2	1	0	0	50	50
w_4	0	$\frac{1}{2}$	$-\frac{1}{2}$	0	-1	1	5	5
P	0	$-\frac{1}{2}$	$\frac{7}{2}$	0	0	M	525	$M+528$
x	1	$\frac{1}{2}$	$\frac{1}{2}$	0	0	0	75	77
w_2	0	1	-2	1	0	0	50	50
w_4	0	(1)	-1	0	-2	2	10	10
P	0	$-\frac{1}{2}$	$\frac{7}{2}$	0	0	M	525	$M+528$
x	1	0	1	0	1	-1	70	72
w_2	0	0	-1	1	2	-2	40	40
y ~~w_4~~	0	1	-1	0	-2	2	10	10
P	0	0	3	0	-1	$M+1$	530	$M+533$

The basic variable w_4 is now replaced by y (the column of the last unit pivot). We now have another negative entry in the index row, so we have to perform the simplex calculation yet again.

For the next round, we get ....

48

Basis	x	y	w_1	w_2	w_3	w_4	b	check
x	1	0	1	0	1	-1	70	72
w_2	0	0	-1	1	2	-2	40	40
y	0	1	-1	0	-2	2	10	10
P	0	0	3	0	-1	$M+1$	530	$M+533$
x	1	0	1	0	1	-1	70	72
w_2	0	0	$-\frac{1}{2}$	$\frac{1}{2}$	(1)	-1	20	20
y	0	1	-1	0	-2	2	10	10
P	0	0	3	0	-1	$M+1$	530	$M+533$
x	1	0	$\frac{3}{2}$	$-\frac{1}{2}$	0	0	50	52
w_3 ~~w_2~~	0	0	$-\frac{1}{2}$	$\frac{1}{2}$	1	-1	20	20
y	0	1	-2	1	0	0	50	50
P	0	0	$\frac{5}{2}$	$\frac{1}{2}$	0	M	550	$M+553$

No further negative entry remains in the index row. The optimal solution has been found i.e.

$$\underline{P_{\max} = 550 \text{ with } x = 50,\ y = 50.}$$

In addition, we see that $w_3 = 20$ while $w_1 = w_2 = w_4 = 0$ since they do not occur in the basic variable column.

Notice, also, that w_4, the artificial variable, does not figure in the optimal solution — as indeed it must not.

Next frame.

49 Here is one for you to do in the same way.

Example 2

$$\begin{aligned} \text{Maximise} \quad & P = 2x + 5y \\ \text{subject to} \quad & x + 4y \leq 60 \\ & 3x + 2y \leq 40 \\ & x + \ y \geq 12 \quad (x, y \geq 0) \end{aligned}$$

Work right through it, just as before. The result is

....

50

$$\boxed{P_{max} = 78 \quad \text{with } x = 4, \ y = 14}$$

$$\begin{aligned} \text{For} \quad & x + 4y + w_1 && = 60 \\ & 3x + 2y + w_2 && = 40 \\ & x + \ y - w_3 + w_4 && = 12 \\ & P - 2x - 5y + Mw_4 && = \ 0 \end{aligned}$$

Basis	x	y	w_1	w_2	w_3	w_4	b	*check*
w_1	1	4	1	0	0	0	60	66
w_2	3	2	0	1	0	0	40	46
w_4	1	(1)	0	0	-1	1	12	14
P	-2	-5	0	0	0	M	0	$M-7$
w_1	-3	0	1	0	4	-4	12	10
w_2	1	0	0	1	2	-2	16	18
y ~~w_4~~	1	1	0	0	-1	1	12	14
P	3	0	0	0	-5	$M+5$	60	$M+63$
w_1	$-\frac{3}{4}$	0	$\frac{1}{4}$	0	(1)	-1	3	$\frac{5}{2}$
w_2	1	0	0	1	2	-2	16	18
y	1	1	0	0	-1	1	12	14
P	3	0	0	0	-5	$M+5$	60	$M+63$
w_3 ~~w_1~~	$-\frac{3}{4}$	0	$\frac{1}{4}$	0	1	-1	3	$\frac{5}{2}$
w_2	$\frac{5}{2}$	0	$-\frac{1}{2}$	1	0	0	10	13
y	$\frac{1}{4}$	1	$\frac{1}{4}$	0	0	0	15	$\frac{33}{2}$
P	$-\frac{3}{4}$	0	$\frac{5}{4}$	0	0	M	75	$M+\frac{151}{2}$
w_3	$-\frac{3}{4}$	0	$\frac{1}{4}$	0	1	-1	3	$\frac{5}{2}$
w_2	(1)	0	$-\frac{1}{5}$	$\frac{2}{5}$	0	0	4	$\frac{26}{2}$
y	$\frac{1}{4}$	1	$\frac{1}{4}$	0	0	0	15	$\frac{33}{2}$
P	$-\frac{3}{4}$	0	$\frac{5}{4}$	0	0	M	75	$M+\frac{151}{2}$
w_3	0	0	$\frac{1}{10}$	$\frac{3}{10}$	1	-1	6	$\frac{32}{5}$
x ~~w_2~~	1	0	$-\frac{1}{5}$	$\frac{2}{5}$	0	0	4	$\frac{26}{5}$
y	0	1	$\frac{3}{10}$	$-\frac{1}{10}$	0	0	14	$\frac{76}{5}$
P	0	0	$\frac{11}{10}$	$\frac{3}{10}$	0	M	78	$M+\frac{397}{5}$

$\therefore\ P_{max} = 78 \quad \text{with} \quad x = 4,\ y = 14.$

On to the next frame.

51 We are not always as lucky as we were in the last two examples and other steps sometimes have to be taken to remove the artificial variable. Consider the following case.

Example 3

$$\begin{aligned} \text{Maximise} \quad & P = 8x + 4y \\ \text{subject to} \quad & 2x + 3y \le 120 \\ & x + y \le 45 \\ & -3x + 5y \ge 25 \qquad (x, y \ge 0) \end{aligned}$$

Inserting the slack variables and artificial variable as required, we have

....

52

$$\begin{aligned} 2x + 3y + w_1 \qquad\qquad\qquad &= 120 \\ x + y \quad + w_2 \qquad\qquad &= 45 \\ -3x + 5y \qquad\quad - w_3 + w_4 &= 25 \\ P - 8x - 4y \qquad\qquad + Mw_4 &= 0 \end{aligned}$$

That is very much as before, so work through it and check with the next frame.

53

Here it is.

Basis	x	y	w_1	w_2	w_3	w_4	b	check
w_1	2	3	1	0	0	0	120	126
w_2	(1)	1	0	1	0	0	45	48
w_4	−3	5	0	0	−1	1	25	27
P	−8	−4	0	0	0	M	0	$M - 12$
w_1	0	1	1	−2	0	0	30	30
x ~~w_2~~	1	1	0	1	0	0	45	48
* → w_4	0	8	0	3	−1	1	160	171
P	0	4	0	8	0	M	360	$M + 372$

There is no further negative entry in the index row, so it looks as though the optimal solution has been attained. However, the artificial variable w_4 still remains in the basic variable column at * and thus must be removed. Therefore, we take the entry at the junction of the y column and the w_4 row as the pivot and proceed to eliminate w_4 by simplifying the tableau a stage further.

If we do that, we get ....

54

Basis	x	y	w_1	w_2	w_3	w_4	b	check
w_1	0	1	1	-2	0	0	30	30
x	1	1	0	1	0	0	45	48
$*\longrightarrow$ w_4	0	$\boxed{8}$	0	3	-1	1	160	171
P	0	4	0	8	0	M	360	$M+372$
w_1	0	1	1	-2	0	0	30	30
x	1	1	0	1	0	0	45	48
$*\longrightarrow$ w_4	0	(1)	0	$\frac{3}{8}$	$-\frac{1}{8}$	$\frac{1}{8}$	20	$\frac{171}{8}$
P	0	4	0	8	0	M	360	$M+372$
w_1	0	0	1	$-\frac{19}{8}$	$\frac{1}{8}$	$-\frac{1}{8}$	10	$\frac{69}{8}$
x	1	0	0	$\frac{5}{8}$	$\frac{1}{8}$	$-\frac{1}{8}$	25	$\frac{213}{8}$
y ~~w_4~~	0	1	0	$\frac{3}{8}$	$-\frac{1}{8}$	$\frac{1}{8}$	50	$\frac{171}{8}$
P	0	0	0	$\frac{13}{2}$	$\frac{1}{2}$	$M-\frac{1}{2}$	280	$M+\frac{573}{2}$

The artificial variable, w_4, is now replaced by y in the basic variable column and the optimal solution has been reached.

$$\therefore P_{\text{max}} = 280 \quad \text{with} \quad x = 25,\ y = 20.$$

Next frame.

55 Now here is one for you to deal with.

Example 4

Maximise $\quad P = 10x + 2y$

subject to

$$-x + 2y \le 60$$
$$5x + 4y \le 260$$
$$-x + 8y \ge 80 \qquad (x, y \ge 0)$$

Work through it as before and see if you agree with the solution in the next frame.

56

$$\begin{aligned} -x + 2y + w_1 \quad & = 60 \\ 5x + 4y \quad + w_2 \quad & = 260 \\ -x + 8y \quad - w_3 + \quad w_4 & = 80 \\ P - 10x - 2y \quad + Mw_4 & = 0 \end{aligned}$$

	Basis	x	y	w_1	w_2	w_3	w_4	b	*check*
	w_1	-1	2	1	0	0	0	60	62
	w_2	5	4	0	1	0	0	260	270
	w_4	-1	8	0	0	-1	1	80	87
	P	-10	-2	0	0	0	M	0	$M - 12$
	w_1	-1	2	1	0	0	0	60	62
	w_2	(1)	$\frac{4}{5}$	0	$\frac{1}{5}$	0	0	52	54
	w_4	-1	8	0	0	-1	1	80	87
	P	-10	-2	0	0	0	M	0	$M - 12$
	w_1	0	$\frac{14}{5}$	1	$\frac{1}{5}$	0	0	112	116
	x ~~w_2~~	1	$\frac{4}{5}$	0	$\frac{1}{5}$	0	0	52	54
* →	w_4	0	$\frac{44}{5}$	0	$\frac{1}{5}$	-1	1	132	141
	P	0	6	0	2	0	M	520	$M + 528$
	w_1	0	$\frac{14}{5}$	1	$\frac{1}{5}$	0	0	112	116
	x	1	$\frac{4}{5}$	0	$\frac{1}{5}$	0	0	52	54
* →	w_4	0	(1)	0	$\frac{1}{44}$	$-\frac{5}{44}$	$\frac{5}{44}$	15	$\frac{705}{44}$
	P	0	6	0	2	0	M	520	$M + 528$
	w_1	0	0	1	$\frac{3}{22}$	$\frac{7}{22}$	$-\frac{7}{22}$	70	$\frac{1565}{22}$
	x	1	0	0	$\frac{2}{11}$	$\frac{1}{11}$	$-\frac{1}{11}$	40	$\frac{453}{11}$
	y ~~w_4~~	0	1	0	$\frac{1}{44}$	$-\frac{5}{44}$	$\frac{5}{44}$	15	$\frac{705}{44}$
	P	0	0	0	$\frac{41}{22}$	$\frac{15}{22}$	$M - \frac{15}{22}$	430	$M + \frac{9501}{22}$

$\therefore\ P_{\text{max}} = 430$ with $x = 40,\ y = 15.$

Now for another example.

57 *Example 5* This one is slightly different, so take note.

$$\begin{aligned}
\text{Maximise} \quad & P = 11x + 15y \\
\text{subject to} \quad & 3x + 5y \le 130 \\
& -4x + 5y \ge 25 \\
& x + 5y \ge 75 \qquad (x, y \ge 0)
\end{aligned}$$

In this problem, notice that there are two 'greater than' inequalities so that there will be two slack variables to be subtracted and two artificial variables to be incorporated. In the objective function, we can use the same factor, $-M$, for both artificial variables, since neither of those two variables will appear in the final optimal solution. So, we have:

$$\begin{aligned}
3x + 5y + w_1 \qquad\qquad\qquad\qquad\qquad &= 130 \\
-4x + 5y \quad - w_2 \quad + w_4 \qquad\quad &= 25 \\
x + 5y \quad\quad - w_3 \quad\quad + w_5 &= 75 \\
P - 11x - 15y \qquad\qquad + Mw_4 + Mw_5 &= 0
\end{aligned}$$

w_1, w_4, w_5 now form the unity matrix from which to start. The method is just the same as in previous examples, so finish it off.

58

Basis	x	y	w_1	w_2	w_3	w_4	w_5	b	*check*
w_1	3	5	1	0	0	0	0	130	139
w_4	-4	5	0	-1	0	1	0	25	26
w_5	1	5	0	0	-1	0	1	75	81
P	-11	-15	0	0	0	M	M	0	$2M-26$
w_1	3	5	1	0	0	0	0	130	139
w_4	$-\frac{4}{5}$	(1)	0	$-\frac{1}{5}$	0	$\frac{1}{5}$	0	5	$\frac{26}{5}$
w_5	1	5	0	0	-1	0	1	75	81
P	-11	-15	0	0	0	M	M	0	$2M-26$
w_1	7	0	1	1	0	-1	0	105	113
y ~~w_4~~	$-\frac{4}{5}$	1	0	$-\frac{1}{5}$	0	$\frac{1}{5}$	0	5	$\frac{26}{5}$
w_5	5	0	0	1	-1	-1	1	50	55
P	-23	0	0	-3	0	$M+3$	M	75	$2M+52$
w_1	7	0	1	1	0	-1	0	105	113
y	$-\frac{4}{5}$	1	0	$-\frac{1}{5}$	0	$\frac{1}{5}$	0	5	$\frac{26}{5}$
w_5	(1)	0	0	$\frac{1}{5}$	$-\frac{1}{5}$	$-\frac{1}{5}$	$\frac{1}{5}$	10	11
P	-23	0	0	-3	0	$M+3$	M	75	$2M+52$
w_1	0	0	1	$-\frac{2}{5}$	$\frac{7}{5}$	$\frac{2}{5}$	$-\frac{7}{5}$	35	36
y	0	1	0	$-\frac{1}{25}$	$-\frac{4}{25}$	$\frac{1}{25}$	$\frac{4}{25}$	13	14
x ~~w_5~~	1	0	0	$\frac{1}{5}$	$-\frac{1}{5}$	$-\frac{1}{5}$	$\frac{1}{5}$	10	11
P	0	0	0	$\frac{8}{5}$	$-\frac{23}{5}$	$M-\frac{8}{5}$	$M+\frac{23}{5}$	305	$2M+305$
w_1	0	0	$\frac{5}{7}$	$-\frac{2}{7}$	(1)	$\frac{2}{7}$	-1	25	$\frac{180}{7}$
y	0	1	0	$-\frac{1}{25}$	$-\frac{4}{25}$	$\frac{1}{25}$	$\frac{4}{25}$	13	14
x	1	0	0	$\frac{1}{5}$	$-\frac{1}{5}$	$-\frac{1}{5}$	$\frac{1}{5}$	10	11
P	0	0	0	$\frac{8}{5}$	$-\frac{23}{5}$	$M-\frac{8}{5}$	$M+\frac{23}{5}$	305	$2M+305$
w_3 ~~w_1~~	0	0	$\frac{5}{7}$	$-\frac{2}{7}$	1	$\frac{2}{7}$	-1	25	$\frac{180}{7}$
y	0	1	$\frac{4}{35}$	$-\frac{3}{35}$	0	$\frac{3}{35}$	0	17	$\frac{634}{35}$
x	1	0	$\frac{1}{7}$	$\frac{1}{7}$	0	$-\frac{1}{7}$	0	15	$\frac{113}{7}$
P	0	0	$\frac{23}{7}$	$\frac{2}{7}$	0	$M-\frac{2}{7}$	M	420	$2M+\frac{2963}{7}$

So there it is. $\underline{P_{max} = 420 \text{ with } x = 15,\ y = 17.}$

Incidentally, also, $w_3 = 25$ and $w_1 = w_2 = w_4 = w_5 = 0$.

---

59 Our examples on the use of artificial variables have so far concerned only two problem variables, x and y. The method, however, is exactly the same when more problem variables are involved, though, naturally, the solution then becomes somewhat longer.

Here is one for you to work through: it brings in most of what we have covered and provides excellent revision. The result is given in the next frame.

Example 6

Maximise $P = 24x + 21y + 30z$

subject to

$$12x + 4y + 8z \leq 240$$
$$8x + 3y + 3z \leq 140$$
$$6x + 2y + 3z \geq 110 \qquad (x,\ y,\ z \geq 0)$$

60

$$P_{max} = 750 \text{ with } x = 10, y = 10, z = 10$$

The simplex technique is designed to maximise a given objective function in the light of stated constraints. However, a problem requiring the minimisation of an objective function (denoting costs, machine idling time, etc.) can easily be converted for solution by the same method.

For this, move on to the next frame.

61

Minimisation

If P denotes the objective function to be minimised, we write Q as the negative of this function. Q_{max} is then determined by the usual simplex method and, finally, the negative value of Q_{max} is the value of the required P_{min}.

i.e. Write $Q = -P$. Determine Q_{max} in the normal way.

Then $\underline{P_{min} = -(Q_{max})}$.

Example 1

$$\begin{aligned} \text{Minimise} \quad & P = -3x + 4y \\ \text{subject to} \quad & x + 3y \leq 54 \\ & 3x + y \leq 34 \\ & -x + 2y \geq 12 \quad (x, y \geq 0) \end{aligned}$$

First write $Q = -P$, i.e. $Q = 3x - 4y$, and maximise Q.

Inserting the usual slack variables and artificial variable as needed, we have ....

62

$$\begin{aligned} x + 3y + w_1 \qquad\qquad\qquad\qquad &= 54 \\ 3x + y \qquad + w_2 \qquad\qquad\quad &= 34 \\ -x + 2y \qquad\qquad - w_3 + w_4 &= 12 \\ Q - 3x + 4y \qquad\qquad\qquad + Mw_4 &= 0 \end{aligned}$$

Now we just carry out the usual simplex routine to evaluate Q_{max} and hence P_{min}, since $P_{min} = -(Q_{max})$.

$P_{min} = ...$

63

$$\boxed{P_{min} = 16 \quad \text{with} \quad x = 8,\ y = 10}$$

For $Q_{max} = -16$ and hence $P_{min} = -(Q_{max}) = 16$.

The full working is available in the next frame, should you need to refer to it.

If not, move straight on to frame 65.

64

Here is the working to Example 1.

Basis	x	y	w_1	w_2	w_3	w_4	b	*check*
w_1	1	3	1	0	0	0	54	59
w_2	3	1	0	1	0	0	34	39
w_4	-1	2	0	0	-1	1	12	13
Q	-3	4	0	0	0	M	0	$M+1$
w_1	1	3	1	0	0	0	54	59
w_2	(1)	$\frac{1}{3}$	0	$\frac{1}{3}$	0	0	$\frac{34}{3}$	13
w_4	-1	2	0	0	-1	1	12	13
Q	-3	4	0	0	0	M	0	$M+1$
w_1	0	$\frac{8}{3}$	1	$-\frac{1}{3}$	0	0	$\frac{128}{3}$	46
x ~~w_2~~	1	$\frac{1}{3}$	0	$\frac{1}{3}$	0	0	$\frac{34}{2}$	13
$* \longrightarrow w_4$	0	$\boxed{\frac{7}{3}}$	0	$\frac{1}{3}$	-1	1	$\frac{70}{3}$	26
Q	0	5	0	1	0	M	34	$M+40$
w_1	0	$\frac{8}{3}$	1	$-\frac{1}{3}$	0	0	$\frac{128}{3}$	46
x	1	$\frac{1}{3}$	0	$\frac{1}{3}$	0	0	$\frac{34}{3}$	13
w_4	0	(1)	0	$\frac{1}{7}$	$-\frac{3}{7}$	$\frac{3}{7}$	10	$\frac{78}{7}$
Q	0	5	0	1	0	M	34	$M+40$
w_1	0	0	1	$-\frac{5}{7}$	$\frac{8}{7}$	$-\frac{8}{7}$	16	$\frac{114}{7}$
x	1	0	0	$\frac{2}{7}$	$\frac{1}{7}$	$-\frac{1}{7}$	8	$\frac{65}{7}$
y ~~w_4~~	0	1	0	$\frac{1}{7}$	$-\frac{3}{7}$	$\frac{3}{7}$	10	$\frac{78}{7}$
Q	0	0	0	$\frac{2}{7}$	$\frac{15}{7}$	$M-\frac{15}{7}$	-16	$M-\frac{110}{7}$

$$Q_{max} = -16 \therefore \underline{P_{min} = 16 \quad \text{with} \quad x = 8, y = 10.}$$

65

Example 2

$$\begin{aligned} &\text{Minimise} && P = -2x + 8y \\ &\text{subject to} && 3x + 4y \leq 80 \\ & && -3x + 4y \geq 8 \\ & && x + 4y \geq 40 \qquad (x, y \geq 0) \end{aligned}$$

Note that we have two constraints that are 'greater than' inequalities, so, beside the slack variables, we shall need two artificial variables.

The three constraints in their new form therefore become

. . . .

66

$$\begin{aligned} 3x + 4y + w_1 \qquad\qquad\qquad\qquad &= 80 \\ -3x + 4y \qquad - w_2 \qquad + w_4 \qquad &= 8 \\ x + 4y \qquad\qquad - w_3 \qquad + w_5 &= 40 \end{aligned}$$

and, in the subsequent manipulation, we must see that w_4 and w_5 disappear from the basic solution before the optimal solution is obtained.

The objective function P is now replaced by Q $(= -P)$ and the new form of Q is written as

. . . .

67

$$Q - 2x + 8y \; + Mw_4 + Mw_5 = 0$$

since $P = -2x + 8y \quad \therefore Q = -P = 2x - 8y$
and with the artificial variables $\; Q = 2x - 8y \; - Mw_4 - Mw_5$.

$$\therefore Q - 2x + 8y \; + Mw_4 + Mw_5 = 0$$

In this example, w_1, w_4, w_5, will form the unity matrix, so work through the solution in the usual way. Simplify the initial tableau and then refer to the next frame.

68

Basis	x	y	w_1	w_2	w_3	w_4	w_5	b	check
w_1	3	4	1	0	0	0	0	80	88
w_4	−3	4	0	−1	0	1	0	8	9
w_5	1	4	0	0	−1	0	1	40	45
Q	−2	8	0	0	0	M	M	0	$2M+6$
w_1	(1)	$\frac{4}{3}$	$\frac{1}{3}$	0	0	0	0	$\frac{80}{3}$	$\frac{88}{3}$
w_4	−3	4	0	−1	0	1	0	8	9
w_5	1	4	0	0	−1	0	1	40	45
Q	−2	8	0	0	0	M	M	0	$2M+6$
x ~~w_1~~	1	$\frac{4}{3}$	$\frac{1}{3}$	0	0	0	0	$\frac{80}{3}$	$\frac{88}{3}$
*→ w_4	0	8	1	−1	0	1	0	88	97
*→ w_5	0	$\frac{8}{3}$	$-\frac{1}{3}$	0	−1	0	1	$\frac{40}{3}$	$\frac{47}{3}$
Q	0	$\frac{32}{3}$	$\frac{2}{3}$	0	0	M	M	$\frac{160}{3}$	$2M+\frac{194}{3}$

At this stage, there is no further negative entry in the index row, but we still must get rid of w_4 and w_5 from the basic variable column. Let us start by dealing with w_5.

We will take the entry $\frac{8}{3}$ at the intersection of the w_5 row and the y column as the next pivot and launch forth on the next stage. Complete the second stage and then again refer to the next frame.

69

Here is the working of stage 2.

Basis	x	y	w_1	w_2	w_3	w_4	w_5	b	check
x	1	$\frac{4}{3}$	$\frac{1}{3}$	0	0	0	0	$\frac{80}{3}$	$\frac{88}{3}$
w_4	0	8	1	-1	0	1	0	88	97
w_5	0	$\boxed{\frac{8}{3}}$	$-\frac{1}{3}$	0	-1	0	1	$\frac{40}{3}$	$\frac{47}{3}$
Q	0	$\frac{32}{3}$	$\frac{2}{3}$	0	0	M	M	$\frac{160}{3}$	$2M+\frac{194}{3}$
x	1	$\frac{4}{3}$	$\frac{1}{3}$	0	0	0	0	$\frac{80}{3}$	$\frac{88}{3}$
w_4	0	8	1	-1	0	1	0	88	97
w_5	0	(1)	$-\frac{1}{8}$	0	$-\frac{3}{8}$	0	$\frac{3}{8}$	5	$\frac{47}{8}$
Q	0	$\frac{32}{3}$	$\frac{2}{3}$	0	0	M	M	$\frac{160}{3}$	$2M+\frac{194}{3}$
x	1	0	$\frac{1}{2}$	0	$\frac{1}{2}$	0	$-\frac{1}{2}$	20	$\frac{43}{2}$
* → w_4	0	0	2	-1	3	1	-3	48	50
y ~~w_5~~	0	1	$-\frac{1}{8}$	0	$-\frac{3}{8}$	0	$\frac{3}{8}$	5	$\frac{47}{8}$
Q	0	0	2	0	4	M	$M-4$	0	$2M+2$

At this point, w_5 is replaced by y in the basic variable column.

Now we deal with w_4 by taking the entry 2 at the junction of the w_4 row and the w_1 column as the next pivot. That should do the trick, so finish off the solution and check with the next frame.

70

$$P_{\min} = 48 \quad \text{with} \quad x = 8, \; y = 8$$

Basis	x	y	w_1	w_2	w_3	w_4	w_5	b	*check*
x	1	0	$\frac{1}{2}$	0	$\frac{1}{2}$	0	$-\frac{1}{2}$	20	$\frac{43}{2}$
* → w_4	0	0	$\boxed{2}$	-1	3	1	-3	48	50
y	0	1	$-\frac{1}{8}$	0	$-\frac{3}{8}$	0	$\frac{3}{8}$	5	$\frac{47}{8}$
Q	0	0	2	0	4	M	$M-4$	0	$2M+2$
x	1	0	$\frac{1}{2}$	0	$\frac{1}{2}$	0	$-\frac{1}{2}$	20	$\frac{43}{2}$
w_4	0	0	(1)	$-\frac{1}{2}$	$\frac{3}{2}$	$\frac{1}{2}$	$-\frac{3}{2}$	24	25
y	0	1	$-\frac{1}{8}$	0	$-\frac{3}{8}$	0	$\frac{3}{8}$	5	$\frac{47}{8}$
Q	0	0	2	0	4	M	$M-4$	0	$2M+2$
x	1	0	0	$\frac{1}{4}$	$-\frac{1}{4}$	$-\frac{1}{4}$	$\frac{1}{4}$	8	9
w_1 ~~w_4~~	0	0	1	$-\frac{1}{2}$	$\frac{3}{2}$	$\frac{1}{2}$	$-\frac{3}{2}$	24	25
y	0	1	0	$-\frac{1}{16}$	$-\frac{3}{16}$	$\frac{1}{16}$	$\frac{3}{16}$	8	9
Q	0	0	0	1	1	$M-1$	$M-1$	-48	$2M-48$

w_4 in the basic variable column is now replaced by w_1, so the conditions are satisfied at last. From the final tableau, we have:

$$Q_{\max} = -48 \qquad \text{But } P_{\min} = -(Q_{\max}) = 48$$

$$\therefore \; \underline{P_{\min} = 48 \quad \text{with} \quad x = 8, \; y = 8.}$$

By this means, then, we can solve minimisation problems by the simplex method and so widen the scope of this valuable technique.

Applications 71

So far we have seen how to solve a typical linear programming problem by the simplex method, when the data is presented as a linear objective function and a number of linear constraints in the form of equations or inequalities. A practical problem, however, must first be interpreted into algebraic form and we conclude this programme with a brief reference to this initial requirement. Let us consider the following example.

Example 1

A firm manufactures two types of couplings, A and B, each of which requires processing time on lathes, grinders and polishers. The machine times needed for each type of coupling are given in the table.

Coupling type	*Time required (hours)*		
	Lathe	*Grinder*	*Polisher*
A	2	8	5
B	5	5	2

The total machine time available is 250 hours on lathes, 310 hours on grinders and 160 hours on polishers. The net profit per coupling of type A is £9 and of type B £10.

Determine (a) the number of each type to be produced to maximise profit

(b) the maximum profit.

If we let $x =$ the number of type A units to be produced

$y =$ the number of type B units to be produced

the objective function to be maximised can be expressed as

....

72

$$P = 9x + 10y$$

Now we have to sort out the constraints from the given data.

Total time available on lathes = 250 hours

$$\therefore 2x + 5y \leq 250 \quad \text{(lathes)}$$

Similar statements for the grinders and polishers are

....

73

$$\begin{aligned} 8x + 5y &\le 310 \quad \text{(grinders)} \\ 5x + 2y &\le 160 \quad \text{(polishers)} \end{aligned}$$

The problem now can be expressed as:

Maximise $\quad P = 9x + 10y$

subject to $\quad 2x + 5y \le 250$

$8x + 5y \le 310$

$5x + 2y \le 160 \quad (x, y \ge 0)$

Then we go through the usual process. Inserting slack variables to convert the inequalities into equations, we have

. . . .

74

$$\begin{array}{llll} 2x + 5y + w_1 & & & = 250 \\ 8x + 5y & + w_2 & & = 310 \\ 5x + 2y & & + w_3 & = 160 \\ P - 9x - 10y & & & = 0 \end{array}$$

and the solution then develops in the usual way. Work through it carefully — it is all good practice — and see if you agree with the result given in the next frame.

The result is . . . .

75

$$P_{\max} = 550 \quad \text{with} \quad x = 10, \; y = 46$$

The maximum profit of £550 occurs with a manufacturing schedule of

10 couplings of type A

and 46 couplings of type B.

Now for another, so move on.

76

Example 2

A firm produces three types of pumps, A, B, C, each of which requires the four processes of turning, drilling, assembling and testing.

Pump type	*Process time (hours) per pump*				*Profit per pump (£)*
	Turning	*Drilling*	*Assembling*	*Testing*	
A	2	1	3	4	84
B	1	1	4	3	72
C	2	1	2	2	52
Total available time (h/week)	98	60	145	160	

From the information given in the table, determine

(a) the weekly output of each type of pump to maximise profit
(b) the maximum profit.

So, if we let $x =$ the number of pumps, type A
$y =$ the number of pumps, type B
$z =$ the number of pumps, type C

we can interpret the problem into its algebraic form, which is

....

77

Maximise $P = 84x + 72y + 52z$
subject to
$$\begin{aligned} 2x + y + 2z &\leq 98 \\ x + y + z &\leq 60 \\ 3x + 4y + 2z &\leq 145 \\ 4x + 3y + 2z &\leq 160 \qquad (x, y, z \geq 0) \end{aligned}$$

Inserting the slack variables and expressing the problem as equations, we have

....

78

$$
\begin{aligned}
2x + y + 2z + w_1 \qquad\qquad &= 98 \\
x + y + z \quad + w_2 \qquad &= 60 \\
3x + 4y + 2z \qquad + w_3 \quad &= 145 \\
4x + 3y + 2z \qquad\qquad + w_4 &= 160 \\
P - 84x - 72y - 52z \qquad\qquad &= 0
\end{aligned}
$$

Now you can proceed to set up the simplex tableau and solve the problem on your own in the usual manner. It is very similar to the other examples you have worked earlier in the programme.
The result you no doubt get is

....

79

$$P_{max} = 3652 \quad \text{with} \quad x = 23,\ y = 8,\ z = 22$$

i.e. by producing 23 pumps, type A
8 pumps, type B
22 pumps, type C
the maximum profit of £3652 is attained.

Care with the calculations and constant use of the check column provide the key to avoiding errors in the working.

That completes the programme. Check down the Revision Summary that comes next before working through the Test Exercise that follows thereafter. As usual, a set of Further Problems provides further necessary practice in these useful techniques.

REVISION SUMMARY

80

1. *Optimisation* — determination of an optimal value (maximum or minimum) of an objective function subject to a set of constraints.
2. *Linear programming (linear optimisation)* — optimisation where the objective function is a linear function and the constraints are linear equations or linear inequalities.
3. *Inequalities* — multiplying or dividing both sides by a negative factor $(-k)$ reverses the inequality, i.e. $\geq$ becomes $\leq$ and $\leq$ becomes $\geq$.
4. *Problem variables* (x, y, z, etc.) are always non-negative.
5. *Feasible solution* — a set of variables that satisfies all the given constraints.
6. *Optimal solution* — a feasible solution for which the objective function becomes a maximum (or minimum) within the constraints.
7. *Basic feasible solution* — a feasible solution for which at least $(n - m)$ of the total variables are zero, where

   n = total number of variables in the constraints

   m = number of constraints.
8. *Basis* — collection of the m variables which are not put equal to zero.
9. *Basic solution* — solution obtained by equating $(n - m)$ variables to zero and solving for the remaining m variables.
10. *Graphical solution*

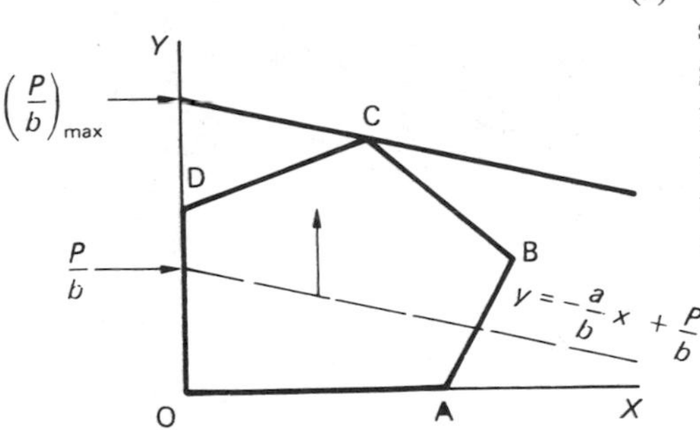

(a) Constraints — graphs of constraints form the feasible polygon or feasible domain.

Feasible point or feasible solution — coordinates of all points within the feasible polygon or on its boundary (OABCD).

(b) Objective function $P = ax + by$ $\therefore y = -\frac{a}{b}x + \frac{P}{b}$ represented by a set of parallel lines, slope $-\frac{a}{b}$, intercept $\frac{P}{b}$. Line through the extreme point C gives P_{max}, the optimal value of P.

11. *Slack variable* — non-negative variable added to, or subtracted from, a linear inequality to form a linear equation.
12. *Simplex method of solution* — computation.

    Refer back to page 985.

    Where necessary, we multiply an inequality by (-1), with consequent reversal of inequality sign, to ensure that the right-hand side constant term $b_i \geq 0$.

13. *Artificial variable* — to convert a 'greater than' inequality to an equation, the slack variable required must be subtracted. To complete the unity matrix in the tableau, a further artificial variable w_i is included to allow the simplex procedure to continue. Such artificial variables must be eliminated before the optimal solution is finally attained.

    The objective function $P = ax + by$ becomes $P = ax + by - Mw_i$.

14. *Minimisation* — If P is the objective function to be minimised:

    (a) write $Q = -P$

    (b) maximise Q by the usual simplex method

    (c) then $Q_{max} = (-P)_{max} = -(P_{min})$
    i.e. $P_{min} = -(Q_{max})$.

81 TEST EXERCISE XX

1. Using a *graphical method*, maximise $P = x + 2y$ subject to the constraints.

$$\begin{aligned} -3x + 4y &\le 8 \\ x + 4y &\le 16 \\ 3x + 2y &\le 18 \\ x, y &\ge 0 \end{aligned}$$

Note: Use the *simplex method* to solve Questions 2 to 6. In each case, all variables are non-negative.

2. Maximise $\quad P = -3x + 4y$
   subject to
$$\begin{aligned} 3x - 2y &\le 15 \\ x + y &\le 10 \\ -x + 4y &\le 15 \\ -2x + y &\le 2 \end{aligned}$$

3. Maximise $\quad P = 8x + 12y + 10z$
   Subject to
$$\begin{aligned} 4x + 3y + 2z &\le 64 \\ 2x + y + 4z &\le 48 \\ x + 2y + z &\le 24 \end{aligned}$$

4. Maximise $\quad P = 44x + 20y$
   subject to
$$\begin{aligned} 12x + 6y &\le 84 \\ 3x + 2y &\ge 24 \end{aligned}$$

5. Minimise $\quad P = 3y - 4x$
   subject to
$$\begin{aligned} x + 4y &\le 60 \\ 2x + y &\le 22 \\ -x + y &\ge 7 \end{aligned}$$

6. A firm makes two types of containers, A and B, each of which requires cutting, assembly and finishing. The maximum available machine capacity in hours per week for each process is: cutting 50, assembly 84, finishing 72.

   The process times for one unit of each type are as follows:

	Time in hours	
Process	A	B
Cutting	2	5
Assembly	4	8
Finishing	4	5

   If the profit margin is £6 per unit A and £10 per unit B, determine

   (a) the optimum weekly output of containers

   (b) the maximum profit.

FURTHER PROBLEMS XX

All variables in the following problems are non-negative.

Graphical Solution

1. Maximise $P = -x + 8y$

   subject to
$$\begin{aligned} -3x + 4y &\le 10 \\ -x + 4y &\le 14 \\ 3x + 2y &\le 21 \\ 3x + y &\le 18. \end{aligned}$$

2. Minimise $P = -4x + 8y$

   subject to
$$\begin{aligned} x + 3y &\le 57 \\ 7x + 4y &\le 110 \\ -x + 5y &\le 40. \end{aligned}$$

3. Maximise $P = 5x + 4y$

   subject to
$$\begin{aligned} x - 2y &\le 2 \\ 3x - 4y &\le 8 \\ 5x + 6y &\le 45 \\ x + 3y &\le 18. \end{aligned}$$

Simplex Solution

4. Maximise $P = 2x + y$
subject to
$$\begin{aligned} x + 4y &\le 24 \\ x + y &\le 9 \\ x - y &\le 3 \\ x - 2y &\le 2. \end{aligned}$$

5. Maximise $P = -3x + 4y$
subject to
$$\begin{aligned} 3x - 4y &\le 12 \\ 5x + 4y &\le 36 \\ -x + 3y &\le 8 \\ -3x + y &\le 0. \end{aligned}$$

6. Maximise $P = x + 2y$
subject to
$$\begin{aligned} -2x + y &\le 1 \\ -x + y &\le 2 \\ x + y &\le 6 \\ 2x - 3y &\le 2. \end{aligned}$$

7. Maximise $P = 4y - 3x$
subject to
$$\begin{aligned} x - 2y &\le 0 \\ x - y &\le 2 \\ x + 2y &\le 14 \\ -x + 2y &\le 6 \\ -3x + 2y &\le 2. \end{aligned}$$

8. Maximise $P = 3x + 4y + 5z$
subject to
$$\begin{aligned} 5x + 4y + 8z &\le 40 \\ 3x + 2y + 12z &\le 30 \\ y &\le 8. \end{aligned}$$

9. Maximise $P = 3x + 4y + 3z$
subject to
$$\begin{aligned} 2x + 3y + 4z &\le 58 \\ 4x + 2y + 3z &\le 51 \\ 3x + 4y + 2z &\le 62. \end{aligned}$$

10. Maximise $P = 4x + 3y + 3z$
subject to
$$\begin{aligned} 4x + y + 2z &\le 40 \\ x + 4y + z &\le 50 \\ 2x + 3y + 4z &\le 60. \end{aligned}$$

11. Maximise $P = 5.3x + 3.6y + 2.0z$
subject to
$$\begin{aligned} 2.1x + 4.3y + 1.5z &\leq 70 \\ 3.2x + 1.4y + 2.2z &\leq 60 \\ 1.6x + 6.2y + 3.1z &\leq 100. \end{aligned}$$

Artificial Variables

12. Maximise $P = 8x + 5y$
subject to
$$\begin{aligned} 2x + y &\leq 80 \\ x + 3y &\leq 90 \\ x + y &\geq 30. \end{aligned}$$

13. Maximise $P = 12x + 8y$
subject to
$$\begin{aligned} x + 2y &\leq 20 \\ 4x - y &\leq 8 \\ -x + y &\geq 1. \end{aligned}$$

14. Maximise $P = 3x + 4y$
subject to
$$\begin{aligned} x + 4y &< 76 \\ -5x + 8y &\geq 40 \\ -x + 4y &\geq 32. \end{aligned}$$

15. Maximise $P = 4x + 5y$
subject to
$$\begin{aligned} x + 2y &\leq 63 \\ 3x + y &\leq 70 \\ 2x + y &\geq 42 \\ x + 4y &\geq 84. \end{aligned}$$

16. Maximise $P = 65x - 23y$
subject to
$$\begin{aligned} 5x - y &\leq 30 \\ 10x + 4y &\geq 150. \end{aligned}$$

17. Maximise $P = 24x - 8y$
subject to
$$\begin{aligned} x + 3y &\leq 360 \\ 2x + y &\leq 850 \\ -5x + 25y &\geq 320. \end{aligned}$$

18. Maximise $P = 4x + 2y$
subject to
$$\begin{aligned} x + 2y &\leq 60 \\ 3x + 2y &\leq 80 \\ -3x + 10y &\geq 40. \end{aligned}$$

19. Maximise subject to

$$P = 18x + 40y + 24z$$
$$5x + 2y + 4z \leq 63$$
$$2x + 4y + 2z \leq 42$$
$$2x + 3y + z \geq 35.$$

20. Maximise subject to

$$P = 60x + 45y + 25z$$
$$4x + 8y + 2z \leq 160$$
$$6x + 3y + 4z \leq 168$$
$$4x + 3y + 3z \geq 128.$$

21. Maximise subject to

$$P = 12x + 8y - 10z$$
$$4x + 2y - 3z \leq 210$$
$$6x + 8y + z \leq 630$$
$$2x - y + 4z \geq 210$$
$$x + y + z \leq 180.$$

Minimisation

22. Minimise subject to

$$P = -4x + 3y$$
$$x + 4y \leq 20$$
$$2x + y \leq 12$$
$$x - y \leq 3.$$

23. Minimise subject to

$$P = -5x + 8y$$
$$x + 2y \leq 40$$
$$3x + 2y \leq 48$$
$$-x + 4y \geq 40.$$

24. Minimise subject to

$$P = -4x + 8y$$
$$-5x + 4y \leq 32$$
$$7x + 4y \leq 80$$
$$-x + 8y \geq 40.$$

25. Minimise subject to

$$P = 2x + 8y$$
$$-x + 2y \leq 24$$
$$7x + 6y \leq 132$$
$$-x + 2y \geq 4$$
$$x + 2y \geq 12.$$

26. Minimise subject to

$$P = 4x - 8y + 5z$$
$$2x + 3y + z \leq 70$$
$$x + 2y + 2z \leq 60$$
$$3x + 4y + z \leq 84$$
$$x + y + z \geq 33.$$

27. Minimise $P = 6x - 5y - 3z$
subject to
$$5x + 8y + 4z \leq 220$$
$$2x + y + 6z \leq 154$$
$$4x + 2y + z \geq 77$$
$$x + y + 2z \geq 55.$$

Applications

28. A firm manufacturing two types of switching module, A and B, is under contract to produce a daily output of at least 35 modules in all. Assembly and testing times for each type of module are as follows:

Module type	*Processing time (hours)*	
	Assembly	*Testing*
A	1.0	2.0
B	2.0	1.0

Available staff resources provide a daily maximum of 80 hours for assembly and 55 hours for testing.

The profit on the sale of each A-module is £4.00 and of each B-module £5.00. Determine

(a) the daily production schedule for maximum profit.

(b) the maximum daily profit.

29. Three different types of coupling units are produced by a firm. The times required for machining, polishing and assembling a unit of each type are included in the information given in the following table.

Type of unit	*Time in hours per unit*			*Profit (£) per unit*
	Machining	*Polishing*	*Assembling*	
A	4	1	2	11
B	2	3	1	10
C	3	2	4	12
Available time (h/week)	320	250	280	

The firm is required to supply a total of at least 100 units of mixed types each week. Determine

(a) the weekly output of each type to maximise profit

(b) the maximum weekly profit.

30. A firm makes three types of wooden cabinets, A, B, C, with profit margins of £35, £30, £24 per unit respectively.

	Time in hours per cabinet		
Process	A	B	C
Preparation	2	5	4
Assembly	2	3	2
Finishing	5	4	3

The manufacturer has 25 men available for preparation, 20 men for assembly and 30 men for polishing, and all staff work a 40 hour week. To remain competitive, at least 300 cabinets in all must be produced each week. Determine

(a) the number of each model to be manufactured each week in order to maximise the profit

(b) the maximum weekly profit.

Appendix

1. Green's theorem

If P and Q are two functions in x and y, finite and continuous inside a region R and on its boundary c in the xy-plane, with continuous first partial derivatives, then Green's theorem states that

$$\iint_R \left(\frac{\partial P}{\partial y} - \frac{\partial Q}{\partial x}\right) \mathrm{d}x\,\mathrm{d}y = -\oint_c \{P\mathrm{d}x + Q\mathrm{d}y\}$$

Proof of Green's theorem

Let the lower boundary of the region be the curve $y_1 = f(x)$ and the upper boundary the curve $y_2 = F(x)$.

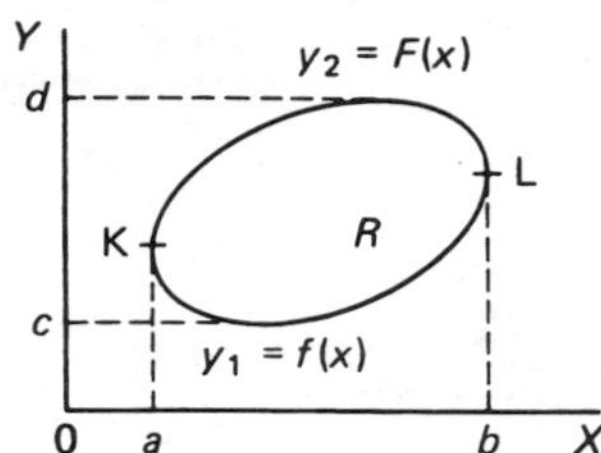

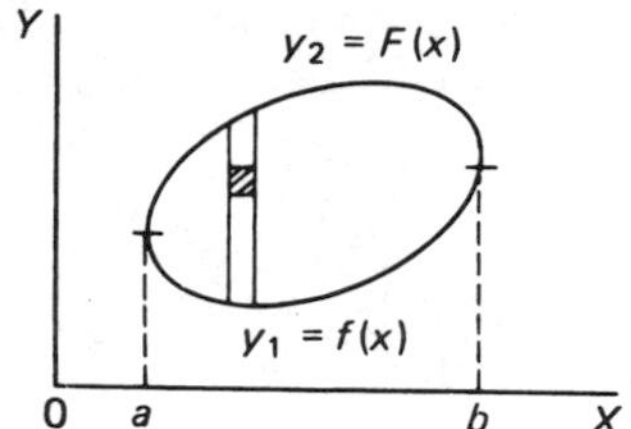

Using vertical strips, we then have

$$\iint_R \frac{\partial P}{\partial y}\,\mathrm{d}x\,\mathrm{d}y = \int_a^b \int_{y_1}^{y_2} \frac{\partial P}{\partial y}\,\mathrm{d}y\,\mathrm{d}x = \int_a^b \Big[P\Big]_{y_1=f(x)}^{y_2=F(x)} \mathrm{d}x$$

$$= \int_a^b \{P(x, y_2) - P(x, y_1)\}\,\mathrm{d}x$$

$$= -\int_a^b P(x, y_1)\,\mathrm{d}x - \int_b^a P(x, y_2)\,\mathrm{d}x$$

$$= -\left\{\int_a^b P(x, y_1)\,\mathrm{d}x + \int_b^a P(x, y_2)\,\mathrm{d}x\right\}$$

$$= -\oint P(x, y)\,\mathrm{d}x \qquad (1)$$

Similarly, using horizontal strips, we have

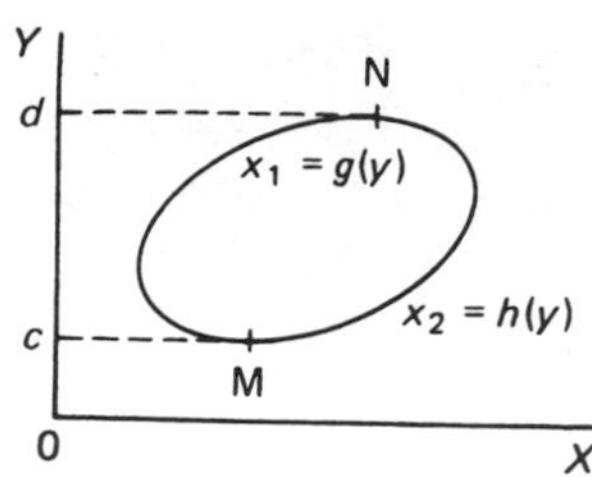

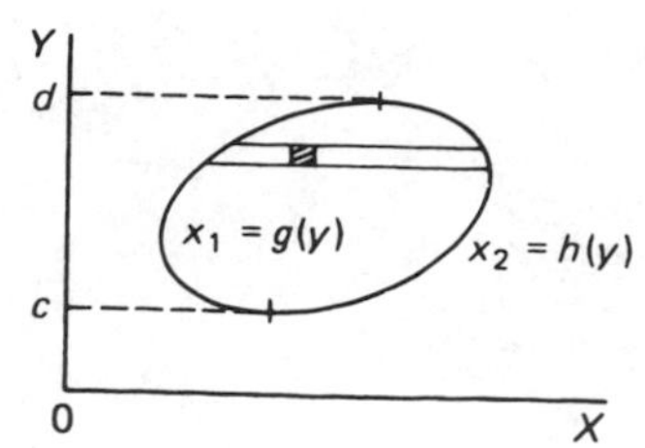

$$\iint_R \frac{\partial Q}{\partial x}\,dx\,dy = \int_c^d\int_{x_1}^{x_2} \frac{\partial Q}{\partial y}\,dx\,dy = \int_c^d \Big[Q\Big]_{x_1=g(y)}^{x_2=h(y)} dy$$

where $x_1 = g(y)$ and $x_2 = h(y)$ are the left-hand and right-hand portions of the boundary curve c.

$$\therefore \iint_R \frac{\partial Q}{\partial x}\,dx\,dy = \int_c^d Q(x_2, y)\,dy - \int_c^d Q(x_1, y)\,dy$$

$$= \int_c^d Q(x_2, y)\,dy + \int_d^c Q(x_1, y)\,dy$$

$$-\oint_c Q(x, y)\,dy \qquad (2)$$

$$\therefore \iint_R \left(\frac{\partial P}{\partial y} - \frac{\partial Q}{\partial x}\right) dx\,dy = -\oint_c P(x, y)\,dx - \oint_c Q(x, y)\,dy$$

$$= -\oint_c \{P\,dx - Q\,dy\}$$

2. **Proof that** $\sec\gamma = \sqrt{1 + \left(\frac{\partial z}{\partial x}\right)^2 + \left(\frac{\partial z}{\partial y}\right)^2}$

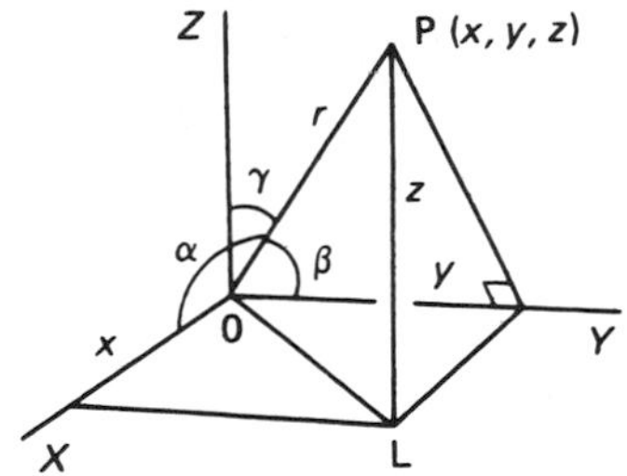

Let α, β, γ be the angles that OP makes with the x, y and z axes respectively.

Then $x = r\cos\alpha$; $y = r\cos\beta$; $z = r\cos\gamma$

Also $x^2 + y^2 + z^2 = r^2$

If $r = 1$ unit, then $x^2 + y^2 + z^2 = 1 \quad \therefore z^2 = 1 - x^2 - y^2$

$$\therefore z = (1 - x^2 - y^2)^{1/2}$$

$$\frac{\partial z}{\partial x} = \frac{1}{2}(1 - x^2 - y^2)^{-1/2}(-2x) = \frac{-x}{\sqrt{1 - x^2 - y^2}}$$

$$\frac{\partial z}{\partial y} = \frac{1}{2}(1 - x^2 - y^2)^{-1/2}(-2y) = \frac{-y}{\sqrt{1 - x^2 - y^2}}$$

$$\therefore 1+\left(\frac{\partial z}{\partial x}\right)^2+\left(\frac{\partial z}{\partial y}\right)^2=1+\frac{x^2}{1-x^2-y^2}+\frac{y^2}{1-x^2-y^2}$$

$$=\frac{1-x^2-y^2+x^2+y^2}{1-x^2-y^2}$$

$$=\frac{1}{1-x^2-y^2}=\frac{1}{z^2}$$

But, with $r=1$, $z=\cos\gamma$ $\quad\therefore \dfrac{1}{z^2}=\sec^2\gamma$

$$\therefore \sec\gamma=\sqrt{1+\left(\frac{\partial z}{\partial x}\right)^2+\left(\frac{\partial z}{\partial y}\right)^2}$$

3. Vector triple products

(a) $\mathbf{A}\times(\mathbf{B}\times\mathbf{C})=(\mathbf{A}\cdot\mathbf{C})\,\mathbf{B}-(\mathbf{A}\cdot\mathbf{B})\,\mathbf{C}$

(b) $(\mathbf{A}\times\mathbf{B})\times\mathbf{C}=(\mathbf{C}\cdot\mathbf{A})\,\mathbf{B}-(\mathbf{C}\cdot\mathbf{B})\,\mathbf{A}$

Let $\mathbf{A}=a_x\mathbf{i}+a_y\mathbf{j}+a_z\mathbf{k}$; $\mathbf{B}=b_x\mathbf{i}+b_y\mathbf{j}+b_z\mathbf{k}$; $\mathbf{C}=c_x\mathbf{i}+c_y\mathbf{j}+c_z\mathbf{k}$

Then $\mathbf{B}\times\mathbf{C}=(b_x\mathbf{i}+b_y\mathbf{j}+b_z\mathbf{k})\times(c_x\mathbf{i}+c_y\mathbf{j}+c_z\mathbf{k})$

$$=\begin{vmatrix}\mathbf{i} & \mathbf{j} & \mathbf{k}\\ b_x & b_y & b_z\\ c_x & c_y & c_z\end{vmatrix}$$

$$=\mathbf{i}\begin{vmatrix}b_y & b_z\\ c_y & c_z\end{vmatrix}-\mathbf{j}\begin{vmatrix}b_x & b_z\\ c_x & c_z\end{vmatrix}+\mathbf{k}\begin{vmatrix}b_x & b_y\\ c_x & c_y\end{vmatrix}$$

Then $\mathbf{A}\times(\mathbf{B}\times\mathbf{C})=\begin{vmatrix}\mathbf{i} & \mathbf{j} & \mathbf{k}\\ a_x & a_y & a_z\\ \begin{vmatrix}b_y & b_z\\ c_y & c_z\end{vmatrix} & \begin{vmatrix}b_z & b_x\\ c_z & c_x\end{vmatrix} & \begin{vmatrix}b_x & b_y\\ c_x & c_y\end{vmatrix}\end{vmatrix}$

$$=\mathbf{i}\left\{a_y\begin{vmatrix}b_x & b_y\\ c_x & c_y\end{vmatrix}-a_z\begin{vmatrix}b_z & b_x\\ c_z & c_x\end{vmatrix}\right\}-\mathbf{j}\left\{a_x\begin{vmatrix}b_x & b_y\\ c_x & c_y\end{vmatrix}-a_z\begin{vmatrix}b_y & b_z\\ c_y & c_z\end{vmatrix}\right\}$$

$$+\mathbf{k}\left\{a_x\begin{vmatrix}b_z & b_x\\ c_z & c_x\end{vmatrix}-a_y\begin{vmatrix}b_y & b_z\\ c_y & c_z\end{vmatrix}\right\}$$

$$= \mathbf{i}\{a_y(b_x c_y - b_y c_x) - a_z(b_z c_x - b_x c_z)\}$$
$$+\mathbf{j}\{a_z(b_y c_z - c_y b_z) - a_x(b_x c_y - b_y c_x)\}$$
$$+\mathbf{k}\{a_x(b_z c_x - b_x c_z) - a_y(b_y c_z - b_z c_y)\}$$

$$= \mathbf{i}\{b_x a_x c_x + b_x a_y c_y + b_x a_z c_z - c_x a_x b_x - c_x a_y b_y - c_x a_z b_z\}$$
$$+\mathbf{j}\{b_y a_x c_x + b_y a_y c_y + b_y a_z c_z - c_y a_x b_x - c_y a_y b_y - c_y a_z b_z\}$$
$$+\mathbf{k}\{b_z a_x c_x + b_z a_y c_y + b_z a_z c_z - c_z a_x b_x - c_z a_y b_y - c_z a_z b_z\}$$

$$= \mathbf{i}\{b_x(a_x c_x + a_y c_y + a_z c_z) - c_x(a_x b_x + a_y b_y + a_z b_z)\}$$
$$+\mathbf{j}\{b_y(a_x c_x + a_y c_y + a_z c_z) - c_y(a_x b_x + a_y b_y + a_z b_z)\}$$
$$+\mathbf{k}\{b_z(a_x c_x + a_y c_y + a_z c_z) - c_z(a_x b_x + a_y b_y + a_z b_z)\}$$

Now $\quad \mathbf{A}\cdot\mathbf{C} = (a_x\mathbf{i} + a_y\mathbf{j} + a_z\mathbf{k})\cdot(c_x\mathbf{i} + c_y\mathbf{j} + c_z\mathbf{k})$
$$= a_x c_x + a_y c_y + a_z c_z$$

and similarly $\quad \mathbf{A}\cdot\mathbf{B} = a_x b_x + a_y b_y + a_z b_z$

$$\therefore\ \mathbf{A}\times(\mathbf{B}\times\mathbf{C}) = \mathbf{i}\{b_x(\mathbf{A}\cdot\mathbf{C}) - c_x(\mathbf{A}\cdot\mathbf{B})\}$$
$$+\mathbf{j}\{b_y(\mathbf{A}\cdot\mathbf{C}) - c_y(\mathbf{A}\cdot\mathbf{B})\}$$
$$\mathbf{k}\{b_z(\mathbf{A}\cdot\mathbf{C}) - c_z(\mathbf{A}\cdot\mathbf{B})\}.$$

$$\therefore\ \mathbf{A}\times(\mathbf{B}\times\mathbf{C}) = (\mathbf{A}\cdot\mathbf{C})\{\mathbf{i}b_x + \mathbf{j}b_y + \mathbf{k}b_z\} - (\mathbf{A}\cdot\mathbf{B})\{\mathbf{i}c_x + \mathbf{j}c_y + \mathbf{k}c_z\}$$

$$\therefore\ \underline{\mathbf{A}\times(\mathbf{B}\times\mathbf{C}) = (\mathbf{A}\cdot\mathbf{C})\,\mathbf{B} - (\mathbf{A}\cdot\mathbf{B})\,\mathbf{C}}$$

In the same way, it can be established that

$$\underline{(\mathbf{A}\times\mathbf{B})\times\mathbf{C} = (\mathbf{C}\cdot\mathbf{A})\,\mathbf{B} - (\mathbf{C}\cdot\mathbf{B})\,\mathbf{A}}$$

4. Divergence theorem (Gauss' theorem)

To prove that $\int_V \operatorname{div}\mathbf{F}\,\mathrm{d}V = \int_S \mathbf{F}\cdot\mathrm{d}\mathbf{S}$ for the region V bounded by the surface S.

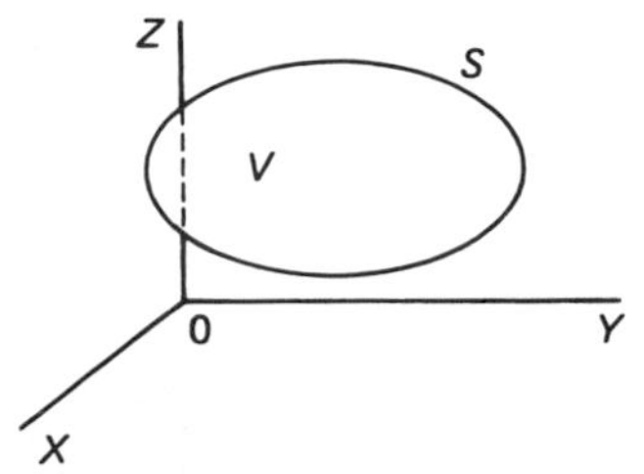

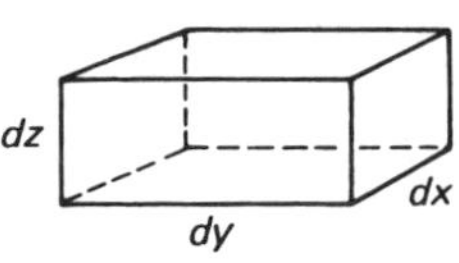

Consider an element of volume $\mathrm{d}V = \mathrm{d}x\,\mathrm{d}y\,\mathrm{d}z$ and let the components of $\mathbf{F}$ in the x, y and z directions be denoted by $F_x\mathbf{i}$, $F_y\mathbf{j}$ and $F_z\mathbf{k}$ respectively at any point P. We then determine $\int \mathbf{F}\cdot\mathrm{d}\mathbf{S}$ over the element $\mathrm{d}V$ and finally sum the results for all such elements throughout the region.

(a) S_1: $\quad \mathrm{d}S_1 = \mathrm{d}y\,\mathrm{d}z; \quad \mathbf{n} = \mathbf{i}$

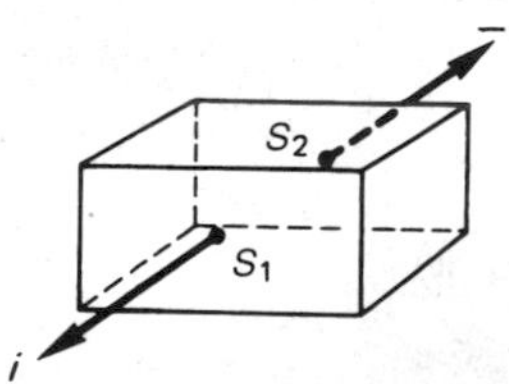

$$\begin{aligned}(\mathbf{F}\cdot\mathrm{d}\mathbf{S})_1 &= (F_x\mathbf{i}+F_y\mathbf{j}+F_z\mathbf{k})\cdot(\mathbf{n})\,\mathrm{d}S_1\\ &= (F_x\mathbf{i}+F_y\mathbf{j}+F_z\mathbf{k})\cdot(\mathbf{i})\,\mathrm{d}S_1\\ &= F_x\,\mathrm{d}S_1\end{aligned}$$

(b) S_2: $\quad \mathrm{d}S_2 = \mathrm{d}y\,\mathrm{d}z; \quad \mathbf{n} = -\mathbf{i}$

$$\begin{aligned}\therefore\ (\mathbf{F}\cdot\mathrm{d}\mathbf{S})_2 &= (F_x\mathbf{i}+F_y\mathbf{j}+F_z\mathbf{k})\cdot(-\mathbf{i})\,\mathrm{d}S_2\\ &= -F_x\,\mathrm{d}S_2\end{aligned}$$

Combining these two results, we have

$$\begin{aligned}(\mathbf{F}\cdot\mathrm{d}\mathbf{S})_1 + (\mathbf{F}\cdot\mathrm{d}\mathbf{S})_2 &= (F_x\,\mathrm{d}S)_1 - (F_x\,\mathrm{d}S)_2\\ &= \frac{\partial}{\partial x}(F_x\,\mathrm{d}S)\,\mathrm{d}x\end{aligned}$$

$$\therefore\ \int_{S_1+S_2} \mathbf{F}\cdot\mathrm{d}\mathbf{S} = \frac{\partial F_x}{\partial x}\,\mathrm{d}S\,\mathrm{d}x = \left(\frac{\partial F_x}{\partial x}\right)\mathrm{d}x\,\mathrm{d}y\,\mathrm{d}z \qquad (1)$$

Similarly, for S_3 and S_4 we have

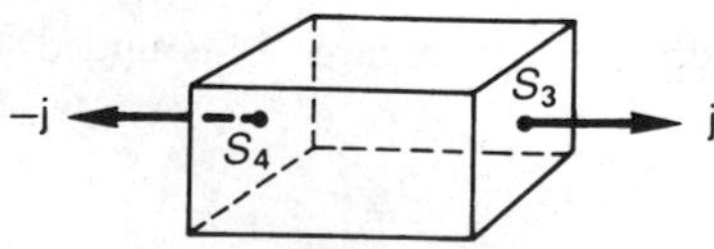

$$\int_{S_3+S_4} \mathbf{F}\cdot\mathrm{d}\mathbf{S} = \left(\frac{\partial F_y}{\partial y}\right)\mathrm{d}x\,\mathrm{d}y\,\mathrm{d}z \qquad (2)$$

and for S_5 and S_6

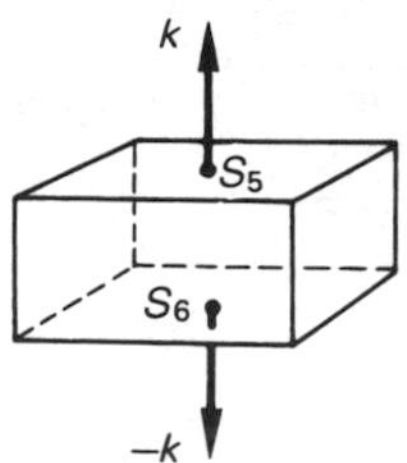

$$\int_{S_5+S_6} \mathbf{F}\cdot\mathrm{d}\mathbf{S} = \left(\frac{\partial F_z}{\partial z}\right)\mathrm{d}x\,\mathrm{d}y\,\mathrm{d}z \qquad (3)$$

These three results together cover the total surface of the element $\mathrm{d}V$.

$$\int_{S_1 \ldots S_6} \mathbf{F} \cdot \mathrm{d}\mathbf{S} = \left(\frac{\partial F_x}{\partial x} + \frac{\partial F_y}{\partial y} + \frac{\partial F_z}{\partial z} \right) \mathrm{d}x\,\mathrm{d}y\,\mathrm{d}z = \operatorname{div} \mathbf{F}\, \mathrm{d}V$$

Finally, summing the results for all such elements throughout the region with $\mathrm{d}V \to 0$ and $\mathrm{d}\mathbf{S} \to 0$, we obtain

$$\int_V \operatorname{div} \mathbf{F}\, \mathrm{d}V = \sum \int \mathbf{F} \cdot \mathrm{d}\mathbf{S} \quad \text{with } \mathrm{d}\mathbf{S} \to 0$$

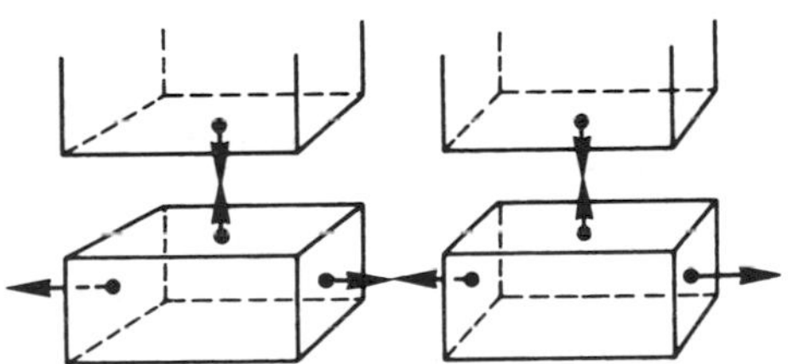

On the common boundaries between adjacent elements, the values of $\int \mathbf{F} \cdot \mathrm{d}\mathbf{S}$ cancel out. On the boundary surface, however, there are no such adjacent faces and the integral $\oint_S \mathbf{F} \cdot \mathrm{d}\mathbf{S}$ remains.

$$\therefore \int_V \operatorname{div} \mathbf{F}\, \mathrm{d}V = \int_S \mathbf{F} \cdot \mathrm{d}\mathbf{S}$$

5. Stokes' theorem

If $\mathbf{F}$ is a single-valued vector field, continuous and differentiable over an open surface S and on the boundary c of the surface, then

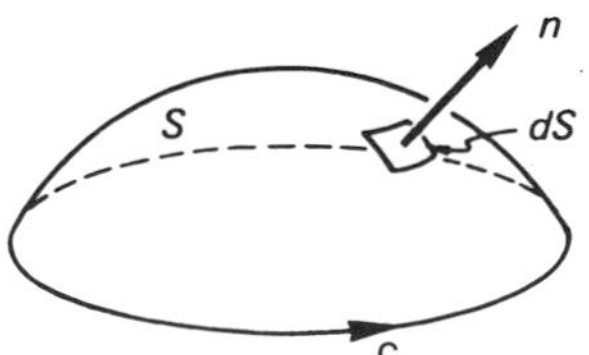

$$\int_S \operatorname{curl} \mathbf{F} \cdot \mathrm{d}\mathbf{S} = \oint_c \mathbf{F} \cdot \mathrm{d}\mathbf{r}$$

Proof of Stokes' theorem

Consider the surface S divided into small rectangular elements and let ABCD be one such element. If axes of reference X and Y be arranged to coincide with AB and AD respectively as shown, a third axis Z will then be normal to the surface at A.

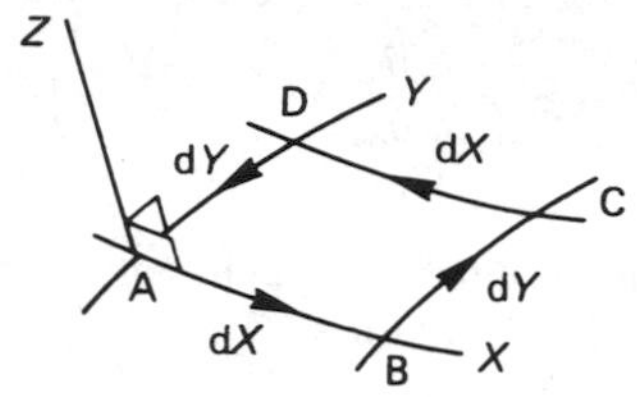

If $AB = dX$, then $\mathbf{dX} = \mathbf{i}\,dX$ and

if $AD = dY$, then $\mathbf{dY} = \mathbf{j}\,dY$.

Let $\mathbf{F}_a$ denote the vector field at A; $\mathbf{F}_b$ that at B; $\mathbf{F}_c$ that at C; and $\mathbf{F}_d$ that at D. Now consider each side in turn.

AB: $\mathbf{F}\cdot\mathbf{dr} = \mathbf{F}_a\cdot\mathbf{dX} = \{F_{ax}\mathbf{i} + F_{ay}\mathbf{j} + F_{az}\mathbf{k}\}\cdot\{\mathbf{i}\,dX\} = F_{ax}\,dX$

BC: $\mathbf{F}\cdot\mathbf{dr} = \mathbf{F}_b\cdot\mathbf{dY} = \{F_{bx}\mathbf{i} + F_{by}\mathbf{j} + F_{bz}\mathbf{k}\}\cdot\{\mathbf{j}\,dY\} = F_{by}\,dY$

CD: $\mathbf{F}\cdot\mathbf{dr} = \mathbf{F}_c\cdot\mathbf{dX} = \{F_{cx}\mathbf{i} + F_{cy}\mathbf{j} + F_{cz}\mathbf{k}\}\cdot\{-\mathbf{i}\,dX\} = -F_{cx}\,dX$

DA: $\mathbf{F}\cdot\mathbf{dr} = \mathbf{F}_d\cdot\mathbf{dY} = \{F_{dx}\mathbf{i} + F_{dy}\mathbf{j} + F_{dz}\mathbf{k}\}\cdot\{-\mathbf{j}\,dY\} = -F_{dy}\,dY$

(a) AB + CD:

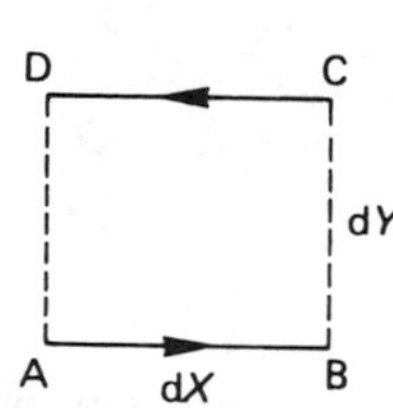

$$\int \mathbf{F}\cdot\mathbf{dr} = F_{ax}\,dX - F_{cx}\,dX = -(F_{cx} - F_{ax})\,dX$$

$$= -\,\delta F_x\,dX = -\frac{\partial F_x}{\partial Y}\,dY\,dX$$

$$\therefore \int_{(AB+CD)} \mathbf{F}\cdot\mathbf{dr} = -\frac{\partial F_x}{\partial Y}\,dX\,dY \qquad (1)$$

(b) BC + DA:

$$\int \mathbf{F}\cdot\mathbf{dr} = F_{by}\,dY - F_{dy}\,dY = (F_{by} - F_{dy})\,dY$$

$$= \delta F_y\,dY = \frac{\partial F_y}{\partial X}\,dX\,dY$$

$$\therefore \int_{(BC+DA)} \mathbf{F}\cdot\mathbf{dr} = \frac{\partial F_y}{\partial X}\,dX\,dY \qquad (2)$$

Adding these two results together for the complete rectangle, we have

$$\int_{(ABCD)} \mathbf{F}\cdot\mathbf{dr} = \left\{\frac{\partial F_y}{\partial X} - \frac{\partial F_x}{\partial Y}\right\} dX\,dY \qquad (3)$$

$$\text{Now curl } \mathbf{F} = \begin{vmatrix} \mathbf{i} & \mathbf{j} & \mathbf{k} \\ \dfrac{\partial}{\partial X} & \dfrac{\partial}{\partial Y} & \dfrac{\partial}{\partial Z} \\ F_x & F_y & F_z \end{vmatrix}$$

$$= \mathbf{i}\left(\frac{\partial F_z}{\partial Y} - \frac{\partial F_y}{\partial Z}\right) - \mathbf{j}\left(\frac{\partial F_z}{\partial X} - \frac{\partial F_x}{\partial Z}\right) + \mathbf{k}\left(\frac{\partial F_y}{\partial X} - \frac{\partial F_x}{\partial Y}\right)$$

$$\therefore \left\{\frac{\partial F_y}{\partial X} - \frac{\partial F_x}{\partial Y}\right\} = (\text{curl } \mathbf{F}) \cdot (\mathbf{k}) \tag{4}$$

From (3) $$\int_{\text{ABCD}} \mathbf{F} \cdot \mathbf{dr} = \text{curl } \mathbf{F} \cdot \mathbf{k} \, \mathrm{d}X \, \mathrm{d}Y = \text{curl } \mathbf{F} \cdot \mathbf{dS}$$

Summing for all such elements over the surface

$$\int_S \text{curl } \mathbf{F} \cdot \mathbf{dS} = \lim_{\mathrm{dr} \to 0} \sum \left\{ \int_{\text{ABCD}} \mathbf{F} \cdot \mathbf{dr} \right\} \tag{5}$$

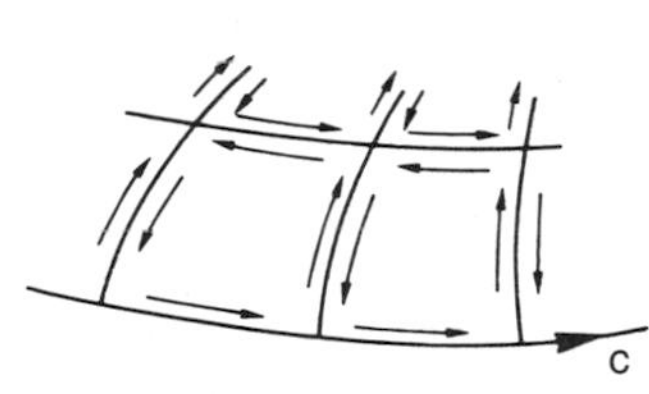

$\int \mathbf{F} \cdot \mathbf{dr}$ on boundary lincs bctwccn adjaccnt rectangular elements will cancel out, expect on the boundary curve c of the surface S. The integral then becomes $\oint_c \mathbf{F} \cdot \mathbf{dr}$.

$$\therefore \int_S \text{curl } \mathbf{F} \cdot \mathbf{dS} = \oint_c \mathbf{F} \cdot \mathbf{dr}$$

ANSWERS

Text Exercise I (page 35)

1. $f(x) = \{[(3x+5)x-4]x+1\}x-2;\ f(2.5) = 170.8$
2. 32.872
3. $2x-1$
4. $x = 1, 6, -2$
5. $y^3 - 5y^2 + 17y - 13 = 0$
6. $x = \frac{1}{3}, 3, -\frac{1}{2}, -2$
7. $k = -2.5$

Further Problems I (page 36)

1. $p = -5, q = -1$
2. $p = 4, q = 9$
3. $x = 2, 3, -3$
4. $x = 1, -3, 9$
5. $y^3 - 13y^2 + 52y - 60 = 0$
6. $x = \frac{1}{2}, \frac{3}{2}, -1$
7. $x = -2, 4, 8$
8. $2y^3 - 15y^2 + 25y = 0$

10. $x = 2.618, 0.382, -0.113, -8.887$
11. $x = -4.791, -1.000, -0.209, 0.382, 2.618$
12. $k = \frac{1}{2};\ x = \frac{1}{2}, \frac{1}{2}, -3$
13. (a) -1.375; (b) 81.104
14. (a) -6.048; (b) 461.500
15. (a) 133; (b) -9.048

Text Exercise II (page 76)

1. 0.8934
2. $x = 2.732, -0.732, -2.000$
3. $y^3 - 3y + 2 = 0;\ x = -4, -1, -1$
4. $x = 1 \pm \sqrt{2}, 3, -1$
5. $y^4 - 2y^2 + 4y = 0;\ x = 1, -1, 2 \pm \mathrm{j}$
6. $x = 1.646$

Further Problems II (page 76)

1. (a) -0.6736; (b) 0.3717

2. (a) $-2.330, 0.2018, 2.1283$

(b) $1, -0.50 \pm \mathrm{j}\,1.66$

(c) $-2.115, 0.254, 1.861$

3. (a) $-4.104, -0.9481 \pm \mathrm{j}\,0.5652$

(b) $0.5, -1.5$ (twice)

(c) $0.25, 1 \pm \mathrm{j}\,3$

4. (a) $1, -5, -2 \pm \mathrm{j}\sqrt{3}$

(b) $1, 2, 4, 5$

(c) $-1, 2, 2, 5$

(d) $1, 3, 2 \pm \mathrm{j}\sqrt{2}$

(e) $2, 4, 3 \pm \mathrm{j}\,2$

(f) $1 \pm \mathrm{j}, 1 \pm \mathrm{j}\,2$

5. (a) $-1, 2, 3, 4$

(b) $-1, 3, 5 \pm \mathrm{j}\,2$

(c) $2, -2, 2 \pm \mathrm{j}\,3$

6. (a) -2.456; (b) 1.765; (c) 0.739;

(d) 1.812; (e) 1.8175; (f) 0.5171;

(g) 0.8449; (h) 0.8806

Text Exercise III (page 121)

2. 145.7 ± 2.6 mm

3. $\dfrac{-2(x+y)}{2x+3y}$; $\dfrac{-2}{(2x+3y)^3}$

4. $\dfrac{x}{2(x^2-y^2)}$; $\dfrac{-y}{4(x^2-y^2)}$; $\dfrac{-y}{2(x^2-y^2)}$; $\dfrac{x}{4(x^2-y^2)}$

5. $(-1, 1)$, saddle; $(-1, -\frac{4}{3})$, min.

6. $1.10\text{ m} \times 1.10\text{ m} \times 0.825$ m high.

Further Problems III (page 122)

1. $(8x\cos x - 6y\sin x)/J$; $-(4x^3\cos y + 6x\sin y)/J$;
$J = 4x\cos x\sin y + 2x^2 y\sin x\cos y$

2. $e^{3y}/2(xe^{3y} + e^{-3y})$; $e^{-3y}/2(xe^{3y} + e^{-3y})$;
$-1/3(xe^{3y} + e^{-3y})$; $x/3(xe^{3y} + e^{-3y})$

5. $(2e^{-x}\sinh 2x\sin 3y + 3ye^{-x}\cosh 2x\cos 3y)/(1 + 3y^2)$;
$\{-4ye^{x}\sinh 2x\sin 3y + 3e^{x}(1 + y^2)\cosh 2x\cos 3y\}/2(1 + 3y^2)$

7. (a) $(4, -4, -11)$, min.; (b) $(1, -2, 4)$, saddle;
(c) $(\frac{10}{7}, \frac{6}{7}, \frac{97}{7})$, max.

8. (0, 0), saddle; (2, 0), min.; $(-2, 0)$, min.

9. (2, 1), max.; $(-\frac{2}{3}, -\frac{1}{3})$, min.

10. (0, 0); (3, 3); $(-3, -3)$, all saddle points

11. (a) (1, 0), saddle; (1, 1), min.; $(-2, \frac{1}{2})$, saddle; $(-\frac{7}{5}, \frac{1}{5})$, max.
(b) (0, 0), max.; (1, 1); $(1, -1)$; $(-1, 1)$; $(-1, -1)$, all four saddle points

12. $x = 66.7$ mm; $\theta = \frac{\pi}{3}$

13. $l = h = \frac{1}{5\pi}\sqrt[3]{60\pi^2 V}$; $d = l\sqrt{5}$

14. $l = 1.00$ cm; $d = 4.48$ cm; $\theta = 48°\ 11'$

15. cube of side $\dfrac{2r}{\sqrt{3}}$; $V_{max} = \dfrac{8r^3}{3\sqrt{3}}$

16. (a) $u = \dfrac{64}{27}$; $x = y = z = \pm\dfrac{2}{\sqrt{3}}$; $u = \dfrac{9}{7}$, $x = y = \pm\dfrac{3}{\sqrt{14}}$
(b) $u = 9$; $x = \pm\dfrac{3}{\sqrt{2}}$, $y = \mp\dfrac{3}{\sqrt{2}}$
(c) $u = \frac{8}{7}$; $x = \frac{6}{7}$, $y = -\frac{4}{7}$, $z = \frac{2}{7}$

Test Exercise IV (page 167)

1. (a) $\dfrac{20}{3}$; (b) $\dfrac{2}{3}$; (c) -2; (d) 120; (e) $\dfrac{15\sqrt{\pi}}{2048}$

2. (a) $\frac{256}{315}$; (b) $\frac{1}{40}$; (c) $\frac{2}{105}$

3. (a) $\dfrac{1}{\sqrt{2}}\cdot K\left(\dfrac{1}{\sqrt{2}}\right)$; (b) $\dfrac{1}{2}\cdot F\left(\dfrac{1}{2}, \dfrac{\sqrt{3}}{2}\right)$

Further Problems IV (page 168)

1. (a) 6; (b) $-\frac{1}{2}$; (c) 0.4; (d) 24; (e) $\frac{315}{4}$

2. (a) 6; (b) $\dfrac{8}{81}$; (c) $\dfrac{\sqrt{2}\pi}{16}$; (d) 4

4. (a) $\dfrac{1}{8960}$; (b) $\dfrac{\sqrt{2}\pi}{64}$; (c) $\dfrac{8}{315}$;

(d) $\frac{2}{7}$; (e) $\frac{1}{63}$; (f) $\dfrac{\pi}{432} = 0.00727$

5. (a) $\sqrt{5}\cdot E\left(\dfrac{2}{\sqrt{5}}\right)$; (b) $\sqrt{2}\cdot K\left(\dfrac{1}{\sqrt{2}}\right) = 2.622$;

(c) $2\cdot E\left(\dfrac{1}{2},\ 1\right) = 2.935$; (d) $\dfrac{1}{4}\cdot F\left(\dfrac{3}{4},\ \dfrac{2}{3}\right) = 0.193$;

(e) $\dfrac{1}{\sqrt{5}}\cdot F\left(\dfrac{2}{\sqrt{5}},\ 1\right)$; (f) $\dfrac{1}{\sqrt{2}}\cdot F\left(\dfrac{1}{\sqrt{2}},\ \dfrac{\pi}{6}\right)$;

(g) $\dfrac{1}{\sqrt{2}}\cdot\left\{F\left(\dfrac{1}{\sqrt{2}},\ \dfrac{\pi}{3}\right) - F\left(\dfrac{1}{\sqrt{2}},\ \dfrac{\pi}{4}\right)\right\}$

6. $\dfrac{1}{2}\cdot\left\{F\left(\dfrac{\sqrt{3}}{2},\ \dfrac{\pi}{2}\right) - F\left(\dfrac{\sqrt{3}}{2},\ \dfrac{\pi}{4}\right)\right\}$

7. (a) $\dfrac{1}{\sqrt{3}}\cdot F\left(\dfrac{1}{\sqrt{3}},\ \dfrac{1}{2}\right) = 0.307$

(b) $\dfrac{1}{\sqrt{3}}\cdot\left\{F\left(\dfrac{1}{\sqrt{3}},\ 1\right) - F\left(\dfrac{1}{\sqrt{3}},\ \dfrac{1}{2}\right)\right\}$

(c) $\dfrac{1}{\sqrt{34}}\cdot K\left(\dfrac{3}{\sqrt{34}}\right) = 0.2905$

(d) $\dfrac{1}{\sqrt{7}}\left\{F\left(\sqrt{\dfrac{3}{7}},\ \dfrac{\pi}{2}\right) - F\left(\sqrt{\dfrac{3}{7}},\ \dfrac{\pi}{6}\right)\right\}$

Text Exercise V (page 227)

2. $$y = a_0\left\{1 + \frac{5x^2}{2} + \frac{15x^4}{8} + \frac{5x^6}{16} + \dots\right\}$$

$$+ a_1\left\{x + \frac{4x^3}{3} + \frac{8x^5}{15} + \dots\right\}$$

3. (a) $y = A\left\{1 - \frac{x}{1\times 2} + \frac{x^2}{(1\times 2)(2\times 5)} - \frac{x^3}{(1\times 2)(2\times 5)(3\times 8)} + \ldots\right\}$

$$+ Bx^{\frac{1}{3}}\left\{1 - \frac{x}{1\times 4} + \frac{x^2}{(1\times 4)(2\times 7)} - \frac{x^3}{(1\times 4)(2\times 7)(3\times 10)} + \ldots\right\}$$

(b) $y = a_0\left\{1 - \frac{x^4}{3\times 4} + \frac{x^8}{(3\times 4)(7\times 8)} + \ldots\right\}$

$$+ a_1\left\{x - \frac{x^5}{4\times 5} + \frac{x^9}{(4\times 5)(8\times 9)} + \ldots\right\}$$

(c) $y_A = A\left\{-\frac{1}{2} - \frac{x}{6} - \ldots\right\}$

$$y_B = B\left\{\ln x\left(-\frac{1}{2} - \frac{x}{6} - \ldots\right) + x^{-2}\left(1 - x + \frac{x^2}{4} + \ldots\right)\right\}$$

Further Problems V (page 227)

1. $y_5 = 64e^{4x}\{16x^3 + 60x^2 + 60x + 15\}$
2. $y_n = (-1)^n e^{-x}\{x^3 - 3nx^2 + n(n-1)3x - n(n-1)(n-2)\},\ n > 3$
3. $y_4 = 480x + 96$
4. $y_6 = -\{(x^4 - 180x^2 + 360)\cos x + (24x^3 - 480x)\sin x\}$
5. $y_4 = -4e^{-x}\sin x$
6. $y_3 = 2x(13 + 12\ln x)$
8. $y_6 = -1018$
10. (a) $y_{2n} = \{x^2 + 2n(2n-1)\}\sinh x + 4nx\cosh x$

    (b) $y_{2n} = \{x^3 + 6n(2n-1)x\}\cosh x$

    $$+\{6nx^2 + 2n(2n-1)(2n-2)\}\sinh x$$
11. $y_6 = 2^5 e^{2x}\{2x^3 + 24x^2 + 81x + 75\}$
12. $y_3 = 2\sqrt{2}a^3 e^{-ax}\{\cos(ax + \pi/4)\}$
14. $y = y_0\left\{1 + \frac{9x^2}{2} + \frac{15x^4}{8} - \frac{7x^6}{16} + \frac{27x^8}{128} + \ldots\right\} + y_1\left\{x + \frac{4x^3}{3}\right\}$
15. $y = A(1 + x^2) + Be^{-x}$

16. $y = y_0\left\{1 + \frac{3^2 \times x^2}{2!} + \frac{3^2 \times 5^2 \times x^4}{4!} + \frac{3^2 \times 5^2 \times 7^2 \times x^6}{6!} + \dots\right\}$

$$+ y_1\left\{x + \frac{4^2 \times x^3}{3!} + \frac{4^2 \times 6^2 \times x^5}{5!} + \dots\right\}$$

17. $y = y_1 x + y_0\left\{1 - x^2 - \frac{x^4}{3} - \frac{x^6}{5} - \frac{x^8}{7} - \dots\right\}$

18. $y = y_0\left\{1 - \frac{2x}{2^2} + \frac{2^2 \times x^4}{2^2 \times 4^2} - \frac{2^3 \times x^6}{2^2 \times 4^2 \times 6^2} + \dots\right\}$

$$+ y_1\left\{x - \frac{2x^3}{3^2} + \frac{2^2 \times x^5}{3^2 \times 5^2} - \frac{2^3 \times x^7}{3^2 \times 5^2 \times 7^2} + \dots\right\}$$

19. $y = A\left\{1 + x + \frac{x^2}{2 \times 4} + \frac{x^3}{(2 \times 3)(4 \times 7)} + \frac{x^4}{(2 \times 3 \times 4)(4 \times 7 \times 10)} + \dots\right\}$

$$+ Bx^{\frac{2}{3}}\left\{1 + \frac{x}{1 \times 5} + \frac{x^2}{(1 \times 2)(5 \times 8)}\right.$$

$$\left. + \frac{x^3}{(1 \times 2 \times 3)(5 \times 8 \times 11)} + \dots\right\}$$

20. $y = a_0\left\{1 - \frac{x^2}{2!} + \frac{x^4}{4!} + \dots\right\} + a_1\left\{x - \frac{x^3}{3!} + \dots\right\}$

21. $y = a_0\left\{1 + \frac{x^3}{2 \times 3} + \frac{x^6}{(2 \times 3)(5 \times 6)} + \dots\right\}$

$$+ a_1\left\{x + \frac{x^4}{3 \times 4} + \frac{x^7}{(3 \times 4)(6 \times 7)} + \dots\right\}$$

22. $y = A\left\{1 - \frac{x}{1 \times 4} + \frac{x^2}{(1 \times 2)(4 \times 7)} - \frac{x^3}{(1 \times 2 \times 3)(4 \times 7 \times 10)} + \dots\right\}$

$$+ Bx^{-\frac{1}{3}}\left\{1 - \frac{x}{1 \times 2} + \frac{x^2}{(1 \times 2)(2 \times 5)}\right.$$

$$\left. - \frac{x^3}{(1 \times 2 \times 3)(2 \times 5 \times 8)} + \dots\right\}$$

23. $y = a_1 x + a_0\left\{1 - \frac{x^2}{2!} - \frac{x^4}{4!} - \frac{3x^6}{6!} - \frac{(3)(5)x^8}{8!} + \dots\right\}$

24. $y = u + v$ where

$$u = A\left\{\frac{-x^4}{4!\,3!} + \frac{x^5}{5!\,3!} - \dots\right\}$$

$$v = B\left\{\ln x\left(\frac{-x^4}{4!\,3!} + \frac{x^5}{5!\,3!} - \dots\right)\right.$$

$$\left. + \left(1 + \frac{x}{1 \times 3} + \frac{x^2}{(1 \times 2)(2 \times 3)} + \dots\right)\right\}$$

25. $y = u + v$ where

$$u = A\left\{1 + \frac{3x}{1^2} + \frac{3^2 \times x^2}{1^2 \times 2^2} + \frac{3^3 \times x^3}{1^2 \times 2^2 \times 3^2} + \ldots\right\}$$

$$v = B\left\{\ln x\left(1 + \frac{3x}{1^2} + \frac{3^2 \times x^2}{1^2 \times 2^2} + \frac{3^3 \times x^3}{1^2 \times 2^2 \times 3^2} + \ldots\right) - \left(\frac{2 \times 3x}{1^2} + \frac{3 \times 3^2 \times x^2}{1^2 \times 2^2} + \frac{11 \times 3^3 \times x^3}{1^2 \times 2^2 \times 3^3} + \ldots\right)\right\}$$

Text Exercise VI (page 278)

1.

x	y
0	1.0
0.1	1.1
0.2	1.211
0.3	1.3352
0.4	1.4753
0.5	1.6343

2.

x	y
1	0
1.2	0.204
1.4	0.4211
1.6	0.6600
1.8	0.9264
2.0	1.2243

3.

x	y
0	1.0
0.1	1.2052
0.2	1.4214
0.3	1.6499
0.4	1.8918
0.5	2.1487

4.

x	y
2.0	3.0
2.1	3.005
2.2	3.0195
2.3	3.0427
2.4	3.0736
2.5	3.1117

5.

x	y
1.0	0
1.1	0.1052
1.2	0.2215
1.3	0.3401
1.4	0.4717
1.5	0.6180

Further Problems VI (page 279)

1.

x	y
0	1.0
0.2	0.8
0.4	0.72
0.6	0.736
0.8	0.8288
1.0	0.9830

2.

x	y
0	1.4
0.1	1.596
0.2	0.8707
0.3	2.2607
0.4	2.8318
0.5	3.7136

3.

x	y
1.0	2.0
1.2	2.0333
1.4	2.1143
1.6	2.2250
1.8	2.3556
2.0	2.5000

4.

x	y
0	0.5
0.1	0.543
0.2	0.5716
0.3	0.5863
0.4	0.5878
0.5	0.5768

5.

x	y
0	1.0
0.1	1.1022
0.2	1.2085
0.3	1.3179
0.4	1.4296
0.5	1.5428

6.

x	y
1.0	1.0
1.1	1.1871
1.2	1.3531
1.3	1.5033
1.4	1.6411
1.5	1.7688

7.

x	y
0	0
0.1	0.1002
0.2	0.2015
0.3	0.3048
0.4	0.4110
0.5	0.5214

8.

x	y
0	1.0
0.2	0.8562
0.4	0.8110
0.6	0.8465
0.8	0.9480
1.0	1.1037

9.

x	y
0	1.0
0.1	0.9138
0.2	0.8512
0.3	0.8076
0.4	0.7798
0.5	0.7653

10.

x	y
0	0.4
0.2	0.4259
0.4	0.4374
0.6	0.4319
0.8	0.4085
1.0	0.3689

11.

x	y
1.0	2.0
1.2	2.4197
1.4	2.8776
1.6	3.3724
1.8	3.9027
2.0	4.4677

12.

x	y
0	1.0
0.2	1.1997
0.4	1.3951
0.6	1.5778
0.8	1.7358
1.0	1.8540

13.

x	y
0	1.0
0.2	1.1679
0.4	1.2902
0.6	1.3817
0.8	1.4497
1.0	1.4983

14.

x	y
0	1.0
0.1	1.11
0.2	1.2422
0.3	1.4013
0.4	1.5937
0.5	1.8271

15.

x	y
0	3.0
0.1	2.88
0.2	2.5224
0.3	1.9368
0.4	1.1424
0.5	0.1683

16.

x	y
0	0
0.2	0.1987
0.4	0.3897
0.6	0.5665
0.8	0.7246
1.0	0.8624

17.

x	y
0	1.0
0.2	1.1972
0.4	1.3771
0.6	1.5220
0.8	1.6161
1.0	1.6487

18.

x	y
0	2.0
0.1	2.0845
0.2	2.1367
0.3	2.1554
0.4	2.1407
0.5	2.0943

19.

x	y
0	1.0
0.2	1.0367
0.4	1.1373
0.6	1.2958
0.8	1.5145
1.0	1.8029

20.

x	y
1.0	0
1.2	0.1833
1.4	0.3428
1.6	0.4875
1.8	0.6222
2.0	0.7500

Test Exercise VII (page 332)

1. (a) $\dfrac{8-2s}{s^2-16}$; (b) $\dfrac{s+4}{s^2+16}$;

(c) $\dfrac{1}{s^4}\{4s^3-s^2+4s+6\}$; (d) $\dfrac{s+2}{s^2+4s+29}$;

(e) $\dfrac{6s}{(s^2+9)^2}$; (f) $\ln\left\{\dfrac{s+2}{s+1}\right\}$

2. (a) $2e^{3t}-e^{4t}$

(b) $2\cos\sqrt{2}t+\dfrac{5}{\sqrt{2}}\sin\sqrt{2}t-e^t$

(c) $e^t(3t+2)-e^{3t}$

(d) $\frac{1}{8}\{e^t(17\cos 2t+9\sin 2t)-e^{3t}\}$

3. (a) $x=e^{-2t}+e^{-3t}$

(b) $x=\frac{1}{12}\{13e^{2t}-\cos 2t-\sin 2t\}$

(c) $x=\frac{1}{6}-\frac{5}{3}e^{3t}+\frac{5}{2}e^{4t}$

(d) $x=e^t\left(1-t+\dfrac{t^3}{6}\right)$

4. $x = \frac{1}{2}\{9\cos t - 7\sin t - e^{-3t}\}$

$y = 3\sin t - 2\cos t + e^{-2t}$

Further Problems VII (page 333)

1. (a) $\dfrac{s-4}{s^2-8s+20}$; (b) $\dfrac{4s}{(s^2+4)^2}$;

(c) $\dfrac{6}{s^4}+\dfrac{8}{s^3}+\dfrac{5}{s}$; (d) $\dfrac{4s^2-24s+38}{(s-3)^3}$;

(e) $\dfrac{2s^3-6s}{(s^2+1)^3}$; (f) $\ln\sqrt{\dfrac{s+2}{s-2}}$

2. (a) $e^{2t}+e^{4t}$; (b) $3e^{4t}+2$;

(c) $e^{2t}\left\{\dfrac{3t^2}{2}+2t+1\right\}$; (d) $e^{-t}\{2\cos t - 5\sin t\} - 2e^{2t}$;

(e) $\frac{1}{3}(\cos t - \cos 2t)$; (f) $e^{-2t}\{\cos 4t - \frac{7}{4}\sin 4t\}$

3. $x = 4e^{4t} - 2$

4. $x = \frac{35}{78}e^{4t/3} - \frac{3}{26}\{\cos 2t + \frac{2}{3}\sin 2t\}$

5. $x = e^t(2t+1) + 2t + 4 + \cos t$

6. $x = \frac{3}{2}e^{4t} - e^{3t} - \frac{1}{2}e^{2t}$

7. $x = \frac{4}{5}\cos 3t + \sin 3t + \frac{1}{5}\cos 2t$

8. $x = \frac{1}{5}\{e^{2t} - e^t(\cos 2t - 2\sin 2t)\}$

9. $x = \frac{1}{8}\{2t^2 - 4t + 3 + e^{-2t}(4t^2 + 6t + 1)\}$

10. $x = \frac{2}{5}\{2(e^{-4t} - 1)\cos 4t + (e^{-4t} + 1)\sin 4t\}$

11. $x = (2t+1)\cos 5t + t\sin 5t$

12. $x = \frac{1}{13}\{2e^{2t} + 3e^{-2t} - 5(\cos 3t - \sin 3t)\}$

$y = \frac{1}{13}\{5(\cos 3t + \sin 3t) - 3e^{2t} - 2e^{-2t}\}$

13. $x = \frac{1}{6}\{7e^{-6t} + 5\}$

$y = \frac{1}{3}\{7e^{-6t} + 5\}$

14. $x = 10e^{-4t} + 2$

$y = 5e^{-4t} + 3$

15. $x = e^{-2t} - e^t + 2t$

$y = 3e^t + \frac{1}{2}e^{-2t} + t - \frac{7}{2}$

16. $x = 5e^{t} + 3e^{-t}$

$y = 4e^{t} - e^{-t}$

17. $x = 4\cos t - 2\sin t - \frac{1}{3}\{8e^{-t} + e^{2t}\}$

$y = 6\cos t + 2\sin t - \frac{4}{3}\{2e^{-t} + e^{2t}\}$

18. $x = \frac{5}{3}\{\cos 2t + \sin 2t - \cosh\sqrt{2}t - \sqrt{2}\sinh\sqrt{2}t\}$

19. $y = \frac{1}{5}\{3\sin 2t - 4\cos 2t + \frac{4}{3}\sin 3t + \frac{48}{7}\cos 3t\} - \frac{4}{7}\cos 4t$

20. $x = \cos\left(t\sqrt{\dfrac{3}{10}}\right) + \dfrac{3}{4}\cos(t\sqrt{6})$

$y = \dfrac{5}{4}\cos\left(t\sqrt{\dfrac{3}{10}}\right) - \dfrac{1}{4}\cos(t\sqrt{6})$

Test Exercise VIII (page 386)

1. (a)

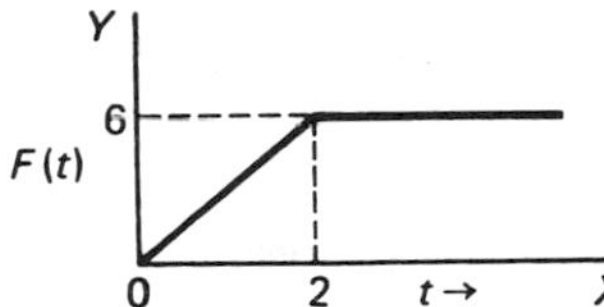

$$f(s) = \frac{3}{s^2}\{1 - e^{-2s}\}$$

(b)

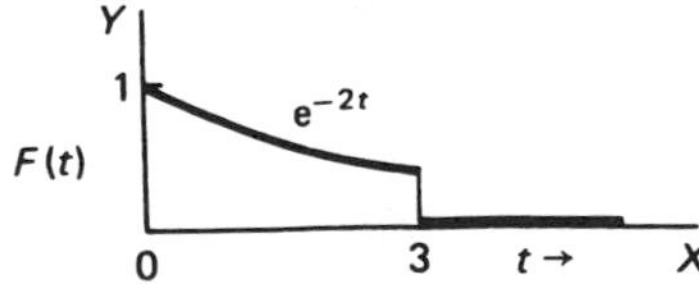

$$f(s) = \frac{1}{s+2}\{1 - e^{-6}\,e^{-3s}\}$$

(c)

$$f(s) = \frac{2}{s^3} - 2e^{-2s}\left\{\frac{1}{s^3} + \frac{2}{s^2} - \frac{1}{s}\right\} + \frac{2}{s}\cdot e^{-3s}$$

(d)

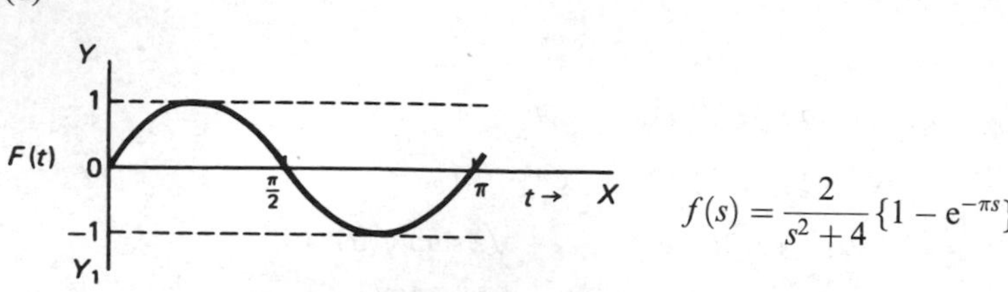

$$f(s) = \frac{2}{s^2 + 4}\{1 - e^{-\pi s}\}$$

2. $F(t) = 2 \cdot H(t) - 5 \cdot H(t-1) + 8 \cdot H(t-3)$

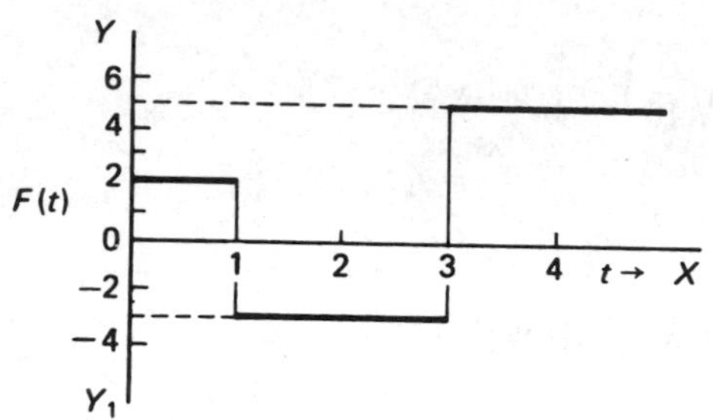

3. $F(t) = t \cdot H(t) + 3(t-2) \cdot H(t-2) - (t-3) \cdot H(t-3) - 3(t-5) \cdot H(t-5)$

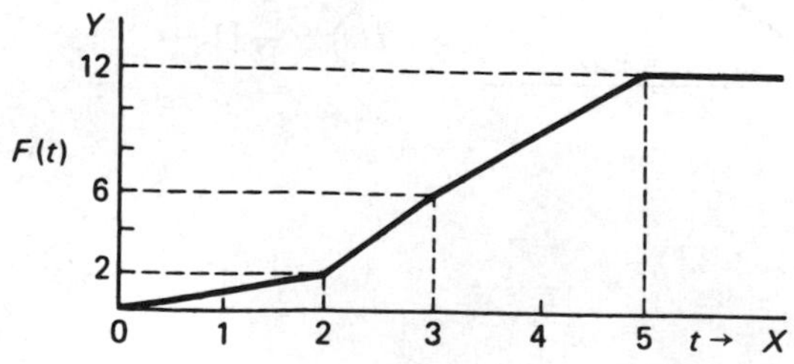

4. $F(t) = 2 \cdot H(t) - 2 \cdot H(t-1) + 2 \cdot H(t-3) - 2 \cdot H(t-4)$

$$+2 \cdot H(t-6) - 2 \cdot H(t-7) + \ldots$$

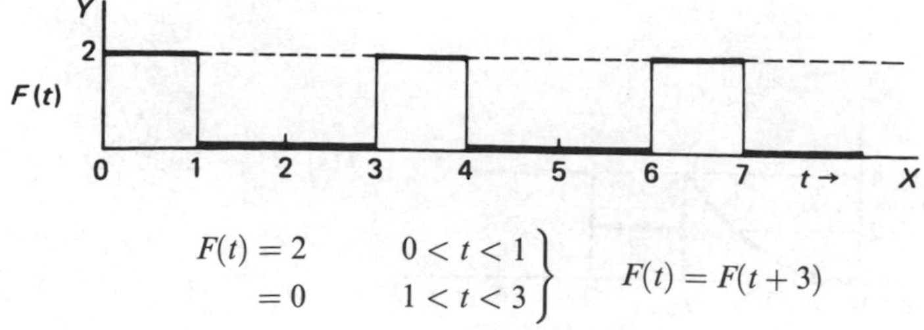

$$\left.\begin{aligned} F(t) &= 2 \qquad 0 < t < 1 \\ &= 0 \qquad 1 < t < 3 \end{aligned}\right\} \quad F(t) = F(t+3)$$

5. $f(s) = \dfrac{2(1 - e^{-2s} - 2s\,e^{-2s})}{1 - e^{-4s}}$

6. (a) e^{-6}; (b) 0; (c) 11

7. (a) $f(s) = 4e^{-3s}$; (b) $f(s) = e^{-2(3+s)}$

8.

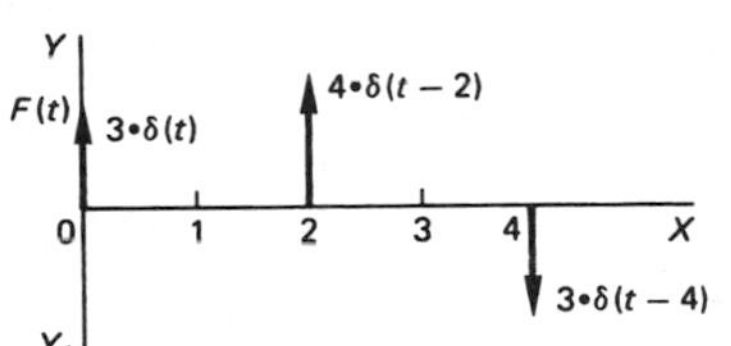

$f(s) = 3 + 4e^{-2s} - 3e^{-4s}$

9. $x = e^{-3t}\{4\sin t - \cos t\}$

10. $x = 3e^4\,e^{-t} \cdot H(t-4) + e^{-2t}\{2 \cdot H(t) - 3e^8 \cdot H(t-4)\}$

Further Problems VIII (page 387)

1. $F(t) = 3 \cdot H(t) + 2(t-2) \cdot H(t-2) - 2(t-5) \cdot H(t-5)$

2. $F(t) = t \cdot H(t) - (t-1) \cdot H(t-1) + (t-2) \cdot H(t-2) - (t-3) \cdot H(t-3)$

3.

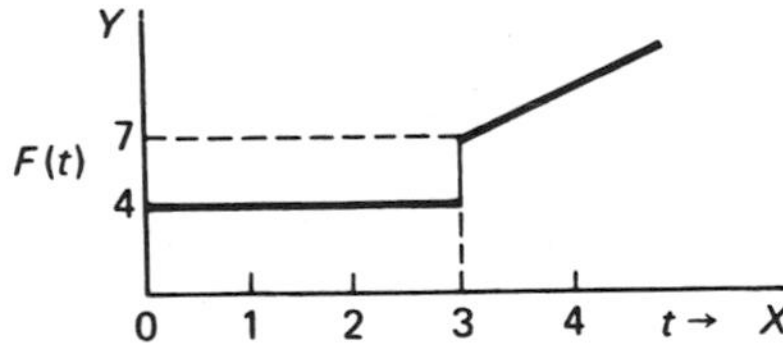

$f(s) = \dfrac{4}{s} + \dfrac{3e^{-3s}}{s} + \dfrac{2e^{-3s}}{s^2}$

4. (a) $F(t) = t^2 \cdot H(t) - (t^2 - 5t) \cdot H(t-3)$

   (b) $F(t) = \cos t \cdot H(t) + (\cos 2t - \cos t) \cdot H(t-\pi)$
   $+(\cos 3t - \cos 2t) \cdot H(t - 2\pi)$

5. $F(t) = e^{-2s}\left\{\dfrac{1}{s^2} + \dfrac{3}{s}\right\} - e^{-3s}\left\{\dfrac{1}{s^2} + \dfrac{4}{s}\right\}$

6. (a) $F(t) = t^2 \cdot H(t) - t^2 \cdot H(t-2) + 4 \cdot H(t-2) - 4 \cdot H(t-5)$

   (b) $f(s) = \dfrac{2}{s^3} - \dfrac{2e^{-2s}}{s^3} - \dfrac{4e^{-2s}}{s^2} - \dfrac{4e^{-5s}}{s}$

8. $\mathscr{L}\{F(t)\} = \dfrac{a(1 + e^{-\pi s})}{(s^2+1)(1 - e^{-\pi s})}$

9. (a) $f(s) = \dfrac{1}{s^2} - \dfrac{w}{s}\left\{\dfrac{e^{-ws}}{1 - e^{-ws}}\right\}$

(b) $f(s) = \dfrac{1 - e^{2(1-s)\pi}}{(s-1)(1 - e^{-2\pi s})}$

(c) $f(s) = \dfrac{1 - e^{-s}(s+1)}{s^2(1 - e^{-2s})}$

(d) $f(s) = \dfrac{1}{1 - e^{-3s}}\left\{\dfrac{2}{s^3} - \dfrac{2e^{-2s}}{s^3} - \dfrac{4e^{-2s}}{s^2} - \dfrac{4e^{-3s}}{s}\right\}$

10. $x = \dfrac{P}{M\omega}\sin\omega t$

11. $i = \dfrac{E}{L}\cos\left(\dfrac{t}{\sqrt{LC}}\right)$

12. $x = 2e^{-2t}\{1 + 10e^8 \cdot H(t-4)\} - 2e^{-3t}\{1 + 10e^{12} \cdot H(t-4)\}$

Test Exercise IX (page 450)

1. (a) $dz = 4x^3\cos 3y\,dx - 3x^4\sin 3y\,dy$

(b) $dz = 2e^{2y}\{2\cos 4x\,dx + \sin 4x\,dy\}$

(c) $dz = xw^2\{2yw\,dx + xw\,dy + 3xy\,dw\}$

2. (a) $z = x^3y^4 + 4x^2 - 5y^3$

(b) $z = x^2\cos 4y + 2\cos 3x + 4y^2$

(c) not exact differential

3. 9 square units

4. (a) 278.6; (b) $\pi/2$; (c) 22.5;

(d) 48; (e) -21; (f) -54π

5. Area $= \frac{5}{12}$ square units

Further Problems IX (page 451)

1. 14; 2. 1.6; 3. $\frac{\pi}{36}\{9 - 4\sqrt{3}\}$;

4. $\frac{1}{2}\{\pi^4 + 4\}$; 5. $\frac{9\pi}{256}$; 6. $\frac{1}{2} \cdot \ln 2$;

7. $2 - \pi/2$; 8. $\frac{1}{8}$; 9. 14;

10. (a) 39.24; (b) 0

Test Exercise X (page 503)

1. $4\sqrt{2}\pi$; **2.** $a(\pi/2)^2$

3. (a) (i) $(4.47, 0.464, 3)$; (ii) $(5.92, 0.564, 0.322)$

(b) (i) $(3.54, 3.54, 3)$; (ii) $(-0.832, 1.82, 3.46)$

4. 12π; **5.** $a^3(8 - 3a)\pi/12$

6. (a) $I = \iint v(1 + u)(1 + u + v)\,\mathrm{d}v\,\mathrm{d}v$

(b) $I = \iiint \dfrac{(2u + v)(v - 4w)}{vw}\,\mathrm{d}u\,\mathrm{d}v\,\mathrm{d}w$

Further Problems X (page 503)

1. $8\sqrt{5}\pi$; **2.** $\left(\dfrac{a}{2}, \dfrac{a}{2}, \dfrac{a}{2}\right)$; **3.** $10\sqrt{61}$

4. $\dfrac{4\sqrt{22}\pi}{3}$; **5.** $\dfrac{\pi}{24}(5\sqrt{5} - 1)$; **6.** $\pi\sqrt{5}$;

7. $16a^2$; **8.** $2a^2(\pi - 2)$; **9.** $4\pi(a + b)\sqrt{a^2 - b^2}$;

10. 45π; **11.** $\dfrac{11}{30}$; **12.** $\dfrac{\pi a^4}{2}$;

13. $2\left(\pi - \dfrac{4}{3}\right)$; **14.** $\bar{x} = \bar{y} = \bar{z} = \dfrac{3a}{8}$; **15.** $\dfrac{\pi a^3}{3}\{4\sqrt{2} - 3\}$;

16. $\dfrac{4\pi abc}{3}$; **17.** $\dfrac{2a^3}{3}$; **18.** $\dfrac{1}{4}\iint (u^2 + v^2)\,\mathrm{d}u\,\mathrm{d}v$;

19. $u^2 v\,\mathrm{d}u\,\mathrm{d}v\,\mathrm{d}w$; **20.** $\bar{z} = -\dfrac{a}{5}$; **21.** $\dfrac{7}{18}$;

22. $2 - \dfrac{\pi}{2}$; **23.** $\dfrac{1}{4}(\sqrt{2} - 1)$

Test Exercise XI (page 565)

1. (a) solutions unique; (b) infinite number of solutions

2. $x_1 = -4,\ x_2 = 2,\ x_3 = -3$

3. $x_1 = -2,\ x_2 = 2,\ x_3 = 3$

4. $x_1 = -3,\ x_2 = 4,\ x_3 = -2$

5. $x_1 = 1,\ x_2 = -2,\ x_3 = 2$

6. $\lambda_1 = 1, \lambda_2 = -2, \lambda_3 = 3$

$$x_1 = \begin{bmatrix} 1 \\ 0 \\ -1 \end{bmatrix}; \quad x_2 = \begin{bmatrix} 1 \\ -1 \\ 3 \end{bmatrix}; \quad x_3 = \begin{bmatrix} 3 \\ 2 \\ -1 \end{bmatrix}$$

7. (a) $\begin{bmatrix} -8 \\ 1 \end{bmatrix}$; (b) $\begin{bmatrix} 7.196 \\ -0.464 \end{bmatrix}$

8. $M = \begin{bmatrix} 1 & 1 \\ -2 & 1 \end{bmatrix}$; $M^{-1} = \begin{bmatrix} 1/3 & -1/3 \\ 2/3 & 1/3 \end{bmatrix}$

$$M^{-1}AM = \begin{bmatrix} -1 & 0 \\ 0 & 5 \end{bmatrix}$$

Further Problems XI (page 566)

1. $x_1 = 1, x_2 = -4, x_3 = 3$
2. (a) $x_1 = 3, x_2 = 1, x_3 = -4$

   (b) $x_1 = 4, x_2 = -2, x_3 = -1$
3. (a) $x_1 = 4, x_2 = 2, x_3 = 5, x_4 = 3$

   (b) $x_1 = 5, x_2 = -4, x_3 = 1, x_4 = 3$

   (c) $x_1 = 3, x_2 = -2, x_3 = 0, x_4 = 5$
4. (a) $x_1 = -3, x_2 = 1, x_3 = 3$

   (b) $x_1 = 5, x_2 = 2, x_3 = -1$

   (c) $x_1 = 4, x_2 = 3, x_3 = -1, x_4 = -2$
5. (a) $\lambda_1 = 2, \lambda_2 = 7; x_1 = \begin{bmatrix} 3 \\ -2 \end{bmatrix}; x_2 = \begin{bmatrix} 1 \\ 1 \end{bmatrix}$

   (b) $\lambda_1 = 1, \lambda_2 = -3; x_1 = \begin{bmatrix} 5 \\ 1 \end{bmatrix}; x_2 = \begin{bmatrix} 1 \\ 1 \end{bmatrix}$

   (c) $\lambda_1 = -8, \lambda_2 = 4; x_1 = \begin{bmatrix} 5 \\ -2 \end{bmatrix}; x_2 = \begin{bmatrix} 1 \\ 2 \end{bmatrix}$

   (d) $\lambda_1 = 4, \lambda_2 = -6; x_1 = \begin{bmatrix} 1 \\ 1 \end{bmatrix}; x_2 = \begin{bmatrix} 9 \\ -1 \end{bmatrix}$

   (e) $\lambda_1 = 1, \lambda_2 = 3; \lambda_3 = 9; x_1 = \begin{bmatrix} 7 \\ -1 \\ -5 \end{bmatrix}; x_2 = \begin{bmatrix} 7 \\ 1 \\ -7 \end{bmatrix}; x_3 = \begin{bmatrix} 1 \\ 1 \\ 5 \end{bmatrix}$

   (f) $\lambda = 1,\ 2,\ 4; x = \begin{bmatrix} 0 \\ 1 \\ 6 \end{bmatrix}, \begin{bmatrix} 1 \\ 1 \\ 3 \end{bmatrix}, \begin{bmatrix} 3 \\ 1 \\ 3 \end{bmatrix}$

(g) $\lambda = -1,\ -3,\ 7;\ x = \begin{bmatrix} 6 \\ -5 \\ 2 \end{bmatrix}, \begin{bmatrix} 2 \\ -1 \\ 0 \end{bmatrix}, \begin{bmatrix} 6 \\ 27 \\ 10 \end{bmatrix}$

(h) $\lambda = -2,\ 4,\ 7;\ x = \begin{bmatrix} 3 \\ -1 \\ 1 \end{bmatrix}, \begin{bmatrix} 3 \\ 4 \\ 5 \end{bmatrix}, \begin{bmatrix} 6 \\ 1 \\ -1 \end{bmatrix}$

6. $\lambda = 0, 7, 13$

7. $I_1 = 2, I_2 = -3, I_3 = 2$

8. $k = \frac{1}{2};\ x_1 = -2,\ x_2 = \frac{1}{2},\ x_3 = 1$

Test Exercise XII (page 623)

1. (a) -15; (b) $-16\mathbf{i} + 10\mathbf{j} + 17\mathbf{k}$

2. (a) 9; (b) $-(47\mathbf{i} + 17\mathbf{j} + 29\mathbf{k})$

3. $\mathbf{A} \cdot (\mathbf{B} \times \mathbf{C}) = 0 \quad \therefore$ vectors coplanar

4. (a) $4\mathbf{i} - 4\mathbf{j} + 24\mathbf{k}$; (b) $2\mathbf{i} - 2\mathbf{j} + 24\mathbf{k}$; (c) 24.66

5. $\mathbf{T} = \dfrac{1}{\sqrt{66}}(4\mathbf{i} + \mathbf{j} + 7\mathbf{k})$

6. $\dfrac{8}{5}(25\mathbf{i} - 6\mathbf{j} - 15\mathbf{k})$ **7.** 5.08

8. $\dfrac{1}{\sqrt{101}}(2\mathbf{i} + 4\mathbf{j} + 9\mathbf{k})$

9. (a) $14\mathbf{i} - 12\mathbf{j} - 30\mathbf{k}$; (b) 8; (c) $5\mathbf{i} - 2\mathbf{j} - 4\mathbf{k}$;
(d) $7\mathbf{i} + 2\mathbf{j} + 3\mathbf{k}$; (e) $3\mathbf{i} + 2\mathbf{j} + \mathbf{k}$

Further Problems XII (page 624)

1. 61; **2.** $29\mathbf{i} - 10\mathbf{j} + 16\mathbf{k}$

3. (a) $22\mathbf{i} + 14\mathbf{j} + 2\mathbf{k}$; (b) $-2\mathbf{i} + 14\mathbf{j} - 22\mathbf{k}$

4. (a) $2x\mathbf{i} + 3\mathbf{j} + \cos x\,\mathbf{k}$; (b) $2\mathbf{i} - \sin x\,\mathbf{k}$;
(c) $(4x^2 + 9 + \cos^2 x)^{1/2}$; (d) $34 + \sin 2$

5. (a) $2 - 2u - 9u^2$; (b) $(3u^2 + 4u + 3)\mathbf{i} + (3u^2 + 6)\mathbf{j} + (1 - 2u)\mathbf{k}$;
(c) $\mathbf{i} - 2\mathbf{j} + (3 - 2u)\mathbf{k}$

6. $\dfrac{1}{5\sqrt{21}}(2\mathbf{i} - 20\mathbf{j} + 11\mathbf{k})$ **7.** $\dfrac{-1}{\sqrt{129}}(10\mathbf{i} + 2\mathbf{j} - 5\mathbf{k})$

8. $\dfrac{-1}{\sqrt{126}}(5\mathbf{i} - \mathbf{j} + 10\mathbf{k})$;

9. $\dfrac{-1}{\sqrt{601}}(12\mathbf{i} + 4\mathbf{j} - 21\mathbf{k})$

10. -8.285;

11. -9.165

12. (a) 15; (b) -33; (c) 7

13. (a) $-6\mathbf{i} + 4\mathbf{j} - 7\mathbf{k}$; (b) $62\mathbf{i} + 10\mathbf{j} - 38\mathbf{k}$;

(c) $18\mathbf{i} - 21\mathbf{j} + 10\mathbf{k}$

14. (a) $12\mathbf{i} - 4\mathbf{j} + 4\mathbf{k}$; (b) $24\mathbf{i} - 4\mathbf{j}$; (c) 144

15. (a) $(2\sin 2)\mathbf{i} + 2e^3\mathbf{j} + (\cos 2 + e^3)\mathbf{k}$

(b) $(4\sin^2 2 + \cos^2 2 + 2e^3\cos 2 + 5e^6)^{1/2}$

16. -5.014

17. $p = \dfrac{1}{\sqrt{29}}(3\mathbf{i} + 2\mathbf{j} - 4\mathbf{k})$; $q = \dfrac{1}{\sqrt{38}}(6\mathbf{i} - \mathbf{j} + \mathbf{k})$; $\theta = 68° \, 48'$

18. (a) $(2t + 3)\mathbf{i} - (6\cos 3t)\mathbf{j} + 6e^{2t}\mathbf{k}$;

(b) $2\mathbf{i} + (18\sin 3t)\mathbf{j} + 12e^{2t}\mathbf{k}$; (c) 12.17

20. $-4x\mathbf{i} + 4z\mathbf{k}$

21. $(2\cos 5.5)\mathbf{i} - (6\sin 5.5)\mathbf{j} - (6\sin 5.5)\mathbf{k}$

22. $p = 6$

23. (a) (i) $p = 15/4$; (ii) $p = -33$;

(b) $\dfrac{1}{7}(3\mathbf{i} - 2\mathbf{j} + 6\mathbf{k})$

Test Exercise XIII (page 686)

1. $3\mathbf{i} + \dfrac{18}{7}\mathbf{j} - \dfrac{81}{8}\mathbf{k}$;

2. 12;

3. $18\pi(2\mathbf{i} + \mathbf{j})$;

4. $24(\mathbf{i} + \mathbf{j})$;

5. $8 + \dfrac{4\pi}{3}$;

6. all conservative;

7. $36\left(\dfrac{\pi}{4} + 1\right)$;

8. 0

Further Problems XIII (page 687)

1. (a) $576\mathbf{k}$; (b) $\dfrac{576}{5}(3\mathbf{i} + \mathbf{j} + 2\mathbf{k})$

2. $1771\mathbf{i} + 1107\mathbf{j} + 830.4\mathbf{k}$

3. $416.1\mathbf{i} + 718.5\mathbf{j} + 5679\mathbf{k}$

4. 46.9; **5.** -4.18; **6.** 8π

7. $\frac{16\pi}{3}(\mathbf{i}+\mathbf{k})$; **8.** $\frac{1}{3}(48\mathbf{i}+64\mathbf{j}-24\mathbf{k})$

9. $64\left(\frac{\pi}{4}-\frac{1}{3}\right)(6\mathbf{i}+4\mathbf{j})$

10. $\frac{9}{2}\{(\pi+2)\mathbf{i}+(\pi+2)\mathbf{j}+4\mathbf{k}\}$

11. $\frac{12}{5}(32\mathbf{j}+15\mathbf{k})$; **12.** -1; **13.** $\frac{250}{3}\pi$

14. $\frac{1}{6}(117\pi+256-28\sqrt{7})=91.58$; **15.** -80

16. 96π; **17.** -2; **18.** 12π

19. $-\frac{a^3}{3}$; **20.** $\frac{81\pi}{4}$

Test Exercise XIV (page 718)

1. Yes, an orthogonal set; **2.** $h_u=1,\ h_v=2v,\ h_\theta=2u$

3. $4\mathbf{I}+\mathbf{K}$

4. (a) $(2\cos\phi+2\cos 2\phi+1)$; (b) $(2\sin 2\phi+\sin\phi)\mathbf{K}$

5. (a) $(\mathrm{d}s)^2=(\mathrm{d}r)^2+r^2(\mathrm{d}\theta)^2+r^2\sin^2\theta\,(\mathrm{d}\phi)^2$

(b) $\mathrm{d}V=r^2\sin\theta\,\mathrm{d}r\,\mathrm{d}\theta\,\mathrm{d}\phi$

6. -10.5

Further Problems XIV (page 719)

1. (a) Yes; (b) No

2. -50.5; **3.** $2\frac{5}{18}$

5. (a) $\nabla^2 V=\frac{\partial^2 V}{\partial\rho^2}+\frac{1}{\rho}\cdot\frac{\partial V}{\partial\rho}+\frac{1}{\rho^2}\cdot\frac{\partial^2 V}{\partial\phi^2}+\frac{\partial^2 V}{\partial z^2}$

(b) $\nabla^2 V=\frac{1}{r^2}\cdot\frac{\partial}{\partial r}\left(r^2\frac{\partial V}{\partial r}\right)+\frac{1}{r^2\sin\theta}\cdot\frac{\partial}{\partial\theta}\left(\sin\theta\frac{\partial V}{\partial\theta}\right)+\frac{1}{r^2\sin^2\theta}\cdot\frac{\partial^2 V}{\partial\phi^2}$

6. (b) $h_u=h_v=\sqrt{u^2+v^2}$; $h_w=1$

(c) $\operatorname{div} F = \dfrac{1}{u^2+v^2}\left\{\dfrac{\partial}{\partial u}\left(\sqrt{u^2+v^2}\cdot\dfrac{\partial F_u}{\partial u}\right)\right.$

$$\left. + \frac{\partial}{\partial v}\left(\sqrt{u^2+v^2}\cdot\frac{\partial F_v}{\partial v}\right)\right\} + \frac{\partial F_w}{\partial w}$$

(d) $\nabla^2 V = \dfrac{1}{u^2+v^2}\left\{\dfrac{\partial^2 V}{\partial u^2}+\dfrac{\partial^2 V}{\partial v^2}\right\}+\dfrac{\partial^2 V}{\partial w^2}$

Test Exercise XV (page 769)

1. (a) $w = 6 - j2$; (b) $w = 3 - j2$;

(c) $w = j3$; (d) $w = 2$

2. Magnification = 2.236; rotation = 63° 26′;

translation = 1 unit to right, 3 units downwards

3.

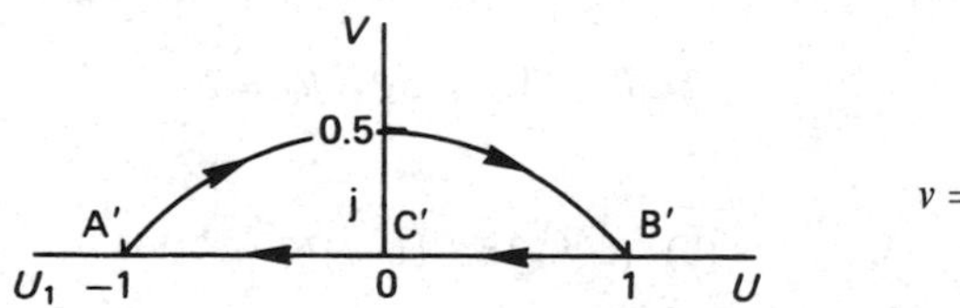

$v = \frac{1}{2}(1 - u^2)$

4.

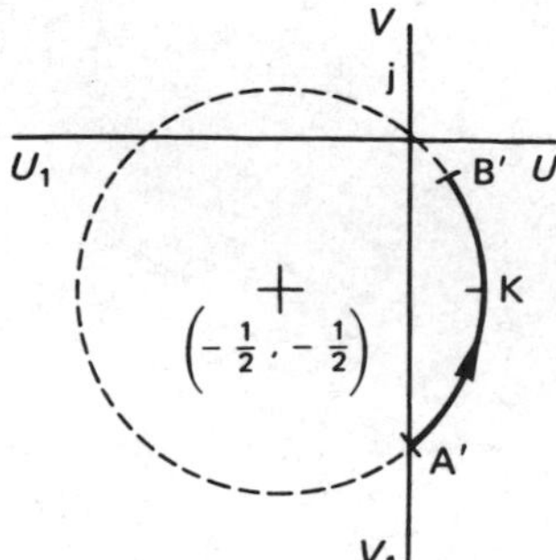

Minor arc of circle, centre $\left(-\frac{1}{2}, -\frac{1}{2}\right)$ radius $\frac{1}{\sqrt{2}} = 0.7071$, between A′ (−j) and B′ (0.12 − j0.16)

5. (a) Centre $\left(u = 0,\ v = \frac{2}{3}\right)$; (b) radius $\frac{1}{3}$

6. Centre $\left(u = \frac{2}{3},\ v = 0\right)$; radius $\frac{2}{3}$

Further Problems XV (page 770)

1. Triangle A′B′C′ with A′ $(-1+j2)$, B′ $(5+j2)$, C′ $(2+j5)$
2. (a) A′ $(-8+j9)$; B′ $(23+j14)$

   (b) Magnification $= \sqrt{29} = 5.385$; rotation $= 68°12'$; translation = nil
3. Straight line joining A′ $(5-j7)$ to B′ $(-3-j)$; magnification $= 3.162$; rotation $= 161°\,34'$ anticlockwise; translation = 2 to right, 4 upwards
4.

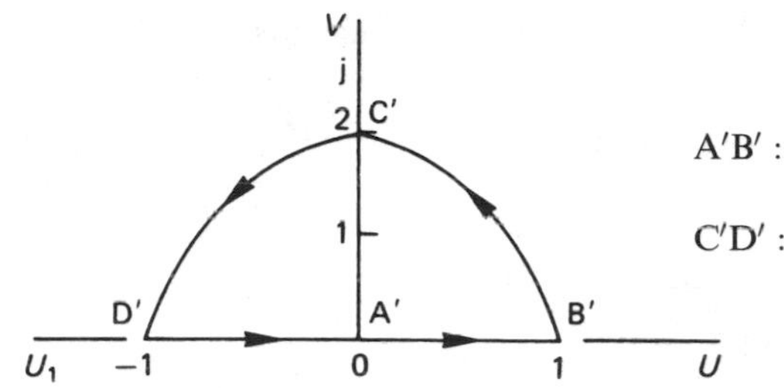

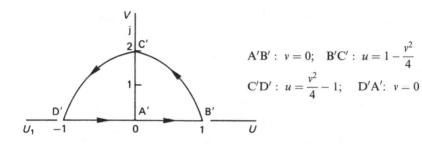

A′B′ : $v = 0$; B′C′ : $u = 1 - \dfrac{v^2}{4}$

C′D′ : $u = \dfrac{v^2}{4} - 1$; D′A′: $v = 0$

5.

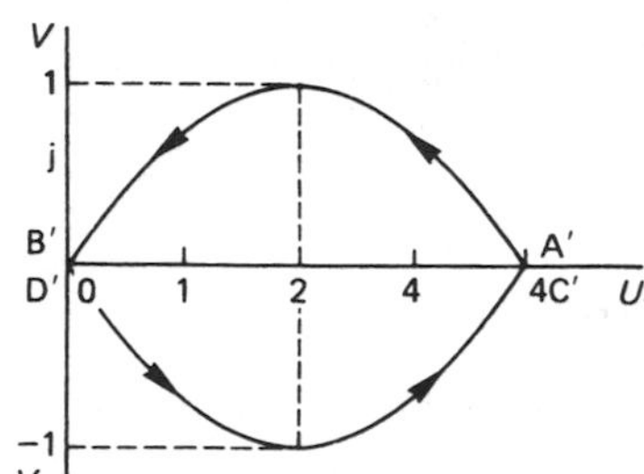

A′B′ and C′D′: $v = \frac{1}{4}(4u - u^2)$

B′C′ and D′A′: $v = \frac{1}{4}(u^2 - 4u)$

6. A′ $(1-j2)$; B′ $(-23+j10)$; C′ $(1-j8)$

   A′B′ : $u = 2 - \dfrac{v^2}{4}$; B′C′ : $v = \dfrac{(u-1)^2}{32} - 8$; C′A′ : $u = 1$
7. Circle, centre $\left(\dfrac{1}{2} - j\dfrac{2}{3}\right)$, radius $\dfrac{7}{6}$
8. (a) Circle, centre $\left(\dfrac{1}{3} - j0\right)$, radius $\dfrac{2}{3}$

   (b) Region outside the circle in (a)

9. Circle, centre $\left(\frac{3}{2}+\mathrm{j}0\right)$, radius 1; clockwise development

10. Circle, $u^2+v^2-\frac{22u}{5}+\frac{8}{5}=0$, centre $\left(\frac{11}{5}+\mathrm{j}0\right)$, radius $\frac{9}{5}$

11. Circle, $u^2+v^2-\frac{u}{2}=0$, centre $\left(\frac{1}{4}+\mathrm{j}0\right)$, radius $\frac{1}{4}$. Region inside this circle

12. Circle, centre $\left(-\frac{7}{3}+\mathrm{j}0\right)$, radius $\frac{5}{3}$

13. (a) Circle, centre $\left(\frac{3}{5}, 0\right)$, radius $\frac{2}{5}$, developed clockwise

(b) Region outside the circle in (a)

14. $v=-\frac{u}{3}$

Test Exercise XVI (page 822)

1. (a) Regular at all points; (b) $z=-5$;
(c) Regular at all points; (d) $z=-1$ and $z=4$;
(e) $z=0$, where $z=x+\mathrm{j}y$

3. $\mathrm{j}4\pi$

4. (a) $z=0$; (b) $z=\pm 1$; (c) no critical point;
(d) $z=\pm\sqrt{2}$; (e) $z=0$; (f) no critical point;

5. $w=\frac{12+\mathrm{j}8z-z^2}{4}$; D′: $w=3$

Further Problems XVI (page 823)

3. Circle, centre (5, −2), radius $\sqrt{2}$

4. Circle, centre $\left(-\frac{1}{3}, 1\right)$, radius $\frac{2}{3}$, anticlockwise

5. (a) $\mathrm{j}10\pi$; (b) $\mathrm{j}6\pi$

6. (a) 0; (b) $\mathrm{j}4\pi$; (c) $\mathrm{j}10\pi$

8. $\mathrm{j}2\pi$; 9. $\mathrm{j}10\pi$

10. (a) (i) $z=0$; (ii) $z=\pm 1$

(b) Ellipse, centre (0, 0), semi major axis $\frac{5}{2}$, semi minor axis $\frac{3}{2}$

11. (a) $u^2+v^2=1$; (b) $u^2+(v-1)^2=2$ $\quad\theta=45°$.

12. Unit circle becomes the real axis on the w-plane. Region within the circle maps on to the upper half plane

13. $w=\sin\dfrac{z\pi}{2a}$

Test Exercise XVII (page 889)

1. (a) yes; (b) yes; (c) no;

(d) yes; (e) no; (f) no

2. $f(x)=2\pi-4\{\sin x+\frac{1}{2}\sin 2x+\frac{1}{3}\sin 3x+\dots\}$

3. (a) odd; (b) odd; (c) even;

(d) neither; (e) neither; (f) even

4. (a) $f(x)=\dfrac{\pi}{2}+\dfrac{4}{\pi}\left\{\cos x+\dfrac{1}{9}\cos 3x+\dfrac{1}{25}\cos 5x+\dots\right\}$

(b) $f(x)=-2\left\{\sin x-\dfrac{1}{2}\sin 2x+\dfrac{1}{3}\sin 3x-\dots\right\}$

5. (a) cosine terms only; (b) sine terms only;

odd harmonics only;

(c) even harmonics only; (d) odd harmonics only

6. $f(t)=\dfrac{1}{2}-\dfrac{1}{\omega^2}\left\{\cos\omega t+\dfrac{1}{9}\cos 3\omega t+\dfrac{1}{25}\cos 5\omega t+\dots\right\}$

$+\dfrac{1}{\omega}\left\{\sin\omega t-\dfrac{1}{2}\sin 2\omega t+\dfrac{1}{3}\sin 3\omega t-\dots\right\}$ where $\omega=\pi/2$

Further Problems XVII (page 891)

1. $f(x)=\dfrac{2}{\pi}\left\{\sin x+\dfrac{1}{2}\sin 2x+\dfrac{1}{3}\sin 3x+\dots\right\}$

2. $f(t)=-1-\dfrac{16}{\pi}\left\{\sin\omega t+\dfrac{1}{3}\sin 3\omega t+\dfrac{1}{5}\sin 5\omega t+\dots\right\}$ where $\omega=\dfrac{\pi}{2}$

3. $f(x)=\dfrac{4}{\pi}\left\{\dfrac{1}{2}-\dfrac{1}{1\times 3}\cos 2x-\dfrac{1}{3\times 5}\cos 4x-\dfrac{1}{5\times 7}\cos 6x-\dots\right\}$

4. $f(x)=\dfrac{\pi}{2}+\dfrac{4}{\pi}\left\{\cos x+\dfrac{1}{9}\cos 3x+\dfrac{1}{25}\cos 5x+\dots\right\}$

5. $f(x) = \dfrac{2A}{\pi}\left\{1 - 2\left(\dfrac{1}{1 \times 3}\cos 2x + \dfrac{1}{3 \times 5}\cos 4x + \ \ldots\right)\right\}$

6. $f(x) = \dfrac{\pi}{4} - \dfrac{2}{\pi}\left\{\cos x + \dfrac{1}{9}\cos 3x + \dfrac{1}{25}\cos 5x + \ \ldots\right\}$

$$+\left\{\sin x - \frac{1}{2}\sin 2x + \frac{1}{3}\sin 3x - \ \ldots\right\};$$

Second harmonic $= 42.2$ per cent

7. $i = f(t) = \dfrac{A}{\pi}\left\{1 + \dfrac{\pi}{2}\sin \omega t - 2\left(\dfrac{1}{1 \times 3}\cos 2\omega t + \dfrac{1}{3 \times 5}\cos 4\omega t\right.\right.$

$$\left.\left. + \frac{1}{5 \times 7}\cos 6\omega t + \ \ldots\right)\right\} \qquad \text{where } \omega = \frac{2\pi}{T}$$

8. $f(x) = \dfrac{3a}{\pi}\left\{\sin 2x + \dfrac{1}{2}\sin 4x + \dfrac{1}{4}\sin 8x + \dfrac{1}{5}\sin 10x + \ \ldots\right\}$

9. (a) $f(x) = \dfrac{\pi^2}{6} - \left(\cos 2x + \dfrac{1}{4}\cos 4x + \dfrac{1}{9}\cos 6x + \ \ldots\right)$

(b) $f(x) = \dfrac{8}{\pi}\left(\sin x + \dfrac{1}{3^3}\sin 3x + \dfrac{1}{5^3}\sin 5x + \ \ldots\right)$

10. $f(x) = \dfrac{2}{\pi}\left\{\dfrac{1}{2} + \dfrac{\pi}{4}\cos x + \dfrac{1}{1 \times 3}\cos 2x - \dfrac{1}{3 \times 5}\cos 4x + \ \ldots\right\}$

11. $f(x) = \ -\dfrac{1}{\pi} + \dfrac{1}{2}\cos x - \dfrac{2}{3\pi}\cos 2x + \dfrac{2}{15\pi}\cos 4x - \ \ldots$

12. $f(x) = \dfrac{4}{\pi}\left\{\sin x - \dfrac{1}{9}\sin 3x + \dfrac{1}{25}\sin 5x - \ \ldots\right\}$

13. $f(t) = -\dfrac{4}{\pi^2}\left\{\cos \pi t + \dfrac{1}{9}\cos 3\pi t + \ \ldots\right\}$

$$+\frac{2}{\pi}\left\{2\sin \pi t - \frac{1}{2}\sin 2\pi t + \ldots\right\}$$

14. $f(x) = \dfrac{\pi^2}{3} - 4\left\{\cos x - \dfrac{1}{4}\cos 2x + \dfrac{1}{9}\cos 3x - \dfrac{1}{16}\cos 4x + \ldots\right\}$

15. $f(x) = 7 - \dfrac{6}{\pi}\left\{\sin x - \dfrac{1}{2}\sin 2x + \dfrac{1}{3}\sin 3x - \dfrac{1}{4}\sin 4x + \ \ldots\right\}$

16. $f(x) = \dfrac{2}{3} + \dfrac{4}{\pi^2}\left\{\cos \pi t - \dfrac{1}{4}\cos 2\pi t + \dfrac{1}{9}\cos 3\pi t - \ \ldots\right\}$

17. $f(x) = -\left\{\sin x + \dfrac{1}{2}\sin 2x + \dfrac{1}{3}\sin 3x + \dfrac{1}{4}\sin 4x + \ \ldots\right\}$

18. $f(t) = -\dfrac{2}{\pi}\left\{\sin \omega t - \sin 2\omega t + \dfrac{1}{3}\sin 3\omega t + \dfrac{1}{5}\sin 5\omega t + \ \ldots\right\}$

where $\omega = \pi/2$

19. (a) $f(x) = \dfrac{4\pi^2}{3} + 4\left\{\cos x + \dfrac{1}{4}\cos 2x + \dfrac{1}{9}\cos 3x + \ldots\right\}$

$$-4\pi\left\{\sin x + \frac{1}{2}\sin 2x + \frac{1}{3}\sin 3x + \ldots\right\}$$

(b) Second harmonic $= 48.2$ per cent

20. $f(t) = 1 - 1.17\cos\omega t + 0.328\cos 2\omega t + 0\cos 3\omega t + \ldots$

$$+\,0.280\sin\omega t + 0.288\sin 2\omega t - 0.318\sin 3\omega t + \ldots$$

where $\omega = \pi/3$;

Third harmonic $= 26.4$ per cent

Test Exercise XVIII (page 921)

1. $f(x) = 2.6 + 0.618\cos x - 0.150\cos 2x - 0.20\cos 3x + \ldots$

$$+\,1.817\sin x + 0.029\sin 2x + 0\sin 3x + \ldots$$

2. $i = f(t) = 5.52 + 1.556\cos x - 1.008\cos 2x + 1.033\cos 3x + \ldots$

$$+\,1.423\sin x + 1.140\sin 2x + 1.033\sin 3x + \ldots$$

where $x = \dfrac{\pi t}{12}$

3. (a) $f(x) = -\,2.943\cos x + 3.000\cos 3x + \ldots$

$$+\,14.383\sin x + 4.333\sin 3x + \ldots$$

(b) Third harmonic $= 35.9$ per cent

Further Problems XVIII (page 922)

1. $f(x) = 3.31 + 0.023\cos x - 0.30\cos 2x + 0.233\cos 3x + \ldots$

$$+\,0.711\sin x + 1.01\sin 2x - 0.250\sin 3x + \ldots$$

2. $f(x) = 6.89 - 1.190\cos x - 0.267\cos 2x + 0.167\cos 3x + \ldots$

$$-\,2.525\sin x - 1.213\sin 2x - 0.067\sin 3x + \ldots$$

3. $f(x) = 5.44 + 1.26\cos x + 0.050\cos 2x + 0.033\cos 3x + \ldots$

$$+\,2.04\sin x - 0.375\sin 2x + 0.833\sin 3x + \ldots$$

4. $f(x) = 2.059 + 1.01\cos x + 0.775\cos 2x + 0.083\cos 3x + \ldots$

$$-\,3.33\sin x - 0.303\sin 2x - 0.300\sin 3x + \ldots$$

5. $f(x) = 5.89 - 0.976\cos x + 1.33\cos 2x - 0.267\cos 3x + \ldots$

$$+\,2.145\sin x + 1.44\sin 2x + 0.083\sin 3x + \ldots$$

6. $f(x) = 5.09 - 2.42\cos x - 1.52\cos 2x + 0.367\cos 3x + \ldots$

$+ 0.989\sin x - 1.15\sin 2x + 0.583\sin 3x + \ldots$

7. $y = 1.242 - 0.062\cos x + 0.217\cos 2x + 0.017\cos 3x + \ldots$

$- 0.545\sin x + 0.290\sin 2x + 0.033\sin 3x + \ldots$

8. (a) $f(x) = 8.96 - 5.26\cos x - 1.79\cos 2x + 1.67\cos 3x + \ldots$

$+ 0.52\sin x + 0.51\sin 2x - 0.25\sin 3x + \ldots$

(b) Second harmonic = 35.2 per cent

9. (a) $f(u) = 12.1 + 0.973\cos x + 0.258\cos 2x + 0.583\cos 3x + \ldots$

$- 4.030\sin x - 5.267\sin 2x + 0.200\sin 3x + \ldots$

where $x = \dfrac{\pi u}{30}$

(b) Third harmonic = 14.87 per cent

10. $f(u) = 5.22 - 1.076\cos x + 2.217\cos 2x - 0.167\cos 3x + \ldots$

$- 1.312\sin x + 1.039\sin 2x - 0.233\sin 3x + \ldots$

where $x = \dfrac{\pi u}{6}$

11. $i = 5.23 + 0.31\cos x + 0.14\cos 2x + 0.05\cos 3x + \ldots$

$+ 3.39\sin x + 0.62\sin 2x - 0.37\sin 3x + \ldots$

where $x = \dfrac{\pi t}{12}$

12. Amplitude $A_2 = 3.62$; phase angle $\phi = 0.987$ radians

13. $f(x) = 1.411\cos x - 1.667\cos 3x + \ldots$

$+ 10.241\sin x + 0.667\sin 3x + \ldots$

$= 10.34\sin(x + 0.138) + 1.80\sin(3x - 1.19) + \ldots$

Test Exercise XIX (page 961)

1. (a) $u = 2x^4(t - 2) + 4xt + e^{2t}$

(b) $u = 2\sin 2x \cdot (e^y - 1) + \sin x + y^2$

2. $u(x, t) = \dfrac{16}{\pi^2}\sum_{r=1}^{\infty}\dfrac{1}{r^2} \cdot \sin\dfrac{r\pi}{2} \cdot \sin\dfrac{r\pi x}{10} \cdot \cos\dfrac{r\pi t}{10}$

3. $u(x,t)=\dfrac{100}{\pi}\sum\limits_{r=1}^{\infty}(-1)^{r+1}\cdot\dfrac{1}{r}\sin\dfrac{\lambda x}{c}\cdot e^{-\lambda^2 t}$

where $\lambda=\dfrac{r\pi c}{2}$

4. $u(x,y)=\sum\limits_{r=1}^{\infty}\dfrac{20}{r\pi}\cdot\operatorname{cosech} r\pi\cdot\sin\dfrac{r\pi x}{2}\cdot\sinh\dfrac{r\pi y}{2}$

$r=1,\ 3,\ 5,\ \ldots$

Further Problems XIX (page 962)

2. $u(x,t)=\dfrac{32}{\pi^3}\sum\limits_{r=1}^{\infty}\dfrac{1}{r^3}\cdot\sin\dfrac{r\pi x}{2}\cdot\cos\dfrac{3r\pi t}{2}$ $(r \text{ odd})$

3. $u(x,t)=\dfrac{2}{25\pi^2}\sum\limits_{r=1}^{\infty}\dfrac{1}{r^2}\cdot\sin\dfrac{r\pi}{2}\cdot\sin\dfrac{r\pi x}{4}\cdot\cos\dfrac{5r\pi t}{2}$

4. $u(x,t)=\dfrac{25}{2\pi^2}\sum\limits_{r=1}^{\infty}\dfrac{1}{r^2}\cdot\sin\dfrac{r\pi}{5}\cdot\sin\dfrac{r\pi x}{10}\cdot\cos\dfrac{cr\pi t}{10}$

5. $u(x,t)=\dfrac{800}{\pi^3}\sum\limits_{r=1}^{\infty}\dfrac{1}{r^3}\cdot\sin\dfrac{r\pi x}{10}\cdot e^{-4\lambda^2 t}$ $\quad r=1,\ 3,\ 5,\ \ldots$

where $\lambda=\dfrac{r\pi}{10}$

6. $u(x,t)=\dfrac{16}{\pi^2}\sum\limits_{r=1}^{\infty}\dfrac{1}{r^2}\cdot\sin\dfrac{r\pi}{2}\cdot\sin\dfrac{r\pi x}{10}\cdot e^{-r^2c^2\pi^2 t/100}$

with $r=1,\ 3,\ 5,\ \ldots$

7. $u(x,y)=\dfrac{128}{\pi^3}\sum\limits_{r=1}^{\infty}\dfrac{1}{r^3}\cdot\operatorname{cosech}\dfrac{r\pi}{2}\cdot\sinh\dfrac{r\pi}{4}(2-y)\cdot\sin\dfrac{r\pi x}{4}$

with $r=1,\ 3,\ 5,\ \ldots$

8. $u(x,y)=\dfrac{200}{\pi^3}\sum\limits_{r=1}^{\infty}\dfrac{1}{r^3}\cdot\operatorname{cosech}\dfrac{2r\pi}{5}\cdot\sin\dfrac{r\pi x}{5}\cdot\sinh\dfrac{r\pi}{5}(y-2)$

with $r=1,\ 3,\ 5,\ \ldots$

Test Exercise XX (page 1016)

1. $P_{max}=10 \quad (x=4,\ y=3)$

2. $P_{max}=13 \quad (x=1,\ y=4)$

3. $P_{max} = 188$ $(x = 10, y = 4, z = 6)$

4. $P_{max} = 296$ $(x = 4, y = 6)$

5. $P_{min} = 16$ $(x = 5, y = 12)$

6. (a) 13 type A + 4 type B; (b) £118

Further Problems XX (page 1017)

1. $P_{max} = 32$ $(x = 4, y = 9/2)$
2. $P_{min} = 40$ $(x = 10, y = 10)$
3. $P_{max} = 40$ $(x = 6, y = 5/2)$
4. $P_{max} = 15$ $(x = 6, y = 3)$
5. $P_{max} = 9$ $(x = 1, y = 3)$
6. $P_{max} = 10$ $(x = 2, y = 4)$
7. $P_{max} = 10$ $(x = 2, y = 4)$
8. $P_{max} = 37$ $(x = 0, y = 8, z = 1)$
9. $P_{max} = 67$ $(x = 4, y = 10, z = 5)$
10. $P_{max} = 65$ $(x = 5, y = 10, z = 5)$
11. $P_{max} = 111$ $(x = 14.8, y = 9.06, z = 0)$ to 3 s.f.
12. $P_{max} = 340$ $(x = 30, y = 20)$
13. $P_{max} = 112$ $(x = 4, y = 8)$
14. $P_{max} = 108$ $(x = 16, y = 15)$
15. $P_{max} = 138$ $(x = 12, y = 18)$
16. $P_{max} = 240$ $(x = 9, y = 15)$
17. $P_{max} = 4400$ $(x = 201, y = 53)$
18. $P_{max} = 100$ $(x = 20, y = 10)$
19. $P_{max} = 410$ $(x = 9, y = 5, z = 2)$
20. $P_{max} = 1560$ $(x = 11, y = 10, z = 18)$
21. $P_{max} = 660$ $(x = 60, y = 30, z = 30)$
22. $P_{min} = -14$ $(x = 5, y = 2)$
23. $P_{min} = 56$ $(x = 8, y = 12)$
24. $P_{min} = 16$ $(x = 8, y = 6)$
25. $P_{min} = 40$ $(x = 4, y = 4)$
26. $P_{min} = -10$ $(x = 6, y = 13, z = 14)$

27. $P_{\text{min}} = -75 \quad (x = 8,\ y = 12,\ z = 21)$

28. (a) 10 type A + 35 type B; (b) £215

29. (a) 22 type A + 44 type B + 48 type C; (b) £1258

30. (a) 125 type A + 50 type B + 125 type C; (b) £8875.

Index